Lecture Notes in Artificial Intelligence   1910

Subseries of Lecture Notes in Computer Science
Edited by J. G. Carbonell and J. Siekmann

Lecture Notes in Computer Science
Edited by G. Goos, J. Hartmanis and J. van Leeuwen

T0189834

Springer
Berlin
Heidelberg
New York
Barcelona
Hong Kong
London
Milan
Paris
Singapore
Tokyo

Djamel A. Zighed   Jan Komorowski
Jan Żytkow (Eds.)

Principles of
Data Mining and
Knowledge Discovery

4th European Conference, PKDD 2000
Lyon, France, September 13-16, 2000
Proceedings

 Springer

Series Editors

Jaime G. Carbonell, Carnegie Mellon University, Pittsburgh, PA, USA
Jörg Siekmann, University of Saarland, Saabrücken, Germany

Volume Editors

Djamel A. Zighed
Université Lyon 2, Laboratoire ERIC
5 avenue Pierre Mendès-France, 69676 Bron, France
E-mail: zighed@univ-lyon2.fr

Jan Komorowski
Norwegian University of Science and Technology
Department of Computer and Information Science
O.S. Bragstads plass 2E, 7491 Trondheim, Norway
E-mail: janko@idi.ntnu.no

Jan Żytkow
University of North Carolina, Department of Computer Science
Charlotte, NC 28223, USA
E-mail: zytkow@uncc.edu

Cataloging-in-Publication Data applied for

Die Deutsche Bibliothek - CIP-Einheitsaufnahme

Principles of data mining and knowledge discovery : 4th European
conference ; proceedings / PKDD 2000, Lyon, France, September 13 - 16,
2000. Djamel A. Zighed ... (ed.). - Berlin ; Heidelberg ; New York ;
Barcelona ; Hong Kong ; London ; Milan ; Paris ; Singapore ; Tokyo;
Springer, 2000
    Lecture notes in computer science ; Vol. 1910 Lecture notes in
    artificial intelligence)
    ISBN 3-540-41066-X

CR Subject Classification (1998): I.2, H.3, H.5, G.3, J.1, F.4.1

ISBN 3-540-41066-X Springer-Verlag Berlin Heidelberg New York

This work is subject to copyright. All rights are reserved, whether the whole or part of the material is
concerned, specifically the rights of translation, reprinting, re-use of illustrations, recitation, broadcasting,
reproduction on microfilms or in any other way, and storage in data banks. Duplication of this publication
or parts thereof is permitted only under the provisions of the German Copyright Law of September 9, 1965,
in its current version, and permission for use must always be obtained from Springer-Verlag. Violations are
liable for prosecution under the German Copyright Law.

Springer-Verlag Berlin Heidelberg New York
a member of BertelsmannSpringer Science+Business Media GmbH
© Springer-Verlag Berlin Heidelberg 2000
Printed in Germany

Typesetting: Camera-ready by author, data conversion by PTP-Berlin, Stefan Sossna
Printed on acid-free paper   SPIN: 10722719   06/3142   5 4 3 2 1 0

Preface

This volume contains papers selected for presentation at PKDD'2000, the Fourth European Conference on Principles and Practice of Knowledge Discovery in Databases. The first meeting was held in Trondheim, Norway, in June 1997, the second in Nantes, France, in September 1998, and the third in Prague, Czech Republic, in September 1999.

PKDD 2000 was organized in Lyon, France, on 13–16 September 2000. The conference was hosted by the Equipe de Recherche en Ingénierie des Connaissances at the Université Lumière Lyon 2. We wish to express our thanks to the sponsors of the Conference, to the University Claude Bernard Lyon 1, the INSA of Lyon, the Conseil général of the Rône, the Région Rhône Alpes, SPSS France, AFIA, and the University of Lyon 2, for their generous support.

Knowledge discovery in databases (KDD), also known as data mining, provides tools for turning large databases into knowledge that can be used in practice. KDD has been able to grow very rapidly since its emergence a decade ago by drawing its techniques and data mining experiences from a combination of many existing research areas: databases, statistics, mathematical logic, machine learning, automated scientific discovery, inductive logic programming, artificial intelligence, visualization, decision science, knowledge management, and high performance computing. The strength of KDD came initially from the value added by the creative combination of techniques from the contributing areas. But in order to establish its identity, KDD has to create its own theoretical principles and to demonstrate how they stimulate KDD research, facilitate communication, and guide practitioners towards successful applications.

Seeking the principles that can guide and strengthen practical applications has been always a part of the European research tradition. Thus "Principles and Practice of KDD" (PKDD) make a suitable focus for annual meetings of the KDD community in Europe. The main long-term interest is in theoretical principles for the emerging discipline of KDD and in practical applications that demonstrate utility of those principles. Other goals of the PKDD series are to provide a European-based forum for interaction among all theoreticians and practitioners interested in data mining and knowledge discovery as well as to foster new directions in interdisciplinary collaboration.

A Discovery Challenge hosted at PKDD 2000 was conducted for the second year in a row. It followed a successful and very broadly attended Discovery Challenge in Prague. Discovery challenge is a new initiative that promotes cooperative research on original and practically important real-world databases. The Challenge requires a broad and unified view of knowledge and methods of discovery, and emphasizes business problems that require an open-minded search for knowledge in data. Two multi-relational databases, in banking and in medicine, were available to all conference participants. The Challenge was born out of the conviction that knowledge discovery in real-world databases requires an open-minded discovery process rather than application of one or another tool limited to one form of knowledge. A discoverer should consider a broad scope of techniques that can reach many forms of knowledge. The discovery process cannot be rigid and selection of techniques must be driven by knowledge hidden in the data, so that the most and the best of knowledge can be reached.

The contributed papers were selected from 157 full papers (48% growth over PKDD'99) by the following program committee:

Magnus L. Andersson (AstraZeneca, Sweden),
Petr Berka (U. Economics, Czech Republic),
Jean-Francois Boulicaut (INSA Lyon, France),
Henri Briand (U. Nantes, France),
Leo Carbonara (American Management System, UK),
Luc De Raedt (U. Freiburg, Germany),
Ronen Feldman (Bar Ilan U., Israel),
Arthur Flexer (Austrian Research Inst. for AI, Austria),
Alex Freitas (PUC-PR, Brazil),
Patrick Gallinari (U. Paris 6, France),
Jean Gabriel Ganascia (U. Paris 6, France),
Mohand-Said Hacid (U. Lyon I, INSA Lyon, France),
Petr Hájek (Acad. Science, Czech Republic),
Howard Hamilton (U. Regina, Canada),
David Hand (Imperial Col. London, UK),
Mika Klemettinen (U. Helsinki, Finland),
Willi Kloesgen (GMD, Germany),
Yves Kodratoff (U. Paris-Sud, France),
Jan Komorowski (Norwegian U. Sci. & Tech.),
Jacek Koronacki (Acad. Science, Poland),
Michel Lamure (U. Lyon I, France),
Heikki Manilla (Nokia, Finland),
Stan Matwin (U. Ottawa, Canada),
Hiroshi Motoda (Osaka U., Japan),
Jan Mrazek (Bank of Montreal, Canada),
Witold Pedrycz (U. Alberta, Canada),
Mohamed Quafafou (U. Nantes, France),
Ricco Rakotomalala (F. Lyon II, France),
Jan Rauch (U. Economics, Czech Republic),
Zbigniew Ras (UNC Charlotte, USA),
Gilbert Ritschard (U. Geneva, Switzerland),
Michele Sebag (Ecole Polytech. Paris, France),
Arno Siebes (CWI, Netherlands),
Andrzej Skowron (U. Warsaw, Poland),
Myra Spiliopoulou (Humboldt U. Berlin, Germany),
Nicolas Spyratos (U. Paris-Sud, France),
Olga Štěpánková (Czech Tech. U., Czech Republic),
Einoshin Suzuki (Yokohama Natl. U., Japan),

Ljupco Todorovski (Josef Stefan Inst., Slovenia),
Shusaku Tsumoto (Shimane Medical U., Japan),
Gilles Venturini (U. Tours, France),
Inkeri Verkamo (U. Helsinki, Finland),
Louis Wehenkel (U. Liege, Belgium),
Gerhard Widmer (U. Vienna, Austria),
Rudiger Wirth (DaimlerChrysler, Germany),
Stefan Wrobel (U. Magdeburg, Germany),
Ning Zhong (Maebashi Inst. Techn., Japan),
Djamel A. Zighed (U. Lyon 2, France), and
Jan Żytkow (UNC Charlotte, USA).

The following colleagues also reviewed papers for the conference and are due our special thanks:

Florence d'Alche buc (U. Paris 6),
Jérôme Darmont (U. Lyon 2),
Stéphane Lallich (U. Lyon 2), and
Christel Vrain (U. Orleans).

Classified according to the first author's nationality, papers submitted to PKDD'2000 came from 34 countries on 5 continents (Europe: 102 papers; Asia: 21; North America: 17; Australia: 3; and South America: 4), including Argentina (1 paper), Australia (3), Austria (3), Belgium (5), Brazil (3), Canada (4), Croatia (1), Czech Republic (4), Finland (3), France (38), Germany (7), Greece (5), Hong Kong (1), Italy (3), Japan (9), Mexico (2), Netherlands (1), Norway (1), Poland (5), Portugal (2), Romania (1), Russia (1), Singapore (4), Slovakia (1), Slovenia (1), South Korea (3), Spain (5), Sweden (1), Switzerland (3), Taiwan (4), Turkey (1), United Kingdom (10), and USA (20).

Many thanks to all who submitted papers for review and for publication in the proceedings. The accepted papers were divided into two categories: 29 oral presentations and 57 poster presentations. In addition to poster sessions each poster paper was available for the web-based discussion forum in the months preceding the conference.

There were three invited presentations:

- Sep.14, **Willi Kloesgen** (GMD, Germany), Multi-relational, statistical, and visualization approaches for spatial knowledge discovery
- Sep.15 **Luc De Raedt** (U.Freiburg, Germany), Data mining in multi-relational databases
- Sep.16 **Arno Siebes** (Utrecht U.), Developing KDD Systems

Six workshops were affiliated with the conference. All were held on 12 September.

- **Workshop 1** Data Mining, Decision Support, Meta-learning, and ILP. Chairs: **Pavel Brazdil and Alipio Jorge, U. Porto, Portugal**;
- **Workshop 2** Temporal, Spatial, and Spatio-temporal Data Mining (TSDM 2000). Chairs: **John F. Roddick, Flinders U., Australia & Kathleen Hornsby, U. Maine, USA**;

- **Workshop 3** Knowledge Discovery in Biology. Chair: **Jan Komorowski, Norwegian U. Sci. & Techn, Norway**;
- **Workshop 4** Machine Learning and Textual Information Access. Chairs: **Hugo Zaragoza, LIP6, U. Paris 6, France Patrick Gallinari, LIP6, U. Paris 6, France Martin Rajman, EPFL, Switzerland**;
- **Workshop 5** Knowledge Management: Theory and Applications. Chair: **Jean-Louis Ermine, CEA Paris, France**;
- **Workshop 6** Symbolic Data Analysis: Theory, Software, and Applications for Knowledge Mining. Chair: **Edwin Diday, U. Paris Dauphine, France**.

Five tutorials were offered to all conference participants on 13 September:

- **Tutorial 1** An Introduction to Distributed Data Mining, presented by **H. Kargupta, Washington State U., USA**;
- **Tutorial 2** Clustering Techniques for Large Data Sets: From the Past to the Future, presented by **A. Hinneburg and D.A. Keim, U. Halle, Germany**;
- **Tutorial 3** Data Analysis for Web Marketing and Merchandizing Applications, presented by **M. Spiliopoulou, Humboldt U. Berlin, Germany**;
- **Tutorial 4** Database Support for Business Intelligence Applications, presented by **W. Lehner, U. Erlangen-Nüremberg, Germany**;
- **Tutorial 5** Text Mining, presented by **Yves Kodratoff, Djamel Zighed, and Serge Di Palma (France)**.

Members of the PKDD'2000 organizing committee did an enormous amount of work and deserve the special gratitude of all participants: **Willi Kloesgen** – Tutorials Chair, **Jan Rauch** – Workshops Chair, **Arno Siebes and Petr Berka** – Discovery Challenge Chairs, **Leonardo Carbonara** – Industrial Program Chair, **Djamel Zighed** – Local Arrangement Chair, and **Jean Hugues Chauchat, Fabrice Muhlenbach, Laure Tougne, Fadila Bentayeb, Omar Boussaid, Salima Hassas, Stéphane Lallich, Céline Agier, David Coeurjolly, Alfredos Mar, Luminita Firanescu Astrid Varaine, Sabine Rabaséda, Fabien Feschet**, and **Nicolas Nicoloyannis**.

Special thanks go to **Alfred Hofmann** of Springer-Verlag for his continuous help and support and to **Karin Henzold** for the preparation of the Proceedings.

July 2000   Djamel Zighed and Jan Komorowski
                                                                              Jan Żytkow

Table of Contents

Session 1A – Towards Broader Foundations ....

Multi-relational Data Mining, Using UML for ILP 1
  Arno J. Knobbe, Arno Siebes, Hendrik Blockeel, and
  Daniël Van Der Wallen

An Apriori-Based Algorithm for Mining Frequent Substructures from
Graph Data 13
  Akihiro Inokuchi, Takashi Washio, and Hiroshi Motoda

Basis of a Fuzzy Knowledge Discovery System 24
  Maurice Bernadet

Session 1B – Rules and Trees ....

Confirmation Rule Sets 34
  Dragan Gamberger and Nada Lavrač

Contribution of Dataset Reduction Techniques to Tree-Simplification and
Knowledge Discovery 44
  Marc Sebban and Richard Nock

Combining Multiple Models with Meta Decision Trees 54
  Ljupčo Todorovski and Sašo Džeroski

Session 2A – Databases and Reward-Based Learning ...

Materialized Data Mining Views 65
  Tadeusz Morzy, Marek Wojciechowski, and Maciej Zakrzewicz

Approximation of Frequency Queries by Means of Free-Sets 75
  Jean-François Boulicaut, Arthur Bykowski, and Christophe Rigotti

Application of Reinforcement Learning to Electrical Power System
Closed-Loop Emergency Control 86
  Christophe Druet, Damien Ernst, and Louis Wehenkel

Efficient Score-Based Learning of Equivalence Classes of Bayesian
Networks 96
  Paul Munteanu and Denis Cau

Session 2B – Classiffication ....

Quantifying the Resilience of Inductive Classification Algorithms......... 106
  Melanie Hilario and Alexandros Kalousis

Bagging and Boosting with Dynamic Integration of Classifiers 116
  Alexey Tsymbal and Seppo Puuronen

Zoomed Ranking: Selection of Classification Algorithms Based on Relevant
Performance Information.................................... 126
  Carlos Soares and Pavel B. Brazdil

Some Enhancements of Decision Tree Bagging 136
  Pierre Geurts

Session 3A – Association Rules and Exceptions..........

Relative Unsupervised Discretization for Association Rule Mining 148
  Marcus-Christopher Ludl and Gerhard Widmer

Mining Association Rules: Deriving a Superior Algorithm by Analyzing
Today's Approaches 159
  Jochen Hipp, Ulrich Güntzer, and Gholamreza Nakhaeizadeh

Unified Algorithm for Undirected Discovery of Exception Rules 169
  Einoshin Suzuki and Jan Żytkow

Sampling Strategies for Targeting Rare Groups from a Bank Customer
Database 181
  Jean-Hugues Chauchat, Ricco Rakotomalala, and Didier Robert

Session 3B – Instance-Based Discovery...................

Instance-Based Classification by Emerging Patterns 191
  Jinyan Li, Guozhu Dong, and Kotagiri Ramamohanarao

Context-Based Similarity Measures for Categorical Databases 201
  Gautam Das and Heikki Mannila

A Mixed Similarity Measure in Near-Linear Computational Complexity
for Distance-Based Methods.................................... 211
  Ngoc Binh Nguyen and Tu Bao Ho

Fast Feature Selection Using Partial Correlation for Multi-valued
Attributes 221
  Stéphane Lallich and Ricco Rakotomalala

Session 4A – Clustering and Classification ....

Fast Hierarchical Clustering Based on Compressed Data and OPTICS 232
  Markus M. Breunig, Hans-Peter Kriegel, and Jörg Sander

Accurate Recasting of Parameter Estimation Algorithms Using Sufficient
Statistics for Efficient Parallel Speed-Up: Demonstrated for Center-Based
Data Clustering Algorithms 243
  Bin Zhang, Meichun Hsu, and George Forman

Predictive Performance of Weighted Relative Accuracy 255
  Ljupčo Todorovski, Peter Flach, and Nada Lavrač

Quality Scheme Assessment in the Clustering Process 265
  Maria Halkidi, M. Vazirgiannis, and Y. Batistakis

Session 5A – Time Series ....

Algorithm for Matching Sets of Time Series 277
  *Iztok Savnik, Georg Lausen, Hans-Peter Kahle, Heinrich Spiecker, and
  Sebastian Hein*

MSTS: A System for Mining Sets of Time Series 289
  Georg Lausen, Iztok Savnik, and Aldar Dougarjapov

Learning First Order Logic Time Series Classifiers: Rules and Boosting ... 299
  Juan J. Rodríguez, Carlos J. Alonso, and Henrik Boström

Posters ....

Learning Right Sized Belief Networks by Means of a Hybrid Methodology . 309
  Sylvia Acid and Luis M. De Campos

Algorithms for Mining Share Frequent Itemsets Containing Infrequent
Subsets 316
  Brock Barber and Howard J. Hamilton

Discovering Task Neighbourhoods through Landmark Learning
Performances 325
  Hilan Bensusan and Christophe Giraud-Carrier

Induction of Multivariate Decision Trees by Using Dipolar Criteria 331
  Leon Bobrowski and Marek Krętowski

Inductive Logic Programming in Clementine 337
  Sam Brewer and Tom Khabaza

A Genetic Algorithm-Based Solution for the Problem of Small Disjuncts .. 345
  Deborah R. Carvalho and Alex A. Freitas

Clustering Large, Multi-level Data Sets: An Approach Based on Kohonen
Self Organizing Maps 353
    Antonio Ciampi and Yves Lechevallier

Trees and Induction Graphs for Multivariate Response 359
    Antonio Ciampi, Djamel A. Zighed, and Jérémy Clech

CEM - Visualisation and Discovery in Email 367
    Richard Cole, Peter Eklund and Gerd Stumme

Image Access and Data Mining: An Approach 375
    Chabane Djeraba

Decision Tree Toolkit: A Component-Based Library of Decision
Tree Algorithms 381
    Nikos Drossos, Athanasios Papagelis, and Dimitris Kalles

Determination of Screening Descriptors for Chemical Reaction Databases . 388
    Laurent Dury, Laurence Leherte, and Daniel P. Vercauteren

Prior Knowledge in Economic Applications of Data Mining 395
    A.J. Feelders

Temporal Machine Learning for Switching Control 401
    Pierre Geurts and Louis Wehenkel

Improving Dissimilarity Functions with Domain Knowledge, Applications
with IKBS System 409
    David Grosser, Jean Diatta, and Noël Conruyt

Mining Weighted Association Rules for Fuzzy Quantitative Items 416
    Attila Gyenesei

Centroid-Based Document Classification: Analysis and Experimental
Results 424
    Eui-Hong (Sam) Han and George Karypis

Applying Objective Interestingness Measures in Data Mining Systems 432
    Robert J. Hilderman and Howard J. Hamilton

Observational Logic Integrates Data Mining Based on Statistics and Neural
Networks 440
    Martin Holeňa

Supporting Discovery in Medicine by Association Rule Mining of
Bibliographic Databases 446
    Dimitar Hristovski, Sašo Džeroski, Borut Peterlin, and
    Anamarija Rozic-Hristovski

Collective Principal Component Analysis from Distributed, Heterogeneous
Data 452
    Hillol Kargupta, Weiyun Huang, Krishnamoorthy Sivakumar,
    Byung-Hoon Park, and Shuren Wang

Hierarchical Document Clustering Based on Tolerance Rough Set Model .. 458
    Saori Kawasaki, Ngoc Binh Nguyen, and Tu Bao Ho

Application of Data-Mining and Knowledge Discovery in Automotive Data
Engineering 464
    Jörg Keller, Valerij Bauer, and Wojciech Kwedlo

Towards Knowledge Discovery from cDNA Microarray Gene Expression
Data 470
    Jan Komorowski, Torgeir R. Hvidsten, Tor-Kristian Jenssen,
    Dyre Tjeldvoll, Eivind Hovig, Arne K. Sanvik, and Astrid Lægreid

Mining with Cover and Extension Operators 476
    Marzena Kryszkiewicz

A User-Driven Process for Mining Association Rules.................... 483
    Pascale Kuntz, Fabrice Guillet, Rémi Lehn, and Henri Briand

Learning from Labeled and Unlabeled Documents: A Comparative Study
on Semi-Supervised Text Classification 490
    Carsten Lanquillon

Schema Mining: Finding Structural Regularity among Semistructured
Data 498
    P.A. Laur, F. Masseglia, and P. Poncelet

Improving an Association Rule Based Classifier 504
    Bing Liu, Yiming Ma, and Ching Kian Wong

Discovery of Generalized Association Rules with Multiple Minimum
Supports.. 510
    Chung-Leung Lui and Fu-Lai Chung

Learning Dynamic Bayesian Belief Networks Using Conditional Phase-
Type Distributions 516
    Adele Marshall, Sally McClean, Mary Shapcott, and Peter Millard

Discovering Differences in Patients with Uveitis through Typical Testors
by Class 524
    José F. Martínez-Trinidad, Miriam Velasco-Sánchez, and
    Edgar E. Contreras-Aravelo

Web Usage Mining: How to Efficiently Manage New Transactions and New
Clients 530
    F. Masseglia, P. Poncelet, and M. Teisseire

Mining Relational Databases 536
    Frédéric Moal, Teddy Turmeaux, and Christel Vrain

Interestingness in Attribute-Oriented Induction (AOI):
Multiple-Level Rule Generation 542
    Maybin K. Muyeba and John A. Keane

Discovery of Characteristic Subgraph Patterns Using Relative Indexing
and the Cascade Model 550
    Takashi Okada and Mayumi Oyama

Transparency and Predictive Power: Explaining Complex Classification
Models 558
    Gerhard Paass and Jörg Kindermann

Clustering Distributed Homogeneous Datasets 566
    Srinivasan Parthasarathy and Mitsunori Ogihara

Empirical Evaluation of Feature Subset Selection Based on a Real-World
Data Set 575
    Petra Perner and Chid Apte

Discovery of Ambiguous Patterns in Sequences: Application to
Bioinformatics 581
    Gerard Ramstein, Pascal Bunelle, and Yannick Jacques

Action-Rules: How to Increase Profit of a Company 587
    Zbigniew W. Ras and Alicja Wieczorkowska

Aggregation and Association in Cross Tables 593
    Gilbert Ritschard and Nicolas Nicoloyannis

An Experimental Study of Partition Quality Indices in Clustering 599
    Céline Robardet, Fabien Feschet, and Nicolas Nicoloyannis

Expert Constrained Clustering: A Symbolic Approach 605
    Fabrice Rossi and Frédérick Vautrain

An Application of Association Rules Discovery to Geographic Information
Systems 613
    Ansaf Salleb and Christel Vrain

Generalized Entropy and Projection Clustering of Categorical Data 619
    Dan A. Simovici, Dana Cristofor, and Laurentiu Cristofor

Supporting Case Acquisition and Labelling in the Context of Web Mining 626
Vojtěch Svátek and Martin Kavalec

Indirect Association: Mining Higher Order Dependencies in Data 632
Pang-Ning Tan, Vipin Kumar, and Jaideep Srivastava

Discovering Association Rules in Large, Dense Databases 638
Tudor Teusan, Gilles Nachouki, Henri Briand, and Jacques Philippe

Providing Advice to Website Designers Towards Effective Websites
Re-organization 646
Peter Tselios, Agapios Platis, and George Vouros

Clinical Knowledge Discovery in Hospital Information Systems: Two Case
Studies 652
Shusaku Tsumoto

Knowledge Discovery Using Least Squares Support Vector Machine
Classifiers: A Direct Marketing Case 657
Stijn Viaene, B. Baesens, T. Van Gestel, J.A.K. Suykens,
D. Van Den Poel, J. Vanthienen, D. De Moor, and G. Dedene

Lightweight Document Clustering 665
Sholom M. Weiss, Brian F. White, and Chidanand V. Apte

Automatic Category Structure Generation and Categorization of Chinese
Text Documents 673
Hsin-Chang Yang and Chung-Hong Lee

Mining Generalized Multiple-Level Association Rules 679
Show-Jane Yen

An Efficient Approach to Discovering Sequential Patterns in Large Databases 685
Show-Jane Yen and Chung-Wen Cho

Using Background Knowledge as a Bias to Control the Rule Discovery
Process 691
Ning Zhong, Juzhen Dong, and Setsuo Ohsuga

Author Index 699

Multi-relational Data Mining, Using UML for ILP

Arno J. Knobbe[1,2], Arno Siebes[2], Hendrik Blockeel[3], and Daniël Van Der Wallen[4]

[1]Perot Systems Nederland B.V., Hoefseweg 1, 3821 AE Amersfoort, The Netherlands,
arno.knobbe@ps.net
[2]CWI, P.O. Box 94079, 1090 GB Amsterdam, The Netherlands
arno@cwi.nl
[3]K.U.Leuven, Dept. of Computer Science, Celestijnenlaan 200A, B-3001 Heverlee, Belgium
hendrik.blockeel@cs.kuleuven.ac.be
[4]Inpact B.V., Nieuwekade 201, 3511 RW Utrecht, The Netherlands
daniel@inpact.nl

Abstract. Although there is a growing need for multi-relational data mining solutions in KDD, the use of obvious candidates from the field of Inductive Logic Programming (ILP) has been limited. In our view this is mainly due to the variation in ILP engines, especially with respect to input specification, as well as the limited attention for relational database issues. In this paper we describe an approach which uses UML as the common specification language for a large range of ILP engines. Having such a common language will enable a wide range of users, including non-experts, to model problems and apply different engines without any extra effort. The process involves transformation of UML into a language called CDBL, that is then translated to a variety of input formats for different engines.

1 Introduction

A central problem in the specification of a multi-relational data mining problem is the definition of a model of the data. Such a model directly determines the type of patterns that will be considered, and thus the direction of the search. Such specifications are usually referred to as declarative or language bias in ILP [14]. Most current systems use logic-based formalisms to specify the language bias (e.g., Progol, S-CART, Claudien, ICL, Tilde, Warmr [13, 12, 6, 7, 3, 8]). Although most of these formalisms are quite similar and make use of the same concepts (e.g., types and modes), there are still differences between the formalisms that make the sharing of the language bias specification between engines a non-trivial task. The main reasons for this are:

- the different formalisms each have their own syntax; the user needs to be familiar with all of them
- many formalisms contain certain constructs, the semantics of which, sometimes in a subtle way, reflect behavioral characteristics of the inductive algorithm.

D.A. Zighed, J. Komorowski, and J. Żytkow (Eds.): PKDD 2000, LNAI 1910, pp. 1–12, 2000.
© Springer-Verlag Berlin Heidelberg 2000

The use of different ILP-systems would be simplified significantly if a common declarative bias language were available. Such a language should have the following characteristics:

- The common language should be usable for a large range of ILP systems, which means that it should be easy to translate a bias specification from the common language to the native language of the ILP system
- It should be easy to learn. This means it should make use of concepts most users are familiar with. In the ideal case, the whole language itself is a language that the intended users are familiar with already
- The bias should not just serve as a necessary prerequisite for running the induction algorithm, but should also be usable as a shared piece of information or documentation about a problem within a team of analysts with varying levels of technical expertise
- It should be easy to judge the complexity of a problem from a single glance at the declarative bias. A graphical representation would be desirable.

In this paper we propose the use of the Unified Modeling Language (UML) [2, 15, 16, 17] as the language of choice for specifying declarative bias of such nature. Over the past few years UML has proven itself as a versatile tool for modeling a large range of applications in various domains. For ILP the Class Diagrams with their usefulness in database modeling are specifically interesting. Our discussion will be based on these diagrams.

Why do we wish to use UML to express bias? First of all, as UML is an intuitive visual language, essentially consisting of annotated graphs, we can easily write down the declarative bias for a particular domain or judge the complexity of a given data model [9]. Another reason for using UML is its widespread use in database (as well as object oriented) modelling. UML has effectively become a standard with thorough support in many commercial tools. Some tools allow the reverse engineering of a data model from a given relational database, directly using the table specifications and foreign key relations. If we can come up with a process of using UML in ILP algorithms, we would then have practically automated the analysis process of a relational database. Finally, UML may serve as a common means of stating declarative bias languages used in the different ILP engines.

Although it is clear that UML is a good candidate for specifying first order declarative bias, it may not be directly clear how the different engines will actually be making use of the UML declarations. Its use in our previously published Multi-Relational Data Mining framework [10, 11] is straightforward, as this framework and the related engine Fiji2 have been designed around the use of UML from the outset. To translate UML bias declarations to logic-based bias languages, we use an intermediate textual representation, called Common Declarative Bias Language (CDBL). CDBL is essentially a set of Prolog predicates, which can be easily processed by the different translation procedures. Translation procedures for the popular engines Tilde, Warmr and Progol are currently available. The whole process of connecting an ILP engine to a relational database now becomes a series of translation steps as is illustrated by the diagram in figure 1. We have implemented and embedded each of these steps into a single GUI.

The investigation of UML as a common declarative bias language for non-experts was motivated by the efforts involved in the Esprit IV project Aladin. This project aims at bringing ILP capabilities to a wider, commercial audience by embedding a range of ILP algorithms into the commercial Data Mining tool, Clementine.

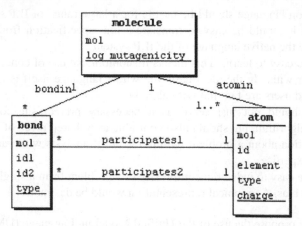

Fig. 1. The complete process of using UML with existing engines.

The outline of this paper is as follows. A section describing UML and its potential as first order declarative bias follows this introduction. We then give a short overview of the syntax of the Common Declarative Bias Language. In **Translating CDBL** we give an algorithm for translating CDBL to ILP input. Next we analyze the usefulness of UML as a declarative bias language compared to other approaches in **Comparing UML to traditional languages**. This section is followed by a **Conclusion**.

2 UML

From the large set of modelling tools provided by UML we will focus on the richest and most commonly used one: Class Diagrams [16]. These diagrams model exactly the concepts relevant for ILP, namely tables and the relation between them. In fact when we write UML in this paper we are referring specifically to Class Diagrams. There are two specific concepts within the Class Diagrams that we will be focusing on. The first is the concept of *class*. A class is a description of a set of objects that share the same features, relationships, and semantics. In a Class Diagram, a class is represented as a rectangle. Typically, a class represents some tangible entity in the problem domain, and maps to a table in the database.

The second concept is that of *association*. An association is a structural relationship that specifies that objects of one class are connected to objects of another. An important aspect of an association is its *multiplicity*. This specifies how many objects in one class are related to a single object in another, and vice versa. The multiplicity can thus be interpreted as the constraints on the association. Associations typically map to (foreign key) relations in the database although they sometimes need to be represented by an extra table, in the case of n-to-m relationships. An association

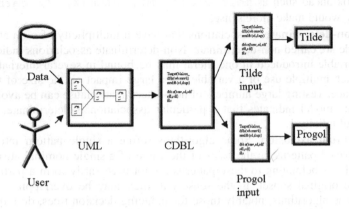

Fig. 2. UML model of the mutagenesis problem

is graphically represented as a line between two classes, which has certain labels (name, multiplicity, etc.) attached to it.

To illustrate these concepts, we show the Class Diagram describing part of the mutagenesis problem, a common benchmark problem within ILP [18]. As can be seen in figure 2, there are three classes: `molecule`, `atom` and `bond`. Four associations between these classes determine how objects in each class relate to objects in another class. As a bond involves 2 atoms (in different roles) there are two associations between `atom` and `bond`. The multiplicity of an association determines how many objects in one class correspond to a single object in another class. For example, a molecule has one or more atoms, but an atom belongs to exactly one molecule.

In order to see how UML may serve as input to an ILP engine we have to examine how the elements of a Class Diagram map to concepts in logic programming. An obvious way of doing this is to make classes and their attributes correspond to predicates and their arguments [8]. We do not specify whether these predicates need to be extensional or intentional. That is, both data originating from relational databases, as well as from Prolog databases including background knowledge can be modelled. Associations between predicates are used as constraints on the sharing of variables between these predicates, much in the same way as foreign links in [19]. Associations join two attributes in one or two predicates, which in logic programming terms means that a single variable may occur in the two positions in the argument lists identified by the two attributes.

Not only the existence of associations, but also the nature thereof in terms of its multiplicity provides bias. We will use this knowledge in three possible ways [10]:

- **Look-ahead**. For some refinements to a clause, the set of objects that support the clause is not actually changed. Therefore, it may be desirable to use a look-ahead in the form of extra refinements [3]. Given the multiplicity of an association involved in a refinement, we can directly decide whether this refinement changes anything to the support of the clause, and therefore whether the look-ahead is necessary. For example, in the mutagenesis domain a clause `molecule(...)` can be extended to form `molecule(X, M)`, `atom(X, Y, N, O, P)`. However, this refinement provides the same support, as all molecules have at least one atom (multiplicity 1..*n*). A look-ahead such as `molecule(X, M)`, `atom(X, Y, carbon, O, P)` would make more sense.

- **Non-determinate**. Associations that have a multiplicity of n on at least one side are called non-determinate. Non-determinate associations indicate that a variable introduced in one literal may be bound in several alternative ways. Such multiple use of a variable has a large impact on the size of the search space. Testing large numbers of non-determinate clause can be avoided if the data model indicates that a particular association is determinate, i.e. has a multiplicity 1.

- **Mutual exclusion**. Some algorithms refine a single pattern into a set of derived patterns on the basis of the value of a single nominal attribute. The subsets belonging to these patterns do not necessarily form a partitioning of the original subset, in the sense that they may be overlapping. However, some algorithms, notably those for inducing decision trees, do require these subsets to be mutually exclusive. A multiplicity of 0..1 or 1 indicates that such a split may be made without the danger of overlap.

3 CDBL

In order to provide a textual format for the information stored in Class Diagrams that is suitable for interpretation by ILP engines, we introduce the Common Declarative Bias Language (CDBL). The aim of CDBL is to act as an intermediate language between UML and the specific declarative bias languages available for each ILP engine. As the translation from the graphical model to CDBL is trivial, we only need to produce engine specific translation procedures for CDBL in order to have an automatic procedure for using UML as declarative bias.

A CDBL definition is built up of a set of statements, ground facts to a Prolog interpreter. Each of these statements provides information about an element of the graphical model. Each table, association or constraint is thus described by a single statement.

One of the ideas behind the design of CDBL is the clear distinction between declarations about the relational bias, which is formed by restrictions from the data model, and search bias, which contains restrictions that are relevant to the current search strategy. The relational bias can usually be derived from the database schema automatically, whereas the search bias depends on the particular interest of the user. In most cases, one will want to try different forms of search bias, but keep the same relational bias.

Relational bias The relational bias is specified by expressions of the sort:

```
table(TableName, AttributeList).
types(TableName, TypeList).
association(AssociationName, Table1, Attribute1,
    Table2, Attribute2, MinMultiplicity1,
    MaxMultiplicity1, MinMultiplicity2,
    MaxMultiplicity2).
```

The types can be one of three predefined types, `numeric`, `nominal` and `binary` and determine which operators can be used in conditions involving the attribute. They are not used to determine how tables can be joined.

Search bias The following statements, among others, are part of the search bias:

```
target(Table, Attribute).
```

This statement indicates that the single table, called `Table`, represents the concept to be learned. The (optional) second argument determines that one of attributes in Table is used as the primary target for prediction. Appointing the target in the set of tables determines what type of object is analysed, and thus for example how support and frequency of patterns are computed. Note that positive and negative examples appear in the same table, and can be identified by the value of `Attribute`.

```
direction(Association, Direction).
```

This statement indicates that refinements may only be made in a certain direction along association `Association` rather than both ways.

```
fix(Table, AttributeList).
```

This statement indicates that if table `Table` is introduced in a pattern, all of the attributes provided in `AttributeList` are fixed by conditions on their values.

The CDBL representation of the example introduced in the previous section looks as follows:

```
table(molecule, [mol, log_mutagenicity]).
table(atom, [mol, id, element, type, charge]).
table(bond, [mol, id1, id2, type]).

types(molecule, [nominal, numeric]).
types(atom, [nominal, nominal, nominal, nominal, numeric]).
types(bond, [nominal, nominal, nominal, nominal]).

association(participates1, atom, id, bond, id1, 1, 1, 0, n).
association(participates2, atom, id, bond, id2, 1, 1, 0, n).
association(atomin, atom, mol, molecule, mol, 1, n, 1, 1).
association(bondin, bond, mol, molecule, mol, 0, n, 1, 1).

target(molecule, log_mutagenicity).

fix(bond, [type]).

direction(atomin, left).
direction(bondin, left).
```

4 Translating CDBL

This section describes an algorithm for translating CDBL specifications into language specifications for one of the currently supported engines, the ILP system Tilde [3, 4]. Our translation algorithm for converting CDBL into Tilde specifications consists of four steps:

1. generate **predict** and **tilde_mode** settings
2. generate types for tables
3. generate types and rmodes for standard operators (comparison etc.)
4. generate rmodes, constraints and look-ahead specifications for tables

Step 1

Step 1 is trivial: based on the **target** specification in the CDBL input, the corresponding **predict** setting is generated; based on the type of the argument to be predicted Tilde is set into classification or regression mode.

Step 2

Step 2 is implemented as follows:

```
2.1 for each table T:
          2.1.1 read its type specification types(T, L)
          2.1.2 change each type t in L into a unary
          term with functor t and an anonymous variable
          as argument
2.2 for each association between T1.A1 and T2.A2 :
          2.2.1 unify the types of the corresponding
          attributes
2.3 ground the type specifications by instantiating
each variable to a different constant
```

For the mutagenesis example given above this yield:

```
Types = { molecule(nominal(1), numeric(2),
atom(nominal(1),
          nominal(3), nominal(4), nominal(5),
          numeric(6)), bond(nominal(1), nominal(3),
          nominal(3), numeric(7)) }
```

With these type specifications it is ensured that the same variable can only occur in the place of arguments that have a chain of associations between them. At the same time, information on the original types is still available; this information will be used to decide, e.g., which variables can be compared using numerical comparison operators.

Step 3

Step 3 consists of generating rmodes, types and possibly look-ahead for a number of standard operators (=, <, ...). Based on the **compare** setting in CDBL, one can allow comparisons between variables that a) have the same type; b) represent the same attribute of different instances; c) represent different attributes of the same instance.

For each allowed operator (the user can specify which operators are allowed), rmode and type specifications for comparisons of variables with the same type can be generated in a trivial manner; e.g.,

```
rmode(+X < +Y).                         % argument of <
predicate are input arguments
type(numeric(X) < numeric(X)). % only comparisons
between same type allowed
```

If we want to allow only comparisons between the same attribute of different instances, respectively different attributes of the same instance, this cannot be specified using only type and mode declarations. Tilde provides the possibility to specify constraints of the form **constraint(A,B)** where A is a literal (or conjunction of literals) that is considered for insertion in a certain node in the tree and B is a condition that has to be true in order for the insertion to be valid. For instance, the following code could be used for these constraints:

```
same_attribute(X,Y) :-
        occurs(Lit1), Lit1 =.. [F|Args1],
        strict_member(X, Args1), occurs(Lit2), Lit2
        \== Lit1, Lit2 =.. [F|Args2], strict_member(Y,
        Args2).
in_same_tuple(X,Y) :-
        occurs(Lit), Lit =.. [F|Args],
        strict_member(X, Args), strict_member(Y,
        Args).

constraint(X<Y, same_attribute(X,Y)).
constraint(X<Y, in_same_tuple(X,Y)).

strict_member(A, [B|C])  :- A == B.
strict_member(A, [B|C])  :- strict_member(A, C).
```

Step 4

Step 4 is the most complicated step. The way our implementation works is as follows: for each association, the algorithm looks from between which tables the association runs, and it adds rmodes for the "to" table (and constraints on these rmodes) in accordance with the CDBL specifications (e.g., which arguments are fixed). At this point the algorithm also inspects the properties of the association and based on these possibly adds look-ahead specifications. The algorithm is shown below.

Its results are 1) a set of rmodes; 2) a set of constraints; 3) a set of look-ahead specifications. The constraints can be seen as an array of constraints indexed on tables. For each table the constraint is initialised to *false* and whenever an association in the CDBL specifications indicates that the table can be added if certain conditions are fulfilled, the currently existing constraint on this table is weakened by extending it disjunctively with the new conditions. The "needs look-ahead" test is based on inspection of multiplicities, as indicated earlier. When adding an rmode or a look-ahead specification, it is implicitly checked whether the rmode or look-ahead already occurs (if it does, it is not added again).

The algorithm can be described as follows.

```
Global variables: rmodes: set of rmodes; initialised to
the empty set.
Constraints: array of constraints indexed on tables

For each association Assoc between an attribute A1 of
some table T1 and an attribute A2 of some table T2 that
goes in the right direction:
    add an rmode R for T2
        the argument corresponding to A2 has mode +
        the arguments occurring in a fix list for T2
        have mode #
        all other arguments have mode -
    constraints(T2) := constraints(T2) OR (the
                        variable occurring as A2 in T2
                        already occurs as A1 in a T1
                        literal)
    if T1 needs look-ahead then create look-ahead(T1,
    T2);
    if T2 needs look-ahead then
        add_rmodes+constraints+look-ahead for
            comparing attributes of T2 with constants
    else
        add_rmodes+constraints for comparing
            attributes of T2 with constants
Add_rmodes+constraints[+look-ahead] for comparing
attributes of T with constants:
For each allowed operator op :
        add rmode(+X op #X)
        add a constraint stating that X must already
        occur inside a T literal in the clause
        [add look-ahead allowing test to be added
        immediately after adding T]
```

The above translation algorithm was tested on a few test cases. The results of these experiments suggest that the bias specifications typically generated by the algorithm are significantly more complex than the ones humans would usually write. This additional complexity is in part attributable to the mismatch between relational concepts and logical concepts, as is explained in the next section. It should be noted, though, that the size of the hypothesis space does not increase due to the more complex specifications and thus the complexity of the specification is not harmful to Tilde's performance.

Although the translator described above was designed with the Tilde system in mind, it is also usable for the Warmr ILP system, which uses the same bias specification language as Tilde except for some minor differences (e.g., no target argument is to be specified for the target table).

5   Comparing UML to Traditional Languages

If we want to compare the usefulness of our approach to existing means of bias specification we have to take different criteria into account. Because one of our goals is to widen the audience of ILP by improving its usability, some of these criteria will be somewhat subjective. We will be looking at the ease of use of UML in the context of ILP as well as the comprehensibility of Class Diagrams for non-experts. However we will start our comparison with a more objective aspect which involves the contrast between the actual bias (i.e. the hypothesis space) produced by UML and more traditional languages.

The main difference in actual bias between UML and traditional declarative bias languages is in how sharing of variables is controlled. In UML associations are used to explicitly allow two arguments to share a variable. In other languages this is less strict and any pair taken from a set of arguments with the same type can share a variable. What effect this difference in restriction has, is best demonstrated by considering the simplest case of three classes, a, b and c, connected by two associations both referring to the same argument in the middle class b. Bias on the basis of types would allow both a (X), b (X), c (X) as well as a (X), c (X), whereas UML would only consider the former.

Assume either of the two associations is compulsory with respect to b (multiplicity 1 or 1..*n* on the side of b). In this case the two expressions are logically equivalent and the bias is effectively the same. The use of UML has a slight preference, because only one of the two alternatives is considered, in fact the one closer to our intuition.

If, on the other hand, both associations are optional (0..1 or 0..*n*), then the two expressions are distinct and UML will only allow a (X), b (X), c (X) . If patterns such as a (X), c (X) should be considered we would have to explicitly introduce a third association between a and c.

From the discussion above we can conclude that the use of associations is more explicit than that of types. Associations can express all bias that can be expressed by types, because each sharing of variables of the same type can be listed explicitly. In the meantime more precise and restrictive forms of bias can be declared by stating only the required associations, something which is hard to achieve by types only.

Because UML is such a versatile and widely-used tool for system modeling it needs little arguing that the language is easy to use and comprehensible for technical roles. Comprehensibility will however be lower for people with a non-technical background. This view is supported by an empirical study at British Airways [1], which reported that comprehensibility of Class Diagrams is high for 'Analyst',

'Designer' and 'Programmer' roles. One of these, the 'Domain Expert', is moderate at understanding these diagrams through its exposure to OO-techniques when co-operating with Analysts. In short we can say that UML is a comprehensible tool for the intended audience, although some users may require a short introduction into the syntax.

Because of its unified nature UML is slightly limited in expressiveness for ILP purposes. The types of possible constraints are limited to relational and search bias, whereas dedicated input languages can offer a wider range of engine-specific constraints. Also, many declarative bias languages allow the expression of general constraints on refinements with the full power of Prolog. Although many of the desirable constraints can be derived from the relational bias, as was shown before, our method is clearly restricted in terms of expressing general constraints.

6 Conclusion

In this paper we have introduced a method that facilitates the use of ILP in a KDD setting. Our approach is centered around the use of UML as a declarative bias language for a wide range of existing ILP engines. UML has a number of advantages over existing languages that are specific to particular engines:

- easy to comprehend for a wide range of people
- close relation to relational database modeling
- industry standard for modeling since 1997
- generic specification language for range of ILP algorithms.

Of course the last advantage only holds with sufficient support from the different engines (currently three). However, UML clearly has inherent features, which make it a good candidate for a common declarative bias language.

Because UML can be seen as a common denominator of a range of bias specification languages, it can not be expected to cover every single construct in a particular language. In order to be able to use the full functionality of an engine the native language will still be necessary. The usage of UML will be attractive in those cases where functionality is required, which is shared by most engines. Typically our approach is preferred for business analysts with a clear understanding of the domain at hand, who have had some minimal training in (database) modeling and inductive techniques and intend to apply a range of tools for comparison.

References

1. Arlow, J., Emmerich, W., Quinn, J. *Literate Modelling – Capturing Business Knowledge with the UML,* In proceedings of UML '98, 1998
2. Blaha, M., Premerlani, W. *Object-Oriented Modeling and Design for Database Applications,* Prentice Hall, 1998
3. Blockeel, H., De Raedt, L. *Top-down induction of first-order logical decision trees,* Artificial Intelligence 101 (1-2), pages 285-297. 1998

4. Blockeel, H., Dehaspe, L. *Tilde and Warmr User Manual*, http://www.cs.kuleuven.ac.be/~ml/PS/TWuser.ps.gz. 1999
5. Blockeel, H., De Raedt, L. *Relational knowledge discovery in databases,* Proceedings of the Sixth International Workshop on Inductive Logic Programming, Volume 314 of Lecture Notes in Artificial Intelligence, 1996, Springer Verlag
6. De Raedt, L., Dehaspe, L. *Clausal Discovery,* Machine Learning 26, pages 99-146. 1997
7. De Raedt, L., Van Laer, W. *Inductive Constraint Logic*, In Proceedings of the 6th International Workshop on Algorithmic Learning Theory. Lecture Notes in Artificial Intelligence, vol. 997, pages 80-94. 1995
8. Dehaspe, L. *Frequent Pattern Discovery in First-Order Logic*, Ph.D. thesis, Katholieke Universiteit Leuven, 1998
9. Knobbe, A.J. *Towards Scalable Industrial Implementations of ILP,* ILPNet2 Seminar on ILP & KDD, Caminha, Portugal, 1998
10. Knobbe, A.J., Blockeel, H., Siebes, A., Van der Wallen, D.M.G. *Multi-Relational Data Mining,* In Proceedings of Benelearn '99, 1999
11. Knobbe, A.J., Siebes, A., Van der Wallen, D.M.G. *Multi-Relational Decision Tree Induction,* In Proceedings of PKDD '99, Prague, Czech Republic, September 1999
12. Kramer, S. *Structural regression trees*, In Proceedings of the 13th National Conference on Artificial Intelligence, pages 812-819. AAAI Press / The MIT Press. 1996
13. Muggleton, S. *Inverse entailment and Progol*, New Generation Computing 13, 1995
14. Nédellec, C., Rouveirol, C., Adé, H., Bergadano, F. and Tausend, B. *Declarative bias in ILP*, Advances in Inductive Logic Programming, pages 82-103. IOS Press, 1996
15. Object Management Group, 492 Old Connecticut Path, Framingham, Mass. *UML Notation Guide,* ad/97-08-05 edition, Nov 1997
16. Rumbaugh, J., Booch, G., Jacobson, I. *Unified Modeling Language Reference Manual,* Addison Wesley, 1998
17. Si Alhir, S. *UML in a Nutshell*, O'Reilly & Associates, Inc., 1998
18. Srinivasan, A., Muggleton, S., Sternberg, M.J.E. and King, R.D. *Theories for mutagenicity: a study in first-order and feature-based induction*, Artificial Intelligence 85, 1996
19. Wrobel, S. *An algorithm for multi-relational discovery of subgroups,* In Proceedings of Principles of Data Mining and Knowledge Discovery (PKDD '97), 78-87, 1997

An Apriori-Based Algorithm for Mining Frequent Substructures from Graph Data

Akihiro Inokuchi*, Takashi Washio, and Hiroshi Motoda

I.S.I.R., Osaka University
8-1, Mihogaoka, Ibarakishi, Osaka, 567-0047, Japan
Phone: +81-6-6879-8541 Fax: +81-6-6879-8544
washio@sanken.osaka-u.ac.jp

Abstract. This paper proposes a novel approach named AGM to efficiently mine the association rules among the frequently appearing substructures in a given graph data set. A graph transaction is represented by an adjacency matrix, and the frequent patterns appearing in the matrices are mined through the extended algorithm of the basket analysis. Its performance has been evaluated for the artificial simulation data and the carcinogenesis data of Oxford University and NTP. Its high efficiency has been confirmed for the size of a real-world problem. . . .

1 Introduction

Mining knowledge from structured data is a major research topic in recent data mining study. "Graph structure" is one of the representatives of the structured data since it frequently appears in real-world data such as web links and chemical compounds structures. In the field of chemistry, CASE and MultiCASE systems have been often used to discover characteristic substructures of chemical compounds [8], [9]. Though these systems can efficiently find the substructures, the class of the substructures is limited to the no-branching atom sequences. Wang and Liu proposed the mining of wider class of substructures which are subtrees called schemas [14]. Though the proposed algorithm is very efficient to mine frequent schemas from massive data, the mining patterns are still limited to acyclic graphs. To mine characteristic patterns having general graph structures, the propositional classification techniques, e.g., C4.5, the regression tree techniques, e.g., M5, and the inductive logic programming (ILP) techniques have been applied in the carcinogenesis predictions of chemical compounds [10], [7]. However, these approaches can discover only limited types of characteristic substructures, because the graph structures must be pre-characterized by some specific features and/or ground instances of predicates.

Recently, a technique to mine the frequent substructures characterizing the carcinogenesis of chemical compounds has been proposed without requiring any conversion of substructures to specific features by Dehaspe et al. [3]. They used

---

* Currently beeing in Tokyo Research Institute, IBM, 1623-14 Shimotsuruma, Yamatoshi, Kanagawa, 242-8502, Japan.

D.A. Zighed, J. Komorowski, and J. Żytkow (Eds.): PKDD 2000, LNAI 1910, pp. 13–23, 2000.
© Springer-Verlag Berlin Heidelberg 2000

the ILP framework combined with levelwise search to minimize the access frequency to the database [11]. Since the efficiency achieved by this approach is much better than the former ILP approaches, some new discovery of substructures characterizing carcinogenesis was expected. However, the full search space was still so large that the search had to be limited within the 6th level where the substructures are represented with 6 predicates at maximum, and they reported that significant substructures have not been obtained within the search level. Some other researches have also developed the techniques to mine the frequent substructures in graph data. The graph-based induction (GBI) is an approach to seek the frequent patterns by iteratively chunking the vertex pairs that frequently appear [12]. SUBDUE is another approach to seek the characteristic graph patterns to efficiently compress the original graph in terms of MDL principle [2]. These approaches do not face the severe computational complexity. However, they may miss some significant patterns, since their search strategies are greedy.

Though the task tackled by these works involves the problem of deciding graph isomorphism which is known to be NP, each work mines some characteristic graph substructures by introducing the limitations on the search space and/or the class of substructures. The objective of this paper is 1) to propose a novel approach named as "Apriori-based Graph Mining", AGM for short, to mine the frequent substructures and the association rules from the general class of graph structured data in a more efficient manner than the preceding work, and 2) to assess the performance of the approach for the artificially simulated data and also for the carcinogenesis data of Oxford University and National Toxicological Program (NTP) [13].

2   Principle of Mining Graph Substructures

The methods studied in the mathematical graph isomorphism problem are not directly applicable to our case, because the methods are only to check if the two given graphs are isomorphic [4]. We introduce the mathematical graph representation of "adjacency matrix" and to combine it with an efficient levelwise search of the frequent canonical matrix code [5]. The levelwise search is based on the extension of the Apriori algorithm of the basket analysis [1].

2.1   Representation of Graph Structures

A graph in which the vertices and edges have labels is mathematically defined as follows.

Definition 1 (Graph having Labels) *Given a set of vertices $V(G) = \{v_1, v_2, ..., v_k\}$, a set of edges connecting some vertex pairs in $V(G)$; $E(G) = \{e_h = (v_i, v_j)|v_i, v_j \in V(G)\}$, a set of vertex labels $L(V(G)) = \{lb(v_i)|\forall v_i \in V(G)\}$ and a set of edge labels $L(E(G)) = \{lb(e_h)|\forall e_h \in E(G)\}$, then a graph G is represented as*

$$G = (V(G), E(G), L(V(G)), L(E(G))).$$

This graph G is represented by an adjacency matrix X which is a very well known representation in mathematical graph theory [4]. This transformation from G to X does not require much computational effort.

Definition 2 (Adjacency Matrix) *Given a graph* $G = V(G), E(G), L(V(G)),$ $L(E(G)))$, *the adjacency matrix* X *has the following* (i, j)-*element,* x_{ij},

$$x_{ij} = \begin{cases} num(lb) \; ; e_h = (v_i, v_j) \in E(G) \text{ and } lb = lb(e_h) \\ 0 \quad\quad\; ; (v_i, v_j) \notin E(G) \end{cases},$$

where $num(lb)$ *is an integer arbitrarily assigned to a label value* lb. *Moreover, a number* $num(lb)$ *is assigned to the* i-*th low* (i-*th column) of the matrix where* $v_i \in V(G)$ *and* $lb = lb(v_i)$.

Definition 3 (Size of a Graph) *The "size" of a graph* G *is the number of vertices in* $V(G)$, *i.e.,* k *in Definition 1.*

Definition 4 (Graph Transaction and Graph Data) *A graph* $G = (V(G),$ $E(G), L(V(G)), L(E(G)))$ *is a transaction, and graph data* GD *is a set of the transactions, where* $GD = \{G_1, G_2, ..., G_n\}$.

Each element of an adjacency matrix in the standard definition is either '0' or '1', whereas each element in Definition 2 can have the number of an edge label. This extended notion of the adjacency matrix gives a compact representation of a graph having labeled edges, and enables an efficient coding of the graph as shown later.

The representation of the adjacency matrix depends on the assignment of each vertex to the i-th row (i-th column). To reduce the variants of the representations and increase the efficiency of the code matching described later, the vertices are sorted according to the numbers of their labels. The adjacency matrix of a graph whose size is k is noted as X_k, and the graph as $G(X_k)$.

Definition 5 (Vertex-sorted Adjacency Matrix) *The adjacency matrix* X_k *of the graph* $G(X_k)$ *is vertex-sorted if*

$$num(lb(v_i)) \leq num(lb(v_{i+1})) \text{ for } i = 1, 2, ..., k - 1.$$

In the standard basket analysis, items within an itemset are kept in lexicographic order [1]. This enables an efficient control of the generation of candidate itemsets. However, the vertex-sorted adjacency matrices do not have such lexicographic order. Thus, a coding method of the adjacency matrices need to be introduced.

Definition 6 (Code of Adjacency Matrix) *In case of an undirected graph, the code* $code(X_k)$ *of a vertex-sorted adjacency matrix* X_k;

$$X_k = \begin{pmatrix} x_{1,1} & x_{1,2} & x_{1,3} & \cdots & x_{1,k} \\ x_{2,1} & x_{2,2} & x_{2,3} & \cdots & x_{2,k} \\ x_{3,1} & x_{3,2} & x_{3,3} & \cdots & x_{3,k} \\ \vdots & \vdots & \vdots & \ddots & \vdots \\ x_{k,1} & x_{k,2} & x_{k,3} & \cdots & x_{k,k} \end{pmatrix},$$

is defined as

$$code(X_k) = x_{1,1}x_{1,2}x_{2,2}x_{1,3}x_{2,3}x_{3,3}x_{1,4}\cdots x_{k-1,k}x_{k,k},$$

where the digits are obtained by scanning the elements along the columns at the upper triangular part of X_k. In case of a directed graph, it is defined as

$$code(X_k) = x_{1,1}x_{1,2}x_{2,1}x_{2,2}x_{1,3}x_{3,1}x_{2,3}x_{3,2}\cdots x_{k-1,k}x_{k,k-1}x_{k,k},$$

where the digits are obtained similarly to the undirected case, but the diagonally symmetric element x_{ji} is added after each x_{ij} when $i \neq j$.

The method proposed in this paper discovers substructures frequently appearing in the graph transaction data GD. The rigorous definition of the substructure is given as follows.

Definition 7 (Induced Subraph) *Given a graph $G = (V(G), E(G), L(V(G)), L(E(G)))$, an induced subgraph of G, $G_s = (V(G_s), E(G_s), L(V(G_s)), L(E(G_s)))$, is a graph satisfying the following conditions.*

$$V(G_s) \subset V(G), E(G_s) \subset E(G),$$

$$\forall u, v \in V(G_s), (u, v) \in E(G_s) \Leftrightarrow (u, v) \in E(G).$$

When G_s is an induced subgraph of G, it is denoted as $G_s \subset G$.

2.2 Algorithm of AGM

Candidate Generation. The two indices which are identical to the definitions of "support" and "confidence" in the basket analysis are introduced.

Definition 8 (Support and Confidence) *Given a graph G_s, the support of G_s is defined as*

$$sup(G_s) = \frac{\text{number of graph transactions } G \text{ where } G_s \subset G \in GD}{\text{total number of graph transactions } G \in GD}.$$

Given two induced subgraphs G_b and G_h, the confidence of the association rule $G_b \Rightarrow G_h$ is defined as

$$conf(G_b \Rightarrow G_h) = \frac{\text{number of graphs } G \text{ where } G_b \cup G_h \subset G \in GD}{\text{number of graphs } G \text{ where } G_b \subset G \in GD}.$$

If the value of $sup(G_s)$ is more than a threshold value $minsup$, G_s is called as a "frequent induced subgraph".

Similarly to the Apriori algorithm, the candidate generation of the frequent induced subgraph is made by the levelwise search in terms of the size of the subgraph. Let X_k and Y_k be vertex-sorted adjacency matrices of two frequent induced graphs $G(X_k)$ and $G(Y_k)$ of size k. If both $G(X_k)$ and $G(Y_k)$ have equal elements of the matrices except for the elements of the k-th row and the k-th column, then they are joined to generate Z_{k+1}.

$$X_k = \begin{pmatrix} X_{k-1} & \boldsymbol{x}_1 \\ \boldsymbol{x}_2^T & x_{kk} \end{pmatrix}, Y_k = \begin{pmatrix} X_{k-1} & \boldsymbol{y}_1 \\ \boldsymbol{y}_2^T & y_{kk} \end{pmatrix},$$

$$Z_{k+1} = \begin{pmatrix} X_{k-1} & \boldsymbol{x}_1 & \boldsymbol{y}_1 \\ \boldsymbol{x}_2^T & x_{kk} & z_{k,k+1} \\ \boldsymbol{y}_2^T & z_{k+1,k} & y_{kk} \end{pmatrix} = \left(\begin{array}{cc|c} & & \boldsymbol{y}_1 \\ & X_k & \\ & & z_{k,k+1} \\ \hline \boldsymbol{y}_2^T & z_{k+1,k} & y_{kk} \end{array} \right), \tag{1}$$

where X_{k-1} is the adjacency matrix representing the graph whose size is $k-1$, $\boldsymbol{x}_i$ and $\boldsymbol{y}_i (i = 1, 2)$ are $(k-1) \times 1$ column vectors. X_k is called the "first matrix" and Y_k the "second matrix". The following relations hold among the vertex-sorted adjacency matrices X_k, Y_k and Z_{k+1}.

$$lb(v_i; v_i \in V(G(X_k))) = lb(v_i; v_i \in V(G(Y_k))) = lb(v_i; v_i \in V(G(Z_{k+1}))),$$

$$lb(v_i; v_i \in V(G(X_k))) \le lb(v_{i+1}; v_{i+1} \in V(G(X_k))),$$

$$lb(v_k; v_k \in V(G(X_k))) = lb(v_k; v_k \in V(G(Z_{k+1}))), \tag{2}$$

$$lb(v_k; v_k \in V(G(Y_k))) = lb(v_{k+1}; v_{k+1} \in V(G(Z_{k+1}))),$$

$$lb(v_k; v_k \in V(G(X_k))) \le lb(v_k; v_k \in V(G(Y_k))).$$

Here, $i = 1, \cdots, k - 1$. $z_{k,k+1}$ and $z_{k+1,k}$ are not determined by X_k and Y_k. Each can take every integer value $num(lb)$ corresponding to each edge label lb or 0 corresponding to the case that no edge exists between v_k and v_{k+1}. In case of an undirected graph, $z_{k,k+1}$ and $z_{k+1,k}$ must have an identical value. This join procedure of X_k and Y_k creates multiple Z_{k+1}s for all possible value pairs of $z_{k,k+1}$ and $z_{k+1,k}$. Note that when the labels of the k-th vertices v_k of $G(X_k)$ and $G(Y_k)$ are the same, exchanging X_k and Y_k (i.e., taking Y_k as the first matrix and X_k as the second matrix), produces redundant adjacent matrices. In order to avoid this redundant generation, the two adjacency matrices are joined only when Eq.(3) is satisfied. The vertex-sorted adjacency matrix generated under this condition is called a "normal form".

$$code(\text{the first matrix}) \le code(\text{the second matrix}) \tag{3}$$

In the standard basket analysis, the $(k + 1)$-itemset becomes a candidate frequent itemset only when all the k-sub-itemsets are confirmed to be frequent itemsets. Similarly, the graph G of size $k + 1$ is a candidate of frequent induced subgraphs only when all adjacency matrices generated by removing from the graph G the i-th vertex v_i $(1 \le i \le k + 1)$ and all its connected links are confirmed to be frequent induced subgraphs of the size k. As this algorithm generates only adjacency matrices of the normal form in the earlier (smaller) k-levels, if the adjacency matrix of the graph generated by removing the i-th vertex v_i is non-normal form, it must be transformed to a normal form to check if it matches one of the normal form matrices found earlier. An adjacency matrix X_k of a non-normal form is transformed into a normal form X_k' by reconstructing the matrix structure in a bottom up manner. First, an adjacency matrix of the size 1×1 is set for each vertex $v_i \in G(X_k)$. Then, the pair of the matrices for the vertices $v_i, v_j \in G(X_k)$ satisfying the constraints of Eq.(2) and (3) are joined by the operation of Eq.(1). At this time, the values of the elements for (v_i, v_j) and (v_j, v_i) in the original X_k are substituted to the non-diagonal elements $z_{1,2}$ and $z_{2,1}$ respectively to reconstruct the structure of $G(X_k)$. Subsequently, the pair of the obtained 2×2 matrices are further joined according to the constraints of Eq.(1), (2) and (3). The values of the elements $z_{2,3}$ and $z_{3,2}$ are determined from X_k in the similar manner. This procedure is repeated until a $k \times k$ matrix X'_k is obtained. Because X'_k precisely reflects the structure of $G(X_k)$, and is construc-ted by following the constraints, X'_k is a normal form of X_k. This reconstruction is called "normalization". In the intermediate levels, the normal forms of all in-duced subgraphs of $G(X_k)$ can be derived. This feature of the normalization is used in the frequency calculation explained latter. The normalization consists of the set of permutations of the rows and the columns of the original matrix X_k. Thus X'_k has the relation of $X'_k = (T_k)^T X_k T_k$ where T_k is the transformation matrix. The details of the normalization can be found in [6].

Canonical From. After all candidate induced subgraphs are derived, the sup-port value of each candidate is counted in the database. However, the normal form representation is in general not unique for a graph. For instance, the follo-wing two matrices which are both normal forms represent an identical undirected graph with a unique link label.

$$X_3 = \begin{pmatrix} 0 & 0 & 1 \\ 0 & 0 & 1 \\ 1 & 1 & 0 \end{pmatrix}, Y_3 = \begin{pmatrix} 0 & 1 & 1 \\ 1 & 0 & 0 \\ 1 & 0 & 0 \end{pmatrix}.$$

If the support value is counted for each representation independently, it has to be summed up to obtain the correct support value for the corresponding graph. To perform this summation efficiently, all normal forms for an identical induced subgraph must be indexed. For this purpose, canonical form is defined for normal forms of adjacency matrices representing an identical induced subgraph, and an efficient method to index each normal form to its canonical form is introduced.

Definition 9 (Canonical Form) *Given a set $NF(G)$ of all normal forms of adjacency matrices representing an identical graph G, its canonical form X_c is defined as X having the minimum code number in $NF(G)$, i.e.,*

$$X_c = arg \min_{X \in NF(G)} code(X).$$

We assume that all the transformation matrices S_{k-1} to the canonical form from the normal forms of every frequent induced subgraph of size $k-1$ are known. Let X^m_{k-1} be the matrix obtained by removing the m-th vertex v_m $(1 \le m \le k)$ from $G(X_k)$. X^m_{k-1} is transformed to one of its normal forms, X'^m_{k-1}, by the afo-rementioned normalization, and thus its transformation matrix T^m_{k-1} is known. Furthermore, let S_{k-1} of X'^m_{k-1} be S^m_{k-1}, then the transformed canonical form is represented by $(T_{k-1}^m S_{k-1}^m)^T X_{k-1}^m T_{k-1}^m S_{k-1}^m$. The canonical form X_{ck} of X_k and the matrices S_k^m, T_k^m to transform X_k to X_{ck} are obtained from S_{k-1}^m, T_{k-1}^m by the following expressions. The detailed proof of this transformation can be found in [6].

$$s_{ij} = \begin{cases} s_{ij}^m & 0 \le i \le k-1 \text{ and } 0 \le j \le k-1, \\ 1 & i = k \text{ and } j = k, \\ 0 & \text{otherwise,} \end{cases}$$

$$t_{ij} = \begin{cases} t_{ij}^m & i < m \text{ and } j \ne k, \\ t_{i-1,j}^m & i > m \text{ and } j \ne k, \\ 1 & i = m \text{ and } j = k, \\ 0 & \text{otherwise,} \end{cases}$$

$$X_{ck} = arg \min_{m=1,\cdots,k} code((T_k^m S_k^m)^T X_k (T_k^m S_k^m)),$$

where s_{ij}, s_{ij}^m, t_{ij} and t_{ij}^m are the elements of matrix S_k^m, S_{k-1}^m, T_k^m and T_{k-1}^m respectively. $T_k^m S_k^m$ which minimize the code is S_k of X_k.

Frequency Calculation. Frequency of each candidate induced subgraph is counted by scanning the database after generating all the candidates of frequent induced subgraphs and obtaining their canonical forms. Every transaction graph G in the database can be represented by an adjacency matrix X_k, but it may not be a normal form in most cases. Since the candidates of frequent induced subgraphs are normal forms, the normalization must be applied to X_k of each transaction G to check if the candidates are contained in G. As previously described, the procedure of the normalization of X_k can derive the normal form of every induced subgraph of G in the intermediate levels. Thus, the frequency of each candidate is counted based on all normal forms of the induced subgraphs of G. When the value of the count exceeds the threshold $minsup$, the subgraph is a frequent induced subgraph. Once all frequent induced subgraphs are found, the association rules among them whose confidence values are more than a given confidence threshold are enumerated by using the algorithm similar to the standard basket analysis.

3 Performance Evaluation

The performance of the proposed AGM was examined using an artificial graph transaction data. The machine used is a PC with 400MHz CPU and 128MB main memory. The size of each transaction is determined by the gaussian distribution with the average $|T|$ and the standard deviation 1. The vertex labels are randomly determined with equal probability. The edges are attached randomly with the probability of p. L basic patterns of the average size $|I|$ are generated, and one of them is randomly overlaid on each transaction. The two groups of the test data, one for the directed graph and the other for the undirected graph, are prepared. The direction of the edges are given randomly in the former group.

Figures 1, 2, 3 and 4 show the results of computation time for different number of transactions, number of vertex labels, minimum support threshold and average transaction size for both directed and undirected graphs, respectively. In every parameter setting, the required computational time and the number of the discovered frequent induced subgraphs are less in the case of directed graph. Because the number of possible subgraph patterns is larger due to existence of edge direction, the frequency of each subgraph pattern is smaller. This also reduces the required computation for search. In short summary, the proposed algorithm does not show intractable computational complexity except the cases for graphs of large size in the database.

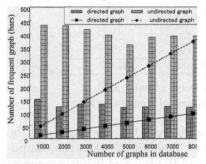

Fig. 1. Complexity v.s. number of transactions.

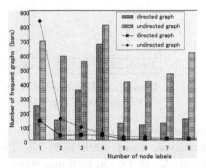

Fig. 2. Complexity v.s. number of vertex labels.

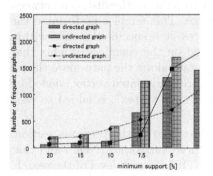

Fig. 3. Complexity v.s. minimum support.

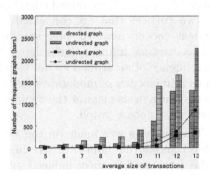

Fig. 4. Complexity v.s. transaction size.

4   Application to Chemical Analysis

AGM was applied to chemical carcinogenesis analysis which is a challenge topic proposed in IJCAI-97 by Srinisavan et al. [13]. The task is to find structures typical to carcinogen of organic chlorides. The objective data were obtained from the website of National Toxicology Program (NTP) and Oxford University. Totally, the 300 compounds were selected for the analysis, of which 185 compounds

Table 1. Results for three *minsup* values.

L	minsup = 20% NOC	NOFS	minsup = 15% NOC	NOFS	minsup = 10% NOC	NOFS
1	24	7	24	8	24	10
2	280	62	360	67	550	108
3	2277	477	2525	640	4558	964
4	6223	2178	9709	3333	18268	5912
5	9767	4806	18740	9372	40744	19568
6	6899	4726	19813	13479	56179	37219
7	2655	2179	11989	9499	52082	41639
8	668	655	4347	4019	33208	29817
9	118	118	1212	1199	15618	15242
10	7	7	220	220	5739	5725
11	-	-	21	21	1455	1455
12	-	-	1	1	23	23
13	-	-	-	-	15	15
Total	28918	15215	68961	41858	228663	157897

L:level(number of vertices included in frequent subgraph)
NOC:number of candidates, NOFS:number of frequent graphs

have positive carcinogenesis and the rests are negative. Thus, the fraction of the carcinogenic compounds is 61.7%. The types of atoms involved in the compounds are C, H, O, Cl, F, S and some cations, and the types of bonds are single, double, aromatic and cation bonds. Each transaction data were preprocessed to add artificial edges from each vertex to every other vertex that is within the distance of 6 edges. Each added edge has a label to indicate the distance between the two vertices that are connected by the edge. This enables us to mine the frequent cooccurrence of some specific structures at a specific distance within 6. The distance limit of 6 was determined based on the chemical insights that the influence of an atom does not usually propagate along the path more than 6 bonds in molecules of moderate sizes. Furthermore, an isolated vertex labeled by the carcinogenesis class of the compound, *i.e.*, "class vertex", is added to each chemical structure graph.

The analysis was made on the same PC described in the previous section. Table 1 shows the number of the candidate induced subgraphs (NOC) and that of the discovered frequent induced subgraphs (NOF) for each level of the search, *i.e.*, the size of the induced subgraphs. In each *minsup* case, all frequent induced subgraphs were exhaustively discovered. The computation time required to complete the search was far longer for the *minsup* of smaller value, and was almost 8 days for 10%, while it was only about 40 minutes for 20%. The size of the largest frequent induced subgraph discovered in the case of 10% was 13.

In Figure 5, the confidence deviation Δ of an association rule $G_b \Rightarrow G_h$ is given as follows.

$$
\Delta = \begin{cases} conf(G_b \Rightarrow G_h) - fr_p & \text{if } G_h \text{ contains a positive class vertex.} \\ conf(G_b \Rightarrow G_h) - fr_n & \text{if } G_h \text{ contains a negative class vertex.} \end{cases}
$$

Here, fr_p is the fraction of positive compounds in the data, *i.e.*, 61.7% in this case, and fr_n is that of negative compounds, i.e., 38.3% (=100%-61.7%). The cover rate CR of a set of association rules is the fraction of chemical compounds whose classes are derived by applying the rule set to the data. Given a value of Δ_{th}, a set of association rules each having Δ more than the Δ_{th} is defined, and CR of the rule set is calculated. As shown in Figure 5, the rule set derived for the 10% threshold contains some rules having significant confidence. Accordingly, the exhaustive search for low support threshold is considered to be very effective to mine valuable rules.

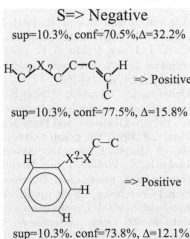

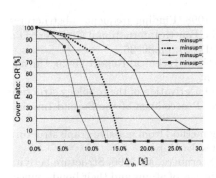

Fig. 5. Relation of Δ_{th} and CR.   **Fig. 6.** Examples of discovered rules.

Figure 6 shows some association rules obtained for the carcinogenesis class under the support threshold 10%. The first rule is very simple, but indicates that a sulfur atom plays an important role to suppress the carcinogenesis. In the second rule, the symbol X of a vertex and ? of an edge indicate that their labels are arbitrary. The third is an example of a less significant but more complex substructure involving a benzene ring. This is consistent with the chemical knowledge that benzene rings frequently have the positive carcinogenesis.

5   Discussion and Conclusion

The largest graphs of the chemical compound discovered by AGM have the size of 13 atoms. In contrast, the approach of ILP in conjunction with a levelwise search proposed by Dehaspe et al. could mine the substructure consisting of 6 predicates at maximum equivalent to the size of a molecule consisting of only 3 atoms or so [3]. This fact shows the practical efficiency of AGM for real world problems. Further investigation on the computational efficiency of AGM in terms of the theoretical aspect remains for the future study.

In conclusion, a novel approach has been developed that can efficiently mine frequently appearing induced subgraphs in a given graph data set and the association rules among the frequent induced subgraphs. Its performance has been evaluated for both the artificial simulation data and the real world chemical carcinogenesis data. The powerful performance of this approach under some practical conditions has been confirmed through these evaluations.

Acknowledgement. The authors wish to thank Prof. Takashi Okada in Center for Information & Media Studies, Kwansei Gakuin University for providing us with the chemical compound data and his expertise in the chemistry domain.

References

1. Agrawal, R. and Srikant, R. 1994. Fast algorithms for mining association rules. In Proc. of the 20th VLDB Conference, pp.487–499.
2. Cook, D.J. and Holder, L.B. 1994. Substructure Discovery Using Minimum Description Length and Background Knowledge, Journal of Artificial Intelligence Research, Vol.1, pp.231-255.
3. Dehaspe, L., Toivonen, H. and King, R.D. 1998. Finding frequent substructures in chemical compounds. In Proceedings of the Fourth International Conference on Knowledge Discovery and Data Mining (KDD-98), pp.30–36.
4. Fortin, S. 1996. The graph isomorphism problem. Technical Report 96-20, University of Alberta, Edomonton, Alberta, Canada.
5. Inokuchi, A., Washio, T. and Motoda, H. 1999. Derivation of the topology structure from massive graph data. Discovery Science: Proceedings of the Second International Conference, DS'99, pp.330–332.
6. Inokuchi, A. 2000. The study on a fast mining method from massive graph structure data. Master thesis (in Japanese), I.S.I.R., Osaka Univ.
7. King, R., Muggleton, S., Srinivasan, A. and Sternberg, M. 1996. Structure-activity relationships derived by machine learning; The use of atoms and their bond connectives to predict mutagenicity by inductive logic programming. In Proceedings of the National Academy of Sciences, Vol.93, pp.438-442.
8. Klopman, G. 1984. Artificial intelligence approach to structure activity studies. J. Amer. Chem. Soc., Vol.106, pp.7315-7321.
9. Klopman, G. 1992. MultiCASE 1. A hierarchical computer automated structure evaluation program, QSAR, Vol.11, pp.176–184.
10. Kramer, S., Pfahringer, B. and Helma, C. 1997. Mining for causes of cancer: Machine learning experiments at various levels of detail. In Proceedings of the Third International Conference on Knowledge Discovery and Data Mining (KDD-97), pp.223–226.
11. Mannila, H. and Toivonen, H. 1997. Levelwise search and borders of theories in knowledge discovery. Data Mining and Knowledge Discovery, Vol.1, No.3, pp.241-258.
12. Matsuda, T., Horiuchi, T., Motoda, H. and Washio, T. 2000. Extension of Graph-Based Induction for General Graph Structured Data. In Proceedings of the Fourth Pacific-Asia Conference of Knowledge Discovery and Data Mining (PAKDD2000), pp.420–431.
13. Srinisavan, A., King, R.D., Muggleton, S.H. and Sternberg, M.J.E. 1997. The predictive toxicology evaluation challenge. In Proceedings of the Fifteenth International Joint Conference on Artificial Intelligence (IJCAI-97), pp.4–9.
14. Wang, K. and Liu, H. 1997. Schema discovery for semistructured data. In Proceedings of the Third International Conference on Knowledge Discovery and Data Mining (KDD-97), pp.271–274.

Basis of a Fuzzy Knowledge Discovery System

Maurice Bernadet

IRIN / Ecole Polytechnique de l'Université de Nantes
Rue Christian Pauc, La Chantrerie
BP 60601, 44306 Nantes Cedex 3, France
Tel: (+33 2)/(02) 40 68 30 00
Fax: (+33 2)/(02) 40 68 30 66
E-mail: Maurice.Bernadet@irin.univ-nantes.fr

Abstract. Considering a fuzzy knowledge discovery system we have realized we describe here the main features of such systems. First, we consider possible methods to define fuzzy partitions on numerical attributes in order to replace continuous or symbolic attributes by fuzzy ones. We explain then how to generalize statistical indexes to evaluate fuzzy rules, detailing a special index, the intensity of implication and its generalization to fuzzy rules. We describe then one algorithm use to extract fuzzy rules. Since many fuzzy operators are available, we propose a method to choose one fuzzy conjunction, one fuzzy implication and one fuzzy aggregation, and we explain how this choice may be validated by comparing the results of the Generalized Modus Ponens applied on the premises of the examples to the effective conclusions in the database. To reduce the important number of fuzzy rules extracted, we consider also some methods to aggregate fuzzy rules, showing that usage of classical reduction schemes requires specific choices of fuzzy operators.

1 Introduction

Knowledge discovery has been defined as "The non-trivial process of identifying valid, novel, potentially useful, and ultimately understandable patterns in data" [9]. To realize a set of knowledge discovery tools, one has some major choices to achieve: the discovery process may be supervised or not, knowledge representation may use decision trees, association rules, neural networks, ... Having to design such tools, we have chosen, for simplicity reasons, a supervised process and a knowledge representation by rules. However, considering that rules using intervals on continuous attributes are often difficult to interpret and that strict thresholds are often too abrupt, we have decided to use fuzzy logics.

Fuzzy logics may be considered as extensions of multivalued logics, allowing usage of intermediate truth-values between false and true [18]. They allow expression of knowledge in a more natural way that classical binary logics, using graduated attributes as in "X is rather high" (for "X is high" is rather true) ... Fuzzy logics offer many logical operators [12], which permits a good expressiveness of various knowledge forms. In this paper, we detail the primary operations needed to extract fuzzy knowledge from a database.

First, usage of fuzzy logics in knowledge discovery needs to convert numerical attributes to their fuzzy representations; for this, it is necessary to define for each

D.A. Zighed, J. Komorowski, and J. Zytkow (Eds.): PKDD 2000, LNAI 1910, pp. 24–33, 2000.
© Springer-Verlag Berlin Heidelberg 2000

classical attribute, a mapping from its possible values to a set of truth-values for each fuzzy attribute. This mapping is often realized by a fuzzy partition, or rather by a fuzzy pseudo partition, and it is then possible to translate classical attributes by valuations on their fuzzy correspondents. These operations are called fuzzification.

To extract rules from a database, one needs to evaluate each possible rule in order to establish which rules must be kept; for this purpose, many indexes are available, of which we have only retained three indexes: the confidence of a rule, its support and a less usual index, called the intensity of implication. After recalling the principles of these indexes, we expose how they can be evaluated in fuzzy logics.

It is then possible to use a knowledge extraction algorithm using the same principles as in classical logics. Our algorithm is an exploratory search in a tree of possible rules, with evaluation of each rule. Fuzzy logics also allow specific methods, based on genetic algorithms, to search the most representative set of weights for a general set of fuzzy rules, but we will not consider this possibility here.

Several evaluations of the fuzzy logical operators are possible. If one only wants to extract rules for a human expert, the nature of the operators does not matter, but if these rules are to be processed by an expert system, a choice of fuzzy operators is necessary. To find the most adequate set of fuzzy operators, we expose a justified restriction to only four possible sets of fuzzy operators and we give a method to find amongst them, the more consistent with a database.

To reduce the huge set of rules that can then be extracted, one may want to use classical reduction schemes. We show that to be valid in a fuzzy logic, classical reduction schemes need specific choices of fuzzy operators. We conclude by recalling the interest to use fuzzy attributes instead of numerical intervals for continuous attributes in the database, and by considering some possible improvements of the systems we have described.

2 The Process of "Fuzzification"

2.1 Definition of Fuzzy Partitions

Let us first recall that fuzzy logics evaluate the truth-value of a fuzzy proposition "X is A", as the degree to which X belongs to the fuzzy set A: $Truth(\text{"}X \text{ is } A\text{"}) = \mu_A(X)$, $\mu_A(X)$ being the membership (or characteristic) function of the fuzzy set A.

Fuzzy sets allow definition of fuzzy C-partitions or "pseudo partitions" in which each value of a continuous attribute may be classified into several fuzzy classes, with a total membership of 1. These fuzzy pseudo partitions allow conversion of continuous attributes into fuzzy ones, giving then the truth-value of fuzzy propositions. For a continuous attribute CA, varying from minCA to maxCA, one can define a fuzzy pseudo partition several ways ([5], [13]).

The simplest method divides the interval [minCA, maxCA] in n sub-intervals, with a small percentage of coverage between two adjacent ones, and to give each sub-interval a symbolic name related to their position; for instance one may divide the interval [minCA, maxCA] in 5 sub-intervals with an overlap of about 20%, giving then 5 fuzzy attributes, attributes such as: *strong negative, rather negative, medium, rather positive and strong positive* (Fig. 1) .

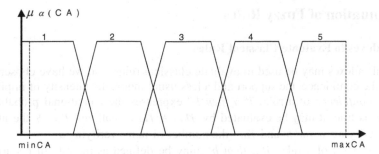

Fig. 1. A fuzzy C-partition (α =1, strong negative; α =2, rather negative; α =3, medium; α =4, rather positive; α =5, strong positive).

The fuzzy classes may also be defined by the experts; otherwise one may propose 3 or 5 classes as standard options. Different numbers of classes may also be used, but a too high number of classes risks to heavily slow down the knowledge discovery process.

Another kind of method extracts the number of classes and defines the fuzzy C-classes from the database. It considers the values of the attributes giving the same conclusion and, when possible, cluster these values in the same fuzzy sets, with a membership value equal to the rate of samples that give this conclusion. These methods often use histograms of the values of the attributes for each possible conclusion. Moreover, it is possible to conceive a more satisfactory method by generalizing to fuzzy logics optimal discretization methods such as those studied in [20].

2.2 Fuzzification of a Database

Once the fuzzy classes have been defined for each continuous attributes, one may convert the related value of each item, by mapping this value to the membership values of each fuzzy class defined for the corresponding classical attribute (Fig. 2).

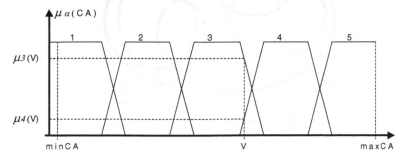

Fig. 2. Mapping from the value V of the continuous attribute CA into membership values of fuzzy attributes (here, only $\mu 3$ and $\mu 4$ are non zero).

3 Evaluation of Fuzzy Rules

3.1 Indexes to Evaluate Classical Rules

Several indexes may be used to evaluate classical rules, and we have chosen three of them: the confidence, the support and a less usual index, the intensity of implication.

The confidence of a rule *"if a then b"* expresses the conditional probability of b when a is true; it may be evaluated by $n_{a \wedge b} / n_a$, calling $n_{a \wedge b}$ the number of items verifying *a and b* and n_a the number of items verifying *a*.

The support of a rule *"if a then b"* may be defined as the rate of occurrences of items verifying *a* and *b* related to all items of the database; so, the support of the rule is $n_{a \wedge b} / n_E$, with $n_{a \wedge b}$ the number of items verifying *a and b* and n_E the total number of items in the database.

The intensity of implication is an index expressing the quality of a rule. This index, defined by R. Gras and A. Larher [8], is based on simple probability concepts: since the cardinalities of two subsets A and B are determined by the objects of the database belonging to A and B, we consider two random subsets X and Y having respectively the same cardinalities as A and B. The implication $a \Rightarrow b$ is characterized by the relation $A \subset B$ and its counter examples are associated to the subset $A \cap \overline{B}$. We compare the cardinality of $A \cap \overline{B}$ (given by the database) with the random variable given by the cardinality of $X \cap \overline{Y}$, supposing that there is no link between X and Y (Fig. 3). If the cardinality of $A \cap \overline{B}$ is unusually small compared to the expectation of the distribution on the cardinalities of $X \cap \overline{Y}$, we accept *"if a then b"* as a rule.

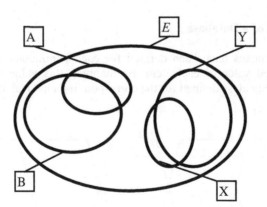

Fig. 3. X and Y vary randomly in E.

The intensity of $a \Rightarrow b$ is therefore the complement to one of the probability for the random variable "cardinality of $X \cap \overline{Y}$" to be smaller than cardinality of $A \cap \overline{B}$. It may be defined by

$$\varphi(a,\overline{b}) = 1 - P\left[Card(X \cap \overline{Y}) \leq Card(A \cap \overline{B}) \right] \tag{1}$$

Calling $\quad n = Card(E)$, $\quad n_a = Card(A)$, $\quad n_{\overline{a}} = Card(\overline{A})$, $\quad n_b = Card(B)$, $n_{\overline{b}} = Card(\overline{B})$, $n_{a \wedge b} = Card(A \cap B)$, $n_{a \wedge \overline{b}} = Card(A \cap \overline{B})$, the random variable $Card(X \cap \overline{Y})$ obeys an hyper geometric distribution [7]:

$$P[Card(A \cap \overline{Y}) = k] = \frac{C_{n_a}^k \cdot C_{n - n_a}^{n - n_b - k}}{C_n^{n - n_b}} = \frac{C_{n_a}^k \cdot C_{n_{\overline{a}}}^{n_{\overline{b}} - k}}{C_n^{n_{\overline{b}}}} . \quad (2)$$

and

$$P[Card(X \cap \overline{Y}) \le Card(A \cap \overline{B})] = \sum_{\substack{i=0 \\ i \ge n_a - n_b}}^{Card(A \cap \overline{B})} \frac{C_{n_a}^i \cdot C_{n_{\overline{a}}}^{n_{\overline{b}} - i}}{C_n^{n_{\overline{b}}}} . \quad (3)$$

The intensity of implication has interesting properties [6]. First, its value increases with the size of the learning set, while other indexes stay constant; some new counter-examples for a strong implication do not change much its value, but progressively doubts come, and finally a few more counter-examples cause its fall. This index is also well adapted to noisy data since a small number of counter-examples does not necessary invalidate the implication; it also proscribes rules such as $a \Rightarrow b$ when proposition b is true for nearly all examples of the learning set (since it is not surprising then that nearly all examples with a true are examples with b true).

3.2 Adaptation of Classical Indexes to Fuzzy Rules

Two categories of indexes evaluate fuzzy rules; indexes based exclusively on fuzzy set theory ([1], [13]) or indexes developed in classical logic and generalized to fuzzy knowledge ([14], [16], [19]). Choosing this second class, we have generalized the three indexes for classical rules by applying a concept from fuzzy probability of fuzzy events [17], for which the number of elements that satisfy a proposition p associated to a fuzzy set P with membership function μ_p, is the crisp cardinality of the fuzzy set P: $Card(P) = \sum_{x \in E} \mu P(x)$.

For a crisp implication $a \Rightarrow b$ on two fuzzy propositions a and b, linked to fuzzy sets A and B with membership functions μ_A and μ_B, we generalize the three indexes by using cardinalities of fuzzy sets instead of cardinalities of crisps sets [2]. With a fuzzy conjunction T (T-norm) and a fuzzy complement $\mu \overline{A}(x) = 1 - \mu A(x)$, one can write:

$$nA = Card(A) = \sum_{x \in E} \mu A(x), \qquad n\overline{A} = Card(\overline{A}) = \sum_{x \in E} \mu \overline{A}(x) = \sum_{x \in E} (1 - \mu A(x)),$$

$$n\overline{B} = Card(\overline{B}) = \sum_{x \in E} \mu \overline{B}(x) = \sum_{x \in E} (1 - \mu B(x)),$$

$$nA \cap B = Card(A \cap B) = \sum_{x \in E} \mu A \cap B(x) = \sum_{x \in E} T(\mu A(x), \mu B(x)))$$

$$nA \cap \overline{B} = Card(A \cap \overline{B}) = \sum_{x \in E} \mu A \cap \overline{B}(x) = \sum_{x \in E} T(\mu A(x), (1 - \mu B(x)))$$

The confidence of a rule, its support and its intensity of implication are then expressed by the same formulas as above, but they use these cardinalities of the fuzzy sets instead of the cardinalities of crisp sets.

4. A Knowledge Extraction Algorithm

Our algorithm uses a depth first strategy to evaluate all the rules that can be constructed from a set of propositions. To limit the number of rules to evaluate, we restrict the number of propositions in the premise of a rule; thus, we use 4 thresholds α, β, γ, δ and one rule is kept if its confidence C is greater than α, its support S is greater than β, its intensity of implication I over γ and if the length L of its premise has at most δ propositions.

Let us call $E = \{e_1, e_2, ..., e_n\}$ the learning set, R the set of rules retained,
$P = \{p_1, p_2, ..., p_n\}$ the set of propositions describing the examples in E,
P' the set of propositions associated to the possible conclusions,
$D = \{a_1, a_2, ..., a_m\}$ the set of attributes in the possible propositions of the premise,
$F_{decision}$ the fuzzy partition associated to the attribute of the classifying decision;
the algorithm is then:

```
Algorithm: Knowledge Extraction
(1) R = ∅
(2) For all values vᵢ ∈ F_decision do
(3)     Let T, the  tree of rules in P with conclusion
        { a_decision = vᵢ }
(4)     Let B, the set of observations in E with proposition
        { a_decision = vᵢ } true
(5)     CurrentNode = Forward(T, R)
(6)     While the tree T has not been totally searched do
(7)         Let r: Premise → pᵢ ∈ P', the rule associated to
            CurrentNode
(8)         Let A, the set of examples in E for which Premise
            is true
(9)         Compute C,S,I from the cardinalities of the sets
            E, A and B
(10)        Let L, the length of Premise
(11)            If (C≥α) and (S≥β) and (I≥γ) and (L≤δ)
(12)                Then R = R ∪ { Premisse → pᵢ }
(13)            End If
(14)            If (S < β) or (L > δ) or (CurrentNode terminal)
(15)                Then CurrentNode = Backward(T, CurrentNode)
(16)                Else CurrentNode = Forward(T, CurrentNode)
(17)            End If
(18)    End While
(19) End For
End.
```

5. The Choice of Fuzzy Operators

The knowledge extraction algorithm highlights a set of interesting rules, but needs another mechanism to evaluate the fuzzy implications. Fuzzy operators accept many possible definitions [12]. A fuzzy conjunction (fuzzy "and") may be chosen inside several classes of T-norms, such as:

Zadeh's minimum: $\mu_{a \cap b}(x) = \min(\mu_a(x), \mu_b(x))$

probabilistic intersection: $\mu_{a \cap b}(x) = \mu_a(x) * \mu_b(x))$

Lukasiewicz's intersection: $\mu_{a \cap b}(x) = \max(\mu_a(x) + \mu_b(x) - 1, 0)$.

As a fuzzy implication, one may use:

Reichenbach's $\mu_{a \Rightarrow b}(x, y) = 1 - \mu_a(x) + \mu_a(x) * \mu_b(y))$

Zadeh's $\mu_{a \Rightarrow b}(x, y) = \max(1 - \mu_a(x), \min(\mu_a(x), \mu_b(y)))$

Kleene-Dienes' $\mu_{a \Rightarrow b}(x, y) = \max(1 - \mu_a(x), \mu_b(y))$

Lukasiewicz's $\mu_{a \Rightarrow b}(x, y) = \min(1, 1 - \mu_a(x) + \mu_b(x))$

Gödel-Brouwer's $\mu_{a \Rightarrow b}(x, y) = 1$ if $\mu_a(x) \leq \mu_b(x)$

$\mu_{a \Rightarrow b}(x, y) = \mu_b(x)$ otherwise

However, since the extracted rules can be used in an expert system by application of the generalized modus ponens (G.M.P.), our choices must be coherent with the operators chosen then. Let us recall that the G.M.P. is the following inference scheme:

If X is A   then   Y is B   And for one T-norm T and the implication
                    X is A' $\mu_{a \Rightarrow b}(\mu_a(x), \mu_b(y)) = I(\mu_a(x), \mu_b(y))$,

────   ──   ──
                    Y is B'   one has $\mu_{b'}(y) = \sup_{x \in A'} T(\mu_{a'}(x), I(\mu_a(x), \mu_b(y)))$.

A comparative study of fuzzy implication operators [11] has shown that the generalized modus ponens gives very good results with four combinations: Lukasiewicz's implication and bold intersection, Kleene-Dienes' implication and minimum, Kleene-Dienes' implication and bold intersection, Gödel-Brouwer's implication and bold intersection. So we have limited our trials to these four combinations. Our algorithms also need an aggregation operator; and since we want an averaging evaluation of the implication and a mechanism to exclude abnormal records, we have chosen the arithmetic mean, which allows usage of standard deviations.

The algorithm to compute the fuzzy implication for each rule highlighted by the knowledge extraction algorithm is given below (more details are given in [3]). Each rule may be composed of one conjunction of fuzzy propositions in premise and one fuzzy proposition in conclusion, and the algorithm prospects a random sample of the database to compute the number of examples and counter examples for each couple (x, y) of the implication, then it evaluates the rate of good examples. If the rate of good examples satisfies the expert, the set of fuzzy operators is kept; otherwise, another set of operators is tried.

```
Algorithm: Evaluation of a set of fuzzy operators
1) For each example:
     Evaluate the premise's fuzzy value using the T-norm,
     Compute the values of the selected implication.
2) Compute the arithmetic mean η(x, y) and the standard
         deviation σ(x, y) for each pair (x, y), calling
               x, one truth-value of the premise,
               y, one truth-value of the conclusion.
3) For each example:
     If the value of its implication for the pair (x, y)
         fits [η(x, y)-2*σ(x, y), η(x, y)+2*σ(x, y)]
         (interval of confidence of 95% for a normal law),
               Note it as one positive example for (x, y),
     Else      Note it as one negative example.
4) The values of the implication are then given by the
   arithmetic means of the good examples for each couple
   (x, y) (x for the premise, y for the conclusion).
5) Evaluate the adequacy of the implication:
     For each positive example of the sample,
         Apply the Generalized Modus Ponens,
         Compute the distance between the truth-values of
             the inferred conclusions and the observation,
         If this distance is greater than a threshold
             chosen by the operator,
               add the example to the set of records for
                   which the rule is inadequate
         Else add the example to the set of records for
                   which the rule is correct
6) The rates of good examples are then given by
```

$$\rho(x,y) = n_+(x,y) / (n_+(x,y) + n_-(x,y))$$

```
End.
```

These algorithms were processed on databases of the UCI KDD Archive; they often gave good results, but sometimes none of the possible fuzzy implication proved really adequate. For these cases, we have proposed a statistical evaluation of the conclusions given the truth-values of the consequent ([4]).

6. Reduction of Fuzzy Rules

In some applications, rules are not extracted in order to build expert systems, but to give human experts a synthetic view of the database records; then, since the number of rules extracted is often high, we have studied methods to aggregate rules.

In classical logics, one may write $(a \Rightarrow c) \wedge (b \Rightarrow c) \vdash (a \vee b) \Rightarrow c$, and we wanted to process similarly with fuzzy rules without having to reevaluate the fuzzy rule $(a \vee b) \Rightarrow c$. So, we wished to write: $\mu_{(a \Rightarrow c) \wedge (b \Rightarrow c)}(x,y) = \mu_{(a \vee b \Rightarrow c)}(x,y)$ or

$\mu_{(a \Rightarrow c)}(x,y) \ T \ \mu_{(b \Rightarrow c)}(x,y) = \mu_{(a \vee b \Rightarrow c)}(x,y)$, using fuzzy operators chosen among the sets defined above. Therefore, we had to find which, if any, of the above operators sets allow to write the condensed form: $\mu_{I(\alpha,\gamma)} \, T \, \mu_{I(\beta,\gamma)} = \mu_{I(\alpha C \beta, \gamma)}$, T being a T-norm, C its complementary T-co-norm, and noting $\mu_{I(\alpha,\gamma)} = \mu_{(a \Rightarrow c)}(x,y)$, $\alpha = \mu_a(x,y)$, $\beta = \mu_b(x,y)$ and $\gamma = \mu_c(x,y)$. We have proved [4] that the Kleene-Dienes' implication, associated with *min* for T-norm and *max* for T-co norm is the only solution in the four sets of fuzzy operators. Similarly, to keep the classical reduction scheme $(a \Rightarrow c) \wedge (a \Rightarrow c) \vdash a \Rightarrow b \wedge c$ or, with the above condensed notations $\mu_{I(\alpha,\gamma)} \, T \, \mu_{I(\alpha,\gamma)} = \mu_{I(\alpha, \beta T \gamma)}$, we have shown that the same set of fuzzy operators must be used.

Adaptations of classical reduction schemes to fuzzy logics must be studied further, but we give these results to show that fuzzy rules often may not be handled as classical rules.

7. Conclusion

We have described here a generalization of knowledge discovery mechanisms to fuzzy logics; first, one needs to fuzzify the continuous attributes of the database. One needs then indexes to evaluate possible rules, and we have generalized to fuzzy logics three indexes: the confidence, the support and a less usual index, the intensity of implication. We have given a KDD algorithm that can be used to extract fuzzy rules.

Since fuzzy logics allow usage of numerous operators, we have justified their restriction to four sets and we have developed a method to choose on set of operators, computing the rates of good examples by applying the Generalized Modus Ponens on a sample extracted from the database.

Finally, to reduce the number of rules proposed to human experts, we have studied methods to cluster fuzzy rules, and we have shown that among the previously retained sets of operators, the association of Kleene-Dienes' implication with *min* as T-norm and *max* as T-co norm is the most interesting choice.

We must remark that the increase in computing complexity induced by usage of fuzzy logics is relatively small for rule extraction, since instead of increasing by one the counters on the number of examples and counter examples, fuzzy logics add membership degrees. The operations of fuzzification, the choice of fuzzy operators and the rules reductions are more complex, but the advantages of using fuzzy logics may compensate for this: intervals on continuous attributes are expressed by more expressive (fuzzy) labels, and abrupt threshold are avoided.

Among future perspectives, we plan to improve automatic configuration of the fuzzy partitions in connection to the samples extracted of the databases; we want to study other possible rules reduction mechanisms, and we also intend to study parallelisation methods of extraction algorithm, using for this a multi-agent platform. A more complete study of fuzzy operators is also desirable, to consider the Goguen implication, and examine cases when the generalized modus ponens is not only an approximate inference but also a valid logical inference [10].

References

1. Aguilar Martin J. and Lopez De Mantaras R.: The Process of Classification and Learning the Meaning of Linguistic Descriptors of Concepts. In: M.M. Gupta et E. Sanchez (eds.) Approximate reasoning in decision analysis, North Holland, pp. 165-175, 1982
2. Bernadet M., Rose G., Briand H.: FIABLE and Fuzzy FIABLE: two learning mechanisms based on a probabilistic evaluation of implications. In: Conference IPMU'96, Granada (Spain), July 1996, pp. 911-916.
3. Bernadet M.: A knowledge discovery mechanism with evaluation of fuzzy implications. In: Conference IPMU'98, Paris (France), July 1998.
4. Bernadet M.: reduction of rules in a fuzzy knowledge discovery system. Research Report RR 00.8, IRIN, Univ. of Nantes, France
5. Bezdek, J.C. and Harris, J.D. (1978): Fuzzy Partitions and Relations: An Axiomatic Basis for Clustering. In: Fuzzy Sets and Systems, 1. pp. 111-127.
6. Briand H., Djeraba C., Fleury L., Masson Y. and Philippe J.: Contribution of the implication intensity in rules evaluations for knowledge discovery in databases. In: ECML'95, Heraklion, Crete, April 1995.
7. Fleury L. and Masson Y.: Intensity of implication: a measurement in learning machine. In: IEAAIE'95, 8th International Conference on Industrial and Engineering applications of AI and Expert Systems, 6-8 June 1995, Melbourne, Australia.
8. Gras R., Larher A.: L'implication statistique, une nouvelle méthode d'analyse de données. In: Mathématiques, Informatique et Sciences Humaines n°120.
9. Frawley W.J., Piatetsky-Shapiro G. and Matheus C.J.: Knowledge Discovery in Databases: An Overview. In: AI Magazine, Fall 1992, pp. 213-228.
10. Hajek P.: Metamathematics of Fuzzy Logic. Kluwer Academic Publisher, 1998.
11. Kerre Etienne E.: A comparative study of the behavior of some popular fuzzy implications on the generalized modus ponens. In: in Fuzzy Logic for the management of uncertainty, L. Zadeh, J Kacprzyk,, Wiley 1992, pp. 281-295.
12. Klir George J, Bo Yuan: Fuzzy Sets, and Fuzzy Logics- Theory and Applications. PrenticeHall. Englewood Cliffs, USA, 1995.
13. Lesmo L., Saitta L. and Torasso P.: Fuzzy production rules : a learning methodology. In : P.P. Wang (ed.) Advances in *Fuzzy Sets, Possibility Theory and Applications*. New York : Plenum Press, pp. 181-198
14. Rives J.: FID3 : Fuzzy Induction Decision Tree. In: ISUMA'90, Maryland, USA, pp. 457-462, December 1990
15. Turksen I.B.: Measurement of membership functions and their acquisition. In: Fuzzy Sets and Systems, 40:5--38, 1991.
16. Weber R.: Fuzzy-ID3: a Class of Methods for Automatic Knowledge Acquisition. In: Proceedings of the 2nd International Conference on Fuzzy Logic and Neural Networks, Iizuka, Japan, pp. 265-268, July 1992
17. Zadeh L.A.: Probability Measures of Fuzzy Events. In: Journal of Mathematical Analysis and Applications, Vol. 23, pp. 421-427, 1968
18. Zadeh L.A.: Fuzzy Logic and its application to approximate reasoning. In: Information Processing, 74:591--594, 1974.
19. Zeidler J., Schlosser M.: Fuzzy Handling of Continuous-Valued Attributes in Decision Trees. In: Proc. ECML-95 Mlnet Familiarization Workshop Statistics, Machine Learning and Knowledge Discovery in Databases, Heraklion, Crete, Greece, April 1995, pp. 41-46.
20. Zighed D.A., Rabaseda S., Rakotomalala R., Feschet F.: Discretization methods in supervised learning. In: Encyclopedia of Computer Science and Technology, vol. 40, pp. 35-50, Marcel Dekker inc., 1999.

Confirmation Rule Sets

Dragan Gamberger[1] and Nada Lavrač[2]

[1] Rudjer Bošković Institute, Bijenička 54,10000 Zagreb, Croatia
gambi@lelhp1.irb.hr
[2] Jožef Stefan Institute, Jamova 39, 1000 Ljubljana, Slovenia
nada.lavrac@ijs.si

Abstract. The concept of confirmation rule sets represents a framework for reliable decision making that combines two principles that are effective for increasing the predictive accuracy: consensus in an ensemble of classifiers and indecisive or probabilistic predictions in cases when reliable decisions are not possible. The confirmation rules concept uses a separate classifier set for every class of the domain. In this decision model different rules can be incorporated: either those obtained by applying one or more inductive learning algorithms or even rules representing human encoded expert domain knowledge. The only conditions for the inclusion of a rule into the confirmation rule set are its high predictive value and relative independence of other rules in the confirmation rule set. This paper introduces the concept of confirmation rule sets, together with an algorithm for selecting relatively independent rules from a set of all acceptable confirmation rules and an algorithm for the systematic construction of a set of confirmation rules.

1 Introduction

The concept of confirmation rule sets represents a framework for reliable decision making that combines two principles that are effective for increasing the predictive accuracy: consensus in an ensemble of classifiers and indecisive or probabilistic predictions in cases when reliable decisions are not possible. It is known that ensembles of classifiers generally demonstrate better results than any of their components [2,3] and that the accuracy and diversity of the components determine the ensemble performance [10]. In most cases classifiers are combined by voting to form a compound classifier. Different classifiers can be obtained either by the application of different learning algorithms on the same training set or by the same learning algorithm on different training (sub)sets. The later approach is used in the well-known *bagging* and *boosting* approaches that employ redundancy to achieve better classification accuracy [4,5,16]. For critical applications the predictive accuracy of compound classifiers can be further increased if, instead of voting, the consensus of classifiers' answers is requested. Regardless of the used combination scheme, the fact that it is difficult or impossible to ensure the independence of the ensemble components has the consequence that high prediction reliability of such classifiers can not be ensured in all situations. Another way for achieving reliable predictions is the systematic construction

D.A. Zighed, J. Komorowski, and J. Żytkow (Eds.): PKDD 2000, LNAI 1910, pp. 34–43, 2000.
© Springer-Verlag Berlin Heidelberg 2000

of redundant rules, whose problem is, however, their algorithm and decision complexity. Independently, there has been significant effort devoted to the development of different other techniques aimed at improving the quality of decision making. Especially in medical domains, some techniques are aimed at the construction of either very sensitive or very specific rules instead of rules with a high overall predictive accuracy.[1] This is however not a general solution because false positive predictions are as inadequate as false negative predictions in many applications. Consequently, some of the techniques applied in medical problems aim at classifiers with high ROC (Receiver Operating Characteristic) curve area (see e.g., [11]).[2] An alternative approach to reliable predictor construction is the introduction of indecisive predictions. In this approach, in the case of a two-class problem, three different predictions are possible: class positive, class negative, and prediction not possible. The approach follows the concept of *reliable, probably almost always useful learning* defined in [17]. In [15] it was shown how existing machine learning algorithms can be transformed into the form which enables indecisive predictions. In [15] a simple voting based decision model that includes indecisive predictions based on a set of concept descriptions constructed by the FOCL system has been introduced. The achieved predictive accuracies measured on a few medical domains show that significant accuracy improvements are possible. The substantial disadvantage of the approach are however indecisive answers, whose amount has to be kept as low as possible. The confirmation rule sets concept presented in this work follows the above paradigm of reliable, probably almost always useful learning, allowing for indecisive predictions. However, it represents a more general, consensus based approach to decision making from a set of rules. Its basic characteristic is that it uses separate rules sets for every target class. In this way it is similar to human decision making processes. In addition, like in association rule learning [1], the presented approach utilizes the minimal support requirement as a rule reliability measure which must be satisfied by every rule in order to be accepted as a confirmation rule and included into the confirmation rule set. The approach is introduced in Section 2. The confirmation rule sets concept does not present a novel rule induction approach but rather a decision model in which different rules can be incorporated: either rules induced by (one or more) learning algorithms or even rules representing human encoded expert domain knowledge. High prediction quality is expected only if, besides the predefined predictive value of every included rule, the whole set is as diverse as possible. A simple and general algorithm for the selection of appropriate confirmation rules is presented in Section 2 together with an ex-

---

[1] Sensitivity measures the fraction of positive cases that are classified as positive, whereas specificity measures the fraction of negative cases classified as negative. Let TP denote true positives, TN true negatives, FP false positives and FN false negative answers, then $Sensitivity = \frac{TP}{TP+FN}$, and $Specificity = \frac{TN}{TN+FP}$.

[2] A ROC curve indicates a tradeoff between the false alarm rate (1 - *Specificity*, plotted on X-axis) that needs to be minimized, and the detection rate (*Sensitivity*, plotted on Y-axis) that needs to be maximized. An appropriate tradeoff, determined by the expert, can be achieved by applying different algorithms, as well as by different parameter settings of a selected data mining algorithm.

haustive search procedure which can be used for systematic confirmation rules construction. This algorithm has been used to construct rule sets for the coronary arthery disease domain. Prediction results of induced rule sets, measured for different acceptance levels of the consensus scheme, are presented in Section 3, illustrating properties of the confirmation rule sets concept in a real medical domain.

2 Confirmation Rules

In the decision model using confirmation rule sets, every diagnostic/prognostic class is treated separately as the target class for which a separate set of confirmation rules is constructed. The basic property of confirmation rules is that they should cover (satisfy) only examples of the given target class. Additionally, in order to ensure high predictive quality of confirmation rule sets, every rule included into the set must be a reliable target class predictor by itself. Given that a conclusion of a confirmation rule is a target class assignment, and a condition is a conjunction of simple literals, confirmation rules are similar to if-then rules as induced by the AQ algorithms [14], and to association rules [1]. In the confirmation rule concept, however, every *complex* (conjunction) of an AQ rule would constitute a separate and independent rule. Moreover, the main difference with association rules is that confirmation rules have only the target class assignment in the conclusion of a rule whereas a conclusion of an association rule is a conjunction of arbitrary attribute values.

2.1 Properties of Confirmation Rules

To summarize, confirmation rules are defined by the following properties:

a) The condition of a confirmation rule has the form of a conjunction of simple literals each being a logical attribute value test. The conclusion of a confirmation rule is a target class assignment.
b) Confirmation rules should cover (satisfy) only examples of their target class; since rules cover only target class examples, prediction quality can be estimated by the number of covered target class examples.
c) Acceptable confirmation rules which may be included into confirmation sets are only those rules that cover a sufficient number of target class examples; the *minimal support level* parameter can be used to define the requested minimal number of covered target class examples.[3]

The number of rules in a confirmation rule set is generally not determined and does not have to be equal for all classes. The sets can include all, or only subset of all, acceptable confirmation rules for the target class. There can be one rule in the set, many of them, but it is also possible that the set is empty if no

---

[3] In our system, the default value of the support level is equal to the second root of the total number of target class examples available.

acceptable confirmation rule is known for the class. In the defined concept it does not matter how confirmation rule have been induced but rather how they cover the problem space. Every confirmation rule may be induced and used independently of other confirmation rules. When confirmation rule sets are used for prediction, the following outcomes are possible:

a) If no confirmation rule fires for the example, class prediction is indecisive (the example is not classified).[4]

b) If a single confirmation rule fires for the example, class prediction is determined by this rule.

c) If two or more confirmation rules from the same set fire for the example, the target class of the set is predicted with increased reliability.

d) If two or more confirmation rules fire for the example and at least two of these rules are from different sets, class prediction is indecisive.

This indicates that the confirmation rule sets do not give decisive predictions in every situation (cases (a) and (d)), and that predictions of increased reliability are made possible (case (c)). In some situations one may decide to accept only predictions of increased reliability as decisive predictions. How many confirmation rules must cover an example in order to make the decisive classification can be determined by the so-called *acceptance level* parameter. Acceptance level 1 denotes that a single rule coverage is sufficient for example classification (case (b)). Typical values for the acceptance level parameter are 1–3.

2.2   Confirmation Rule Subset Selection Algorithm

Typically, there are many acceptable confirmation rules for every class, satisfying the requested minimal number of covered target class examples (defined by the *minimal support level* parameter). Inclusion of all these rules into the rule set is generally not desired because (a) it is difficult to make decisions based on very large sets of rules, and (b) experiments demonstrated that there are subsets of very similar rules which use almost the same attribute values and have similar prediction properties. The second characteristic is especially undesirable in cases when the *acceptance level* greater than one is used, intended at increasing the prediction reliability because in this case very similar rules will cover an example more than once. A solution to this problem is to reduce confirmation rule sets so that they include only a relatively small number of confirmation rules which are as diverse as possible. It must be noted that a simple increase of the required support level can reduce the number of acceptable confirmation rules, however probably the remaining rules will still be from the same subset of similar rules. This fact was experimentally detected. A better solution is to leave the required support level unchanged so that there are many acceptable confirmation rules

---

[4] Alternatively, a probabilistic classification could be proposed for this case, e.g., using a simple Bayesian classifier scheme. In our system this approach has not yet been implemented and tested.

and then select among them a small subset of relatively independent rules. Selecting a subset of independent classifiers for the same target class is known as a complex task which occurs in most multiclassifier decision systems [9]. The problem is difficult because there are many combinations which can make different rules statistically dependent, including (even for domain expert unknown) relations among attribute values reflecting inherent domain properties. The confirmation rule sets concept is intended to be able to include and combine all the available knowledge without other restrictions than the requested prediction quality of individual confirmation rules. Consequently, our approach accepts as diverse those rules that cover as different sets of target class examples as possible. Obviously the approach can not guarantee statistical independence of the selected rules, but its advantages are the simplicity and robustness concerning all the known and unknown dependences among attribute values.   Algorithm

Algorithm 1: CONFIRMATION RULE SUBSET SELECTION
Input: A set of all acceptable confirmation rules for the target class
              P target class examples
Parameter: *number* (number of rules in the selected subset)
Output: subset of *number* relative independent confirmation rules
            for the target class
(1)   **for** every $e \in P$ **do** $c(e) \leftarrow 1$
(2)   **repeat** *number* times
(3)   **select** from A the rule with
              the highest weight $\sum 1/c(e)$ where summation is over the set
              $P' \subseteq P$ of target class examples covered by the rule
(4)   **add** the selected rule into the output confirmation rule set
(5)   **for** every $e \in P'$ of the selected rule **do** $c(e) \leftarrow c(e) + 1$
(6)   **eliminate** the selected rule from A
(7)   **end repeat**
(8)   **exit** with *number* of selected confirmation rules

1 presents an approach to selecting the subset of a *number* of relative independent confirmation rules. Input is the set of all acceptable confirmation rules A and the set of all target class examples P. For every example $e \in P$ there is a counter $c(e)$. Initially the output set of selected rules is empty and all counter values are set to 1 (step 1). After that in each iteration of the loop (steps 2 to 7) one confirmation rule is added into the output set (step 4). From set A the rule with the highest *weight* value is selected. For each rule, *weight* is computed such that $1/c(e)$ values are added for all target class examples covered by this rule (step 3). After rule selection, the rule is eliminated from the set A (step 6) and $c(e)$ values for all target class examples covered by the selected rule are incremented by 1 (step 5). This is the central part of the algorithm which ensures that in the first iteration all target class examples contribute the same value $1/c(e) = 1$ to the *weight*, while in following iterations the contributions of examples are inverse proportional to their coverage by previously selected rules. In this way the examples already covered by one or more selected rules can contribute substantially less to the *weight* and the rules covering many yet

uncovered target class examples have a greater chance to be selected in the following iterations. For noisy domains the condition that confirmation rules should
not cover any of the non-target class examples may be too strong since it may
result in a small total number of target class examples covered by every possible
confirmation rule. The requirement may be relaxed by accepting also confirmation rules covering a few non-target class examples as well. Such a modification
offers a simple and practical noise handling approach, but it may lead to the
reduction of predictive accuracy of induced confirmation rules. Instead of such
noise handling, experiments presented in Section 3 are done using a procedure
for explicit noise detection and elimination [7]. This procedure is based on the
consensus of saturation filters, performing reliable filtering of noisy examples in
preprocessing. The characteristic of this approach is that only a small number of
examples with high probability of actually being noisy are detected and eliminated from the training set before confirmation rule induction. This is important
for the confirmation rules concept which should provide for a high reliability of
decisive predictions.

2.3   Confirmation Rule Set Construction

Algorithm 1 builds the set of relative independent confirmation rules by selecting the rules from the input set A, consisting of all acceptable confirmation
rules. In some cases there are neither expert knowledge nor rules generated by
inductive learning systems that can be used as acceptable confirmation rules.
In such situation Algorithm 1 can be modified so that instead of its steps 3
and 6, Procedure 1 is used in every iteration to construct a confirmation rule
with highest weight $\sum 1/c(e)$. This procedure is actually an exhaustive search
algorithm which uses a set of literals defined for the domain. Such set can be
constructed and potentially optimized by algorithms described in [12]. The procedure builds the confirmation rule in the form of logical conjunction of literals
so that: **a)** the rule does not cover examples of the non-target class, **b)** the rule
covers more than *minimal support level* of target class examples, and **c)** the rule
has maximal possible weight $\sum 1/c(e)$. Algorithm 1 with included Procedure 1
is used in experiments presented in Section 3. Computational complexity of the
exhaustive search in Procedure 1 is hight what restricts its applicability to problems of with up to few hundred examples. Procedure 1 needs as its inputs the
complete training set E, the appropriate literal set L, and $c(e)$ values imported
from Algorithm 1 for all positive training examples. The procedure additionally
requires that the parameter *min_support* is defined which restricts the space
of acceptable confirmation rules. In case when internal variable *best_weight* is
greater than zero then procedure output is the best acceptable confirmation rule
that could be constructed with available literals. If the procedure could not find
any acceptable solution then *best_weight* = 0.

Procedure 1: CONFIRMATION RULE CONSTRUCTION

Input: $E = P \cup N$ (E training set, P positive or target class examples,
$\qquad$ N negative or non-target class examples).
$\qquad$ L (set of literals, $l \in L$ covers a positive example $e \in P$ if l is true for e,
$\qquad$ l covers a negative example $e \in N$ if l is false for e).
$\qquad$ $c(e)$ values imported from Algorithm 1.

Parameter: *min_support* (minimal support level for rule acceptance with
$\qquad$ default value equal to the second root of $|P|$)

Output: selected confirmation rule in best solution if *best_weight* > 0

(1) $\quad$ set *best_weight* $= 0$, loop level pointer $V = 0$
(2) $\quad$ **repeat** literal selection loop
(3) $\qquad$ **if** a literal from the literal set at level V covering an uncovered negative
$\qquad\qquad$ example exists **then**
(4) $\qquad\qquad$ **include** this literal into the present solution at level V
(5) $\qquad\qquad$ **compute** coverage of positive and negative examples at level V
(6) $\qquad\qquad$ **compute** $weight = \sum 1/c(e)$, $e \in P$ and covered at level V
(7) $\qquad\qquad$ **if** $weight \le best\_weight$ or total number of covered positive
$\qquad\qquad\qquad$ examples $<$ *min_support* **then** forget this literal at level V
$\qquad\qquad\qquad$ and continue the loop at level V
(8) $\qquad\qquad$ **if** all negative examples are covered at level V **then** copy the
$\qquad\qquad\qquad$ present solution into the best solution, *best_weight* $= weight$,
$\qquad\qquad\qquad$ forget this literal at level V, and continue the loop at level V
(9) $\qquad\qquad$ **continue** the loop at level $V + 1$
(10) $\quad$ **else**
(11) $\qquad\qquad$ **if** level $V = 0$ **then** exit the loop
(12) $\qquad\qquad$ **else** continue the loop at level $V = V - 1$
(13) $\quad$ **end repeat**

3 Summary of Confirmation Rule Sets Application on a Medical Domain

Application characteristics of confirmation rule sets are illustrated in this section with a summary of measured prediction results for the coronary artery disease diagnosis dataset, collected at the University Medical Center, Ljubljana, Slovenia. Details about the domain and some other machine learning results can be found in [8,6]. Independently, in [7] the same domain was used to test our noise handling algorithm (the so-called consensus saturation filter). The results were good because the system detected in total 15 noisy examples (out of 327 patient records) out of which the medical doctor who collected the data recognized 14 as being real outliers, either being errors or possibly noisy examples with coronary angiography tests very close to the borderline between the two classes. In accordance with the standard 10-fold cross-validation procedure, for every training set, 5 confirmation rules were generated for the class *positive* and 5 for the class *negative*. Such experimental setting enabled the testing of generated confirmation rules sets with different acceptance levels. The prediction

Table 1. Results of 10-fold cross-validation presenting the percentage of correct predictions, measured error rate, measured relative error rate and real relative error rate for **a)** without and **b)** with noise elimination in preprocessing. For each fold with about 294 training examples and 33 test examples, 5 **confirmation rules** for the class *positive* and 5 **confirmation rules** for the class *negative* were generated. Results are presented for acceptance levels 1–3, where level 3 means that the example must satisfy at least 3 out of 5 rules for decisive prediction. The percentage of correct predictions represents the total number of correct predictions divided by the total number of predictions (327), while the measured error rate is the total number of erroneous predictions divided by the total number of all predictions. The measured relative error rate is equal to the ratio of the number of erroneous predictions and the number of decisive predictions. The real relative error rate is computed so that the number of erroneous predictions is, at first, reduced so that it does not include expert-evaluated domain outliers, and then it is divided by the number of decisive predictions.

accept. level	correct predictions	measured error rate	meas. relative error rate	real relative error rate
a) without noise elimination in preprocessing				
1	72.48%	7.65%	9.54%	4.2%
2	47.71%	2.75%	5.45%	1.8%
3	28.44%	1.22%	4.12%	1.0%
b) with noise elimination in preprocessing				
1	76.15%	5.81%	7.09%	3.2%
2	60.86%	3.06%	4.78%	2.0%
3	47.40%	1.83%	3.73%	0.6%

is correct (successful) if the example is classified into a single class, which has to be the same as the expert classification. The prediction is erroneous if the example is classified into a single class which is different from the expert classification. Experimental results are presented in Table 1. The table has two parts: the first presents results obtained *without* and the second *with* noise elimination in preprocessing. In both cases results for three different acceptance levels are reported. The first column of every row is the acceptance level, follows the percentage of correct predictions and the percentage of erroneous predictions. Difference between the numbers is the percentage of indecisive predictions in the corresponding experiment. In the fourth column is the measured relative error computed as the ratio of the number of erroneous predictions and the number of decisive predictions. The numbers in this column are greater than those in the third column because decisive predictions are only a part of all predictions. In this sense values in column four are more realistic from the users point of view. But values in column four (and column three) include noisy cases already detected and evaluated by domain expert in [7]. Misprediction of these cases are not actual errors but expected result of good rules. In order to estimate the real relative error rate, such cases (14 of them for the domain) were eliminated from the measured error sets in which they occur and then the real relative error rate was computed. The values are presented in the last column of Table 1. Measured error rates in this domain are between 3.7% and 9.5% (column 4) while estimated real error rates are about 0.6% – 4.2% (column 5) what are better results than those obtained both by other machine learning algorithms and medical experts [8]. It must be noted that the elimination of the 'expected' domain noise was extremely conservative, based on the consensus of the saturation filter preprocessor and the domain expert, potentially resulting in overestimation of the real error rate. The least estimated real error rate is detected with acceptance level 3 and noise detection in preprocessing. In this case the number of indecisive predictions is about 50% with only one really wrong prediction in about 150 decisive classifications. This result proves high reliability of the induced confirmation rules. Results in Table 1 demonstrate the differences in prediction quality of various acceptance levels. As expected, the increased acceptance level reduces the number of correct predictions but it also significantly reduces the number of erroneous predictions, especially the real predictive errors. The observation holds with and without noise elimination in preprocessing. Noise elimination itself is very useful. The comparison of the number of correct predictions for confirmation rules generated without and with noise detection and elimination in preprocessing demonstrates the importance of the use of this (or a similar) noise handling mechanism for effective confirmation rule induction. For example, for acceptance level 3 the increase is from 28% to 47%.

4   Conclusion

This work stresses the importance of reliable decision making and for this purpose the paper elaborates the concept of confirmation rule sets. It is shown that in critical applications where decision errors need to be minimized, confirmation rule sets provide a simple, useful and reliable decision model. The proposed confirmation rule sets framework is general because it enables the incorporation of results of different machine learning algorithms, as well as the existing expert knowledge. The induced structure of an unordered list of simple rules and the possibility of providing predictions of increased reliability are its main advantages. The main disadvantage of the approach are indecisive answers. In presented experiments the number of indecisive predictions has been high, always greater than 20% with a maximum greater than 70%. In the case of indecisive predictions, a probabilistic classification could be proposed, e.g., using a simple Bayesian classifier scheme. In our system this approach has not yet been implemented and tested. This is planned in further work.

Acknowledgement

This work has been supported in part by Croatian Ministry of Science and Technology, Slovenian Ministry of Science and Technology, and the EU funded project Data Mining and Decision Support for Business Competitiveness: A European Virtual Enterprize (IST-1999-11495).

References

1. R. Agrawal, H. Mannila, R. Srikant, H. Toivonen and A.I. Verkamo (1996) Fast discovery of association rules. In U.M. Fayyad, G. Piatetski-Shapiro, P. Smyth and R. Uthurusamy (Eds.) *Advances in Knowledge Discovery and Data Mining*, pp. 307–328. AAAI Press.
2. K.M. Ali and M.J. Pazzani (1996) Error reduction through learning multiple descriptions. *Machine Learning*, 24:173–206.
3. P.J.Boland (1989) Majority systems and the Concorcet jury theorem. *Statistician*, 38:181–189.
4. L. Breiman (1996) Bagging predictors. *Machine Learning* 24(2): 123–140.
5. Y. Freund and R.E. Shapire (1996) Experiments with a new boosting algorithm. In *Proc. Thirteenth International Machine Learning Conference ICML'96*, 148–156, Morgan Kaufmann.
6. D. Gamberger, N. Lavrač, C. Grošelj (1999) Diagnostic rules of increased reliability for critical medical applications. In *Proc. Joint European Conference on Artificial Intelligence in Medicine and Medical Decision Making AIMDM-99*, 361–365, Springer Lecture Notes in AI 1620.
7. D. Gamberger, N. Lavrač, C. Grošelj (1999) Experiments with noise filtering in a medical domain. In *Proc. of International Conference on Machine Learning ICML-99*, 143–151. Morgan Kaufmann.
8. C. Grošelj, M. Kukar, J.J. Fetich and I. Kononenko (1997) Machine learning improves the accuracy of coronary artery disease diagnostic methods. *Computers in Cardiology*, 24:57–60.
9. T.Kagan and J. Ghosh (1996) Error correlation and error reduction in ensemble classifiers. *Connection Science*, 8:385–404.
10. J. Kittler, M. Hatef, R.P.W. Duin, and J.Matas (1998) On combining classifiers. *IEEE Trans. on Pattern Analysis and Machine Intelligence* 20:226–239.
11. M. Kukar, I. Kononenko, C. Grošelj, K. Kralj, J.J. Fettich (1998) Analysing and improving the diagnosis of ischaemic heart disease with machine learning. *Artificial Intelligence in Medicine*, special issue on *Data Mining Techniques and Applications in Medicine*, Elsevier.
12. N. Lavrač, D.Gamberger, and P. Turney (1997) A relevancy filter for constructive induction. *IEEE Intelligent Systems & Their Applications*, 13:50–56.
13. B. Liu, W. Hsu and Y. Ma (1998) Integrating classification and association rule mining. In *Proc. Fourth International Conference on knowledge Discovery and Data Mining*, KDD-98, New York, USA, 1998.
14. R.S. Michalski, I. Mozetič, J. Hong, and N. Lavrač (1986) The multi-purpose incremental learning system AQ15 and its testing application on three medical domains. In *Proc. Fifth National Conference on Artificial Intelligence*, pp. 1041–1045, Morgan Kaufmann.
15. M. Pazzani, P. Murphy, K. Ali, and D. Schulenburg (1994) Trading off coverage for accuracy in forecasts: Applications to clinical data analysis. In *Proceedings of the AAAI Symposium on AI in Medicine*, pp. 106–110.
16. J.R. Quinlan (1996) Boosting, bagging, and C4.5 . In *Proc. Thirteenth National Conference on Artificial Intelligence*, 725–730, AAAI Press.
17. R.L. Rivest and R. Sloan (1988) Learning complicated concepts reliably and usefully. In *Proc. Workshop on Computational Learning Theory*, 69–79, Morgan Kaufman.

Contribution of Dataset Reduction Techniques to Tree-Simplification and Knowledge Discovery

Marc Sebban and Richard Nock

TRIVIA Research Team, French West Indies and Guiana University
Department of Mathematics and Computer Science, 95159 - Pointe-à-Pitre (France)
{msebban,rnock}@univ-ag.fr

Abstract. In the Knowledge Discovery in Databases (KDD) field, the human comprehensibility of models is as important as the accuracy optimization. To address this problem, many methods have been proposed to simplify decision trees and improve their understandability. Among different classes of methods, we find strategies which deal with this problem by *a priori* reducing the database, either through feature selection or case selection. At the same time, many other efficient selection algorithms have been developed in order to reduce storage requirements of case-based learning algorithms. Therefore, their original aim is not the tree simplification. Surprisingly, as far as we know, few works have attempted to exploit this wealth of efficient algorithms in favor of knowledge discovery. This is the aim of this paper. We analyze through large experiments and discussions the contribution of the state-of-the-art reduction techniques to tree simplification. Moreover, we propose an original mixed procedure which deals with the selection problem by jointly removing features and instances. We show that in some cases, this algorithm is very efficient to improve the standard post-pruning performances, used to combat the overfitting problem.

1 Introduction

While numerous predictive models have the only main objective to maximize the generalization accuracy (such as neural networks or nearest-neighbor algorithms), the aim of decision trees seems to be more complicated. In the field of Knowledge Discovery in Databases (KDD), even if the predictive accuracy is obviously important, the human comprehensibility of the model is actually also crucial. Unfortunately, with the huge development of the modern databases, this second goal is rarely reached, often resulting in deep incomprehensible trees. This large size is often explained by the two following reasons: (i) some target concepts can not be represented by trees, (ii) large modern databases can have a lot of noise, that causes overfitting. In such a context, researchers in statistics and computer science have proposed many methods for simplifying trees. Breslow and Aha [3] have proposed a clustering of methods in machine learning. They draw up the five following categories according to five simplification strategies:

D.A. Zighed, J. Komorowski, and J. Zytkow (Eds.): PKDD 2000, LNAI 1910, pp. 44–53, 2000.
© Springer-Verlag Berlin Heidelberg 2000

control tree size, modify space of tests, modify search for tests, reduce dataset, transform into an alternative data structure.

In this paper, we focus only on dataset reduction techniques which belong to the fourth category. In this class of methods, two reduction strategies can be applied. The first one consists in removing irrelevant features before the tree induction process in order to build smaller trees. A study in [16] shows that such a strategy can be very efficient. Using C4.5 [12], the non deletion of weakly relevant features generates actually deeper decision trees with lower performances. *Ad hoc* algorithms has been proposed to deal with this phenomenon on decision trees [1, 5]. These methods belong to a wider family of reduction algorithms, called wrapper models [9], which do not always use a decision tree representation.

The second strategy consists in deleting irrelevant instances before the construction of the tree. Such algorithms have been presented during the last decades in [11] (with a windowing strategy), [17] (with a sampling strategy) and [8] (with *Robust C4.5* decision trees). The Quinlan's strategy [11] consists in incrementally adding misclassified instances to an original learning sample and inducing a new decision tree until no cases are misclassified. To avoid to completely re-induce the tree, Utgoff [17] updates trees incrementally as each new case is added. John [8] chooses an opposite strategy by removing all instances misclassified by the current decision tree. Note that the tree is re-induced at each stage in the original algorithm, even if John claims that *RC4.5* could be modified with a scheme such as that proposed in [17]. This previous algorithm is one of the most efficient techniques, that is why we will take it as a reference in this paper (figure 1 recalls the pseudocode of *RC4.5*).

```
ROBUSTC45(TRAININGDATA)
Repeat
    T ← C45BuildTree(TrainingData)
    T ← C45PruneTree(T)
    foreach Record in TrainingData
        if T misclassified Record then
        remove record from TrainingData
Until T correctly classifies all Records in TrainingData
```

Fig. 1. Pseudocode of the *ROBUST-C*4.5 algorithm

At the same time, many other case-reduction techniques have been proposed, not for reducing tree size and improving human comprehensibility, but rather to limit the size required to store databases in memory [4, 6, 7, 15, 18]. These algorithms (often called Prototype Selection algorithms) have been developed in order to compensate for the drawback of case-based algorithms, such as the nearest-neighbor classifiers, well known to be very efficient, but also computationally expensive. As far as we know, no recent study has attempted to establish a close link between the best prototype selection procedures and tree simplification. No study has tried to compare algorithms for tree simplification, such as the John's *Robust C4.5 Decision Trees* with such algorithms. This behavior can be in fact doubly justified:

- The first reason (already mentioned above) comes from the main goal of the prototype selection algorithms: they are commonly used to reduce storage requirements imposed by expensive algorithms such as the k Nearest-Neighbors (kNN). Therefore, their reduction strategies are not *a priori* directed towards the construction of smaller trees.
- The second reason directly ensues from the first. Since most of these algorithms usually use a kNN classifier during the selection, their adaptation to tree simplification is *a priori* not natural.

Nevertheless, despite this established fact, three arguments prompt us to test them in order to simplify trees. First, while repeated tree inductions during the selection can be computationally costly on huge databases, some of prototype selection methods are known to be very fast. Even if incremental tree updates have been proposed [17], a local modification of a neighborhood graph is faster. Secondly, all the tree-simplification procedures optimize the accuracy during the selection by the induction algorithm. Recent theoretical [10] and empirical results [15] show that accuracy is not always the most suitable criterion to optimize. Since some of prototype selection algorithms take into account this phenomenon [13, 15], it would be interesting to analyze their effect for simplifying trees. Thirdly, the best prototype selection algorithms allow to dramatically reduce datasets, while not compromising the generalization accuracy. We think that some of them can be as efficient for simplifying trees.

In the first part of this paper, we will analyze the contribution of four state-of-the-art prototype selection algorithms: the Gates's *Reduced Neighbor Rule* [6], *the consensus filters* of Brodley and Friedl [4], the Wilson & Martinez's *RT3* [18] and the Sebban and Nock's *PSRCG* [15]. We think that they are representative of the evolution of the selection strategies in this field. After recalling the principles of these methods, we study their specificities in the field of tree simplification, and we statistically analyze their performances. Our goal is to answer to the two following questions:

- Which effect on tree simplification entails the nature of the optimized criterion during the selection, and the type of the learning algorithm used?
- What is the efficiency of these algorithms in comparison with procedures such as the *Robust C4.5 decision trees* in terms of *predictive accuracy, tree size, noise tolerance and learning speed*?

Among the tested algorithms, we show that *PSRCG* [15], which does not optimize the accuracy, present performances statistically comparable to *RC4.5*, with a possible adaptation to other tree induction algorithms, such as CART

[2], or ID3 [11]. The information criterion optimized in *PSRCG* is actually independent on a *ad hoc* algorithm, and allows us to generalize its use for other decision trees. From this criterion, we propose in the second part a new mixed procedure which attempts to jointly delete irrelevant features and instances. A statistical study shows that our approach is suited to dramatically reduce tree sizes, while controlling the predictive accuracy, even in the presence of noise.

2 Prototype Selection as a Tree Simplification Procedure

2.1 Prototype Selection Strategies

In the wide family of prototype selection algorithms, we can distinguish three categories according to their selection strategy. For each category, we present the main methods (for more details see mentioned articles).

The first one only removes *mislabeled* instances or *outliers*. In such a category, the most efficient algorithm is probably the *Consensus Filters* [4]. This approach constructs a set of base-level detectors and uses them to identify mislabeled instances by consensus vote, *i.e.* all classifiers must agree to eliminate an instance. The base-level detectors can come from varied fields (*neural network, induction tree, kNN, genetic algorithm*, etc.).

The second category keeps only relevant instances, by removing not only outliers, but also useless instances, and examples at the class boundaries. Two examples of such algorithms are *RT3* [18] and *PSRCG* [15]. They are known to belong to the most efficient prototype selection algorithms. In *RT3* an instance ω_i is removed if its removal does not hurt the classification of the instances remaining in the sample set, notably instances that have ω_i in their neighborhood (called *associates*). The algorithm is based on 2 principles: it uses a noise-filtering pass, which removes instances misclassified by their kNN, that avoids "overfitting" the data; it removes instances in the center of clusters before border points, by sorting instances according to the distance from their nearest *enemy* (*i.e.* of a different class). In *PSRCG*, authors investigate the problem of selection as an information preserving problem. Rather than optimizing the accuracy of a classifier, they build a statistical information criterion based on a quadratic entropy computed from the nearest-neighbor topology. From neighbors linked by an edge to a given instance (including *associates*), *PSRCG* computes a quadratic entropy by taking into account the label of each neighbors. From this entropy (which conveys a local uncertainty), it deduces a global uncertainty of the learning sample. While an instance deletion is statistically significant, *PSRCG* eliminates uninformative examples (see its statistical properties in [15]).

The last category presents an opposite strategy by keeping only misclassified instances. *CNN* [7] and *RNN* [6] were the pioneer works in this field. Their principle is based on the search for a consistent (*CNN*) or a reduced (*RNN*) subset, which correctly classifies all of the remaining instances in the sample set.

Table 1. Predictive accuracy (Acc.), tree size (number of nodes # N) and number of prototypes (# pr.) on 21 benchmarks

Dataset	C4.5			RNN+C4.5			Brodley+C4.5			RT3+C4.5			PSRCG+C4.5		
	LS	# N	Acc.	# pr.	# N	Acc.	# pr.	# N	Acc.	# pr.	# N	Acc.	# pr.	# N	Acc.
Australian	400	31	84.1	209	29	81.0	344	60	84.8	37	9	83.0	315	52	86.6
Balance	400	106	74.0	138	0	-	328	75	62.0	28	0	-	185	91	72.0
Breast W.	400	64	96.0	50	14	84.0	391	32	97.7	11	0	-	79	5	92.0
Brighton	500	34	86.0	95	5	77.0	491	65	94.0	108	34	85.4	188	34	84.7
Dermato.	200	20	92.0	59	20	92.0	180	20	92.0	50	17	71.0	134	19	92.0
Echocardio	70	5	65.6	58	0	-	52	2	63.0	6	2	63.0	40	2	73.0
German	500	5	69.6	411	27	69.0	411	27	69.0	29	2	64.5	357	9	69.0
Glass	150	73	68.0	76	9	60.0	107	82	67.0	28	19	50.0	122	64	70.0
Hepatitis	100	5	67.3	29	0	-	89	5	67.3	3	0	-	52	2	67.3
Horse	200	49	82.1	92	36	85.0	160	49	81.5	17	2	83.3	186	45	81.5
Ionosphere	185	124	83.6	40	27	66.0	163	25	91.4	19	2	87.9	140	46	88.8
Iris	100	9	98.3	14	2	64.0	97	5	94.0	18	5	78.0	49	5	94.0
LED+17	300	5	78.5	141	14	78.0	242	59	79.5	35	5	50.5	251	14	78.5
LED2	300	20	89.0	114	12	75.0	266	38	85.0	24	14	75.0	193	9	87.5
Pima	468	5	74.3	245	0	-	369	27	75.0	25	5	73.7	438	5	74.7
Tic tac toe	600	70	75.7	218	2	60.0	506	76	79.1	164	17	62.0	533	98	81.6
Vehicle	400	9	69.7	178	2	61.0	335	65	74.0	52	23	70.6	375	20	70.6
Waves	300	291	73.0	127	99	63.0	260	152	73.0	52	12	67.0	238	218	74.0
White H.	235	9	90.5	25	5	87.0	226	2	94.0	15	2	94.0	42	2	94.0
Wine	100	12	93.6	25	5	78.0	91	12	92.3	14	5	73.1	84	2	94.9
Xd6	400	96	83.5	157	0	-	350	168	85.5	56	50	78.5	344	110	84.0
Average	300.4	49.6	80.7	119.1	14.7	73.8	259.9	49.8	81.0	37.7	10.7	72.5	206.9	41.0	81.5

2.2 Comparative Study

The interest of this section consists in testing the listed methods and analyzing their effect on the tree size *a posteriori* built with the *C4.5* algorithm. Deliberately, each algorithm has been applied using its original specificities (*i.e.* to reduce the storage requirement), without any methodological adaptation to tree simplification. Therefore, *RT3* and *RNN* has been run using a *k*NN classifier (here *k*=5) during the selection (note that *CNN* is not tested here because it contains the subset deduced by *RNN*). *PSRCG* has been applied from a *5*NN graph. The problem for the Consensus Filters (*CF*) was more complicated, because it requires to *a priori* provide the number of base-detectors and their nature. For this experimental study, we decided to take two base-detectors: a *5*NN classifier and *C4.5*. These two classifiers must agree to remove an instance with *CF*.

We tested here a large panel of data bases, the large majority coming from the UCI Repository. The experimental set-up is the following: each data base is divided into a learning sample LS (2/3 of instances), and a validation set VS (the remaining third). In our concern to control the time complexity, we did not used a cross-validation procedure. While it is not too expensive to compute distances for searching nearest-neighbors, the time requirement for building decision trees is actually higher. In our study, each prototype selection algorithm is applied on LS. *C4.5* is then used to build the decision tree from the selected subset. VS is finally used to assess the predictive accuracy of the model. According to the tested association, several remarks can be made from table 1:

RNN+C4.5 vs C4.5: The main specificity of the *RNN* rule is the keeping of misclassified instances. This way to proceed results in a difficulty to build efficient trees because only overlapping and border points are used to identify relevant features. The results of *RNN+C4.5* confirm this drawback: while the average tree size is highly reduced (14.7 vs 49.6 nodes on average), that improves the model comprehensibility, the predictive accuracy is not controlled and decreases a lot (73.8 vs 80.7%). Therefore, we can not claim that *RNN+C4.5* is an efficient tree simplification procedure.

RT3+C4.5 vs C4.5: While *RT3* is particularly suited to reduce storage requirements of a kNN classifier without reducing the kNN predictive accuracy [15, 18], it seems also to be inefficient for simplifying trees. *RT3+C4.5* does not provide actually a good balance between the tree size reduction, *i.e.* the comprehensibility of the model (10.7 nodes on average vs 49.6 for *C4.5*), and the control of the predictive accuracy which falls much (from 80.7% to 72.5). A Student paired t-test proves the significant superiority of *C4.5*. This noting is in fact not amazing because *RT3* is originally optimized for the kNN classifier. The specificities of *RT3* based on distance calculations (noise-filtering pass removing instances misclassified by their kNN; instance sorting according to the distance from their nearest enemy) make its adaptation for tree simplification difficult.

CF+C4.5 vs C4.5: *CF* are known to be very efficient to improve kNN performances because such classifiers are very sensitive to noise. On the other hand, its use for simplifying trees is a little of interest. Since *CF* are only intended to remove mislabeled instances, the reduction of the data base size is limited (here, only 13.5% on average), resulting in near similar tree sizes on average (49.8 vs 49.6), with same predictive accuracies (81.0% vs 80.7, that is not statistically significant with a critical risk near 40%). One solution to improve *CF* in such a context would consist in using only *C4.5* as base-level detector. But this way to proceed would amount in fact to applying the *Robust-C4.5* algorithm with only one pass.

PSRCG+C4.5 vs C4.5: Incontestably, *PSRCG* seems to be the most efficient algorithm for simplifying trees. Several criteria are actually improved in comparison with the standard *C4.5* algorithm: (i) only 69% of the total number of instances are used for building the tree, that reduces the learning process; (ii) the number of nodes is 17.3% smaller than with the standard *C4.5*, that allows a higher model comprehensibility; (iii) the predictive accuracy is improved (81.5% vs 80.7); using a Student paired t test over accuracies, we found that it is statistically significant with a critical risk near 12%.

Table 2. *Accuracies for C4.5 and RC4.5 on 21 databases*

Accuracy C4.5	Accuracy RC4.5	Tree Size C4.5	Tree Size RC4.5
84.42	84.88	46.2	32.9

PSRCG+C4.5 vs Robust-C4.5: Since *PSRCG* seems to be the most efficient procedure, we have to compare its performances to the *Robust-C4.5* algorithm. We recall in table 2 the average results on accuracies and tree sizes presented in [8]. Note that even if benchmarks are not the same (8 among 21 are the same in the two experiments) and that John used a cross-validation procedure to assess the predictive accuracy, we can reasonably compare performances of the two methods. Even if the main advantage of *RC4.5* comes from a higher tree size reduction (28.8% vs 17.3% with *PSRCG*), we can note that *PSRCG* has four assets:

- using a Student paired t test over accuracies, and testing the superiority to *C4.5*, we found that the critical risk with *RC4.5* is a little higher than with *PSRCG* (14.5% vs 12%),
- While *RC4.5* requires the global update of the tree at each deletion stage, *PSRCG* is based on fast local modifications of the kNN graph. The complexity of *PSRCG* is then smaller than *RC4.5*.
- our information criterion has statistical properties which establish a theoretical framework for halting selection,
- *PSRCG* does not depend on a given learning algorithm (as *C4.5* for *RC4.5*). The kNN are actually used in *PSRCG* only as an information graph, and not as a classifier.

3   Joint Dataset Reduction for Tree Simplification

We have shown in the previous section that thanks its theoretical properties, *PSRCG* can constitute a good solution for simplifying trees. Until now, only instances have been at the center of our selection strategy to reduce tree size. However, we have proposed in [14] an original mixed procedure to take into account at the same time the two degrees of freedom of the selection problem: the number of features and the number of instances. This was the first attempt to take the reduction problem as a whole (features + instances) for a use in a kNN classification, and to establish a functional link between feature selection and instance selection. As far as we know, the endeavour to reduce tree size by removing features and instances in a single algorithm has not be achieved. We propose here to test our strategy for simplifying trees.

We recall in figure 2 the pseudocode of our algorithm, called *(FS+PS)RCG*. Note that *RCG* assesses the gain of uncertainty due to the deletion of an instance

Table 3. Contribution of our mixed procedure for simplifying different decision trees. The tree size is here presented by the number of induced rules # R

Dataset	C4.5		(FS+PS)RCG+C4.5		CART		(FS+PS)RCG+CART		ID3		(FS+PS)RCG+ID3	
	# R	Acc.	# R	Acc.	# R	Acc.	# R	Acc.	# R	Acc.	# R	Acc.
Australian	4	82.0	4	81.0	2	82.7	4	82.7	92	78.4	6	80.0
Bigpole	3	69.5	3	69.5	11	63.5	6	64.0	60	56.0	59	61.8
Breast W.	13	94.9	5	96.7	8	97.0	4	96.7	66	92.3	9	96.4
Brighton	21	94.2	13	96.2	15	95.5	13	97.5	76	87.9	19	94.9
Car	11	96.0	5	86.2	11	92.0	5	86.2	37	91.9	7	86.2
Dermato.	8	92.0	6	92.0	6	91.0	6	89.0	34	84.4	11	90.0
Echocardio	4	65.0	3	70.5	5	68.8	2	70.0	19	58.2	4	69.7
German	3	69.6	4	70.8	7	67.0	7	68.8	145	64.0	7	68.8
Hepatitis	8	70.0	2	70.0	2	70.0	2	70.0	15	70.0	3	64.5
Horse	5	83.0	8	72.6	3	83.0	5	82.1	39	70.8	11	78.6
Iris	3	96.0	3	84.0	3	96.0	3	84.0	16	91.0	3	84.0
LED+17	8	78.0	10	72.5	4	78.5	3	70.0	70	61.0	17	72.8
LED2	4	87.0	4	87.5	4	87.5	2	83.0	37	86.8	4	87.5
Pima	3	74.6	7	70.5	8	72.0	14	65.7	98	66.8	18	62.0
Tic tac toe	10	79.0	10	76.8	30	84.0	9	77.1	128	82.8	105	79.0
Vehicle	7	73.0	7	71.7	5	72.8	6	72.6	88	68.6	14	73.1
Waves	32	68.0	11	67.5	15	77.0	4	71.0	50	60.3	13	65.5
White H.	3	94.0	2	94.0	2	94.0	2	94.0	41	89.8	2	94.0
Wine	8	79.5	3	92.3	4	79.5	3	92.3	11	89.6	4	92.3
Average	**8.9**	**81.0**	**6.2**	**79.6**	**8.3**	**81.1**	**5.8**	**79.4**	**58.8**	**76.2**	**17.1**	**78.6**

or a feature and $U_{loc}(\omega)$ is the local uncertainty as presented in the previous section (for more details see [14]).

The *RCG* specificities should allow us to use this procedure as a pre-process before the induction of decision trees by different algorithms, such as ID3 [11] or CART [2]. To test these other algorithms, we used the SIPINA software, developed in the computer science ERIC_Lyon laboratory[1], which collects the state-of-the-art tree induction algorithms. We tested our selection procedure on 20 datasets. Even if we kept the same experimental set-up, we decided here to assess the sensitivity of our algorithm to noise by adding artificially noise in data (by changing the output class of 10% of the learning instances). Results are presented in the table 3, and discussed below:

(FS+PS)RCG+C4.5 vs C4.5: the effect of our procedure as a pre-process to *C4.5* is interesting. Even if the predictive accuracy of *C4.5* is 1.4% reduced, this reduction is counterbalanced by an average reduction in the number of rules of about 30%. In a field where the comprehensibility is crucial, such a size reduction keeping an excellent predictive accuracy would be very relevant. Using a Student paired t test, we found that *C4.5* is significantly higher than *(FS+PS)RCG+C4.5* with a risk near 15%.

---

[1] http://eric.univ-lyon2.fr

(FS+PS)RCG+CART vs CART: the rule base with *CART* is 30% smaller than with *(FS+PS)RCG+CART*. However, the predictive accuracy of the induced tree is statistically reduced with a smaller critical risk near 7.4%.

(FS+PS)RCG+ID3 vs ID3: the positive contribution of *(FS+PS)RCG* is greatly presented by results on *ID3*. Not only the number of rules is considerably reduced (-71%), but also the generalization accuracy is significantly improved (78.6% vs 76.2% on average) that is highly significant with a risk near 3%. These results are not amazing, because *ID3* does not use a post-pruning (as *C4.5* and *CART*), and is then very sensitive to noise because of the overfitting problem. By applying our mixed selection, it results in fact in deleting these noisy data and improving *ID3* performances.

```
(FS+PS)RCG ALGORITHM
E = {X₁, ..., Xₚ}; Compute RCG in E; Stop=False
Repeat
    For each Xᵢ ∈ E Compute RCGᵢ in E − {Xᵢ}
    Select Xₘᵢₙ with RCGₘᵢₙ = Min{RCGⱼ}
    If RCGₘᵢₙ >> RCG then
        E = E − {Xₘᵢₙ}
        RCG ← RCGₘᵢₙ
        Remove mislabeled + center instances
        that have U_loc(ω) = 0
    else Stop ← true
Until Stop=true
Repeat
    Compute RCG₀ in E
    Delete ω that have a maximum U_loc(ω)
    Compute RCG₁ after the deletion of ω
Until (RCG₁<RCG₀) or not(RCG₁>>0)
```

Fig. 2. Pseudocode of $(FS + PS)RCG$

4   Conclusion

With the development of modern data bases, dataset reduction becomes a major problem in many research fields such as in statistics or in computer science. Many reduction methods have been proposed for knowledge discovery in databases or for reducing computational costs of instance-based learning algorithms, without attempting to establish relations between these subfields. In this paper, we have tested some state-of-the-art prototype selection methods in order to simplify decision tree size. Experiments show that their contribution is not inconsiderable, and that future works will have to take into account these efficient algorithms for improving understandability of models. We also proposed an original procedure

(FS+PS)RCG to jointly reduce the number of features and instances. By this way to proceed, we emphasize the idea that future reduction algorithms will probably have to theoretically analyze the ways in which instance selection can aid the feature selection process. Moreover, we emphasize the need for studies designed to help understand and quantify the relationship that relates feature and prototype selections.

References

1. H. ALMUALLIM and T.G. DIETTERICH. Learning with many irrelevant features. In *Ninth National Conference on Artificial Intelligence*, pages 547–552, 1991.
2. L. BREIMAN, J.H. FRIEDMAN, R.A. OLSHEN, and C.J STONE. *Classification And Regression Trees*. Chapman & Hall, 1984.
3. L.A. BRESLOW and D.W. AHA. Simplifying decision trees: A survey. *to appear in Knowledge Engineering Review*, 1997.
4. C.E. BRODLEY and M.A. FRIEDL. Identifying and eliminating mislabeled training instances. In *Thirteen National Conference on Artificial Intelligence*, 1996.
5. K.J. CHERKAUER and J.W. SHAVLIK. Growing simpler decision trees to facilitate knowledge discovery. In *Second International Conference on Knowledge Discovery and Data Mining*, 1996.
6. G.W. GATES. The reduced nearest neighbor rule. *IEEE Trans. Inform. Theory*, pages 431–433, 1972.
7. P.E. HART. The condensed nearest neighbor rule. *IEEE Trans. Inform. Theory*, pages 515–516, 1968.
8. G.H. JOHN. Robust decision trees: Removing outliers from databases. In *First International Conference on Knowledge Discovery and Data Mining*, pages 174–179, 1995.
9. G.H. JOHN, R. KOHAVI, and K. PFLEGER. Irrelevant features and the subset selection problem. In *Eleventh International Conference on Machine Learning*, pages 121–129, 1994.
10. M.J. KEARNS and Y. MANSOUR. On the boosting ability of top-down decision tree learning algorithms. *Proceedings of the Twenty-Eighth Annual ACM Symposium on the Theory of Computing*, pages 459–468, 1996.
11. J.R. QUINLAN. Induction of decision trees. *Machine Learning*, 1:81–106, 1986.
12. J.R. QUINLAN. *C4.5 : Programs for Machine Learning*. Morgan Kaufmann, 1993.
13. R. E. SCHAPIRE and Y. SINGER. Improved boosting algorithms using confidence-rated predictions. In *Proceedings of the Eleventh Annual ACM Conference on Computational Learning Theory*, pages 80–91, 1998.
14. M. SEBBAN and R. NOCK. Combining feature and example pruning by uncertainty minimization. In *Sixteenth Conference on Uncertainty in Artificial Intelligence*, 2000.
15. M. SEBBAN and R. NOCK. Instance pruning as an information preserving problem. In *Seventeenth International Conference on Machine Learning*, 2000.
16. THRUN ET AL. The monk's problem: a performance comparison of different learning algorithms. *Technical report CMU-CS 91-197-Carnegie Mellon University*, 1991.
17. P.E. UTGOFF. An improved algorithm for incremental induction of decision trees. In *Eleventh Iternationla Conference on Machine Learning*, pages 318–325, 1994.
18. D.R. WILSON and T.R. MARTINEZ. Instance pruning techniques. In *Fourteenth International Conference on Machine Learning*, pages 404–411, 1997.

Combining Multiple Models
with Meta Decision Trees

Ljupčo Todorovski and Sašo Džeroski

Department of Intelligent Systems, Jožef Stefan Institute
Jamova 39, 1000 Ljubljana, Slovenia
Ljupco.Todorovski@ijs.si, Saso.Dzeroski@ijs.si

Abstract. The paper introduces meta decision trees (MDTs), a novel method for combining multiple models. Instead of giving a prediction, MDT leaves specify which model should be used to obtain a prediction. We present an algorithm for learning MDTs based on the C4.5 algorithm for learning ordinary decision trees (ODTs). An extensive experimental evaluation of the new algorithm is performed on twenty-one data sets, combining models generated by five learning algorithms: two algorithms for learning decision trees, a rule learning algorithm, a nearest neighbor algorithm and a naive Bayes algorithm. In terms of performance, MDTs combine models better than voting and stacking with ODTs. In addition, MDTs are much more concise than ODTs used for stacking and are thus a step towards comprehensible combination of multiple models.

1 Introduction

The task of combining multiple models can be broken down into two subtasks. The first is the *generation* of a diverse set of base-level models. Once the base-level models have been generated, the issue of *combination* of their predictions arises. This is the second subtask. The task of multiple model combination is the focus of this paper.

Several approaches for generating base-level models have been developed. One way is to generate multiple models with different learning algorithms for heterogeneous model representations as in [10]. Another way is to use a single learning algorithm with different initial settings. Multiple models can also be generated by applying a single base-level learning algorithm to different versions of learning data. Different methods for manipulating the set of learning examples can be used, such as random sampling with replacement (also called bootstrap aggregation) in bagging [2] or re-weighting misclassified training examples in boosting [6].

The techniques for combining the predictions obtained from the base-level models can be clustered in three combining paradigms: voting (used in bagging and boosting), stacking [15] and cascading [7]. In a voting scheme, each base-level model gives a vote for its prediction. The prediction receiving the most votes is the final prediction. In stacking, a learning algorithm is used to induce a meta-level model for combining the predictions of the base-level models. Cascading is

D.A. Zighed, J. Komorowski, and J. Żytkow (Eds.): PKDD 2000, LNAI 1910, pp. 54–64, 2000.
© Springer-Verlag Berlin Heidelberg 2000

an iterative process of combining classifiers: at each iteration, the learning data set is extended with the predictions obtained in the previous iteration.

This paper introduces meta decision trees (MDTs), a novel method for combining multiple models. The difference between meta and ordinary decision trees (ODTs) is that MDT leaves specify which base-level model should be used, instead of predicting the class value directly. The decisions are made based on the class probability distributions for the given example predicted by the base-level models. The method is general in the sense that it can be used to combine a set of predictions of base-level models, independently of how they are generated. We developed MLC4.5, a modification of C4.5 [12], for inducing meta decision trees. The description of the method is given in Section 2.

The performance of the proposed method is evaluated on a collection of twenty-one data sets. We combine models generated by five learning algorithms: two tree-learning algorithms C4.5 [12] and LTree [8], the rule-learning algorithm CN2 [4], the k-nearest neighbor (k-NN) algorithm [14] and a modification of the naive Bayes algorithm [9]. In the experiments, we compare the performance of stacking with MDTs to the performance of stacking with ODTs. We also compare MDTs with two voting schemes. Section 3 reports on the experimental methodology and results. The presented work is put in the context of previous work on combining multiple models in Section 4. Section 5 presents conclusions based on the empirical evaluation along with directions for further work.

2 Meta Decision Trees

2.1 What Are Meta Decision Trees

The structure of a meta decision tree is identical to the structure of an ordinary decision tree. A decision (inner) node specifies a test to be carried out on a single attribute value and each outcome of the test has its own branch leading to the appropriate subtree. In a leaf node, a MDT predicts which model is to be used for classification of an example, instead of predicting the class value of the example directly like an ODT.

In the process of inducing meta decision trees two types of attributes are used. *Ordinary* attributes are used in the decision (inner) nodes of the MDT. The role of these attributes is identical to the role of the attributes used for inducing ordinary decision trees. *Class* attributes are used in the leaf nodes only. Each base-level model has its class attribute: the values of the class attribute are equal to the predictions obtained by the base-level model. Thus, the class attribute assigned to the leaf node of the MDT decides which base-level model should be used for prediction.

Attributes in MDTs are properties of the class probability distributions predicted for a given example by the base-level models. Namely, the most general form of a prediction returned by a classification model for a given example is a probability distribution over the possible classes. Let the base-level classifier $\mathcal{C}_\mathcal{L}$ generated with learning algorithm $\mathcal{L}$ return the probability distribution $\mathbf{p}_\mathcal{L}(e)$, when applied to example e: $\mathbf{p}_{\mathcal{L}}(e) = (p_{\mathcal{L}}^{(1)}(e), p_{\mathcal{L}}^{(2)}(e), \ldots, p_{\mathcal{L}}^{(m)}(e))$, where m is the number of classes. The k-th element in this vector denotes the probability that the example e belongs to class c_k as estimated by the model $\mathcal{C}_{\mathcal{L}}$. The class c_* with the highest class probability $p_{\mathcal{L}}^{(*)}$ is predicted by base-level classifier $\mathcal{C}_{\mathcal{L}}$.

The following three properties of class probability distributions are used as attributes in MDTs. First, $\mathcal{L}$_maxprob is the highest class probability (i.e. the probability of the predicted class):

$$\mathcal{L}\text{\_maxprob} = \max_{k=1}^{m} p_{\mathcal{L}}^{(k)}(e).$$

Next, $\mathcal{L}$_entropy is the entropy of the class probability distribution:

$$\mathcal{L}\text{\_entropy} = -\sum_{k=1}^{m} p_{\mathcal{L}}^{(k)}(e) \cdot \log_2 p_{\mathcal{L}}^{(k)}(e).$$

Finally, the $\mathcal{L}_{\text{weight}}$ is the fraction of the training examples used to estimate the class distribution for example e. For decision trees, it is the weight of the examples in the leaf node used to classify the example. For rules, it is the weight of the examples covered by the rule(s) which was used to classify the example. This property does not apply to the nearest neighbor and naive Bayes classifiers.

Table 1. A meta decision tree learned in the balance domain.

```
ltree_entropy <= 0.37699:
|        knn_maxprob <= 0.75079: LTREE (*)
|        knn_maxprob > 0.75079: KNN
ltree_entropy > 0.37699:
|        knn_entropy > 1.49841: KNN
|        knn_entropy <= 1.49841:
|        |        c45_weight <= 0.11388: LTREE
|        |        c45_weight > 0.11388:
|        |        |        c45_maxprob <= 0.95: LTREE
|        |        |        c45_maxprob > 0.95: C45
```

Both the entropy and the maximum probability of a probability distribution can be interpreted as estimates of the confidence of the model in its prediction. If the probability distribution returned is highly spread, the maximum probability will be low and the entropy will be high, indicating the model is not very confident in its prediction. On the other hand, if the probability distribution returned is highly focussed, the maximum probability is high and the entropy low, thus indicating the model is confident in its prediction. Finally, the weight quantifies how reliable the model's estimate of its own confidence is: the higher the weight, the more reliable the estimate.

An example MDT is given in Table 1. The leaf denoted by (*) specifies that the LTree model is to be used to classify an example, if the entropy of the probability distribution returned by it is smaller than 0.38 and the maximum probability in the probability distribution returned by the k-NN model is smaller than 0.75. In sum, if the LTree model is confident in its prediction and the k-NN model is not so confident in its prediction, the leaf recommends using the LTree prediction, which is consistent with common sense in the domain of model combination. The other branch of the tree is based on a much smaller number of examples and is thus much less reliable. It also doesn't make much sense.

2.2   MLC4.5 – A Modification of C4.5 for Learning MDTs

C4.5 is a greedy divide and conquer algorithm for building classification trees [12]. On each step, the best split according to the gain (or gain ratio) criterion is chosen from the set of all possible splits for all attributes. The gain criterion is based on the entropy of the class probability distribution of the examples in the current subset S of training examples:

$$info(S) = -\sum_{i=1}^{m} p(c_i, S) \cdot \log_2 p(c_i, S)$$

where $p(c_i, S)$ denotes the relative frequency of examples in S that belong to class c_i. The gain criterion selects the split that maximizes the decrement of the *info* measure.

When adapting C4.5 for learning meta decision trees, we are interested in the accuracies of each of the individual models $\mathcal{C_L}$ on the examples in S, i.e., the proportion of the examples in S that have a class equal to the corresponding class attribute. The newly introduced measure, used in MLC4.5, is defined as $A = \max_{\mathcal{L} \in LearningAlgorithms} \text{accuracy}(\mathcal{C_L}, S)$, where $\text{accuracy}(\mathcal{C_L}, S)$ denotes the relative frequency of examples in S that are correctly classified by base-level classifier $\mathcal{C_L}$. The vector of accuracies does not have probability distribution properties (its elements do not sum to 1), so the entropy can not be calculated. That is the reason for replacing the entropy based measure with an accuracy based one:

$$info_{\mathcal{ML}}(S) = -\frac{1+A}{2} \cdot \log_2 \frac{1+A}{2} - \frac{1-A}{2} \cdot \log_2 \frac{1-A}{2}.$$

At present, we do not post-prune meta decision trees. The rest of the MLC4.5 algorithm is equivalent to the original C4.5 algorithm for building ordinary classification trees. In order to compare MDTs with ODTs in a principled fashion, we also developed a intermediate version of C4.5 (called AC4.5) that induces ODTs using the accuracy based $info_A$ measure, where $A = \max_{i=1}^{m} p(c_i, S)$, and $info_A$ calculated from A using the same formula as for $info_{\mathcal{ML}}$.

3   Experiments

3.1   Experimental Methodology

In order to evaluate the performance of meta decision trees, we performed experiments on a collection of twenty-one data sets from the UCI Repository of Machine Learning Databases and Domain Theories [11]. These data sets have been widely used in other comparative studies.

Five learning algorithms were used in the base-level experiments: two tree-learning algorithms C4.5 [12] and LTree [8], the rule-learning algorithm CN2 [4], the k-nearest neighbor (k-NN) algorithm [14] and a modification of the naive Bayes algorithm [9]. All algorithms were used with their default settings.

We used five different algorithms for combining classifiers. Two of them are voting schemes, and three are based on stacking:

P-VOTE is a simple plurality vote algorithm. According to this voting scheme, the example is classified in the class that is most frequently predicted by the base-level classifiers.

CD-VOTE is a refinement of the plurality vote algorithm for the case where class probability distributions are given by the base-level classifiers [5]. The probability distribution vectors returned by the base-level classifiers can be summed to obtain the class probability distribution of the meta-level voting classifier. The class predicted is the class with the highest class probability in the summed class distribution.

S-MLC4.5 is a stacking algorithm with meta-level classification trees built using MLC4.5.

S-AC4.5 is a stacking algorithm with ordinary classification trees built using AC4.5, a version of C4.5 with the accuracy measure presented in Section 2.

S-C4.5 is a stacking algorithm with ordinary classification trees built with C4.5.

The same set of attributes was used for all three stacking methods. Class attributes are treated as ordinary ones when inducing ordinary decision trees.

The stacking algorithm we use is similar to the standard stacking technique described in [15]. For time complexity reasons, we use stratified 10-fold cross validation for base-level classification instead of the leave-one-out method. By performing 10-fold cross validation, the class attributes in the meta-level training set are obtained by classifying testing examples that were not used for building the base-level classifiers. Another difference from the standard stacking technique is the use of class probability distributions obtained by the base-level classifiers. The output of each base-level classifier for each example in the test set consist of at least two components: the predicted class and the class probability distribution. All the base level algorithms used in this study calculate the class probability distribution for classified examples, but two of them (k-NN and naive Bayes) do not calculate the weight of the examples used for classification (see Section 2). The code of the other three of them (C4.5, CN2 and LTree) was adapted to output the class probability distribution as well as the weight of the examples used for classification.

Classification errors were measured using 10-fold stratified cross validation. Cross validation is repeated 10 times using a different random reordering of the examples in the data set. The same set of re-orderings were used for all experiments. The average and standard deviation (over the ten cross validations) of the classification error on the test examples are reported.

Table 2. Classification errors (in %) of different meta-level classifiers

Data set	P-VOTE	CD-VOTE	S-MLC4.5	S-AC4.5	S-C4.5
australian	13.96 ±0.71	13.81 ±0.57	14.14 ±0.83	15.45 ±1.19	14.90 ±0.68
balance	14.54 ±1.17	12.35 ±0.79	6.74 ±0.52	8.14 ±0.58	7.91 ±1.17
breast-w	3.86 ±0.19	3.49 ±0.15	2.97 ±0.14	3.30 ±0.32	2.94 ±0.28
bridges-td	13.66 ±0.65	14.35 ±0.80	14.09 ±1.16	17.30 ±2.55	17.06 ±2.27
car	6.02 ±0.22	6.13 ±0.27	3.95 ±0.39	3.80 ±0.33	3.17 ±0.40
chess	0.97 ±0.06	0.68 ±0.08	0.57 ±0.10	0.60 ±0.09	0.46 ±0.07
diabetes	23.13 ±0.65	23.41 ±0.59	24.04 ±0.92	25.08 ±1.25	25.53 ±0.91
echo	30.59 ±1.66	30.97 ±1.83	32.60 ±2.61	36.65 ±3.93	35.63 ±2.84
german	25.43 ±0.55	25.21 ±0.59	25.68 ±0.45	26.79 ±1.16	27.50 ±0.89
glass	27.92 ±0.77	26.65 ±1.51	29.35 ±1.65	36.17 ±2.84	35.10 ±2.43
heart	16.19 ±1.17	17.74 ±1.25	17.80 ±1.18	19.43 ±1.96	18.13 ±1.20
hepatitis	17.55 ±1.14	17.99 ±1.47	16.30 ±1.69	18.11 ±1.96	19.98 ±3.47
hypothyroid	1.01 ±0.05	0.97 ±0.04	0.82 ±0.06	0.95 ±0.08	0.75 ±0.06
image	2.49 ±0.18	2.16 ±0.21	2.51 ±0.17	2.81 ±0.22	2.80 ±0.12
ionosphere	8.20 ±0.79	7.98 ±0.66	9.18 ±0.96	9.81 ±1.32	9.66 ±1.44
iris	3.67 ±0.57	4.20 ±0.45	3.14 ±0.83	4.08 ±1.19	3.15 ±0.55
soya	6.96 ±0.31	6.74 ±0.35	6.11 ±0.60	7.11 ±0.84	6.94 ±0.49
tic-tac-toe	10.21 ±0.87	7.95 ±0.84	0.51 ±0.21	0.35 ±0.20	0.11 ±0.16
vote	3.93 ±0.35	3.86 ±0.37	4.12 ±0.50	4.47 ±0.80	4.24 ±0.54
waveform	14.38 ±0.26	14.73 ±0.32	14.30 ±0.15	15.60 ±0.34	16.11 ±0.45
wine	1.36 ±0.60	1.58 ±0.64	2.64 ±0.64	1.64 ±0.49	1.64 ±0.49
Average	**11.72 ±0.62**	**11.57 ±0.66**	**11.03 ±0.75**	**12.27 ±1.13**	**12.08 ±1.00**

3.2   Experimental Results

Table 2 presents the classification errors of the five meta-level classifiers. The lowest average error is achieved using stacking with meta decision trees. A comparative analysis of the performance of **S-MLC4.5** versus the other four meta-level classifiers and the best base-level classifier (LTree) is given in Table 3.

The figures in Table 3 represent the relative error reduction achieved by using the **S-MLC4.5** algorithm as compared to each of the other algorithms, calculated as $1 - s\_mlc4.5\_error/other\_method\_error$. Positive/negative figures denote better/worse performance of **S-MLC4.5**. The statistical significance of the differences is tested using paired t-tests with significance level of 95%: $+/-$ to the right of a figure in the table means that **S-MLC4.5** is signi-

ficantly better/worse. The average here is calculated as $1 - \mathtt{GeometricMean}$
$(s\_mlc4.5\_error/other\_method\_error)$.

Table 3. Classification error reduction (in %) achieved with stacking using meta decision trees as compared to other meta-level classifiers and the base-level classifier with
smallest average error. The + and - signs indicate the significance of the difference.

Data set	P-VOTE		CD-VOTE		S-AC4.5		S-C4.5		LTree	
australian	-1.29		-2.39		8.48	+	5.10	+	-2.24	
balance	53.65	+	45.43	+	17.20	+	14.79	+	0.15	
breast-w	23.06	+	14.90	+	10.00	+	-1.02		50.00	+
bridges-td	-3.15		1.81		18.55	+	17.41	+	3.76	
car	34.39	+	35.56	+	-3.95		-24.61	-	64.67	+
chess	41.24	+	16.18	+	5.00		-23.91	-	26.92	+
diabetes	-3.93	-	-2.69		4.15		5.84	+	4.38	+
echo	-6.57	-	-5.26		11.05	+	8.50	+	8.09	+
german	-0.98		-1.86	-	4.14	+	6.62	+	4.75	+
glass	-5.12	-	-10.13	-	18.86	+	16.38	+	7.30	+
heart	-9.94	-	-0.34		8.39	+	1.82		-1.19	
hepatitis	7.12		9.39		9.99	+	18.42	+	12.55	+
hypothyroid	18.81	+	15.46	+	13.68	+	-9.33	-	16.33	+
image	-0.80		-16.20	-	10.68	+	10.36	+	23.01	+
ionosphere	-11.95	-	-15.04	-	6.42		4.97		20.79	+
iris	14.44		25.24	+	23.04		0.32		-0.32	
soya	12.21	+	9.35	+	14.06	+	11.96	+	75.03	+
tic-tac-toe	95.00	+	93.58	+	-45.71		-363.64	-	97.16	+
vote	-4.83		-6.74		7.83		2.83		2.14	
waveform	0.56		2.92	+	8.33	+	11.24	+	2.32	+
wine	-94.12	-	-67.09	-	-60.98	-	-60.98	-	9.59	
Average	**18.99**		**16.33**		**5.82**		**-5.29**		**32.34**	

When compared to the best base-level classifier (last column in Table 3),
we can clearly see the improvement obtained with meta decision trees used for
stacking. Significant improvement is achieved in 14 out of 21 data sets, the overall
improvement being 32%.

Stacking with meta decision trees is significantly better than plurality vote
in 7 domains and significantly worse in 6. However, the significant improvements
are much higher than the significant drops of accuracy, giving overall accuracy
improvement of 19%. Since **CD-VOTE** performs slightly better than plurality
vote, smaller overall improvement of 16% is achieved. **S-MLC4.5** is significantly
better in 9 data sets and significantly worse in 5.

To compare stacking with MDTs and ODTs, we first look at the relative
performance of **S-MLC4.5** and **S-C4.5**. **S-MLC4.5** performs significantly better in 11 and worse in 5 data sets. There is a 5% overall decrease of accuracy
(this is a geometric mean), but this is entirely due to result in the tic-tac-toe do-

main, where all stacking methods perform very well. If we exclude the tic-tac-toe domain, a 3% overall increase is observed. We can thus say that **S-MLC4.5** performs slightly better in terms of accuracy. However, the MDTs are much smaller (the size reduction factor being 4, see Table 4), despite the fact that ODTs induced with C4.5 are post-pruned and MDTs are neither pre- nor post-pruned.

Table 4. Sizes (in number of nodes) of meta decision trees and ordinary decision trees used for stacking

Data set	S-MLC4.5	S-AC4.5	S-C4.5
australian	17.16 ±2.67	88.48 ±3.98	35.00 ±3.04
balance	6.48 ±1.19	121.95 ±3.79	32.89 ±1.78
breast-w	5.82 ±1.30	30.76 ±3.00	6.64 ±2.39
bridges-td	4.72 ±1.06	21.04 ±1.83	7.52 ±1.19
car	27.04 ±3.71	181.20 ±3.93	43.90 ±2.71
chess	11.79 ±2.00	34.66 ±3.67	9.50 ±1.32
diabetes	18.92 ±3.71	123.00 ±4.73	79.20 ±5.36
echo	6.54 ±0.90	59.04 ±3.79	22.54 ±3.20
german	20.76 ±3.88	132.68 ±3.60	101.58 ±5.20
glass	7.44 ±0.74	226.12 ±6.63	49.98 ±4.56
heart	7.54 ±1.73	59.56 ±3.77	18.78 ±3.17
hepatitis	7.08 ±0.73	42.22 ±2.26	14.76 ±1.47
hypothyroid	7.68 ±1.72	40.46 ±4.14	4.50 ±0.92
image	19.44 ±2.37	320.14 ±10.59	63.97 ±3.27
ionosphere	12.48 ±2.07	51.30 ±2.84	19.48 ±2.34
iris	3.62 ±0.63	23.11 ±2.19	5.45 ±0.49
soya	8.38 ±1.06	436.57 ±15.08	81.43 ±11.29
tic-tac-toe	6.04 ±1.13	16.74 ±1.46	7.54 ±0.30
vote	9.62 ±0.93	38.98 ±2.17	8.90 ±1.44
waveform	37.84 ±5.66	479.55 ±8.26	353.31 ±21.65
wine	3.98 ±0.61	13.22 ±1.97	4.60 ±0.35
Average	**11.92** ±1.90	**120.99** ±4.46	**46.26** ±3.69

To get a clearer picture of the performance differences due to the increased representation power of MDTs as compared to ODTs, we compare **S-MLC4.5** and **S-AC4.5**. Both MLC4.5 and AC4.5 use the same learning algorithm. The only difference between them is the types of trees they induce: MLC4.5 induces meta decision trees and AC4.5 induces ordinary ones. The comparison clearly shows that MDTs outperform ODTs for stacking. The overall accuracy improvement is only 6%, but **S-MLC4.5** is significantly better than **S-AC4.5** in 13 out of 21 data sets and is significantly worse in only one. Furthermore, the MDTs are, on average, ten times smaller than the ODTs (see Table 4). The reduction of the tree size improves the comprehensibility of meta decision trees. For example, we were able to interpret and comment on the MDT in Table 1 (Section 2.1).

4   Related Work

An overview of methods for constructing ensembles of classifiers can be found in [5]. Two recent studies are closely related to our work: the SCANN method, based on stacking using correspondence analysis of the classifications of the base-level classifiers [10] and local cascade generalization [7]. The SCANN method outperforms the plurality vote scheme, especially in the case when the base-level classifiers are highly correlated. The SCANN method does not use any class probability distribution properties of the predictions by the base-level classifiers. Therefore, no comparison with the CD voting scheme is included in their study. The set of base-level classifiers used is similar to ours: C4.5 and another algorithm for trees, CN2 for rules, two nearest neighbor algorithms and a naive Bayesian classifier.

In local cascading generalization, the base-level classifiers are used in every node of the decision tree. New attributes, based on the class probability distribution of the example obtained by the base-level classifiers are generated at each step of the divide and conquer algorithm for building decision trees. The base-level classifiers used in this study are naive Bayes and Linear Discriminant. The integration of base-level classifiers is much tighter than in stacking. The similarity to our approach is that class probability distributions are used.

Ordinary decision trees have already been used for combining multiple models in [3]. However, the emphasis of their study is more on partitioning techniques for massive data sets and combining multiple models trained on different subsets of massive data sets. Our study focuses on combining multiple models generated on the same data set. Therefore, the obtained results are not directly comparable to theirs.

The present study is also related to our previous work on the topic of meta-level learning [13]. There we introduced an inductive logic programming (ILP) framework for learning the relation between data set characteristics and the performance of different (base-level) learning algorithms. MDTs use a representation language that is slightly more expressive than propositional decision trees, but much less than ILP.

5   Conclusions and Further Work

We have presented a new technique for combining classifiers based on meta decision trees (MDTs). MDTs increase the expressiveness of propositional decision trees to make them more suitable for stacking. The empirical evaluation of MDTs showed that they outperform ordinary decision trees in terms of accuracy and are much more concise. MDTs are usually so small that they can easily be inspected: we regard this as a step towards a comprehensible model of combining multiple models. In contrast, most existing work uses non-symbolic learning methods (e.g. neural networks) to stack classifiers [10].

There are several obvious directions for further work. For ordinary classification trees, it is already known that post-pruning gives better results than pre-pruning. Preliminary experiments show that pre-pruning degrades the classification accuracy of MDTs. Thus, one of the priorities for further work is the development of a post-pruning method for meta decision trees and its implementation in MLC4.5.

An interesting aspect of our work is that we use class-distribution properties for meta-level learning. Most of the work on combining classifiers only uses the predicted classes and not the corresponding probability distributions (with the notable exception of Gama). It would be interesting to use other learning algorithms (neural networks, Bayesian classification) to combine models based on the probability distributions returned by them. A comparison of stacking with predictions of models only vs. with predictions and probability distribution properties would be also worthwhile.

The consistency of meta decision trees with common sense model combination knowledge, as briefly discussed in Section 2.1, opens another question for further research. The process of inducing meta-level classifiers should be biased to produce only meta-level classifiers consistent with existing knowledge. This can be achieved using strong language bias within MLC4.5 or, probably more easily, within a framework of meta decision rules, where rule templates could be used.

Note that meta decision trees are, in principle, transferable across domains, in the sense that a MDT built on one data set can be used on any other data set (since it uses the same set of attributes). MDTs can be also built using examples from data sets originating from different domains. Combining data from different domains for learning MDTs is an interesting avenue for further work that would bring together the present study with meta-level learning work on selecting appropriate classifiers for a given domain [1]. In this case, attributes describing individual data set properties can be added to the class distribution properties in the meta-level learning data set.

Acknowledgements

The work reported was supported in part by the Slovenian Ministry of Science and Technology and by the EU-funded project Data Mining and Decision Support for Business Competitiveness: A European Virtual Enterprise (IST-1999-11495).

References

1. Brazdil, P. B. and Henery, R. J. (1994) Analysis of Results. In Michie, D., Spiegelhalter, D. J., and Taylor, C. C., editors: *Machine learning, neural and statistical classification*. Ellis Horwood.
2. Breiman, L. (1996) Bagging Predictors. *Machine Learning* 24(2): 123–140.
3. Chan, P. K. and Stolfo, S. J. (1997) On the Accuracy of Meta-learning for Scalable Data Mining. *Journal of Intelligent Information Systems* 8(1): 5–28.

4. Clark, P. and Boswell, R. (1991) Rule induction with CN2: Some recent improvements. In *Proceedings of the Fifth European Working Session on Learning*: 151–163. Springer-Werlag.

5. Dietterich, T. G. (1997) Machine-Learning Research: Four Current Directions. *AI Magazine* 18(4): 97–136.

6. Freund, Y. and Schapire, R. E. (1996) Experiments with a New Boosting Algorithm. In *Proceedings of the Thirteenth International Conference on Machine Learning*. Morgan Kaufmann.

7. Gama, J. (1998) Combining Classifiers by Constructive Induction. In *Proceedings of the Ninth European Conference on Machine Learning*.

8. Gama, J. (1999) Discriminant trees. In *Proceedings of the Sixteenth International Conference on Machine Learning*: 134-142. Morgan Kaufmann.

9. Gama, J. (2000) A Linear-Bayes Classifier. Technical Report. Artificial Intelligence and Computer Science Laboratory, University of Porto.

10. Merz, C. J. (1999) Using Correspondence Analysis to Combine Classifiers. *Machine Learning* 36(1/2): 33–58. Kluwer Academic Publishers.

11. Murphy, P. M. and Aha, D. W. (1994) *UCI repository of machine learning databases* [http://www.ics.uci.edu/~mlearn/MLRepository.html]. Irvine, CA: University of California, Department of Information and Computer Science.

12. Quinlan, J. R. (1993) *C4.5: Programs for Machine Learning*. Morgan Kaufmann.

13. Todorovski, L. and Džeroski, S. (1999) Experiments in Meta-Level Learning with ILP. In *Proceedings of the Third European Conference on Principles of Data Mining and Knowledge Discovery*: 98–106. Springer-Werlag.

14. Wettschereck, D. (1994) *A study of distance-based machine learning algorithms*. PhD Thesis, Department of Computer Science, Oregon State University, Corvallis.

15. Wolpert, D. (1992) Stacked Generalization. *Neural Networks* 5(2): 241–260.

Materialized Data Mining Views*

Tadeusz Morzy, Marek Wojciechowski, and Maciej Zakrzewicz

Poznan University of Technology
Institute of Computing Science
ul. Piotrowo 3a, 60-965 Poznan, Poland
tel. +48 61 6652378, fax +48 61 8771525
{morzy,marek,mzakrz}@cs.put.poznan.pl

Abstract. Data mining is a useful decision support technique, which can be used to find trends and regularities in warehouses of corporate data. A serious problem of its practical applications is long processing time required by data mining algorithms. Current systems consume minutes or hours to answer simple queries. In this paper we present the concept of materialized data mining views. Materialized data mining views store selected patterns discovered in a portion of a database, and are used for query rewriting, which transforms a data mining query into a query accessing a materialized view. Since the transformation is transparent to a user, materialized data mining views can be created and used like indexes.

1 Introduction

Data mining, also referred to as database mining or knowledge discovery in databases (KDD), aims at discovery of useful patterns from large databases or warehouses. Currently we are observing the evolution of data mining environments from specialized tools to multi-purpose data mining systems offering some level of integration with existing database management systems. From a user's point of view data mining can be seen as advanced querying: a user specifies the source data set and the requested class of patterns, the system chooses the right data mining algorithm and returns discovered patterns to the user. The most serious problem concerning data mining queries is a long response time. Current systems consume minutes or hours to answer simple queries.

Another important feature of data mining is that it is an iterative and interactive process. Users very often periodically perform the same data mining tasks to get the up-to-date information. We claim that data mining systems should provide support for such repetitive queries. It is desirable to store the results of a data mining query that will be repeated after some changes to the database because there are known algorithms for incremental data mining. In this paper we propose using periodically refreshed *materialized data mining views* (MDMVs) for repetitive data mining

---

* This work was partially supported by the grant no. KBN 43-1309 from the State Committee for Scientific Research (KBN), Poland.

D.A. Zighed, J. Komorowski, and J. Zytkow (Eds.): PKDD 2000, LNAI 1910, pp. 65-74, 2000.
© Springer-Verlag Berlin Heidelberg 2000

queries in the same manner as materialized views are used in relational database management systems to store results of complex and time consuming queries.

Benefits of using MDMVs to answer data mining queries where the query to be answered is the same as the query defining an existing MDMV are obvious. The question we try to answer in this paper is: can we use MDMVs to efficiently answer a data mining query that is not equal but only similar to the query defining some MDMV? We consider two data mining queries similar, if they have the same schema of source datasets and resulting patterns, and differ in selection predicates applied to the query on the source dataset and/or constraints concerning statistical strength and contents of patterns.

In this paper we present the concept of MDMVs and their application in the discovery of frequent itemsets and association rules. Since it is straightforward to generate association rules from frequent itemsets, we focus on the frequent itemsets only. We illustrate our optimization rules with many examples expressed in *MineSQL*, which is a declarative, multi-purpose *SQL*-like language for interactive and iterative data mining in relational databases, developed by us over the last couple of years [8][9].

1.1 Basic Definitions

Frequent itemsets. Let $L=\{l_1, l_2, ..., l_m\}$ be a set of literals, called items. Let a non-empty set of items T be called an *itemset*. Let D be a set of variable length itemsets, where each itemset $T \subseteq L$. We say that an itemset T *supports* an item $x \in L$ if x is in T. We say that an itemset T *supports* an itemset $X \subseteq L$ if T supports every item in the set X. The *support* of the itemset X is the percentage of T in D that support X. The problem of mining frequent itemsets in D consists in discovering all itemsets whose support is above a user-defined support threshold.

Association rules. An *association rule* is an implication of the form $X \rightarrow Y$, where $X \subseteq L$, $Y \subseteq L$, $X \cap Y = \varnothing$. Each rule has associated measures of its statistical significance and strength, called *support* and *confidence*. The rule $X \rightarrow Y$ holds in the set D with support s if $s\%$ of itemsets in D support $X \cup Y$. The rule $X \rightarrow Y$ has confidence c if $c\%$ of itemsets in D that support X also support Y. The problem of mining association rules in D consists in discovering all associations rules whose support and confidence are above user-defined support thresholds for support and confidence respectively.

1.2 Data Mining Queries

We have proposed a declarative language, called *MineSQL*, for expressing data mining problems by means of *data mining queries*. *MineSQL* is a *SQL*-based interface between a client application and a data mining system. It plays similar role to data mining applications as *SQL* does to traditional database applications. *MineSQL* is *declarative* - the client application is separated from the data mining algorithm being used. Any modifications and improvements done to the algorithm do not influence the applications. *MineSQL* follows the syntax philosophy of *SQL* language – data mining queries can be combined with *SQL* queries, i.e. *SQL* results can be mined and *MineSQL* results can be queried. Thus, existing database applications can be easily modified to use data mining methods. In this section we present elements of *MineSQL* that are used later in the paper, where MDMVs and their usage are discussed. The detailed syntax of *MineSQL* can be found in [8].

MineSQL language defines a set of new *SQL* data types, which are used to store and manage association rules and itemsets. The *SET OF* data types family (*SET OF NUMBER, SET OF CHAR*, etc.) is used to represent sets of items, e.g. a shopping cart contents. In order to convert single item values into a *SET OF* value, we use a new *SQL* group function called *SET*.

The *ITEMSET OF* data types family is used to represent frequent itemsets. For an itemset its support is stored together with the set of items it contains. We define a set of *SQL* functions and operators that operate on rules: SIZE(x) returns the number of items in the itemset x, s CONTAINS q returns *TRUE* if the itemset s contains the set q, SUPPORT(x) returns support of the itemset x.

The *RULE OF* data types family is used to represent association rules, containing body, head, support and confidence values. We define a set of *SQL* functions and operators that operate on rules: BODY(x) returns the SET OF value representing the body of the rule x, HEAD(x) returns the SET OF value representing the head of the rule x, SUPPORT(x) returns support of the rule x, CONFIDENCE(x) returns confidence of the rule x.

The central statement of the *MineSQL* language is *MINE*. *MINE* is used to discover frequent itemsets or association rules from the database. *MINE* also specifies a set of predicates to be satisfied by the returned rules or patterns.

The following *MINE* statement uses the *PURCHASED_ITEMS* table to discover all frequent itemsets, whose support is greater than 0.1. We display the itemsets and their supports.

```
MINE ITEMSET, SUPPORT(ITEMSET)
FOR X FROM (SELECT SET(ITEM) AS X
                FROM PURCHASED_ITEMS GROUP BY T_ID)
WHERE SUPPORT(ITEMSET)>0.1;
```

1.3   Related Work

The problem of mining association rules was first introduced in [1] and an algorithm called *AIS* was proposed. In [2], two new algorithms were presented, called *Apriori* and *AprioriTid* that are fundamentally different from the previous ones. The algorithms achieved significant improvements over *AIS* and became the core of many new algorithms for mining association rules. *Apriori* and its variants first generate all frequent itemsets (sets of items appearing together in a number of database records meeting the user-specified support threshold) and then use them to generate rules. *Apriori* and its variants rely on the property that an itemset can only be frequent if all of its subsets are frequent. It leads to a level-wise procedure. First, all possible 1-itemsets (itemsets containing 1 item) are counted in the database to determine *frequent 1-itemsets*. Then, frequent 1-itemsets are combined to form potentially frequent 2-itemsets, called *candidate 2-itemsets*. Candidate 2-itemsets are counted in the database to determine *frequent 2-itemsets*. The procedure is continued by combining the frequent 2-itemsets to form *candidate 3-itemsets* and so forth. A disadvantage of the algorithm is that it requires K or $K+1$ passes over the database to discover all frequent itemsets, where K is the size of the greatest frequent itemset found.

In [4], an algorithm called *FUP* (Fast Update Algorithm) was proposed for finding the frequent itemsets in the expanded database using the old frequent itemsets. The major idea of *FUP* algorithm is to reuse the information of the old frequent itemsets and to integrate the support information of the new frequent itemsets in order to reduce the pool of candidate itemsets to be re-examined. Another approach to incremental mining of frequent itemsets was presented in [11]. The algorithm introduced there required only one database pass and was applicable not only for expanded but also for reduced database. Along with the itemsets, a *negative border* [12] was maintained.

In [10] the issue of interactive mining of association rules was addressed and the concept of *knowledge cache* was introduced. The cache was designed to hold frequent itemsets that were discovered while processing other queries. Several cache management schemas were proposed and their integration with the *Apriori* algorithm was analyzed. An important contribution was an algorithm which used itemsets discovered for higher support thresholds in the discovery process for the same task, but with a lower support threshold.

The idea of precomputing frequent itemsets in a partitioned database and using them while discovering association rules in the whole database or parts of it was discussed in [13]. The itemsets were materialized and store in a compact form. The proposed method exploited the property that an itemset can be frequent in the union of a number of partitions if and only if it is frequent in at least one of the partitions. Thus itemsets that were frequent in at least one of the partitions of the mined dataset, formed the set of candidates for one verifying database pass.

The notion of data mining queries (or *KDD* queries) was introduced in [6]. The need for Knowledge and Data Management Systems (KDDMS) as second generation data mining tools was expressed. The ideas of application programming interfaces and data mining query optimizers were also mentioned. Several data mining query languages that are extensions of *SQL* were proposed [3][5][7][8][9].

2 Data Mining Views

Relational databases provide users with a possibility of creating views and materialized views. A view is a virtual table presenting the results of the *SQL* query hidden in the definition of the view. Views are used mainly to simplify access to frequently used data sets that are results of complex queries. When a user selects data from a view, its defining query has to be executed but the user does not have to be familiar with its syntax.

Since data mining tasks are repetitive in nature and the syntax of data mining queries may be complicated, we propose to extend the usage of views to handle both *SQL* queries and *MineSQL* queries. The following statement creates the data mining view presenting the results of the data mining task discussed earlier.

```
CREATE VIEW BASKET_ITEMSETS
```

```
AS MINE ITEMSET
FOR X FROM (SELECT SET(ITEM) AS X FROM PURCHASED_ITEMS GROUP BY T_ID)
WHERE SUPPORT(ITEMSET)>0.1;
```

In the defining statement of a data mining view, there are two classes of constraints: database constraints and mining constraints. *Database constraints* are located within the *SELECT* statement in the *FROM* clause of the *MINE* statement. Database constraints are used to apply selection conditions on the source dataset that is being mined. *Mining constraints* are located in the *WHERE* clause of the *MINE* statement and are used to specify selection conditions on the set of patterns to be discovered.

An important advantage of data mining views is separation of applications processing results of data mining queries from predicates defining parameters of data mining algorithms. If applications access frequent patterns by means of data mining views, they do not have to be changed when only selection predicates (database or mining predicates) are changed in a data mining query. In such case only views have to be modified.

Any *SQL* query concerning the view presented above involves performing the data mining task according to the data mining query that defines the view. This guarantees access to up-to-date patterns but leads to long response times, since data mining algorithms are time consuming.

3 Materialized Data Mining Views

In database systems it is possible to create materialized views that materialize the results of the defining query to shorten response times. Of course, data presented by a materialized view may become invalid as the source data changes. One of the solutions minimizing effects of this problem is periodic refreshing of materialized views. In fact, in the area of data mining, changes to the source database should not be considered to be a serious problem because data mining tasks are usually performed on data warehouses rather than on operational databases. In data warehouses, changes are applied in bulks and materialized data mining views should be refreshed only after a series of changes, together with other views existing in the data warehouse.

We introduce materialized data mining views (MDMVs) with the option of automatic periodic refreshing. A materialized data mining view is a database object containing patterns (association rules or frequent itemsets) discovered as a result of a data mining query. It contains rules and patterns that were valid at a certain point of time. MDMVs can be used for further selective analysis of discovered patterns with no need to re-run mining algorithms. They can be automatically refreshed according to a user-defined time interval. This might be useful when a user is interested in a set of rules or itemsets, whose specification does not change in time, but he or she always wants to have access to relatively recent information.

The following statement creates a MDMV containing all frequent itemsets with support greater than 0.1, discovered in the set of transactions from *PURCHASED_ITEMS* table. The view is to be refreshed once a week.

```
CREATE MATERIALIZED VIEW BASKET_ITEMSETS
REFRESH 7 AS MINE ITEMSET
FOR X FROM (SELECT SET(ITEM) AS X FROM PURCHASED_ITEMS GROUP BY T_ID)
WHERE SUPPORT(ITEMSET)>0.1;
```

In most cases when a MDMV is being refreshed, it can be refreshed efficiently with one of the algorithms for incremental mining. Moreover, it is desirable to store information about the time of last changes applied to the source objects, in order to detect situations when refreshing is not necessary, since the source dataset has not changed.

4 Data Mining Query Rewriting with Materialized Data Mining Views

MDMVs can be also used to reduce execution time of data mining queries, which are not identical to those, on which the views were built. Consider the following example: we are given a MDMV defined over the following data mining query.

```
CREATE MATERIALIZED VIEW V1
AS MINE ITEMSET
FOR ITEMS FROM (SELECT SET(ITEM) AS ITEMS
                FROM PURCHASED_ITEMS GROUP BY T_ID)
WHERE SUPPORT(ITEMSET)>0.2;
```

Assume that a user wants to discover frequent itemsets with the following data mining query.

```
MINE ITEMSET
FOR ITEMS FROM (SELECT SET(ITEM) AS ITEMS
                FROM PURCHASED_ITEMS GROUP BY T_ID)
WHERE SUPPORT(ITEMSET)>0.2
AND ITEMSET CONTAINS TO_SET('A,D');
```

Notice that in order to execute the query, we can simply filter the actual contents of the materialized data mining view *V1*, without running a data mining algorithm. Thus, MDMVs can play a similar role to data mining queries, as indexes or materialized views do to database queries. Application developers can create MDMVs to transparently decrease execution times of their applications' data mining queries.

We need formal methods for determining data mining query execution plans, which use MDMVs to reduce time complexity. First, we define four relations, which may occur between two data mining queries, DMQ_1 and DMQ_2. We say that:

1. *DMQ_1 extends database constraints* of *DMQ_2*, if *DMQ_1* does one of the following:
- appends a *WHERE* or *HAVING* clause of database constraints of *DMQ_2*
- appends an additional *ANDed* condition to a *WHERE* or *HAVING* clause of database constraints of *DMQ_2*
- removes an *ORed* condition from a *WHERE* or *HAVING* clause of database constraints of *DMQ_2*

Example. The following data mining query *DMQ_1* extends database constraints of the data mining query *DMQ_2*.

DMQ_1:
```
MINE ITEMSET
FOR ITEMS
FROM (SELECT SET(ITEM) AS ITEMS
      FROM PURCHASED_ITEMS
```

DMQ_2:
```
MINE ITEMSET
FOR ITEMS
FROM (SELECT SET(ITEM) AS ITEMS
      FROM PURCHASED_ITEMS
```

```
    WHERE ITEM!='D' AND T_ID>100          WHERE ITEM!='D'
    GROUP BY T_ID)                         GROUP BY T_ID)
 WHERE SUPPORT(ITEMSET)>0.2;            WHERE SUPPORT(ITEMSET)>0.2;
```

Intuitively, extension of database constraints means narrowing the mined data set.

2. *DMQ₁ reduces database constraints* of *DMQ₂*, if *DMQ₁* does one of the following:
- removes a *WHERE* or *HAVING* clause of database constraints of *DMQ₂*
- appends an additional *ORed* condition to a *WHERE* or *HAVING* clause of database constraints of *DMQ₂*
- removes an *ANDed* condition from a *WHERE* or *HAVING* clause of database constraints of *DMQ₂*

Example. The following data mining query *DMQ₁* reduces database constraints of the data mining query *DMQ₂*.

DMQ₁:

```
MINE ITEMSET
FOR ITEMS
FROM (SELECT SET(ITEM) AS ITEMS
    FROM PURCHASED_ITEMS
    GROUP BY T_ID)
WHERE SUPPORT(ITEMSET)>0.2;
```

DMQ₂:

```
MINE ITEMSET
FOR ITEMS
FROM (SELECT SET(ITEM) AS ITEMS
    FROM PURCHASED_ITEMS
    GROUP BY T_ID
    HAVING COUNT(*)>10)
WHERE SUPPORT(ITEMSET)>0.2;
```

Intuitively, reduction of database constraints means extending the mined data set.

3. *DMQ₁ extends mining constraints* of *DMQ₂*, if *DMQ₁* does one of the following:
- replaces SUPPORT(ITEMSET)>x with SUPPORT(ITEMSET)>y in *DMQ₂*, where $x<y$
- replaces SUPPORT(ITEMSET)<x with SUPPORT(ITEMSET)<y in *DMQ₂*, where $x>y$
- replaces ITEMSET CONTAINS X with ITEMSET CONTAINS Y in *DMQ₂*, where $X \subset Y$
- replaces ITEMSET NOT CONTAINS X with ITEMSET NOT CONTAINS Y in *DMQ₂*, where $Y \subset X$
- replaces SIZE(ITEMSET)>x with SIZE(ITEMSET)>y in *DMQ₂*, where $x<y$
- replaces SIZE(ITEMSET)<x with SIZE(ITEMSET)<y in *DMQ₂*, where $x>y$
- appends a *WHERE* or *HAVING* clause of mining predicates of *DMQ₂*
- appends an additional *ANDed* condition to a *WHERE* or *HAVING* clause of mining constraints of *DMQ₂*
- removes an *ORed* condition from a *WHERE* or *HAVING* clause of mining constraints of *DMQ₂*

Example. The following data mining query *DMQ₁* extends mining constraints of the data mining query *DMQ₂*.

DMQ₁:

```
MINE ITEMSET
FOR ITEMS
FROM (SELECT SET(ITEM) AS ITEMS
    FROM PURCHASED_ITEMS
    GROUP BY T_ID)
WHERE SUPPORT(ITEMSET)>0.4;
```

DMQ₂:

```
MINE ITEMSET
FOR ITEMS
FROM (SELECT SET(ITEM) AS ITEMS
    FROM PURCHASED_ITEMS
    GROUP BY T_ID)
WHERE SUPPORT(ITEMSET)>0.2;
```

Intuitively, extension of mining constraints means narrowing the resulting set of discovered patterns.

4. DMQ_1 *reduces mining constraints* of DMQ_2, if DMQ_1 does one of the following:
- replaces SUPPORT(ITEMSET)>x with SUPPORT(ITEMSET)>y in DMQ_2, where $x>y$
- replaces SUPPORT(ITEMSET)<x with SUPPORT(ITEMSET)<y in DMQ_2, where $x<y$
- replaces ITEMSET CONTAINS X with ITEMSET CONTAINS Y in DMQ_2, where $Y{\subset}X$
- replaces ITEMSET NOT CONTAINS X with ITEMSET NOT CONTAINS Y in DMQ_2, where $X{\subset}Y$
- replaces SIZE(ITEMSET)>x with SIZE(ITEMSET)>y in DMQ_2, where $x>y$
- replaces SIZE(ITEMSET)<x with SIZE(ITEMSET)<y in DMQ_2, where $x<y$
- removes a *WHERE* or *HAVING* clause from mining constraints of DMQ_2
- appends an additional *ORed* condition to a *WHERE* or *HAVING* clause of mining constraints of DMQ_2
- removes an *ANDed* condition from a *WHERE* or *HAVING* clause of mining constraints of DMQ_2

Example. The following data mining query DMQ_1 reduces mining constraints of the data mining query DMQ_2.

DMQ_1:	*DMQ_2:*
`MINE ITEMSET`	`MINE ITEMSET`
`FOR ITEMS`	`FOR ITEMS`
`FROM (SELECT SET(ITEM) AS ITEMS`	`FROM (SELECT SET(ITEM) AS ITEMS`
`   FROM PURCHASED_ITEMS`	`   FROM PURCHASED_ITEMS`
`   GROUP BY T_ID)`	`   GROUP BY T_ID)`
`WHERE SUPPORT(ITEMSET)>0.4;`	`WHERE SUPPORT(ITEMSET)>0.2;`

Intuitively, reduction of mining constraints means expanding the resulting set of discovered patterns.

We also define four different mining methods, which will be used to execute data mining queries over MDMVs: full mining, incremental mining, complementary mining, and verifying mining. *Full mining* (FM) refers to executing a complete algorithm for discovering frequent itemsets (e.g. [2]). This method is used if MDMV contents is unusable to execute the data mining query. *Incremental mining* (IM) refers to discovering frequent itemsets in an incremented data set (e.g. [4]). It can be used for data mining queries which reduce database constraints. *Complementary mining* (CM) refers to discovering frequent itemsets based on currently materialized itemsets, which will remain frequent (e.g. [10]). This method can be used for data mining queries which reduce mining constraints. Finally, we have *verifying mining* (VM), that simply consists in pruning those materialized itemsets, which do not satisfy mining constraints. It is used for data mining queries, which extend mining constraints.

If two relations occur between a data mining query and a data mining query on which a MDMV is based, then we use the compatibility table (see Table 1) to decide what mining method to use.

Table 1. Compatibility table for using materialized data mining views

	reduction of database constraints	extension of database constraints	-
reduction of mining constraints	CM, IM	FM	CM
extension of mining constraints	VM, IM	FM	VM
-	IM	FM	-

Example. We are given the following data mining query DMQ_1 and the materialized data mining view $MDMV_1$.

DMQ_1:
```
MINE ITEMSET
FOR ITEMS
FROM (SELECT SET(ITEM) AS ITEMS
    FROM PURCHASED_ITEMS
    GROUP BY T_ID)
WHERE SUPPORT(ITEMSET)>0.4;
```

$MDMV_1$:
```
MINE ITEMSET
FOR ITEMS
FROM (SELECT SET(ITEM) AS ITEMS
    FROM PURCHASED_ITEMS
    GROUP BY T_ID
    HAVING COUNT(*)>10)
WHERE SUPPORT(ITEMSET)>0.2;
```

Since DMQ_1 extends mining constraints (higher minimum support) and reduces database constraints (removed HAVING clause) of the data mining query of $MDMV_1$, we perform verifying mining (*VM*), and then incremental mining (*IM*). The verifying mining prunes all materialized itemsets, whose support value is not above 0.4, while the incremental mining discovers frequent itemsets using the information on frequent itemsets discovered in a subset of the mined data set. It was proven in the literature that the execution time of the above mining algorithms will be shorter than when performing full mining.

Example. We are given the following data mining query DMQ_1 and the materialized data mining view $MDMV_1$.

DMQ_1:
```
MINE ITEMSET
FOR ITEMS
FROM (SELECT SET(ITEM) AS ITEMS
    FROM PURCHASED_ITEMS
    GROUP BY T_ID)
WHERE SUPPORT(ITEMSET)>0.3
AND ITEMSET CONTAINS TO_SET('A,B');
```

$MDMV_1$:
```
MINE ITEMSET
FOR ITEMS
FROM (SELECT SET(ITEM) AS ITEMS
    FROM PURCHASED_ITEMS
    GROUP BY T_ID)
WHERE SUPPORT(ITEMSET)>0.2;
```

Since DMQ_1 extends mining constraints (higher minimum support as well as additional *AND*ed condition) of the data mining query of $MDMV_1$, we perform verifying mining (*VM*). The verifying mining prunes all materialized itemsets, whose support value is not above 0.3 or which do not contain the subset {A,B}.

5 Conclusions and Future Work

In this paper we have presented the concept of materialized data mining views. We have proposed several rules for optimization of data mining queries in environments, where MDMVs, containing results of other data mining queries are available. These rules can serve as a basis for rule-based data mining query optimizers. An important advantage of the solutions we propose is that the algorithms required to implement our optimization framework have already been introduced and verified.

In the future we plan to address the problem of cost-based data mining query optimization, especially concentrating on situations when there are several MDMVs that can be used to optimize the processing of a given data mining query.

Another topic that we plan to discuss is concurrent refreshing of several MDMVs. We believe that in such case, sometimes it might be desirable to combine mining tasks associated with several MDMVs to optimize the global performance of the refresh operation.

In the paper we focused on discovery of frequent itemsets and association rules. In the future we plan to analyze possibilities of using data mining views for optimizing queries concerning other data mining tasks such as discovery of sequential patterns.

References

1. Agrawal R., Imielinski T., Swami A.: Mining Association Rules Between Sets of Items in Large Databases. Proc. of the 1993 ACM SIGMOD Conf. on Management of Data (1993)
2. Agrawal R., Srikant R.: Fast Algorithms for Mining Association Rules. Proc. of the 20th Int'l Conf. on Very Large Data Bases (1994)
3. Ceri S., Meo R., Psaila G.: A New SQL-like Operator for Mining Association Rules. Proc. of the 22nd Int'l Conference on Very Large Data Bases (1996)
4. Cheung D.W., Han J., Ng V., Wong C.Y.: Maintenance of Discovered Association Rules in Large Databases: An Incremental Updating Technique. Proc. of the 12th ICDE (1996)
5. Han J., Fu Y., Wang W., Chiang J., Gong W., Koperski K., Li D., Lu Y., Rajan A., Stefanovic N., Xia B., Zaiane O.R.: DBMiner: A System for Mining Knowledge in Large Relational Databases. Proc. of the 2nd KDD Conference (1996)
6. Imielinski T., Mannila H.: A Database Perspective on Knowledge Discovery. Communications of the ACM, Vol. 39, No. 11 (1996)
7. Imielinski T., Virmani A., Abdulghani A.: Datamine: Application programming interface and query language for data mining. Proc. of the 2nd KDD Conference (1996)
8. Morzy T., Wojciechowski M., Zakrzewicz M.: Data Mining Support in Database Management Systems. Proc. of the 2nd DaWaK Conference (2000)
9. Morzy T., Zakrzewicz M.: SQL-like Language for Database Mining. ADBIS'97 Symposium (1997)
10. Nag B., Deshpande P.M., DeWitt D.J.: Using a Knowledge Cache for Interactive Discovery of Association Rules. Proc. of the 5th KDD Conference (1999)
11. Thomas S., Bodagala S., Alsabti K., Ranka S.: An Efficient Algorithm for the Incremental Updation of Association Rules in Large Databases. Proc. of the 3rd KDD Conference (1997)
12. Toivonen H.: Sampling Large Databases for Association Rules. Proc. of the 22nd Int'l Conference on Very Large Data Bases (1996)
13. Wojciechowski M., Zakrzewicz M.: Itemset Materializing for Fast Mining of Association Rules. Proc. of the 2nd ADBIS Conference (1998)

Approximation of Frequency Queries by Means of Free-Sets

Jean-François Boulicaut, Artur Bykowski, and Christophe Rigotti

Laboratoire d'Ingénierie des Systèmes d'Information
INSA Lyon, Bâtiment 501
F-69621 Villeurbanne Cedex, France
{Jean-François.Boulicaut,Artur.Bykowski,Christophe.Rigotti}@insa-lyon.fr

Abstract. Given a large collection of transactions containing items, a basic common data mining problem is to extract the so-called frequent itemsets (i.e., set of items appearing in at least a given number of transactions). In this paper, we propose a structure called free-sets, from which we can approximate any itemset support (i.e., the number of transactions containing the itemset) and we formalize this notion in the framework of ε-adequate representation [10]. We show that frequent free-sets can be efficiently extracted using pruning strategies developed for frequent itemset discovery, and that they can be used to approximate the support of any frequent itemset. Experiments run on real dense data sets show a significant reduction of the size of the output when compared with standard frequent itemsets extraction. Furthermore, the experiments show that the extraction of frequent free-sets is still possible when the extraction of frequent itemsets becomes intractable. Finally, we show that the error made when approximating frequent itemset support remains very low in practice.

1 Introduction

Several data mining tasks (e.g., association rule mining [1]) are based on the evaluation of frequency queries to determine how often a particular pattern occurs in a large data set. We consider the problem of frequency query evaluation, when patterns are itemsets, in dense data sets[1] like, for instance in the context of census data analysis [4] or log analysis [8]. In these important but difficult cases, there is a combinatorial explosion of the number of frequent itemsets and computing the frequency of all of them turns out to be intractable. In this paper, we present an efficient technique to approximate closely the result of the frequency queries, and formalize it within the ε-adequate representation framework [10]. Intuitively an ε-adequate representation is a representation of data that can be substituted to another representation to answer the same kind of queries, but eventually with some lost of precision (bound by the ε parameter). First evidences of the practical interest of such representations has been given in [10,5].

---

[1] e.g., data sets containing many strong correlations.

D.A. Zighed, J. Komorowski, and J. Żytkow (Eds.): PKDD 2000, LNAI 1910, pp. 75–85, 2000.
© Springer-Verlag Berlin Heidelberg 2000

In this paper, we propose a new ϵ-adequate representation for the frequency queries. This representation called *free-sets*, is more condensed than the ϵ-adequate representation based on itemsets [10]. The key intuition of the free-set representation is the following. Let A, B, C, D represent binary attributes in a database. If we know that the association rule $A, B, C \Rightarrow D$ is nearly an exact rule (i.e., it has only a few exceptions), then we can approximate the frequency of itemset $\{A, B, C, D\}$ using the frequency of $\{A, B, C\}$. Moreover, we can approximate the frequency of any itemset X such that $\{A, B, C, D\} \subseteq X$ by the frequency of $X \setminus \{D\}$. We call free-set an itemset Y such that the items in Y can not be used to form a nearly exact association rule. We show that *frequent* free-sets are an ϵ-adequate representation for frequency queries that can be extracted efficiently, even on dense data sets. We also show that the error made when approximating itemset frequency using frequent free-sets remains very low in practice.

Organization of the Paper. In the next section we introduce preliminary definitions used in this paper. In Section 3, we present the notion of free-set, and show that it can be used as an ϵ-adequate representation for the frequency queries. Due to space limitation proofs are omitted. In section 4, we present an algorithm to extract the frequent free-sets. In Section 5, we give practical evidence that frequent free-sets can be extracted efficiently and that the estimation of the supports of frequent itemsets using frequent free-sets leads in practice to very low errors. We review related work in Section 6. Finally, we conclude with a summary and directions for future work.

2   Preliminary Definitions

When applicable, we use the notational conventions and definitions from [10,11].

2.1   Frequent Sets and Association Rules

In this section we recall standard definitions.

Definition 1 (binary database). Let R be a set of symbols called items. A *row* (also called *transaction*) is a subset of R. A *binary database* r *over* R is a multiset of transactions.

Definition 2 (support and frequency). We note $\mathcal{M}(r, X) = \{t \in r | X \subseteq t\}$ the multiset of rows matched by the itemset X and $Sup(r, X) = |\mathcal{M}(r, X)|$ the *support* of X in r, i.e., the number of rows matched by X. The *frequency* of X in r is $Sup(r, X)/|r|$. Let σ be a frequency threshold, $Freq(r, \sigma) = \{X | X \subseteq R$ and $Sup(r, X)/|r| \geq \sigma\}$ is the set of all σ-frequent itemsets in r.

For notational convenience, we also need the following specific definition.

Definition 3 (frequent sets). $FreqSup(r, \sigma)$ is the set of all pairs containing a frequent itemset and its support, i.e., $FreqSup(r, \sigma) = \{\langle X, Sup(r, X)\rangle | X \subseteq R$ and $Sup(r, X)/|r| \geq \sigma\}$.

2.2 ϵ-Adequate Representation

Definition 4 (ϵ-adequate representation [10]). Let $\mathcal{S}$ be a class of structures. Let $\mathcal{Q}$ be a class of queries for $\mathcal{S}$. The value of a query $Q \in \mathcal{Q}$ on a structure $s \in \mathcal{S}$ is assumed to be a real number in $[0,1]$ and is denoted $Q(s)$. An ϵ-adequate representation for $\mathcal{S}$ w.r.t. a class of queries $\mathcal{Q}$, is a class of structures $\mathcal{C}$, a representation mapping $rep : \mathcal{S} \rightarrow \mathcal{C}$ and a query evaluation function $m : \mathcal{Q} \times \mathcal{C} \rightarrow [0,1]$ such that $\forall Q \in \mathcal{Q}, \forall s \in \mathcal{S}, |Q(s) - m(Q, rep(s))| \leq \epsilon$.

Example 1. An example of a class of structures is the set noted $\mathcal{DB}_R$ of all possible binary databases over a set of items R. An interesting query class is $\mathcal{Q}_R$, the set of all queries retrieving the frequency of an itemset $\subseteq R$. If we denote Q_X the query in $\mathcal{Q}_R$ asking for the frequency of itemset X then $\mathcal{Q}_R = \{Q_X | X \subseteq R\}$ and the value of Q_X on a database instance $r \in \mathcal{DB}_R$ is defined by $Q_X(r) = Sup(r, X)/|r|$.

An example of ϵ-adequate representation for $\mathcal{DB}_R$ w.r.t. $\mathcal{Q}_R$ is the representation of $r \in \mathcal{DB}_R$ by means of $Freq(r, \epsilon)$. The corresponding $rep, \mathcal{C}$ and m are as follows. $\forall r \in \mathcal{DB}_R, rep(r) = FreqSup(r, \epsilon)$, $\mathcal{C} = \{rep(r) | r \in \mathcal{DB}_R\}$, $\forall Q_X \in \mathcal{Q}_R, \forall c \in \mathcal{C}$, if $\exists \langle X, \alpha \rangle \in rep(r)$ then $m(Q_X, c) = \alpha/|r|$ else $m(Q_X, c) = 0$. It is straightforward to see that this is an ϵ-adequate representation for $\mathcal{DB}_R$ w.r.t. $\mathcal{Q}_R$ since $\forall Q_X \in \mathcal{Q}_R, \forall r \in \mathcal{DB}_R, |Q_X(r) - m(Q_X, rep(r))| \leq \epsilon$.

Interesting ϵ-adequate representations are *condensed representations*, i.e., ϵ-adequate representations where structures have a smaller size than the original structures.

3   The Free-Sets as a Condensed Representation

Even though this paper do not concerned directly with *association rules*, we need the following definitions to introduce the concept of free-set in a concise way.

Definition 5 (δ-strong rule). Let R be a set of items, an *association rule* based on R is an expression of the form $X \Rightarrow Y$, where $X, Y \subseteq R$ and $X \cap Y = \emptyset$. A *δ-strong rule*[2] in a binary database r over R is an association rule $X \Rightarrow Y$ such that $Sup(r, X) - Sup(r, X \cup Y) \leq \delta$, i.e., the rule is violated in no more than δ rows.

In this definition, δ is supposed to have a small value, so a δ-strong rule is intended to be a rule with very few exceptions.

---

[2] Stemming from the notion of *strong rule* of [15]

3.1   Free-Sets

Definition 6 (δ-free-set). Let r be a binary database over R, $X \subseteq R$ is a δ-*free-set* w.r.t. r if and only if there is no δ-strong rule based on X in r. The set of all δ-free-sets w.r.t. r is noted $Free(r, \delta)$.

Since δ is supposed to be rather small, informally, a free-set is a set of items such that its subsets (seen as conjunction of properties) are not related by any very strong positive correlation.
One of the most interesting properties of *freeness* is its *anti-monotonicity* w.r.t. itemset inclusion.

Definition 7 (anti-monotonicity). A property ρ is *anti-monotone* if and only if for all itemsets X and Y, $\rho(X)$ and $Y \subseteq X$ implies $\rho(Y)$.

The anti-monotonicity has been identified as a key property for efficient pattern mining [11,12,6], since it is the formal basis of a safe pruning criterion. Indeed, efficient frequent set mining algorithms like APRIORI [1] make use of the (anti-monotone) property *"to be frequent"* for pruning.
The anti-monotonicity of freeness follows directly from the definition of free-set and is stated by the following theorem.

Theorem 1. *Let X be an itemset. For all $Y \subseteq X$ if $X \in Free(r, \delta)$ then $Y \in Free(r, \delta)$.*

3.2   Free-Sets as an ϵ-Adequate Representation

We show now that δ-free-sets can be used to answer frequency queries with a bounded error. The following lemma states that the support of any itemset can be approximated using the support of one of the free-sets.

Lemma 1. *Let r be a binary database over a set of items R, $X \subseteq R$ and $\delta \in [0, |r|]$, then there exists $Y \subseteq X$ such that $Y \in Free(r, \delta)$ and $Sup(r, Y) \geq Sup(r, X) \geq Sup(r, Y) - \delta |X|$.*

This lemma states that the support of an itemset X can be approximated using the support of one of the free-sets, but it remains to determine which free-set to use. We now show that this can be done by simply choosing among the free-sets included in X any free-set with a minimal support value. This is stated more formally by the following theorem.

Theorem 2. *Let r be a binary database over a set of items R, $X \subseteq R$ and $\delta \in [0, |r|]$, then for any $Y \subseteq X$ such that $Y \in Free(r, \delta)$ and $Sup(r, Y) = min(\{Sup(r, Z)|Z \subseteq X \text{ and } Z \in Free(r, \delta)\})$ we have $Sup(r, Y) \geq Sup(r, X) \geq Sup(r, Y) - \delta |X|$.*

In practice, computing the whole collection of δ-free-sets is often intractable. We show now that such an exhaustive mining can be avoided since an ϵ-adequate representation to answer frequency queries can be obtained if we extract only *frequent* free-sets together with a subset of the corresponding negative border [11].

Definition 8 (frequent free-set). Let r be a binary database over a set of items R, we noted $FreqFree(r, \sigma, \delta) = Freq(r, \sigma) \cap Free(r, \delta)$ the set of σ-frequent δ-free-sets w.r.t. r.

Let us adapt the concept of negative border from [11] to our context.

Definition 9 (negative border of frequent free-sets). Let r be a binary database over a set of items R, the negative border of $FreqFree(r, \sigma, \delta)$ is noted $\mathcal{B}d^-(r, \sigma, \delta)$ and is defined as follows: $\mathcal{B}d^-(r, \sigma, \delta) = \{X | X \subseteq R, X \notin FreqFree(r, \sigma, \delta) \wedge (\forall Y \subset X, Y \in FreqFree(r, \sigma, \delta))\}$.

Informally, the negative border $\mathcal{B}d^-(r, \sigma, \delta)$ consists of the smallest itemsets (w.r.t. set inclusion) that are not σ-frequent δ-free. Our approximation technique only needs a subset of the negative border $\mathcal{B}d^-(r, \sigma, \delta)$. This subset, noted $Free\mathcal{B}d^-(r, \sigma, \delta)$, is the set of all free-sets in $\mathcal{B}d^-(r, \sigma, \delta)$.

Definition 10. $Free\mathcal{B}d^-(r, \sigma, \delta) = \mathcal{B}d^-(r, \sigma, \delta) \cap Free(r, \delta)$

As in the case of an ϵ-adequate representation for $\mathcal{DB}_R$ w.r.t. $\mathcal{Q}_R$ using frequent itemsets (see Section 2.2), we need the free-sets and their supports.

Definition 11. $FreqFreeSup(r, \sigma, \delta)$ is the set of all pairs containing a frequent free-set and its support, i.e., $FreqFreeSup(r, \sigma, \delta) = \{\langle X, Sup(r, X) \rangle | X \in FreqFree(r, \sigma, \delta)\}$.

We can now define the ϵ-adequate representation w.r.t. the frequency queries.

Definition 12. The *frequent free-sets representation* w.r.t. σ, δ and a query class $\mathcal{Q} \subseteq \mathcal{Q}_R$, is defined by a class of structures $\mathcal{C}$, a representation mapping rep and a query evaluation function m, where $\forall r \in \mathcal{DB}_R$,
$rep(r) = \langle FreqFreeSup(r, \sigma, \delta), Free\mathcal{B}d^-(r, \sigma, \delta) \rangle$,
$\mathcal{C} = \{rep(r) | r \in \mathcal{DB}_R\}$,
$\forall Q_X \in \mathcal{Q}, \forall c \in \mathcal{C}$, if $\exists Y \in Free\mathcal{B}d^-(r, \sigma, \delta), Y \subseteq X$ then $m(Q_X, c) = 0$ else $m(Q_X, c) = min(\{\alpha | \exists Z \subseteq X, \langle Z, \alpha \rangle \in FreqFreeSup(r, \sigma, \delta)\})/|r|$.

Using this representation, the frequency of an itemset X is approximated as follows. If X has a subset Y which is free but not frequent then the frequency of X is considered to be 0. Otherwise we take the smallest support value among the supports of the subsets of X that are free and frequent.

We now establish that this representation is an ϵ-adequate representation for the following database class and query class.

Definition 13. $\mathcal{DB}_{R,s} = \{r | r \in \mathcal{DB}_R \text{ and } |r| \leq s\}$, i.e., the set of all binary databases having no more than s rows. $\mathcal{Q}_{R,n} = \{Q_X | X \subseteq R \text{ and } |X| \leq n\}$, i.e., the set of frequency queries on itemsets having no more than n items.

Theorem 3.
A frequent free-sets representation *w.r.t.* σ, δ *and a query class* $\mathcal{Q}_{R,n}$ *is an* ϵ-*adequate representation for* $\mathcal{DB}_{R,s}$ *w.r.t.* $\mathcal{Q}_{R,n}$ *where* $\epsilon = max(\sigma, n\delta/s)$.

4   Discovering All Frequent Free-Sets

In this section, we describe an algorithm, called MINEX , generating all frequent free-sets. For clarity, we omit the fact that it outputs their supports as well. Implementation issues are presented in Section 4.2.

4.1   The Algorithm – An Abstract Version

MINEX can be seen as an instance of the levelwise search algorithm presented in [11]. It explores the itemset lattice (w.r.t. set inclusion) levelwise, starting from singletons and stopping at the level of the largest frequent free-sets. More precisely, the collection of candidates is initialized to the collection of all sets of size 1 and then the algorithm iterates on candidate evaluation and larger candidate generation. At the ith iteration of this loop, it scans the database to find out which candidates are frequent free-sets. Then, it generates candidates for the $i+1^{th}$ iteration, taking every set of size $i+1$ such that all proper subsets are frequent free-sets. The algorithm finishes when there is no more candidates.

Algorithm 1
Input: r a *binary database over a set of items* R, σ and δ *two thresholds.*
Output: $FreqFree(r, \sigma, \delta)$

1. $\mathcal{C}_1 := \{\{A\}|A \in R\};$
2. $i := 1;$
3.   **while** $\mathcal{C}_i \neq \emptyset$ **do**
4. $\mathcal{F}req\mathcal{F}ree_i := \{X|X \in \mathcal{C}_i \text{ and } X \text{ is a } \sigma\text{-frequent } \delta\text{-free-set}\};$
5. $\mathcal{C}_{i+1} := \{X|X \subseteq R \text{ and } \forall\, Y \subset X, Y \in \bigcup_{j \leq i} \mathcal{F}req\mathcal{F}ree_j\} \setminus \bigcup_{j \leq i} \mathcal{C}_j;$
6. $i := i+1;$
7.   **od;**
8.   **output** $\bigcup_{j < i} \mathcal{F}req\mathcal{F}ree_j;$

Using the correctness result of the levelwise search algorithm given in [11] the following theorem is straightforward.

Theorem 4 (Correctness). Algorithm MINEX computes the sets of all σ-frequent δ-free-sets.

4.2   Implementation Issues

We used techniques similar to the ones described in [2] for frequent itemsets mining. The candidate generation is made using a join-based function, and the itemset support counters are updated w.r.t. a row of the database using a *prefix-tree* data structure.
The key point that needs a new specific technique is the freeness test in step 4 of the algorithm. An efficient computation of this test can be done, based on the

following remark: Z is not a δ-free-set iff there exists $A \in Z$ and $X = Z \setminus \{A\}$ such that X is not δ-free or X is δ-free and $X \Rightarrow \{A\}$ is a δ-strong rule. Furthermore, the step 5 of the algorithm guarantees that if Z is a candidate, X, which is a subset of Z, must be δ-free. Therefore, during the ith iteration, we might first compute the δ-strong rules of the form $X \Rightarrow \{A\}$ where $X \in \mathcal{F}req\mathcal{F}ree_i$ and $A \in R \setminus X$ and then use them to remove candidates in $\mathcal{C}_{i+1}$ that are not δ-free. Thus, at the begining of an iteration, only free-sets are candidates.

This is incorporated in the algorithm by replacing steps 4 and 5 with the following steps:

4.1 $\mathcal{F}req\mathcal{F}ree_i := \{X | X \in \mathcal{C}_i$ and X is a σ-frequent$\}$;

4.2 $\mathcal{N}ot\mathcal{F}ree_{i+1} := \{Z | Z = X \cup \{A\}$ where $X \in \mathcal{F}req\mathcal{F}ree_i, A \in R \setminus X$
                and $X \Rightarrow \{A\}$ is a δ-strong rule $\}$;

5.1 $\mathcal{C}_{i+1}^g := \{X | X \subseteq R$ and $\forall Y \subset X, Y \in \bigcup_{j \leq i} \mathcal{F}req\mathcal{F}ree_j\} \setminus \bigcup_{j \leq i} \mathcal{C}_j$;

5.2 $\mathcal{C}_{i+1} := \mathcal{C}_{i+1}^g \setminus \mathcal{N}ot\mathcal{F}ree_{i+1}$;

and the step 1 with the following initialization step:

1. $\mathcal{C}_1 := \{\{A\} | A \in R$ and $\emptyset \Rightarrow \{A\}$ is not a δ-strong rule $\}$;

More details on this technique can be found in [7], where it is shown in particular that steps 4.1 and 4.2 can be computed efficiently within the same database scan.

5   Experiments

The running prototype is implemented in C++. We use a PC with 512 MB of memory and a 500 MHz Pentium III processor under Linux operating system.

For an experimental evaluation, we chose the pumsb* data set, a PUMS census data set[3] preprocessed by researchers from IBM Almaden Research Center. The particularity of PUMS data sets is that they are very dense and make the mining of all frequent itemsets together with their supports intractable for low frequency thresholds, because of the combinatorial explosion of the number of frequent itemsets [4].

5.1   Frequent Free-Sets vs. Frequent Sets Condensation

Table 1 shows a comparison of the extraction of frequent sets and frequent free-sets for different frequency thresholds and different values of δ. The collections $FreqFree(r, \sigma, \delta)$ are significantly smaller than the corresponding $Freq(r, \sigma)$. For frequency thresholds of 15% and 20% $Freq(r, \sigma)$ is so large that it is clearly impossible to provide it on our platform, while the extraction of $FreqFree(r, \sigma, \delta)$ remains tractable. For this two frequency thresholds of 15% and 20%, we use lower-bound estimations of $|Freq(r, \sigma)|$. These lower-bounds are computed using the δ-strong rules collected by MinEx (see Section 4.2) to find the size of the

---

[3] http://www.almaden.ibm.com/cs/quest/data/long_patterns.bin.tar

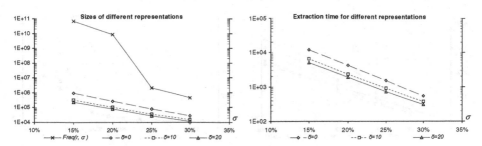

Fig. 1. Extraction time and sizes of different representations.

largest frequent itemset. If this size is m then there is a least 2^m frequent itemsets. Figure 1 (left) emphasizes, using logarithmically scaled axes, the difference of the size of the various representations.

Using also logarithmically scaled axes, Figure 1 (right) shows that the extraction time for MINEX grows up exponentially when the frequency threshold is reduced. This is due to the combinatorial explosion of the number of frequent free-sets. APRIORI-based algorithms have the same global exponential evolution of the extraction time, due in this case to the combinatorial explosion of the number of frequent sets.

Table 1. Comparison of different representations at various frequency thresholds.

σ	15%			20%			25%			30%		
δ	0	10	20	0	10	20	0	10	20	0	10	20
Max frequent free-set size ($\approx$MIN-EX DB scans)	12	11	10	12	10	9	11	9	9	10	9	8
$|FreqFree(r, \sigma, \delta)|$	909 806	324 743	232 887	253 107	105 615	76 413	78 220	36 310	27 137	26 972	14 631	11 079
$FreqFree(r, \sigma, \delta)$ extraction time (sec.)	11 977	6 590	5 126	4 233	2 342	1 890	1 540	905	731	533	373	302
Max frequent set size ($\approx$APRIORI DB scans)	35			32			18			16		
$|Freq(r, \sigma)|$	>2^{35}			>2^{32}			2 064 946			432 699		

5.2   Scale-Up Experiment

On figure 2 we report the extraction time (for $\sigma = 20\%$) when changing the number of rows or the number of items in the data set. We observe an exponential complexity w.r.t. the number of items and a linear complexity w.r.t. number of rows in the data set if the value of δ follows the number of tuples (e.g., if we double the number of rows then we double the value of δ). This is emphasized by a superimposed straight line on figure 2 (left).

5.3   Approximation Error in Practice

In this experiment, we report the practical error made on σ-frequent itemset supports when using the approximation based on σ-frequent δ-free-sets. The data

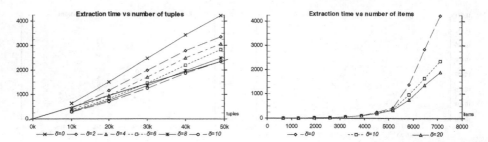

Fig. 2. Behavior of MINEX w.r.t. the number of rows and the number of items.

set is a PUMS data set of Kansas state[4]. We use a version of this data set that has been preprocessed at the University of Clermont-Ferrand (France) in Prof. L. Lakhal's research group. We have reduced this data set to 10000 rows and 317 items to be able to extract all σ-frequent itemsets at a low frequency threshold. For $\sigma = 0.05$ (500 rows), there are 90755 σ-frequent sets and the largest has $n = 13$ items. As a condensed representation, we computed $FreqFreeSup(r, 0.05, 6)$ which contains 4174 elements.

Theoretical error bounds on the frequent set support can be determined using Theorem 2 as follows. In this experiment, the maximal absolute support error is $\delta * n = 6 * 13 = 78$ rows. The maximal relative support error can be obtained assuming that the maximal theoretical absolute error occurs on the σ-frequent set of minimal frequency (i.e. σ). So, the maximal relative support error is $\delta * n/(N * \sigma) = 15.6\%$ ($N = 10000$ rows in the experiment).

The support of each of the 90755 σ-frequent itemsets is approximated using the collection $FreqFreeSup(r, 0.05, 6)$ and Theorem 2 and then compared to the exact support. The maximal absolute support error is 18 rows, and the maximal relative support error is 3.1%. The average absolute support error is 2.12 rows and the average relative support error is 0.28%. Table 2 shows that this error remains very low even for frequent sets containing a lot of items.

Table 2. Error observed on σ-frequent itemset supports by itemset size.

itemset size	1	2	3	4	5	6	7	8	9	10	11	12	13
average abs. sup. error	0	0.24	0.65	1.10	1.53	1.92	2.31	2.75	3.28	3.9	4.58	5.2	5.5
average rel. sup. error	0	0.03%	0.07%	0.13%	0.18%	0.24%	0.31%	0.38%	0.47%	0.58%	0.71%	0.83%	0.88%

---

[4] ftp://ftp2.cc.ukans.edu/pub/ippbr/census/pums/pums90ks.zip

6   Related Work

Using incomplete information about itemset frequencies for some mining task, e.g., rule mining, has been proposed in [10], and formalized in the general framework of ϵ-adequate representations. Probabilistic approaches to the problem of frequency queries have also been investigated (see [14]).

Several search space reductions based on nearly exact (or exact) association rules have been proposed. The use of the nearly exact association rules to estimate the confidence of other rules and then to prune the search space has been suggested in [3] but not investigated nor experimented. Efficient mining of nearly exact rules (more specifically rules with at most δ exceptions) with a single attribute in both the left and the right hand sides has been proposed in [9]. Search space pruning using exact association rules has been experimented in [3] in the context of rule mining and developed independently in the context of frequent itemsets mining in [13]. [13] implicitely proposes a kind of condensed representation called *closed* itemsets which is strongly related to the notion of δ-free-set when $\delta = 0$.

7   Conclusion and Future Work

We proposed a structure called free-sets that can be extracted efficiently, even on dense data sets, and that can be used to approximate closely the support of frequent itemsets. We formalized this approximation in the framework of ϵ-adequate representations [10] and gave a correct extraction algorithm formulated as an instance of the levelwise search algorithm presented in [11].

We reported experiments showing that frequent free-sets can be extracted even when the extraction of frequent itemsets turns out to be intractable. The experiments also show that the error made when approximating the support of frequent itemsets using the support of frequent free-sets remains very low in practice.

Interesting future work includes applications of the notion of free-set to the discovery of association rules with approximated confidence and support, and to the approximation of boolean formula support as investigated in [10].

References

1. R. Agrawal, H. Mannila, R. Srikant, H. Toivonen, and A. I. Verkamo. Fast discovery of association rules. In *Advances in Knowledge Discovery and Data Mining*, pages 307–328. AAAI Press, 1996.
2. R. Agrawal and R. Srikant. Fast algorithms for mining association rules in large databases. In *Proc. VLDB'94*, pages 487 – 499, 1994.
3. R. J. Bayardo. Brute-force mining of high-confidence classification rules. In *Proceedings KDD'97*, pages 123–126, 1997.

4. R. J. Bayardo. Efficiently mining long patterns from databases. In *Proceedings of the 1998 ACM SIGMOD International Conference on Management of Data*, pages 85–93. ACM Press, 1998.

5. J.-F. Boulicaut and A. Bykowski. Frequent closures as a concise representation for binary data mining. In *Proc. PAKDD'00*, volume 1805 of *LNAI*, pages 62–73, Kyoto, JP, 2000. Springer-Verlag.

6. J.-F. Boulicaut and B. Jeudy. Using constraints during itemset mining: a generic approach. Technical Report 2000-01, INSA Lyon, LISI, F-69621 Villeurbanne, Mar. 2000.

7. A. Bykowski. Frequent set discovery in highly-correlated data. Technical Report July 1999, Master of Science thesis, INSA Lyon, LISI, F-69621 Villeurbanne, 1999.

8. A. Bykowski and L. Gomez-Chantada. Frequent itemset extraction in highly-correlated data: a web usage mining application. In *Proc. WKDDM'00*, pages 27–42, Kyoto, JP, Apr. 2000.

9. S. Fujiwara, J. D. Ullman, and R. Motwani. Dynamic miss-counting algorithms: Finding implication and similarity rules with confidence pruning. In *Proc. ICDE'00*, pages 501–511, San Diego, USA, 2000.

10. H. Mannila and H. Toivonen. Multiple uses of frequent sets and condensed representations. In *Proceedings KDD'96*, pages 189–194, Portland, USA, 1996.

11. H. Mannila and H. Toivonen. Levelwise search and borders of theories in knowledge discovery. *Data Mining and Knowledge Discovery*, 1(3):241–258, 1997.

12. R. Ng, L. V. Lakshmanan, J. Han, and A. Pang. Exploratory mining and pruning optimization of constrained association rules. In *Proc. ACM SIGMOD'98*, pages 13–24, Seattle, USA, 1998.

13. N. Pasquier, Y. Bastide, R. Taouil, and L. Lakhal. Efficient mining of association rules using closed itemset lattices. *Information Systems*, 24(1):25–46, 1999.

14. D. Pavlov, H. Mannila, and P. Smyth. Probalistic models for query approximation with large data sets. Technical Report 2000-07, Univsersity of California, Department of Information and Computer Science, Irvine, CA-92697-3425, Feb. 2000.

15. G. Piatetsky-Shapiro. Discovery, analysis, and presentation of strong rules. In *Knowledge Discovery in Databases*, pages 229 – 248. AAAI Press, Menlo Park, CA, 1991.

Application of Reinforcement Learning to Electrical Power System Closed-Loop Emergency Control

C. Druet, D. Ernst, and L. Wehenkel

Department of Electrical and Computer Engineering - Institut Montefiore
University of Liège - Sart-Tilman B28 - B4000 Liège - Belgium

Phone: +32-4-3662645 - Fax: +32-4-3662984
{druet,ernst,lwh}@montefiore.ulg.ac.be

Abstract. This paper investigates the use of reinforcement learning in electric power system emergency control. The approach consists of using numerical simulations together with on-policy Monte Carlo control to determine a discrete switching control law to trip generators so as to avoid loss of synchronism. The proposed approach is tested on a model of a real large scale power system and results are compared with a quasi-optimal control law designed by a brute force approach for this system.

1   Introduction

Reinforcement learning techniques are currently being investigated for suitability of use in a wide variety of environments. These range from game playing environments such as backgammon where these machine learning techniques have been successfully applied to develop systems capable of Master-level play ([8]) to self-adjusting algorithms for packet routing in computer networks ([1]).

In the field of automatic control *Reinforcement Learning* starts also to be well-known. Its principle is to learn how to control by associating a certain benefit to being in a particular state and taking a particular action in that state. We can divide such control learner (agent) in two categories. First we can plug the agent and wait for him to know enough about the system to control it. Unfortunately this is not always possible. Let us imagine a car driven by an agent which doesn't now anything about driving rules. We will have to buy lots of cars before having a *capable* agent. It leads to the second category, the agent which learns from simulated experience before being used in real-life.

In the electric power system community, Dynamic Security Assessment (DSA) has long been recognized to be an issue of great practical concern ([2]). The recent deregulated practices make the need for effective DSA methods more urgent than ever. What holds true for predictive DSA holds even more true for emergency DSA. Predictive DSA concerns what can be done in study mode and then used in the control center to enhance security. Emergency DSA concerns

D.A. Zighed, J. Komorowski, and J. Żytkow (Eds.): PKDD 2000, LNAI 1910, pp. 86–95, 2000.
© Springer-Verlag Berlin Heidelberg 2000

implementation of control systems to deal with emergency situations. It becomes a necessity, given the trend to operate the systems increasingly closer to their limits and given the difficulties in predicting the operating conditions and the troublesome contingencies likely to occur.

This paper deals with during transient emergency TSA (Transient Stability Assessment) and control. The purpose is to demonstrate that an agent using reinforcement learning techniques can appraise and trigger control actions so as to prevent the electrical power system from serious degradation.

2   Reinforcement Learning

2.1   Basics

The idea that we learn by interacting with our environment is probably the first to occur to us when we think about the nature of learning. Reinforcement learning is learning what to do, i.e. how to map situations to actions, so as to maximize a reward signal ([7]). The goal is to discover the actions which yield the most reward, by trying them out. In the most interesting and challenging case, actions may affect not only the immediate reward but also the next situation and all the subsequent rewards.

One of the challenges in reinforcement learning is the trade-off between exploration and exploitation. To obtain a high reward, an agent must prefer actions that it has tried before and found to be rewarding. But some other actions could be better, i.e. actions not yet taken, it must then try them out. The first behavior is called *exploitation*, the second *exploration*. The challenge is that neither exploration nor exploitation can be pursued exclusively without failing the task ([9]).

As we will use the terms *policy*, *reward function* and *value function*, we will define them. A **policy** (π) defines the agent's way of behaving at a given time. A policy is thus a mapping from perceived states to actions. A **reward function** defines the goal in a reinforcement learning problem. This function defines what are the good and the bad events for the agent. The **value function** is more complicated. It specifies what is good in the long run. The value of a state is the reward (R) an agent can expect to accumulate over the future, starting from that state.

Two kinds of value functions exist: the state-value function $V(s)$, where s denotes the state, and the action-value function[1] $Q(s,a)$ where a denotes the action. The first says how good it is to be in a particular state, the second how good it is to take a particular action in a particular state.

The agent makes its decisions as a function of a signal from the environment (state). Certainly, state signal should include the immediate sensations, e.g. measurements, but it has to contain more than that. This state signal is of course not expected to inform the agent of everything about the environment, or even everything that would be useful to it in making decision. Ideally it must sum-

---

[1] Also called state-action function.

marize past sensations compactly, in such a way that all relevant information is retained. A state signal that succeeds in retaining all relevant information is said to have the *Markov property*[2]. Of course this is very restrictive, but even when the state signal doesn't have the Markov property, one can consider it as an approximation of a Markov state, i.e. the more the state signal approaches the Markov property, the better the performance from reinforcement learning systems will be.

2.2   On-Policy Monte Carlo Control

Monte Carlo methods require only experience-sample sequences of states, actions and rewards from on-line or simulated interaction with an environment. Learning from *on-line* experience is striking because it requires no prior knowledge of the environment's dynamics, yet can still reach an optimal behavior in the long run. Learning from simulated experience is also powerful. Although a model is required, the model only generates sample transitions, not all the possible transitions.

We assume experience is divided into episodes. An episode is a sequence of state-action pairs and rewards. All episodes terminate no matter what actions are selected. Monte Carlo methods are incremental in an episode-by-episode sense.

The idea of the Monte Carlo Control method is to maintain both approximate policy and approximate value function. In this scheme, the value function is repeatedly altered to more closely approximate the value function for the current policy, and the policy is repeatedly improved with respect to the current value function. These two changes work against each other as each creates a target for

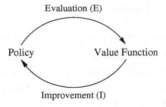

Fig. 1. Approach Scheme

the other, but together they cause both policy and value function to approach optimality.

$$Q^0 \xrightarrow{I} \pi_0 \xrightarrow{E} Q^1 \xrightarrow{I} \pi_1 \xrightarrow{E} Q^2 \xrightarrow{I} \pi_2 \xrightarrow{E} \cdots \xrightarrow{E} Q^* \xrightarrow{I} \pi^*$$

Policy evaluation (E) is done in the following manner: for each state-action pair (s, a) appearing in the episode k,

$$Q^{k+1}(s, a) = Q^k + \alpha[R - Q^k(s, a)]$$

---

[2] Or to be Markov

where R is the reward following that state-action pair. Two techniques have been tested in this paper:

1. $\alpha = \frac{1}{n_{s,a}}$, where $n_{s,a}$ is the number of times the agent has passed through the state-action pair $(s,a)^3$, which corresponds to estimate the state-action values as a sample average of observed rewards;
2. $\alpha = C^{st}$ indicating that the estimates never completely converge but continue to vary in response to the most recently received rewards[4].

Policy improvement (I) can be done making the policy greedy, i.e. a policy that selects the action which has the higher expected reward, with respect to the current action-value function.

$$\pi_k(s) = \arg\max_a Q^k(s,a)$$

This procedure is modified by introducing the exploration part in it. Instead of using the greedy policy described above, one can use an ϵ-greedy policy, i.e. a policy that selects another action than the one with the higher expected reward with a probability of ϵ.

The general algorithm used to implement the Monte Carlo control is thus based on the following scheme.

1. Initialize $Q^0(s,a) \to \pi_0$.
2. Generate an episode using π_k.
3. Evaluate the policy and update the state-action function $\to Q^{k+1}(s,a)$.
4. Improve the policy $\to \pi_{k+1}$.
5. Return to point 2 (loop forever).

The method is called *on-policy* Monte Carlo control because it attempts to estimate the value of a policy while using it. In opposition is the *off-policy* Monte Carlo control method which uses separate policies: one policy is used to generate the episodes (*behavior* policy) which can be unrelated to the second policy that is evaluated and improved, the *estimation* policy.

3   The Practical Problem

Transient stability concerns the ability of an electrical power system to maintain synchronism when subjected to a severe transient disturbance, e.g. a three phase short-circuit cleared by opening a line. The resulting system response involves large excursions of generator rotor angles and is influenced by the nonlinear power angle relationship. When the excursions becomes too large, a loss of synchronism may occur: the system is driven to instability.

---

[3] This method is called *every-visit Monte-Carlo method* by opposition to the *first-visit Monte-Carlo method* which consists in using only the rewards associated to the first visit to (s,a) in each episode.

[4] This is desirable in a nonstationary environment.

The selective tripping of generating units for a severe disturbance which weakens the system transfer capabilities can be used as a method of improving system stability. The rejection of an appropriate amount of generation in the system reduces power to be transferred over the critical transmission interface. Since generating units can be tripped rapidly, the method constitutes a very effective means of improving transient stability ([6]).

The illustration is based on a lightly modified Brasilian power system. The resulting system comprises 63 machines, 1180 busses and 1968 lines and is modeled in its usual detailed way. The generation shedding scheme is applied to the Itaipu transmission system (figure 2): 8 machines of 700 WM, at 60Hz side of Itaipu).

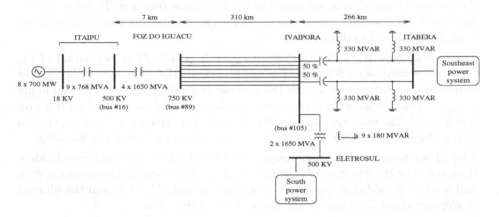

Fig. 2. Topology of the Itaipu transmission system (60 Hz)

What this experience attempts to demonstrate is that it is possible to use reinforcement learning method to control the shedding of the units at Itaipu to maintain stability in terms of synchronism. The control center Itaipu receives measurements of the δ (relative electrical angle, *deg*: $\delta - \delta_{initial}$), ω (angular speed, *rad/s*), P_m (mechanical power, *MW*), P_e (electrical power, *MW*) and the status of the 8 units (on, stopping, off). This is clearly one of the possible descriptions of the state of the system. At each time step, one can decide to shed units. In our experiments one can only shed one machine by time step. As the units are identical, two control actions are possible: to shed one of the 8 units or to wait (*shed-one-unit* or *no-shedding*). Once the control action is decided, a delay of 50 *ms* is introduced before the transition actually occurs, so as to better suit the real-life case[5]. A loss of synchronism will correspond to a bad event (negative reward). A positive reward will be associated to a stable system (system stable x seconds after the fault).

The Measurements. For want of real-world measurements, the illustration is based on time-domain simulation using the ST600 program of Hydro-Québec

---

[5] The *stopping* status of a unit: the control action is decided but the action will be effective after the transmission delay.

([10]). Real-time measurements are thus artificially created. The acquisition of these measurements is supposed to have an observation rate of 5 ms.

The Contingencies. A simulation, i.e. an episode, consists of a three-phase short-circuit applied at bus # 89. The starting time t_s is randomly chosen between $t_{s\,min}$ and $t_{s\,max}$ and the clearing time t_e is randomly chosen between $t_{e\,min}$ and $t_{e\,max}$ to cover a wide variety of cases as a light warranty of generality. The non-zero t_s and the use of a contingency without short-circuit (*no-fault* contingency) prevent the systematic shedding of units. Several post-fault configurations (*x-lines-tripped* contingencies) are also used.

The Agent. Starting from a state s and taking action a one cannot determine the resulting state s', thus we must use a state-action function. The state-action function is an associative table $((s, a) \leftrightarrow r)$; the continuous variation spaces of δ, ω, P_m, P_e are thus discretized.

The policy followed by the agent is an ϵ-greedy policy. After evaluation of the expected reward of each action, the agent chooses the action with the highest reward. If several actions match the highest reward, it chooses one randomly among them. To maintain exploration, a random action is selected with a probability of ϵ. The choice of this ϵ is difficult as we want to keep as much generation as possible. This leads to a very small ϵ [6], so as to avoid excessive shedding.

The Discretizations are very simple. As the eight generators are exactly identical, only one combination of δ, ω and P_a $(P_m - P_e)$ is enough to represent each unit's state. In addition, the number of *running* units U_r (0-8) and the number of *stopping* units U_s complete the description of the state.

The space of δ, ω and P_a are respectively divided into 41, 21 and 15 ranges (respectively 40, 20 and 15 of identical length for each variable plus an additional category to represent the *no-unit-running* case).

The Rewards. The goal of the agent is not only to save the system but also to preserve as much generation as possible. When the system is stable, the reward is positive and corresponds to the total amount of electrical power produced at the end of the simulation so as to distinguish good stable situations (6 units producing) from less good stable situations (e.g. 4 units only). When the system is unstable, the reward is negative and corresponds to the loss of the starting generation, i.e. -5600 (*MW*).

4   The Simulation Results

Our experiments demonstrate that one can apply successfully reinforcement learning techniques to the transient stability emergency problem. On the figure 3 (Contingency: *4-lines-tripped*, $50 < t_e < 60\,ms$, $0 < t_s < 100\,ms$, $\alpha = \frac{1}{n_{s,a}}$, $\epsilon = 0.005$), one can see the evolution of the stability of the system (*dots* are stable episodes and *crosses* unstable episodes) and the evolution of the final production P_f when the number of simulations increases k.

---

[6] $\epsilon < \dfrac{\text{number of units}}{\text{maximum number of time steps}}$.

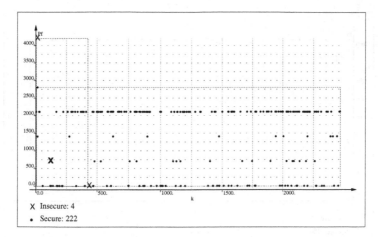

Fig. 3. MW not rejected **versus** evolution of learning

The convergence is very fast in terms of saving the system (a small number of unstable cases with low values of k). This agent learns also very fast to shed as few generation as possible (P_f already high after 250 episodes). The goal is of course to have higher MW not rejected points when the number of episodes increases (later we will see that ideally the value of P_f should still increase up to 2800 MW).

4.1 $\alpha = C^{st}$ Versus $\alpha = \frac{1}{n_{s,a}}$

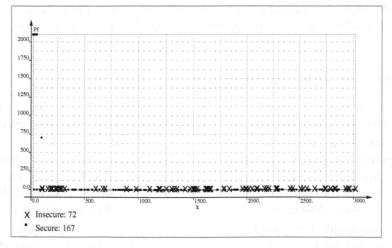

Fig. 4. MW not rejected **versus** evolution of learning for $\alpha = C^{st} (= 0.2)$

We have observed that the second technique ($\alpha = \frac{1}{n_{s,a}}$) converges better and faster than the first technique ($\alpha = C^{st}$). The reason for that is the way we defined the problem. As the state of the system is always the same at the begin-

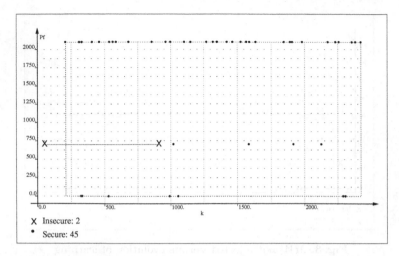

Fig. 5. MW not rejected **versus** evolution of learning for $\alpha = \frac{1}{n_{s,a}}$

ning of the simulation, as the contingencies are always the same, as the models behind the simulator are fixed, the problem can be considered as a stationary problem. Thus there is no reason to use a constant α which is particularly useful in a non-stationary environment.

Looking on the figures 4 and 5 (Contingency: *3-lines-tripped*, $40 < t_e < 50\,ms$, $0 < t_s < 100\,ms$, $\epsilon = 0.005$), one can see that the technique using $\alpha = C^{st}$ (= 0.2) does not even converge to save the system (a lot of crosses during the whole learning). but that the one using $\alpha = \frac{1}{n_{s,a}}$ converges very fast (small number of unstable cases with low values of k). Moreover it successfully succeeds in saving as much production as possible (a already high P_f for low k) despite $\alpha = C^{st}$ cannot.

4.2 Quality of the Control

For this experiment, we use two contingencies, the *no-fault* contingency and the *4-lines-tripped* contingency. The fault duration varies between $40\,ms$ and $100\,ms$.

Analysis of the reinforcement learning control. The state-action values are equal to the sample average of the observed rewards ($\alpha = \frac{1}{n_{s,a}}$). According to the procedure previously established, 2500 simulations have been run. After these simulations, the reward is not corrected anymore and the ϵ-greedy decision process is changed to greedy. The decision process is thus deterministic[7].

These two modifications having been done, we can study the evolution of the controlled system to a *no-fault* contingency. The simulation shows that the method has learned that it was not necessary to exclude machines to stabilize the system: the best reward associated to the current state s is always bound to the *no-action*. Note that in such a scenario, if the decision to take an action is

---

[7] Except when several actions have the same reward in a state s.

not taken at the first instant, no action will be taken at all. This is the direct consequence that the state $s = (\delta, \omega, P_a, U_r, U_s)$ is here constant if any action is decided. 8 machines are thus still in activity at the end of the simulation, the reward is maximum and equal to 5600 ($8 * 700\ MW$).

For the *4-lines-tripped* contingency each state of the period preceding the fault is constant and the same as the constant state of a *no-fault* contingency. Such an observation suggests that the instant of appearance of the fault does not influence the control process of the system. The only relevant parameter is the fault duration (t_e). The following table represents the number of shed machines (U_n) and the production (P_f) at the end of the simulation for different values of t_e. Even if all the machines are sometimes shed (8), no unstable simulations occurs: the control process has always been able to stabilize the system. Note that except for $t_e = 85\ ms$, the number of shed machines is an increasing function of the fault duration.

$t_e(ms)$	U_n	P_f	$t_e(ms)$	U_n	P_f	$t_e(ms)$	U_n	P_f	$t_e(ms)$	U_n	P_f
40	5	2100	55	5	2100	70	7	700	85	7	700
45	5	2100	60	5	2100	75	7	700	90	7	700
50	5	2100	65	7	700	80	8	0	95	8	0

To summarize, we can say that for a *no-fault* contingency the control of the system is optimum because no machines are shed while avoiding loss of stability. For the *4-lines-tripped* contingency, the control also stabilizes the system but the price to pay (the number of shed machines) is sometimes very high. Nothing guarantees that the control designed here is for the *4-lines-tripped* contingencies the optimum or even close to the optimum.

Design of the optimal control. The reinforcement learning procedure that we have applied here to the problem of generation shedding control is self-reliant in the sense that to establish the control law, no human knowledge of the dynamics of the system was required. Exploiting the knowledge of the system we have ([3] and [5]) we are able to establish a near to optimum control law. We described its design in an internal report ([4]).

The following table summarizes the results obtained with simulations carried out using this near optimum control procedure. A comparison with the previous table highlights better quality, but we have to keep in mind that it was only possible to establish this law thanks to the perfect knowledge that we had of the system dynamics and the restricted set of contingencies considered. The comparison is just aimed to give an idea of how far we are from the optimum when we use the reinforcement learning procedure to design a control.

$t_e(ms)$	U_n	P_f	$t_e(ms)$	U_n	P_f	$t_e(ms)$	U_n	P_f	$t_e(ms)$	U_n	P_f
40	3	3500	55	4	2800	70	4	2800	85	5	2100
45	4	2800	60	4	2800	75	4	2800	90	5	2100
50	4	2800	65	4	2800	80	5	2100	95	5	2100

The disappointing performances of the reinforcement learning control law originate from the state discretization which decreases the degree of observability of the system and thus the ability to associate to each state an optimal control action.

5 Discussion

The interest of reinforcement learning for electrical power system emergency control has been demonstrated. To improve the results, we believe that further research is necessary, in particular in terms of choosing a better reward signal.

Problem enhancement. To be applied in Itaipu this procedure has of course to be improved. First, one must consider noise and other uncertainties in the state signal. Second, the experimentation domain should be enlarged by including other starting states, other topologies and other perturbations.

Method enhancement. The first thing to improve is the discretization of δ, ω and P_a. An alternative to that is the use of continuous variable spaces using more sophisticated machine learning techniques such as regression trees or neural networks (multi-layer perceptron) to generalize the state-action function approximation.

References

1. J.A. Boyan and M.L. Littman. Packet routing in dynamically changing networks: A reinforcement learning approach. In J.D. Cowan, G. Tesauro, and J. Alspector, editors, *Advances in Neural Information Processing Systems*, volume 6. Morgan Kaufmann, San Francisco CA, 1993.
2. T. Dy Liacco. Security functions in power system control centers. In *IFAC Symp. on Power System Control, New Delhi*, 1979.
3. D. Ernst, A. Bettiol, Y. Zhang, L. Wehenkel, and M. Pavella. Real time transient stability emergency control of the south-southeast brazilian system. *SEPOPE, Salvador, Brazil*, May 1998.
4. D. Ernst and C. Druet. Design of an optimal control law for emergency transient stability, 2000. Internal report.
5. D. Ernst and M. Pavella. Closed-loop transient stability emergency control. *Presented at the On-line Transient Stability Assessment and Control, panel session at the IEEE/PES Winter Meeting 2000, Singapore*, 2000.
6. P. Kundur and G.K. Morison. Techniques for emergency control of power systems and their implementation. In *Proceedings of the IFAC-CIGRE Symp. on Control of Power Plants and Power Systems, Beijing*, pages 679–684, 1997.
7. R.S. Sutton and A.G. Barto. *Reinforcement Learning, an introduction.* The MIT Press, 1998.
8. G.J. Tesauro. Td-gammon, a self-teaching backgammon program, achieves master-level play. In *Proceedings of the AAAI Fall Symp. on Games: Planning and Learning*, pages 19–23. AAAI Press Technical Report FS93-02, 1993.
9. S.B. Thrun. Efficient exploration in reinforcement learning. Technical Report CMU-CS-92-102, Shool of Computer Science, Carnegie-Mellon University, Pittsburgh, Pennsylvania 15213-3890, January 1992.
10. A. Vallette, F. Lafrance, S. Lefebvre, and L. Radakovitz. *ST600 programme de stabilité : manuel d'utilisation version 701.* Hydro-Québec, Vice-présidence technologie et IREQ, 1987.

Efficient Score-Based Learning of Equivalence Classes of Bayesian Networks

Paul Munteanu and Denis Cau

Centre de Recherche du Groupe ESIEA
38 rue des Docteurs Calmette et Guérin
53 000 Laval, France
{munteanu, cau}@esiea-ouest.fr

Abstract. The use of bayesian networks for knowledge discovery requires learning algorithms which emphasize not only the predictive power but also the structural fidelity of the discovered networks.

Previous work on score-based methods for learning equivalence classes of bayesian networks showed that they generally provide better results than classical algorithms, that explore the space of bayesian networks. However, they are considerably slower, mainly because they use more complicated search operators and because they have to build instances of the equivalence classes in order to check their consistency and in order to calculate their score.

We propose here a new greedy learning algorithm that explores the space of equivalence classes with a reduced set of operators and realizes the verification of the consistency and the computation of the score without any need for instantiation. We show on five experimental tasks that this algorithm is rather efficient, obtains better scores and discovers structures closer to the "gold-standard" than classical greedy and tabu search in the space of bayesian networks.

1 Introduction

Learning bayesian networks from data is one of the most ambitious approaches to Knowledge Discovery in Databases. Unlike most other data mining techniques, it does not focus its search on a particular kind of knowledge but aims to found all the (probabilistic) relations which hold between the considered variables.

From a statistical viewpoint, a bayesian network efficiently encodes the joint probability distribution of the variables describing an application domain. This kind of knowledge allows making rational decisions involving any arbitrary subset of these variables on the basis of the available knowledge about another arbitrary subset of variables.

Moreover, bayesian networks may be represented in a graphical annotated form which seems quite natural to human experts for a large variety of applications. The nodes of a bayesian network correspond to domain variables and the edges which connect the nodes correspond to direct probabilistic relations between these variables. Under certain assumptions [1], these relations have causal

D.A. Zighed, J. Komorowski, and J. Żytkow (Eds.): PKDD 2000, LNAI 1910, pp. 96–105, 2000.
© Springer-Verlag Berlin Heidelberg 2000

semantics (a directed edge $A \to B$ may be interpreted as A *is a direct cause of* B), while most other data mining approaches deal exclusively with correlation.

There are two main approaches to learning bayesian networks with unknown structure. The first one is to build the network according to the conditional independence relations found in data (*e.g.,* [1]). Traditionally, these methods aim at discovering causal relations between the variables and, therefore, emphasize the structural fidelity of the bayesian networks they learn. Unfortunately, they suffer from the lack of reliability of high-dimensional conditional independence tests.

The other approach to learning bayesian networks is to define an evaluation function (or score) which accounts for the quality of candidate networks with respect to the available data and to use some kind of search algorithm in order to find, in a "reasonable" amount of time, a network with an "acceptable" score (we use the terms "reasonable" and "acceptable" because this learning task have been proven to be NP-hard for the evaluation functions mentioned in the following section). These algorithms are less sensitive to the quality of the available data and their results can be successfully used in various decision making tasks.

However, as we will see in the following section, the exploration of the space of bayesian network structures by a greedy search algorithm may end with a structure which fails to reveal some independence relations between the variables and, therefore, may be rather different from the true one. The space of equivalence classes of bayesian network structures seems to be better suited for this kind of search. Learning algorithms which explore this space have already been proposed by some authors, as described in section 3. Unfortunately, these algorithms are considerably slower than classical ones, mainly because they use more complicated search operators, and because they have to build instances of the equivalence classes in order to check their consistency and in order to calculate their score.

Section 4 introduces a new algorithm, EQ1, that explores the space of equivalence classes with a reduced set of operators and realizes the verification of the consistency and the computation of the score without instantiating the equivalence classes constructed during the search. The experimental results presented in section 5 confirm the fact that EQ1 is able to efficiently produce better results than classical greedy and tabu search in the space of bayesian networks.

2   Heuristic Search in the Space of Bayesian Networks

Like in many other machine learning approaches, the quality of a bayesian network with respect to the available data is evaluated on the basis of a score issued from the information theory or from bayesian inference. Some of these scores, like MDL [2] and BDe [3] have been proven to be asymptotically correct and have some nice mathematical properties that can be exploited by the search algorithms:

- the score of a bayesian network may be expressed as a sum of local scores involving only a node and its parents;

- the bayesian networks belonging to the same equivalence class (*i.e.*, representing the same conditional independence relations) have the same score.

Since the unconstrained learning of bayesian networks is NP-hard, most authors propose the use of heuristic search algorithms, which explore the space of bayesian network structures. The transformation operators are the addition, the suppression, and the reversal of edges, submitted to the constraint that the resulting network contains no cycle (on the basis of score decomposability, these operators ensure that no more than two local scores have to be re-evaluated in order to evaluate the resulting network). The search strategy is based on some general-purpose method (greedy search, simulated annealing or tabu search).

Unfortunately, there are many local optima in the space of bayesian networks and heuristic search algorithms may easily be trapped in one of them. The main reason for this difficulty is the equality of the score of equivalent networks. We illustrate this statement by a learning task which is very simple but nevertheless confusing for greedy search (fig. 1). In this example, we have three variables

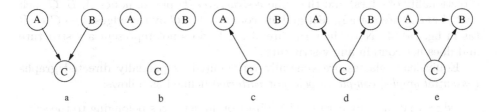

a   b   c   d   e

Fig. 1. Greedy search

distributed according to the bayesian network 1a (which is the only instance of its equivalence class). Let us consider that there is enough data, such as the evaluation function we use (which is asymptotically correct) assigns the best score to the network 1a, among all possible network structures.

Suppose the search starts with the totally unconnected network of fig. 1b. Hopefully, the search algorithm will immediately find that adding edges between A and C and between B and C improves the score of the network. Suppose, for instance, that the addition of an edge between A and C produces the greatest improvement of the score. Since the structures $(A \rightarrow C \quad B)$ and $(A \leftarrow C \quad B)$ belong to the same equivalence class (as described in the next section), the addition of the edge $A \rightarrow C$ and the addition of the edge $A \leftarrow C$ have the same impact on score at the beginning of the search. Therefore, there is a 50 % chance[1] that the search algorithm adds the first edge in the wrong direction $(A \leftarrow C)$. If

---

[1] From our experience with publicly available software, it appears that most programmers seem to neglect this issue and simply apply the first best operator found. Since the order of evaluation of edge additions generally depends on node order and publicly available networks often declare nodes in their topological order, the results of

it does so, the direction of the edge between B and C also becomes indifferent from the score viewpoint, as the structures $(A \leftarrow C \rightarrow B)$ and $(A \leftarrow C \leftarrow B)$ belong to the same equivalence class. Globally, there is a 25 % chance that the network found after the second iteration is $(A \leftarrow C \rightarrow B)$. In this case, since A and B become dependent when C is known (as shown by fig. 1a), an edge will be added between these nodes (*e.g.*, $A \rightarrow B$). This makes the search stop in an incorrect state, because no further single edge reversal or removal can improve the score. In larger networks, this kind of early wrong decisions are very probable and their effects can cumulate and make the final network very different from the ideal one.

3   Heuristic Search in the Space of Equivalence Classes of Bayesian Network Structures

Bayesian networks that represent the same conditional independence relations form an equivalence class. All bayesian networks belonging to the same equivalence class have the same *skeleton* (undirected graph resulting from ignoring the directionality of edges) and the same *v-structures* (triples of nodes A, B, C such that A and B are not adjacent and are connected to C by the edges $A \rightarrow C \leftarrow B$ (as in fig. 1a) [4]. Note the structure of fig. 1e does not represent a v-structure and does not contain any v-structure.

Equivalence classes are generally represented as partially directed graphs (*essential graphs*, *completed pdags* or *patterns*) defined as follows:

- edges that may appear in either direction in networks belonging to the same equivalence class are represented as undirected edges;
- the other edges are represented as directed edges.

These conditions define a unique representation for equivalence classes but do not ensure that the equivalence classes represented this way are legal (*i.e.*, can be instantiated).

In order to overcome the difficulties presented in the previous section, we can realize the search in the space of equivalence classes. Intuitively, this approach consists in allowing the addition of undirected edges when no direction is preferred by the score. Edge orientation is delayed until the interactions between edges make possible the choice of a direction on the basis of the score. Since the obtained partially directed graphs may be interpreted as equivalence classes, this solution consists in a modification of the search space: the search algorithm explores the space of equivalence classes of bayesian networks instead of the space of bayesian networks.

This kind of solution has already been studied in [5]. The conclusion of this work was that the search in the space of equivalence classes generally provides better results than the search in the space of bayesian networks but, unfortunately, it is much more time consuming.

---

these programs on "gold-standard" benchmarks are over-optimistic. Reversing the order of nodes declarations can lead to serious degradations of their performance.

One of the difficulties Chickering met in his work was the evaluation of equivalence classes. Since available evaluation functions have been designed to score bayesian networks (and not equivalence classes), his solution consists in the generation of an arbitrary instance of the equivalence class to be evaluated and the evaluation of this instance according to the classical formulas. Additionally, Chickering's algorithm relies on this procedure in order to prevent the construction of illegal equivalence classes (without instances).

This procedure is much more time-consuming than the evaluation of a bayesian network because it needs some additional time to generate an instance from the equivalence class and because more than two local scores may have to be evaluated in order to evaluate the generated instance. Furthermore, Chickering's algorithm uses a complex set of transformation operators, that produce a great number of candidate graphs at each iteration of the search.

4   The EQ1 Algorithm

EQ1 is a learning algorithm that explores the space of equivalence classes with a reduced set of operators and realizes the verification of the consistency and the computation of the score without instantiating the equivalence classes constructed during the search.

EQ1 greedily uses the following transformation operators:

- *Operator 1 :* addition of a directed edge $X \to Y$ between two nodes which are not adjacent;
- *Operator 2 :* addition of a v-structure $X \to Y \leftarrow Z$ between three nodes in configuration $X \quad Y - Z$, where X and Z are not adjacent (addition of a directed edge together with the orientation of a previously undirected edge);
- *Operator 3 :* addition of an undirected edge $X - Y$ between two nodes which are not adjacent.

Since "repair" operators (edge deletions and reversals), generally used in traditional learning algorithms, are mainly needed for correcting the errors of edge orientation made in the early phases of the search, we decided not to use them in EQ1.

4.1   Constraints on the Transformation Operators

As discussed earlier, not all partially directed graphs represent legal equivalence classes. A partially directed graph G represents a legal equivalence class if and only if it satisfies the following conditions [6]:

1. G is a chain graph (*i.e.*, it contains no directed or partially directed cycle);
2. every chain component of G is chordal (*i.e.*, on every undirected cycle of length ≥ 4 there are two non-consecutive nodes connected by an undirected edge - a *chord*);

3. the configuration $A \to B - C$ does not occur as an induced subgraph of G;
4. every edge $A \to B$ must occur in at least one of the four configurations of fig. 2 as an induced subgraph of G.

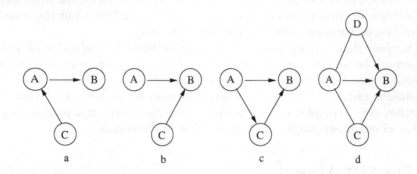

Fig. 2. Possible configurations for directed edges

It can be easily verified that this characterization entails the following constraints on the transformation operators of EQ1 :

- *Operator 1 :* X and Y must have different sets of parents. $X \to Y$ may not introduce any directed or partially directed cycle.
- *Operator 2 :* $X \to Y$ may not introduce any directed or partially directed cycle (after the orientation of $Y \leftarrow Z$).
- *Operator 3 :* X and Y must have the same (possibly empty) sets of parents. If $X - Y$ introduces an undirected cycle, it must also introduce an undirected triangle.

Furthermore, the addition of a v-structure may require the orientation of some previously undirected edges, in order to satisfy the third condition of the above theorem. Similarly, the addition of a directed edge (*e.g.*, $C \to B$ in fig. 2a), or of an undirected edge (*e.g.*, $A - C$ in fig. 2b or $C - D$ in fig. 2d) may request the removal of the orientation of some previously directed edges ($A \to B$ in fig. 2a, $A \to B$ and $C \to B$ in fig. 2b, $A \to B$, $C \to B$ and $D \to B$ in fig. 2d) in order to satisfy the fourth condition of the theorem. The addition or the removal of an edge orientation may have cascading effects.

Since these additions or removals of edge orientations do not have any influence on the evaluation of the transformations (presented in the next subsection), they are implemented as post-processing operations, performed by EQ1 only for the best found transformation, which is really applied on the current graph.

4.2   Scoring the Transformations

Since an equivalence class contains instances with the same score, we can define the score of an equivalence class as being the score of any of its instances. In the following, $score(G)$ denotes the score of the (directed or partially directed) graph G. Remember the score of a bayesian network is decomposable on nodes. $score(X \mid \mathcal{P})$ denotes the score of node X with the set of nodes $\mathcal{P}$ as parents. $Pa_G(X)$ ("parents" of X in G) denotes the set of nodes Y such that $X \leftarrow Y$ belongs to graph G. $Br_G(X, Y)$ ("common brothers" of X and Y in G) denotes the set of nodes Z such that $X - Z$ and $Y - Z$ belong to G.

All formulas used by EQ1 for scoring the transformations can be derived on the basis of the following argument: Let G and G' be the partially oriented graphs corresponding to the current equivalence class and to the transformed equivalence class. We will see in the following paragraphs that, for all transformation operators, we can build some instances I of G and I' of G' such that all nodes have the same parents in I and I', except for a single node, Y. It follows that the difference between the scores of G' and G is equal to the difference between the score of Y in I' and the score of Y in I. All transformations can therefore be evaluated by computing the score of a single node in two different configurations.

First consider the addition of a directed edge $X \rightarrow Y$ (*Operator 1*). We can build I such that all undirected edges adjacent to Y in G, $Y - Z$, are oriented as $Y \rightarrow Z$ in I. Let I' be the directed graph obtained by adding $X \rightarrow Y$ to I. It can be easily verified that I' is an in instance of G'. We have, therefore :

$$\Delta score(G', G) = score(Y \mid Pa_G(Y), X) - score(Y \mid Pa_G(Y))$$

If the addition of the edge $X \rightarrow Y$ is done together with the orientation of a previously undirected edge $Y - Z$, which becomes $Y \leftarrow Z$ (*Operator 2*), we build I such that $Y - Z$ is already oriented as $Y \leftarrow Z$ and all the other undirected edges adjacent to Y in G, $Y - W$, are oriented as $Y \rightarrow W$. An instance of G', I', can be obtained from I by adding the edge $X \rightarrow Y$. We have, therefore :

$$\Delta score(G', G) = score(Y \mid Pa_G(Y), Z, X) - score(Y \mid Pa_G(Y), Z)$$

Let us now consider the addition of an undirected edge $X - Y$ (*Operator 3*). Since instances have only directed edges, I' must be produced from I by adding a directed edge, for instance $X \rightarrow Y$. In order to avoid directed cycles in I', we have to orient all edges $Y - Z$, where Z is a common brother of X and Y, as $Y \leftarrow Z$. This will not introduce any spurious v-structure in I and I', because X and Y have the same parents in G and all their common brothers are interconnected (*cf.* constraints on *Operator 2*). All other undirected edges adjacent to Y in G, $Y - W$, are oriented as $Y \rightarrow W$. The formula used for scoring this transformation is, therefore:

$$\Delta score(G', G) = score(Y \mid Pa_G(Y), Br_G(X, Y), X)$$
$$-score(Y \mid Pa_G(Y), Br_G(X, Y))$$

5   Experimental Results

In order to evaluate the performances of EQ1, we have compared it experimentally to greedy search and tabu search in the space of bayesian networks

(GreedyBN and TabuBN). TabuBN uses a tabu list of 10 states and stops after 10 consecutive iterations without score improvement.

All algorithms use the MDL score (see [2] for more details and justification):

$$score(G) = \sum_{i=1}^{n} (\log(n) + \log \binom{n}{|Pa_i|} + \|Pa_i\|(\|X_i\| - 1)\frac{\log(N)}{2} + NH(X_i|Pa_i))$$

where $X_1, \cdots, X_n$ are the nodes of G, $Pa_i = Pa_G(X_i)$ are their sets of parents, $|Pa_i|$ is the number of parents of X_i, $\|X_i\|$ is the number of values of X_i, $\|Pa_i\|$ is the number of different instantiations of the set of variables Pa_i, N is the number of examples and $H(X_i|Pa_i)$ is the conditional entropy of X_i, given Pa_i.

The comparison has been realized on learning tasks involving five publicly available bayesian networks of various sizes: *Cancer* (5 nodes, 5 edges), *Asia* (8 nodes, 8 edges), *CarStarts* (18 nodes, 17 edges), *Alarm* (37 nodes, 46 edges), *Hailfinder* (56 nodes, 66 edges).

In order to improve the statistical significance of the experimental results, we have compared the algorithms on thirty different data sets for each network (1,000 examples for the small networks *Cancer* and *Asia*, and 10,000 for the others, generated according to the probability distributions modeled by the networks).

The first criterion we have evaluated is the score. Tables 1 and 2 present:

- the network used to generate the examples;
- the means of the score of the compared algorithms[2];
- the number of data sets on which the first algorithm obtained a score inferior, equal or superior to the score of the second one (the MDL score has to be *minimized*);
- the probability p that the means of the score of the two algorithms are equal (paired t-test);
- the algorithm which obtained the better mean of the score, if $p < 0.01$.

Table 1. GreedyBN vs. EQ1

Network	GreedyBN	EQ1	<	=	>	p	Best
Cancer	3266.29	3261.57	0	15	15	1.12E-05	EQ1
Asia	3343.20	3335.82	0	14	16	2.06E-04	EQ1
CarStarts	33563.80	33517.19	0	4	26	2.69E-07	EQ1
Alarm	139719.52	139198.80	5	0	25	2.76E-06	EQ1
Hailfinder	720712.31	720038.42	0	0	30	3.07E-11	EQ1

---

[2] Since this work deals exclusively with the optimization of evaluation functions proposed elsewhere, only the scores obtained on train sets are reported. Nevertheless, we have also evaluated the scores on independent test sets and observed the same patterns of relative behavior of the algorithms as those reported in tables 1 and 2.

Table 2. TabuBN vs. EQ1

Network	TabuBN	EQ1	<	=	>	p	Best
Cancer	3262.61	3261.57	0	27	3	8.32E-02	
Asia	3336.69	3335.82	1	24	5	1.05E-01	
CarStarts	33553.79	33517.19	0	11	19	1.32E-05	EQ1
Alarm	139558.87	139198.80	6	0	24	2.72E-04	EQ1
Hailfinder	720383.23	720038.42	4	2	24	9.94E-06	EQ1

These results clearly show that EQ1 is statistically more successful than GreedyBN, and overpasses even TabuBN on non-trivial tasks.

The differences between scores do not seem very important but they represent different local optima that may correspond to rather different structures. In order to appreciate the fidelity of the discovered structures, we have compared them with the networks used for generating the data sets ("gold-standards").

Table 3 present these comparisons on the basis of:

- the number of edges of the skeleton of the learned network that do not exist in the skeleton of the "gold-standard" (A+);
- the number of edges of the skeleton of the "gold-standard" that do not exist in the skeleton of the learned network (A-);
- the number of v-structures of the learned network that do not exist in the "gold-standard" (VS+);
- the number of v-structures of the "gold-standard" that do not exist in the skeleton of the learned network (VS-);

In order to make easier the interpretation of this table, the best results for each of these criteria are presented in bold face.

Table 3. Structural differences

Network	GreedyBN				TabuBN				EQ1			
	A+	A-	VS+	VS-	A+	A-	VS+	VS-	A+	A-	VS+	VS-
Cancer	**0.40**	0.50	0.50	**0.00**	**0.40**	0.10	0.10	0.03	**0.40**	**0.00**	**0.00**	0.03
Asia	1.50	0.83	0.80	0.43	1.40	0.47	0.27	0.27	**1.23**	**0.23**	**0.17**	**0.17**
CarStarts	3.40	2.80	10.27	1.67	3.40	1.97	9.10	1.57	**2.90**	**0.10**	**7.33**	**0.00**
Alarm	2.73	14.10	17.60	5.43	2.67	12.53	15.83	5.00	**2.07**	**4.13**	**5.90**	**1.27**
Hailfinder	21.90	25.73	23.70	5.33	21.70	20.87	19.03	5.30	**20.83**	**17.23**	**16.37**	**4.60**

Once again, EQ1 is clearly more successful than the other two algorithms, notably on the *Alarm* network.

The last table present the comparison of the average execution times of the three algorithms. They are all programmed in Java, using the same base classes, the same methods for computing scores and the same caching schemas. The

Table 4. Execution times

Network	Greedy	Tabu	EQ1
Cancer	1.21	1.27	0.96
Asia	2.32	2.81	2.15
CarStarts	53.48	62.29	64.25
Alarm	319.55	496.47	364.80
Hailfinder	695.00	1490.02	811.84

tabu list of TabuBN is implemented as a hash table. The comparison has been realized on a PIII 500Mhz CPU. The results are given in seconds.

The execution times of EQ1 are comparable to those of GreedyBN and globally smaller than those of TabuBN. These results are very contrasting with those reported by Chickering [5] (his equivalence class learning algorithm was 10 to 20 times slower than greedy search in the space of bayesian networks).

6   Conclusion

The main result presented in this paper is that the efficiency of the search in the space of equivalence classes of bayesian networks can be considerably improved if the verification of the consistency and the computation of the score of candidate structures can be made locally, without instantiation.

This paper also confirm the interest of exploring the space of equivalence classes, even with a reduced set of transformation operators, both from the viewpoint of the score and of the structural fidelity of the learned networks.

References

1. P. Spirtes, C. Glymour, and R. Scheines. *Causation, Prediction and Search.* Springer-Verlag, 1993.
2. N. Friedman and M. Goldszmidt. Learning bayesian networks with local structure. In *Proceedings of the Twelfth Intenational Conference on Uncertainty in Artificial Intelligence.* Morgan Kaufmann, 1996.
3. D. Heckerman, D. Geiger, and D.M. Chickering. Learning bayesian networks: The combination of knowledge and statistical data. *Machine Learning*, 20:197–243, 1995.
4. T. Verma and J. Pearl. Equivalence and synthesis of causal models. In *Proceedings of the Sixth Conference on Uncertainty in Artificial Intelligence.* Elsevier, 1990.
5. D.M. Chickering. Learning equivalence classes of bayesian-network structures. In *Proceedings of the Twelfth Conference on Uncertainty in Artificial Intelligence.* Morgan Kaufmann, 1996.
6. S.A. Andersson, D. Madigan, and M.D. Perlman. A characterization of markov equivalence classes for acyclic digraphs. *Annals of Statistics*, 25:505–541, 1997.

Quantifying the Resilience of Inductive Classification Algorithms

Melanie Hilario and Alexandros Kalousis

CSD - University of Geneva, CH-1211 Geneva 4, Switzerland
hilario|kalousis@cui.unige.ch

Abstract. Selecting the most appropriate learning algorithm for a given task has become a crucial research issue since the advent of multi-paradigm data mining tool suites. To address this issue, researchers have tried to extract dataset characteristics which might provide clues as to the most appropriate learning algorithm. We propose to extend this research by extracting inducer profiles, i.e., sets of metalevel features which characterize learning algorithms from t of view of their representation and functionality, efficiency, practicality, and resilience. Values for these features can be determined on the basis of author specifications, expert consensus or previous case studies. However, there is a need to characterize learning algorithms in more quantitative terms on the basis of extensive, controlled experiments. This paper illustrates the proposed approach and reports empirical findings on one resilience-related characteristic of learning algorithms for classification, namely their tolerance to irrelevant variables in training data.

1 Background and motivation

It is by now a matter of consensus that there are no universally superior models and methods for induction; the no-free-lunch theorems [19] and the conservation law of generalization performance [18] express basically the same thing, i.e., that no learning algorithm can systematically outperform others across the entire range of application tasks. The key question then is not whether a learning method is superior to others, but under which conditions a particular method can significantly outperform others on a given application problem/dataset. To define these conditions, researchers have attempted to isolate a set of data characteristics which impact the performance of learning algorithms as measured by a given evaluation metric [15][12]. While significant progress has been made on data characterization, the complementary task of extracting inducer characteristics has been relatively neglected. Characterizing a learning algorithm is usually reduced to classifying it as applicable or not to a given dataset [7]. We propose to extend research in this area by building profiles of learning algorithms, i.e., by extracting salient characteristics which allow for meaningful mappings between classes of algorithms and datasets. Rules expressing such mappings can be constructed either manually or automatically (via meta-learning). Such rules can then be used in a prior model selection phase to restrict the space of candidate

D.A. Zighed, J. Komorowski, and J. Zytkow (Eds.): PKDD 2000, LNAI 1910, pp. 106–115, 2000.
© Springer-Verlag Berlin Heidelberg 2000

learning algorithms. This paper focuses on learning algorithms for classification. Section 2 gives an overview of characteristics that can be used to build algorithm profiles and highlights inadequacies of characterizations gleaned from the literature. In particular, it focuses on characteristics that comprise a learning algorithm's resilience and proposes an experiment-based approach to quantifying these. Section 3 illustrates this approach via a study of ten learning algorithms from the point of view of their sensitivity or tolerance to irrelevant variables. The experimental setup is described and major findings are reported and discussed in the light of related work. Section 4 summarizes and gives a preview of ongoing and future work.

2   Characterizing learning algorithms

Our knowledge of model characteristics comes from three different sources. First, certain characteristics are given explicitly in algorithm specifications; they concern the basic requirements, capabilities or limitations of an algorithm. Examples of such author-specified characteristics are the the types of data supported by the algorithm. A second source is observed consensus of experts in machine learning and data mining; however, when experts disagree or are simply in doubt, one can turn to controlled experimentation. The goal of our work is to complete the first two sources of knowledge by devising experimental strategies for characterizing learning algorithms. For the purposes of this paper, we group these characteristics along four dimensions: representation and functionality, efficiency, robustness and practicality.

2.1   Dimensions of algorithm characteristics

The first dimension along which learning algorithms can be described is representational power and functionality; this subsumes the types of data that can be processed by an algorithm, its incrementality, its ability to handle (mis)classification costs, and its bias-variance profile. Standard statistical methods typically don't handle symbolic representations while certain machine learning methods such as AQ cannot handle real-valued data. Most classifier inducers are non incremental and unable to handle (mis)classification costs, though researchers are actively exploring ways of overcoming these limitations for certain algorithms. The bias/variance profile of a learning method is a rough, qualitative indication of the direction in which the algorithm tends to resolve the trade-off between bias and variance [9]. High-bias learners generate simple, highly constrained models which are quite insensitive to data fluctuations, so that variance is low (e.g., perceptrons, Naive Bayes). Algorithms with a high-variance profile can generate arbitrarily complex models which fit data variations more readily (e.g., decision trees, neural networks). Characteristics which reflect a learning algorithm's efficiency are its average training and execution time as well as space demands; this dimension has been given increased attention with the advent of data mining applications, where scalability of learning algorithms is of primary importance.

An algorithm's practicality is the ease with which a user can use and understand it as well as its results. Examples of characteristics along this dimension are runtime parameter handling, the comprehensibility of the learning method, and the interpretability of the learned classifier. Assessment of an algorithm's practicality depends very much on user preferences and priorities. Most of the characteristics related to practicality can be described only by reporting users' subjective evaluations. The resilience of a learning algorithm refers to its capability of ensuring reliable performance despite variations in training conditions and especially in the training data. Resiliency characteristics express the sensitivity or tolerance of an algorithm to data characteristics or pathologies that are liable to affect performance adversely. Examples are an algorithm's scalability and tolerance to noise, missing values, and irrelevant or redundant features. The rest of this paper will focus on this group of inducer characteristics.

2.2   Quantifying inducer resilience

There have been previous attempts to evaluate, compare, or rank learning algorithms along the feature dimensions described in the preceding section. Often, such characterizations are tentative approximations; they express qualitative generalizations over broad classes of models or algorithms. For instance, while it is generally recognized that neural networks typically take much more time to train than decision trees, little is known about the average magnitude of this difference, or about any potential benefit that might be put in the balance against this added cost. Predictive accuracy has been the dominant metric for evaluating inductive algorithms, and only recently have efficiency-related criteria been taken into account [14]. The Statlog project included an attempt to quantify learning algorithms' practicality, in particular the comprehensibility of the underlying learning principle and the interpretability of results. However, a gap remains to be filled in our understanding of their resilience as defined above. To fill this gap, we have undertaken an extensive experimental study aimed at devising a quantiative scale for evaluating and comparing the resilience-related characteristics of classifier inducers. In the rest of this paper we focus on one characteristic which illustrates most clearly the need for quantitative metrics with precise semantics, as opposed to previous binary or other categorizations or rankings. We studied the impact of irrelevant attributes on the behavior of ten learning algorithms: C5.0-tree, C5.0-rules, C5.0-boost [17], Naive Bayes and instance-based learning from the MLC++ library [10], linear discriminants and Ltree [8], multilayer perceptrons and radial basis function networks as implemented in Clementine [4], and Ripper [5]. For each series of experiments, we used 43 datasets from the UCI Repository [13].

3   Tolerance to irrelevant attributes

3.1   Experimental setup

To evaluate and compare the impact of irrelevant variables on classifier inducers, we adopted the following experimental setup using the ten learning algorithms and 43 datasets mentioned in Section 2.2. For each dataset, we generated corrupt versions with 10%, 20%, 30%, 40%, and 50% irrelevant variables. To produce a version with p% irrelevant variables from a dataset containing No original variables, we added $N_i = int(.01p * (N_0/(1 - .01p)))$ new variables whose values were generated randomly following a uniform distribution. A couple of precautionary measures were taken for the sake of experimental soundness. First, to ensure that variation in performance is due only to the irrelevant variables and not to side effects such as changes in the balance of variable types, we strove to maintain the original distribution of numeric and symbolic variables when adding irrelevant variables. Second, to mitigate fears that the random-valued features might be serendipitously correlated or associated in any way with the classes, we ascertained that the addition of such features effectively increased the quantity of irrelevant information in the datasets. To do this, we measured the mutual information between the class and each predictive (original or artificially added) variable for each dataset after discretizing all continuous variables using Fayyad and Irani's method [6]. Mutual information measures were averaged over all predictors to produce a single measure for each dataset, then averaged over all datasets with the same percentage of (additional) irrelevant variables to yield a single average measure of mutual information over all datasets at each level of corruption. The resulting curve (Fig. 1) shows that the average mutual information between predictive variables and classes decreases monotonically with the addition of random-valued features, thus confirming their irrelevance. The ten

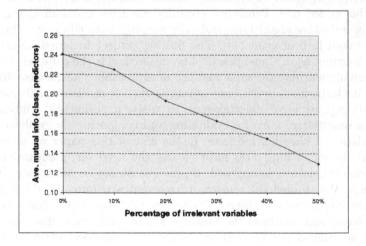

Fig. 1. Average mutual information as a measure of relevance

learning algorithms under study were run on each original dataset and its 5 corrupted variants using stratified 10-fold cross-validation. Each algorithm was run with its default settings. The performance of these algorithms on the original versions provided the baseline against which to study degradation of learning ability as the proportion of irrelevant variables increased. Performance measures used were test-set error and total processing (training plus test) time.

3.2   Results

The degradation in predictive accuracy is summarized in Figure 2.

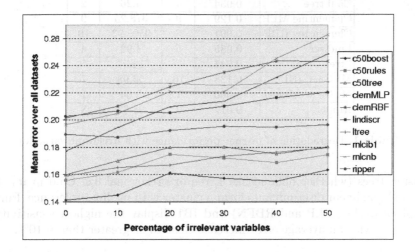

Fig. 2. Generalization error vs percentage of irrelevant variables

As a first cut at interpreting these results, note that an algorithm's predictive accuracy on the given datasets are quite independent of their resilience to irrelevant variables. For instance, boosted C5.0 tree attains the highest predictive accuracy at all levels, but its error curve rises more rapidly than most of the other algorithms in the study. On the other hand, the mean error of Naive Bayes remains among the highest at all levels of irrelevance, yet it seems unaffected by the proportion of irrelevant variables. To quantify sensitivity to irrelevance, it is thus important to decouple the magnitude of the mean error from its variation in response to irrelevant variables. We considered three candidate measures: the error increase $error_i - error_0$ for each level of irrelevance i was eliminated because it does not sufficiently abstract away the magnitude of the error. An alternative is the relative error increase $(error_i - error_0)/error_0$, whose obvious shortcoming is that for equal error increases, the sensitivity score increases with lower values of $error_0$, thus penalizing algorithms with high predictive accuracy. The third measure is simply the slope of the error curve which not only is exempt from the previously mentioned drawbacks but has the additional advantage of a clear semantics: sensitivity to irrelevance is defined quantitatively as the average rate of increase in error (or some other performance metric) with increase in the proportion of irrelevant variables. The resulting error-sensitivity scores and ranks are shown in Table 1(a) and the corresponding regressed curves in Fig. 3.

A closer look at the error sensitivity scores reveals several distinct clusters among the 10 algorithms. Naive Bayes is by far the most resistant to irrelevant variables, maintaining a comfortable distance from Ripper, its closest runner-up.

Algorithm	(a)		(b)	
	Error slope	Rank	Time slope	Rank
Boosted C5.0	0.039	6	33.58	6
C5.0 rules	0.032	3	4.76	3
C5.0 tree	0.034	4	4.26	2
Clementine MLP	0.129	9	378.25	9
Clementine RBF	0.091	8	9595.93	10
Lindiscr	0.035	5	4.92	4
Ltree	0.041	7	14.57	5
MLC++ IB1	0.135	10	56.98	7
MLC++ NBayes	0.000	1	3.64	1
Ripper	0.017	2	98.67	8

Table 1. Degradation of learner performance with irrelevant variables

Decision trees (whether oblique like Ltree or orthogonal like C5.0 in its three variants) and linear discriminants form a cluster with medium tolerance. Finally, neural networks (MLP and RBFN) and IB1 display the highest sensitivity to irrelevance with an average error increase close to or greater than 0.10%.

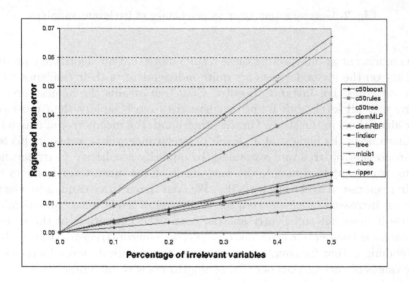

Fig. 3. Regressed error increase with irrelevant variables

The same approach was adopted using total training and test time as the performance criterion. Time measures were standardized across different machines and expressed in Sun Sparc Ultra 10-equivalent CPU seconds (for 124 MB main memory). The derivative of the run time curves was similarly used to quantify the time-sensitivity of learning algorithms to irrelevant variables; the resulting scores are shown in Table 1(b) and the regressed runtime curves in Fig. 4.

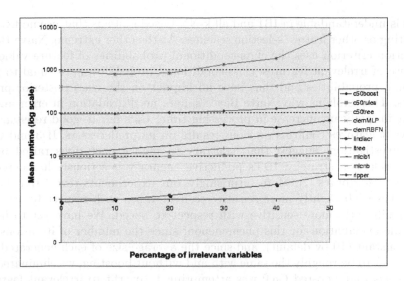

Fig. 4. Regressed run time increase with irrelevant variables

The ranking of the different algorithms undergoes a few changes with the shift to the runtime metric, though a clear overall picture subsists after combining insights based on both accuracy and computational cost. Naive Bayes is still a clear winner with a mean degradation in speed of around 3 CPU seconds. Linear discriminants and single-classifier C5.0 trees and rules trail closely behind, but boosting increases C5.0's speed degradation considerably. Ripper appears to stay close to its baseline accuracy at a much higher computational cost. As with predictive accuracy, the worst degradation is exhibited by the neural networks, MLP and RBF [1]. The runtime curves are represented using a logarithmic scale in Figure 4 and show that Clementine-RBFN's training and test time grows faster by an order of magnitude than Clementine-MLP and by several orders of magnitude than the 8 other algorithms.

3.3   Discussion and related work

Results concerning IB1 confirm expert consensus that instance-based learning (and standard k- NN in general) is highly sensitive to irrelevant variables [1]. Previous work has shown that k-NN's 90% sample complexity (the number of training examples needed to achieve 90% accuracy) grows exponentially with the number of irrelevant features while that of Naive Bayes grows only linearly [11]. This difference can be explained by the underlying inductive principles; instance-based classification identifies nearest neighbors via distance measures based on feature values, so that irrelevant features have as great an effect on classifier as the relevant ones. Thus a dramatic deterioration in both accuracy and

---

[1] Of the 43 datasets used, the monks problems were left out in computing MLP runtime since a bug in Clementine provokes an indefinite loop in default mode.

speed is understandable in IB1 and all k-NN based systems that use no feature-weighting or other feature selection schemes. At the other extreme, Naive Bayes' prediction criterion uses the class-conditional probabilities of feature values; in the case of irrelevant features, class-conditional probabilities are equal to their marginal probabilities and thus have no impact on the class posterior probabilities. This explains why Naive Bayes suffers no degradation in error and an almost insignificant increase in processing time. C5's fair-to-good resistance to irrelevant features confirms previous results on its predecessors, ID3 and C4.5. Past investigations suggest that, contrary to high expectations raised by its feature selection strategy, ID3's predictive accuracy is seriously hampered by irrelevant attributes [2]. More surprisingly, our experiments reveal that boosting makes C5.0 slightly more sensitive to irrelevance with respect to accuracy and significantly more sensitive with respect to speed. We have yet to find a plausible explanation for this phenomenon; since the number of iterations was kept constant (10 by default), and since the average size of each generated tree turned out to be roughly the same with and without boosting, we eliminated the hypothesis that boosted C5.0 was attempting to overfit to irrelevant features. Another surprise is the extreme sensitivity to irrelevance of the neural networks used in this study. For Clementine-MLP, this contradicts predictions based on the theory. MLPs are expected to be relatively unaffected by irrelevant features since the hidden layer projects inputs into a subspace of much lower dimensionality within which approximation can take place [16]. Concretely, connection weights from irrelevant inputs are expected to gradually tend towards zero as the training process converges; thus the only adverse effect should be a significant increase in training time. Unfortunately our experiments tend to show that accuracy deteriorates significantly with irrelevant inputs despite the dramatic increase in computational costs. For Clementine-MLP, this could be explained by overfitting to irrelevant inputs. In default mode, the number of hidden units is determined automatically, and we observed that the average number of hidden units increased monotonically with the percentage of irrelevant variables, doubling between the baseline and the 50% level (Table 2).

% of irrelevant variables	0%	10%	20%	30%	40%	50%
Number of hidden units	12.4	13.9	15.3	17.2	20.7	24.4

Table 2. Growth of MLP network complexity with irrelevant variables

On the other hand, Clementine-RBFN's sensitivity to irrelevance is predictable from the underlying approach [3]. First, in RBFNs hidden units represent basis functions whose parameters (e.g. centres) are selected using unsupervised training methods such as K- means clustering. This implies reliance on a distance measure which, as in the case of k-NN, gives equal importance to all variables, whether relevant or not. In addition, each hidden unit has a local receptive field, i.e., it influences network output only in the neighborhood of its center. While predictive accuracy gains from locality when the hidden units reflect actual clusters in the data, it deteriorates significantly when these are built/activated using distance measures distorted by irrelevancies. This degradation could be alleviated by increasing the number of hidden units, which is however set to 20 in Clementine default mode. In short, two essential features of the RBFNs used—unsupervised K-means training and local receptive fields of hidden units plus a particularity of the Clementine default implementation— explain the observed lack of resilience to irrelevant variables. While the ranks of tested algorithms help to visualize overall trends and patterns, we decided to keep the original quantifications of sensitivity to irrelevance in the algorithm profiles. There were several reasons for this. The original sensitivity measures have a precise semantics, as explained in Section 3. Converting them to ranks would, first, result in loss of information and, second, complicate the task of adding a new learning algorithm to the profiled set. Finally, like other (meta-)datasets, algorithm characterizations can be pre-processed and continuous attributes reencoded in ordinal or qualitative form as the need arises. To conclude this section, it should be emphasized that the findings reported on the above algorithms cannot be generalized beyond the specific implementations used, even when they are confirmed by the underlying theory.

4   Summary and future work

This paper argued for the need to build learning algorithm profiles which can be used in prior model selection for data mining. We distinguished four groups of characteristics depending on whether they concern a learning algorithm's representation and functionality, efficiency, resilience, or practicality. We gave a quick overview of each group of characteristics and proposed an experimental setup for quantifying characteristics related to an inductive algorithm's resilience. We illustrated the proposed approach on a study of tolerance to irrelevant variables. Other studies (e.g., on scalability, bias/variance trade-off, resistance to other data idiosyncrasies such as missing values and redundant features) have been conducted and will be presented in forthcoming publications. Algorithm profiles built from such studies will be used, together with dataset characterizations, to discover mappings between broad classes of learning tasks/data and models/algorithms. We shall attempt such mappings both manually or automatically, i.e., via (static or batch) meta-learning. In a subsequent phase, the knowledge thus built will be deployed to support the user in the model selection task; in return, feedback from novel applications/ environments will be assimilated via incremental meta-learning to dynamically refine or augment this experimentally cumulated expertise.

Acknowledgements

This work was partially supported by the Swiss Office for Education and Science (OFES) in the framework of ESPRIT IV LTR Project METAL-26357. We cannot thank Johann Petrak (OeFAI, Vienna) enough for his Perl scripts which greatly facilitated our experimentation.

References

1. D. W. Aha, D. Kibler, and M. K. Albert. Instance-based learning algorithms. *Machine Learning*, 6(1):37–66, 1991.
2. H. Almuallim and T. Dietterich. Efficient algorithms for identifying relevant features. In *Proceedings of the Ninth Canadian Conference on Artificial Intelligence*, pages 38–45, Vancouver, BC, 1992. Morgan Kaufmann.
3. C. M. Bishop. *Neural Networks for Pattern Recognition*. Oxford University Press, 1995.
4. Clementine. http://www.spss.com.
5. W. W. Cohen. Fast effective rule induction. In A. Prieditis and S. Russell, editors, *Proc. of the 11th International Conference on Machine Learning*, pages 115–123, Tahoe City, CA, 1995. Morgan Kaufmann.
6. U. Fayyad and K. Irani. Multi-interval discretization of continuous-valued attributes for classification learning. In *Proc. of the 13th IJCAI*, pages 1022–1027, Chambéry, France, 1993. Morgan Kaufmann.
7. J. Gama and P. Brazdil. Characterization of classification algorithms. In E. Pinto-Ferreira and N. Mamede, editors, *Progress in Artificial Intelligence. 7th Portuguese Conference on Artificial Intelligence (EPIA-95)*, pages 189–200. Springer-Verlag, 1995.
8. J. Gama and P. Brazdil. Linear tree. *Intelligent Data Analysis*, 3:1–22, 1999.
9. S. Geman, E. Bienenstock, and R. Doursat. Neural networks and the bias/variance dilemma. *Neural Computation*, 4:1–58, 1992.
10. R. Kohavi, D. Sommerfeld, and J. Dougherty. Data mining using mlc++. In *International Conference on Tools with AI*, pages 234–245, 1996.
11. P. Langley and S. Sage. Scaling to domains with irrelevant features. In R. Greiner et al., editor, *Computational Learning Theory and Natural Learning Systems*, volume 4, Cambridge, MA, 1996. MIT Press.
12. D. Michie, D. J. Spiegelhalter, and C. C. Taylor, editors. *Machine learning, neural and statistical classification*. Prentice-Hall, 1994.
13. P. M. Murphy and D.W. Aha. UCI machine learning repository. http://www.ics.uci.edu/ mlearn/MLRepository.html, 1991. Irvine, CA: University of California, Dept. of Information and Computer Science.
14. G. Nakhaeizadeh and A. Schnabl. Development of multi-criteria metrics for evaluation of data mining algorithms. In *Proc. Third International Conference on Knowledge Discovery and Data Mining*, pages 37–42, Newport Beach, CA, 1997. AAAI Press.
15. L. Rendell and E. Cho. Empirical learning as a function of concept character. *Machine Learning*, 5:267–298, 1990.
16. B.D. Ripley. *Pattern Recognition and Neural Networks*. Cambridge U. Press, 1996.
17. http://www.rulequest.com.
18. C. Schaffer. A conservation law for generalization performance. In W. W. Cohen and H. Hirsh, editors, *Proc. of the 11th International Conference on Machine Learning*, pages 259–265, Rutgers, NJ, 1994. Morgan Kaufmann.
19. D. Wolpert. The lack of a priori distinctions between learning algorithms. *Neural Computation*, 8(7):1381–1390, 1996.

Bagging and Boosting with Dynamic Integration of Classifiers

Alexey Tsymbal and Seppo Puuronen

Department of Computer Science and Information Systems,
University of Jyväskylä, P.O.Box 35, FIN-40351 Jyväskylä, Finland
{Alexey,sepi}@cs.jyu.fi

Abstract. One approach in classification tasks is to use machine learning techniques to derive classifiers using learning instances. The cooperation of several base classifiers as a decision committee has succeeded to reduce classification error. The main current decision committee learning approaches boosting and bagging use resampling with
the training set and they can be used with different machine learning
techniques which derive base classifiers. Boosting uses a kind of
weighted voting and bagging uses equal weight voting as a combining
method. Both do not take into account the local aspects that the base
classifiers may have inside the problem space. We have proposed a dynamic integration technique to be used with ensembles of classifiers. In
this paper, the proposed dynamic integration technique is applied with
AdaBoost and bagging. The comparison results using several datasets of
the UCI machine learning repository show that boosting and bagging
with dynamic integration of classifiers results often better accuracy than
boosting and bagging result with their original voting techniques.

1 Introduction

There are several approaches in data mining which tries to find previously unknown
and potentially interesting patterns and relations in large databases [6]. One typical
data mining task is to predict a value of an attribute of a new instance when the values
of the other attributes of the instance are known. A common approach is to use a machine learning technique. A collection of instances with known values of all the attributes is treated as a training set for a machine learning algorithm that derives a classifier. This classifier is later used to predict the unknown value of the attribute for a new
instance.

Decision committee learning has succeeded to reduce classification error
[1,3,4,7,9,16,17,18,24]. A committee (final classifier) is composed of base classifiers
(committee members), each of which makes its own classifications that are combined
to create a single classification result of the whole committee. Two decision committee learning approaches, boosting [16,17] and bagging [3], have received extensive
attention recently and they have been deeply analyzed [4,9,18]. Both generate by re

D.A. Zighed, J. Komorowski, and J. Zytkow (Eds.): PKDD 2000, LNAI 1910, pp. 116-125, 2000.
© Springer-Verlag Berlin Heidelberg 2000

sampling training sets from the original data set to the learning algorithm [24] which builds up a base classifier for each training set. There are two major differences between bagging and boosting. First, boosting changes adaptively the distribution of the training set based on the performance of previously created classifiers while bagging changes the distribution of the training set stochastically [24]. Second, boosting uses a function of the performance of a classifier as a weight for voting, while bagging uses equal weight voting [24]. With both techniques error decreases when the size of the committee increases, but the marginal error reduction of an additional member tends to decrease [24]. In general, bagging is more consistent, increasing the error of the base learner less frequently than boosting. However, boosting has greater average effect, leading to substantially larger error reductions than bagging on average [4]. The error reduction has been a little mysterious also to researchers and several analysis to reveal the real background of the phenomenon have been suggested [1,9,18]. One explanation is based on that boosting tends to reduce both the bias and variance terms of error while bagging tends to reduce the variance term only [1,4]. Another is using the concept of margin distribution [18], and another recent one establishes connection between boosting and additive models [9] but as comments, as [8] to the last paper show there is still much work to do before the behaviour of methods are deeply understood.

Both bagging and boosting uses a voting technique which is unable to take into account the heterogeneity of the instance space. When majority of the base classifiers give a wrong prediction for a new instance then the majority vote will result in a wrong prediction. The problem may consist in discarding base classifiers (by assigning small weights) that are highly accurate in a restricted region of the instance space because this accuracy is swamped by their inaccuracy outside the restricted area. It may also consist in the use of classifiers that are accurate in most of the space but still unnecessarily confuse the whole classification committee in some restricted areas of the space. To overcome this problem we have suggested a dynamic integration approach [13]. In this paper we compare bagging and boosting with and without our dynamic integration approaches using using several data sets from the UCI machine learning repository [2]. Some preliminary results with small committee sizes were presented in [20].

In chapter 2 the bagging and boosting algorithms are reviewed. Chpater 3 includes a short describtion of our dynamic integration methods. Chapter 4 includes the results of comparisons and chapter 5 concludes with a short summary and some further research topics.

2   Bagging and Boosting

In this chapter bagging and boosting algorithms to be used in comparisons are reviewed.

The bagging algorithm (**B**ootstrap **agg**regat**ing**) [3] uses bootstrap samples to build the base classifiers. Each bootstrap sample of m instances is formed by uniformly sampling m instances from the training set with replacement. The amount of the different instances in the bootstrap sample is for large m about $1-1/e=63.2\%$. This results that dissimilar base classifiers are built with unstable learning algorithms (e.g., neural networks, decision trees) [1] and the performance of the committee can become better. However, bagging may slightly degrade the performance of stable algorithms (e.g., k-nearest neighbor) because effectively smaller training sets are used to train each base classifier [3]. The bagging algorithm generates T bootstrap samples $B_1, B_2, ..., B_T$ and then the corresponding T base classifiers $C_1, C_2, ..., C_T$. The final classification produced by the committee of these base classifiers is received using equal weight voting where ties are broken arbitrarily

Boosting was developed as a method for boosting the performance of any weak learning algorithm [17], which needs only to be a little bit better than random guessing [16]. The AdaBoost algorithm (**Ada**ptive **Boost**ing) [7] was introduced as an improvement of the initial boosting algorithm and several further variants are presented. The concept used to correspond a generation of one base classifier is a trial. AdaBoost changes the weights of the training instances after each trial based on the misclassifications made by the resulting base classifier trying to force the learning algorithm to minimize the expected error over different input distributions [1]. Similarly, the correctly classified instances will have a lower total weight.Thus AdaBoost generates during the T trials T training sets $S_1, S_2, ..., S_T$ where instances have weights and thus T base classifiers $C_1, C_2, ..., C_T$ are built. A final classification produced by the committee of these base classifiers is received using a weighted voting scheme where the weight of each classifier depends on its performance on the training set used to build it.

The AdaBoost algorithm requires a weak learning algorithm whose error is bounded by a constant strictly less than ½. In practice, the learning algorithm that we use provide no such guarantee. The original algorithm aborted when the error bound was breached [17], but this case is fairly frequent for some multiclass problems [1]. In this paper, a bootstrap sample from the original data S is generated in such case, and the boosting continues up to a limit of 25 such samples at a given trial, as proposed in [1]. The same is done when the error is equal to zero at some trial. Some implementations of AdaBoost use boosting by resampling because the inducers used were unable to support weighted instances [1]. In this work, boosting by reweighting is implemented with the C4.5 decision tree learning algorithm [15], which is a more direct implementation of the theory. Some evidence exists that reweighting works better in practice [1].

3   Dynamic Integration of Classifiers

The challenge of integration is to decide which classifier to rely on or how to combine classifications produced by several classifiers. The integration approaches can be divided into static and dynamic ones. Recently, two main approaches to integration have been used: first, combination of the classifications produced by the base classifiers, and, second, selection of the best classifier from the base classifiers. The most popular method of combining classifiers is voting [1]. More sophisticated selection approaches use estimates of the local accuracy of the base classifiers by considering errors made in similar instances [12] or the meta-level classifiers ("referees"), which predict the correctness of the base classifiers for a new instance [11].

We have presented a dynamic integration technique that estimates the local accuracy of the base classifiers by analyzing the accuracy in near-by instances [13]. Knowledge is collected about the errors that the base classifiers make with the training instances and this knowledge is taken benefit during the classification of new instances. The goal is to use each base classifier just in that subarea for which it is the most reliable one. The C4.5 decision tree induction algorithm is used to predict the errors of the base classifiers [15] and in the dynamic integration the weighted nearest neighbor classification (WNN) is used [5].

The dynamic integration approach contains two phases. In the learning phase, information is collected about how each classifier succeeds with each training set instance. With bagging and boosting this is done directly using the base classifiers produced by C4.5. In the original version of our algorithm we have used cross-validation technique to estimate the errors of the base classifiers on the training set [13]. In the application phase, the final classification of the committee is formed so that for a new instance the seven nearest neighbor instances of the learning set are found out. The error information of each classifier with these instances is collected and then depending on the dynamic integration approach the final classification is made using this information. Two of these three different functions implementing the application phase were considered [13]. The first function DS implements Dynamic Selection. In the DS application phase the classification error is predicted for each base classifier using the WNN procedure and a classifier with the smallest error (with the least global error in the case of ties) is selected to make the final classification. The second function DV implements Dynamic Voting. In the DV application phase each component classifier receives a weight that depends on the local classifier's performance and the final classification is conducted by voting classifier predictions with their weights. The third function DVS was first presented in [21] and it applies mixed approach. Using the collected error information first about half of classifiers (the better half) is selected and then the final classification is derived using weighted voting among them.

Previously an application of the dynamic classifier integration in medical diagnostics was considered in [19,20,22]. A number of experiments comparing the dynamic integration with such widely used integration approaches as CVM, and weighted voting were also conducted [13,14,22,23]. The comparison results show that the dynamic integration technique outperforms often weighted voting and CVM. In [21] the dynamic classifier integration was applied to decision committee learning, combining the generated classifiers in a more sophisticated manner than voting. This paper continues this research considering more experiments comparing bagging and boosting with and without the three dynamic integration functions.

4 Experiments

In this chapter we present experiments where the dynamic classifier integration algorithm is used with classifiers generated by AdaBoost and bagging. The experimental setting is described before the results of the experiments. We have used nine datasets from the UCI machine learning repository [2]. Previously the dynamic classifier integration was experimentally evaluated in [13] and preliminary comparisons with AdaBoost and bagging were made in [21].

The main characteristics of the eight datasets are presented in Table 1. The table includes the name of the dataset, number of instances included in the dataset, number of different classes of instances, and numbers of different kind of features included in the instances.

Table 1. Characteristics of the datasets

Dataset	Instances	Classes	Features	
			Discrete	Continuous
Breast	286	2	9	0
Diabetes	768	2	0	8
Glass	214	6	0	9
Heart	270	2	8	5
Iris	150	3	0	4
Liver	345	2	0	6
MONK-1	432	2	6	0
MONK-2	432	2	6	0

For each dataset 30 test runs are made. First, in each run 30 percent of the instances of the data set are randomly picked up to the test set. Then the rest 70 percent of the instances are passed to bagging and boosting algorithm. These manipulate the learning set as described in Chapter 2 producing required amount of base classifiers using C4.5 decision tree algorithm with pruning [15]. We make experiments with 5, 10, and 25 base classifiers in the committees.

We calculate average accuracy numbers for base classifiers, for ordinary bagging and boosting, and for bagging and boosting with the three dynamic integration method described in Chapter 3 separately. In the learning phase of the dynamic integration part information about the errors made by each base classifier is collected for all the instances included in the 70 percent of the original data set, selected in the beginning of each run. In the application phase of the dynamic integration the instances included in the test set is used. For each of them the classification errors of all the base classifiers of the committee is collected for seven nearest neighbors of the new instance. The selection of the number of nearest neighbors has been discussed in [13]. Based on the comparisons between different distance functions for dynamic integration presented in [14] we decided to use the Heterogeneous Euclidean-Overlap Metric, which produced good test results earlier. The test environment was implemented within the MLC++ framework (the machine learning library in C++) [10].

Table 2 presents accuracy values for Bagging with different voting methods. The first column includes the name of the corresponding data set. The second column tells how many base classifiers were produced to the committee. The next three columns include the average of the minimum accuracies of the base classifiers (min), the average of average accuracies of the base classifiers (aver), and the average of the maximum accuracies of the base classifiers (max) over the 30 runs.

Table 2. Accuracy values for Bagging with different voting methods

DB	# of base classifiers	C4.5 base classifiers			Bagging with different voting			
		min	aver	max	EWV	DS	DV	DVS
Breast	5	0.649	0.691	0.739	0.710	0.697	0.710	0.706
	10	0.625	0.686	0.746	0.717	0.693	0.712	0.702
	25	0.605	0.689	0.763	0.717	0.692	0.720	0.714
Diabetes	5	0.675	0.706	0.737	0.740	0.704	0.739	0.724
	10	0.668	0.708	0.745	0.752	0.707	0.751	0.740
	25	0.659	0.710	0.760	0.755	0.710	0.755	0.751
Glass	5	0.544	0.608	0.670	0.656	0.663	0.675	0.677
	10	0.522	0.608	0.687	0.669	0.666	0.683	0.689
	25	0.500	0.610	0.711	0.682	0.681	0.696	0.705
Heart	5	0.677	0.731	0.788	0.777	0.747	0.777	0.765
	10	0.659	0.732	0.804	0.788	0.748	0.790	0.779
	25	0.641	0.735	0.818	0.799	0.751	0.798	0.786
Iris	5	0.921	0.942	0.957	0.950	0.944	0.950	0.944
	10	0.914	0.943	0.963	0.947	0.944	0.949	0.943
	25	0.904	0.942	0.965	0.947	0.941	0.948	0.947
Liver	5	0.557	0.610	0.657	0.645	0.621	0.653	0.637
	10	0.541	0.606	0.669	0.649	0.622	0.663	0.657
	25	0.526	0.610	0.695	0.669	0.621	0.682	0.675
MONK-1	5	0.727	0.827	0.928	0.894	0.975	0.900	0.964
	10	0.699	0.834	0.947	0.895	0.988	0.931	0.975
	25	0.666	0.826	0.976	0.913	0.995	0.942	0.976
MONK-2	5	0.510	0.550	0.588	0.567	0.534	0.565	0.547
	10	0.499	0.551	0.600	0.600	0.536	0.566	0.540
	25	0.487	0.551	0.610	0.582	0.539	0.571	0.537
Average	5	0.658	0.708	0.758	0.742	0.736	0.746	0.746
	10	0.641	0.709	0.770	0.752	0.738	0.756	0.753
	25	0.624	0.709	0.787	0.758	0.741	0.764	0.761

The last four columns on the right-hand side of Table 2 include average accuracies for different voting methods. These are: equal weight voting (EWV), dynamic selection (DS), dynamic voting (DS), and dynamic voting with selection (DVS). DS, DV, and DVS were considered in chapter 3. Previous experiments with these two integration strategies [13] have shown that the accuracies of the strategies usually differ significantly; however, it depends on the dataset, what a strategy is preferable. In this paper we apply also a combination of the DS and DV strategies that were in [21] expected to be more stable than the other two.

The accuracies in the table 2 confirm the known result that raising the number of committee members makes bagging more accurate. In this experiment the average increase of accuracy over the 8 data sets was 1.3% and 2.1% when the number of the members was raised from 5 to 10 and 25 correspondingly. The increase was almost same for DV (1.3% and 2.4%) and DVS (1.0% and 2.1%) but much lower for DS (0.3% and 0.8%). It seems to be that bagging with dynamic selection is not able to take as efficiently benefit of the increase of the base classifiers, at least with these data sets. When individual data sets are looked it is noticed that with ordinary equal weight voting and dynamic voting the accuracy decreases only with Iris dataset, with dynamic selection with Iris and Breast datasets, and with DVS with MONK-2 dataset and with Breast and Iris datasets when the number is raised from 5 to 10.

When the different voting techniques are compared it is noticed that in average over the 8 datasets DV and DVS produce only about 0.5% higher accuracies than the ordinary voting. Dynamic selection, on the other hand gives in average about 2% smaller accuracy for bigger committees. There is big difference between individual data sets in the average accuracies achieved.   The biggest ones are that both DVS and DS produce over 10% smaller accuracy with MONK-2 dataset with 10 base classifiers, and DS over 10% higher accuracy with MONK-1 dataset with 10 base classifiers than the ordinary voting.   As a summary it can be said that in average over these 8 datasets there seems not to be great benefit about the dynamic integration with bagging but because there are many dataset related difference it needs further systematic research before being able to make any firm conclusions.

Table 3 presents accuracy values for AdaBoost with different voting methods. The columns are same as in Table 2 except that the ordinary voting with boosting is weighted voting (WV) and again the accuracies are averages over 30 runs.

The accuracies in the table 3 confirm the known result that raising the number of committee members makes also boosting more accurate. In this experiment the average increase of accuracy over the 8 data sets was 0.7% and 1.3% when the number of the members was raised from 5 to 10 and 25 correspondingly. The increase was a little bit higher for DV (1.3% and 1.7%) and DVS (0.9% and 1.7%) but DS suffered from additional base classifiers (-0.0% and -0.7%). It seems to be that also boosting with dynamic selection is not able to take as efficiently benefit of the increase of the base classifiers, at least with these data sets. When individual data sets are looked it is noticed that with ordinary weighted voting and dynamic voting the accuracy decreases only with MONK-2 dataset and with Liver dataset when the amount is raised from 5 to 10. The accuracy of dynamic voting with selection decreases only with the MONK-2 dataset but the dynamic selection has problems with almost every dataset.

When the different voting techniques are compared it is noticed that in average over the 8 datasets DVS produces 0.9% higher accuracy with 25 base classifiers, 0.8% higher accuracy with 10 base classifiers, and 06% higher accuracy with 5 base classifiers than the ordinary weighted voting. The average numbers of DV over the datasets are also a little bit better than the ordinary weigted voting. Instead, the dynamic selection produces lower accuracies, in average about 1.0%, 1.7%, and 3.0% for 5, 10, and 25 base classifiers, correspondingly.

Table 3. Accuracy values for AdaBoost with different voting methods

DB	# of base classifiers	C4.5 base classifiers			AdaBoost with different voting			
		min	aver	max	WV	DS	DV	DVS
Breast	5	0.577	0.637	0.703	0.684	0.664	0.686	0.689
	10	0.548	0.628	0.708	0.686	0.658	0.693	0.689
	25	0.504	0.618	0.717	0.687	0.655	0.697	0.697
Diabetes	5	0.640	0.680	0.720	0.722	0.690	0.722	0.710
	10	0.618	0.674	0.726	0.727	0.685	0.729	0.723
	25	0.528	0.658	0.732	0.738	0.685	0.741	0.742
Glass	5	0.524	0.595	0.666	0.683	0.667	0.685	0.685
	10	0.457	0.572	0.676	0.697	0.668	0.702	0.705
	25	0.371	0.533	0.683	0.703	0.661	0.704	0.708
Heart	5	0.634	0.705	0.773	0.763	0.727	0.763	0.751
	10	0.589	0.685	0.776	0.770	0.719	0.771	0.763
	25	0.495	0.655	0.785	0.769	0.719	0.773	0.776
Iris	5	0.911	0.939	0.965	0.945	0.943	0.945	0.945
	10	0.895	0.939	0.967	0.949	0.946	0.949	0.949
	25	0.846	0.933	0.974	0.951	0.945	0.949	0.948
Liver	5	0.550	0.601	0.650	0.642	0.609	0.642	0.624
	10	0.526	0.593	0.656	0.639	0.606	0.640	0.634
	25	0.474	0.581	0.668	0.645	0.593	0.646	0.647
MONK-1	5	0.740	0.849	0.936	0.949	0.966	0.948	0.967
	10	0.705	0.843	0.950	0.968	0.980	0.974	0.984
	25	0.570	0.803	0.965	0.989	0.983	0.990	0.993
MONK-2	5	0.473	0.535	0.594	0.509	0.572	0.509	0.559
	10	0.458	0.535	0.607	0.503	0.575	0.517	0.538
	25	0.442	0.534	0.625	0.493	0.558	0.501	0.520
Average	5	0.631	0.693	0.751	0.737	0.730	0.738	0.741
	10	0.600	0.684	0.758	0.742	0.730	0.747	0.748
	25	0.529	0.664	0.769	0.747	0.725	0.750	0.754

There is big difference between individual data sets in the average accuracies only with the dynamic selection method and MONK-2 dataset with the dynamic voting with selection. As a summary it can be said that in average over these 8 datasets there might be benefit about the dynamic integration with boosting when dynamic voting especially with selection is used, but further systematic research is needed to confirm this.

5 Conclusion

Decision committee learning has demonstrated spectacular success in reducing classification error with learned classifiers. These techniques a committee of base classifiers which produce the final classification using some voting method. Ordinary voting methods has however an important shortcoming, that they do not take into account local context related things.

In this paper a technique for dynamic integration of classifiers was experiment as a method instead of ordinary voting methods with bagging and boosting. The technique for dynamic integration of classifiers is based on the assumption that each base classifier is the best inside certain sub areas of the whole feature space. The considered algorithm for dynamic integration of classifiers is a new variation of stacked generalization, which uses a distance metric to locally estimate the errors of base classifiers.

The proposed dynamic integration technique was evaluated with AdaBoost and Bagging, the decision committee approaches which have received extensive attention recently, on eight datasets from the UCI machine learning repository. The results achieved are promising and show that especially boosting but in some contexts also bagging might give better accuracy with dynamic integration of classifiers than with simple voting.

Further analysis and experiments are needed to make deeper analysis of combining the dynamic integration of classifiers with different approaches to decision committee learning.

Acknowledgments: This research is partly supported by the COMAS Graduate School of the University of Jyväskylä. We would like to thank the UCI machine learning repository of databases, domain theories and data generators for the datasets, and the machine learning library in C++ for the source code used in this study. We want to thank also the anonymous reviewers for their positive and helpful criticisms.

References

1. Bauer, E., Kohavi, R.: An Empirical Comparison of Voting Classification Algorithms: Bagging, Boosting, and Variants. Machine Learning, Vol.36 (1999) 105-139.
2. Blake, C.L., Merz, C.J.: UCI Repository of Machine Learning Databases [http://www.ics.uci.edu/ ~mlearn/ MLRepository.html]. Dep-t of Information and CS, Un-ty of California, Irvine CA (1998).
3. Breiman, L.: Bagging Predictors. Machine Learning, Vol. 24 (1996) 123-140.
4. Breiman, L.: Arcing classifiers. The Annals of Statistics, Vol. 26, No. 3 (1998) 801-823.
5. Cost, S., Salzberg, S.: A Weighted Nearest Neighbor Algorithm for Learning with Symbolic Features. Machine Learning, Vol. 10, No. 1 (1993) 57-78.
6. Fayyad, U., Piatetsky-Shapiro, G., Smyth, P., Uthurusamy, R.: Advances in Knowledge Discovery and Data Mining. AAAI/ MIT Press (1997).
7. Freund, Y., Schapire, R.E.: A Decision-Theoretic Generalization of On-Line Learning and an Application to Boosting. In: Proc. 2nd European Conf. on Computational Learning Theory, Springer-Verlag (1995) 23-37.

8. Freund, Y., Schapire, R.E.: Discussion of the Paper "Additive Logistic Regression: a Statistical View of Boosting by Jerome Friedman, Trevor Hastie and Robert Tibshirani. The Annals of Statistics, Vol. 28, No. 2 (2000).
9. Friedman, J., Hastie, T, Tibshirani, R.: Additive Logistic Regression: a Statistical View of Boosting. The Annals of Statistics, Vol. 28, No. 2 (2000).
10. Kohavi, R., Sommerfield, D., Dougherty, J.: Data Mining Using MLC++: A Machine Learning Library in C++. Tools with Artificial Intelligence, IEEE CS Press (1996) 234-245.
11. Koppel, M., Engelson, S.P.: Integrating Multiple Classifiers by Finding their Areas of Expertise. In: AAAI-96 Workshop On Integrating Multiple Learning Models (1996) 53-58.
12. Merz, C.: Dynamical Selection of Learning Algorithms. In: D.Fisher, H.-J.Lenz (eds.), Learning from Data, Artificial Intelligence and Statistics, Springer-Verlag, NY (1996).
13. Puuronen, S., Terziyan, V., Tsymbal, A.: A Dynamic Integration Algorithm for an Ensemble of Classifiers. In: Z.W. Ras, A. Skowron (eds.), Foundations of Intelligent Systems: ISMIS'99, Lecture Notes in AI, Vol. 1609, Springer-Verlag, Warsaw (1999) 592-600.
14. Puuronen, S., Tsymbal, A., Terziyan, V.: Distance Functions in Dynamic Integration of Data Mining Techniques. In: B.V. Dasarathy (ed.), Data Mining and Knowledge Discovery: Theory, Tools, and Techniques, SPIE Press, USA (2000) to appear.
15. Quinlan, J.R.: C4.5 Programs for Machine Learning. Morgan Kaufmann, San Mateo, CA (1993).
16. Schapire, R.E.: A Brief Introduction to Boosting. In: Proc. 16th Int. Joint Conf. AI (1999).
17. Schapire, R.E.: The Strength of Weak Learnability. Machine Learning, Vol. 5, No. 2 (1990) 197-227.
18. Schapire, E., Freaund, Y., Bartlett, P., Lee, W.S.: Boosting the Margin: A New Explanation for the Effectiveness of Voting Methods. The Annals of Statistics, Vol. 26, No 5 (1998) 1651-1686.
19. Skrypnik, I., Terziyan, V., Puuronen, S., Tsymbal, A.: Learning Feature Selection for Medical Databases. In: Proc. 12th IEEE Symp. on Computer-Based Medical Systems CBMS'99, IEEE CS Press, Stamford, CT (1999) 53-58.
20. Terziyan, V., Tsymbal, A., Puuronen, S.: The Decision Support System for Telemedicine Based on Multiple Expertise. Int. J. of Medical Informatics, Vol. 49, No. 2 (1998) 217-229.
21. Tsymbal, A.: Decision Committee Learning with Dynamic Integration of Classifiers. In: Proc. 2000 ADBIS-DASFAA Symposium on Advances in Databases and Information Systems, Prague, September 2000, Lecture Notes in Computer Science, Springer-Verlag (to appear).
22. Tsymbal, A., Puuronen, S., Terziyan, V.: Advanced Dynamic Selection of Diagnostic Methods. In: Proceedings 11th IEEE Symp. on Computer-Based Medical Systems CBMS'98, IEEE CS Press, Lubbock, Texas, June (1998) 50-54.
23. Tsymbal, A., Puuronen, S., Terziyan, V.: Arbiter Meta-Learning with Dynamic Selection of Classifiers and its Experimental Investigation. In: J.Eder, I.Rozman, T.Welzer (eds.), Advances in Databases and Information Systems: 3rd East European Conference ADBIS'99, Lecture Notes in CS, Vol. 1691, Springer-Verlag, Maribor (1999) 205-217.
24. Webb, G.I.: MultiBoosting: A Technique for Combining Boosting and Wagging. Machine Learning (2000) in press.

Zoomed Ranking: Selection of Classification Algorithms Based on Relevant Performance Information

Carlos Soares and Pavel B. Brazdil

LIACC/FEP, University of Porto, R. Campo Alegre 823, 4150-180 Porto, Portugal
{csoares,pbrazdil}@ncc.up.pt

Abstract. Given the wide variety of available classification algorithms and the volume of data today's organizations need to analyze, the selection of the right algorithm to use on a new problem is an important issue. In this paper we present a combination of techniques to address this problem. The first one, *zooming*, analyzes a given dataset and selects relevant (similar) datasets that were processed by the candidate algoritms in the past. This process is based on the concept of "distance", calculated on the basis of several dataset characteristics. The information about the performance of the candidate algorithms on the selected datasets is then processed by a second technique, a ranking method. Such a method uses performance information to generate advice in the form of a ranking, indicating which algorithms should be applied in which order. Here we propose the *adjusted ratio of ratios* ranking method. This method takes into account not only accuracy but also the time performance of the candidate algorithms. The generalization power of this ranking method is analyzed. For this purpose, an appropriate methodology is defined. The experimental results indicate that on average better results are obtained with zooming than without it.

1 Introduction

The need for methods which would assist the user in selecting classification algorithms for a new problem has frequently been recognized as an important issue in the fields of Machine Learning (ML) [13, 5] and Knowledge Discovery in Databases (KDD) [3].

Previous meta-learning approaches to algorithm selection consist of suggesting one algorithm or a small group of algorithms that are expected to perform well on the given problem [4, 21, 10]. We believe that a more informative and flexible solution is to provide rankings of the candidate algorithms [15, 19, 5]. A ranking can be used to select just one algorithm, i.e. the one for which the best results are expected. However, if enough resources are available, more than one algorithms may be applied on the given problem.

The problem of constructing rankings can be seen as an alternative to other ML methods, such as classification and regression. Therefore, we must develop

D.A. Zighed, J. Komorowski, and J. Zytkow (Eds.): PKDD 2000, LNAI 1910, pp. 126–135, 2000.
© Springer-Verlag Berlin Heidelberg 2000

methods to generate rankings and also methodologies to evaluate and compare such methods [5].

Recently, several methods that generate rankings of algorithms based on their past performance have been developed with promising results. Some are based only on accuracy [5], others on accuracy and time [19]. So far, these methods were used without taking the dataset which the ranking was intended for into account. That is, given a new dataset, a ranking was generated by processing all available performance information. However, considering the NFL theorem we cannot expect that all that information is relevant for the problem at hand. Therefore, we do not expect that rankings generated this way accurately represent the relative performance of the algorithms on the new problem.

We, therefore, address the problem of algorithm selection by dividing it into two distinct phases. In the first one we identify a subset of relevant datasets. For that purpose we present a technique called *zooming*. It employs the k-Nearest Neighbor algorithm with a distance function based on a set of statistical, information theoretic and other dataset characterization measures to identify datasets that are similar to the one at hand. More details concerning this are in Section 2.

In the second phase we proceed to construct a ranking on the basis of the performance information of the candidate algorithms on the selected datasets. In Section 3 we present the *adjusted ratio of ratios* ranking method. This method processes performance information on accuracy and time. In Section 4 we evaluate this approach by assessing the gains that can be attributed to zooming. In this analysis we assess the effect of varying the number of neighbors and adopting different compromises between the importance of accuracy and time. Finally, we describe some related work (Section 6) and present conclusions (Section 7).

2   Selection of Relevant Datasets

As explained earlier, the ranking of the candidate algorithms is preceded by selecting, from a set of previously processed datasets, those whose performance information is expected to be relevant for the dataset at hand. The ranking is based on that information. We refer to the selection process as *zooming*, because, given the space of all previously processed datasets, it enables us to focus on the "neighborhood" of the new one.

The relevance of a processed dataset to the one at hand is defined in terms of similarity between them, according to a set of measures (meta-attributes). It is given by function $dist(d_i, d_j) = \sum_x \delta(v_{x,d_i}, v_{x,d_j})$ where d_i and d_j are datasets, v_{x,d_i} is the value of meta-attribute x for dataset d_i, and $\delta(v_{x,d_i}, v_{x,d_j})$ is the distance between the values of meta-attribute x for datasets d_i and d_j. In order to give all meta-attributes the same weight, they are normalized in the following way: $\delta(v_{x,d_i}, v_{x,d_j}) = \frac{|v_{x,d_i} - v_{x,d_j}|}{\max_{k \neq i}(v_{x,d_k}) - \min_{k \neq i}(v_{x,d_k})}$, where $\max_{k \neq i}(v_{x,d_k})$ calculates the maximum value of meta-attribute x for all datasets except d_i and $\min_{k \neq i}(v_{x,d_k})$ calculates the corresponding minimum. Note that, it may be the case that a meta-attribute is not applicable on a dataset. For instance, if dataset d_i has no numerical attributes then it makes no sense to calculate mean skew, which is a statistical meta-attribute. It seems reasonable to say that, with respect to this attribute, dataset d_i is very close to dataset d_j if d_j does not have any numerical attributes either. We have, thus, determined that δ is 0 in this case. Furthermore, we can say that dataset d_i is quite different from dataset d_k if the latter has some numerical attributes. In this case δ is assigned the maximum distance, 1.

The meta-attributes used were obtained with the Data Characterization Tool (DCT) [11]. They can be grouped into three categories: general, statistical and information theoretic measures. Examples of general measures used are number of attributes and number of cases. As for the statistical measures, we included mean skew and number of attributes with outliers, among others. Finally, some of the information theoretic measures are class entropy and noise-signal ratio. A full listing of the measures used is given in the appendix.

The meta-attributes used were chosen simply because they are provided by DCT and because they were used before for the same purpose [11]. We do not investigate whether they are appropriate or not, and if different weights should be assigned to them in the distance function, although these are questions that we plan to address in the future.

The distance function defined is used as part of the k-Nearest Neighbor (kNN) algorithm to identify the datasets that are most similar to the one at hand. The kNN algorithm is a simple instance-based learner [13]. Given a case, this algorithm simply selects k cases which are nearest to it according to some distance function.

Performance information for the given candidate algorithms on the selected datasets is then used to construct a ranking. Several methods can be used for that purpose [19, 5]. Details of one of them are given in the next section.

3   Ranking Based on Accuracy and Time

In the previous section we have explained how to select performance information that is relevant to the problem at hand. Here we explain how that information can be used to generate a ranking of the corresponding algorithms. Since the datasets selected are similar to the one at hand, it is expected that algorithms perform similarly. In other words, the method should provide us with a good advice for the selection of algorithms to apply on the dataset at hand.

The ranking method presented here is referred to as the *adjusted ratio of ratios* (ARR) ranking method [19]. This method uses information about accuracy and total execution time to rank the given classification algorithms. We start by defining the measure underlying the method and the parameter that determines the relative importance of time and accuracy. Next we describe how the method works. Finally we describe the experimental setup and give an example.

Weighing Success Rates and Time: The ARR method is based on the ratio of success rate ratio and an adjusted time ratio:

$$ARR_{a_p,a_q}^{d_i} = \frac{\dfrac{SR_{a_p}^{d_i}}{SR_{a_q}^{d_i}}}{1 + \dfrac{\log\left(\dfrac{T_{a_p}^{d_i}}{T_{a_q}^{d_i}}\right)}{K_T}} \tag{1}$$

where $SR_{a_p}^{d_i}$ and $T_{a_p}^{d_i}$ are the success rate and time of algorithm a_p on dataset d_i, respectively, and K_T[1] is a user-defined value that determines the relative importance of time.

The formula may seem ad-hoc at first glance, but its form can be related to the ones used in other areas of science. We can look at the ratio of success rates, $SR_{a_p}^{d_i}/SR_{a_q}^{d_i}$, as a measure of the advantage, and the ratio of times, $T_{a_p}^{d_i}/T_{a_q}^{d_i}$, as a measure of the disadvantage of algorithm a_p relative to algorithm a_q on dataset d_i. The former can be considered a *benefit* while the latter a *cost*. Thus, by dividing a measure of the benefit by a measure of the cost, we assess the overall quality of an algorithm. A similar philosophy underlies the efficiency measure of Data Envelopment Analysis (DEA) that has been proposed for multicriteria evaluation of data mining algorithms [15].

Furthermore, the use of ratios of a measure, namely success rate, has been shown earlier to lead to competitive rankings overall when compared to other ways of aggregating performance information [19,5]. A parallel can be established between the ratio of success rates and performance scatterplots that have been used in some empirical studies involving comparisons of classification algorithms [17].

Relative Importance of Accuracy and Time: The reason behind the adjustment of the time ratio is concerned with the fact that time ratios have, in general, a much wider range of possible values than success rate ratios. Therefore, if a simple time ratio were used, it would dominate the ratio of ratios. By using $\log\left(T_{a_p}^{d_i}/T_{a_q}^{d_i}\right)$, i.e. the order of magnitude of the difference between the times of algorithms a_p and a_q, this effect is minimized. We, thus, obtain values that vary around 1, as happens with the success rate ratio. The parameter K_T enables us to determine the relative importance of the two criteria, which is expected to vary for different applications.

However, the use of the K_T parameter is not very intuitive and would present an obstacle if the method were to be used by non-expert users. We have therefore devised a way to obtain K_T in a way that is more user-friendly. We need an estimate of how much accuracy we are willing to trade for a 10 times speedup or slowdown. The defined setting is represented as 10x≅X%. The parameter K_T is then approximated by $1/X\%$. For instance, if the user is willing to trade 10% of accuracy for 10 times speedup/slowdown (10x≅10%), then $K_T = 1/10\% = 10$.

---

[1] Here, to avoid confusion with the number of nearest-neighbors (k), we refer to the compromise between time and accuracy as K_T, rather then K, as in [19].

Aggregating ARR Information: The method aggregates the given performance information as follows. First, we create an *adjusted ratio of ratios table* for each dataset. The table for dataset d_i is filled with the corresponding values of adjusted ratio of ratios, $ARR^{d_i}_{a_p,a_q}$. Next, we calculate a *pairwise mean adjusted ratio of ratios* for each pair of algorithms, $ARR_{a_p,a_q} = \left(\sum_{d_i} ARR^{d_i}_{a_p,a_q} \right) / n$ where n is the number of datasets. This represents an estimate of the general advantage/disadvantage of algorithm a_p over algorithm a_q. Finally, we derive the *overall mean adjusted ratio of ratios* for each algorithm, $ARR_{a_p} = \left(\sum_{a_q} ARR_{a_p,a_q} \right) / (m-1)$ where m is the number of algorithms. The ranking is derived directly from this measure. The higher the value an algorithm obtains, the higher the corresponding rank.

Experimental Setup: Before presenting an example, we describe the experimental setting. We have used three decision tree classifiers, C5.0, C5.0 with boosting [18] and Ltree, which is a decision tree that can introduce oblique decision surfaces [8]. We have also used an instance based classifier, TiMBL [7], a linear discriminant and a naive bayes classifier [12]. We will refer to these algorithms as c5, c5boost, ltree, timbl, discrim and nbayes, respectively. We ran these algorithms with default parameters on 16 datasets. Seven of those (australian, diabetes, german, heart, letter, segment and vehicle) are from the StatLog repository[2] and the rest (balance-scale, breast-cancer-wisconsin, glass, hepatitis, house-votes-84, ionosphere, iris, waveform and wine) are from the UCI repository[3] [2]. The error rate and time were estimated using 10-fold cross-validation.[4]

Example: Supposing that we want to obtain a ranking of the given algorithms to use on the segment dataset (*test dataset*), without having tested any of them on that dataset. We must, thus, exclude the dataset in question from consideration and use only the remaining datasets (*training datasets*) in the process. In Table 1 we present two rankings. The first is generated by ARR based on all training datasets while the second is based only on the two datasets that are most similar to segment. Here we refer to zooming with a given k followed by the application of ARR on the selected datasets as $Z_k(ARR)$. We note that ARR can be considered as a special case of $Z_k(ARR)$, where k spans across all training datasets. In our meta-data, the two datasets that are most similar to segment are ionosphere (*dist* = 4.99) and glass (*dist* = 8.28). The results presented are obtained with 10x≅1% or $K_T = 100$.

---

[2] See http://www.liacc.up.pt/ML/statlog/.

[3] Some preparation was necessary in some cases, so some of the datasets may not be exactly the same as the ones used in other experimental work.

[4] It must be noted that this is not a comparative study of the algorithms involved. Not all of them were executed on the same machine. However, this does not conflict with the purpose of this work because, in a real-world setting, not all algorithms may be available on the same machine.

Table 1. Recommended rankings for the segment dataset based on all the other datasets and on its two nearest neighbors (*left*). The ideal ranking and part of the calculation of Spearman's correlation for the latter recommended ranking (*right*)

| | Recommended Ranking | | | | Ideal | | Spearman |
| | ARR | | $Z_2(ARR)$ | | Ranking | | |
Rank	a_p	ARR_{a_p}	a_p	ARR_{a_p}	a_p	ARR_{a_p}	$D^2_{a_p}$
1	ltree	1.066	c5boost	1.151	c5boost	1.151	0
2	c5boost	1.057	c5	1.075	c5	1.088	0
3	discrim	1.046	ltree	1.049	ltree	1.088	0
4	c5	1.009	discrim	0.991	discrim	1.031	0
5	nbayes	0.974	timbl	0.902	nbayes	1.008	1
6	timbl	0.919	nbayes	0.900	timbl	0.769	1

We observe that the rankings generated are quite different. The obvious question is which one is the best, i.e. the one that most accurately reflects the actual performance of the algorithms on the test dataset? We try to answer it in the next section.

4   Assessment of Generalization Power

A ranking should naturally be evaluated by comparing it to the actual performance of the algorithms on the dataset the ranking is generated for. Our approach consists of using that performance information to generate an *ideal ranking* [5]. The quality of the ranking being evaluated (*recommended ranking*) is assessed by measuring the distance to the ideal ranking.

The ideal ranking represents the correct ordering of the algorithms on a test dataset. Here, it is based on the assumption that the ARR measure (Eq. 1) appropriately represents the criteria to be used to evaluate the results and that the measured accuracies and times are good estimates of the corresponding true accuracies and times.

The distance between two rankings is best calculated using correlation. Here we use Spearman's rank correlation coefficient [16]. To illustrate this measure, we show how we evaluate the ranking recommended by $Z_2(ARR)$ for the segment dataset with 10x$\cong$1% (Table 1). First we calculate the squared differences, $D^2_{a_p}$, between the recommended and the ideal ranks for algorithm a_p. Then we calculate $D^2 = \sum_{a_p} D^2_{a_p}$. The score of the recommended ranking is the correlation coefficient, $r_s = 1 - \frac{6D^2}{n^3-n}$, where n is the number of algorithms. In the example used $D^2 = 2$ and $r_s = 0.943$, while the correlation for the ranking recommended by ARR is 0.714.

It is not possible to draw any conclusion based on one dataset only. We have, therefore, carried a leave-one-out procedure. As the name suggests, in each iteration one dataset is selected as the test dataset. The rankings generated based on the corresponding training datasets are then evaluated in the way described in the previous paragraphs. The methods compared were $Z_2(ARR)$, $Z_4(ARR)$

Table 2. Mean correlations obtained by $Z_2(ARR)$, $Z_4(ARR)$ and ARR for different values of K_T. The +/- column indicates the number of datasets where the corresponding method has higher/lower correlation than ARR

10x≅	$Z_2(ARR)$		$Z_4(ARR)$		**ARR**
	mean r_S	+/-	mean r_S	+/-	mean r_S
10%	0.64	9/6	0.68	9/2	0.64
1%	0.63	10/6	0.56	9/6	0.54
0.1%	0.64	10/6	0.57	8/6	0.54

and also ARR without zooming. The later method serves as a baseline to assess whether zooming is really advantageous and if it is, quantify that advantage. The values for k were determined in order for the rankings to be built based on 10% and 25%, respectively, of the datasets available. These seem to be sensible values. As for K_T, values were chosen for the time to have large, medium and small importance, respectively 10x≅10%, 1% and 0.1% ($K_T = 10, 100$ and 1000).

If we analyze the mean correlations in Table 2, we observe that when time is predominant (10x≅10%), $Z_4(ARR)$ performs better than ARR while the mean correlations of $Z_2(ARR)$ and ARR are equal. The situation changes when time is given less importance. The advantage of $Z_4(ARR)$ over ARR remains, although in a smaller scale. On the other hand, $Z_2(ARR)$ is now considerably better than ARR, exceeding also the performance of $Z_4(ARR)$. Note that the performance of $Z_2(ARR)$ seems to be quite robust with respect to the variation of K_T.

5   Discussion

According to the previous section, it appears that zooming improves the quality of the rankings generated. However, we would like to obtain statistical support for this claim. For that purpose, we have applied Friedman's test to the results obtained [16]. The values obtained for M, after correction due to the occurrence of ties, are 3.37, 1.34 and 1.24, for 10x≅10%, 1% and 0.1%, respectively. The critical value is 6.5 for $k = 3^5$ and $n = 16$ at a 5% significance level, so we do not reject the null hypothesis that their performance is not significantly different. We, thus, have no statistical evidence of the difference in mean correlation of the methods compared. We must note, however, that, although Friedman's test was used before for the same purpose [19,5] with a conclusive result, it is a distribution-free test, which implies that it is not expected to be very powerful. Also we have restricted our experiments to 6 algorithms and 16 datasets. We expect that by increasing the number of datasets and algorithms used, we are able to obtain statistical evidence of the improvement brought by zooming. The extended study is currently being carried out.

One drawback of the ideal ranking used is that it is built with average accuracies and times. Given that these are only estimates, the ranking generated may

---

[5] This k is a parameter of Friedman's test representing the number of methods being compared.

not be reliable. To minimize this problem, the ideal ranking can be generated as a set of n orderings, one for the results in each fold of the cross-validation procedure used to estimate the performance of the algorithms. A similar procedure as been used before with satisfactory results [20, 5].

As for the measure of distance between rankings used here, it has been shown that correlation is appropriate for that purpose [20]. One drawback is, though, the lack of distinction between rank importance. For instance, it is obvious that the switch made between the 5th and 6th algorithm by the $Z_2 (ARR)$ on the segment dataset (Table 1) is less important than if it would involve the 1st and the 2nd (c5boost and c5). We have previously developed a measure to solve this problem, *weighted correlation* [20]. However, it has not yet been thoroughly analyzed, and, thus was not used here. An alternative to Spearman's correlation coefficient that could be tried is Kendall's tau [16].

6   Related Work

Meta-knowledge as been used before for the purpose of algorithm selection. This knowledge can be either of theoretical or of experimental origin, or a mixture of both. The rules described by Brodley [6] for instance, captured the knowledge of experts concerning the applicability of certain classification algorithms. The meta-knowledge of [1], [4] and [9] was of experimental origin and was obtained by *meta-learning* on past performance information of the algorithms. Its objective is to capture certain relationships between the measured dataset characteristics and the relative performance of the algorithms. As was demonstrated, meta-knowledge can be used to predict the errors of individual algorithms or construct a ranking with a certain degree of success.

Not much work exists in the areas of Machine Learning or KDD concerning multicriteria ranking and evaluation. A noteworthy exception is the work of Nakhaeizadeh et al. [15, 14], who have applied a technique that originated in the area of Operations Research, Data Envelopment Analysis (DEA). It remains to be seen how this approach compares with the method described here.

7   Conclusions

We have presented a combination of techniques that uses past performance information to assist the user in the selection of a classification algorithm for a given problem. The first technique, zooming, works by selecting datasets and associated performance information that is relevant to the problem at hand. This process is based on the distance between datasets, according to a set of statistical, information theoretic and other measures. Here, it is performed using the k-Nearest Neighbor algorithm. We have selected dataset measures that were previously used for the same purpose. Work is under way to select the most predictive subset of those measures.

The ranking method used here is the Adjusted Ratio of Ratios (ARR) method. This is a multicriteria method that takes into account both accuracy and total execution time information. It has a parameter that enables us to determine the relative importance of each criteria. One of the main advantages is its intuitiveness, which is essential to enable its use by non-experts.

We have reported experiments varying the number of neighbors and the relative importance of accuracy and time. The results obtained are compared to results obtained by ARR without zooming. It appears that zooming improves the quality of the generated rankings, although the results obtained are not significantly different according to the Friedman's test.

In summary, our contributions are (1) exploiting rankings rather then classification or regression, (2) providing a general evaluation methodology for ranking, (3) providing a way of combining success rate and time and (4) exploiting dataset characteristics to select relevant performance information prior to ranking.

Acknowledgments We would like to thank the METAL partners for useful discussions. Also thanks to João Gama for providing his implementations of Linear Discriminant and Naive Bayes and to Rui Pereira for implementing an important part of the methods. The financial support from ESPRIT project METAL, project ECO under PRAXIS XXI, FEDER, Programa de Financiamento Pluri-anual de Unidades de I&D and Faculty of Economics is gratefully acknowledged.

References

1. D.W. Aha. Generalizing from case studies: A case study. In D. Sleeman and P. Edwards, editors, *Proceedings of the Ninth International Workshop on Machine Learning (ML92)*, pages 1–10. Morgan Kaufmann, 1992.
2. C. Blake, E. Keogh, and C.J. Merz. Repository of machine learning databases, 1998. http:/www.ics.uci.edu/~mlearn/MLRepository.html.
3. R.J. Brachman and T. Anand. The process of knowledge discovery in databases. In U.M. Fayyad, G. Piatetsky-Shapiro, P. Smyth, and R. Uthurusamy, editors, *Advances in Knowledge Discovery and Data Mining*, chapter 2, pages 37–57. AAAI Press/The MIT Press, 1996.
4. P. Brazdil, J. Gama, and B. Henery. Characterizing the applicability of classification algorithms using meta-level learning. In F. Bergadano and L. de Raedt, editors, *Proceedings of the European Conference on Machine Learning (ECML-94)*, pages 83–102. Springer-Verlag, 1994.
5. P. Brazdil and C. Soares. A comparison of ranking methods for classification algorithm selection. In R.L. de Mántaras and E. Plaza, editors, *Machine Learning: Proceedings of the 11th European Conference on Machine Learning ECML2000*, pages 63–74. Springer, 2000.
6. C.E. Brodley. Addressing the selective superiority problem: Automatic Algorithm/Model class selection. In P. Utgoff, editor, *Proceedings of the 10th International Conference on Machine Learning*, pages 17–24. Morgan Kaufmann, 1993.
7. W. Daelemans, J. Zavrel, K. Van der Sloot, and A. Van Den Bosch. TiMBL: Tilburg memory based learner v2.0 guide. Technical Report 99-01, ILK, 1999.
8. J. Gama. Probabilistic linear tree. In D. Fisher, editor, *Proceedings of the 14th International Machine Learning Conference (ICML97)*, pages 134–142. Morgan Kaufmann, 1997.

9. J. Gama and P. Brazdil. Characterization of classification algorithms. In C. Pinto-Ferreira and N.J. Mamede, editors, *Progress in Artificial Intelligence*, pages 189–200. Springer-Verlag, 1995.
10. A. Kalousis and T. Theoharis. NOEMON: Design, implementation and performance results of an intelligent assistant for classifier selection. *Intelligent Data Analysis*, 3(5):319–337, November 1999.
11. G. Lindner and R. Studer. AST: Support for algorithm selection with a CBR approach. In C. Giraud-Carrier and B. Pfahringer, editors, *Recent Advances in Meta-Learning and Future Work*, pages 38–47. J. Stefan Institute, 1999.
12. D. Michie, D.J. Spiegelhalter, and C.C. Taylor. *Machine Learning, Neural and Statistical Classification*. Ellis Horwood, 1994.
13. T.M. Mitchell. *Machine Learning*. McGraw-Hill, 1997.
14. G. Nakhaeizadeh and A. Schnabl. Towards the personalization of algorithms evaluation in data mining. In R. Agrawal and P. Stolorz, editors, *Proceedings of the Third International Conference on Knowledge Discovery & Data Mining*, pages 289–293. AAAI Press, 1997.
15. G. Nakhaeizadeh and A. Schnabl. Development of multi-criteria metrics for evaluation of data mining algorithms. In D. Heckerman, H. Mannila, D. Pregibon, and R. Uthurusamy, editors, *Proceedings of the Fourth International Conference on Knowledge Discovery in Databases & Data Mining*, pages 37–42. AAAI Press, 1998.
16. H.R. Neave and P.L. Worthington. *Distribution-Free Tests*. Routledge, 1992.
17. F. Provost and D. Jensen. Evaluating knowledge discovery and data mining. Tutorial Notes, Fourth International Conference on Knowledge Discovery and Data Mining, 1998.
18. R. Quinlan. *C5.0: An Informal Tutorial*. RuleQuest, 1998. http://www.rulequest.com/see5-unix.html.
19. C. Soares. Ranking classification algorithms on past performance. Master's thesis, Faculty of Economics, University of Porto, 1999. http://www.ncc.up.pt/~csoares/miac/thesis_revised.zip.
20. C. Soares, P. Brazdil, and J. Costa. Measures to compare rankings of classification algorithms. In *Proceedings of the Seventh Conference of the International Federation of Classification Societies IFCS (to Be Published)*, 2000.
21. L. Todorovski and S. Dzeroski. Experiments in meta-level learning with ILP. In *Proceedings of the Third European Conference on Principles and Practice of Knowledge Discovery in Databases (PKDD99)*, pages 98–106, 1999.

Appendix The dataset characterization measures used in this study were obtained with the DCT program. They consist of simple (number of attributes, number of symbolic and numerical attributes, number of cases and classes, default accuracy, standard-deviation of classes, number of missing values and cases with missing values), statistical (mean skew and kurtosis, number of attributes with outliers, M statistic, degrees of freedom of the M statistic, chi-square M statistic, SD ratio, relative importance of the most important eigenvalue, canonical correlation for the most discriminant function, Wilks Lambda and Bartlett's V statistics, chi square V statistic and number of discriminant functions) and information theoretic measures (minimum, maximum and average symbolic attributes, class entropy, attributes entropy, average mutual information, joint entropy, equivalent number of attributes and noise signal ratio). More details can be found in [11, 12].

Some Enhancements of Decision Tree Bagging

Pierre Geurts

University of Liège, Department of Electrical and Computer Engineering
Institut Montefiore, Sart-Tilman B28, B4000 Liège, Belgium
geurts@montefiore.ulg.ac.be

Abstract. This paper investigates enhancements of decision tree bagging which mainly aim at improving computation times, but also accuracy. The three questions which are reconsidered are: discretization of continuous attributes, tree pruning, and sampling schemes. A very simple discretization procedure is proposed, resulting in a dramatic speedup without significant decrease in accuracy. Then a new method is proposed to prune an ensemble of trees in a combined fashion, which is significantly more effective than individual pruning. Finally, different resampling schemes are considered leading to different CPU time/accuracy tradeoffs. Combining all these enhancements makes it possible to apply tree bagging to very large datasets, with computational performances similar to single tree induction. Simulations are carried out on two synthetic databases and four real-life datasets.

1 Introduction

The bias/variance tradeoff is a well known problem in machine learning. Bias relates to the systematic error component, whereas variance relates to the variability resulting from the randomness of the learning sample and both contribute to prediction errors. Decision tree induction [5] is among the machine learning methods which present the higher variance. This variance is mainly due to the recursive partitioning of the input space, which is highly unstable with respect to small perturbations of the learning set. Bagging [2] consists in aggregating predictions produced by several classifiers generated from different bootstrap samples drawn from the original learning set. By doing so, it reduces mainly variance and indirectly bias, and hence leads to spectacular improvements in accuracy when applied to decision tree learners. Unfortunately, it destroys also the two main attractive features of decision trees, namely computational efficiency and interpretability.

This paper approaches three topics on which improvements can be obtained with tree bagging either in terms of computation time or in terms of accuracy: discretization of continuous attributes, tree pruning, and sampling schemes. A very simple discretization procedure is proposed, resulting in a dramatic speedup without significant decrease in accuracy. Then a new method is proposed to prune an ensemble of trees in a combined fashion, which is significantly more effective than individual pruning. Finally, different resampling schemes are considered leading to different CPU time/accuracy tradeoffs.

D.A. Zighed, J. Komorowski, and J. Żytkow (Eds.): PKDD 2000, LNAI 1910, pp. 136–147, 2000.
© Springer-Verlag Berlin Heidelberg 2000

The paper is organized as follows. Section 2 introduces bagging. Section 3 describes the proposed enhancements and Section 4 is devoted to their empirical study, reporting results in terms of accuracy, variance, tree complexity and computation time.

2   Bootstrap Aggregating (Bagging)

Let us denote by X the random input variable (attribute vector) and Y the (scalar) output variable, and by $P(X,Y)$ the probability distribution from which the data are sampled. We assume that the learning sample is a sequence ($LS = \{(x_1,y_1),...,(x_N,y_N)\}$) of independent and identically distributed observations drawn from $P(X,Y)$. Let us denote by $f_N(x)$ the (random) function which is produced by a learning algorithm in response to such a sample and by $\overline{f_N}(x) = E_{LS}\{f_N(x)\}$ the averaged model over the set of all learning sets of size N. Bias denotes the discrepancy between the best model (the one which minimizes a given loss criteria, also called the bayes model) and the averaged model while variance denotes the variability of the predictions with respect to the learning set randomness. Both, bias and variance, lead to prediction errors and thus should be minimized.

The averaged model $\overline{f_N}(x)$, by definition, has no variance (as it does depend on a particular learning set) and the same bias as the original model. So, if we could compute it, it would certainly have smaller prediction errors than a single model. In this context, bagging [2] has been suggested as a way to approximate this averaged model. As we do not have access to an infinite source of learning sets, the process of sampling from nature is approximated by bootstrap sampling from the original learning set. More precisely, starting from a learning set (LS) of size N, bagging consists in randomly drawing n subsamples of size N with replacement from LS. Let us denote by $ls_1,...,ls_n$ these subsamples. Then from each ls_i, a model is learned denoted by $f_{ls_i}(x)$. Finally, the bagged model $f_{LS}(x)$ is obtained by aggregating the $f_{ls_i}(x)$. When output Y is discrete (classification), the final prediction is determined either by voting:

$$f_{LS}(x) = \{C_k | k = \arg\max_j (\sum_{i=1}^{n} \delta(f_{ls_i}(x), C_j))\}, \tag{1}$$

or by averaging class-conditional probability estimates if they are available:

$$f_{LS}(x) = \{C_k | k = \arg\max_j (\hat{P}_{LS}(C_j|x)) = \arg\max_j (\frac{1}{n}\sum_{i=1}^{n} \hat{P}_{ls_i}(C_j|x))\}, \tag{2}$$

where C_k denotes one of the classes and $\hat{P}_{ls_i}(C_k|x)$ the class-conditional probability estimates given by the ith model. The two approaches have been shown to give very similar results.

For a fixed individual model complexity, this way of doing indeed reduces significantly variance while having little effect on the bias term. So bagging is mostly effective in conjunction with unstable predictors like decision trees which present high variance.

3 Proposed Enhancements of Decision Tree Bagging

While very effective, bagging in conjunction of decision trees learners destroys also the two main attractive features of decision trees, namely computational efficiency and interpretability. In this section we propose three enhancements of the tree bagging algorithm which try to improve its performance in terms of computation time or prediction accuracy only, not taking into account interpretability.

3.1 Median Discretization

In another paper [8], we have investigated different ways to reduce the variance due to the discretization of continuous attributes in the context of top down induction of decision trees. It turns out from this paper that a very simple discretization algorithm which always chooses the median to split a local learning subset gives at least comparable results to the classical discretization algorithm. At the same time, the use of the median allows to reduce significantly variance of the probability estimates of the trees and computation times. However, we point out that the median comes with a loss of interpretability as the threshold is not even related to the class in the learning subset. While this loss of interpretability is a drawback in the context of single decision tree induction, it has no importance in the context of bagging.

Usually node splitting is carried out in two stages: the first stage selects for each input attribute an optimal test and the second stage selects the optimal attribute. In the case of numerical attributes, the first stage consists in selecting a discretization threshold for each attribute. Denoting by a an attribute and by $a(o)$ its value for a given sample o, this amounts to selecting a threshold value a_{th} in order to split the node according to the test $T(o) \equiv [a(o) < a_{\text{th}}]$. To determine a_{th}, normally a search procedure is used so as to maximize a score measure evaluated using the subset $ls = \{o_1, o_2, ..., o_n\}$ of learning samples which reach the node to split. Supposing that the ls is already sorted by increasing values of a, most discretization techniques exhaustively enumerate all thresholds $\frac{a(o_i)+a(o_{i+1})}{2} (i = 1...n-1)$. Denoting the observed classes by $C(o_i), (i = 1, ..., n)$, the score measures how well the test $T(o)$ correlates with the class $C(o)$ on the sample ls. In the literature, many different score measures have been proposed. In our experiments we use the following normalization of Shannon information (see [13] for a discussion):

$$C_C^T = \frac{2I_C^T}{H_C + H_T}, \tag{3}$$

where H_C denotes class entropy, H_T test entropy (also called split information by Quinlan), and I_C^T their mutual information.

Our modification of this classical discretization algorithm simply consists in evaluating for each numerical attribute only one threshold value, the one which splits the learning set into two sets of the same size. According to the previous notation, we can compute this threshold as $\frac{a(o_{n/2})+a(o_{n/2+1})}{2}$ if n is even or $\frac{a(o_{(n+1)/2})+a(o_{(n+3)/2})}{2}$ if n is odd. We then split the node according to the pair attribute-threshold which gives the test receiving the best score.

The median discretization is of course faster than the classical one. First, we only have to compute the score for one threshold value when we need to do this computation n times in the classical algorithm. Second, we do not have to sort the local learning set for each numerical attribute. Indeed, there exist algorithms linear in the number of samples which is obviously better than the $N\log(N)$ order needed for sorting. Actually this second argument is not always relevant as it is necessary to sort the learning set only once for each attribute before any splits are made. However this pre-sorting has the disadvantage of needing a lot of memory space to store the sorted learning sets (pointers) and thus is not always possible to implement in the case of very large databases. In our implementation, we use pre-sorting to compute the median and so the difference between the two discretization algorithms is essentially due to the number of score computations. To give a first idea of the time which can be saved, we discretized 50 random samples of size 1000 from the Waveform database. Classical discretization took about 5000 ms while the median discretization only took 300 ms.

3.2   Combined Pruning

Decision tree pruning. Like bagging, pruning is a way to handle the bias-variance tradeoff in decision tree induction. Pruning aims at cuting useless part of decision trees. By doing so, it reduces complexity significantly and variance to some extent, but it also increases bias. Thus, it generally improves only slightly interpretability and accuracy.

There are two ways to prune a tree : stop-splitting and post-pruning. The former approach consists in evaluating best test significance on the current learning set and decide whether to split or not the node according to this measure. One problem of this approach is that it often relies on a parameter, the significance threshold, and the optimum value of this parameter could be application dependant and difficult to appreciate. The latter approach consists first in building a full tree and then use cross-validation to prune useless parts of the tree in a bottom-up fashion. This approach is not parametric anymore but it is necessary to save some samples to form the validation set.

In our implementation of DT induction, stop-splitting is based on a hypothesis test. Indeed, assuming the independence of the test with respect to the classification in the learning sample, Kvålseth ([9]) has shown that the following quantity:

$$G^2 \stackrel{\triangle}{=} 2N.\ln 2.\hat{I}_C^T,$$

follows a χ-square distribution with $(m-1)(p-1)$ degrees of freedom, where m is the number of classes and p is the number of test issues. Thus we choose not to split a node if the probability of G^2 being greater than the observed value is larger than an a priori fixed value α. To $\alpha = 1.0$ correspond fully developed trees and to $\alpha = 0.0$ correspond only trivial trees.

Post-pruning usually relies on a quality measure of trees where a parameter β weights complexity versus reliability of a tree. The quality measure we used is:

$$Q_\beta(\mathcal{T}, LS) \triangleq N \hat{I}_C^{\mathcal{T}}(LS) - \beta.\delta(\mathcal{T}),$$

where N is the number of learning states in the learning set LS, $\hat{I}_C^{\mathcal{T}}(LS)$ the information provided by the tree $\mathcal{T}$ about the classification in the learning set, $\delta(\mathcal{T})$ the tree complexity which is defined as the number of terminal nodes of the tree minus one. Then for increasing values of β from 0 to ∞, we compute the optimally pruned tree $\mathcal{T}_\beta$ which maximizes $Q_\beta(\mathcal{T}_\beta, LS)$. From the additive nature of the quality criterion, it follows that the β-optimal pruned trees form a nested sequence for increasing β (and thus there are at most k of them, if k is the number of test nodes) and that the sequence of trees can be computed by a simple recursive bottom up algorithm (see [13] for a complete discussion). This yields the sequence $S = \{(\beta_1, \mathcal{T}_1), (\beta_2, \mathcal{T}_2), \dots, (\beta_k, \mathcal{T}_k)\}$ where $\beta_i < \beta_{i+1}$, $\mathcal{T}_1$ is the full tree, $\mathcal{T}_k$ the trivial tree and $\mathcal{T}_i$ $(1 < i \leq k)$ is the best pruned tree for $\beta_{i-1} < \beta \leq \beta_i$. Among this sequence of trees, we extract the one which gives the best result on the pruning cross-validation sample.

Pruning with bagging. In the context of DT ensembles, we could imagine two ways of decreasing the model complexity: first by removing some trees from the ensemble, second by pruning the constituting trees. The first approach has been successfully applied to boosted ensembles in [10] but we believe it is not likely to give good results with bagging as models are i.i.d. and generalization errors have been shown to be monotone functions of the number of aggregated models. To decrease constituting trees complexity, we could first apply the above pruning techniques individually to each tree of the ensemble, i.e. not taking into account their future averaging. However, because bagging reduces variance, the optimal complexity of averaged trees should be higher than the optimal complexity of a single tree induced from the same learning set. So individual pruning should yield trees which are not complex enough given the new bias/variance tradeoff resulting from averaging. So we propose here a new method to post-prune the trees from a bagged ensemble in a combined way.

The algorithm proceeds as follows. First, we compute for each tree $\mathcal{T}_j$ of the ensemble the sequence $S_j = \{(\beta_{j,1}, \mathcal{T}_{j,1}), (\beta_{j,2}, \mathcal{T}_{j,2}), \dots, (\beta_{j,k_j}, \mathcal{T}_{j,k_j})\}$ according to the previously defined single tree post-pruning algorithm. We then let increase β from 0 to ∞ and get from each individual pruning sequence S_j the tree $\mathcal{T}_{j,i}$ such that $\beta_{j,i-1} < \beta \leq \beta_{j,i}$. This yields a sequence of pruned trees ensembles. Among this sequence, we select the ensemble which yields the best error rates on the validation set. Of course, only critical values of β (i.e. corresponding to a $\beta_{j,i}$) need to be considered and if there are n trees with maximum k test nodes each, the length of the sequence will be less than kn.

In our experiments, we will compare stop-splitting and individual postpruning in the context of bagging to combined pruning. We expect that pruning the ensemble in a combined way would give better results than individual pruning and also, because of the non zero variance, will be better than no pruning at all.

Table 1. Datasets

Dataset	#Variables	#Classes	#Samples	#LS	#PS	#TS	domain
Omib	6	2	20000	16000	2000	2000	Power system [13]
Waveform	21	3	5000	3000	1000	1000	artificial [5]
Satellite	36	6	6435	3000	1435	2000	soil type recognition [11]
Pendigits	16	10	10992	5000	2494	3498	digit recognition
Dig44	16	10	18000	6000	3000	9000	digit recognition [11]
Vst	136	2	4041	2430	815	796	voltage stability [13]

3.3   Resampling Method

In average, bootstrap samples leave out about 37% of the examples. So, a question which could be raised is what happens if we replace this "with replacement, full size" sampling with a "without replacement, smaller size" sampling. Theoretical studies of bagging (in [12] for bagging of linear models and more recently in [6] for non-linear models) have shown that without replacement sampling could give similar or better results than bootstrap sampling. In the mean time, the computation times could be reduced significantly as we build trees from smaller learning sets. In our experiments, we propose to study in an empirical way the effect of various resampling scheme on real problems in terms of accuracy and computation times improvements.

4   Empirical Studies

In this section, we evaluate the proposed enhancements. We first consider two artificial problems where thorough experiments are pursued and then look at four real-life datasets. A description of the databases is given in Table 1. All input variables are numerical and the datasets were selected to provide large enough samples.

4.1   Artificial Problems (Omib and Waveform)

Experimentation protocol. To evaluate variants of decision tree induction and tree bagging and be able to assess their variance, we carried out experiments in the following way. First, the database is split into three disjoint parts: a set used to pick random samples for tree growing (LS), a set used for cross-validation during tree pruning (PS) and a set used for testing (TS) (the divisions for each database are shown in table 1). Then 50 random subsets of size 1000 are drawn without replacement from the pool LS, yielding $LS_1, LS_2, ..., LS_{50}$. For each method:

- A model $\mathcal{M}_i$ (either a single tree or a set of trees) is grown from each LS_i according to the studied variant.
- Average test set error rate $\overline{P}_e$ and complexity $\overline{C}$ of the 50 models are computed.

Table 2. Effect of median discretization

method	Omib				Waveform			
	$P_e\%$	C	$v\hat{a}r$	CPU	$P_e\%$	C	$v\hat{a}r$	CPU
Single trees (post-pruned)								
classic	11.20	67.6	0.0572	6	27.30	45.96	0.0434	16
median	10.39	103.92	0.0383	4	27.30	66.04	0.0382	7
Bagging, 10 trees (unpruned)								
classic	7.71	967.0	0.0158	21	20.39	1618.4	0.0172	87
median	7.55	2026.0	0.0134	8	20.61	3032.9	0.0160	27

– As proposed in [7], we estimate classification model variance by computing the variance of the class-probability estimates given by the model. To this end, we determine the conditional class-probability estimates $\hat{P}_{\mathcal{M}_i}(C|o)$ for each case o belonging to the test sample and the following quantity is computed:

$$\hat{Var}(\hat{P}_{\mathcal{M}_i}(C|.)) = \frac{1}{\#TS} \sum_{o \in TS} \{\frac{1}{50} \sum_{i=1}^{50} (\hat{P}_{\mathcal{M}_i}(C|o))^2 - (\frac{1}{50} \sum_{i=1}^{50} \hat{P}_{\mathcal{M}_i}(C|o))^2\}.$$

We also give the average CPU time[1] in seconds needed to build one model of each type. The complexity of a single tree is the total number of nodes and the complexity of a bagged ensemble is the sum of the complexities of the component trees. The conditional class-probability estimates for bagging are computed by averaging the conditional class-probability estimates of each tree:

$$\hat{P}_{\mathcal{M}_i}(C|o) = \frac{1}{n} \sum_{j=1}^{n} \hat{P}_{T_{i,j}}(C|o),$$

where $T_{i,j}, j = 1, ..., n$ are the trees constituting the model $\mathcal{M}_i$. The predictions on test samples are done according to these estimates (attributing the maximum probability class to each case).

Results

Discretization: median vs classical score. Table 2 presents results obtained with classical discretization and median discretization, on one hand with single decision tree induction and pruning, and, on the other hand with bagging of full trees. The median does not decrease or increase the error rates significantly but it decreases the variance especially with single trees. On the other hand, the complexity of the resulting trees are multiplied by a factor 2 (when we do not use pruning). Note here also the positive effect of tree bagging which decreases dramatically the variance and significatively the error rates. In terms of computation times, we observe that given our (non optimal) implementation, median

---

[1] The system is implemented in Common Lisp and runs on a SUN Ultrasparc 2 - $300MHz$ microprocessor.

Table 3. Effect of pruning

method	Omib				Waveform			
	$P_e\%$	C	vâr	CPU	$P_e\%$	C	vâr	CPU
$\alpha = 1.0$	7.71	967.0	0.0158	21	20.39	1618.4	0.0172	87
$\alpha = 0.05$	7.77	873.4	0.0156	21	20.21	1523.9	0.0167	86
$\alpha = 0.01$	7.96	660.9	0.0149	21	20.15	1208.4	0.0156	79
$\alpha = 0.005$	8.21	589.0	0.0145	18	20.05	1057.9	0.0153	76
$\alpha = 0.001$	8.37	469.0	0.0137	17	20.11	780.4	0.0137	70
$\alpha = 0.0005$	8.54	429.3	0.0139	17	20.15	688.4	0.0127	66
$\alpha = 0.0001$	8.87	353.6	0.0134	16	20.35	535.6	0.0115	59
individual pruning	7.90	659.9	0.0143	30	20.03	668.4	0.0122	91
combined (classic)	7.67	831.2	0.0153	32	20.02	777.4	0.0134	93
combined (median)	7.51	1557.5	0.0123	17	19.79	1296.4	0.0113	32

Table 4. Effect of resampling method (with median and combined pruning)

method	Omib				Waveform			
	$P_e\%$	C	vâr	CPU	$P_e\%$	C	vâr	CPU
replacement	7.51	1557.5	0.0123	17	19.79	1296.4	0.0113	32
$p = 0.1$	10.27	344.9	0.0112	8	20.96	205.4	0.0103	8
$p = 0.3$	8.19	807.1	0.0108	12	20.17	568.3	0.0102	18
$p = 0.5$	7.91	1138.5	0.0120	15	20.11	817.3	0.0099	29
$p = 0.7$	7.65	1506.8	0.0140	18	20.30	1284.2	0.0122	35

makes bagging about three times faster than classical discretization although the resulting models are two times bigger.

Stop-splitting vs individual pruning vs combined pruning. To assess the effect and usefulness of pruning in the context of bagging, we made experiments using stop-splitting with different values of α, with individual tree pruning, and two versions of combined tree pruning (classical discretization and median discretization). These results are summarized in Table 3.

Results are different from one dataset to the other. On the Omib dataset, it appears that the more complex the trees are, the better are the error rates. On the Waveform database, we observe an optimal value of α smaller than 1.0 which justifies pruning in this case. However, the interest of combined pruning with respect to individual pruning does not yet appear from these experiments, as it gives more complex trees without really improving error rates. Nevertheless, together with median discretization it allows to decrease error rates and variance.

With replacement vs without replacement. Table 4 summarizes several experiments with different resampling schemes. The one called *replacement* is the classical bootstrapping method. The other ones are sampling without replacement and with learning sets of size p times the original learning set size. All experiments use median discretization and combined pruning.

Results are very interesting. First we note a decrease of variance as we reduce the fraction p. Of course, this decrease of variance comes with an increase of

Table 5. Results on real-life datasets

method	Satellite			Pendigits			Dig44			Vst		
	$P_e\%$	C	CPU	$P_e\%$	C	CPU	$P_e\%$	C	CPU	$P_e\%$	C	CPU
Single trees (post-pruned)												
classic	13.95	163	27	7.52	363	19	16.12	455	127	11.81	21	178
median	15.5	269	19	11.29	893	17	18.378	601	45	14.32	331	59
Bagging, 25 trees, classical dis.												
$\alpha = 1.0$	11.24	9465	465	5.36	7413	305	9.71	21746	2010	8.48	4868	2332
ind. prun.	12.54	3074	492	5.47	6593	358	10.41	11398	2081	8.67	1325	2362
comb. prun.	11.44	6689	501	5.38	7042	343	9.71	19030	2109	8.12	3022	2386
Bagging, 25 trees, median dis.												
comb. prun.	11.13	11392	318	6.14	15465	271	10.47	39862	767	9.27	5476	916
Without replacement, combined pruning, classical dis.												
$p = 0.1, n = 10$	15.04	624	42	9.56	877	46	14.6	1581	135	12.59	282	125
$p = 0.1, n = 25$	13.54	1631	106	8.89	2247	111	12.75	4360	330	10.84	721	293
$p = 0.3, n = 25$	12.10	3579	261	6.38	4275	216	10.52	10057	956	8.89	1915	972
$p = 0.5, n = 25$	11.73	5741	397	5.82	6054	296	10.06	15219	1588	8.40	2513	1730

bias as the trees are much smaller. We thus observe a bias/variance tradeoff regulated this time by p. From this tradeoff results an optimal value of p of 0.5 on the Waveform database while on the Omib database the optimal value seems to be the greatest. These experiments show that it is possible to tune this parameter, on one hand to get better results and on the other hand to decrease the computation times. One other interesting fact is that, whatever the dataset, using 10 times smaller learning sets to build 10 decision trees is better (slightly for Omib, significantly for Waveform) than building a big one using the whole learning set. While CPU times are similar, building a big tree should be more demanding in terms of resources.

4.2   Real-Life Datasets

We also have pursued experiments on four real-life datasets (the last four datasets of Table 1). Each database was split into three disjoint parts: a learning set (LS), a pruning set (PS) and a test set (TS). Results are summarized in Table 5 (from top to bottom):

- A single decision tree was built from LS, then post-pruned using PS with classical and median discretization.
- Using the same 25 bootstrap samples, we built 25 fully developed trees $(\alpha = 1.0)$ with classical discretization, we pruned them individually, then in a combined way and tested the three bagged sets of trees using class-probability estimates averaging. We also report the result of the same experiment with median discretization in the context of combined pruning only.
- We made as well experiments with sampling without replacement using $p = 1.0$ and 10 samples and $p = 0.1, 0.3, 0.5$ and 25 samples, each time with combined pruning and classical discretization.

Table 6. Random splitting

method	Satellite			Pendigits			Dig44			Vst		
	$P_e\%$	C	CPU	$P_e\%$	C	CPU	$P_e\%$	C	CPU	$P_e\%$	C	CPU
Single trees (post-pruned)												
classic	19.85	374	6	14.17	987	6	24.14	1187	36	15.61	146	30
median	17.51	593	5	15.89	2232	8	29.66	2995	18	16.57	176	10
Bagging, 25 trees, combined pruning												
classic	10.68	12552	101	4.71	17122	130	7.33	35409	384	9.18	4014	335
median	11.59	23489	102	4.95	43810	178	8.82	80787	368	10.68	5498	132

Results on each dataset are summarized in terms of error rates, complexity and CPU times in seconds (which include building and pruning). When bagging was used, displayed values are the average of ten experiments with different sets of bootstrap samples. We comment here these results separately in terms of the three enhancements.

Median discretization. On all datasets, median discretization gives significantly worse error rates than classical discretization while building single trees. On the other hand, in conjunction with bagging, median still allows to obtain competitive results with respect to the classical discretization. Anyway the main interest of median discretization is the saving of time which is already significant on the Satellite and Pendigits datasets but becomes very impressive on the other two datasets (we gain a factor three). The smaller gain in computation time with the first two examples can be explained by the fact that on these datasets numerical attributes are integer valued between 0 and 256 and so the number of discretization thresholds to evaluate is much smaller than (and not related to) the number of samples in the learning set.

Since the naive median discretization works well with bagging, we pursued some more experiments with an even more naive splitting procedure in order to see how far we can go. To select a test, we first draw randomly an attribute, then compute the threshold. To ensure not to choose attribute too much unrelated to the classification, this procedure is repeated until we find an attribute which yields a score greater than an a priori fixed threshold (here 0.1). Table 6 shows results obtained with this random splitting on the four datasets, first with single trees, then with bagging of 25 trees (10 experiments). While random splitting deteriorates significantly error rates in the case of single trees, we obtain better results with bagging on the first three domains with a strong reduction of computation times but an increase of complexity. Our feeling at this point is that, in the context of bagging, the optimality of node splitting can be very much relaxed, thus allowing to decrease very strongly computation times.

Combined pruning. From Table 5, it is clear that individual pruning tends to produce trees which are not complex enough given the reduction of variance due to bagging. On the other hand, combined pruning always decreases complexity (by 20% on average) without notable change in accuracy with respect to full trees. However, individual pruning still could be preferred as it decreases much more complexity (50% on average) without very important decreases in error rates.

Resampling. without replacement bagging with $p = 0.5$ gives similar results than with replacement bagging but is better in terms of CPU times because learning sets and thus trees are a bit smaller. On the other hand, the sampling fraction p allows to make a tradeoff between ensembles accuracy (and ensembles complexity) and computation times. As on the two artificial problems, averaging ten small trees sometimes is better than building a single big one (as on Dig44) but we see here that it can also significantly reduces accuracy (as on Pendigits). On the other hand, even if building ten small trees is done in the same amount of time as building a big one, the latter is more demanding in terms of memory space and real computation times should be higher in this case.

5   Discussion

In this paper we have investigated three enhancements of decision tree bagging on which we give the following conclusions and future work directions:

Discretization. While the "median" discretization method does not work generally as good as the classical one when building single decision trees, the loss it introduces vanishes when we use it in conjunction with bagging. Our feeling is that bagging has the effect of counterbalancing the sub-optimality of the node-splitting procedure used to build the individual trees. Thus, there is definitely an opportunity here to reduce computing times of tree bagging, without sacrificing accuracy, especially when dealing with large datasets. Although we did mainly focus in this paper on the discretization part of node splitting, our results with random-splitting suggest that further simplification of the node-splitting algorithm can be obtained without notable degradations of accuracy. This opens a new field of investigations to design ad hoc tree growing procedures in the context of bagging.

Pruning. Breiman [4] and others [1] have come to the conclusion that unpruned trees were better than individually pruned trees in the context of bagging, since bagging reduces variance only. However, our experiments have shown that in some problems (e.g. the Waveform database) pruning the trees in a combined way taking into account the model aggregation step, may lead to more accurate models than fully grown trees. We thus believe it is a good idea to use it in conjunction with tree bagging since it never decreases significantly performances in terms of error rates, and it is still very efficient in terms of computing time.

Resampling scheme. Our experiments with different resampling schemes agree with the theoretical developments in [6]. Sampling without replacement can give similar results than sampling with replacement but with decreased computation times. From these experiments comes the idea of alternative uses of bagging already proposed in [3], for example as a way to handle very large datasets or to learn incrementally a model. In the first goal, we could randomly partition a large database into several small parts (say ten times smaller than the whole set), build a tree from each part and then average the resulting models. Our experiments and the one in [3] show that this procedure would work at least as well as building only one tree. On the other hand, the appropriate use of the scheme in an incremental and adaptive way to handle time-varying data has still to be investigated, but seems to be a very promising direction of complementary work.

Allover, the various improvements suggested in this paper can be combined to make tree bagging a very attractive procedure in the context of data mining of very large databases, more accurate than classical tree induction and of the same or even higher computational efficiency and scalability.

References

1. E. Bauer and R. Kohavi. An empirical comparison of voting classification algorithms : Bagging, boosting, and variants. *Machine Learning*, 36:105–139, 1999.
2. L. Breiman. Bagging predictors. Technical report, University of California, Department of Statistics, September 1994.
3. L. Breiman. Pasting small votes for classification in large databases and on-line. *Machine Learning*, 36:85–103, 1999.
4. L. Breiman. Using adaptive bagging to debias regressions. Technical report, Statistics Department, University of California, Berkeley, February 1999.
5. L. Breiman, J.H. Friedman, R.A. Olsen, and C.J. Stone. *Classification and Regression Trees*. Wadsworth International (California), 1984.
6. J. H. Friedman and P. Hall. On bagging and nonlinear estimation. Technical report, Statistics Department, Standford University, January 2000.
7. J.H. Friedman. On bias, variance, 0/1-loss, and the curse-of-dimensionality. *Data Mining and Knowledge Discovery*, 1:55–77, 1997.
8. P. Geurts and L. Wehenkel. Investigation and reduction of discretization variance in decision tree induction. In *Proc. of the 11th European Conference on Machine Learning (ECML-2000), Barcelona*, pages 162–170, May 2000.
9. T.O. Kvålseth. Entropy and correlation : Some comments. *IEEE Trans. on Systems, Man and Cybernetics*, SMC-17(3):517–519, 1987.
10. Dragos D. Margineantu and Thomas G. Dietterich. Pruning adaptive boosting. In Morgan Kaufmann, editor, *Proc. of Fourteenth International Conference on Machine Learning (ICML-97)*, 1997.
11. D. Michie and D.J. Spiegelhalter, editors. *Machine learning, neural and statistical classification*. Ellis Horwood, 1994.
12. Peter Sollich and Anders Krogh. Learning with ensembles : How over-fitting can be useful. In D.S. Touretzky, M.C. Mozer, and M.E. Hasselmo, editors, *Advances in Neural Information Processing Systems*, volume 8, pages 190–196. MIT Press, 1996.
13. L. Wehenkel. *Automatic learning techniques in power systems*. Kluwer Academic, Boston, 1998.

Relative Unsupervised Discretization for Association Rule Mining

Marcus-Christopher Ludl[1] and Gerhard Widmer[1,2]

[1] Austrian Research Institute for Artificial Intelligence, Vienna,
[2] Department of Medical Cybernetics and Artificial Intelligence,
University of Vienna, Austria

Abstract. The paper describes a context-sensitive discretization algorithm that can be used to completely discretize a numeric or mixed numeric-categorical dataset. The algorithm combines aspects of unsupervised (class-blind) and supervised methods. It was designed with a view to the problem of finding association rules or functional dependencies in complex, partly numerical data. The paper describes the algorithm and presents systematic experiments with a synthetic data set that contains a number of rather complex associations. Experiments with varying degrees of noise and "fuzziness" demonstrate the robustness of the method. An application to a large real-world dataset produced interesting preliminary results, which are currently the topic of specialized investigations.

1 Introduction

Association rules have become one of the most popular objects of study in data mining research. Following the seminal work of e.g. [2], there has been a wealth of research on improvements and extensions of the original algorithms.

One limitation of classical association rule mining algorithms that has been addressed only fairly recently is the fact that they require *categorical data*, i.e., they cannot directly deal with numeric attributes. As numeric information is abundant in real-world databases, this is a problem of practical importance. There have been some attempts at remedying this situation in recent years; section 4 summarizes the most important ones. However, there is still room for improvement, and we hope to show an interesting direction in this paper.

What we present in this paper is a *discretization algorithm* named RUDE that can be used to completely discretize a numeric or mixed numeric-categorical dataset. The algorithm was designed especially with a view to the association rule mining task (see section 2.1). It can be used as a pre-processor to "standard" association rule mining algorithms like Apriori [2]. What distinguishes our algorithm is that it attempts to construct a discretization that as much as possible reflects all the interdependencies between attributes in the database. Experimental results with synthetic data show that RUDE is not only effective, but also quite robust with respect to inaccuracies and noise in the data. We have also applied the algorithm to a large real-world dataset, with interesting preliminary results, which are currently the topic of specialized investigations.

D.A. Zighed, J. Komorowski, and J. Żytkow (Eds.): PKDD 2000, LNAI 1910, pp. 148–158, 2000.
© Springer-Verlag Berlin Heidelberg 2000

2 RUDE – Relative Unsupervised Discretization

2.1 Discretization for Association Rules

Association rules describe systematic dependencies between fields (attributes) in a relational database. Unlike in classification problems, there is no designated class attribute; rather, any combination of attributes may be related through associations, and many different dependencies involving different sets of attributes may be hidden in the data. The goal is to find all of these. A good discretizer should thus produce discretizations that enable a rule finder both to express and to find all dependencies between attribute ranges, where any attribute may potentially occur in the left-hand or the right-hand side of some rules.

Of course, one could simply use some standard "unsupervised" discretization method, like equal-width, equal-frequency, or k-means [4], and apply it to every numeric attribute. However, these methods simply discretize each attribute in isolation, ignoring any patterns of correlation between attributes; the intervals they construct are thus likely to be irrelevant and unsuitable for describing the hidden dependencies. In association rule mining, some researchers have investigated more sophisticated pre-discretization methods (e.g.,[9]), but again, the attributes are discretized in isolation, the discretization step does not pay attention to systematic dependencies between attributes.

In the field of classification, there has been some research on "supervised" discretization, where, given a numeric attribute to be discretized, one looks at correlations between this attribute and the class attribute in order to create intervals that maximize the numeric attribute's predictivity of the class (see [5] for an overview). While this is not directly applicable for association rule mining, it has inspired the approach to be presented in this paper.

What we propose is a *global* discretization strategy that attempts to construct a discretization of all the numeric attributes which reflects, as much as possible, all the potential dependency patterns between all the attributes simultaneously. The algorithm RUDE (**R**elative **U**nsupervised **Discr**E**tization) combines aspects of both unsupervised and supervised discretization. It is an unsupervised method in that it does not require a dedicated class attribute; nonetheless, the split points are not constructed independently of the "other" attributes (hence "relative").

2.2 RUDE – The Top-Level

The basic idea when discretizing a particular attribute (the *target*) is to use information about the value distribution of all attributes other than the target (the *source attributes*). Intuitively, a "good" discretization would be one that creates split points that correlate strongly with changes in the value distributions of the source attributes. The process that tries to accomplish this (the central component of RUDE) is called *structure projection*.

Here is the top level of the algorithm:

1. **Preprocessing:** Discretize (via some unsupervised method – see below) all source attributes that are continuous;
2. **Structure Projection:** Project the structure of each source attribute a_i onto the target attribute t:
   a) Filter the dataset by the different values of attribute a_i.
   b) For each such filtering perform a **clustering** procedure on values of the target attribute (see section 2.4) and gather the split points created.
3. **Postprocessing:** Merge the split points found.

Note that source attributes that are themselves numeric are first subjected to a simple pre-discretization step to turn them into nominal attributes for the purpose of structure projection.

The above algorithm is repeated for every numeric attribute in the database, that is, every numeric attribute in turn gets to act as target and is discretized based on the other attributes. The time required for discretizing one continuous attribute is $O(nm \log m)$, with n the number of attributes and m the number of tuples. A complete discretization of all continuous attributes can therefore be performed in time $O(n^2 m \log m)$.

2.3   The Main Step: Structure Projection

The intuition behind structure projection is best illustrated with an example (see Figure 1). Suppose we are to discretize a target attribute t with a range of, say, $[0..1]$, which happens to be uniformly distributed in our case. The values of t in the tuples of our database have been drawn along the lowest line in Figure 1. The two lines above indicate the same tuples when filtered for the values 1 and 2, respectively, of some particular binary source attribute a. ("Filtering a database for a value v of an attribute a" means retaining only those tuples that have value v for attribute a.) Given the distribution of t, any unsupervised discretizer would return a rather arbitrary segmentation of t that would not reflect the (to us) obvious distribution changes in the source attribute a. The idea of structure projection is to find points where the distribution of the values of a changes drastically, and then to map these "edges" onto the target t. We

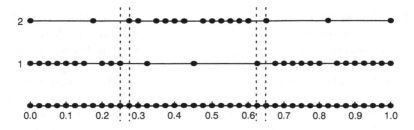

Fig. 1. Structure Projection: An Example.

achieve that by filtering t for each possible value of a separately and clustering the resulting filtered values; the boundaries between clusters are candidates for useful discretization points.

Given:
- a database consisting of m tuples;
- a set of (possibly continuous) source attributes $a_1, \ldots, a_n$
- information on what attribute should be discretized (the target t);

The algorithm:
1. Sort the database in ascending order according to attribute t.
2. For each attribute a_i with $a_i \neq t$ do the following:
   a) If continuous, discretize attribute a_i by *equal-width*
   b) For each symbolic value (interval) v of a_i do the following:
      i. Filter the database for value v in attribute a_i.
      ii. Perform *clustering* on the corresponding values of t in the filtered database.
      iii. Gather the split points thereby created in a *split point list* for attribute t.

Fig. 2. RUDE – The basic discretization algorithm

The basic discretization algorithm is given in fig. 2. RUDE successively maps the "structure" of all source attributes onto the sequence of t's values, creating split points only at positions where some significant distribution changes occur in some of the a_i. For pre-discretizing continuous source attributes in 2(a) above, we have decided to use *equal-width discretization*, because it not only provides a most efficient (linear) method, but also has some desirable statistical properties.

The critical component in all this is the clustering algorithm that groups values of the target t into segments that are characterized by more or less common values of some source attribute a_i. Such segments correspond to relatively densely populated areas in the range of t when filtered for some value of a_i (see Figure 1). Thus, an essential property of this algorithm must be that it tightly delimits such dense areas in a given sequence of values.

2.4 A Characterizing Clustering Algorithm

The algorithm we devised for this purpose has its roots in the concept of *edge detection* in grayscale image processing [7]. The central problem there is to find boundaries between areas of markedly different degrees of darkness. The analogy to our clustering problem is fairly obvious and has led us to develop an algorithm that basically works by opening a "window" of a fixed size around each of the values in an ordered sequence and determining whether this value lies at an "edge", i.e. whether one half of the window is "rather empty" and the other half is "rather full". The notions of "rather full" and "rather empty" are operationalized by two user-defined parameters. For lack of space, we omit the details here.

Given:
- A split point list $s_1, s_2, \ldots$
- A merging parameter (minimal difference s)

The algorithm:

1. Sort the sequence of split points in ascending order.
2. Run through the sequence until you find two splits s_i and s_{i+1} with $s_{i+1} - s_i \leq s$.
3. Starting at $i + 1$ run through the sequence until you find two split points s_j and s_{j+1} with $s_{j+1} - s_j > s$.
4. Calculate the *median* m of $[s_i, \ldots, s_j]$.
   - If $s_j - s_i \leq s$ **merge** all split points in $[s_i, \ldots, s_j]$ to m.
   - If $s_j - s_i > s$ **triple** the set of split points in $[s_i, \ldots, s_j]$ to $\{s_i, m, s_j\}$.
5. Start at s_{j+1} and go back to step 2.

Fig. 3. Merging the split points.

One advantage of the clustering algorithm is that it autonomously determines the appropriate number of clusters/splits, which is in contrast to simpler clustering methods like, e.g., k-means clustering [4]. In fact, the algorithm may in some cases refuse to produce any clusters if it cannot find any justification for splitting. Some numeric attributes may thus be lumped into just one big interval if they do not seem to be correlated with other attributes.

2.5   Post-processing: Merging the Split Points

Of course, due to the fact that RUDE projects multiple source attributes onto a single target attribute, usually many "similar" split points will be formed during the projections. It is therefore necessary to *merge* the split points in a post-processing phase. Figure 3 shows an algorithm for doing that. In step 3, we find a subset of split points with successive differences lower than or equal to a certain pre-defined value s. If all these split points lie closer than s (very dense), they are merged down to only one point (the median). If not, the region is characterized by the median and the two outer borders. (Admittedly, this is a rather *ad hoc* procedure that lacks a strong theoretical foundation.)

3   Experiments

We have evaluated our discretization algorithm as a pre-processing step to the well-known Apriori association rule mining algorithm [2]. In section 3.1, we present a systematic study on a synthetic database that contains a number of rather complex associations. Section 3.2 then briefly hints at some interesting preliminary results on a large, real-world dataset.

3.1 Systematic Experiments with Synthetic Data

As a first testbed for systematic experiments, we used the data and dependencies originally described in [1] and also used in experiments in [10]. In this hypothetical problem, there are six numeric attributes (*salary, commission, age, hvalue, hyears, loan*) and 3 categorical attributes (*elevel, car, zipcode*). All attributes except *commission* and *hvalue* take values randomly from a uniform distribution over certain numeric or nominal ranges. *Commission* and *hvalue* are made to depend on other attributes: *commission* is uniformly distributed between $10K$ and $75K$ if *salary* $< 75K$, otherwise *commission* := 0. *Hvalue* is made to lie in different numeric ranges depending on the value of *zipcode*.

In addition to these attribute dependencies, Agrawal et al. [1] defined a set of 10 binary functions of these attributes, of increasing complexity. We used the first four of these for our experiments:

$f1 = true \Leftrightarrow$ (age < 40) $\vee$ ($60 \leq$ age)

$f2 = true \Leftrightarrow$ ((age < 40) $\wedge$ (($50K \leq$ salary $\leq 100K$)) $\vee$
$\quad$ (($40 \leq$ age < 60) $\wedge$ ($75K \leq$ salary $\leq 125K$)) $\vee$
$\quad$ ((age ≥ 60) $\wedge$ ($25K \leq$ salary $\leq 75K$))

$f3 = true \Leftrightarrow$ ((age < 40) $\wedge$ (elevel $\in [0..1]$)) $\vee$
$\quad$ (($40 \leq$ age < 60) $\wedge$ (elevel $\in [1..3]$)) $\vee$
$\quad$ ((age ≥ 60) $\wedge$ (elevel $\in [2..4]$))

$f4 = true \Leftrightarrow$ ((age < 40) $\wedge$ (**if** elevel $\in [0..1]$
$\quad$ **then** ($25K \leq$ salary $\leq 75K$) **else** ($50K \leq$ salary $\leq 100K$))) $\vee$
$\quad$ (($40 \leq$ age < 60) $\wedge$ (**if** elevel $\in [1..3]$
$\quad$ **then** ($50K \leq$ salary $\leq 100K$) **else** ($75K \leq$ salary $\leq 125K$))) $\vee$
$\quad$ ((age ≥ 60) $\wedge$ (**if** elevel $\in [2..4]$
$\quad$ **then** ($50K \leq$ salary $\leq 100K$) **else** ($25K \leq$ salary $\leq 75K$)))

In all our experiments, we generated a predefined number of tuples, described by the 9 attributes. Then the binary functions were computed and the corresponding labels (*true* or *false*) added to the dataset as additional columns.

Qualitative Results: In a first run, RUDE+Apriori were applied to a database of 10.000 tuples; RUDE generated the following 11 split points:

salary:	**50.5**, **75**, 88.5, **101**, 113.5, **125.5**
commission:	**9.5**
age:	**40, 59.5**
hvalue:	518.5
hyears:	–
loan:	245.5

It is obvious that RUDE has found all the relevant splits but one. These are printed in bold. Based on this discretization, Apriori succeeded in finding near perfect definitions of functions $f1$ and $f3$ and good approximations of $f2$ and $f4$ (see also Table 1 below). For instance, here are the rules that relate to $f1$:

{age $\in$ [19.5...40)} $\Rightarrow$ $f1 = true$
{age $\in$ [40...59.5)} $\Rightarrow$ $f1 = false$
{age $\in$ [59.5...81.5)} $\Rightarrow$ $f1 = true$

It should be pointed out that RUDE+Apriori also discovered the dependencies between the numeric attributes *commission* and *salary*, and, at least partly, between *hvalue* and *zipcode*. Here are the rules for *commission*:

{salary $\in$ [19.5...50.5)} $\Rightarrow$ *commission* $\in$ [9.5...76.5)
{salary $\in$ [50.5...75)} $\Rightarrow$ *commission* $\in$ [9.5...76.5)
{salary $\in$ [75...88.5)} $\Rightarrow$ *commission* $\in$ [0...9.5)
{salary $\in$ [88.5...101)} $\Rightarrow$ *commission* $\in$ [0...9.5)
{salary $\in$ [101...113.5)} $\Rightarrow$ *commission* $\in$ [0...9.5)
{salary $\in$ [113.5...125.5)} $\Rightarrow$ *commission* $\in$ [0...9.5)
{salary $\in$ [125.5...151.5)} $\Rightarrow$ *commission* $\in$ [0...9.5)

For the more complex dependency between *hvalue* and *zipcode*, RUDE+Apriori found a rather crude approximation that maps a subset of the *zipcodes* to an interval representing low *hvalues*, and another subset to a range of high *hvalues*.

Quantitative Results of Systematic Study: Now we study how imprecision and noise affect the algorithm's ability to extract and represent the hidden dependencies. We vary two factors systematically:

1. *"Fuzziness"*: To model fuzzy boundaries between numeric ranges, the values of numeric attributes are perturbed (see also [1]): if the value of attribute A_i for a tuple t is v and the range of A_i is a, then the value of A_i after perturbation becomes $v + r \times \alpha \times a$, where r is a uniform random variable between -0.5 and $+0.5$, and α is our fuzziness parameter.
2. *Noise*: At a given noise level β, the value of each attribute (numeric or nominal) is replaced by a random value from its domain with probability β.

To get objective quantitative measures, we run RUDE+Apriori on the data, extract all the association rules that involve the attributes in the definitions of the hidden functions (remember our primary interest is in whether the algorithm can find the dependencies), and regard these rules as a *classifier*. Let p be the total number of positive examples (i.e., of class *true*) in the dataset, n the number of negative examples, p_c the number of positive examples covered by the rules, and n_c the number of negative examples erroneously covered. Then we measure the classifier's *coverage* $C = p_c/p$ and the fraction of *false positives* $FP = n_c/n$. This gives a more detailed picture than simply the total error of the classifier.

Table 1 shows the results for three different levels of fuzziness ($\alpha \in \{0, 0.05, 0.1\}$) and noise ($\beta \in \{0, 0.05, 0.1\}$), respectively. In each case, we used a database of size 10.000. To save on experimentation time, APriori was then run on a random sample of 200 tuples from the discretized version of the database.

We see that RUDE is quite effective in finding a useful discretization. Generally, Apriori reaches high levels of coverage and low levels of error in the

Table 1. Results for different fuzziness and noise levels (size of database = 10.000).

dependency	C|	FP| 0% fuzziness	C|	FP| 5% fuzziness	C|	FP| 10% fuzziness
$[age] \rightarrow [f1]$	100.00%	2.77%	98.60%	3.51%	97.09%	8.06%
$[salary, age] \rightarrow [f2]$	92.83%	3.08%	81.06%	3.00%	78.92%	2.42%
$[age, elevel] \rightarrow [f3]$	99.08%	0.00%	90.05%	2.24%	90.46%	2.18%
$[s, a, e] \rightarrow [f4]$	90.80%	3.20%	69.37%	2.56%	61.05%	0.81%
	0% noise		**5% noise**		**10% noise**	
$[age] \rightarrow [f1]$	100.00%	2.77%	96.21%	7.69%	99.30%	8.78%
$[salary, age] \rightarrow [f2]$	92.83%	3.08%	80.00%	3.50%	71.25%	1.83%
$[age, elevel] \rightarrow [f3]$	99.08%	0.00%	89.62%	4.24%	84.68%	14.60%
$[s, a, e] \rightarrow [f4]$	90.80%	3.20%	63.51%	0.79%	51.93%	0.50%

noise-free case, using RUDE's discretization. The more difficult functions seem to be $f2$ and $f4$, because they depend on the correct identification of more numeric split points than $f1$ and $f3$ (7 for $f2$ and $f4$ vs. 2 for $f1$ and $f3$). As expected, increasing fuzziness and noise leads to a decrease in the coverage of the rules that Apriori can find, and to an increase in the errors of commission (false positives FP). Noise has a stronger detrimental effect than fuzziness; this is because in simulating noise, we replaced attribute values by *random* values, as opposed to values in the *vicinity*. The degradation with increasing fuzziness and noise seems graceful, except for function $f4$; but $f4$ is a complex dependency indeed.

We also performed experiments with datasets of varying sizes, i.e. sets with 1.000, 5.000, 10.000 and 100.000 tuples. There it turned out that the number of intervals constructed by RUDE does not grow with the size of the dataset. On the contrary, there is a slight *decrease* — with more data, RUDE tends to generate fewer spurious splits.

To summarize, the experiments indicate that our discretization algorithm is both effective and robust. It does find most of the relevant split points that are needed to discover the dependencies in the data. Of course, it also generates some unnecessary splits. What should be acknowledged is that RUDE constructs a *globally good* discretization; that is, it defines intervals that allow the association rule learner to uncover all or most of the hidden dependencies *simultaneously*. This is in strong contrast to the results described in [10], where each function was dealt with separately, and where both a template for the type of association rule and the correct number of intervals had to be given for the system to be able to rediscover the function.

3.2 Preliminary Results on Real-World Data

We have also started to apply our algorithm to a large, real-world dataset from a long-term research project being carried out at our institute (see [11]). The data describes expressive performances of pieces of music by a concert pianist. Every single note is documented and described by various music-theoretic attributes; in addition, for each note we have exact numeric values that specify the *tempo*, *loudness*, and *relative duration* (i.e., degree of *staccato* vs. *legato*) with which the note was played. The goal of the project is to discover general principles of *expressive performance*; we want to find rules that explain and predict how a performer will usually shape aspects like tempo, dynamics, and articulation when playing a piece. Note that these factors are inherently continuous.

When applied to a database representing the performed melodies of 13 complete piano sonatas by W.A. Mozart (more than 44.000 notes), RUDE+Apriori discovered a number of interesting relations between different expression dimensions and other factors. For instance, there seems to be a roughly inverse relationship (at least in terms of ranges) between performed relative tempo and loudness (dynamics), as indicated by association rules like the following:

{duration $\in$ [0.012...0.34), tempo $\in$ [0.26...1.03)} $\Rightarrow$ dynamics $\in$ [0.99...1.41)
     (*support:* 10.63%, *confidence:* 72.73%)
{tempo $\in$ [1.03...5.27), articulation $\in$ [0.69...1.01)} $\Rightarrow$ dynamics $\in$ [0.59...0.99)
     (*support:* 7.13%, *confidence:* 70.40%)

In other words (and simplified), when speeding up and playing notes faster than average, the pianist tends to play notes more softly, and vice versa (at least in some cases). Such associations point to very interesting interrelations between different expression dimensions; these and other discoveries are currently the topic of further specialized investigations (in cooperation with musicologists).

4   Related Work

Extending association rule mining to numeric data has been the topic of quite some research recently. Most related to our approach is the method by Srikant & Agrawal [9]. They first perform a pre-discretization of the numeric attributes into a fixed number of intervals (determined by a "partial completeness" parameter) and then combine adjacent intervals as long as their support in the database is less than a user-specified value. The difference to our approach is that attributes are discretized in isolation; no information about other attributes is taken into account. In terms of results, [9] only report on the effect of the parameters on the number of rules found, and on time complexity.

A second type of methods might be called "template-based" [8,10]. These require a template of an association rule (with right-hand and part of left-hand side instantiated and one or more uninstantiated numeric attributes in the left-hand side) to be given, and then find intervals or hyperrectangles for the numeric attributes that maximize support, confidence, or some other interest measure.

A very different approach is taken by Aumann & Lindell [3]. They do not discretize numeric attributes at all. Rather, they use simple measures describing the distribution of continuous attributes, specifically mean and variance, directly in association rules. That is, their algorithm can discover rules like $sex = female \Rightarrow wage : mean = 7.90/h$. One could easily imagine combining our approach with theirs.

5   Conclusion

This paper has presented RUDE, an algorithm for discretizing numeric or mixed numeric-categorical datasets. The central ideas underlying the algorithm are *mutual structure projection* — using information about the value distributions of other attributes in deciding how many split points to create, and where — and the use of a novel *clustering algorithm* for finding points of significant changes. The algorithm can be viewed as a combination of supervised and unsupervised discretization. Our experimental results show that RUDE constructs discretizations that effectively reflect several associations simultaneously present in the data, and that perturbations in the data do not affect it in an unreasonable way.

The applicability of RUDE is by no means confined to association rule mining. In [6] we evaluate RUDE in a regression-by-classification setting, and it turns out that discretization by mutual structure projection can produce substantial improvements in terms of classifier size (though not in terms of accuracy).

One of the main problems with the current system is that the user-defined parameters still need to be fine-tuned when dealing with a new dataset. Also, some of the parameters represent absolute values; the problem of defining relative threshold measures (like percentages) is also a current research topic.

Acknowledgments

This research is supported by the Austrian *Fonds zur Förderung der Wissenschaftlichen Forschung* under grants P12645-INF and Y99-INF (START).

References

1. Agrawal, R., Imielinski, T., and Swami, A. (1993). Database Mining: A Performance Perspective. *IEEE Transactions on Knowledge and Data Engineering* 5(6) (Special issue on Learning and Discovery in Knowledge-Based Databases), 914-925.
2. Agrawal, R. and Srikant, R. (1994). Fast Algorithms for Mining Association Rules. In *Proc. of the 20th Int.l Conference on Very Large Databases*, Santiago, Chile.
3. Aumann, Y. and Lindell, Y. (1999). A Statistical Theory for Quantitative Association Rules. In *Proceedings of the Fifth International Conference on Knowledge Discovery and Data Mining (KDD-99)*. Menlo Park, CA: AAAI Press.
4. Dillon, W., and Goldstein, M. (1984). *Multivariate Analysis*. New York: Wiley.
5. Dougherty, J., Kohavi, R., and Sahami, M. (1995). Supervised and Unsupervised Discretization of Continuous Features. In *Proceedings of the 12th International Conference on Machine Learning (ML95)*. San Francisco, CA: Morgan Kaufmann.
6. Ludl, M.-C. and Widmer, G. (2000). Relative Unsupervised Discretization for Regression Problems. In *Proceedings of the 11th European Conference on Machine Learning (ECML'2000)*. Berlin: Springer Verlag.

7. Pavlidis T. (1982). *Algorithms for Graphics and Image Processing*. Rockville, MD: Computer Science Press.
8. Rastogi, R. and Shim, K. (1999). Mining Optimized Support Rules for Numeric Attributes. In *Proc. of the International Conference on Data Engineering* 1999.
9. Srikant, R. and Agrawal, R. (1996). Mining Quantitative Association Rules in Large Relational Tables. In *Proceedings of the ACM-SIGMOD Conference on Management of Data*, Montreal.
10. Wang, K., Tay, S., and Liu, B. (1998). Interestingness-Based Interval Merger for Numeric Association Rules. In *Proceedings of the 4th International Conference on Knowledge Discovery and Data Mining (KDD-98)*. Menlo Park: AAAI Press.
11. Widmer, G. (1998). Applications of Machine Learning to Music Research: Empirical Investigations into the Phenomenon of Musical Expression. In R.S.Michalski, I.Bratko and M.Kubat (eds.), *Machine Learning and Data Mining: Methods and Applications*. Chichester, UK: Wiley.

Mining Association Rules: Deriving a Superior Algorithm by Analyzing Today's Approaches

Jochen Hipp[1], Ulrich Güntzer[1], and Gholamreza Nakhaeizadeh[2]

[1] Wilhelm Schickard-Institute, University of Tübingen, 72076 Tübingen, Germany
{jochen.hipp, guentzer}@informatik.uni-tuebingen.de

[2] DaimlerChrysler AG, Research & Technology FT3/AD, 89081 Ulm, Germany
rheza.nakhaeizadeh@daimlerchrysler.com

Abstract. Since the introduction of association rules, many algorithms have been developed to perform the computationally very intensive task of association rule mining. During recent years there has been the tendency in research to concentrate on developing algorithms for specialized tasks, e.g. for mining optimized rules or incrementally updating rule sets. Here we return to the "classic" problem, namely the efficient generation of *all* association rules that exist in a given set of transactions with respect to minimum support and minimum confidence. From our point of view, the performance problem concerning this task is still not adequately solved.

In this paper we address two topics: First of all, today there is no satisfying comparison of the common algorithms. Therefore we identify the fundamental strategies of association rule mining and present a general framework that is independent of any particular approach and its implementation. Based on this we carefully analyze the algorithms. We explain differences and similarities in performance behavior and complete our theoretic insights by runtime experiments. Second, the results are quite surprising and enable us to derive a new algorithm. This approach avoids the identified pitfalls and at the same time profits from the strengths of known approaches. It turns out that it achieves remarkably better runtimes than the previous algorithms.

1 Introduction

Association rules were introduced in [1]. The intuitive meaning of such rules $X \Rightarrow Y$, where X, Y are sets of items, is that a transaction containing X is likely to also contain Y. The prototypical application is the analysis of basket data where rules like "$p\%$ of all customers who buy $\{x_1, x_2, \ldots\}$ also buy $\{y_1, y_2, \ldots\}$" are found. Our paper deals with the "classic" mining of associations. That is, the mining of *all* rules that exist in a given database with respect to thresholds on support and confidence. We assume transactions that are typical for retail applications, i.e. transactions containing between 10 to 20 items on average and items out of $\mathcal{I}$ with $|\mathcal{I}| \approx 1,000 - 100,000$. Several algorithms for association rule mining have been developed, e.g. Apriori [2], Partition [9], DIC [4], and

D.A. Zighed, J. Komorowski, and J. Żytkow (Eds.): PKDD 2000, LNAI 1910, pp. 159–168, 2000.
© Springer-Verlag Berlin Heidelberg 2000

Eclat [11], but finally from our point of view the performance problem is still not satisfyingly solved. Especially in the context of an interactive KDD-process, c.f. [5], performance is still a remaining problem.

Although the algorithms share basic ideas they fundamentally differ in certain aspects and this is also true for their runtime performance. Unfortunately today there is no exhaustive comparison of at least the most important of the algorithms concerning the fundamental ideas behind them or their different runtime behavior. Both topics are addressed in this paper. Moreover based on the insights of our study we derive a new algorithm that remarkably reduces runtimes compared to the previous approaches.

The paper is structured as follows: In Section 2 we identify the general strategies of association rule mining. The overview given is independent of any particular algorithm and its implementation. In Section 3, we systematize the most common algorithms by putting them into the general framework developed in the section before. In addition we show several performance evaluations that, from our point of view, are quite surprising. After that we come to the main part of this section, the detailed explanation of the different runtimes. In Section 4, we deploy the insights from the previous sections and introduce a hybrid algorithm. This algorithm avoids the identified pitfalls and at the same time profits from the recognized strengths of the fundamental strategies. We conclude with performance studies that show the efficiency of this new algorithm. Finally we give a short summary and close with interesting topics for future research.

2   Identifying Fundamental Strategies of Association Rule Mining

2.1   Problem Description

Let $\mathcal{I} = \{x_1, \ldots, x_n\}$ be a set of distinct items. A set $X \subseteq \mathcal{I}$ with $|X| = k$ is said to be a k-itemset or just an itemset. Let $\mathcal{D}$ be a multi-set of transactions T, $T \subseteq \mathcal{I}$. A transaction T supports an itemset X if $X \subseteq T$. The fraction of transactions from $\mathcal{D}$ that support X is called the support of X, denoted by $\mathsf{supp}(X)$. An association rule is an implication $X \Rightarrow Y$, where $X, Y \subseteq \mathcal{I}$ and $X \cap Y = \emptyset$. In addition to $\mathsf{supp}(X \Rightarrow Y) = \mathsf{supp}(X \cup Y)$ every rule is assigned a confidence $\mathsf{conf}(X \Rightarrow Y) = \mathsf{supp}(X \cup Y)/\mathsf{supp}(X)$, c.f. [2]. An itemset X is called frequent if $\mathsf{supp}(X) \geq \mathsf{minsupp}$. For the purpose of association rule generation it suffices to find all frequent itemsets, c.f. [2].

2.2   Search Strategies

Except the empty set the lattice of all $2^{|\mathcal{I}|}$ subsets of the itemset $\mathcal{I}$ are shown in Figure 1(a). The bold line is an example of actual itemset support and separates the frequent itemsets in the upper part from the infrequent ones in the lower part. Obviously the search space is exponentially growing with $|\mathcal{I}|$. It is therefore not practicable to determine the support of each of the subsets of $\mathcal{I}$ in order to

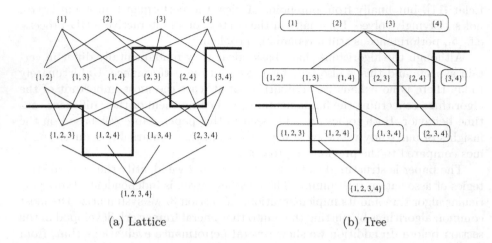

(a) Lattice   (b) Tree

Fig. 1. Subsets of $\mathcal{I} = \{1, 2, 3, 4\}$

decide whether it is frequent or not. Instead, the idea is to traverse the lattice in such a way that *all* frequent itemsets are found but as few infrequent itemsets as possible are visited. To achieve this the downward closure property of itemset support is employed: All subsets of a frequent itemset are also frequent.

In Figure 1(b) a tree on the itemsets is shown: The nodes are the classes $E(P), P \subseteq \mathcal{I}$, $E(P) = \{H \subseteq \mathcal{I} \mid |H| = |P| + 1$ and P is a "prefix" of $H\}$, where the sets are represented as ordered lists, c.f. [6] for precise formalization. Two nodes are connected by an edge, if all itemsets of the lower class can be generated by joining two itemsets of the upper class. A class E and its descendants only contain frequent itemsets if the parent class E' of E contains at least two frequent itemsets. That is, whenever we encounter a class E that contains less than two frequent itemsets than we are allowed to prune all branches starting in node E without accidentally missing any of the frequent itemsets.

With that pruning, we drastically reduce the search space and moreover we are free to choose the strategy – typically breadth-first search (BFS) or depth-first search (DFS) – to traverse the classes of the lattice, respectively the nodes of the tree.

2.3   Counting Strategies

When traversing the search space, i.e. the subsets of $\mathcal{I}$, we always have to check whether or not potential frequent itemsets achieve minsupp. We call these itemsets candidates and counting their support is done in two fundamentally different ways:

The first strategy is to *count the occurrences* of the candidates by setting up counters and then pass over all transactions. Whenever one of the candidates is contained in a transaction, we increment its counter.

The second strategy indirectly counts the support of candidates by *intersecting sets*. This requires for every itemset that the set of transactions containing this itemset is provided. Such tid-sets are denoted by X.tids. The support of a candidate $C = X \cup Y$ is obtained by the intersection C.tids $= X$.tids $\cap Y$.tids and evaluating $|C$.tids$|$.

3   Analysis of the Algorithms

3.1   Algorithms

In this subsection we put the most common algorithms into the general framework that we identified in the previous section:

Apriori [2] is the very first efficient algorithm to mine association rules. Basically it combines BFS with counting of occurrences of candidates. In addition, it employs the following: When using BFS all frequent itemsets at level s of the tree (c.f. Figure 1(b)) are known when starting to count candidates at level $s+1$. Together with the downward closure property of itemset support, this enables Apriori to do additional pruning: Look at all the subsets of size $|C| - 1$ of candidate C and whenever there is at least one of those infrequent, then prune C without counting its support.

Partition [9] combines the Apriori approach with set intersections instead of counting occurrences. That is, Partition also checks the subsets of each candidate for frequency before actually determining its support. In addition, Partition "partitions" the database in several chunks. This is necessary because otherwise the memory usage of the tid-sets to be held simultaneously in main memory would easily grow beyond the physical limitations of common machines.

DIC [4] is an extension of Apriori that aims at minimizing the number of database passes. The idea is to relax the strict separation between generating and counting of candidates: During counting DIC looks for candidates that already achieve minsupp though their final support may still not be determined. Based on these DIC generates new candidates and immediately starts counting them.

In contrast, **Eclat** [11] relies on DFS instead of BFS. With that, Eclat does not need to partition even huge databases although it counts the support values by intersections. The reason is that only the tid-sets of the itemsets on one path from the root down to one of the leaves have to be kept in memory simultaneously. But there is a draw back not described in [11]: DFS implies that in general Eclat cannot prune candidates by looking at their subsets. That is, Eclat does not fully realize the so called apriori_gen()-function. The reason is that basic DFS descends from the root to the leaves of the tree without caring about any subset relation among the itemsets. [11] introduces an important optimization they call "fast intersections". In brief only tid-sets that achieve a size greater than minimum support are of relevance. The idea is to immediately stop an intersection as soon as it is foreseeable that it will never reach this threshold. Eclat as introduced in [11] presumes the frequent itemsets of size 1 and 2 to be known. I.e. Eclat starts at level 3 in the tree. In order to make a fair runtime comparison of Eclat with the other algorithms, we modify Eclat to also start at level 1. This is straight forward to implement by simply calling Eclat on the class consisting of all frequent 1-itemsets.

None of the above algorithms combines counting occurrences with DFS. The reason is evident when considering how counting occurrences actually works: First, a set of candidates is set up, second, the algorithm passes all transactions and increments the counters of the candidates. That is, for each set of candidates one pass over all transactions is necessary. With BFS such a pass is needed for each of level s of the tree as long as there exists at least one candidate of size s. In contrast, DFS would make a separate pass necessary for each class that contains at least one candidate. The costs for each pass over the whole database would contrast with the relatively small number of candidates counted in the pass.

3.2   Performance Studies

We performed the experiments on an Pentium III Linux machine, running at 500Mhz, and C++-implementations of the algorithms. The experiments and datasets were taken from [2,9]. The naming convention of the datasets reflects their basic characteristics, e.g. "T20.I4.D100K" is a dataset with average transaction size of 20, average frequent itemset size of 4, and consisting of 100,000 transactions. The test sets were generated with the dataset generator from [8]. The number of items was always set to 1,000 and the number patterns always to 2,000. In addition to [2,9], we experimented with different restrictions on the maximal size of the generated frequent itemsets based on the dataset "T20.I4.D100K" at minsupp 0.33%.

3.3   Comparing the Algorithms

In Figure 2 the algorithms show quite similar runtime behaviors. At least there is no algorithm fundamentally beating out the other ones. This is quite surprising, especially with regard to former publications. To explain this, we start with some general thoughts on performance issues concerning the strategies described in Section 2.

The influence of the search strategy is relatively small. Only the fact that DFS does not allow proper candidate pruning by subset checking makes BFS somewhat superior to DFS. But we must keep in mind that checking subsets might be costly, especially for larger itemsets. In addition, it makes only sense for itemsets of a size greater than 2. But as Figure 2(d) and [7] show, the time spent with the itemsets of size 2 may dominate the whole generation process.

Counting occurrences is usually done by using a hashtree, c.f. [2]. Counting a candidate that occurs rather infrequently is quite cheap. Costs are only caused by the actual occurrences of the candidate in the transactions. In contrast, whenever incrementing the candidate size by one the hashtree grows one level. I.e., especially for larger candidate sizes – caused by the characteristics of the dataset or by smaller values for minsupp – counting can get fairly expensive. At the

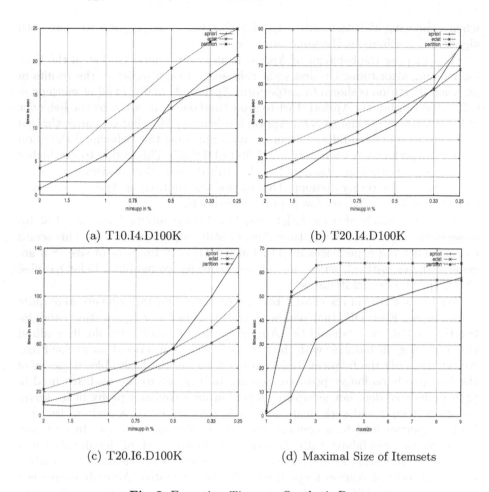

(a) T10.I4.D100K

(b) T20.I4.D100K

(c) T20.I6.D100K

(d) Maximal Size of Itemsets

Fig. 2. Execution Times on Synthetic Data

same time, the size of the candidates has no influence when using intersections. No matter how long the candidates are only the sizes of their tid-sets count. Of course, there is also a drawback: The costs for an intersection are at least $\min\{|X.\text{tids}|, |Y.\text{tids}|\}$ operations regardless of the actual number of occurrences of the candidate $X \bigcup Y$.[1]

As to be expected Partition[2] and Eclat show very similar runtimes, with Eclat beating Partition by a fairly constant factor. The reason is that Partition does not employ "fast intersections". The effect of the additional candidate pruning employed by Partition is not able to compensate this disadvantage. In fact,

---

[1] "Fast intersections" reduce the costs but are also not directly bound by the number of occurrences of a candidate.

[2] We were always able to skip the partitioning step, because our machine is equipped with sufficient main memory.

when enhancing Partition with "fast intersections" we experienced that both algorithms reach about the same runtimes.

As a surprise the behavior of Apriori is also very similar to that of the tids-intersecting algorithms. On first sight this seems to contradict to the results in [9] where Partition is shown to outperform Apriori. Actually, in [9] at minsupp $\approx$ 0.75% the runtimes of Apriori start to grow fundamentally whereas the behavior of Partition does not change. Our Apriori implementation, that uses an optimized structure to count candidate 2-itemsets does not show this behavior.[3] Partition clearly outperforms Apriori only on "T20.I6.D100K". This is due to the higher average size of candidate itemsets found in this dataset. Higher average sizes are also caused by lowering minsupp. Due to this the runtime of Apriori compared to Partition suffers at very low support thresholds.

The above holds also for Eclat that in addition profits from the "fast intersections". But most of the time Apriori still outperforms Eclat. This seems to contradict the experiences in [11], but in [11] only itemsets of size ≥ 3 are generated. As justified before, for our experiments we modified Eclat to mine also frequent 1- and 2-itemsets.

We left out DIC in the charts, because our very first experiments were quite discouraging. Even DIC that passes all transactions before generating candidates, that is DIC that "should be" Apriori, performed badly. We finally realized that replacing the hashtree with the structure from [4] has two draw backs: First, considering only the frequent items when setting up the hashtables in nodes of the hashtree is no longer possible. Second, in contrast to the hashtree used in [2] a prefix tree does not group itemsets sharing a common prefix in its leaves but each itemset and each of its subsets is represented by a node of its own. Both properties lead to a tremendous growth of memory usage. In addition, when counting candidates with the modified hashtree each of the already counted frequent itemsets causes overhead no matter whether it yields to a candidate or not. In contrast Apriori keeps the candidates separated. Actually even when overcoming both, DIC only showed an improvement of less than $\approx 30\%$ over Apriori but no fundamental different behavior on basket data in [4].

The surprisingly similar behavior of the considered algorithms shows that the advantages and disadvantages identified in the beginning of this subsection balance out on market basket-like data. This is also supported by Figure 2(d).

4   New Approach

4.1   Hybrid Approach

The performance studies and explanations of the results in the previous section suggest the development of a hybrid approach. The idea is to count occurrences whenever determining the support values of relatively small candidates and to rely on tid-set intersections for the remaining candidates. Of course this implies

---

[3] Replacing the hashtree for candidates of size 2 with an array is suggested in [10]

additional costs for generating the tid-sets when switching between the two counting strategies. For this purpose we use a hashtree-like structure that contains
pointers to tid-sets instead of counters.

On the one hand with BFS the hybrid algorithm would suffer from memory
problems when using tid-sets intersections. At least a costly mechanism like partitioning the database would be needed. On the other hand using DFS would
elegantly solve this problem but when starting to count occurrences the runtime of our algorithm would suffer substantially, c.f. Section 3.1. The solution
is to switch from BFS to DFS when switching from counting occurrences to
intersections.

```
(1)  algorithm hybrid(transactions, sw, minsupp)
(2)  {
(3)      frequent_itemsets[1] = get_frequent_items(transactions, minsupp);
(4)      // BFS together with counting occurrences:
(5)      for(s = 2; s ≤ sw;++s)
(6)      {
(7)          candidates = generate_candidates(frequent_itemsets[s − 1]);
(8)          count_candidates(candidates, transactions);
(9)          frequent_itemsets[s] = get_frequent_itemsets(candidates, minsupp);
(10)     }
(11)     // DFS together with tid-set intersections:
(12)     for each class ∈ frequent_itemsets[s-1] do
(13)     {
(14)         class_with_tid-sets = generate_tid-sets(class, transactions);
(15)         depth-first_search(class_with_tid-sets, minsupp);
(16)     }
(17) }
```

Fig. 3. Hybrid Algorithm

The finally resulting algorithm **Hybrid** is sketched in Figure 3. The argument
sw determines when to change the counting strategy. As explained basic DFS
does not allow candidate pruning by infrequent subsets. To overcome this we
employ right-most DFS from [6].

4.2   Evaluation

We repeated our experiments with two versions of our hybrid algorithm. One
switching at candidate size 2 and the other at size 3, c.f. Figure 4. Moreover
we made experiments on real-world data from a supermarket with about 60, 000
items in roughly 70, 000 transactions. The new algorithm performs best in nearly
all cases and shows the anticipated behavior. This is most obvious in Figure 4(e)
where Hybrid shows the smooth rise for the 2-, 3- and to some extend 4-itemsets
of Apriori combined with the ability of Eclat to mine frequentent itemsets of
size ≥ 4 at hardly any additional effort. The generation of the tid-sets that takes
place at sw is efficiently solved by the modified hashtree. Nevertheless a very
small average size of frequent itemsets let the algorithm suffer.

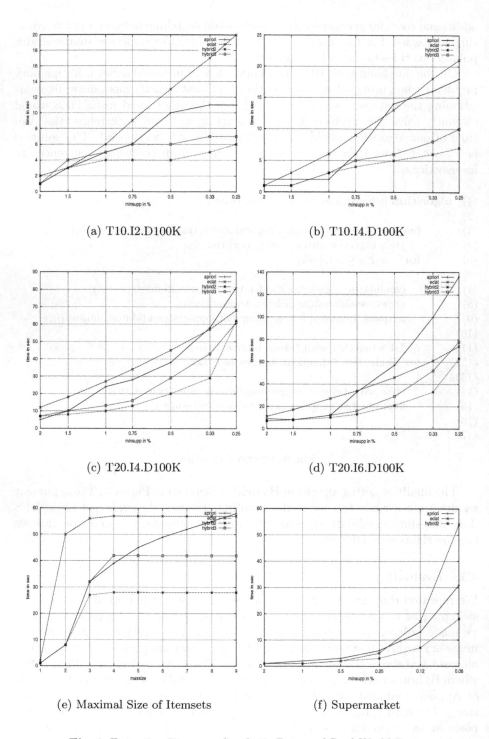

(a) T10.I2.D100K

(b) T10.I4.D100K

(c) T20.I4.D100K

(d) T20.I6.D100K

(e) Maximal Size of Itemsets

(f) Supermarket

Fig. 4. Execution Times on Synthetic Data and Real-World Data

5 Summary

In this paper we addressed the "classic" association rule problem, i.e. the generation of *all* association rules that exist in a given set of transactions with regard to minsupp and minconf. We identified the fundamental strategies of association rule mining and derived a general framework that is independent of any particular algorithm. Based on this we analyzed the performance of todays approaches both theoretically and by carrying out experiments. The results were quite surprising. In addition, our insights lead to the development of a new approach. The resulting algorithm Hybrid exploits the strengths of the known approaches and at the same time avoids their weaknesses. It turns out that in general for the "classic" association rule problem our algorithm achieves remarkably better runtimes than the previous approaches.

Our future research we will focus on the following: We want to explore how the approaches behave when mining databases that fundamentally differ from retail databases, e.g. dense datasets [3]. In addition, the aspects of memory usage are still not exhaustively studied. In fact memory usage is closely related to runtime because all the described algorithms suffer substantially when physical memory is exhausted and parts of the memory are paged out to disk.

References

1. R. Agrawal, T. Imielinski, and A. Swami. Mining association rules between sets of items in large databases. In *Proc. of the ACM SIGMOD '93*, USA, May 1993.
2. R. Agrawal and R. Srikant. Fast algorithms for mining association rules. In *Proc. of the 20th Int'l Conf. on Very Large Databases (VLDB '94)*, Chile, June 1994.
3. R. J. Bayardo Jr., R. Agrawal, and D. Gunopulos. Constraint-based rule mining in large, dense databases. In *Proc. of the 15th Int'l Conf. on Data Engineering*, Sydney, Australia, March 1999.
4. S. Brin, R. Motwani, J. D. Ullman, and S. Tsur. Dynamic itemset counting and implication rules for market basket data. In *Proc. of the ACM SIGMOD '97*, 1997.
5. CRISP. The CRISP-DM process model. http://www.crisp-dm.org.
6. J. Hipp, A. Myka, R. Wirth, and U. Güntzer. A new algorithm for faster mining of generalized association rules. In *Proc. of the 2nd European Symp. on Principles of Data Mining and Knowledge Discovery (PKDD '98)*, France, Sept. 1998.
7. M. Holsheimer, M. Kersten, H. Mannila, and H. Toivonen. A perspective on databases and data mining. In *Proc. of the 1st Int'l Conf. on Knowlegde Discovery and Data Mining (KDD '95)*, Montreal, Canada, August 1995.
8. IBM. QUEST Data Mining Project. http://www.almaden.ibm.com/cs/quest.
9. A. Savasere, E. Omiecinski, and S. Navathe. An efficient algorithm for mining association rules in large databases. In *Proc. of the 21st Conf. on Very Large Databases (VLDB '95)*, Switzerland, September 1995.
10. R. Srikant and R. Agrawal. Mining generalized association rules. In *Proc. of the 21st Conf. on Very Large Databases (VLDB '95)*, Switzerland, September 1995.
11. M. J. Zaki, S. Parthasarathy, M. Ogihara, and W. Li. New algorithms for fast discovery of association rules. In *Proc. of the 3rd Int'l Conf. on KDD and Data Mining (KDD '97)*, Newport Beach, California, August 1997.

Unified Algorithm for Undirected Discovery of Exception Rules

Einoshin Suzuki[1] and Jan M. Żytkow[2]

[1] Electrical and Computer Engineering, Yokohama National University,
79-5 Tokiwadai, Hodogaya, Yokohama 240-8501, Japan.
suzuki@dnj.ynu.ac.jp
[2] Computer Science Department, UNC Charlotte, Charlotte, N.C. 28223.
zytkow@uncc.edu

Abstract. This paper presents an algorithm that seeks every possible exception rule which violates a common sense rule and satisfies several assumptions of simplicity. Exception rules, which represent systematic deviation from common sense rules, are often found interesting. Discovery of pairs that consist of a common sense rule and an exception rule, resulting from undirected search for unexpected exception rules, was successful in various domains. In the past, however, an exception rule represented a change of conclusion caused by adding an extra condition to the premise of a common sense rule. That approach formalized only one type of exceptions, and failed to represent other types. In order to provide a systematic treatment of exceptions, we categorize exception rules into eleven categories, and we propose a unified algorithm for discovering all of them. Preliminary results on fifteen real-world data sets provide an empirical proof of effectiveness of our algorithm in discovering interesting knowledge. The empirical results also match our theoretical analysis of exceptions, showing that the eleven types can be partitioned in three classes according to the frequency with which they occur in data.

Keywords: Exception/Deviation Detection, Rule Discovery, Exception Rule, Rule Triplet

1   Introduction

Exceptions and/or deviations, which focus on a very small portion of a data set, have long been ignored or mistaken as noise in machine learning. The goal of data mining is broader, however. Exceptions were always interesting to discoverers, as they challenged the existing knowledge and often led to the growth of knowledge in new directions. In addition to predictions, decision optimization is important in data mining [5]. We strongly believe that exception and/or deviation can improve the quality of decisions, and their detection deserves more attention.

An increasing number of studies is devoted to exception/deviation detection. Examples of such studies are outlier discovery [4], OLAP operator for explaining

D.A. Zighed, J. Komorowski, and J. Zytkow (Eds.): PKDD 2000, LNAI 1910, pp. 169–180, 2000.
© Springer-Verlag Berlin Heidelberg 2000

increase/decrease of a continuous attribute [7], and exception rule discovery [6, 8, 10–13].

Exception rules are typically represented as deviational patterns to common sense rules of high generality and accuracy. Exception rule discovery can be classified as either directed [6, 8] or undirected [10–13]. A directed method finds a set of exception rules which deviate from the given common sense rules. In distinction, common sense rules are not given to an undirected method, which finds a set of pairs of a common sense rule and an exception rule. The advantage of an undirected method is that it can discover highly unexpected patterns since it also discovers common sense rules [14].

In undirected discovery of exception rules two natural questions arise: "what other kinds of exception rules can be defined?" and "is there an efficient algorithm to discover all of them?" In this paper we give a constructive answer to both questions. We first define all kinds of exception rules that occur in situations described with three literals. Next we present an efficient algorithm, which is not restricted to binary attributes and thus can be applied to a broad range of ordinary data sets. We finally demonstrate the effectiveness of our approach with experiments using fifteen data sets.

2   Categories of Exception Rules

2.1   A Rule and a Negative Rule

Let a data set contain n examples, each expressed by m attributes. Let a literal be a conjunction of atoms, while an atom is either a value assignment for a nominal attribute or a range assignment for a continuous attribute. An atom can be also a missing value assignment for any attribute.

In this paper, we define a rule as $u \rightarrow v$, where u and v are literals. We follow the definition of ITRULE [9] and define the generality condition and the accuracy condition of $u \rightarrow v$ as the right-hand sides of (1), where θ_S and θ_F are thresholds given by the user, and $\widehat{\Pr}(u)$ is the ratio of examples that satisfy u in the data set:

$$u \rightarrow v \Leftrightarrow \widehat{\Pr}(u) \geq \theta_S \text{ (generality) } \& \widehat{\Pr}(v|u) \geq \theta_F \text{ (accuracy)} \qquad (1)$$

In association rule discovery[1], support $\widehat{\Pr}(uv)$ is used instead of $\widehat{\Pr}(u)$, where uv represents $u \wedge v$, but the idea of generality is essentially the same.

In this paper, we also introduce a negative rule as $u \not\rightarrow v$, where θ_I is a threshold given by the user.

$$u \not\rightarrow v \Leftrightarrow \widehat{\Pr}(u) \geq \theta_S \& \widehat{\Pr}(v|u) \leq \theta_I \qquad (2)$$

2.2   Rule Triplets

Let y and z be literals, and x and x' be atoms with the same attribute but with a different value. Suzuki [11] has considered the discovery of rule triplets that consists of a common sense rule $y \rightarrow x$, an exception rule $yz \rightarrow x'$ and a reference rule $z \nrightarrow x'$. Representation of discovered rules is given by

$$(y \rightarrow x, \ yz \rightarrow x', \ z \nrightarrow x') \tag{3}$$

This pattern can be interpreted as "If y and z then x' and not x. This is an interesting exception since usually if y then x, and if z then not frequently x'."

This kind of pattern holds together several pieces of knowledge, including exception rules as unexpected, surprising, anomalous and thus interesting additions to other rules.

We use the term "common sense rule" because it well-represents a user-given belief in the direct method of search for exceptions, where common sense rules are given in the input. In undirected search, however, there may be little common sense in "common sense rules," as we are concerned with relations between rules, and we do not deal with user's knowledge.

We will now seek a systematic generation of patterns similar to (3), starting from the simplest cases. We formalize the situation as a rule triplet, where a common sense rule is represented by $y \rightarrow x$, and an exception rule and a reference rule are represented by a negative rule $\alpha \nrightarrow \beta$ and a rule $\gamma \rightarrow \delta$ respectively.

$$t(y, x, \alpha, \beta, \gamma, \delta) = (y \rightarrow x, \ \alpha \nrightarrow \beta, \ \gamma \rightarrow \delta) \tag{4}$$

Four meta-level variables α, β, γ, and δ can be instantiated in different ways, with the use of literals x, y, and z, to form different specific patterns of exceptions, analogous to (3). This rule triplet has a generic reading "it is common that if y then x, but we have exceptions that if α then not frequently β. This is surprising since if γ then δ". Note that this terminology differs from (3) in that an exception rule and a reference rule are represented by a negative rule and a rule respectively. Of course, that meta-level reading does not communicate any specific exception, anomaly and surprise, but we will seek those features in different instances of the pattern. Here we only justify this definition by stating that a violation (exception rule) of two rules (common sense rule and reference rule) can be interesting.

In order to systematically categorize exception rules, we restrict our attention to rule triplets with three free literals x, y, z. Number three is chosen since it represents the simplest situation which makes sense: a rule triplet with two literals would likely to be meaningless since it tends to be overconstrained, and a rule triplet with more than three literals would likely to be more difficult to interpret.

We assume that a conjunction of two literals can appear only in the premise of an exception rule. This restriction is justified because a conjunction of two literals in the premise makes a good candidate for an exception to a rule that holds one of those literals in the premise.

$$\beta, \gamma, \delta \in \{x, y, z\} \tag{5}$$

$$\alpha \in \{x, y, z, xy, yz, zx\} \tag{6}$$

Since a literal z is not contained in a common sense rule, it must occur both in an exception rule and in a reference rule, otherwise a triplet situation will reduce to the case of two literals x and y. Possible candidates for a reference rule are then restricted to $(\gamma, \delta) \in \{(y, z), (z, y), (z, x), (x, z)\}$. If we pair each of these four with the common sense rule $y \to x$, we realize that the case of $(\gamma, \delta) = (x, z)$ is equivalent (isomorphic) to $(\gamma, \delta) = (z, y)$, one case can be produced from the other by renaming the variables. Therefore, we neglect $(\gamma, \delta) = (x, z)$ in this paper, so that

$$(\gamma, \delta) \in \{(y, z), (z, y), (z, x)\}. \tag{7}$$

2.3   Rule Triplets without Conjunctions

In this section, we consider the simplest triplets. We categorize rule triplets that do not contain conjunctions of literals. In such a case, possible candidates for an exception rule are restricted to the following four:

$$(\alpha, \beta) \in \{(y, z), (z, y), (z, x), (x, z)\}. \tag{8}$$

Since there are three possible candidates for a reference rule from (7), there are twelve candidates for rule triplets without conjunctions. Note that first, however, three candidates that satisfy $(\alpha, \beta) = (\gamma, \delta)$ show a contradictory relation, i.e. $(\alpha \not\to \beta, \alpha \to \beta)$, and must be removed. Second, consider the three candidates that satisfy $(\alpha, \beta) = (\delta, \gamma)$. In such a case, the exception rule $\alpha \not\to \beta$ and its reference rule $\beta \to \alpha$ jointly mean proper inclusion of α in β, which is not sufficient for an exception from a common sense rule $y \to x$. We therefore remove these three candidates. Third, as shown in figure 1, $t(y, x, x, z, y, z)$ is equivalent to $t(y, x, z, x, y, z)$ if we exchange x and z. Similarly, $t(y, x, z, y, z, x)$ is equivalent to $t(y, x, y, z, z, x)$ if we exchange z and y. We conclude that there are $12 - 3 - 3 - 2 = 4$ kinds of rule triplets without conjunctions. We define type 1, 2, 3, and 4 as shown in figure 1. They are defined by the following instantiations of α, β, γ, and δ:

$$(\alpha, \beta, \gamma, \delta) \in \{(z, x, y, z), (z, x, z, y), (x, z, z, y), (y, z, z, x)\}. \tag{9}$$

Considering the remaining four types of triplets in figure 1, we can argue that all are interesting as each of them represents a kind of violation. For example, type 1 represents a surprise ($z \not\to x$) for an expected overlap relation between z and x which can be naturally derived from two rules ($y \to x$ and $y \to z$). Further, we can interpret type 3 and 4 as showing mild exceptions. Type 2, however, shows a violation of transitivity, and can thus be regarded as demonstrating the strongest deviations in these rule triplets.

2.4   Rule Triplets with a Conjunction

In this section we categorize rule triplets of which premise of an exception rule is a conjunction of two literals. In such a case, possible candidates for an exception

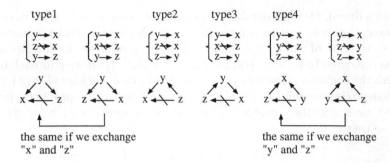

Fig. 1. Rule triplets without conjunctions.

rule are restricted to the following three:

$$(\alpha, \beta) \in \{(xy, z), (xz, y), (yz, x)\}. \tag{10}$$

Since there are three possible candidates for a reference rule from (7), there are nine candidates for rule triplets with a conjunction. Note that, as shown in figure 2, $t(y, x, yz, x, y, z)$ is equivalent to $t(y, x, xy, z, y, z)$ if we exchange x and z. In addition, $t(y, x, xz, y, z, x)$ is equivalent to $t(y, x, xy, z, z, x)$ if we exchange y and z. We here conclude that there are $9 - 2 = 7$ kinds of rule triplets with a conjunction. We define type 5, 6, $\cdots$, and 11 as shown in figure 2:

$$(\alpha, \beta, \gamma, \delta) \in \{(xy, z, y, z), (xz, y, y, z), (xy, z, z, y), (xz, y, z, y),$$
$$(yz, x, z, y), (xy, z, z, x), (yz, x, z, x)\}. \tag{11}$$

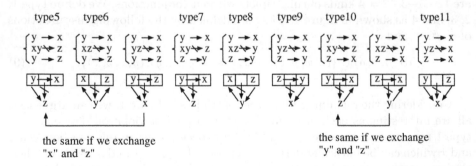

Fig. 2. Rule triplets with a conjunction. A rectangle on the top center for each triplet represents a conjunction of literals in the top right and left.

By examining Figure 2, we can interpret type 6, 7, and 10 as interesting triplets, but we can hardly see them as exceptions. Type 5, 8, 9, and 11, however, are different. In each of the four cases the exception rule logically contradicts at least one of the other rules. The situations of type 5, 8, 9, and 11 can only occur if those records that make exceptions to each of the rules in the triplet are distributed in a very specific way, so that the thresholds set by θ_F and θ_I can be met. It is an interesting empirical questions whether the triplets derived from data are going to match those expectations.

3   Unified Algorithm for All Exception Rules

In this section we propose an algorithm for discovering all the rule triplets which we categorized in the previous section.

3.1   Depth-First Search for Rule Triplets

Association rule discovery [1] assumes a transaction data set, which has only binary attributes. Each attribute can take either "y" or "n" as its value, and most of the attribute values are "n". The sparseness of the data set allows to employ breadth-first search. Note that if the breadth-first search was employed for an ordinary data set, the number of large item-sets would be huge. In such a case, space efficiency would be so poor that any breadth-first algorithm would be impractical.

On the other hand, ITRULE [9] assumes an ordinary data set, which can have an attribute with more than two values, and have no assumption on the distribution of attribute values. Since sparseness of the data set is not assumed, it employs a depth-first search.

In this paper, we assume an ordinary data set and propose an algorithm which performs depth-first search for triplets of literals a, b, c. We leave an algorithm for a transaction data set for future work. Let d be a candidate for a rule triplet and ! be a logical negation. A threshold vector θ represents $(\theta_S, \theta_F, \theta_I)$, and $|a|$ is the number of atoms in a literal a. Without loss of generality, we assume that every attribute is allocated a unique positive integer i(a). The main routine of the algorithm is given below, followed by the supporting procedures. Some of them are presented in the next sections.

Algorithm: rule_triplet_discovery(θ, M, D, R).
Input: threshold vector θ, maximum length of literals M, data set D.
Output: a set of discovered rule-triplets R.
begin
$R, a, b, c := \phi$. //initialization
foreach $a, b, c \in$ the atom set of D **such that** i(a) < i(b) < i(c) //search
  begin
    $d := (a, b, c)$. //generation of an initial rule triplet
    evaluateRt(d, θ, D, R). //evaluation
    if(! Prune(d, θ, D, R)) //pruning
      extend(d, *true*, *true*, θ, M, D, R) //search for depth ≥ 2
  end
end

The routine of search for depth ≥ 2 is given as follows.

Procedure: extend(d, f_a, f_b, $\boldsymbol{\theta}$, M, D, R).
Input: literal triplet d, flag f_a, flag f_b, $\boldsymbol{\theta}$, M, D.
Output: R.
begin
if $(|a| > M)$ **or** $(|b| > M)$ **or** $(|c| > M)$ //limit of search depth
    return
if (f_a) //can extend literal a_i
    foreach $a_i \in$ the atom set of D **such that** a_i does not appear in d
        begin
            $d' = (aa_i, b, c)$ //add an atom to a_i
            extendSub(d', *true*, *true*, $\boldsymbol{\theta}$, M, D, R)
        end
if (f_b) //can extend literal b_i
    foreach $b_i \in$ the atom set of D **such that** b_i does not appear in d
        begin
            $d' = (a, bb_i, c)$ //add an atom to b_i
            extendSub(d', *false*, *true*, $\boldsymbol{\theta}$, M, D, R)
        end
foreach $c_i \in$ the atom set of D **such that** c_i does not appear in d
    begin
        $d' = (a, b, cc_i)$ //add an atom to c_i
        extendSub(d', *false*, *false*, $\boldsymbol{\theta}$, M, D, R)
    end
end

Procedure: extendSub(d, f_a, f_b, $\boldsymbol{\theta}$, M, D, R).
Input: d, f_a, f_b, $\boldsymbol{\theta}$, M, D.
Output: R.
begin
evaluateRt(d). //evaluation
if(! Prune(d)) //pruning
    extend(d, f_a, f_b) //search for depth+1
end

Recall that the number of atoms in a literal a is $|a|$. We consider $|a|, |b|, |c| \leq M$ as the search restriction in the above algorithm. For illustration, figure 3 shows the traversal order in the search tree when $M = 2$. Time efficiency of this algorithm is $O(m^{3M})$, where m is the number of attribute in D. This is justified since this algorithm is complete in the sense that it discovers all rule triplets. This inefficiency is remedied by the pruning procedure, which will be described in section 3.4.

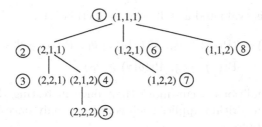

Fig. 3. Traversal order in a search tree when $M = 2$, where each circled figure represents the order, and a node represents $(|a|, |b|, |c|)$.

3.2   Selection of an Atom for a Continuous Attribute

In this section, we explain how to select an atom given an attribute x [13]. If x is nominal, the algorithm considers all single-value assignments to x as atoms. On the other hand, if x is continuous, the algorithm considers single-range assignments to x as atoms, and these ranges are obtained by discretizing x.

Discretization of a continuous attribute can be classified as either global (done before hypothesis construction) or local (done during hypothesis construction) [3]. It can be also classified as either supervised (considers class information) or unsupervised (not using class information) [3]. In this paper, we employ a global unsupervised method due to its time efficiency.

The equal frequency method, when the number of intervals is k, divides n examples so that each bin contains n/k (possibly duplicated) adjacent values. The equal frequency method belongs to unsupervised methods and is widely used [3]. We obtain the minimum value and the maximum value of a range by the global equal-frequency method.

In order to exploit the stopping conditions presented in the previous section, ranges are selected as follows. Let the k intervals be $\pi_1, \pi_2, \cdots, \pi_k$ in ascending order. Note that there are two possible ranges, π_1 to π_{k-1} and π_2 to π_k, as a range which consists of $k - 1$ adjacent intervals. Our algorithm first selects these two as ranges, then selects the ranges of $k - 2$ adjacent intervals, i.e. π_1 to π_{k-2}, π_2 to π_{k-1}, and π_3 to π_k. Pruning conditions (13) which will be presented in section 3.4 are employed in ignoring unnecessary ranges. This procedure is iterated by decrementing the number of adjacent intervals in a range until no ranges are left for consideration.

3.3   Evaluation of a Literal Triplet

The algorithm in section 3.1 searches for triplets of literals, and rule triplets made of those literal triplets. Note that there are six one-to-one correspondence from a set $\{A, B, C\}$ to a set $\{x, y, z\}$, and there are eleven categories of rule triplets from section 2.3 and 2.4. In the procedure "evaluateRt(d)", our algorithm considers these $6 * 11 = 66$ possibilities for a literal triplet, and outputs if it satisfies the conditions for a rule triplet.

A rule triplet is evaluated as follows (see section 2.1):

$$t(y, x, \alpha, \beta, \gamma, \delta) \Leftrightarrow (\widehat{\Pr}(y) \geq \theta_S, \ \widehat{\Pr}(x|y) \geq \theta_F, \ \widehat{\Pr}(\alpha) \geq \theta_S, \ \widehat{\Pr}(\beta|\alpha) \leq \theta_I,$$
$$\widehat{\Pr}(\gamma) \geq \theta_S, \ \widehat{\Pr}(\delta|\gamma) \geq \theta_F) \qquad (12)$$

Some of these conditions occur more than once in testing different types of rule triplets. Our algorithm applies such conditions only once in order to avoid redundant calculations.

3.4   Pruning

Assume the current literal triplet represents a rule triplet $t(y', x', \alpha', \beta', \gamma', \delta')$. It is straightforward to prove that if at least one of (13) holds, no rule triplets $t(y, x, \alpha, \beta, \gamma, \delta)$ in the children nodes satisfy all conditions in the right-hand side of (12). Note that here y expands y'.

$$\widehat{\Pr}(y') < \theta_S, \ \widehat{\Pr}(x'y') < \theta_S\theta_F, \ \widehat{\Pr}(\alpha') < \theta_S, \ \widehat{\Pr}(\gamma') < \theta_S, \ \widehat{\Pr}(\gamma'\delta') < \theta_S\theta_F (13)$$

As described in section 2.3 and 2.4, there are eleven types of rule triplets. Our algorithm checks the above conditions for all these types. Similarly as in the previous section, our algorithm checks the above conditions for six one-to-one mappings from a set $\{A, B, C\}$ to a set $\{x, y, z\}$ in the procedure "Prune(d)".

4   Experimental Evaluation

We here analyze empirically the statistics of the searched nodes and the discovered rule-triplets. We have chosen UCI data sets [2] since they have served for a long time as benchmark data sets in the machine learning community. In our experiments, we deleted attributes that have only one value in a data set. Table 1 shows characteristics of the data sets employed in the experiments.

In applying our algorithm, the number k of discretization bins was set to 4. Other parameters were set to $\theta_S = 0.025$, $\theta_F = 0.7$, $\theta_I = 0.6$, and $M = 2$. Table 2 shows the results of experiments. Data sets are sorted on the "rule triplets" column, i.e. with respect to the number of discovered rule triplets.

From this table we see a rough correlation between the number of searched nodes and the number of discovered rule triplets. By inspecting Table 1 we see that the number of discovered rule triplets typically increases as the number of attributes, continuous attributes, and values of nominal attributes increase. Data sets "vote", "mushroom", "credit", and "shuttle" are exceptions, and we consider that this is due to the distribution of attribute values.

Table 2 shows that pruning is effective, since without pruning the nodes increase by 5 % ("nursery" and "diabetes") to 285 % ("mushroom"). This is due to the fact that a considerable number of nodes in a search tree tend to have small probabilities for their literals and are thus pruned.

Numbers of discovered rule triplets per types reveal interesting tendencies. From Table 2, we see that type 3, 4, 6, 7, 10 are extremely numerous: they are

Table 1. Characteristics of the data sets employed in the experiments, where " ex.", " att.", "c.", and " val." represent the number of examples, the number of attributes, the number of continuous attributes and the number of possible values for the nominal attributes respectively.

data set	ex.	att. (c.)	val.	data set	ex.	att. (c.)	val.
car	1,728	7 (0)	3 - 4	australian	690	15 (6)	2 - 14
nursery	12,960	9 (0)	2 - 5	credit	690	16 (6)	2 - 15
postoperative	90	9 (1)	2 - 3	vote	435	17 (0)	2 - 3
yeast	1,484	9 (6)	2 - 10	hepatitis	155	20 (6)	2 - 3
diabetes	768	9 (8)	2	german	1,000	21 (6)	2 - 10
abalone	4,177	9 (8)	3	mushroom	8,124	22 (0)	2 - 12
breastcancer	699	10 (9)	2	thyroid	7,200	22 (6)	2 - 3
shuttle	58,000	10 (9)	7				

more than $1 * 10^5$ in 11 data sets. Type 1 is also numerous since it is more than $1 * 10^5$ in 3 data sets. On the other hand, type 2 and 9 are modest in number: they never exceed $1 * 10^5$ in any data sets, and exceed $1 * 10^4$ in 9 data sets. Finally, type 11, 8, and 5 are rare in this order: type 11 exceeds $1 * 10^4$ in 1 data set, and type 8 and 5 never exceed $1 * 10^4$ in any data sets. Similar tendencies were observed for $M = 1$. Interestingly, we anticipated the exceptionality of type 2, 5, 8, 9, and 11 as stronger than the other types in section 2.3 and 2.4. We are currently investigating this tendency analytically.

5   Conclusion

In this paper, we formalized discovery of interesting exception rules as rule-triplet discovery, and we categorized rule triplets with three literals into eleven types. We also analyzed these eleven types according to their interestingness. Moreover, we proposed an efficient algorithm for simultaneous discovery of all these types based on literal-triplet search and sound pruning.

Our algorithm has been applied to fifteen data sets, and confirmed the analysis on the interestingness of the eleven types. Experimental results clearly show the effectiveness of pruning in reducing the number of searched nodes. The ongoing work focuses on relating these rule-triplet types with domain-dependent interestingness in collaboration with experts in various domains. A general interestingness measure for rule triplets represents a promising avenue for future research.

References

1. R. Agrawal, H. Mannila, R. Srikant *et al.*: Fast Discovery of Association Rules, *Advances in Knowledge Discovery and Data Mining*, AAAI Press, Menlo Park, Calif., pp. 307–328 (1996)

Table 2. Number of searched nodes and discovered rule triplets for each data set. Here, "unp." represents the number of nodes without pruning divided by the number of nodes with pruning.

data set	nodes (unp.)	rule triplets	type1 type7	type2 type8	type3 type9	type4 type10	type5 type11	type6
car	1.06E5	0	0	0	0	0	0	0
	(1.11)		0	0	0	0	0	
nursery	1.12E5	71	31	0	0	0	0	40
	(1.05)		0	0	0	0	0	
postoperative	4.75E4	2.63E4	238	131	1.44E3	1.06E4	2	1.57E3
	(1.42)		1.44E3	2	122	1.05E4	126	
vote	6.41E5	1.90E5	5.94E3	2.05E3	3.10E4	5.11E4	0	2.96E4
	(1.69)		2.53E4	2	845	4.39E4	10	
breastcancer	1.35E6	1.65E6	3.50E4	2.19E4	1.98E5	5.14E5	45	1.93E5
	(1.14)		1.81E5	1	1.64E4	4.93E5	120	
mushroom	8.64E6	2.92E6	9.68E4	7.50E3	1.22E5	1.18E6	28	2.18E5
	(3.85)		1.15E5	226	6.11E3	1.16E6	4.92E3	
credit	6.53E6	4.25E6	4.87E4	1.33E4	2.27E5	1.75E6	45	2.51E5
	(2.09)		2.21E5	67	1.18E4	1.71E6	2.43E3	
abalone	3.25E6	4.41E6	2.62E5	8.53E3	4.87E5	1.32E6	29	6.49E5
	(1.08)		3.86E5	11	5.76E3	1.29E6	1.72E3	
diabetes	3.97E6	4.78E6	3.42E4	2.24E4	3.43E5	1.86E6	91	3.23E5
	(1.05)		3.38E5	54	2.20E4	1.83E6	4.39E3	
yeast	3.44E6	5.09E6	2.66E4	1.42E4	4.00E5	1.93E6	0	3.88E5
	(1.41)		3.99E5	9	1.31E4	1.91E6	1.25E3	
australian	6.81E6	5.54E6	5.74E4	1.58E4	3.28E5	2.24E6	61	3.57E5
	(1.45)		3.22E5	96	1.44E4	2.19E6	3.28E3	
shuttle	5.47E6	7.39E6	1.36E5	6.64E4	6.27E5	2.63E6	233	6.88E5
	(1.15)		6.05E5	972	5.64E4	2.57E6	7.91E3	
hepatitis	1.78E7	1.65E7	1.97E5	8.59E4	1.07E6	6.46E6	1.50E3	1.21E6
	(1.51)		1.06E6	2.66E3	7.28E4	6.33E6	7.22E4	
german	2.05E7	1.76E7	7.03E4	3.39E4	8.28E5	7.55E6	35	8.52E5
	(1.32)		8.24E5	102	3.08E4	7.40E6	8.48E3	
thyroid	7.00E6	1.88E7	6.12E4	5.78E4	1.98E6	6.43E6	30	1.92E6
	(1.56)		1.96E6	150	5.45E4	6.34E6	1.96E3	

2. C.L. Blake and C.J. Merz: "UCI Repository of Machine Learning Databases", *http://www.ics.uci.edu/~mlearn/MLRepository.html*, Dept. of Information and Computer Sci., Univ. of California Irvine (1998).
3. J. Dougherty, R. Kohavi, and M. Sahami: Supervised and Unsupervised Discretization of Continuous Features, in *Proc. Twelfth Int'l Conf. Machine Learning (ICML)*, pp. 194–202 (1995).
4. E.M. Knorr and R.T. Ng: Algorithms for Mining Distance-Based Outliers in Large Datasets, in *Proc. 24th Ann. Int'l Conf. Very Large Data Bases (VLDB)*, pp. 392–403 (1998).
5. T.M. Mitchell: "Machine Learning and Data Mining", *CACM*, Vol. 42, No. 11, pp. 31–36 (1999).
6. B. Padmanabhan and A. Tuzhilin: "A Belief-Driven Method for Discovering Unexpected Patterns", *Proc. Fourth Int'l Conf. Knowledge Discovery and Data Mining (KDD)*, AAAI Press, Menlo Park, Calif., pp. 94–100 (1998).
7. S. Sarawagi: Explaining Differences in Multidimensional Aggregates, in *Proc. 25th Int'l Conf. Very Large Data Bases (VLDB)*, pp. 42–53 (1999).
8. A. Silberschatz and A. Tuzhilin: "What Makes Patterns Interesting in Knowledge Discovery Systems", *IEEE Trans. Knowledge and Data Eng.*, Vol. 8, No. 6, pp. 970–974 (1996).
9. P. Smyth and R.M. Goodman: "An Information Theoretic Approach to Rule Induction from Databases", *IEEE Trans. Knowledge and Data Eng.*, Vol. 4, No. 4, pp. 301–316 (1992).
10. E. Suzuki and M. Shimura : Exceptional Knowledge Discovery in Databases Based on Information Theory, *Proc. Second Int'l Conf. Knowledge Discovery and Data Mining (KDD)*, AAAI Press, Menlo Park, Calif., pp. 275–278 (1996).
11. E. Suzuki: "Autonomous Discovery of Reliable Exception Rules", *Proc. Third Int'l Conf. Knowledge Discovery and Data Mining (KDD)* , AAAI Press, Menlo Park, Calif., pp. 259–262 (1997).
12. E. Suzuki and Y. Kodratoff: "Discovery of Surprising Exception Rules based on Intensity of Implication", *Principles of Data Mining and Knowledge Discovery, LNAI 1510 (PKDD)*, Springer, Berlin, pp. 10–18 (1998).
13. E. Suzuki: "Scheduled Discovery of Exception Rules", *Discovery Science, LNAI 1721 (DS)*, Springer, Berlin, pp. 184–195 (1999).
14. E. Suzuki and S. Tsumoto: "Evaluating Hypothesis-Driven Exception-Rule Discovery with Medical Data Sets", *Knowledge Discovery and Data Mining, LNAI 1805 (PAKDD)*, Springer, Berlin, pp. 86–97 (2000).

Sampling Strategies for Targeting Rare Groups from a Bank Customer Database

J-H. Chauchat[1], R. Rakotomalala[1], and D. Robert[2]

[1] ERIC Laboratory - University of Lyon 2
5, av. Pierre Mendes-France
F-69676 Bron - FRANCE

[2] Crédit Agricole Centre-Est
1, rue Pierre de Truchis
F-69410 Champagne aux Monts d'Or - FRANCE

Abstract. This paper presents various balanced sampling strategies for building decision trees in order to target rare groups. A new coefficient to compare targeting performances of various learning strategies is introduced. A real life application of targeting specific bank customer group for marketing actions is described. Results shows that local sampling on the nodes while constructing the tree requires small samples to achieve the performance of processing the complete base, with dramatically reduced computing times.

Keywords: sampling, customer targeting, targeting quality coefficient, imbalanced database, decision tree, application

1 Introduction

This paper studies supervised learning using a real life application of targeting for the "Crédit Agricole Centre-Est" bank. More specifically, the use of decision trees [1][2] or induction graphs [19] on large databases to learn discriminating between two unequal size classes is of interest.

Crédit Agricole manages a several hundred of thousands customers data-base for whom some 200 attributes are known, 95% of them continuous. The class attribute is whether a client connected to some remote service. The study should identify those clients most susceptible to connect in the future; these types of clients shall be targeted by remote services marketing campaigns.

Learning form imbalanced classes is known to be difficult, yet it is quite common in practice: detection of rare diseases in epidemiology; detection of bank card frauds; process breakdown forecasting in industry; targeting specific client groups for marketing actions.

Moreover, if the database is large, the computing time is long especially if continuous attributes must be optimally discretized at each step [18][7]. Then, learning must be done on a sample [5], with efficiency gains if the sample is balanced [4].

The paper is organized as follows: the two sampling strategies that were implemented are presented in the next section. A new coefficient to compare

D.A. Zighed, J. Komorowski, and J. Żytkow (Eds.): PKDD 2000, LNAI 1910, pp. 181–190, 2000.
© Springer-Verlag Berlin Heidelberg 2000

client-targeting performances is introduced in the third section. This coefficient is used to compare the quality of the decision trees derived from the various sampling methods. The fourth section presents numerical results (computing time and targeting quality coefficient) from the bank customers database. Conclusion and future work are in the fifth section.

2   Sampling Strategies and Probability Distributions in Decision Trees

The focus here is on balanced sampling; indeed, the detection of rare classes with the classical induction tree on imbalanced training set works poorly [10]. Sampling for a decision tree can be executed in one of the two following ways:

1. either a sample is drawn from the original database, and the tree is built from the sample;
2. or a random sample is drawn on each node of the tree as it is being constructed.

Each method has advantages and disadvantages. The former is quicker, as it accesses the database only once and builds a learning set from the sample. On the other hand, as the tree grows, the leaves become smaller and smaller, making estimation of the probabilities less reliable while a wealth of data is available to comfortably make those estimations. If those probabilities are to be estimated for the initial population (or the complete base taken as the population), then they must be adjusted using Bayes theorem to obtain correct distributions on each node [12].

The latter is not hampered by data fragmentation. As the tree is constructed, on each node, the needed sample is drawn. There is, however, a severe drawback to this method for multiple accesses to the database are required. Even with fast algorithms [16], the method remains computer intensive. On the other hand, at each pass, exact probability distributions can be computed.

2.1   Building a Global Sample Before the Learning Process

A random sample of size $n_{k.}$ is to be drawn from the original database for each of the values y_k of the class attribute Y ($k = 1, \ldots, K$). The size of the sample file is n ($n = \sum_{k=1}^{K} n_{k.}$). If the $n_{k.}$ are equal, the sample is said to be balanced. This sampling scheme is a K-sample retrospective sampling [3], the $n_{k.}$ are not random and cannot be used to estimate the $\pi_{k.} = P(Y = y_k)$, the prior probabilities of obtaining one of the values for the class attribute. Here, the $\pi_{k.}$ are considered as computed from the complete database.

Let ℓ be a leaf on the decision tree. This leaf can be described by a statement such as $(X_1 = x_1, \ldots, X_p = x_p)$, and correct estimates of the conditional probabilities $P(Y = y_k/l) = P(Y = y_k/X_1 = x_1, \ldots, X_p = x_p)$ can be obtained; the later can be derived from the n_{kl}, the observed empirical frequencies on ℓ, the leaf of interest (Figure 1).

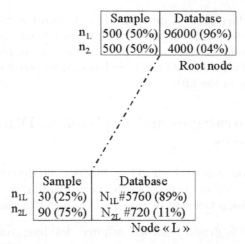

Fig. 1. Estimated sample size and conditional distribution on a node using global sampling

Posterior probabilities can be obtained using Bayes theorem:

$$\pi_{k/l} = P(Y = y_k/l) \tag{1}$$

$$= \frac{P(Y = y_k) \times P(l/Y = y_k)}{P(l)} \tag{2}$$

$$= \frac{P(Y = y_k) \times P(l/Y = y_k)}{\sum_{j=1}^{K} P(Y = y_j) \times P(l/Y = y_j)} \tag{3}$$

The estimates from the learning sample are readily obtained as:

$$\widehat{\pi}_{k/l} = \frac{\pi_{k.} \times \frac{n_{kl}}{n_{k.}}}{\sum_{j=1}^{K} \pi_{j.} \times \frac{n_{jl}}{n_{j.}}} \tag{4}$$

If the population size is noted N, the number of individuals accounted for by the leaf ℓ is given by:

$$\widehat{N}_{.l} = N \times \left[\sum_{k=1}^{K} \pi_{k.} \times \frac{n_{kl}}{n_{k.}} \right] \tag{5}$$

The main advantage of this method is that a single pass is required to obtain N and the $\pi_{k.}$. The induction tree and probability estimates are obtained from the sample, which can be a separate file created once and for all before the learning process (if the sample size is reasonable, it can fit in memory). The reliability of the estimates of the conditional probabilities depends on the sample size [14].

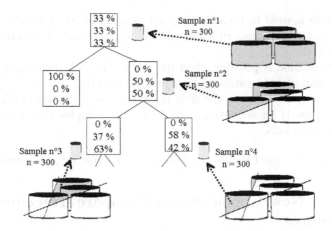

Fig. 2. Steps of building decision tree using local sampling

2.2 Local Sampling while Constructing the Tree

This approach follows work developed in [5]. On each leaf, while constructing the tree, a sample is drawn from the section of the base outlined by the rules defined by the path leading to the leaf (Figure 2). Each time, the sample is full size as long as the database contains enough individuals for the leaf; otherwise the available individuals are selected. Thus, little information is spoiled: at first, information is superabundant and a sample is enough to set the correct rules in a reasonable time; by the end, when information becomes scarce, a larger fraction of what is available is drawn, even all of it.

Computing time is less than that in learning from the complete base, especially when the database contains many continuous attributes that need to be sorted and discretized.

The property of decreasing global entropy may be lost when selecting a new sample on each node, but this is of little consequence. When a tree is built on a fixed set of examples, the global entropy can only decrease at each step [13], but this is an artefact of the learning set. In general, this property does not translate well to another set on which the tree would be applied, for example, a test sample. And, truly, the dataminer is especially interested in the generalization properties.

Compared to a global sample drawn prior to learning, the need to go back to the base to sample for each node allows the determination of the size of the population concerned and of the exact probabilities.

In practice, build a decision tree with local sampling is as follows:

1. first, a complete list of examples on the base is drawn;
2. the first sample is selected while the base is being read; an array of records associated with each attribute is kept in memory;

3. this sample is used to identify the best segmentation attribute, if it exists; otherwise, the stopping rule has played its role and the node becomes a terminal leaf;

4. if a segmentation is possible, then the list in (1.) is broken up into sub-lists corresponding to the various leaves just obtained;

5. step (4.) requires passing through the database to update each example's leaf; this pass is an opportunity to select the samples that will be used in later computations.

Steps (3.) to (5.) are iterated until all nodes are converted to terminal leaves.

3   TQC, a New Coefficient to Compare Tracking Procedures

In this section, TQC (Targeting Quality Coefficient), a coefficient to compare two tracking procedures is introduced. The coefficient is similar to a Gini coefficient in statistics [8], lift charts used in marketing, ROC curves from signal theory [6] or medicine [15]. The coefficient can help comparing trees derived from different sampling processes and that constructed on the complete database.

In general, classifiers are compared using the "test error rate", that is the proportion of "misclassified" among a sample independent of the learning sample [11]. For the situation at hand (tracking rare groups), the usual error rate is ill adapted. Rather than looking for the most likely class of an individual given his characteristics, the probability of having a rare characteristic is estimated: disease, fraud, breakdown, tele-purchase...

Individuals with a predicted probability of belonging to the rare group of at least $x\%$ are tracked; by varying $x\%$ with respect to cost and expected benefits ensuing actions, a larger or smaller set of individuals "at risk" is selected.

Hence, the quality coefficient must depend on the predicted probabilities given by the classifier: it ranges from 0 for a random classification (i.e. all predicted probabilities are equal to p, the global probability of having the rare characteristic), to 1 if the classifier recognizes perfectly the members of both classes (in this case, the predicted probability is 1 for members having the rare characteristic, and is set to 0 for the other ones).

Table 1 shows how the TQC coefficient is constructed. Individuals are sorted by decreasing predicted probabilities; then, two cumulative functions of the relative frequencies are computed:

1. the cumulative proportion of individuals in the population,
2. the cumulative proportion of individuals with the rare characteristic.

Computations from the decision tree built on a validation file of size $N = 1000$ individuals, with $A = 100$ bearing the rare characteristic, are displayed in Table 1. For a given individual, the predicted probability is the proportion of "rare" individuals among the individuals on the same leaf of the decision tree. For example, selecting the 4 individuals with the largest predicted probabilities (that

Table 1. Building TQC, the Quality Targeting Coefficient, on an artificial example

Rank $= i$	% Total Population	Pred. Prob $= P_i$	Class	% Cumulative Class "1"$= F_i$	Surface element $= (1/N) * (F_{i-1} + F_i)/2$
1	$1/N = 1/1000$	100 %	1	$1/ A = 1/100$	$(1/N)*(1/A)/2$
2	$2/N = 2/1000$	100 %	0	$1 / A = 1/100$	$(1/N)*(1/100+1/100)/2$
3	$3/N = 3/1000$	70 %	0	$1 / A = 1/100$	$(1/N)*(1/100+1/100)/2$
4	$4/N=4/1000$	70 %	1	$2 / A = 2/100$	$(1/N)*(2/100+1/100)/2$
...	...	...	...	...	...
N	$N/N = 100\%$	0 %	0	$A/A=100$ %	$(1/N)*(F_{N-1}+1)/2$
SUM =	—	—	A	—	Area

is $x\% = 4/1000$ of the population), $F_i = 2/100$ of the "rare" individuals are expected to be covered.

The two cumulative distributions are linked in Figure 3 : the proportion of selected population on the horizontal axis and the estimated proportion of targeted individuals on the vertical axis. The true curve must lie between two extremes:

1. Perfect targeting, displayed as two straight segments joining three points: $(0; 0)$ where no one is selected and no one is covered, $(\frac{A}{N}; 100\%)$ exactly $\frac{A}{N}$ of the population is selected and it is the whole targeted group, and $(100\%; 100\%)$ where every one is selected hence the target is attained;
2. Random targeting, displayed as the diagonal: selecting $x\%$ of the population covers $x\%$ of the targeted group.

The coefficient TQC is defined as the ratio of two areas: the "$Area$" between the real curve and the diagonal, and the area between perfect targeting and the diagonal. From Table 1,

$$TQC = \frac{2 \times Area - 1}{1 - \frac{A}{N}}$$

Hence, $TQC = 0$ for random targeting (no one selected), and $TQC = 1$ when targeting is perfect.

TQC may be negative if a very bad targeting procedure is used: few targeted instances would be selected first, and most of them at the end.

4   Results from Crédit Agricole Client Database

4.1   Characteristics of the Client Database - Sampling Strategies

The Crédit Agricole client base contains several hundreds of thousands of individuals, with some 200 attributes (95% of them continuous). Given the computer available to us, a master sample of 200,000 was drawn to represent the complete database because we want to fit all databases in memory to speed up computing. The attribute of interest is quite skewed, with a prior distribution of 4% "positive" and 96% "negative".

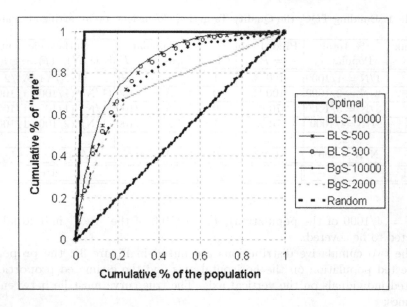

Fig. 3. Comparing various sampling strategies: cumulative proportions scatter-plot for targeting clients most susceptible to connect to some remote service

Three strategies were laid out:

BgS(n) - Balanced global sampling : extract a balanced learning size n sample, and the rules will be applied to the 200, 000 individuals of the master sample; sample sizes of $n = 2000$, $n = 10000$ and $n = 20000$ were tested;

BLS(n) - Balanced local sampling : extract a balanced sample at each node, then apply the classification rules; sizes n=500, 1000, 1500, 2500, 5000 and 10000 were tested;

ALL - All database : work on the full database and apply the rules.

We use the ChAID algorithm [9], the experimentation protocol was as follows: construct a tree according to the suggested strategy (BgS, BLS, ALL), apply the ensuing classification rules to the master sample to obtain predicted probabilities for every individual [17]. Each procedure was replicated ten times.

4.2   Computing Time

Changes in computing time are as expected (Table 2):

- creating a decision tree from the complete database is rather long compared to processing samples;
- computing time for BgS(n) and BLS(n) grows with the sample size;
- learning from a prior sample is quicker than from sample drawn at each node, partly because the number of examples processed at each node diminishes with the growth of the tree;

Table 2. Computing times (in seconds) according to various sampling strategies and sample size

Size	100	200	300	500	1,000	2,000	10,000	20,000	ALL
BgS	-	-	-	-	-	6	42	92	1,381
BLS	2	8	20	45	93	212	311	-	-

Table 3. Quality of targeting coefficient TQC according to various sampling strategies and sample size

Size	100	200	300	500	1,000	2,000	10,000	20,000	ALL
BgS	-	-	-	-	-	0.498	0.600	0.628	0.737
BLS	0.524	0.619	0.664	0.694	0.711	0.722	0.722	-	-

- for comparable computing times [BgS(10,000)-BLS(500), or BgS(20,000)-BLS(1,000)], the quality of prediction for local sampling surpasses that for global sampling (Table 3). This last point is further developed in the next section.

4.3   Quality of Targeting

Using the complete base as our yardstick, for which TQC=0.737 (ALL), the alternative sampling strategies are ranked (Table 3):

- for global sampling (BgS), all possible file sizes were exhausted, yet performances can never approach those achieved by working on the complete file. Indeed, the number of targeted individuals in the master sample does not exceed 8,000 (4% of 200,000); so the largest size of a balanced sample is 16,000 (8,000 positives and 8,000 others). In the sample of 20,000, the balanced sample had to be packed with others (8,000 positives and 12,000 others). Relatively bad targeting quality results of data fragmentation as the tree grows : stoping rules are activated on small sets of individuals; then, test powers are low and no more significant segmentation is find.
- local sampling approaches maximum performance as soon as the local sample size reaches 2,000 on each node. It is remarkable that n=300 seems enough to beat the best performance of global sampling [BgS(20,000)]. This result conforms with earlier empirical and theoretical work [5] on sample sizes for a classic learning problem.

5   Conclusion

The work described here aimed at building an efficient client targeting tool for Crédit Agricole Centre-Est. A number of sampling strategies were developed, well adapted to tracking rare target groups with decision trees. A new quality coefficient was introduced to assess the quality of a tracking strategy. This co-efficient is better suited to our study as recognizing individuals is not the goal; isolating those of interest is.

The study shows that local sampling on the nodes while constructing the tree requires small samples to achieve the performance of processing the complete base, with dramatically reduced computing times.

Acknowledgements

This work was done in collaboration with the marketing service of Crédit Agricole Centre-Est. The authors thank Mrs. Morel and M. Chapuisot for their support. We also thank M. Jean Dumais, Section Chief, Methodology Branch, Statistics Canada and guest lecturer at Université Lumière for his careful proof reading of the manuscript and his suggestions.

References

1. L. Breiman, J.H. Friedman, R.A. Olshen, and C.J. Stone. *Classification and Regression Trees*. California : Wadsworth International, 1984.
2. L. A. Breslow and D. W. Aha. Simplifying decision trees: a survey. *Knowledge Engineering Review*, 12(1):1–40, 1997.
3. G. Celeux and A. Mkhadri. Méthodes dérivées du modèle multinomial. In G. Celeux and J.P. Nakache, editors, *Analyse Discriminante Sur Variables Qualitatives*, chapter 2. Polytechnica, 1994.
4. J.H. Chauchat, O. Boussaid, and L. Amoura. Optimization sampling in a large database for induction trees. In *Proceedings of the JCIS'98-Association for Intelligent Machinery*, pages 28–31, 1998.
5. J.H. Chauchat and R. Rakotomalala. A new sampling strategy for building decision trees from large databases. In *Proceedings of the 7th Conference of the International Federation of Classification Societies, IFCS'2000*, pages 45–50, 2000.
6. J.P. Egan. *Signal Detection Theory and ROC Analysis*. Series in Cognition and Perception. Academic Press, New York, 1975.
7. Eibe Frank and Ian H. Witten. Making better use of global discretization. In *Proc. 16th International Conf. on Machine Learning*, pages 115–123. Morgan Kaufmann, San Francisco, CA, 1999.
8. C.W. Gini. Variabilita e mutabilita, contributo allo studio delle distribuzioni e relazioni statiche. Technical report, Studi Economico-Giuridici della R. Universita di Caligiari, 1938.
9. G.V. Kass. An exploratory technique for investigating large quantities of categorical data. *Applied Statistics*, 29(2):119–127, 1980.
10. Miroslav Kubat and Stan Matwin. Addressing the curse of imbalanced training sets: one-sided selection. In *Proc. 14th International Conference on Machine Learning*, pages 179–186. Morgan Kaufmann, 1997.
11. T.M. Mitchell. *Machine learning*. McGraw Hill, 1997.
12. Y.H. Pao. *Adaptive pattern recognition and neural networks*. Addison Wesley, 1989.
13. J.R. Quinlan. Discovering rules by induction from large collections of examples. In D. Michie, editor, *Expert Systems in the Microelectronic Age*, pages 168–201, Edinburgh, 1979. Edinburgh University Press.

14. R. Rakotomalala. *Graphes d'Induction.* PhD thesis, University Claude Bernard - Lyon 1, December 1997.
15. J. Swets. Measuring the accuracy of diagnostic systems. *Science*, 240:1285–1293, 1988.
16. J.S. Vitter. Faster methods for random sampling. In *Communications of ACM*, volume 27, pages 703–718, 1984.
17. I.H. Witten and E. Frank. *Data Mining: practical machine learning tools and techniques with JAVA implementations.* Morgan Kaufmann, 2000.
18. D.A. Zighed, S. Rabaseda, R. Rakotomalala, and F. Feschet. Discretization methods in supervised learning. In A. Kent and J.G. Williams, editors, *Encyclopedia of Computer Science and Technology*, volume 40, pages 35–50. Marcel Dekker, Inc., 1999.
19. D.A. Zighed and R. Rakotomalala. *Graphes d'Induction - Apprentissage et Data Mining.* Hermes, 2000.

Instance-Based Classification by Emerging Patterns

Jinyan Li[1], Guozhu Dong[2], and Kotagiri Ramamohanarao[1]

[1] Dept. of CSSE, The University of Melbourne, Vic. 3010, Australia.
{jyli, rao}@cs.mu.oz.au
[2] Dept. of CSE, Wright State University, Dayton OH 45435, USA. gdong@cs.wright.edu

Abstract. Emerging patterns (EPs), namely itemsets whose supports change significantly from one class to another, capture discriminating features that sharply contrast instances between the classes. Recently, EP-based classifiers have been proposed, which first mine as many EPs as possible (called eager-learning) from the training data and then aggregate the discriminating power of the mined EPs for classifying new instances. We propose here a new, instance-based classifier using EPs, called DeEPs, to achieve much better accuracy and efficiency than the previously proposed EP-based classifiers. High accuracy is achieved because the instance-based approach enables DeEPs to pinpoint all EPs relevant to a test instance, some of which are missed by the eager-learning approaches. High efficiency is obtained using a series of data reduction and concise data-representation techniques. Experiments show that DeEPs' decision time is linearly scalable over the number of training instances and nearly linearly over the number of attributes. Experiments on 40 datasets also show that DeEPs is superior to other classifiers on accuracy.

1 Introduction

The problem of classification has been studied extensively, using eager-learning approaches or lazy instance-based approaches, in machine learning, pattern recognition, and recently also in the data mining community. In this paper we introduce *DeEPs*, a new instance-based classifier which makes *De*cisions through *E*merging *P*atterns. The notion of emerging patterns (EPs) was proposed in [5] and is defined as multivariate features (i.e., *itemsets*) whose *supports* (or frequencies) change significantly from one class to another. Because of sharp changes in support, EPs have strong discriminating power. Two eager-learning classifiers based on the concept of EPs, CAEP [6] and the JEP-Classifier [9], have been proposed and developed, using the novel idea of aggregating the discriminating power of pre-mined EPs for classification. The newly proposed DeEPs classifier has considerable advantages on accuracy, speed, and dimensional scalability over CAEP and the JEP-Classifier, because of its efficient new ways to select sharp and relevant EPs, its new ways to aggregate the discriminating power of individual EPs, and most importantly the use of instance-based approach which creates a remarkable reduction on both the volume (i.e., the number of instances) and the dimension (i.e., the number of attributes) of the training data. Another advantage is that DeEPs can handle new training data without the need to re-train the classifier which is, however, commonly required by the eager-learning based classifiers. This feature is extremely useful for practical applications where the training data must be frequently updated. DeEPs

D.A. Zighed, J. Komorowski, and J. Zytkow (Eds.): PKDD 2000, LNAI 1910, pp. 191–200, 2000.
© Springer-Verlag Berlin Heidelberg 2000

can handle both numeric and discrete attributes and DeEPs is nicely scalable over the number of training instances.

Given two classes of data $\mathcal{D}_1$ and $\mathcal{D}_2$ and a test instance T, the basic idea of DeEPs is to discover those subsets of T which are emerging patterns between $\mathcal{D}_1$ and $\mathcal{D}_2$, and then use the supports of the discovered EPs for prediction. The test instance T may not contain any pre-mined EPs if the eager-learning approaches are used; in contrast, using the instance-based approach, DeEPs will be able to efficiently pinpoint all relevant EPs for classifying T. The basic idea of DeEPs is made practical and scalable to high dimension data because we use a series of data reduction and concise data represensation techniques. We use the following example to illustrate the ideas behind DeEPs.

Example 1. Table 1, taken from [14], contains a training set, for predicting whether the weather is good for some "Saturday morning" activity. The instances, each described by four attributes, are divided into two classes: class $\mathcal{P}$ and class $\mathcal{N}$.

Table 1. Weather conditions and Saturday Morning activity

| Class $\mathcal{P}$ (suitable for activity) | | | | Class $\mathcal{N}$ (not suitable) | | | |
outlook	temperature	humidity	windy	outlook	temperature	humidity	windy
overcast	hot	high	false	sunny	hot	high	false
rain	mild	high	false	sunny	hot	high	true
rain	cool	normal	false	rain	cool	normal	true
overcast	cool	normal	true	sunny	mild	high	false
sunny	cool	normal	false	rain	mild	high	true
rain	mild	normal	false				
sunny	mild	normal	true				
overcast	mild	high	true				
overcast	hot	normal	false				

Now, given the test instance $T=\{sunny, mild, high, true\}$, which class label should it take? Basically, DeEPs calculates the supports (in both classes) of the proper subsets of T in its first step. The proper subsets of T and their supports are organized as the following three groups:

1. those that only occur in Class $\mathcal{N}$ but not in Class $\mathcal{P}$, namely, {sunny, high}, {sunny, mild, high}, and {sunny, high, true}; their supports in Class $\mathcal{N}$ are 60%, 20%, and 20% respectively.
2. those that only occur in Class $\mathcal{P}$ but not in Class $\mathcal{N}$, namely, {sunny, mild, true}; its support in Class $\mathcal{P}$ is 11%.
3. those that occur in both classes, namely, $\emptyset$, {mild}, {sunny}, {high}, {true}, {sunny, mild}, {mild, high}, {sunny, true}, {high, true}, {mild, true}, and {mild, high, true}. Except for the patterns $\emptyset$ and {mild}, all these subsets have larger supports in Class $\mathcal{N}$ than in Class $\mathcal{P}$.

Obviously, the first group of subsets — which are indeed EPs of Class $\mathcal{N}$ because they do not appear in Class $\mathcal{P}$ at all — favors the prediction that T should be classified as Class $\mathcal{N}$. However, the second group of subsets gives us a contrasting indication that T should be classified as Class $\mathcal{P}$, although this indication is not as strong as that of the first group. The third group also strongly suggests that we should favor Class $\mathcal{N}$

as T's label, although the pattern {mild} contradicts this mildly. Using these EPs in a *collective* way, not separately, DeEPs would decide that T's label is Class $\mathcal{N}$ since the "aggregation" of EPs occurring in Class $\mathcal{N}$ is much stronger than that in Class $\mathcal{P}$.

In practice, an instance may contain many (e.g., 100 or more) attributes. To examine all subsets and discover the relevant EPs contained in such instances by naive enumeration is too expensive (e.g., checking 2^{100} or more sets). We make DeEPs efficient and scalable to high dimensional data by the following data reduction and concise data representation techniques.

- We reduce the training datasets firstly by removing those items that do not occur in the test instance and then by selecting the *maximal* ones from the processed training instances. (Set X is *maximal* in collection $\mathcal{S}$ if there are no proper supersets of X in $\mathcal{S}$.) This data reduction process makes the training data sparser in both horizontal and vertical directions.
- We use *border* [5], a two-bound structure like $<\mathcal{L}, \mathcal{R}>$, to succinctly represent all EPs contained in a test instance. Importantly, we use efficient border-based algorithms to derive EP borders from the reduced training datasets.
- We select *boundary* EPs, those in $\mathcal{L}$ (typically small in number, e.g., 81 in mushroom), for DeEPs' decision making. These selected EPs are "good" representatives of all EPs occurring in a test instance. This selection method also significantly reduces the number of EPs that are used for classification.

Detailed discussions of these points will be given in the next three sections. Table 2 illustrates the first stage of the sparsifying effect on both the volume and dimension of $\mathcal{D}_\mathcal{P}$ and $\mathcal{D}_\mathcal{N}$, by removing all items that do not occur in T. Observe that the transformed $\mathcal{D}_\mathcal{P}$ and $\mathcal{D}_\mathcal{N}$ are sparse, whereas the original $\mathcal{D}_\mathcal{P}$ and $\mathcal{D}_\mathcal{N}$ are *dense* since there is a value for every attribute of any instance. Sections 3 and 4 will discuss formally how to select the maximal ones of the reduced training instances and how to utilize the reduced training data to gain more efficiency with the use of borders.

Table 2. Reduced training data after removing items irrelevant to the instance {sunny, mild, high, true}. A "*" indicates that an item is discarded.

Reduced Class $\mathcal{P}$				Reduced Class $\mathcal{N}$			
outlook	temperature	humidity	windy	outlook	temperature	humidity	windy
*	*	high	*	sunny	*	high	*
*	mild	high	*	sunny	*	high	true
*	*	*	*	*	*	*	true
*	*	*	true	sunny	mild	high	*
sunny	*	*	*	*	mild	high	true
*	mild	*	*				
sunny	mild	*	true				
*	mild	high	true				
*	*	*	*				

Since EPs usually have very low supports, they are not suitable to be used individually for classification. We will use the *compact summation* method to aggregate the discriminating power contributed by all selected EPs to form classification scores.

Importantly, the *compact summation* method avoids duplicate contribution of training instances.

For continuous attributes, we introduce a new method, called *neighborhood-based intersection*. This allows DeEPs to determine which continuous attribute values are relevant to a given test instance, without the need to pre-discretize data.

The remainder of this paper is organized as follows. Section 2 reviews the notions of EPs and borders. Section 3 formally introduces the DeEPs classifier. Section 4 presents border-based algorithms to implement the main ideas of DeEPs. Section 5 presents the experimental results of DeEPs on 40 datasets, taken from the UCI Repository on Machine Learning [2]. The experiments demonstrate the scalability and high accuracy of DeEPs. Section 6 compares our DeEPs classifier with other classification models. Section 7 concludes this paper.

2   Preliminaries: EPs and Borders

In our discussion, the basic elements are the *items*. Relational data in the form of vectors (or tuples) are first translated as follows. For each relational attribute A with a continuous domain and a given interval $[l, h)$, "$A \in [l, h)$" is an item. If A is a discrete attribute and a is in the domain of A, then "$A = a$" is an item. Vectors or tuples are now represented as sets of items. An *instance* is a set of items, and a *dataset* $\mathcal{D}$ is a set of instances. A set X of items is also called an *itemset*. We say an instance S *contains* an itemset X, if $X \subseteq S$. The *support* of an itemset X in a dataset $\mathcal{D}$, $supp_{\mathcal{D}}(X)$, is $\frac{count_{\mathcal{D}}(X)}{|\mathcal{D}|}$, where $count_{\mathcal{D}}(X)$ is the number of instances in $\mathcal{D}$ containing X.

The notion of *emerging patterns* (EPs) [5] and a special type, *jumping emerging patterns* (JEPs), were proposed to capture differences between classes.

Definition 1. *[5] (**EP** and **JEP**) Given a real number $\rho > 1$ and two datasets $\mathcal{D}_1$ and $\mathcal{D}_2$, an itemset X is called an ρ-emerging pattern (EP) from $\mathcal{D}_1$ to $\mathcal{D}_2$ if the support ratio $\frac{supp_{\mathcal{D}_2}(X)}{supp_{\mathcal{D}_1}(X)} \geq \rho$. (Define $\frac{0}{0} = 0$ and $\frac{\neq 0}{0} = \infty$.) Specially, a jumping EP (JEP) of $\mathcal{D}_2$ is such an EP which occurs in $\mathcal{D}_2$ (or, whose support in $\mathcal{D}_2$ is non-zero) but does not occur in $\mathcal{D}_1$ (or, whose support in $\mathcal{D}_1$ is zero).*

We have already seen three JEPs of $\mathcal{D}_{\mathcal{N}}$ and one JEP of $\mathcal{D}_{\mathcal{P}}$ in Example 1.

Informally, a border is a concise structure used to describe a large collection of sets. For example, the border $<\{\{1\}\}, \{\{1, 2\}, \{1, 2, 3, 4, 5, 6\}\}>$ is bounded by its *left bound* $\{\{1\}\}$ and its *right bound* $\{\{1, 2\}, \{1, 2, 3, 4, 5, 6\}\}$. This border represents the collection of all sets which are supersets of $\{\{1\}\}$ and are subsets of either $\{1, 2\}$ or $\{1, 2, 3, 4, 5, 6\}$. This collection of sets is denoted $[\{\{1\}\}, \{\{1, 2\}, \{1, 2, 3, 4, 5, 6\}\}]$. More details about the definitions of borders can be found in [5].

3   The DeEPs Classifier

Our DeEPs classifier needs three main steps to determine the class of a test instance: (i) Discovering border representations of EPs; (ii) Selecting the more discriminating EPs; (iii) Determining collective scores based on the selected EPs for classification.

We present algorithms for each of the three steps in the subsequent subsections. We begin by presenting algorithms to handle datasets with only two classes and then generalize DeEPs in Section 3.4 to handle datasets with more classes.

3.1   Discovering Border Representations of EPs

This step aims to learn discriminating knowledge from training data and represent them concisely, by first reducing the data and then discovering all JEPs (and optionally other EPs). Assume we are given a classification problem, having a set $\mathcal{D}_p = \{P_1, \cdots, P_m\}$ of positive training instances, a set $\mathcal{D}_n = \{N_1, \cdots, N_n\}$ of negative training instances, and a set of test instances.

For each test instance T, the DeEPs classifier uses the three procedures below to discover border representations of the EPs from the training data.

1. **Intersecting the training data with** T: $T \cap P_1, \cdots, T \cap P_m$ **and** $T \cap N_1, \cdots, T \cap N_n$. We will discuss how to conduct intersection operation, using *neighborhood-based intersection* method, when continuous attributes are present.
2. **Selecting the maximal itemsets from** $T \cap P_1, \cdots, T \cap P_m$, **and similarly from** $T \cap N_1, \cdots, T \cap N_n$. Denote the former collection of maximal itemsets as $\mathcal{R}_p$ and the latter as $\mathcal{R}_n$.
3.a **Discovery of jumping emerging patterns**. Mining those subsets of T which occur in $\mathcal{D}_p$ but not in $\mathcal{D}_n$, i.e., *all* the JEPs in $\mathcal{D}_p$ (positive class), by taking *border difference* operation $[\{\emptyset\}, \mathcal{R}_p] - [\{\emptyset\}, \mathcal{R}_n]$. On the other hand, mining those subsets of T which occur in $\mathcal{D}_n$ but not in $\mathcal{D}_p$, i.e., *all* the JEPs in $\mathcal{D}_n$ (negative class), by similarly taking *border difference* operation $[\{\emptyset\}, \mathcal{R}_n] - [\{\emptyset\}, \mathcal{R}_p]$.
3.b **Discovery of common emerging patterns**. Mining those subsets of T which occur in both $\mathcal{D}_p$ and $\mathcal{D}_n$, namely, $commonT = [\{\emptyset\}, \mathcal{R}_p] \cap [\{\emptyset\}, \mathcal{R}_n]$, and then selecting those itemsets whose supports change significantly from $\mathcal{D}_p$ to $\mathcal{D}_n$ or from $\mathcal{D}_n$ to $\mathcal{D}_p$. This step is optional, and is omitted if decision speed is important.

Using intersection and maximal itemsets to reduce volume and dimension of training data. First, with the intersection operation in step 1, the dimension of the training data is substantially reduced, because many values of the original training data do not occur in the test instance T. Second, with the maximal itemset selection step, the volume of the training data is also substantially reduced since itemsets $T \cap P_i$ are frequently contained in some other itemsets $T \cap P_j$. Then, $\mathcal{R}_p$ can be viewed as a compressed $\mathcal{D}_p$, and $\mathcal{R}_n$ a compressed $\mathcal{D}_n$. We use the mushroom dataset to demonstrate this point. The original mushroom data has a volume of 3788 edible training instances, with 22 attributes per instance. The average number of items (or *length*) of the 3788 processed instances (by intersection with a test instance) is 11, and these processed instances are further compressed into 7 maximal itemsets. Thus we have achieved a reduction from 3788 to 7 in volume and from 22 to 11 in dimension in the mushroom data. Table 2 of Section 1 also briefly illustrated this 2-directional sparsifying effect. Therefore, this data reduction mechanism narrows our search space. This compression effect is made possible by the instance-based learning approach and it lays a foundation for the high accuracy and high efficiency of DeEPs.

Using border algorithms to efficiently discover EPs. Step 3.a is used to efficiently discover the JEP border representations. Note that $<\{\emptyset\}, \mathcal{R}_p>$ or $<\{\emptyset\}, \mathcal{R}_n>$ can still represent large collections of sets, despite the tremendous effect of reduction as discussed above. To enumerate all itemsets covered by these borders is costly. The *border difference* operation avoids the expensive enumeration, by manipulating the boundary elements $\mathcal{R}_p$ and $\mathcal{R}_n$, to output $<jump\mathcal{L}_p, jump\mathcal{R}_p>$ or $<jump\mathcal{L}_n, jump\mathcal{R}_n>$ as a succinct border representation of the discriminating features (JEPs) of T in the positive class or in the negative class. The border difference operation itself is reviewed later in Section 4. Similarly, by manipulating the boundary elements in $\mathcal{R}_p$ and $\mathcal{R}_n$, we can represent $commonT$ by the border $<\{\emptyset\}, commonR>$. The above three borders are border representations of the EPs that DeEPs needs.

Neighborhood-based intersection of continuous attributes. For datasets containing continuous attributes, we need a new way to intersect two instances. We introduce a method called *neighborhood-based intersection*. It helps select the most relevant information from the training instances for classifying the test instance T, and avoids pre-discretizing the training instances. Suppose `attri_A` is a continuous attribute and its domain is $[0, 1]$. (We can normalize all `attri_A` values in the training instances to the range of $[0, 1]$ if its domain is not $[0, 1]$.) Given a test instance T and a training instance S, $T \cap S$ will contain the `attri_A` value of T, if the `attri_A` value of S is in the neighborhood $[a_1 - \alpha\%, a_1 + \alpha\%]$, where a_1 is the normalized `attri_A` value for T. The parameter α is called *neighborhood factor*, which can be used to adjust the length of the neighborhood.

We observed that different neighborhood factors α can cause accuracy variance, although slight, on the test data. When α is too large, it may happen that many originally different instances from different classes can be transformed into an identical binary instance; consequently, the inherent discriminating features among these instances disappear. When α is too small, nearly identical attribute values may be considered different, and thus useful discrimination information in the training data might be missed. Briefly, we select a suitable α for each dataset by using part of training data as a guide; the details are given in [10]. This is also a research topic in our future work.

3.2 Selecting the More Discriminating EPs

We observed that the number of JEPs (or optional EP candidates) that occur in a test instance is usually large (e.g., of the order of 10^6 in mushroom, waveform, ionosphere, and sonar data). It is expensive to aggregate the contributions of all those EPs. To solve this problem, we select more discriminating EPs by taking advantage of border representation of EPs: We select the most *general* JEPs among all JEPs, namely those in $jump\mathcal{L}_p$ and $jump\mathcal{L}_n$, as the necessary EPs, and select itemsets in the right bound of $commonT$, namely $commonR$, as optional EPs. By the most *general* JEPs, we mean that their proper subsets are not JEPs any more. Partial reasons of this selection are given in [9]. Whether the optional EPs are used depends on how much time a user is willing to spend in classifying instances. With less time, we only select those EPs that must be included; with more time, we can consider the optional EPs for inclusion.

3.3 Determining Collective Scores for Classification

We determine the collective score of T for any specific class C by aggregating the supports of the selected EPs in class C, using our *compact summation* method.

Definition 2. *The compact summation of the supports in $\mathcal{D}_C$ of a collection of selected EPs is defined as the percentage of instances in $\mathcal{D}_C$ that contain one or more of the selected EPs; this percentage is called the* compact score *of T for C, that is,*

$compactScore(C) = \frac{count_{\mathcal{D}_C}(SEP)}{|\mathcal{D}_C|}$, *where SEP is the collection of selected EPs and*
$count_{\mathcal{D}_C}(SEP)$ *is the number of instances in $\mathcal{D}_C$ that contain one or more EPs in*
SEP.

The main purpose of this aggregation is to avoid counting duplicate contribution of training instances.

DeEPs makes the final decision only when the compact scores, formed by compact summation, for all different classes are available. Then the DeEPs classifier will simply assign to T the class for which T's score is largest. We use majority rule to break ties.

3.4 Handling Datasets Containing More than Two Classes

We have already discussed how DeEPs is applied to the problems with two classes. In the following, the DeEPs classifier is generalized to handle more classes. For example, given a test instance T and a training database containing 3 classes of data $\mathcal{D}_1$, $\mathcal{D}_2$, and $\mathcal{D}_3$, we only need to discover the border representation of the JEPs (and optionally EPs) from $(\mathcal{D}_2 \cup \mathcal{D}_3)$ to $\mathcal{D}_1$, those from $(\mathcal{D}_1 \cup \mathcal{D}_3)$ to $\mathcal{D}_2$, and those from $(\mathcal{D}_1 \cup \mathcal{D}_2)$ to $\mathcal{D}_3$. The compact scores for these three classes can be calculated based on three groups of the boundary EPs. We choose the class in which the biggest compact score is obtained as T's class label.

4 Algorithms for DeEPs

We need three algorithms, MAXSELECTOR, INTERSECOPERATION, and DIFFOPERATION [11], for DeEPs to find the desired EPs. MAXSELECTOR is used to select the maximal ones from a collection of sets. INTERSECOPERATION and DIFFOPERATION are used to conduct *border intersection* and *border difference* respectively; they are needed to discover border representations of *commonT* and of JEPs. As the latter two algorithms only manipulate the bounds of borders to handle huge collections, they are highly efficient and scalable in practice.

The DIFFOPERATION algorithm [11] is essential to the efficiency of DeEPs. It is used to discover the border representation of the JEPs of T in each class. Given two borders $<\{\emptyset\}, \mathcal{R}_1>$ and $<\{\emptyset\}, \mathcal{R}_2>$, the *border difference* operation, implemented by DIFFOPERATION, is used to derive the border of $[\{\emptyset\}, \mathcal{R}_1] - [\{\emptyset\}, \mathcal{R}_2]$.

More details about these three algorithms can be found in [11,5] and in our technical report [10].

5 Performance Evaluation: Accuracy, Speed, and Scalability

We now present the experimental results to demonstrate the accuracy, speed, and scalability of DeEPs. We have run DeEPs on 40 datasets taken from the UCI Machine Learning Repository [2]. The accuracy results were obtained using the methodology of ten-fold cross-validation (CV-10). These experiments were carried out on a 500MHz PentiumIII PC, with 512M bytes of RAM.

5.1 Pre-processes

The experiment's pre-processes include: (i) download original datasets, say $\mathcal{D}$, from the sources; (ii) partition $\mathcal{D}$ into class datasets $\mathcal{D}_1, \mathcal{D}_2, \cdots, \mathcal{D}_q$, where q is the number of classes in $\mathcal{D}$; (iii) randomly shuffle each $\mathcal{D}_i, i = 1, \cdots, q$, using the function random_shuffle() in Standard Template Library [4]; (iv) for each $\mathcal{D}_i$, do CV-10 partition; (v) if there exist continuous attributes, scale all the values in the training datasets of each attribute into the values in the range of $[0, 1]$, and then use the same parameters to scale the values in the testing datasets. This step is used to prepare for *neighborhood-based intersection of continuous attributes* and to prepare for the conversion of training datasets into binary ones. In this paper, we use the formula $\frac{x-min}{max-min}$ to scale every value x of the attribute attri_A, where max and min are respectively the biggest and the smallest values of attri_A in the *training* data.

5.2 Accuracy, Speed, and Scalability

We compare DeEPs with five other state-of-the-art classifiers: C4.5 [15], Naive Bayes (NB) [7], TAN [8], CBA [12], and LB [13].

Table 3 reports results of the experiments. Column 1 lists the name of the datasets; column 2 lists the numbers of instances, attributes, and classes; columns 3 and 4 present the average accuracy of DeEPs, when the neighborhood factor α is fixed as 12 for all datasets, and respectively when α is dynamically selected within each dataset (as explained in [10]). Note that for the datasets such as chess, flare, nursery, splice, mushroom, voting, soybean-l, t-t-t, and zoo which do not contain any continuous attributes, DeEPs does not need α. Columns 5, 6, 7, 8, and 9 give the accuracies of CBA, C4.5, LB, NB, and TAN respectively; these results are exactly copied from Table 1 in [13] for the first 21 datasets. For the remaining datasets, we select for CBA and C4.5 the *best* result from Table 1 of [12]. A dash indicates that we were not able to find previous reported results and "N/A" means the classifier is not applicable to the datasets. The last column gives the average time used by DeEPs to test one instance.

The results for DeEPs were obtained by selecting only the left bounds of JEPs, without selecting any optional EPs. Our experiments also show that selecting optional EPs can increase accuracy but degrade speed.

We now discuss the case when the neighborhood factor is fixed as 12 for all datasets. We highlight some interesting points as follows:

- Among the first 21 datasets where results of the other five classifiers are available, DeEPs achieves the best accuracy (the numbers in bold font) on 9 datasets. C4.5, LB, NB, and TAN achieve the best accuracy on 3, 5, 4 and 3 datasets respectively. It can be seen that DeEPs in general outperforms the other classifiers.
- For the remaining 16 datasets where results of CBA and C4.5 are available, DeEPs achieves the best accuracy on 7 datasets. CBA and C4.5 achieve the best accuracy on 6 and 3 datasets respectively. In addition, DeEPs can reach 100% accuracy on mushroom, 99.04% on nursery, and 98.21% on pendigits.

Discussions on speed and scalability of DeEPs can be found in our technical report [10] and omitted here for space reasons.

6 Related Work

CAEP [6] and the JEP-Classifier [9] are two relatives to DeEPs. The former two are eager-learning based approaches, but DeEPs is instance-based. For more comparison, the readers are refered to [6,9,10].

Table 3. Accuracy Comparison.

Datasets	#inst, attri, class	DeEPs		CBA	C4.5	LB	NB	TAN	time (sec.)
		$\alpha = 12$	dynamical α						
australian	690, 14, 2	84.78	88.41* (5)	85.51	84.28	**85.65**	**85.65**	85.22	0.054
breast-w	699, 10, 2	96.42	96.42 (12)	95.28	95.42	96.86	**97.00**	N/A	0.055
census-inc	**30162, 16, 2**	**85.93**	85.93* (12)	85.67	85.4	85.11	84.12	N/A	2.081
chess	3196, 36, 2	97.81	97.81	98.12	**99.5**	90.24	87.15	92.12	0.472
cleve	303, 13, 2	81.17	84.21* (15)	77.24	72.19	82.19	**82.78**	N/A	0.032
diabete	768, 8, 2	**76.82**	76.82* (12)	72.9	71.73	76.69	75.13	76.56	0.051
flare	1066,10,2	**83.50**	83.50	83.11	81.16	81.52	79.46	82.64	0.028
german	1000, 20, 2	74.40	74.40 (12)	73.2	71.7	**74.8**	74.1	72.7	0.207
heart	270, 13, 2	81.11	82.22 (15)	81.87	76.69	82.22	82.22	**83.33**	0.025
hepatitis	155, 19, 2	81.18	82.52 (11)	80.20	80.00	**84.5**	83.92	N/A	0.018
letter	**20000, 16, 26**	**93.60**	93.60* (12)	51.76	77.7	76.4	74.94	85.7	3.267
lymph	148, 18, 4	75.42	75.42 (10)	77.33	78.39	**84.57**	81.86	83.76	0.019
pima	768, 8, 2	**76.82**	77.08* (14)	73.1	72.5	75.77	75.9	75.77	0.051
satimage	**6435, 36, 6**	**88.47**	88.47* (12)	84.85	85.2	83.9	81.8	87.2	2.821
segment	2310, 19, 7	94.98	95.97* (5)	93.51	**95.8**	94.16	91.82	93.51	0.382
shuttle-small	5800, 9, 7	97.02	99.62* (1)	99.48	99.5	99.38	98.7	**99.64**	0.438
splice	3175, 60, 3	69.71	69.71	70.03	93.3	**94.64**	**94.64**	94.63	0.893
vehicle	846, 18, 4	**70.95**	74.56* (15)	68.78	69.82	68.8	61.12	70.92	0.134
voting	433, 16, 2	95.17	95.17	93.54	**95.66**	94.72	90.34	93.32	0.025
waveform	**5000, 21, 3**	**84.36**	84.36* (12)	75.34	70.4	79.43	78.51	79.13	2.522
yeast	1484, 8, 10	**59.78**	60.24* (10)	55.1	55.73	58.16	58.05	57.21	0.096
anneal	998, 38, 6	94.41	95.01 (6)	**98.1**	94.8	–	–	–	0.122
automobile	205, 25, 7	67.65	72.68 (3.5)	79.00	**80.1**	–	–	–	0.045
crx	690, 15, 2	84.18	88.11* (3.5)	**85.9**	84.9	–	–	–	0.055
glass	214, 9, 7	58.49	67.39 (10)	**72.6**	72.5	–	–	–	0.021
horse	368, 28, 2	84.21	85.31* (3.5)	82.1	83.7	–	–	–	0.052
hypo	3163, 25, 2	97.19	98.26 (5)	98.4	**99.2**	–	–	–	0.275
ionosphere	351, 34, 2	86.23	91.24 (5)	**92.1**	92.00	–	–	–	0.147
iris	150, 4, 3	**96.00**	96.67* (10)	92.9	95.3	–	–	–	0.007
labor	57, 16, 2	**87.67**	87.67* (10)	83.00	79.3	–	–	–	0.009
mushroom	**8124, 22, 2**	**100.0**	100.0	–	–	–	–	–	0.436
nursery	**12960, 8, 5**	**99.04**	99.04	–	–	–	–	–	0.290
pendigits	**10992, 16, 10**	**98.21**	98.44 (18)	–	–	–	–	–	1.912
sick	4744, 29, 2	94.03	96.63 (5)	97.3	**98.5**	–	–	–	0.284
sonar	208, 60, 2	**84.16**	86.97* (11)	78.3	72.2	–	–	–	0.193
soybean-small	47, 34, 4	**100.0**	100.0* (10)	98.00	98.00	–	–	–	0.022
soybean-large	683, 35, 19	90.08	90.08	**92.23**	92.1	–	–	–	0.072
tic-tac-toe	958, 9, 2	99.06	99.06	**100.0**	99.4	–	–	–	0.032
wine	178, 13, 3	95.58	96.08* (11)	91.6	92.7	–	–	–	0.028
zoo	101, 16, 7	**97.19**	97.19	94.6	92.2	–	–	–	0.007

Both being instance-based, DeEPs and the k-nearest-neighbor (k-NNR) classifier [3] are closely related. Their fundamental differences are: k-NNR uses the distances of the k nearest neighbors of a test instance T to determine the class of T; however, DeEPs uses the support change of some selected subsets of T.

7   Conclusions

In this paper, we have proposed a new classifier, DeEPs. The DeEPs classifier uses the instance-based, lazy approach to the mining of discriminating knowledge in the form of EPs. This strategy ensures that good representative EPs that are present in a new instance can be efficiently found and can be effectively used for classifying the new instance. Our experimental results have shown the classification accuracy achieved by DeEPs is very high. Our experimental results also have shown DeEPs' decision is quick and DeEPs is scalable over the number of training instances and nearly scalable over the number of attributes in large datasets.

References

1. D. W. Aha, D. Kibler, and M. K. Albert. Instance-based learning algorithms. *Machine Learning*, 6:37–66, 1991.
2. C.L. Blake and P.M. Murphy.   UCI Repository of machine learning database. [http://www.cs.uci.edu/ mlearn/mlrepository.html]. 1998.
3. T. M. Cover and P. E. Hart. Nearest neighbor pattern classification. *IEEE Transactions on Information Theory*, 13:21–27, 1967.
4. H. M. Deitel and P. J. Deitel. *C++ how to program, second edition*. Prentice Hall, Upper Saddle River, New Jersey, USA, 1998.
5. G. Dong and J. Li. Efficient mining of emerging patterns: Discovering trends and differences. In *Proceedings of the Fifth International Conference on Knowledge Discovery and Data Mining, San Diego, USA* (SIGKDD'99), pages 43–52, 1999.
6. G. Dong, X. Zhang, L. Wong, and J. Li. CAEP: Classification by aggregating emerging patterns. In *Proceedings of the Second International Conference on Discovery Science, Tokyo, Japan*, pages 30-42, 1999.
7. R. Duda and P. Hart. *Pattern Classification and Scene Analysis*. New York: John Wiley & Sons, 1973.
8. N. Friedman, D. Geiger, and M. Goldszmidt. Bayesian network classifiers. In *Machine Learning*, 29: 131–163, 1997.
9. J. Li, G. Dong, and K. Ramamohanarao. Making use of the most expressive jumping emerging patterns for classification. In *Proceedings of the Fourth Pacific-Asia Conference on Knowledge Discovery and Data Mining, Kyoto, Japan*, pages 220–232, 2000.
10. J. Li, G. Dong, and K. Ramamohanarao. DeEPs: Instance-based classification by emerging patterns. Technical Report, Dept of CSSE, University of Melbourne, 2000.
11. J. Li, K. Ramamohanarao, and G. Dong. The space of jumping emerging patterns and its incremental maintenance algorithms. In *Proceedings of the Seventeenth International Conference on Machine Learning, Stanford, CA* June, 2000.
12. B. Liu, W. Hsu, and Y. Ma. Integrating classification and association rule mining. In *Proceedings of the Fourth International Conference on Knowledge Discovery in Databases and Data Mining, KDD'98 New York, USA*, 1998.
13. D. Meretakis and B. Wuthrich. Extending naive bayes classifiers using long itemsets. In *Proceedings of the Fifth ACM SIGKDD International Conference on Knowledge Discovery and Data Mining, San Diego*, pages 165–174, 1999.
14. J. R. Quinlan. Induction of decision trees. *Machine Learning*, 1:81–106, 1986.
15. J.R. Quinlan. *C4.5: Programs for machine learning*. Morgan Kaufmann, 1993.

Context-Based Similarity Measures for Categorical Databases

Gautam Das[1] and Heikki Mannila[2]

[1] Microsoft Research
gautamd@microsoft.com
[2] Nokia Research
Heikki.Mannila@nokia.com

Abstract. Similarity between complex data objects is one of the central notions in data mining. We propose certain similarity (or distance) measures between various components of a 0/1 relation. We define measures between attributes, between rows, and between subrelations of the database. They find important applications in clustering, classification, and several other data mining processes. Our measures are based on the contexts of individual components. For example, two products (i.e., attributes) are deemed similar if their respective sets of customers (i.e., subrelations) are similar. This reveals more subtle relationships between components, something that is usually missing in simpler measures. Our problem of finding distance measures can be formulated as a system of nonlinear equations. We present an iterative algorithm which, when seeded with random initial values, converges quickly to stable distances in practice (typically requiring less than five iterations). The algorithm requires only one database scan. Results on artificial and real data show that our method is efficient, and produces results with intuitive appeal.

1   Introduction

Similarity between complex data objects is a crucial notion in data mining and information retrieval. In order to look for patterns or regularities in a database, it is often necessary to be able to quantify how far from each other two objects in the database are. Once we have a natural notion for similarity between objects in a database, we can for example use distance-based clustering or nearest neighbor techniques to search for interesting information from the data set. Recently, there has been considerable interest in defining intuitive and easily computable measures of similarity between complex objects and in using abstract similarity notions in querying databases [1, 2, 10, 14, 17, 19, 22, 4, 12, 15].

Ideally, the similarity notion is defined by the user, who understands the domain concepts well and is able to explicate the notions needed for similarity computations. However, in many applications the domain expertise is not available. The users do not understand the interconnections between objects well enough to formulate exact definitions of similarity or distance. In such cases it is quite useful to have (semi)automatic methods for finding out the similarity (or distance) between objects.

D.A. Zighed, J. Komorowski, and J. Zytkow (Eds.): PKDD 2000, LNAI 1910, pp. 201–210, 2000.
© Springer-Verlag Berlin Heidelberg 2000

In this paper we focus on defining similarity between various components of *categorical databases*. Such databases typically do not contain numeric data. Instead, the domains of the attributes are small, unordered sets of values. An important example is a *market basket database* which is typically a supermarket's record of customer purchases. Conceptually this is a binary table where columns (or attributes) represent products, and rows represent customers. Such databases also frequently occur in domains outside retail (for example, in this paper we consider a TV viewership dataset, where the columns are TV shows and the rows are viewers). We consider the problem of defining distance measures between various components of such databases, such as between attributes, between rows, and between subrelations. Since the data is nonnumeric, simple distance measures such as Euclidean distances are inappropriate in this context.

We develop measures that are *context-based*, i.e. similarity between components is determined by examining the contexts in which the components appear. For example, two products (i.e. attributes) are deemed similar if their respective sets of customers (i.e. subrelations) are similar. This reveals more subtle relationships between components, something that is usually missing in simpler, context insensitive measures. A highlight of our methods is that they are based on plausible probabilistic arguments: data components are assumed to be samples of underlying probability distributions, and computing the similarity between two components is reduced to computing the similarity between two respective probability distributions. While other context-based approaches to similarity and clustering problems have been investigated by other researchers (see discussion at the end of this section), our proposal is particularly appropriate for the kind of applications we consider.

In the rest of the paper we describe our results only for market basket databases, i.e. binary tables. Our results can be extended to general categorical databases by mapping them into binary relations; we omit details from this version of the paper. In what follows, we use the words *attribute, column* and *product* interchangeably; likewise we also use *row* and *customer* interchangeably.

Consider the problem of defining (dis)similarities between attributes. This has several applications, such as in forming product hierarchies or clusters of attributes [13, 20]. Typically, one assumes that the hierarchy is given by a domain expert. However, in the absence of such knowledge, we could produce a product hierarchy that is derived from the data. This can be done through standard similarity measures such as Euclidean distance, correlations, etc. However, if we use them in a typical supermarket scenario, we might conclude that two products such as Coke and Pepsi are quite dissimilar because they do not have many common customers, whereas in reality both are soft drinks and should thus be more similar than say, Pepsi and Mustard. Secondly, consider a high volume product such as Coke, and a relatively low volume product such as RC Cola. Correlations or Euclidean distances would not detect much similarity between the two. We propose the following measure: two products are similar if the *buying patterns* of the customers of each are similar. Thus Coke and Pepsi can be deemed similar attributes, if the buying behavior of buyers of Coke and buyers of Pepsi are similar. But buyers of Coke (resp. Pepsi) are simply subrelations of the database. Thus we relate similarity between attributes to similarity between certain subrelations.

Consider the problem of defining similarities between rows. This is affected by the fact that not all attributes are equidistant from each other (for example, a Coke customer is closer to a Pepsi customer than to a Mustard customer). In other words, our row similarity measure depends on our attribute similarity measure. So we see that various different similarity notions are interdependent on one another: row similarity depends on attribute similarity, which depends on subrelation similarity, which in turn depends on row similarity (since a subrelation is basically a set of rows). The main result of this paper is that the problem of simultaneously solving for all distance measures is formulated as a system of nonlinear equations. We present ICD (*Iterated Contextual Distances*), an iterative algorithm which, when seeded with random initial values, converges quickly to stable distances in practice (typically requiring less than five iterations). The algorithm requires only one database scan.

We complete this section with a discussion of related research. Recently, several iterative procedures (quite different from ours) have been developed for other problems, such as document retrieval systems [5, 15], extraction of information in hyperlinked environments [9, 16], and clustering of categorical data [8]. Much of the emphasis in document retrieval is on grouping/clustering documents based on co-occurances of keywords. The problem of clustering values of a categorical database has attracted much attention recently. A variety of approaches have been developed, such as dynamical systems formulations, probabilistic mixture models, combinatorial methods, etc. [8, 3, 12, 7, 11].

However clustering only gives an indirect sense of inter-attribute distances, and in some applications it is important to have the direct distance measure. It is not clear whether the above clustering methods will easily extend to defining and computing the kind of similarities that we desire (such as defining product similarity based on buying patterns of customers, etc.). Attribute distances of a binary database has been investigated in [12, 4]. The idea of using contexts to define attribute distances has been used in [4], but there the distance between subrelations is defined by looking at the marginal distributions for certain *probe* attributes. Also, the paper did not take into account issues such as the dependencies between attribute distances and row distances.

The rest of the paper is organized as follows. We first describe our Iterated Contextual Distances algorithm. We then describe a variety of experiments on artificial and real data. We conclude with some open problems.

2   The Iterated Contextual Distances (ICD) Algorithm

Let r be a $0/1$ relation containing n rows, over the set of attributes R where $|R| = m$. As mentioned earlier, given distances between attributes we will define distances between rows, given distances between rows we will define distances between subrelations, and, given distances between subrelations we will define distances between attributes. This way of defining distances between attributes is circular, and therefore we use an iterative approach. The basic idea is to initially start with an arbitrary distance function d_0 between attributes of R and use that to derive a vector representation for the rows. From this we go to a vector representation for subrelations. This representation gives us a distance metric Δ_0 between subrelations of r. Then the value of $\Delta_0(r_A, r_B)$ is used to get a new distance value $d_1(A, B)$ for attributes. A few iterations of these steps quickly produces a stable set of distances between attributes. We denote the resulting distance function by d^* and call it *the iterated contextual distance* between attributes. From these attribute distances, we can easily compute other stable distances such as row distances and subrelation distances.

For attributes A and B, let $r_A = \{t \in r \mid t[A] = 1\}$ and $r_B = \{t \in r \mid t[B] = 1\}$ be two corresponding subrelations. We shall relate the distance between A and B to the distance between r_A and r_B. Defining distance between attributes by distance between relations might seem a step backwards; we want to define distance between two objects of size $n \times 1$ by reducing this to distance between objects of dimensions $n_A \times m$ and $n_B \times m$, where $n_A = |r_A|$ and $n_B = |r_B|$. However, we will see later that we can rely on some well-established probabilistic notions for defining distance between subrelations.

2.1   From Attribute Distances to Row Representations

Assume we have a distance function d between attributes. We will use this to define distances between rows.

A row in r can also be viewed as an m-dimensional vector with values from $\{0, 1\}$. In what follows, we will often speak of real-valued rows over R as vectors over R. We map each row $t \in r$ to a vector $f(t)$ over R as follows. We build the vector $f(t)$ for row t by combining vectors g_A for each $A \in t$ (i.e., $A \in R$ such that $t[A] = 1$).

For each attribute $A \in R$ define the function g_A from R to $[0, 1]$ by

$$g_A(B) = \frac{K(d(A, B))}{\sum_{C \in R} K(d(A, C))}$$

where K is a *kernel smoothing* function. We only require that K is monotonically decreasing (for example, $K(x) = 1/(1 + x)$).

Note that $\sum_{B \in R} g_A(B) = 1$. It is useful to think of g_A as a probability distribution that represents the amount of *equivalence* of other products with product A. Thus, if a customer bought A, then $g_A(B)$ represents the probability that the customer would have been satisfied buying B instead. If products A and B are completely similar, i.e., $d(A, B) = 0$, then $g_A(A) = g_A(B)$.

Let $t \in r$, and let $A_1, \ldots, A_k$ be the attributes for which t has value 1. Define a vector $f(t)$ where

$$f(t)(C) = c(g_{A_1}(C), \ldots, g_{A_k}(C))$$

where c is a combination function defined as

$$c(g_1, \ldots, g_k) = 1 - \prod_{i=1}^{k} (1 - g_i),$$

i.e., we have

$$f(t)(C) = 1 - \prod_{i=1}^{k} (1 - g_{A_i}(C))$$

corresponding to a disjunction of the different pieces of evidence $g_{A_i}(C)$ for the presence of C.

Using this mapping f, we define the row distance $d(t, u)$ between rows t and u as the L_1 distance between the two vectors, i.e. $L_1(f(t), f(u))$.

Remarks: There are several possible choices for the kernel smoothing function, $K(x)$. If we run the ICD algorithm with different kernel functions, we get different final distances. However, the actual choice of a does not seem to affect the relative *ranks* of the distances by much.

2.2   Distances between Subrelations

Consider two subrelations r_A and r_B of r (say the Coke customers and the Pepsi customers). We would like to define a notion of distance between r_A and r_B. We construct the following sets of vectors for r_A and r_B: $V(r_A) = \{f(u)|u \in r_A\}$ and $V(r_B) = \{f(v)|v \in r_B\}$. Then the task is reduced to defining the distance between two sets of vectors in m-space.

We could view each relation as a point set and use one of a variety of combinatorial methods for defining the distance between them, such as Hausdorff distance, closest pair, etc. [6]. However, we choose to view the subrelations as samples of corresponding underlying probability distributions (e.g. the distributions of Coke and Pepsi customers), and reduce the problem to computing distance between the two distributions. A comprehensive way of doing this would be to fit distributions (such as gaussians) to each sample and then compute the Kulback-Leibler distance between the two. In the interest of simplicity, we simply choose to compute the distance between the two corresponding centroids c_A and c_B of the sets $V(r_A)$ and $V(r_B)$. Thus,

$$\Delta(r_A, r_B) = L_1(c_A, c_B)$$

2.3   Details of the ICD Algorithm

In detail, the process of finding the attribute distances is as follows. Let r be a $0/1$ relation over attributes R.

1. Initialize with random seed: Set $d(A, A) = 0$ for all $A \in R$, and let $d(A, B) = d(B, A) = rand()$ for all $A, B \in R$ with $A \neq B$.

2. Normalize distances: Multiply all distances $d(A, B)$ by a constant c so that the average pairwise distance between attributes is 1.

3. Compute attribute vectors: For each $A \in R$, compute the vector $g_A : R \to [0, 1]$ by

$$g_A(B) = \frac{K(d(A, B))}{\sum_{C \in R} K(d(A, C))}.$$

4. Compute row vectors: For each $t \in r$, let $A_1, \ldots, A_k$ be the attributes for which t has value 1. Define a vector $f(t)$ as

$$f(t)(C) = c(g_{A_1}(C), \ldots, g_{A_k}(C)),$$

where $c(g_1, \ldots, g_k) = 1 - \prod_{i=1}^{k}(1 - g_i)$.

5. Form subrelation centers: For each $A \in R$, let c_A be the vector on R defined as the average of the vectors $f(t)$, where $t(A) = 1$.

6. Compute distances between subrelation centers: For each pair of attributes $A, B \in R$, let

$$\Delta(r_A, r_B) = L_1(c_A, c_B).$$

7. Iterate: for each pair of attributes $A, B \in R$, let $d(A, B) = \Delta(r_A, r_B)$. If the method has converged, stop; now the distance function d is the iterated contextual distance function d^*. Otherwise, go to step 2.

2.4   Analysis of the ICD Method

It can be shown that the distances produced by the ICD algorithm satisfy the triangle inequality. We omit details from this version of the paper.

The ICD algorithm seems to always converge very quickly in practice, typically not requiring more than five iterations. A theoretical analysis of the convergence seems quite difficult, as the ICD algorithm is essentially trying to compute the fixed points of a nonlinear dynamical system. In practice, different starting points sometimes (but rarely) lead to different fixed points, and when that occurs, we have to select the solution(s) that satisfies our requirements best. Our approach is to select the solutions with large variances in attribute distances, the rationale being that solutions which "spread" attributes apart are more interesting.

A naive implementation requires I database scans, where I is the number of iterations required for convergence (usually less than five). A much better implementation which requires only one database scan is possible for sparse relations. Here the first scan can be used to prepare certain main memory data structures (such as the distance matrix, and for each centroid the formula which determines how it is updated after every iteration). The remaining iterations can proceed without having to access the database.

3 Experiments

We experimented with three types of data: (a) small illustrative examples, (b) TV shows viewing data, and (c) data on pages visited at the MSN website.

3.1 An Illustrative Example

Consider the following relation:

	A	B	C	D	E	F	G
t_1	1	0	0	0	0	0	0
t_2	1	1	0	0	0	0	0
t_3	1	0	1	0	0	0	0
t_4	0	1	1	0	0	0	0
t_5	0	1	0	0	0	0	0
t_6	1	0	0	1	0	0	0
t_7	0	1	0	0	1	0	0
t_8	0	0	0	0	0	1	0
t_9	0	0	0	0	0	0	1

If we run the ICD algorithm (with kernel smoothing function $K(x) = 1/(1+x)$), it converges quickly to the following attribute distances:

	A	B	C	D	E	F	G
A	0.000	0.032	0.338	0.338	0.338	2.065	2.065
B	0.032	0.000	0.338	0.338	0.338	2.065	2.065
C	0.338	0.338	0.000	0.058	0.058	2.322	2.322
D	0.338	0.338	0.058	0.000	0.076	2.320	2.320
E	0.338	0.338	0.058	0.076	0.000	2.320	0.320
F	2.065	2.065	2.322	2.320	2.320	0.000	0.063
G	2.065	2.065	2.322	2.320	2.320	0.063	0.000

Upon examining these distances (especially close to the main diagonal), we notice that there are essentially three clusters of attributes: A and B are similar to each other, C, D, and E are similar to each other, and F and G are similar to each other.

We give some intuitive reasoning for the above. Let us try to determine the attribute distance between A and B by examining the buying behavior of their customers. The customers buying A buy the same proportion of C as the customers buying B. Also, the proportion of D bought by A's customers is the same as the proportion of E bought by B's customers. If we knew that D and E are themselves similar, then we can conclude that the customers of A and B are very similar, and thus A and B are themselves very similar. But how do we know anything about the similarity between D and E? If we compare the customers of D with the customers of E, we see that they buy A and B respectively. Thus by a circular argument, we see that A and B must be similar, and D and E also must be similar.

We next consider C. The customers buying C buy the same proportion of A (or B) as the customers buying D. Thus C is similar to D (and thus to E).

As for F and G, since their customers have nothing in common with the others, they are far from the other attributes.

3.2   TV Shows Data

We experimented using data about which TV shows people watched. The data set contained 2272 observations (rows) over about 15 shows (columns)[3] Our objective was to compute distances between TV shows based on viewership patterns.

If we examine the ICD attribute distances for the TV shows in sorted order, they make intuitive sense. For example, the most distant/similar pairs are shown in Figure 1.

Most distant pairs	
AMER.FUNNIEST-HM VIDEOS	FRIENDS
ABC WORLD NEWS TONIGHT	FRIENDS
FRIENDS	CBS EVE NEWS-RATHER/CHUN
MAD ABOUT YOU-THU.	CBS EVE NEWS-RATHER/CHUN
Most similar pairs	
FRIENDS	MAD ABOUT YOU-THU
ABC WORLD NEWS TONIGHT	GOOD MORNING, AMERICA
AMER.FUNNIEST-HM VIDEOS	NBC NIGHTLY NEWS
MURDER, SHE WROTE	60 MINUTES

Fig. 1. Distance between shows

One interesting result is that ABC NEWS:NIGHTLINE is close to both NBC NIGHTLY NEWS and CBS EVE NEWS-RATHER/CHUN. Also, ABC WORLD NEWS TONIGHT and GOOD MORNING, AMERICA are close to each other, but tend to be far from most of the other shows. These relationships were not discovered by other methods, such as by computing correlations between the respective shows.

During the experiments, we tried out the following family of kernel smoothing functions: $K(x) = 1/(1 + ax)$, for $a > 0$. The precise value of a did not seem to affect the relative ranks of the attribute distances by much. Also, the ICD algorithm seems to converge to fixed points very quickly from random starting points (typically within five iterations).

---

[3] The original data had about 130 shows, but it was pruned to contain only those shows that were watched by at least 300 persons.

3.3   MSN Data

We have started an extensive examination of data on the pages visited at the MSN website. The data has been collected over a six months period, and sampled to extract information about the most frequent 5000 visitors. Since MSN is a vast site, we focus on the activities at the top of the hierarchy, i.e. at the top 50 (in number of hits) DNSids. After some aggregation, we prepared a 50 × 5000 binary table (in sparse form) where the columns are the DNSid's and the rows are visitors. An entry of 1 indicates that the visitor has visited that DNSid during those six months. Our objective is to compute similarities between DNSids based on similarity between their respective visitor sets.

Due to lack of space, the details of the experiments will appear in the full version of this paper. We only mention here that the initial results appear encouraging. As an example, the pair of DNSids *women.msn.com* and *underwire.msn.com* are found similar by our method, but do not appear similar if we simply compute correlations.

4   Open problems

How reasonable and natural are the resulting distance measures? Our preliminary investigations with real data seem to indicate that the ICD method has intuitive appeal. We are carefully investigating if some of the other context-senstive methods can be unified with ICD to produce even better results in our application domains.

Even though it seems to work well in practice, the ICD method needs further theoretical study. The criteria (if any) for convergence, estimates of the number of fixed points, etc., need to be investigated.

References

1. R. Agrawal, C. Faloutsos and A. Swami. Efficient similarity search in sequence databases. In *Proc. of the 4th Intl. Conf. on Foundations of Data Organization and Algorithms (FODO'93)*, 1993.
2. R. Agrawal, K.-I. Lin, H. S. Sawhney and K. Shim. Fast similarity search in the presence of noise, scaling, and translation in time-series databases. In *Proc. of the 21st Intl. Conf. on Very Large Data Bases (VLDB)*, 1995, pp 490–501.
3. P. Cheeseman and J. Stutz. Bayesian classification (Autoclass): theory and results. In *Advances in Knowledge Discovery and Data Mining*, MIT Press, 1996, pp. 153–180.
4. G. Das, H. Mannila, and P. Ronkainen. Similarity of attributes by external probes. In *Proc. of the 4th Intl. Conf. on Knowledge Discovery and Data Mining (KDD)*, 1998, pp 23–29.
5. S. Deerwester, S. T. Dumais, T. K. Landauer, G. W. Furnas, and R. A. Harshman. Indexing by latent semantic analysis. *Journal of the Society of Information Science*, 41(6), 1990, 391–407.
6. T. Eiter and H. Mannila. Distance measures for point sets and their computation. *Acta Informatica*, 34(2), 1997, pp. 109–133.

7. Venkatesh Ganti, J. E. Gehrke, and Raghu Ramakrishnan. CACTUS–clustering categorical data Using summaries. In *Proc. of the 5th Intl. Conf on Knowledge Discovery and Data Mining (KDD)*, 1999.
8. D. Gibson, J. Kleinberg, and P. Raghavan. Clustering categorical data: an approach based on dynamical systems. In *Proc. of VLDB*, 1998, pp. 311–322.
9. D. Gibson, J. Kleinberg, and P. Raghavan. Inferring Web communities from link topology. In *Proc. 9th ACM Conference on Hypertext and Hypermedia*, 1998.
10. D. Q. Goldin and P.Kanellakis. On similarity queries for time-series data: Constraint Specification and Implementation. In *Intl. Conf. on Principles and Practices of Constraint Programming*, 1995.
11. S. Guha, R. Rastogi and K. Shim. ROCK: A Robust Clustering Algorithm for Categorical Attributes. In *Proc. of ICDE*, 1999, pp. 512–521.
12. E.-H. Han, G. Karypis, V. Kumar and B. Mobasher. Clustering Based On Association Rule Hypergraphs. In *Workshop on Research Issues on Data Mining and Knowledge Discovery*, 1997.
13. J. Han, Y. Cai and N. Cercone. Knowledge discovery in databases: an attribute oriented approach. In *Proc. of the 18th Conf. on Very Large Data Bases (VLDB)*, 1992, pp. 547–559.
14. H.V. Jagadish, A. O. Mendelzon and T. Milo. Similarity-based queries. In *Proc. of 14th Symp. on Principles of Database Systems (PODS)*, 1995, pp. 36–45.
15. Y. Karov and S. Edelman. Similarity-based word sense disambiguation. *Computational Linguistics*, 24(1), 1998, pp. 41-59.
16. J. Kleinberg. Authoritative sources in a hyperlinked environment. In *Proc. 9th ACM-SIAM Symposium on Discrete Algorithms (SODA)*, 1998.
17. A. J. Knobbe and P. W. Adriaans. Analyzing binary associations. In *Proc. of the 2nd Intl. Conf. on Knowledge Discovery and Data Mining (KDD)*, 1996, pp. 311-314.
18. R. T. Ng and J. Han. Efficient and effective clustering methods for spatial data mining. In *Proc. of VLDB*, 1994, pp. 144–155.
19. D. Rafiei and A. Mendelzon. Similarity-based queries for time series data. *SIGMOD Record*, 26(2), 1997, pp. 13–25.
20. R. Srikant and R. Agrawal. Mining generalized association rules. In *Proc. of the 21st Intl. Conf. on Very Large Data Bases (VLDB)*, 1995, pp. 407–419.
21. G. Strang. Linear algebra and its applications. Harcourt Brace International Edition, 1988.
22. D. A. White and R. Jain. Algorithms and strategies for similarity retrieval. Technical Report VCL-96-101, Visual Computing Laboratory, UC Davis, 1996.
23. T. Zhang, R. Ramakrishnan, and M. Livny. BIRCH: An efficient data clustering method for very large databases. In *Proc. of SIGMOD*, 1996, pp. 103–114.

A Mixed Similarity Measure in Near-Linear Computational Complexity for Distance-Based Methods

Ngoc Binh Nguyen * and Tu Bao Ho

Graduate School of Knowledge Science
Japan Advanced Institute of Science and Technology
Tatsunokuchi, Ishikawa, 923-1292 JAPAN
{binh,bao}@jaist.ac.jp

Abstract. Many methods of knowledge discovery and data mining are distance-based such as nearest neighbor classification or clustering where similarity measures between objects play an essential role. While real-world databases are often heterogeneous with mixed numeric and symbolic attributes, most available similarity measures can only be applied to either symbolic or numeric data. In such cases, data mining methods often require transforming numeric data into symbolic ones by discretization techniques. Mixed similarity measures (MSMs) without discretization of numeric values are desirable alternatives for objects with mixed symbolic and numeric data. However, the time and space complexities of computing available MSMs are often very high that make MSMs not applicable to large datasets. In the framework of Goodall's MSM inspired by biological taxonomy, computing methods have been done but their time and space complexities so far are at least $O(n^2 \log n^2)$ and $O(n^2)$, respectively. In this work, we propose a new and efficient method for computing this MSM with $O(n \log n)$ time and $O(n)$ space complexities. We demonstrate experimentally the applicability of new method to large datasets and suggest meta-knowledge on the use of this MSM. Practically, the experimental results show that only the near-linear time and space MSM could be applicable to mining large heterogeneous datasets.
Keywords: mixed similarity measure (MSM), computational complexity, very large databases, distance-based methods.

1 Introduction

Traditionally, distance-based methods can be applied to datasets which contain either symbolic or numeric data as most similarity measures are available for homogeneous data. However, real-world data are often heterogeneous with mixed numeric and symbolic attributes. In such cases, most distance-based methods require transforming numeric data into symbolic ones by discretization techniques.

---

* Currently with the Faculty of Information Technology, Hanoi University of Technology. Dai Co Viet, Hai Ba Trung, Hanoi, VIETNAM.

D.A. Zighed, J. Komorowski, and J. Żytkow (Eds.): PKDD 2000, LNAI 1910, pp. 211–220, 2000.
© Springer-Verlag Berlin Heidelberg 2000

Discretization however may cause a loss of information and a difficulty in interpreting results. There have been an interest in finding *mixed similarity measures*
(MSMs) for processing mixed symbolic and numeric data [2], [4], [5], [7]. In [2],
the authors use a Cartesian join operator whenever a mutual pair of symbolic
objects is selected for agglomeration based on minimum dissimilarity. In [5], the
author extended the Minkowski metric on multidimensional space containing
mixed data. Thought having good properties the authors do not provide any experiment on large real-world data or an analysis of computational complexity of
their methods. In [1], Goodall proposed an MSM for biological taxonomy. This
MSM provides a uniform framework for processing mixed numeric and symbolic
data. However, in this framework the computation for numeric attributes may
take the time of $O(n^4)$ where n is the maximum number of unique values met in
the database for any numeric attributes. Recently, in [7] the authors introduced a
computing method for Goodall's MSM and showed that their computing method
(with direct implementation of the MSM) is more efficient than the original one.
However, with computation time of $O(n^2 \log n^2)$ and the space of $O(n^2)$, this
computing method for MSM has still not been able to be applied to large datasets.
We propose a new method for calculating the same MSM of Goodall but with
time $O(n \log n)$ and linear space $O(n)$. We show experimentally the applicability of the new method to large datasets. The efficiency of the new computing
method makes the MSM applicable to many KDD distance-based methods. Section 2 of this paper summaries the basics of computation for Goodall's MSM.
Section 3 provides a computation complexity analysis for MSM as the basics of
our computing method, and the description of a linear-time algorithm. Section
4 reports experimental results on efficiency of the proposed method.

2   Basics of Mixed Similarity Measure Computation

The computation of the mixed similarity measure in [1] consists of two phases: (1) computation of the similarity in respect of single numeric and symbolic
attributes, and (2) combination of the obtained similarities into a unique similarity value. The key idea of similarity computation is *uncommon attribute values
make greater contributions to the overall similarity between two objects*. For any
particular attribute, similarity will be defined in such a way that it generates
an ordering relationship between pairs of values so that, for two pairs of values
(a, b) and (c, d), either (a, b) are more similar than (c, d), or the reverse is true,
or the two pairs are equally similar. The definition of similarity depends on the
type of attribute in question – whether metrical (numeric) or non-metrical (symbolic). Once all pairs of values for an attribute have been ordered in respect of
similarity, *the similarity for each pair of instances is defined as the complement
of the probability that a random pair of instances will have a similarity equal to,
or greater than, the pair in question.*
Similarity between symbolic values. Pairs of differing values are all regarded as equally dissimilar, but pairs of values which agree are ordered according
to the rule: agreement between two instances in possessing an uncommon value

of the attribute is considered as indicating closer similarity than agreement in possessing a commoner value. The possible pairs of values having been ordered in this way, their probabilities are then calculated, and the measure of similarity for any pair is then the complement of the sum of the probabilities for all pairs of values equal or greater in similarity. In practical situation, of course, the p_i will not be known but need to be estimated from the sample of m instances. **Proposition 1** *From a sample of m instances, if f_i is the number of instances in m which have V_i as the value of the attribute in question, the appropriate estimate of similarity S_{ij} between two symbolic values can be*

$$\hat{S}_{ij} = \begin{cases} 1 - \sum_{f_k \leq f_i} \frac{f_k(f_k-1)}{m(m-1)}, & if \ \ i = j \\ 0, & if \ \ i \neq j \end{cases} \tag{1}$$

Similarity between numeric values. Let an attribute take a set of ordered numeric values $V_1 < V_2 < \cdots < V_n$. The numeric values as metric data: they have the order and the difference between any two values. Pairs with identical values are more similar than those which different values, those with a small difference are more similar than those with a larger difference. It appears desirable to take into account the size of the groups encompassed by the two values. Again, once these ordering relations have been established, the degree of similarity is expressed by the complement of the probability of the observed pair or any more similar. Given a pair of numeric values (V_i, V_j), the probability P_{ij} that a random pair of values will be as similar as or more similar than (V_i, V_j) is defined by

$$P_{ij} = \sum_{k \in Q_{ij}} \left(p_k^2 + 2 \sum_{t \in T_k} p_k p_t \right) \tag{2}$$

$$T_k = \left\{ t \ \Big| \ \left[(|V_t - V_k| < |V_j - V_i|) \vee \left[(|V_t - V_k| = |V_j - V_i|) \right. \right. \right.$$
$$\left. \left. \wedge \left(\sum_{u=i}^{j} p_u \geq \sum_{u=k}^{t} p_u \right) \right] \right] \wedge [k < t \leq n] \right\} \tag{3}$$

$$Q_{ij} = \left\{ k \ \Big| \ [(i \neq j) \vee (p_k \leq p_i)] \wedge [1 \leq k \leq n] \right\} \tag{4}$$

Proposition 2 *The similarity S_{ij} between two numeric values can be estimated from sample data by*

$$\hat{S}_{ij} = 1 - \hat{P}_{ij} = 1 - \frac{1}{m(m-1)} \sum_{f_k \leq f_i} \left(f_k(f_k - 1) + 2 \sum_{t \in T_k} f_k f_t \right) \tag{5}$$

where $\hat{P}_{ij}$ is the estimation of the probability P_{ij} by frequencies. **Combining similarities of numeric and symbolic values.** Goodall in [1], Li and Biswas in [7] have described a method to combine similarities in respect of different attributes by using Fisher's transformation and Lancaster's transformation [6] as follows. Denote $(P_{ij})_k$ the probability P_{ij} that a random pair of values at attribute k will be as similar as, or more similar than, the pair (V_i, V_j) at attribute k, then such a probability for all t_c numeric attributes is combined by Fisher's χ^2 transformation:

$$(\chi_c)_{ij}^2 \;=\; -2 \sum_{k=1}^{t_c} \ln((P_{ij})_k) \tag{6}$$

where t_c is the number of numeric attributes in the data. Similarly, the probabilities from symbolic attributes are combined using Lancaster's *mean value* χ^2 transformation [6]:

$$(\chi_d)_{ij}^2 \;=\; 2 \sum_{k=1}^{t_d} \left(1 - \frac{(P_{ij})_k \ln(P_{ij})_k - (P_{ij})_k' \ln(P_{ij})_k'}{(P_{ij})_k - (P_{ij})_k'} \right) \tag{7}$$

where t_d is the number of symbolic attributes in the data, $(P_{ij})_k$ is the probability for symbolic attribute value pair $((V_i)_k, (V_j)_k)$ at attribute k, $(P_{ij})_k'$ is the next smaller probability in the symbolic set. The significance value of this χ^2 distribution can be looked up in standard tables or approximated from the expression:

$$P_{ij} \;=\; e^{-\frac{\chi_{ij}^2}{2}} \sum_{k=0}^{(t_d+t_c-1)} \frac{(\frac{1}{2}\chi_{ij}^2)^k}{k!} \tag{8}$$

where $\chi_{ij}^2 = (\chi_c)_{ij}^2 + (\chi_d)_{ij}^2$. The overall similarity representing the set of $(t_c + t_d)$ independent similarity measures is $S_{ij} = 1 - P_{ij}$. The similarity measures obtained from individual attributes are combined to give the overall similarity between pairs of objects.

3   An MSM in Linear Time and Space Computation

3.1   New Efficient Computation

Because computing the similarities for symbolic features requires the $O(n)$ time complexity, therefore our efforts are to speed up computing the similarities for numeric features. **Proposition 3** *The following properties of P_{ij} hold regarding*

1. $P_{ij} = 1$ *iff* $d_{ij} = |V_n - V_1|$ *(i.e. $\{i,j\} = \{1,n\}$)*
2. $d_{ij} < d_{kl} \implies P_{ij} < P_{kl}$
$d_{ij} = |V_j - V_i|$:   3. $0 < P_{ij} \le 1$ *for* $\forall i, j$; *but* $\hat{P}_{ij} = 0$ *iff* $(i = j) \wedge (f_i = 1)$ Ob-
4. $P_{ij} > \sum_{k=1}^{n} p_k^2$ *for* $\forall i \ne j$
5. $T_k = \phi$ *when* $i = j$ *in Eqs.(3) and (4)*

serving these properties, the essential idea of our method is to compute the quantity $\hat{\overline{P}}_{ij}$ as an estimate of $\overline{P}_{ij} = 1 - P_{ij}$ instead of computing $\hat{P}_{ij}$. The following sequential analyses will lead us to the key result formulated in Proposition 4. Note that T_k is a set of indices t $(k < t \le n)$ such that either

$$|V_t - V_k| < |V_j - V_i| \tag{9}$$

or

$$(|V_t - V_k| = |V_j - V_i|) \wedge (\sum_{u=i}^{j} p_u \geq \sum_{u=k}^{t} p_u) \tag{10}$$

Thus, $\overline{P}_{ij}$ is related to the complement of Eq.(9) and Eq.(10), that is

$$|V_t - V_k| > |V_j - V_i| \tag{11}$$

and

$$(|V_t - V_k| = |V_j - V_i|) \wedge (\sum_{u=i}^{j} p_u < \sum_{u=k}^{t} p_u) \tag{12}$$

In Eq.(3) when $i = j$ we have $T_k = \phi$. Hence

$$P_{ij} = \sum_{k \in \{k | p_k \leq p_i\}} p_k^2, \quad i = j \tag{13}$$

and in this case, the computation of P_{ii} takes the $O(n)$ time. From the definition $\overline{P}_{ij} = 1 - P_{ij}$, in Eq.(4) when $i \neq j$: $Q = \{1, 2, ..., n\}$. Using Eq.(2) we have

$$\overline{P}_{ij} = 2 \sum_{k=1}^{n} \left(p_k \sum_{t \in \overline{T}_k} p_t \right), \quad i \neq j \tag{14}$$

$$\overline{T}_k = \left\{ t \mid \left[(|V_t - V_k| > |V_j - V_i|) \vee \left[(|V_t - V_k| = |V_j - V_i|) \right. \right. \right.$$
$$\left. \left. \left. \wedge \left(\sum_{u=i}^{j} p_u < \sum_{u=k}^{t} p_u \right) \right] \right] \wedge [k < t \leq n] \right\} \tag{15}$$

We can write $\overline{T}_k$ i a simpler form:

$$\overline{T}_k = \{t \mid t_k \leq t \leq n\} \cup T_k^0 \tag{16}$$

where t_k is the smallest t satisfying Eq.(11), and T_k^0 contains values t satisfying Eq.(12), if any. Note that if there exists $t \in T_k^0$, then t is equal to $t_k \pm 1$. To show the correctness of computing $\overline{P}_{ij}$ in Eq.(14), using Eq.(3) and Eq.(15) to get

$$T_k \cup \overline{T}_k = \{t \mid k < t \leq n\} \tag{17}$$

then summing Eq.(2) and Eq.(14), for $i \neq j$, we have

$$P_{ij} + \overline{P}_{ij} = \sum_{k=1}^{n} \left(p_k^2 + 2 \sum_{t>k} p_k p_t \right) = \left(\sum_{k=1}^{n} p_k \right)^2 = 1^2 = 1 \tag{18}$$

Using Eq.(14) and Eq.(16) together, we can rewrite $\overline{P}_{ij}$ in the form:

$$\overline{P}_{ij} = 2 \sum_{k=1}^{n} \left(p_k \sum_{t=t_k-|T_k^0|}^{n} p_t \right), \quad i \neq j \tag{19}$$

or

$$\overline{P}_{ij} = 2\sum_{k=1}^{n}\left(p_k(1 - \sum_{t=1}^{t_k-|T_k^0|-1} p_t)\right) = 2\sum_{k=1}^{n}(p_k(1 - s_k)), \ i \neq j \qquad (20)$$

where

$$s_k = \sum_{t=1}^{t_k-|T_k^0|-1} p_t \qquad (21)$$

Proposition 4 *The probability P_{ij} in Eq.(2) can be estimated by $\hat{P}_{ij} = 1 - \overline{\hat{P}}_{ij}$ where*

$$\overline{\hat{P}}_{ij} = \frac{2}{m(m-1)}\sum_{k=1}^{n}(f_k(1-\hat{s}_k)), \ i \neq j \ , \ \hat{s}_k = \sum_{t=1}^{t_k-|T_k^0|-1} f_t \qquad (22)$$

For given i, j at step k, all t_k, T_k^0, and s_k could be computed in the $O(1)$ time by using the accumulated information from the previous steps. Therefore, computing $\overline{\hat{P}}_{ij}$ by Eq.(22) takes the $O(n)$ time complexity. In the next subsection we propose such an algorithm to estimate P_{ij} via computing $\overline{\hat{P}}_{ij}$ by Eq.(22). In fact, by the definition of S_{ij}, we have $\overline{P}_{ij} = 1 - P_{ij} = S_{ij}$, or $\overline{\hat{P}}_{ij} = 1 - \hat{P}_{ij} = \hat{S}_{ij}$. In sum, we have the following result:

> The probability $\hat{P}_{ij}$ in Eq.(5), used to be computed with the $O(n^2 \log n^2)$ time and $O(n^2)$ space complexities, could be computed by the complement $\overline{\hat{P}}_{ij}$ of its estimation $\hat{P}_{ij}$ in Eq.(22) with $O(n)$ time and $O(n)$ space complexities.

Example. Consider a set of $n = 5$ values with occurrence probabilities as follows

i	1	2	3	4	5
V_i	5.5	6.0	7.5	9.0	10.5
p_i	0.1	0.2	0.4	0.2	0.1

Let take (V_3, V_5), then $d_{3,5} = |10.5 - 7.5| = 3.0$. Consider step by step $k = 1 : t_k = 4, T_k^0 = \phi, s_k = 0.7, \overline{P}_{3,5} = 2*0.1*(1.0-0.7) = 0.06$,
$k = 2 : t_k = 5, T_k^0 = \{4\}, s_k = 0.9, \overline{P}_{3,5} = 0.06 + 2*0.2*(1-0.9) + 2*0.2*0.2 = 0.18$,
$k = 3 : t_k = 6, T_k^0 = \phi, s_k = 1.0, \overline{P}_{3,5} = 0.18 + 2*0.1*(1.0-1.0) = 0.18$,
$k = 4$ and 5 are not considered because $s_k = 1.0$. Hence, $P_{3,5} = 1 - \overline{P}_{3,5} = \mathbf{0.82}$.
For direct computation: $T_1 = \{2,3\}; T_2 = \{3\}; T_3 = \{4,5\}; T_4 = \{5\}; T_5 = \phi, Q = \{1,2,3,4,5\}$, so $P_{3,5} = 0.1^2 + 2*0.1*(0.2+0.4) + 0.2^2 + 2*0.2*0.4 + 0.4^2 + 2*0.4*(0.2+0.1) + 0.2^2 + 2*0.2*0.1 + 0.1^2 = \mathbf{0.82}$, exactly as $1 - \overline{P}_{3,5}$.   □

3.2   Algorithm

Fig.1 describes an algorithm for computing P_{ij} via $\overline{P}_{ij}$. The inputs to the algorithm are: $\{V_i\}_{i=1}^{n}$ with $V_i < V_{i+1}, \forall i$; probabilities p_i (or frequencies f_i) of occurrence of V_i; and two indices i, j of the pair of values (V_i, V_j) to be considered for computing their similarity. The output is the value of P_{ij} for (V_i, V_j).

1. $u = 0.0$; $s = 0.0$; $d = |V_j - V_i|$; $t = 1$; $P = 0.0$; $\overline{P} = 0.0$;
2. **if** $i = j$ **then**
3.   **begin**
4.   **for** $k = 1$ **to** n **do**
5.   **if** $p_k \leq p_i$ **then** $P \leftarrow P + p_k$
6.   **end**
7. **else**
8.   **begin**
9.   **if** $i > j$ **then**
10.   **begin**
11. $k \leftarrow i$; $i \leftarrow j$; $j \leftarrow k$
12.   **end**
13. $q = 0.0$; **for** $k = i$ **to** j **do** $q \leftarrow q + p_k$
14.   **for** $k = 1$ **to** n **do**
15.   **if** $V_k + d \leq V_n$ **and** $s < 1$ **then**
16.   **begin**
17.   **if** $k > 1$ **then** $u \leftarrow u + p_{k-1}$
18.   **while** $t \leq n$ **and** $|V_t - V_k| \leq d$ **do**
19.   **begin**
20. $s \leftarrow s + p_t$; $t \leftarrow t + 1$
21.   **end**
22. $\overline{P} \leftarrow \overline{P} + 2p_k(1 - s)$
23.   **if** $|V_{t-1} - V_k| = d$ **and** $s - u > q$ **then**
24. $\overline{P} \leftarrow \overline{P} + 2p_k p_{t-1}$
25.   **end**
26.   **else break**
27. $P \leftarrow 1 - \overline{P}$
28.   **end**
29. **return** P;

Fig. 1. An $O(n)$ time and $O(n)$ space complexity algorithm for computing P_{ij}

Table 1. Comparisons of efficiency by experiments on large datasets: Census Bureau

Specification	Datasets						
	SS1	SS2	SS3	SS4	SS5	SS6	Full Set
# instances (m)	500	1,000	1,500	2,000	5,000	10,000	199,523
Dataset size (MB)	0.2	0.5	0.9	1.1	2.6	5.2	103
# diff. num. values (n)	497	992	1,486	1,973	4,858	9,651	97,799
Time of LIBISWAS	67.3s	26m6.2s	1h46m31s	6h59m45s	>60h	n/a	n/a
Time of OURS	**0.1s**	**0.2s**	**0.3s**	**0.5s**	**2.8s**	**9.2s**	**36m26s**
Memory LIBISWAS (MB)	5.3	20.0	44.0	77.0	455.0	n/a	n/a
Memory OURS (MB)	**0.5**	**0.7**	**0.9**	**1.1**	**2.1**	**3.4**	**64.0**
Preprocessing time	0.1s	0.1s	0.2s	0.5s	0.9s	6.2s	127.2s

Note that the body of the **while**-loop is executed in the total of n times on whole n steps of its outer **for**-loop, hence, the average number of executions of the **while**-loop's body on each step k is of $O(1)$, therefore, any **for**-loop of the algorithm is of $O(n)$. There are no nested **for**-loops, thus the algorithm is with the $O(n)$ time. The space is also of $O(n)$ because there are only 1-dimension arrays with indices from 1 to n in the algorithm.

4   Experimental Evaluation

Our experimental evaluation aims to two objectives: (i) to show that the proposed MSM can work on large datasets; and (ii) to investigate for which kinds of data the MSM can be used in distance-based methods to achieve good results. The proposed algorithm has been implemented in C and evaluated with large mixed datasets. We have also implemented Li and Biswas's algorithm to do comparisons of efficiency of the algorithms. All experiments were carried out on a workstation Alpha 21264 (OS: Digital UNIX V4.0E; Clock frequency: 500 MHz; RAM: 2 GB). Two datasets have been used to evaluate new MSM efficiency:

1. The U.S. Census Bureau (<http://www.census.gov/main/www/access.html>) of 199,523 instances with 33 symbolic attributes and 8 numeric attributes, and the size of the dataset is of 103MB. It is found that the maximal number of unique numeric values for the numeric attribute *"instance weight"* nearly equals the number of instances (i.e., $n \approx m$).
2. KDD Cup 1999 (<http://kdd.ics.uci.edu/databases/kddcup99/kddcup99.html>) with 485,790 instances (73MB) as 10% of full dataset described by 6 symbolic attributes and 35 numeric attributes.

We carried out an experimental comparative evaluation on efficiency of the proposed method (called the OURS) and the previous one of Li and Biswas (called the LIBISWAS) in [7]. We randomly took from each of these two datasets six subsets SS1; SS2; ...; SS6 of 500; 1,000; 1,500; 2,000; 5,000; and 10,000 instances, respectively. The experiment was designed as follows: from each of these subsets and the full dataset, find the nearest neighbor for an instance randomly taken from outside of the subset; do compare the memory size used and execution time needed for the OURS and the LIBISWAS to do this task. That is, for each running on the dataset of m instances, it was necessary to compute MSM between the new instance and each of m instances of the subset. Experiment results on these two datasets are shown in Table 1 and Table 2. These results make clearly the advantage of our proposed method. Note that for SS6 with $m = 10,000$ and the full dataset Census Bureau with $m = 199,523$ and $m = 485,790$ for 10% of KDD-Cup 99, the LIBISWAS cannot fit the program with the data into the RAM of 2 GB (with "n/a": not applicable), but the OURS needs only 64 MB and 276 MB due to its requirement of $O(n)$ of memory size. Therefore, the OURS could be used for very large databases. Note that the OURS works under an assumption that $V_1, V_2, \cdots, V_n$ have been sorted (i.e., $V_1 < V < V_2 < \cdots < V_n$). For large databases and for the first time reading, it is necessary to collect the

Table 2. Comparisons of efficiency by experiments on large datasets: KDD-cup99

Specification	Datasets						
	SS1	SS2	SS3	SS4	SS5	SS6	10% Full Set
# instances (m)	500	1,000	1,500	2,000	5,000	10,000	485,790
Data size (MB)	0.1	0.1	0.2	0.3	0.7	1.5	73.6
# diff. num. values (n)	276	557	867	1,147	2,238	2,889	36,185
Time of LIBISWAS	6.2s	1m28.4s	12m54.2s	1h23m32s	>10h	n/a	n/a
Time of OURS	**0.1s**	**0.2s**	**0.3s**	**0.4s**	**2.0s**	**4.8s**	**11m21.3s**
Memory LIBISWAS (MB)	5.1	10.0	19.0	30.0	225.0	486.0	n/a
Memory OURS (MB)	**0.5**	**1.1**	**1.7**	**1.9**	**4.8**	**6.9**	**276.0**
Preprocessing time	0.1s	0.1s	0.1s	0.3s	0.5s	4.3s	20.1s

values of each attribute, to count their frequencies, to encode and sort the values, and so on. The slowest in this processing is sorting, which is with the $O(n \log n)$ time. When $n \approx m$ the number of t satisfying Eq.(10) is of $O(n)$, hence, the total time from reading a new database until the final results of the NNR program with the OURS is of $O(m + mn + n \log n) = O(mn) = O(n^2)$, the LIBIS is with $O(m + n^3) = O(n^3)$. To evaluate the usefulness of MSM, the decision tree induction program C4.5 with discretization of numeric data [8]) and k-nearest neighbor method with this MSM have been compared on 18 datasets containing mixed symbolic and numeric data from the UCI repository of databases. These datasets are divided into two groups: one with datasets having more numeric attributes than symbolic ones, and vice-versa. The preliminary results reported in [3] suggest some hypotheses about meta-knowledge of the use of this MSM and discretization for a given task of classification:

1. When a database contains more numeric attributes than symbolic ones, it is better to use MSM than discretization. The predictive accuracy of k-NNR with this MSM is often higher than that of C4.5 using discretization. However, k-NNR requires much more time than C4.5 to calculate MSM in these cases;
2. When a database contains more symbolic attributes than numeric ones, it maybe better to do discretization and use classifiers with symbolic data, vice-versa. The predictive accuracy of of k-NNR with this MSM is often lower than that of C4.5 using discretization. However, the computational time for this MSM is much lower than that of cases of many numeric attributes.

5 Conclusions

We studied a mixed similarity measure (MSM), initially proposed by Goodall [1], and the improved computation by Li and Biswas [7]. This MSM has been shown to work well with mixed datasets, but its computational costs are $O(n^2 \log n^2)$ time complexity and $O(n^2)$ space complexity, therefore, it was not suitable for the large databases with large number n of unique values for a numeric feature/attribute. We developed a mathematical basis for computing the same

MSM, but with a new and fast algorithm of only $O(n)$ time and $O(n)$ space complexities ($O(n \log n)$ time for the fisrt use of each dataset). The new method makes MSM, used to be impractical for large datasets, applicable to very large databases. The experimental results on some large datasets have demonstrated the efficiency of the proposed computing method and the algorithm, and suggested some meta-knowledge on the use of MSM and discretization. Further investigating the MSM with the proposed computation method by applying it to KDD distance-based methods such as NNR, clustering, classification, and so on, for large databases is very promising future work.

References

1. Goodall, D.W.: A New Similarity Index Based On Probability. *Biometrics*, Vol. 22 (1966) 882–907.
2. Gowda, K.C., Diday, E.: Symbolic Clustering Using a New Similarity Measure. *IEEE Transactions on Systems, Man, and Cybernetics*, Vol. 22, No. 2 (1992) 368–378.
3. Ho, T.B., Nguyen, N.B., Morita, T.: Study of a Mixed Similarity Measure for Classification and Clustering. *3th Pacific-Asia Conf. on Knowledge Discovery and Data Mining PAKDD'99*. Lecture Notes in Artificial Intelligence 1574. Springer-Verlag (1999) 375-379.
4. Huang, Z.: Clustering Large Data Sets With Mixed Numeric and Categorical Values. *KDD: Techniques and Application*. World Scientific (1997) 21–34.
5. Ichino, M., Yaguchi, H.: Generalized Minkowski Metrics for Mixed Feature-Type Data Analysis. *IEEE Trans. Systems, Man and Cybernetics*, Vol. 24 (1994) 679–709.
6. Lancaster, H.O.: The Combining of Probabilities Arising from Data in Discrete Distributions. *Biometrika*. Vol. 36 (1949) 370-382.
7. Li, C., Biswas, G.: Unsupervised Clustering with Mixed Numeric and Nominal Data - A New Similarity Based Agglomerative System. *KDD: Techniques and Application*. World Scientific (1997) 33–48.
8. Quinlan, J.R. *C4.5: Programs for Machine Learning*, Morgan Kaufmann, 1993.

Fast Feature Selection using Partial Correlation for Multi-valued Attributes

S. Lallich and R. Rakotomalala

ERIC Laboratory - University of Lyon 2
5, av Pierre Mendes-France
F-69676 Bron
FRANCE

Abstract. We propose a fast feature selection method in supervised learning for multi-valued attributes. The main idea is to rewrite the multi-valued problem in the space of examples into a boolean problem in the space of pairwise examples. On basis of this approach, we can use point correlation coefficient which is null in the case of conditional independence, and verifies a formula connecting partial coefficients with marginal coefficients. This property allows to reduce considerably the computing times because a single pass over the database is necessary to compute all coefficients. We test our algorithm on benchmark databases.
Keywords: feature selection, partial association, marginal association

1 Introduction

Feature selection is a key step in any machine learning process. In supervised learning, only the relevant variables are selected, reducing the volume of computation and making the classifier more efficient in generalization, as shown in a variety of papers on sensitivity of noisy attribute classifiers (nearest neighbour [1], naive bayes classifier [11]).

With the development of searches in very large databases [7], preselection -even reduction of- the variables becomes more crucial and more resource consuming [4]. Fast feature selection methods that are general enough to deal with multi-valued categorical variables are thus needed. Typically, there are two kinds of variable selection methods [17]: stepwise filtering [16] and wrapper strategies [12].

Wrapper strategies explicitly use the classifier to select the subset of predictive attributes which minimizes the generalization error rate obtained using cross validation. The main difficulty is choosing between exploring all solutions and the greedy elementary strategy while maintaining the generalization error rate and a reasonable computation time. For this very reason, studies have often relied on so-called rapid learning methods such as decision trees and naive bayes models [11]. Large volumes of data make this strategy harder to apply.

A filtering-type method is suggested here; this quicker method takes advantage of a measure of association that allows the derivation of partial association directly from the marginal pairwise associations. With Boolean variables, we

D.A. Zighed, J. Komorowski, and J. Zytkow (Eds.): PKDD 2000, LNAI 1910, pp. 221–231, 2000.
© Springer-Verlag Berlin Heidelberg 2000

have shown that such a filtering method can be built on point correlation coefficients [21]. For categorical attributes, we suggest in this paper to reduce them into boolean attributes, by rewriting the problem in the space spanned by the co-labels of the original attributes.

In section 2, the principle of our pairwise correlation measure and the linking formula from marginal coefficient to partial coefficient are laid out. The corresponding feature selection algorithm is given in section 3. Results of experiments on real and artificial datasets using a naive bayes classifier are given in section 4. Related works are described in Section 5, while a conclusion is given in Section 6.

2   Principles

Let's first review stepwise selection using learning set. At the first step, the variable showing the strongest predictive association with the class attribute, say Y, is selected. At each following step, the variable adding the most to the quality of the prediction is selected and added to the set of variables already selected. A stopping rule is needed, and a measure of predictive association in marginal form (first step) or partial form (following steps) that can show the marginal gain brought about by each of the added remaining variables. Thus, we seek a measure of (partial) association between X and Y, given Z, with two important features:

P_1 : the partial measure can be written as a function of the marginal measures between all of the variables taken two at a time (linking formula);

P_2 : if X and Y are independent given Z, than the partial measure is null (conditional independence).

2.1   A linking formula

With a linking formula, the partial coefficients can be computed gradually from the mere marginal coefficients, which represents an important reduction in the need for computational resources [24].

Pearson's correlation coefficient, defined for continuous attributes, verifies such a linking formula:

$$r(Y; X/Z) = \frac{r(Y; X) - r(Y; Z)r(X; Z)}{\sqrt{(1 - r^2(Y; Z))(1 - r^2(X; Z))}} \tag{1}$$

Kendall's rank correlation for ordinal variables verifies a similar equation. Conversely, Saporta [24] has proposed a partial coefficient related to Tschuprow's coefficient[1] that is formally derived from the linking formula.

---

[1] a Chi-2 coefficient standardized to account for the sample size and the table dimensions

2.2   Conditional independence

Conditional independence means that when X and Y are independent, given Z, then the partial measure is null. This is an important feature for stepwise procedures, as adding a predictor X, independent of Y given Z, to a predictor Z will add nothing to the quality of the model.

Using combinatorics, Lerman [15] studied measures of association for categorical variables, giving, for each type of cross-classification, a formulation for the null hypothesis and an expression for the partial correlation coefficient that guarantees nullity under the null hypothesis. Unfortunately, to our knowledge, there is no linking formula.

As Lerman [15] points out, partial Tschuprow's coefficient proposed by Saporta [24] is not null under conditional independence, except, noted Daudin [5], if the conditioning variable has only 2 values, which is indeed the case for boolean attributes !

Coefficients similar to Proportional Reduction in Error [9] could also be used; these have been generalized as the Proportional Reduction in Entropy, be it Daroczy's-type entropy or rank entropy [13],[22]. The partial association coefficient is then defined as the weighted mean of the marginal associations given Z; weights can be the probabilities of Z [9], or better yet, those probabilities times the conditional entropy of Y given Z [20]. All these coefficients are null under conditional independence as long as the marginal coefficients are null under independence, but they do not yield to a linking formula !

2.3   Boolean attributes

Boolean attributes, whether binary discretized continuous variables or categorical variables rewritten as set of Boolean variables, can be treated using a point correlation coefficient, as we have proposed [21]; this coefficient satisfies P1 and P2 above. Saporta's [24] partial coefficient after Tschuprow's, while null for conditionally independent Boolean variables, can be negative for certain $2 \times 2 \times 2$ tables.

Cross tables of Boolean variables show some interesting properties linked to the fact that those variables can be regarded, without loss of generality, as 0-1 variables since any other coding can be deduced by a linear transformation. Let Y and X be two Boolean variables; let the joint proportion of 1 be p_{11}, the marginal proportions be p_{1+} and p_{+1}; then the expected value and variance of X are p_{+1} and $p_{+1}(1 - p_{+1})$ respectively; those of Y are p_{1+} and $p_{1+}(1 - p_{1+})$ respectively; and the covariance is $p_{11} - p_{+1}p_{1+}$. The linear correlation coefficient (in this very case called the point correlation coefficient), invariant under linear transformation, is obtained as [2]:

$$r(Y;X) = \varphi = \frac{p_{00}p_{11} - p_{01}p_{10}}{\sqrt{p_{0+}p_{1+}p_{+0}p_{+1}}} \tag{2}$$

_____

[2] In a 2×2 table, all standardizations of the Chi-2 are equivalent : r^2, ϕ^2, $\frac{\chi^2}{n}$, Cramer's V, Kendall's tau, Tschuprow, Goodman and Kruskal's tau.

Examples	Y	X
1	1	1
2	1	2
3	2	2
4	3	3

Pairwise	I_Y	I_X
1,1	1	1
1,2	1	0
1,3	0	0
1,4	0	0
2,1	1	0
2,2	1	1
2,3	0	1
2,4	0	0
3,1	0	0
3,2	0	1
3,3	1	1
3,4	0	0
4,1	0	0
4,2	0	0
4,3	0	0
4,4	1	1

Table 1. From original dataset to pairwise co-labeled dataset

There are many advantages to reason with the point correlation coefficient: it is naturally signed, it verifies a linking formula being a special case of Pearson's correlation, and it is null under conditional independence. This last point can be shown easily as any regression on a Boolean regressor is linear. As with Tschuprow's coefficient, $r(Y; X/Z)$ can be null even if X and Y are not independent. This is true for all coefficients verifying the linking equation. Consider a $2\times2\times2$ table for which a new predictor X is uncorrelated with a former predictor Z and uncorrelated with the class attribute.

2.4 Multi-valued Attributes

For multi-valued attributes, there is no method that satisfies P_1 and P_2, as stated above. Categorical attributes can be reduced to Boolean attributes by rewriting the problem in the space spanned by the co-labels[3] of the original attributes, and thus the selection method described earlier can be applied. Hence, step by step, co-labels associated with the regressors that explain the best the class attribute will be selected using the point correlation coefficient. The $n \times (p+1)$ matrix of data (n is the number of examples in the learning set, $(p+1)$ the number of multi-valued attributes including the class) changes to a $n^2 \times (p+1)$ Boolean matrix (e.g. Table 1).

Formally, we will prefer to define individuals pairs with replacement taking order into account. Indeed, in this case, one can demonstrate that independence of categorical variables Y and X implies independence between co-labels associated with each variable [18] and then the point correlation coefficient $r(I_Y, I_X)$

---

[3] A co-label I_X associated with an attribute X is set to 1 if both individuals have the same value for the attribute of interest, 0 otherwise.

has zero value. If the pairs are without replacement and non ordered, we have demonstrated that under Y and X independence hypothesis, $r(I_Y, I_X)$ has a negative value.

In order to compute the point correlation coefficient between the co-labels, it is simpler to use a contingency-type formula based on the original Y and X, rather than the customary "r" based on I_Y and I_X flags. Consequently, it is not necessary to explicitly form the I_Y and I_X co-labels attributes. The computational cost of the point correlation $r(I_Y, I_X)$ remains in $O(n)$, which is an essential condition of the quickness of the algorithm.

Frequencies of the cross-classified flags, say $g_{ij}, i = 0, 1; j = 0, 1$, are written as functions of the frequencies of the cross-classified attributes Y and X, say $n_{kl}, k = 1, 2, ..., K; l = 1, 2..., L$.

Y\X	$\neq$ (different label on X)	= (same label on X)
$\neq$	$g_{00} = \sum_{k=1}^{K} \sum_{l=1}^{L} n_{kl} [n - n_{k+} - n_{+l} + n_{kl}]$	$g_{01} = \sum_{k=1}^{K} \sum_{l=1}^{L} n_{kl} [n_{+l} - n_{kl}]$
=	$g_{10} = \sum_{k=1}^{K} \sum_{l=1}^{L} n_{kl} [n_{k+} - n_{kl}]$	$g_{11} = \sum_{k=1}^{K} \sum_{l=1}^{L} n_{kl}^2$

$$r(I_Y, I_X) = \frac{g_{11}g_{00} - g_{10}g_{01}}{\sqrt{g_{1+}g_{0+}g_{+1}g_{+0}}}$$

Once the table of the point correlation coefficients is completed, partial correlations are derived, thus selecting attributes step by step. The transition from the space of examples to the space of pairs changes the predictions: now, whether two examples have the same Y-labels given their X-labels are identical or not is predicted, rather than the Y-label given the X-labels. This Boolean set-up is particular as the roles of the labels are not interchangeable. Here, "r", not "r^2", must be maximized, since a strong negative correlation is a sign that X is ill-adapted at predicting Y. Similarly, "r_{part}" and not "r_{part}^2" will be maximized in the following steps. Let's consider the second step of the procedure, when the best X is sought, given the Z selected at the first step. For a given Z, either all pairs are concordant in Z or all pairs are discordant in Z; in either case, concordance in X should correspond to the concordance in Y, that is maximizing "r_{part}".

A rigorous stopping rule is still needed. Value zero is not convenient because we work on a sample. For the time being, we use an empirical stopping rule which operates when "r_{part}" is less than $\frac{2.5}{n}$. This would correspond approximately to the upper 0.5% point of the distribution of "r" under independence, namely, $r \sim N(0, \frac{1}{n})$. A theoretically sound critical value for "r_{part}" that accounts for the number of tests is needed.

3   A Greedy Algorithm for Feature Selection

Building the partial correlations from the marginal correlation without any additional passes though the data set is a key feature of our algorithm. Only the marginal correlations require passing through the data set, a $(p + 1) \times (p + 1)$

$S = \emptyset,\; k = 0$
Compute marginal coefficients $T_{r_0(Y,X)}$
Repeat
    Find $X^* = \arg \max_{X} r_k(Y, X\,/\,S)$
    If $r_k(Y, X^*\,/\,S) > \frac{2.5}{n}$
      Then $S = S \cup \{X^*\}$
        $k = k + 1$
          Compute $T_{r_k(Y,X)}$ from $T_{r_{k-1}(Y,X)}$
   End if
Until "Last add refused"
Return S

Table 2. G3 greedy algorithm for feature selection

table containing the marginal correlations $r_{Y,X_j} (j = 1, \ldots, p)$ and $r_{X_i,X_j}(i, j = 1, \ldots, p)$, say $T_{r(Y,X)}$, being then constructed (p is the number of descriptors on the data set). As correlations are symmetrical, only the upper triangle of $T_{r(Y,X)}$ need be computed.

First, S, the set of selected attributes, is empty. The attribute X^* the most correlated with Y, as indicated in $T_{r(Y,X)}$, is sought. If this search yields a significant solution, then X^* is selected and inserted in S, and the table T is refreshed with the partial correlations $r(Y, X/X^*)$ using the linking formula. In the next step, X^{**}, the attribute showing the strongest correlation with Y given X^*, is selected. This goes on until the stopping rule is activated. Thus, with a very simple greedy algorithm, T is updated each time an attribute is selected. Pseudo-code for the corresponding algorithm $G3$ is shown in Table 2.

The one and only pass through the data set is of magnitude $O(p^2 \times n)$, where n is the number of individuals on the dataset. All other computations can be derived from the coefficients computed first. The maximum complexity, if all attributes were selected, is of magnitude $O(p^2)$.

4 Experiments

Whether our algorithm indeed selects the "right" attributes, and the impact of the selection on a classifier have to be assessed. The naive bayes classifier will be used for two reasons: its complexity is easy to compute $[O(p \times n)]$ and hence the reduction of computing time due to the reduction of attributes is easily seen; it is sensitive to noisy attributes [11], hence eliminating irrelevant attributes should improve its performances.

Databases extracted from the UCI Irvine [3] server were used. We used a very diversified set of databases so that the performance of the algorithm could be assessed in a variety of situations. Some are real-life cases (adult, auto, dermatology, heart, iris, lung cancer), some are artificial (monks1, monks3, mushroom);

Base	Examples	Err. init	Naive bayes	G3	MIFS
Adult	48842	0.24	14 (0.161)	7 (0.144)	8 (0.146)
Autos	205	0.55	25 (0.248)	12 (0.249)	23 (0.243)
Breast noisy	699	0.345	18 (0.037)	8 (0.036)	9 (0.040)
Dermatology	366	0.70	34 (0.102)	23 (0.085)	34 (0.101)
Heart	270	0.44	13 (0.169)	10 (0.175)	12 (0.174)
Iris	150	0.66	4 (0.060)	2 (0.028)	2 (0.027)
Led noisy	10000	0.90	24 (0.264)	7 (0.264)	24 (0.264)
Lung-cancer	32	0.60	56 (0.742)	3 (0.308)	38 (0.700)
Monks-1	556	0.50	6 (0.254)	1 (0.254)	4 (0.254)
Monks-3	554	0.48	6 (0.036)	2 (0.036)	3 (0.036)
Mushroom	8416	0.47	22 (0.004)	13 (0.007)	7 (0.023)
Segmentation	2310	0.86	11 (0.163)	11 (0.163)	4 (0.080)
Wave noisy	5000	0.67	40 (0.224)	14 (0.219)	30 (0.221)

Table 3. Databases characteristics and results - Number of selected attributes (Error rate)

the last set are sets to which random noise was added (wave noisy, breast cancer noisy, led noisy).

Some continuous attributes were discretized using FUSINTER [25]. This is a supervised discretization method which has the advantage of suggesting partitions maximizing the link of each attribute with the class attribute. Compared to less sophisticated methods (mainly unsupervised strategies) [6], we could think that this approach gives the advantage to discretized continuous attributes compared to other categorical one. Experiments show that this undesirable effect does not appear in practice. If we refer to the analogy with decision trees which we will explain in more details below, doing a global discretization before learning process is a practicable strategy [8].

Data bases characteristics and results are shown in Table 3; 10 cross-validations were used to measure the error rate. Results for G3 were compared to results for MIFS [2], an alternative greedy algorithm that will be reviewed in the next section. Table 3 shows the number of cases, the default classifier error rate (the default classifier always predicts the class with the highest frequency), the number of selected attributes and the error rate for the naive bayes algorithm, G3, and finally MIFS. Some observations follow:

- where noise was deliberately added (wave noisy, breast cancer noisy, led noisy), all random attributes were eliminated; thus the method is effective when a large number of attributes are present, not all relevant;
- more interesting yet, for the bases where the "right" attributes were known, these were among the first selected (iris, wave, breast, led); the test error rate should not be severely affected by the reduction of attributes that would follow the introduction of a more stringent stopping rule;

- on the other bases, even UCI Irvine's that have been "worked at", with few irrelevant attributes [10], the selection almost always drastically reduces the number of attributes;
- for the monks datasets (*monks-1, monks-3*), where the concepts to learn are disjunctions of conjunctions, the greedy algorithm failed to select the right attributes; in both cases, only one set of the disjunction was identified. This also happens when a strong interaction is present among the variables (mushroom);.
- error rates are insensitive to reductions in the number of attributes, with two exceptions (dermatology and especially lung cancer); in the lung cancer case, curse of dimensionality disturbs, and a reduction of the number of dimensions improved learning;
- how well MIFS operates depends on how well β, the parameter that rules how many attributes are to be selected, is set. In practice, it appears that β varies with the problem at hand; a constant $\beta=1$ was used, following [14], to obtain a stable comparison. In some instances (auto, dermatology, led, lung cancer, wave) this value seems insufficient; in other cases (mushroom) it appears too restrictive; it appears appropriate for iris and segmentation. From this point of view, G3 is more stable. Tuning G3 is done by adjusting the stopping rule threshold, and even when it was set to 0, results were comparable to those displayed in Table 3;
- as regards calculation times, $G3$ and $MIFS$ are theoretically equivalent since they have the same computational complexity. However, in our experiments, we can note that the $MIFS$ method is a little slower, simply because it selects a greater number of attributes.

5   Related works

In this paper, we are mainly concerned with filtering feature selection methods. The basic idea is to identify the most discriminating subset of attributes before learning starts [17]. Contrary to the wrapper method, filtering methods do not use the classifier's characteristics to build the classifier. There is no guarantee that the best set of attributes will be selected for a given learning algorithm. For example, in the case of the Boolean XOR, a filtering method could select the right attributes, but the decision tree used later on to learn the concept, short-sightedness, fails to see the correct model [19]. This apparent flaw can become an advantage. As it does not depend on any learning algorithm, the filtering aims at outlining the right representation space. Then, a suitable learning algorithm can be found for the definitive set of attributes. Using a multilayer perceptron in the example above, the solution is trivial. The second appealing feature of filtering algorithms is their computational speed. In data mining situations [7], the huge size of the database is such that direct processing of all the data is hardly ever possible. The data base must be reduced in length (by sampling cases) and in width (by selecting variables) to make computations possible and affordable [4]. Of course, this requires a fast selection algorithm.

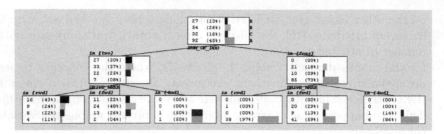

Fig. 1. Constrained Decision Tree with two level on AUTOS dataset

The best representation space is determined by a stepwise selection of the predictors that exhibit the strongest correlation with the class attribute. Hence, exploring the solution space ought to be greedy. In this sense, FGMIFS [14] is similar to our algorithm. FGMIFS constructs a decision tree under the constraint that, for each level, the leaves be split using the same attribute (Figure 1). Here, the measure for the attribute selection is akin to information gain [26]. Typically, it is the description of a stepwise regression using discrete explicative variables. G3 implicitly does the same, but differs from FGMIFS on two essential points: (1) at each node, the partial correlations are derived from previously computed parameters and do not require an additional pass over the databases; (2) the tree not being explicitly constructed, it is not penalized by data fragmentation. Indeed, with FGMIFS, for a large tree (assume 10 Boolean attributes, hence 2^{10} leaves), the number of individuals on each leave is too small for reliable estimation of probabilities.

In our opinion, MIFS [2] is closer to G3 than FGMIFS: the additional information about the class attribute Y brought by an additional X, given the already selected attributes, is computed. Here, the main difference with G3 is that the author uses mutual information, and while partial association is derived from marginal, it is defined empirically. Let S be the set of the already selected attributes, the additional information about Y brought by X is given by:

$$I(Y, X/S) = I(Y, X) - \beta \times \sum_{Z \in S} \frac{I(X, Z)}{card(S)}$$

This is in fact a valid procedure whatever the number of categories. But the success of the procedure relies heavily on the choice of β. In practice, many tests are required before a suitable β is found, and the expected gain in time is lost.

Lastly, work in [23] is older, yet close to ours. The main idea is to derive the partial coefficients from the marginals. While we seek coefficients satisfying some properties, namely the iterative derivation of the partial coefficients, coefficients in [23] are defined after the linking equation. Such coefficients need not be null under conditional independence, and may lie outside the domain of the marginal coefficient (Tschuprow's is bounded by 0 and 1; the partial Tschuprow, as defined by its author, may be negative) which makes interpretation difficult.

6   Conclusion and Future Work

In this paper, a fast multi-valued attribute selection method based on the recurrent computation of partial correlations is developed. It is quite fast as a single pass through the data is necessary, and can thus be used as a starting point far an induction process.

Tests on real and artificial data showed that this approach is pragmatic, and have identified situations where it is quite advantageous. Where the method appears to fail, namely in disjunction problems or with data showing large interaction, the greediness of the method seems at fault. Similarities with decision trees could open new venues: borrowing sophisticated algorithms such as lookahead search, post pruning, or synthetic attributes [26].

Acknowledgements

We are very grateful to the two anonymous referees who provided helpful comments, and to M. Jean Dumais, Section Chief, Methodology Branch, Statistics Canada and guest lecturer at Université Lumière for his careful proof reading of the manuscript and his suggestions.

References

1. D. Aha. Tolerating noisy, irrelelvant and novel attributes in instance-based algorithms. *International Journal of Man-Machine Studies*, 36:267–287, 1992.
2. R. Battiti. Using mutual information for selecting features in supervised neural net learning. *IEEE Transactions on Neural Networks*, 5(4):537–550, 1994.
3. S.D. Bay. The uci kdd archive [http://kdd.ics.uci.edu]. Irvine, CA: University of California, Department of Computer Science, 1999.
4. A. Blum and P. Langley. Selection of relevant feature and examples in machine learning. *Artificial Intelligence*, pages 245–271, 1997.
5. J.J. Daudin. Analyse factorielle des dependances partielles. *Revue de Statistique Appliquee*, 29(2):15–29, 1981.
6. J. Dougherty, R. Kohavi, and M. Sahami. Supervised and unsupervised discretization of continuous attributes. In Morgan Kaufmann, editor, *Machine Learning : Proceedings of the 12th International Conference (ICML-95)*, pages 194–202, 1995.
7. U. Fayyad, G. Piatetsky-Shapiro, and P. Smyth. Knowledge discovey and data mining : Towards an unifying framework. In *Proceedings of the 2nd International Conference on Knowledge Discovery and Data Mining*, 1996.
8. Eibe Frank and Ian H. Witten. Making better use of global discretization. In *Proc. 16th International Conf. on Machine Learning*, pages 115–123. Morgan Kaufmann, San Francisco, CA, 1999.
9. L.A. Goodman and W.H. Kruskall. Measures of association for cross classifications. *Journal of American Statistical Association*, 49:732–764, 1954.
10. R.C. Holte. Very simple classification rules perform well on most commonly used datasets. *Machine Learning*, 11:63–91, 1993.
11. G. John and P. Langley. Static versus dynamic sampling for data mining. In *Proceedings of the 2nd International Conference on Knowledge Discovery in Databases and Data Mining*. AAAI/MIT Press, 1996.

12. R. Kohavi and G. John. Wrappers for feature subset selection. *Journal of Artificial Intelligence, Special issue on Relevance*, 1997.
13. S. Lallich and R. Rakotomalala. Les entropies de rangs généralisés en induction par arbres. In *Proceedings of 7mes Journées de la Société Francophone de Classification - SFC'99*, pages 101–107, September 1999.
14. K-C. Lee. A technique of dynamic feature selection using the feature group mutual information. In *Proceedings of the Third PAKDD-99*, pages 138–142, 1999.
15. I.C. Lerman. Correlation partielle dans le cas qualitatif. Technical Report 111, INRIA, 1982.
16. H. Liu and H. Motoda. *Feature Extraction, Construction and Selection : A Data Mining Perspective*. Kluwer Academic Publishers, 1998.
17. H. Liu and H. Motoda. *Feature selection for knowledge discovery and data mining*, volume 454 of *The kluwer international series in engineering and computer science*. Kluwer, 1998.
18. F. Marcotorchino. Utilisation des comparaisons par paires en statistique des contingences - partie iii. Technical Report F 081, Centre Scientifique IBM-France, 1985.
19. D. Michaud. *Filtrage et Selection D'attributs En Apprentissage*. PhD thesis, Universite de Franche-Comte, 1999.
20. M. Olszak and G. Ritschard. The behaviour of nominal and ordinal partial association measures. *The statistician*, 44(2):195–212, 1995.
21. R. Rakotomalala and S. Lallich. Sélection rapide de variables booléennes en apprentissage supervisé. In *Proceedings of 2nd Conférence Apprentissage - CAP'2000*, pages 225–234, 2000.
22. R. Rakotomalala, S. Lallich, and S. Di Palma. Studying the behavior of generalized entropy in induction trees using a m-of-n concept. In *Proceedings of the Third European Conference PKDD'99*, pages 510–517, 1999.
23. G. Saporta. *Liaisons entre plusieurs ensembles de variables et codage de données qualitatives*. PhD thesis, 1975.
24. G. Saporta. Quelques applications des operateurs d'Escouffier au traitement des variables qualitatives. *Statistique et Analyse de Donnees*, (1):38–46, 1976.
25. D.A. Zighed, S. Rabaseda, and R. Rakotomalala. Fusinter : a method for discretization of continuous attributes for supervised learning. *International Journal of Uncertainty, Fuzziness and Knowledge-Based Systems*, 6(33):307–326, 1998.
26. D.A. Zighed and R. Rakotomalala. *Graphes d'Induction - Apprentissage et Data Mining*. Hermes, 2000.

Fast Hierarchical Clustering Based on Compressed Data and OPTICS

Markus M. Breunig, Hans-Peter Kriegel, Jörg Sander

Institute for Computer Science, University of Munich
Oettingenstr. 67, D-80538 Munich, Germany
{breunig | kriegel | sander}@dbs.informatik.uni-muenchen.de
phone: +49-89-2178-2225
fax: +49-89-2178-2192

Abstract: One way to scale up clustering algorithms is to squash the data by some intelligent compression technique and cluster only the compressed data records. Such compressed data records can e.g. be produced by the BIRCH algorithm. Typically they consist of the sufficient statistics of the form (N, X, X^2) where N is the number of points, X is the (vector-)sum, and X^2 is the square sum of the points. They can be used directly to speed up k-means type of clustering algorithms, but it is not obvious how to use them in a hierarchical clustering algorithm. Applying a hierarchical clustering algorithm e.g. to the centers of compressed subclusters produces a very weak result. The reason is that hierarchical clustering algorithms are based on the distances between data points and that the interpretaion of the result relies heavily on a correct graphical representation of these distances. In this paper, we introduce a method by which the sufficient statistics (N, X, X^2) of subclusters can be utilized in the hierarchical clustering method OPTICS. We show how to generate appropriate distance information about compressed data points, and how to adapt the graphical representation of the clustering result. A performance evaluation using OPTICS in combination with BIRCH demonstrates that our approach is extremely efficient (speed-up factors up to 1700) and produces high quality results.

1   Introduction

Knowledge discovery in databases (KDD) is known as the non-trivial process of identifying valid, novel, potentially useful, and understandable patterns in large amounts of data. One of the primary data analysis tasks which should be applicable in this process is cluster analysis. Therefore, improving clustering algorithms with respect to efficiency and the quality of the results has received a lot of attention in the last few years in the research community..

The goal of a clustering algorithm is to group objects into meaningful subclasses. Applications of clustering are, e.g., the creation of thematic maps in geographic information systems, the detection of clusters of objects in geographic information systems and to explain them by other objects in their neighborhood, or the clustering of a Web-log database to discover groups of similar access patterns corresponding to different user profiles.

There are different types of clustering algorithms suitable for different types of applications. The most commonly known distinction is between *partitioning* and *hierar-*

D.A. Zighed, J. Komorowski, and J. Zytkow (Eds.): PKDD 2000, LNAI 1910, pp. 232–242, 2000.
© Springer-Verlag Berlin Heidelberg 2000

chical clustering algorithms (see e.g. [8]). Partitioning algorithms construct a partition of a database D of n objects into a set of k clusters. Typical examples are the k-means [9] and the k-medoids [8] algorithms. Most hierarchical clustering algorithms such as the single link method [10] and OPTICS [1] do not construct a clustering of the database explicitly. Instead, these methods compute a representation of the data set (single link: dendrogram, OPTICS: reachability plot) which reflects its clustering structure. It depends on the application context whether or not the data set is decomposed into definite clusters. Heuristics to find a decomposition of a hierarchical representation of the clustering structure exist, but typically the results are analyzed interactively by a user.

Clustering algorithms, in general, do not scale well with the size and/or dimension of the data set. One way to overcome this problem is to use *random sampling* in combination with a clustering algorithm (see e.g. [6]). The idea is to apply a clustering algorithm only to a randomly chosen subset of the whole database. The clustering for the whole database is then "inferred" from the clustering of the subset. This usually leads to a significant inaccuracy of the clustering, introduced by sampling variance.

Another approach is to use more intelligent *data compression* techniques to squash the data into a manageable amount, and cluster only the compressed data records. This approach has been pioneered in the BIRCH algorithm [11]. More general compression techniques to support clustering algorithms are e.g. presented in [2], [4]. Compressed data items can be produced by any of the above compression methods in linear time. They are, however, tailored to k-means type of clustering algorithms. On the other hand, it is not obvious how to use compressed data items in a hierarchical clustering algorithm without unacceptably deteriorating the quality of the result.

In this paper, we show how compressed data items of the above form can be used in the hierarchical clustering method OPTICS. For this purpose, we adapt the OPTICS algorithm so that it can profit maximally from the information generated in the compression step. A detailed qualitative and performance analysis for real and synthetic data sets is conducted using the BIRCH method for the compression step.

The rest of the paper is organized as follows. In section 2, we give the problem statement in more detail. Section 3 elaborates on how OPTICS is extended to make optimal use of the information generated by a common data compression step. In section 4, we present basic experimental results and discuss the tradeoff between quality of the results and improvement of the runtime for OPTICS when using compressed data. Finally, section 7 concludes the paper.

2   Problem Statement

Specialized data compression methods have been developed to scale up clustering algorithms. The sufficient statistics intended to support clustering algorithms are basically the same for all these compression methods. As an example, we give a short description of the method BIRCH and only discuss the major differences and the common features for the other methods in this section. Then, we will show why a hierarchical clustering algorithm cannot directly benefit from the compressed data while k-means type of clustering algorithms are well supported.

The clustering method *BIRCH* [11] uses a highly specialized tree-structure for the purpose of clustering very large sets of d-dimensional vectors. It incrementally computes compact descriptions of subclusters, called Clustering Features.

Definition 1: (Clustering Feature, *CF*)

Given a set of n d-dimensional data points $\{X_i\}$, $1 \le i \le n$. The *Clustering Feature* (*CF*) for the set $\{X_i\}$ is defined as the triple $CF = (n, LS, ss)$, where

$$LS = \sum_{i\,=\,1\ldots n} X_i \quad \text{is the linear sum and } ss = \sum_{i\,=\,1\ldots n} X_i^2 \quad \text{the square sum of the points.}$$

The *CF*-values are sufficient to compute information about subclusters like centroid, radius and diameter. They satisfy an important additivity condition, i.e. if $CF_1 = (n_1, LS_1, ss_1)$ and $CF_2 = (n_2, LS_2, ss_2)$ are the clustering features for sets of points S_1 and S_2 respectively, then $CF_1 + CF_2 = (n_1 + n_2, LS_1 + LS_2, ss_1 + ss_2)$ is the clustering feature for the set $S_1 \cup S_2$.

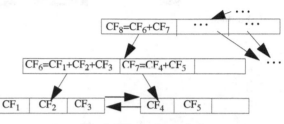

Fig. 1. CF-tree structure

The CFs are organized in a balanced tree with branching factor B and a threshold T (see figure 1). A non-leaf node represents a subcluster consisting of all the subclusters represented by its entries. A leaf node has to contain at most L entries and the diameter of each entry in a leaf node has to be less than T.

BIRCH performs a sequential scan over all data points and builds a *CF*-tree similar to the construction of B$^+$-trees. A point is inserted by inserting the corresponding *CF*-value into the closest leaf. If an entry in the leaf can absorb the new point without violating the threshold condition, its *CF* is updated. Otherwise, a new entry is created in the leaf node, and, if the leaf node then contains more than L entries, it and maybe its ancestors are split. A clustering algorithm can then be applied to the entries in the leaf nodes of the *CF*-tree.

Bradley et al. [2] propose another compression technique for scaling up clustering algorithms. Their method produces basically the same type of compressed data items as BIRCH, i.e. triples of the form (n, LS, ss) as defined above. The method is, however, more specialized to k-means type of clustering algorithms than BIRCH in the sense that the authors distinguish different sets of data items. A very general framework for compressing data has been introduced recently by DuMouchel et al. [4]. Their technique is intended to scale up a large collection of data mining methods.

The application of k-means type clustering algorithms to compressed data items is rather straightforward. The k-means clustering algorithm represents clusters by the mean of the points contained in that cluster. It starts with an assignment of data points to k initial cluster centers, resulting in k clusters. It then iteratively performs the following steps while the cluster centers change: 1) Compute the mean for each cluster. 2) Re-assign each data point to the closest of the new cluster centers. When using CFs, the algorithm just has to be extended so that it treats the triplets (n, LS, ss) as data points LS/n with a weight of n when computing cluster means.

When we want to apply a hierarchical clustering algorithm to compressed data items, however, it is not clear whether we can follow a similar approach, i.e. treat clustering features as data points LS/n. Since hierarchical clustering algorithms do not compute any cluster centers but use only distances between points and between clusters, the

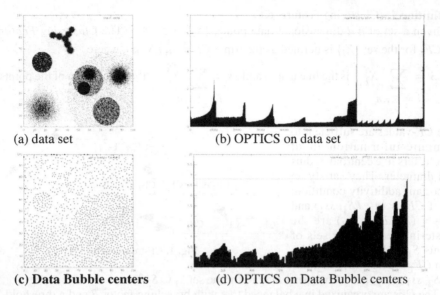

(a) data set   (b) OPTICS on data set

(c) **Data Bubble centers**   (d) OPTICS on Data Bubble centers

Fig. 2. 2-d example data set and its reachability plot

main problem with this approach is that we need a new concept of how to utilize the weight n of a clustering feature in a hierarchical clustering algorithm. Applying hierarchical clustering only to the set of centers LS/n of compressed data items (n, LS, ss) will in general produce only very inferior results.

Figure 2 (a) shows the 2-dimensional example used throughout the paper. The data set consists of 100,000 points grouped into several nested clusters of different densities and distributions (uniform and gaussian). Figure 2 (b) is the "reachability plot" for this data set produced by OPTICS. The "dents" in the plot represent the clusters, clearly showing the hierarchical structure. For a detailed explanation of reachability plots and their interpretation see [1]. Figure 2 (c) shows the distribution of the 1,484 clustering feature centers produced by BIRCH and (d) the corresponding reachability plot.

The failure observed in the example results from the following two problems when applying a hierarchical clustering algorithm to compressed data items:

- Hierarchical clustering algorithms need a notion of distance between data points which is obviously not represented well by only the distance between cluster feature centers. If we cannot infer some information about the distances between the points compressed into a single clustering feature, a hierarchical clustering algorithm cannot determine the correct clustering structure of the data set.
- Utilization of the output of a hierarchical clustering algorithm relies heavily on the graphical representation of the result. If we cannot integrate an appropriate representation of the number of points compressed into a single clustering feature, we loose much of the information about the clustering structure of a data set.

3   Extending OPTICS to Process Compressed Input Data

In this section we review the hierarchical clustering method OPTICS [1], and define the concept of *Data Bubbles* which contain information about compressed input data. We then extend OPTICS to work on Data Bubbles instead of data points.

3.1   Review of OPTICS

First, we define the basic concepts of neighborhood and nearest neighbors.

Definition 2: (ε-neighborhood and k-distance of an object P)

Let P be an object from a database D, let ε be a distance value, let k be a natural number and let d be a distance metric on D. Then:

- the ε-neighborhood $N_\varepsilon(P)$ is a set of objects X in D with $d(P,X) \leq \varepsilon$:
    $N_\varepsilon(P) = \{\, X \in D \mid d(P,X) \leq \varepsilon \,\}$,
- the k-distance of P, $k\text{-}dist(P)$, is the distance $d(P, O)$ between P and an object $O \in D$ such that at least for k objects $O' \in D$ it holds that $d(P, O') \leq d(P, O)$, and for at most k-1 objects $O' \in D$ it holds that $d(P, O') < d(P, O)$. Note that $k\text{-}dist(P)$ is unique, although the object O which is called 'the' k-nearest neighbor of P may not be unique. When clear from the context, we write $N_k(P)$ as a shorthand for $N_{k\text{-}dist(P)}(P)$, i.e. $N_k(P) = \{\, X \in D \mid d(P, X) \leq k\text{-}dist(P) \}$.

The objects in the set $N_k(P)$ are called the "k-nearest-neighbors of P" (although there may be more than k objects contained in the set if the k-nearest neighbor of P is not unique).

In [5] a density-based notion of clusters is introduced. The basic idea is that for each object of a cluster, the ε-neighborhood has to contain at least a minimum number of objects. Such an object is a *core object*. Clusters are defined as maximal sets of density-connected objects. An object P is density-connected to Q if there exists an object O such that both P and Q are density-reachable from O (directly or transitively). P is directly density-reachable from O if $P \in N_\varepsilon(O)$ and O is a core object. Thus, a flat partitioning of a data set into a set of clusters is defined, using *global* density parameters. Very different local densities may be needed to reveal and describe clusters in different regions of the data space. In [1] the density-based clustering approach is extended to create an augmented *ordering* of the database representing its density-based clustering structure. This cluster-ordering contains information which is equivalent to the density-based clusterings corresponding to a broad range of parameter settings. This cluster-ordering of a data set is based on the following notions of "core-distance" and "(density-)reachability-distance".

Definition 3: (core-distance of an object P)

Let P be an object from a database D, let ε be a distance value and let *MinPts* be a natural number. Then, the *core-distance* of P is defined as

$$core\text{-}dist_{\varepsilon,MinPts}(P) = \begin{cases} \text{UNDEFINED, if } |N_\varepsilon(P)| < MinPts \\ MinPts\text{-}dist(P), \text{ otherwise} \end{cases}.$$

The core-distance of an object P is the smallest distance $\varepsilon' \leq \varepsilon$ such that P is a core object with respect to ε' and *MinPts* - if such a distance exists, i.e. if there are at least *MinPts* objects within the ε-neighborhood of P. Otherwise, the core-distance is UNDEFINED.

Definition 4: (reachability-distance of an object P w.r.t. object O)

Let P and O be objects, $P \in N_\varepsilon(O)$, let ε be a distance value and let *MinPts* be a natural number. Then, the *reachability-distance* of P with respect to O is defined as

$$reach\text{-}dist_{\varepsilon,MinPts}(P,\,O) = \begin{cases} \text{UNDEFINED, if } |N_\varepsilon(O)| < MinPts \\ max(core\text{-}dist_{\varepsilon,\,MinPts}(O),\,d(O,P)), \text{otherwise} \end{cases}$$

Intuitively, *reach-dist(P,O)* is the smallest distance such that P is directly density-reachable from O if O is a core object. Therefore *reach-dist(P,O)* cannot be smaller than *core-dist(O)* because for smaller distances no object is directly density-reachable from O. Otherwise, if O is not a core object, *reach-dist(P,O)* is UNDEFINED. Figure 3 illustrates these notions.

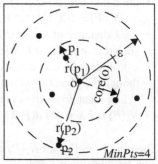

Fig. 3. core-dist(O), reach-dists r(P_1,O), r(P_2,O)

Using the core- and reachability-distances, OPTICS computes a "walk" through the data set, and assigns to each object O its core-distance and the smallest reachability-distance with respect to an object considered *before O* in the walk (see [1] for details). This walk satisfies the following condition: Whenever a set of objects C is a density-based cluster with respect to *MinPts* and a value ε' smaller than the value ε used in the OPTICS algorithm, then a permutation of C (possibly without a few border objects) is a subsequence in the walk. Therefore, the *reachability-plot*, which consists of the reachability values of all objects, plotted in the OPTICS ordering, yields an easy to understand visualization of the clustering structure of the data set. Roughly speaking, a low reachability-distance indicates an object within a cluster, and a high reachability-distance indicates a noise object or a jump from one cluster to another cluster.

3.2   From Sufficient Statistics to Data Bubbles

In section 2, we have seen different methods to compute subsets of the input data and sufficient statistics for these sets containing the number of points, the linear sum and the square sum. Based on these statistical information we define Data Bubbles as a convenient abstraction on which density-based hierarchical clustering can be done.

Definition 5: (Data Bubble)

Let $X=\{X_i\}$, $1 \le i \le n$ be a set of n d-dimensional data points.

Then, a *Data Bubble B* is defined as a triple $B = (n, M, e)$, where

$$M = \left(\sum_{i=1...n} X_i \right) \bigg/ n \text{ is the } center \text{ of } X, \text{ and } e = \sqrt{\frac{\sum\limits_{i=1..n} \sum\limits_{j=1..n} (X_i - X_j)^2}{n \cdot (n-1)}}$$

is called the *extent* of X (i.e. the average pairwise distance between the points in X).

Data Bubbles are a compressed representation of the sets of points they describe. If the points are approximately uniformly distributed around the *center*, a sphere of radius *extent* around the *center* will contain most of the points described by the Data Bubble. The following lemma shows how to compute Data Bubbles from sufficient statistics.

Corollary 1:

Let $X=\{X_i\}$, $1 \le i \le n$ be a set of n d-dimensional data points. Let LS be the linear sum and ss the square sum of the points in X as defined in definition 1. Then, the Data

Bubble B describing X is equal to $B = \left(n, \dfrac{LS}{n}, \sqrt{\dfrac{2 \cdot n \cdot ss - 2 \cdot LS^2}{n \cdot (n-1)}} \right)$.

3.3   Basing OPTICS on Data Bubbles by Modifying the core-dist and reach-dist

In any state, the OPTICS algorithm needs to know which data object is closest to the ones considered so far. To extend OPTICS to work on Data Bubbles, we therefore need a suitable measure for the distance between Data Bubbles. Given a distance between Data Bubbles, it is possible to extend the OPTICS algorithm by defining a suitable core- and reachability-distance for Data Bubbles.

If we assume that a Data Bubble B is a good description of its points, i.e. that all (or almost all) points it describes are inside a sphere of the extent of B around the center of B, we can compute the expected k-nearest neighbor distance of the points in B in the following way:

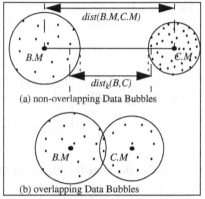

(a) non-overlapping Data Bubbles

(b) overlapping Data Bubbles

Fig. 4. distance of Data Bubbles

Lemma 1: (expected k-nearest neighbor distance inside a Data Bubble B)

Let $B = (n, M, e)$ be a Data Bubble of dimension d. If the n points described by B are uniformly distributed inside a sphere with center M and radius e, then the *expected k-nearest neighbor distance of B* is then equal to $nndist(k, B) = \left(\dfrac{k}{n}\right)^{1/d} \cdot e$.

Proof: The volume of a d-dimensional sphere of radius e is $V_S(e) = \dfrac{\sqrt{\pi}^d}{\Gamma\left(\dfrac{d}{2}+1\right)} \cdot e^d$

(where Γ is the Gamma-Function). If the n points are uniformly distributed in such a sphere, we expect exactly one point in the volume $V_S(e) / n$ and k points in the volume $k\, V_S(e) / n$, which is exactly the volume of a sphere of radius $nndist(k, B)$.

Using the radius and the expected k-nearest neighbor distance, we can now define a distance measure between two Data Bubbles that is suitable for OPTICS.

Definition 6: (distance between two Data Bubbles)

Let $B=(n_1, M_1, e_1)$ and $C=(n_2, M_2, e_2)$ be two Data Bubbles and k a natural number. Then, the k-*distance* between B and C is defined as

$$dist_k(B, C) = \begin{cases} dist(M_1, M_2) - (e_1 + e_2) + nndist(k, B) + nndist(k, C) \\ \qquad \text{if } dist(M_1, M_2) - (e_1 + e_2) \ge 0 \\ max(nndist(k, B), nndist(k, C)) \qquad \text{otherwise} \end{cases}.$$

We have to distinguish two cases (c.f. figure 4). The distance between two non overlapping Data Bubbles is the distance of their centers, minus their radii plus their expected k-nearest neighbor distances. If the Data Bubbles overlap, their distance is the maximum of their expected k-nearest neighbor distances. Intuitively, this distance definition is intended to approximate the distance of the two closest points in the Data Bubbles. Using this distance, we can define the notion of a core-distance and a reachability-distance.

Definition 7: (core-distance of a Data Bubble B)

Let B be a Data Bubble, let ε be a distance value and let $MinPts$ be a natural number. Then, the *core-distance* of B is defined as

$$
core\text{-}dist_{\varepsilon, MinPts}(B) = \begin{cases} \text{UNDEFINED if} \left(\displaystyle\sum_{X = (n, M, e)\, \in\, N_\varepsilon(B)} n \right) < MinPts \\[2em] dist(B, C) \quad \text{otherwise} \end{cases} ,
$$

where C is the Data Bubble in $N_\varepsilon(B)$ with minimal $dist(B, C)$ such that

$$
\left(\sum_{X = (n, M, e)\, \in\, N_\varepsilon(B)\, \wedge\, dist(B, X) < dist(B, C)} n \right) \geq MinPts \text{ holds.}
$$

Note that $core\text{-}dist_{\varepsilon, MinPts}(B=(n,M,e))=0$ if $n \geq MinPts$.

This definition is based on a similar notion as the core-distance for data points. For points, the core-distance is undefined if the number of points in the ε-neighborhood is smaller than $MinPts$. Analogously, the core-distance for Data Bubbles is undefined if the sum of the numbers of points represented by the Data Bubbles in the ε-neighborhood is smaller than $MinPts$. For points, the core-distance (if defined) is the distance to the $MinPts$-neighbor. For Data Bubbles, it is the distance to the Data Bubble containing the $MinPts$-neighbor.

Given the core-distance, the reachability-distance for Data Bubbles is defined in the same way as the reachability-distances on data points.

Definition 8: (reachability-distance of a Data Bubble B w.r.t. Data Bubble C)

Let B and C be Data Bubbles, $B \in N_\varepsilon(C)$, let ε be a distance value and let $MinPts$ be a natural number. Then, the *reachability-distance* of B with respect to C is defined as

$$
reach\text{-}dist_{\varepsilon, MinPts}(B, C) = \begin{cases} \text{UNDEFINED, if} \left(\displaystyle\sum_{X = (n, M, e)\, \in\, N_\varepsilon(B)} n \right) < MinPts \\[2em] max(core\text{-}dist_{\varepsilon,\, MinPts}(C), dist(C, B)) \quad \text{otherwise} \end{cases} .
$$

3.4   Modifying the Output

In order to handle the output correctly, we need to define a *virtual* reachability for the data points described by a Data Bubble $B=(n, M, e)$. If we assume that the points described by B are uniformly distributed in a sphere of radius e around the center M, and B describes at least *MinPts* points, we expect the reachability of most of these points to be close to the *MinPts*-nearest neighbor distance. If, on the other hand, B contains less than *MinPts* points, we expect the core-distance of any of these points to be close to the core-distance of B itself. We use this heuristics to define the virtual reachability as an approximation of the true, but unknown, reachability of the points described by B.

Definition 9: (virtual reachability of a Data Bubble B w.r.t Data Bubble C)
  Let $B=(n, M, e)$ and C be Data Bubbles and *MinPts* a natural number. The *virtual reachability* of the N points described by B w.r.t. C is then defined as

$$virtual\text{-}reachability(B, C) = \begin{cases} nndist(MinPts, B) & \text{if } n \geq MinPts \\ reach\text{-}dist(B, C) & \text{otherwise} \end{cases}.$$

As Data Bubbles can contain strongly varying numbers of data points, we need to weigh each Data Bubble by the number its data points, if we want to maintain the proportional sizes of the clusters in the reachability plot. The output of the original OPTICS algorithm is generated by appending the reachability of the next chosen data point to the output file. When outputting a Data Bubble B, we first append the reachability of B to the output file (marking the jump to B) followed by n-times the virtual reachability of B (marking the n points that B describes).

4   Experimental Evaluation

4.1   Efficiency

In figure 5 the speed-up factors of OPTICS on Data Bubbles over "pure" OPTICS for different dimensional data are shown for a constant compression to approx. 1000 Data Bubbles. The size of the input data sets ranges from 50,000 points to 400,000 points. All data sets contain 30% noise and 10 clusters of random locations and sizes. The speed-up ranges from a factor of about 50 for small

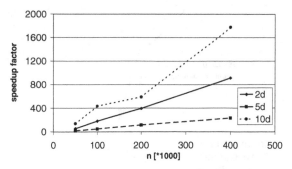

Fig. 5. Speed-up Factors

data sets to more than 1,700 for large, high-dimensional data. This is the wall-clock speed-up, i.e. it includes both CPU and I/O-time.

For "pure" OPTICS, an index structure [3] supported the range-queries (the index build time is not included; if it were, the speed-up would be even higher!), while OPTICS on the Data Bubbles used the sequential scan. We expect the scale-up to be approx. constant.

4.2   Understanding the Trade-off between Quality and Efficiency

We have seen so far that we gain large performance improvements by trading in relatively small amounts of quality. To get a better understanding of the trade-off involved see figure 6, which shows different compressed representations. For each of these, the runtime is depicted in the graph on the left side, followed by the number of Data Bubbles (DB) and the percentage of Data Bubbles relative to the number of original data points. Furthermore, each compression is visualized as a 2-dimensional plot, where the Data Bubbles B=(*n, M, e*) are shown as circles with center *M* and radius *e*. Finally, on the far right side we see the reachability plots generated from these Data Bubbles. Note that sometimes the order of the clusters in the reachability switches; this is a positive effect of the non-deterministic choices in the walk that OPTICS allows for.

The compression rate increases from top to bottom, i.e. the Data Bubbles get larger and the reachability plots become coarser. The top-most experiment, in which the data is compressed by 86.4%, shows no visible deterioration of quality at all. Compression rates of 94.6% and 98.5% show almost none. With a compression rate of 99.5%, the general clustering structure is still visible, with the one exception that the star-formed cluster consisting of 7 small, gaussian subclusters, starts to loose its internal structure. And even for the most extreme case on the bottom, in which the 100,000 data points are compressed into just 228 Data Bubbles, the general clustering structure is still preserved rather well.

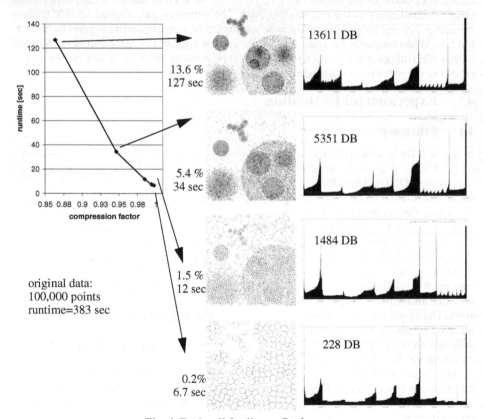

Fig. 6. Trade-off Quality vs. Performance

5   Conclusions

In this paper, we adapt the hierarchical clustering algorithm OPTICS to work on compressed data. First, we generate Data Bubbles consisting of essential information about compressed data points. Second, we adapt OPTICS to take Data Bubbles as input and make maximal use of the contained information. Third, we modify the graphical representation of the OPTICS result accordingly. Combining OPTICS with the compression technique BIRCH yields performance speed-up-factors between 50 and more than 1,700 for large, high-dimensional data sets, while at the same time maintaining the high quality of the clustering results.

Currently, we are evaluating the quality of our approach on high-dimensional data. In the future, we plan to compare data compression with sampling and to utilize other data compression methods to further improve the quality of the results, e.g. by using higher-order moments. We are also investigating how to extend the method to categorical data sets.

Acknowledgments

We thank the authors of BIRCH for making their code available to us and Ekkehard Krämer for his help with the implementation.

References

1. Ankerst M., Breunig M. M., Kriegel H.-P., Sander J.: *"OPTICS: Ordering Points To Identify the Clustering Structure"*, Proc. ACM SIGMOD'99 Int. Conf. on Management of Data, Philadelphia, 1999, pp. 49-60.
2. Bradley P. S., Fayyad U., Reina C.: *"Scaling Clustering Algorithms to Large Databases"*, Proc. 4th Int. Conf. on Knowledge Discovery and Data Mining, New York, NY, AAAI Press, 1998, pp. 9-15.
3. Berchthold S., Keim D., Kriegel H.-P.: *"The X-Tree: An Index Structure for High-Dimensional Data"*, 22nd Conf. on Very Large Data Bases, Bombay,India,1996, pp. 28-39.
4. DuMouchel W., Volinsky C., Johnson T., Cortez C., Pregibon D.: *"Sqashing Flat Files Flatter"*, Proc. 5th Int. Conf. on Knowledge Discovery and Data Mining, San Diego, CA, AAAI Press, 1999, pp. 6-15.
5. Ester M., Kriegel H.-P., Sander J., Xu X.: *"A Density-Based Algorithm for Discovering Clusters in Large Spatial Databases with Noise"*, Proc. 2nd Int. Conf. on Knowledge Discovery and Data Mining, Portland, OR, AAAI Press, 1996, pp. 226-231.
6. Ester M., Kriegel H.-P., Xu X.: *"Knowledge Discovery in Large Spatial Databases: Focusing Techniques for Efficient Class Identification"*, Proc. 4th Int. Symp. on Large Spatial Databases (SSD'95), Portland, ME, 1995, LNCS 591, Springer, 1995, pp. 67-82.
7. Fayyad U., Piatetsky-Shapiro G., Smyth P.: *"Knowledge Discovery and Data Mining: Towards a Unifying Framework"*. Proc. 2nd Int. Conf. on Knowledge Discovery and Data Mining, Portland, OR, 1996, pp. 82-88.
8. Kaufman L., Rousseeuw P. J.: *"Finding Groups in Data: An Introduction to Cluster Analysis"*, John Wiley & Sons, 1990.
9. MacQueen, J.: *"Some Methods for Classification and Analysis of Multivariate Observations"*, Proc. 5th Berkeley Symp. on Math. Statist. and Prob., Vol. 1, 1965, pp. 281-297.
10. Sibson R.: *"SLINK: An Optimally Efficient Algorithm for the Single-Link Cluster Method"*, The Computer Journal, Vol. 16, No. 1, 1973, pp. 30-34.
11. Zhang T., Ramakrishnan R., Linvy M.: *"BIRCH: An Efficient Data Clustering Method for Very Large Databases"*, Proc. ACM SIGMOD Int. Conf. on Management of Data, ACM Press, New York, 1996, pp. 103-114.

Accurate Recasting of Parameter Estimation Algorithms Using Sufficient Statistics for Efficient Parallel Speed-Up

Demonstrated for Center-Based Data Clustering Algorithms

Bin Zhang, Meichun Hsu, and George Forman

Data Mining Solutions Department, Hewlett-Packard Laboratories
{bzhang,mhsu,gforman}@hpl.hp.com

Abstract. Fueled by advances in computer technology and online business, data collection is rapidly accelerating, as well as the importance of its analysis—data mining. Increasing database sizes strain the scalability of many data mining algorithms. Data clustering is one of the fundamental techniques in data mining solutions. The many clustering algorithms developed face new challenges with growing data sets. Algorithms with quadratic or higher computational complexity, such as agglomerative algorithms, drop out quickly. More efficient algorithms, such as K-Means EM with linear cost per iteration, still need work to scale up to large data sets. This paper shows that many parameter estimation algorithms, including K-Means, K-Harmonic Means and EM, can be recast without approximation in terms of Sufficient Statistics, yielding an superior speed-up efficiency. Estimates using today's workstations and local area network technology suggest efficient speed-up to several hundred computers, leading to effective scale-up for clustering hundreds of gigabytes of data. Implementation of parallel clustering has been done in a parallel programming language, ZPL. Experimental results show above 90% utilization.

Keywords: Parallel Algorithms, Data Mining, Data Clustering, K-Means, K-Harmonic Means, Expectation-Maximization, Speed-up, Scale-up, Sufficient Statistics.

1 Introduction

Clustering is one of the principle workhorse techniques in the field of data mining. Its purpose is to organize a dataset into a set of groups, or clusters, which contain "similar" data items, as measured by some distance function. Example applications of clustering include document categorization, scientific data analysis, and customer/market segmentation.

Many algorithms for clustering data have been developed in recent decades, however, they all face a major challenge in scaling up to very large database sizes, an accelerating trend brought on by advances in computer technology, the Internet, and electronic commerce. Clustering algorithms with quadratic (or higher order) computational complexity, such as agglomerative algorithms, scale poorly. Even for more efficient algorithms, such as K-Means and Expectation-Maximization (EM), which have linear cost per iteration, research is needed to improve their scalability for

D.A. Żighed, J. Komorowski, and J. Zytkow (Eds.): PKDD 2000, LNAI 1910, pp. 243–254, 2000.
© Springer-Verlag Berlin Heidelberg 2000

very large and ever increasing data sets. In this paper, we develop a class of iterative parallel parameter estimation algorithms—covering K-Means, K-Harmonic Means, and EM algorithms—that are both efficient and accurate.

There have been several recent publications in scaling up K-Means and EM by approximation. For example, in BIRCH [19] and Microsoft-TR by Bradley et. al. [4], a single scan of the data and subsequent aggregation of each local cluster into Sufficient Statistics (SS) enables a data set to be pared down to fit the available memory. Such algorithms provide an approximation to the original algorithm and have been successfully applied to very large datasets. However, the higher the aggregation ratio, the less accurate the result is in general. As reported in the BIRCH paper that the quality of the clustering depends on the original scanning order.

There is also recent work on non-approximated, parallel versions of K-Means [10]. The Kantabutra and Couch algorithm requires re-broadcasting the data set to all computers each iteration, which leads to heavy communication overhead. Their analytical and empirical analysis estimates 50% utilization of the processors. The technology trend is for processors to improve faster than networks are improving, making the network a greater bottleneck in the future. Finally, the number of slave computing units in their algorithm is limited to the number of clusters to be found.

In this paper, we produce a parallel K-Means algorithm that limits inter-processor communication to SS only, reducing the network bottleneck. The dataset is partitioned across the memory of the processors and does not need to be communicated between iterations. The number of computing units is not limited in any way by the number of clusters sought. The method applies not only to K-Means, but also to other iterative clustering methods, such as K-Harmonic Means and EM.

There is no approximation introduced by the method; the results are exactly as if the original algorithm were run on a single computer. Further, it is also complementary to aggregation techniques (as in BIRCH), and by combining the two approaches in a hybrid, even larger data sets can be handled or better accuracy can be achieved with less aggregation.

There has also been work done parallel clustering using special hardware [1] [2]. The parallel algorithms we present require no special hardware or special networking, even though special networking may help further scale up the number of computers that can be deployed with high utilization. We emphasize running the parallel clustering algorithm on existing networking structures (LAN) because of practical and economic considerations. The total computing resources in a collection of "small" computers, modern "PCs" or desktop workstations, easily exceeds the total computing resources available in a supercomputer. Small computers, which are already everywhere, are much more accessible and can even be considered a free resource if they can be utilized when they would otherwise be idle.

An example application of this idea is in on-line commercial product recommendations where clustering is involved in calculating the recommendations (see collaborative filtering and recommender systems [14][15]). Large numbers of PCs that are used during business hours for processing orders can be used at night for updating the clustering of customers and products based on the daily revised sales information (customer buying patterns).

The parallel algorithm presented in this paper is not limited to data clustering algorithms. We use the class of center-based clustering algorithms, which includes K-Means [11], K-Harmonic Means [18] and EM [5][13], to illustrate the parallel algorithm for iterative parameter estimations.   By finding K centers (local high densities of data), $M = \{m_i \mid i=1,...,K\}$, the data set, $S = \{x_i \mid i=1,...,N\}$, can be partitioned either into discrete clusters using the Voronoi partition (each data item belongs to the center that it is closest to, as in K-Means) or into fuzzy clusters given by the local density functions (as in K-Harmonic Means or EM).

The problem is formulated as an optimization of a performance function, *Perf(S, M)*, depending on both the data items and the center locations.   A popular performance function for measuring the goodness of a clustering is the sum of the mean-square error (MSE) of each data point to its center.   The popular K-Means algorithm attempts to find a local optimum for this performance function.   The K-Harmonic Means (KHM) algorithm optimizes the harmonic average of these distances. KHM is very insensitive to the initialization of the centers, a major problem for K-Means. The Expectation-Maximization (EM) algorithm, in addition to the centers, optimizes a covariance matrix and a set of mixing probabilities.

The next section presents an abstraction of a class of center-based algorithms and the following section presents the parallel algorithm.   Section 4 applies this to three examples: K-Means, K-Harmonic Means and EM.   Section 5 gives an analysis of the utilization of the computing units, and we conclude in section 6.

2   A Class of Center-Based Algorithms

Let R^{dim} be the Euclidean space of dimension dim; $S \subseteq R^{dim}$ be a finite subset of data of size $N = |S|$; and $M = \{m_k \mid k=1,...,K\}$, the set of parameters to be optimized. (The parameter set M consists of K centroids for K-Means, K centers for K-Harmonic Means, and K centers with co-variance matrices and mixing probabilities for EM.) We write the performance function and the parameter optimization step for this class of algorithms in terms of SS.   The performance function is decomposed as follows:

Performance Function: $\quad F(S,M) = f_0(\sum_{x \in S} f_1(x,M), \sum_{x \in S} f_2(x,M),...\sum_{x \in S} f_R(x,M))$.   (1)

What is essential here is that f_0 depends only on the SS, represented by the sums, whereas the remaining f_i functions can be computed independently for each data point.   The detailed form of f_i, $i=1,...,R$, depend on the particular performance function considered.   It will become clear when examples of K-Means, K-Harmonic Means and EM are given in later sections.

We write the center-based algorithm, which minimizes the value of the performance function over M, as an iterative algorithm in the form of Q SS ($I()$ stands for the iterative algorithm, and $\sum_{x \in S} g_j$, $j=1,...,Q$, stands for SS.):

$$M^{(u+1)} = I(\sum_{x \in S} g_1(x,M^{(u)}), \sum_{x \in S} g_2(x,M^{(u)}),......, \sum_{x \in S} g_Q(x,M^{(u)})).$$ (2)

$M^{(u)}$ is the parameter vector after the uth iteration.   We are only interested in algorithms that converge: $M^{(u)} \rightarrow M$.   The values of the parameters for the 0th iteration, $M^{(0)}$, are by initialization.   One method often used is to randomly initialize the parameters (centers, covariance matrices and/or mixing probabilities).   There are many different ways of initializing the parameters for particular types of center-based algorithms in the literature [3]. The computation carried out here will be identical to the traditional, sequential equivalent. The set of quantities

$$Suff = \{ \sum_{x \in S} f_r(x, M) \mid r = 1, \ldots, R\} \bigcup \{ \sum_{x \in S} g_q(x, M) \mid q = 1, \ldots, Q\}. \qquad (3)$$

is called the global SS of the problem (1)+(2). As long as these quantities are available, the performance function and the new parameter values can be calculated and the algorithm can be carried out to the next iteration. We will show in Section 4 that K-Means, K-Harmonic Means and Expectation-Maximization clustering algorithms all belong to this class defined in (1)+(2).

3 Parallelism of Center-Based Algorithms

This decomposition of center-based algorithms (and many other iterative parameter estimation algorithms) leads to a natural parallel structure with minimal need for communication. Let L be the number of computing units (with a CPU and local memory – PCs, workstations or multi-processor computers, shared memory is not required). To utilize all L units for the calculation of (1)+(2), the data set is partitioned into L subsets, $S = D_1 \cup D_2 \cup \ldots \cup D_L$, and the lth subset, D_l, resides on the lth unit. It is important not to confuse this partition with the clustering in the data:
a) This partition is arbitrary and has nothing to do with the clustering in the data.
b) This partition is static. Data points in D_l, after being loaded into the memory of the lth computing unit, need not be moved from one computer to another. (Except for the purpose of load balancing among units, whose only effect is on the execution time of the processors and does not affect the algorithm. See Section 5.)

We need not assume homogeneous processing units. The sizes of the partitions, $|D_l|$, besides being constrained by the storage of the individual units, are ideally set to be proportional to the speed of the computing units. Partitioned thus, it will take about the same amount of time for each unit to finish its computation on each iteration, improving the utilization of all the units. A scaled-down test could be carried out on each computing unit in advance to measure the actual speed (do not include the time of loading data because it is only loaded once at the beginning) or a load balancing over all units could be done at the end of first iteration.

The calculation is carried out on all L units in parallel. Each subset, D_l, contributes to the refinement of the parameters in M in exactly the same way as the algorithm would have been run on a single computer. Each unit independently computes its partial sum of the SS over its data partition. The SS of the lth partition are

$$Suff_l = \{ \sum_{x \in D_l} f_r(x, M) \mid r = 1, \ldots, R\} \bigcup \{ \sum_{x \in D_l} g_q(x, M) \mid q = 1, \ldots, Q\}. \qquad (4)$$

One of the computing units is chosen to be the Integrator. It is responsible for sums up the SS from all partitions (4), to get the global SS (3); calculates the new parameter values, M, from the global SS; evaluates the performance function on the new parameter values, (2); checks the stopping conditions; and informs all units to stop or sends the new parameters to all computing units to start the next iteration. The duties of the Integrator may be assigned as a part time job to one of the regular units. There may also be more than one computer used as Integrators, possibly organized in a hierarchy if the degree of parallelism is sufficiently high. Special networking support is also an option. If broadcast is supported efficiently, it may be effective to have every node be an Integrator, eliminating one direction of communication. Studies of this sort can be found in the parallel computing literature [6][9].

The Parallel Clustering Algorithm:

Step 0:   Initialization: Partition the data set and load the lth partition to the memory of the lth computing unit. Use any preferred algorithm to initialize the parameters, $\{m_k\}$, on the Integrator.

Step 1:   Broadcast the integrated parameter values to all computing units.

Step 2:   Compute at each unit independently the SS of the local data based on (4).

Step 3:   Send SS from all units to the Integrator

Step 4:   Sum up the SS from each unit to get the global SS, calculate the new parameter values based on the global SS, and evaluate the performance function. If the Stopping condition[1] is not met, goto Step 1 for the next iteration, else inform all computing units to stop.

4   Examples

This section demonstrates the decomposition for three examples: K-Means, K-Harmonic Means, and EM.

4.1   K-Means Clustering Algorithm

K-Means is one of the most popular clustering algorithms ([11][17] and other references therein).   The algorithm partitions the data set into K clusters, $S = (S_1,\ldots,S_K)$, by putting each data point into the cluster represented by the center nearest to the data point.   K-Means algorithm finds a local optimal set of centers that minimizes the total within cluster variance, which is K-Means performance function:

$$Perf_{KM}(X,M) = \sum_{k=1}^{K} \sum_{x \in S_i} \| x - m_k \|^2,$$ (5)

where the kth center, m_k, is the centroid of the kth partition.   The double summation in (5) can instead be expressed as a single summation over all data points, adding only the distance to the nearest center expressed by the MIN function below:

$$Perf_{KM}(X,M) = \sum_{i=1}^{N} MIN\{\| x_i - m_l \|^2 | l = 1,\ldots,K\}.$$ (6)

The K-Means algorithm starts with an initial set of centers and then iterates through the following steps: For each data item, find the closest m_k and assign the data item to the kth cluster.   The current m_k's are not updated until the next phase (Step 2).   A proof of optimality can be found in [8].

---

[1] Typically test for sufficient convergence or the number of iterations.

1. Recalculate all the centers. The kth center becomes the centroid of the kth cluster. A proof can be found in [8] that this phase gives the optimal center locations for the given partition of data.
2. Iterate through 1 & 2 until the clusters no longer change significantly.

After each phase, the performance value never increases and the algorithm converges to a local optimum. More precisely, the algorithm will reach a stable partition in a finite number of steps for finite datasets. The cost per iteration is $O(K \cdot dim \cdot N)$.

   a)   *The functions for calculating both global and local SS for K-Means are the 0th, 1st and 2nd moments of the data on the unit belonging to each cluster as shown in (7). Both the K-Means performance function and the new center locations can be calculated from these three moments.*

$$\begin{cases} g_1(x,M) = f_1(x,M) = (\delta_1(x), \delta_2(x), \ldots, \delta_K(x)), \\ \quad g_2(x,M) = f_2(x,M) = f_1(x,M) \cdot x, \\ \quad g_3(x,M) = f_3(x,M) = f_1(x,M) \cdot x^2. \end{cases} \tag{7}$$

$\delta_k(x) = 1$ if x is closest to m_k, otherwise $\delta_k(x) = 0$ (resolve ties arbitrarily). The summation of these functions over a data set (see (3) and (4)) residing on the lth unit gives the count, $n_{k,l}$, first moment, $\Sigma_{k,l}$, and the second moment, $s_{k,l}$, of the clusters (this is called the CF vector in the BIRCH paper [19]). The vector $\{n_{k,l}, \Sigma_{k,l}, s_{k,l}, |$ $k=1,\ldots,K\}$, has dimensionality $2 \cdot K + K \cdot dim$, which is the size of the SS that have to be communicated between the Integrator and each computing unit.

The set of SS presented here is more than sufficient for the simple version of K-Means algorithm. The aggregated quantity, $\Sigma_k s_{k,l}$, could be sent instead of the individual $s_{k,l}$. But there are other variations of K-Means performance functions that require individual $s_{k,l}$ for evaluating the performance functions. Besides, the quantities that dominate the communication cost are $\Sigma_{k,l}$.

The lth computing unit collects the SS, $\{ n_{k,l}, \Sigma_{k,l}, s_{k,l} | k=1,\ldots,K \}$, on the data in its own memory, and then sends them to the Integrator. The Integrator simply adds up the SS from each unit to get the global SS,

$$n_k = \sum_{l=1}^{L} n_{k,l}, \quad \Sigma_k = \sum_{l=1}^{L} \Sigma_{k,l}, \quad s_k = \sum_{l=1}^{L} s_{k,l}.$$

The leading cost of integration is $O(K \cdot dim \cdot L)$, where L is the number of computing units. The new location of the kth center is given by $m_k = \Sigma_k/n_k$ from the global SS (this is the $I()$ function in (2)), which is the only information all the computing units need to start the next iteration. The performance function is calculated by (proof by direct verification), $Perf_{KM} = \sum_{l=1}^{L} s_k.$

The parallel version of the K-Means algorithm gives exactly the same result as the original centralized K-Means because both the parallel version and the sequential version are based on the same global SS except on how the global SS are collected.

4.2 K-Harmonic Means Clustering Algorithm

K-Harmonic Means is a clustering algorithm designed by the author [18]. A major strength is its insensitivity to the initialization of the centers (cf. K-Means, where the dependence on the initialization has been the major problem and many authors have tried to address the problem by finding good initializations).

The iteration step of the K-Harmonic Means algorithm adjusts the new center locations to be a weighted average of all x, where the weights are given by

$$1 \bigg/ d_{x,k}{}^s (\sum_{x \in S} \frac{1}{d_{x,l}^2})^2 . \tag{8}$$

(K-Means is similar, except its weights are the nearest-center membership functions, making its centers centroids of the cluster.) Overall then, the recursion equation is given by

$$m_k = \sum_{x \in S} \frac{1}{d_{x,k}{}^s (\sum_{l=1}^{K} \frac{1}{d_{x,l}})^2} \; x \bigg/ \sum_{x \in S} \frac{1}{d_{x,k}{}^s (\sum_{l=1}^{K} \frac{1}{d_{x,l}})^2} . \tag{9}$$

where $d_{x,k} = \|x - m_k\|$ and s is a constant ≤ 4. The decomposed functions for calculating SS (see (3) and (4)) are then

$$\begin{cases} g_1(x,M) = 1 \bigg/ \sum_{k=1}^{K} \frac{1}{d_{x,k}^2} \\ g_2(x,M) = g_1^2(x,M) \cdot (\frac{1}{d_{x,1}^s}, \frac{1}{d_{x,2}^s}, \ldots, \frac{1}{d_{x,K}^s}) . \\ g_3(x,M) = g_1^2(x,M) \cdot (\frac{1}{d_{x,1}^s}, \frac{1}{d_{x,2}^s}, \ldots, \frac{1}{d_{x,K}^s}) x \end{cases} \tag{10}$$

Each computing unit collects the SS,

$$Suff_l = \{ \sum_{x \in D_l} g_2(x,M), \sum_{x \in D_l} g_3(x,M) \} \tag{11}$$

on the data in its own memory, and then sends it to the Integrator. The size of the SS vector is $K+K \cdot dim$ (g_3 is a matrix). The Integrator simply adds up the SS from each unit to get the global SS. The new centers are given by the *component-wise* quotient:

$$(m_1, m_2, \ldots, m_K) = \sum_{x \in S} g_3(x,M) \bigg/ \sum_{x \in S} g_2(x,M) = \sum_{l=1}^{L} \sum_{x \in D_l} g_3(x,M) \bigg/ \sum_{l=1}^{L} \sum_{x \in D_l} g_2(x,M) , \tag{12}$$

which is the only information the units need to start the next iteration. This calculation costs $O(K \cdot dim \cdot L)$. The updated global centers are sent to each unit for the next iteration. If broadcasting is an option, this is the total cost in time. If the Integrator finds the centers stop moving significantly, the clustering is considered to have converged to an optimum and the units are stopped.

4.3 Expectation-Maximization (EM) Clustering Algorithm

We limit ourselves to the *EM* algorithm with linear mixing of K bell-shape (Gaussian) functions. Unlike K-Means and K-Harmonic Means in which only the centers are to be estimated, the EM algorithm estimates the centers, the co-variance matrices, Σ_k, and the mixing probabilities, $p(m_k)$. The performance function of the EM algorithm is

$$Perf_{EM}(X,M,\Sigma,p) = -\log\left\{\prod_{x\in S}\left[\sum_{k=1}^{K}p_k\cdot\frac{1}{\sqrt{(2\pi)^D\det(\Sigma_k)}}\cdot EXP(-(x-m_k)\Sigma_k^{-1}(x-m_k)^T)\right]\right\}. \quad (13)$$

where the vector $p =(p_1, p_2, ..., p_K)$ is the mixing probability. *EM* algorithm is a recursive algorithm with the following two steps:

E-Step: Estimating "the percentage of x belonging to the kth cluster",

$$p(m_k\,|\,x) = p(x\,|\,m_k)\cdot p(m_k)\Big/\sum_{x\in S}p(x\,|\,m_k)\cdot p(m_k), \quad (14)$$

where $p(x|m)$ is the prior probability with Gaussian distribution, and $p(m_k)$ is the mixing probability.

$$p(x\,|\,m_k) = \frac{1}{\sqrt{(2\pi)^D\det(\Sigma_k)}}\cdot EXP(-(x-m_k)\Sigma_k^{-1}(x-m_k)^T) \quad (15)$$

M-Step: With the fuzzy membership function from the E-Step, find the new center locations new co-variance matrices and new mixing probabilities that maximize the performance function.

$$m_k = \frac{\sum_{x\in S}p(m_k\,|\,x)\cdot x}{\sum_{x\in S}p(m_k\,|\,x)},\ \Sigma_k = \frac{\sum_{x\in S}p(m_k\,|\,x)\cdot(x-m_k)^T(x-m_k)}{\sum_{x\in S}p(m_k\,|\,x)},\ p(m_k) = \frac{1}{|S|}\sum_{x\in S}p(m_k\,|\,x). \quad (16\text{-}18)$$

For details see [5][13]. The functions for calculating the SS are:

$$f_1(x,M,\Sigma,p) = -\log\left[\sum_{l=1}^{K}p(x\,|\,m_k)p(m_k)\right]$$

$$g_1(x,M,\Sigma,p) = (p(m_1\,|\,x), p(m_2\,|\,x),......, p(m_K\,|\,x))$$

$$g_2(x,M,\Sigma,p) = (p(m_1\,|\,x)x, p(m_2\,|\,x)x,......, p(m_K\,|\,x)x)$$

$$g_3(x,M,\Sigma,p) = (p(m_1\,|\,x)x^Tx, p(m_2\,|\,x)x^Tx,......, p(m_K\,|\,x)x^Tx)$$

The vector length (in number of scalars) of the SS is $1+K+K\cdot dim +K\cdot dim^2$. The global SS is also the sum of the SS from all units. The performance function value is given by the first global sufficient statistic. The global centers are from the component-wise "ratio" of the third and the second global SS (see (16)), the co-variance matrices from (17) and the mixing probability from (18). All these quantities, $\{m_k, \Sigma_k, p(m_k)\,|\,k = 1,......,K\}$, have to be propagated to all the units at the beginning of each iteration. The vector length is $K+K\cdot dim +K\cdot dim^2$.

5   Time/Space Complexities

The storage required at each node is $O(dim\cdot(N/L + K))$—the data set S is partitioned across the processors and the list of centers M is replicated at each processor. In contrast, the algorithm in [10] partitions the centers across the processors. Utilization is determined by the percentage of time each unit works on its own data to adjust the centers and collect SS, vs. the time waiting between sending out the local SS to the Integrator and receiving the new global parameters (see Fig. 1).

When the data size on each unit is far greater than the number of computing units ($N>>L$, which is true in general), the amount of work each unit has to do, $O(K{\cdot}dim{\cdot}N/L)$, is far greater than the amount of work the Integrator has to do, $O(K{\cdot}dim{\cdot}L)$. Therefore, the integration time is marginal compared to the time for collecting SS. The main source of waiting comes from network communication of the SS and global parameters. The ratio between the speed of network communication and the speed of the computing units is an important factor for determining the utilization (see (19) later). Since the size of the SS is very small compared with the original data size on each unit (see the following table), we will give an example to show that the speed ratio between the network transmission and CPU processing of current technology is sufficient to support high utilization. If the network is slower (relative to the unit's CPU speed), the number of units, L, has to be smaller to maintain high utilization. The faster the network, the higher degree of parallelism can be achieved for a given utilization target. Fig. 2 shows the breakdown of four periods in each iteration. Since the parameters sent from the Integrator to all units are the same, broadcasting, if supported by the network, may shorten the waiting time of the units. Let t_d be the unit of time to process one datum, t_n be the unit of time to transmit one datum, and t_c the communication latency for sending an empty message. The utilization U of the (slave) units can be estimated by (using K-Means as an example),

$$Utilization = \frac{C{\cdot}K{\cdot}\dim{\cdot}(N/L){\cdot}t_d}{C{\cdot}K{\cdot}\dim{\cdot}(N/L){\cdot}t_d + C^2K{\cdot}\dim{\cdot}L{\cdot}t_d + L{\cdot}(C''{\cdot}K{\cdot}\dim{\cdot}t_n + t_c)} \quad (19)$$

$$\approx \frac{1}{1+(C'/C){\cdot}(L^2/N)+(C''/C){\cdot}(L^2/N){\cdot}(t_n/t_d)}.$$

where C, C', and C'' are constants. The communication latency is shared by all $O(K{\cdot}dim)$ data items that are transmitted together. When $K{\cdot}dim$ is so large that the latency t_c becomes negligible compared with the transmission time $C''{\cdot}K{\cdot}\dim{\cdot}t_n$, the term $t_c / K{\cdot}\dim$ on the second line of (19) may be ignored. Both the computational cost at each unit and the communication costs grow linearly in dim and K. Therefore, the ratio of the two costs, which is close to the utilization of the units if the starting cost of communication is ignored, does not depend on dim and K.

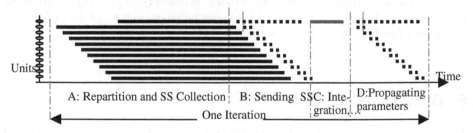

Units | A: Repartition and SS Collection | B: Sending SS | C: Integration | D: Propagating parameters

One Iteration

Fig 1. The timeline of all the units in one iteration. Network has no broadcasting feature. The top unit is the Integrator given a lighter load of data. The slope in period B and D are not same in general. Communication time from the Integrator to the units is drawn as sequential supposing of no support for broadcasting. Period A and Period D overlap.

The length of each period is determined by the following factors:

Table 1. The Analysis of Computational Costs (*N>>L*).

	The size of SS from each unit (in # floating points).	The size of data on each unit.	The cost of collecting the SS on each unit/iteration.	The cost of integration per iteration.
KM	2·K+K·dim	N/L	O(K·dim·N/L)	O(K·dim·L)
KHM	K+K·dim	N/L	O(K·dim·N/L)	O(K·dim·L)
EM	1+K+K·dim+K·dim2	N/L	O(K·dim2 ·N/L)	O(K·dim2 ·L)

6 Experimental Results – Speedup Curves

The parallel clustering algorithms presented in this paper have been implemented in the parallel programming language ZPL [16]. The plots in Fig. 2 show the speed-ups as the number of processors change from 1 to 8. Four different data sizes are used, N = *50k, 100k, 500k,* and *10M*. At *N=50,000* points, it takes only 0.24 seconds per iteration to cluster on 8 processors.   The dotted line represents 90% speed-up efficiency. The data used in all experiments are randomly generated with uniform distribution, which could be any other distribution and has no effect on speed-up for the clustering algorithms.   The number of clusters *K=100*.   The dimensionality *dim=2*.   We have shown analytically in Section 5 that the speed-ups are very insensitive to *K*, and *dim*.   The measurements are taken on 8 identical HP-UX 9000/735's each with 208 MB memory sharing a normal 10Mbps Ethernet. The HP 9000/735 has roughly the same floating point SPECmarks as the 233 MHz PentiumMMX, for comparison.

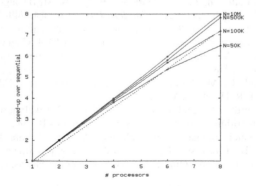

Fig 2. The speedup curves under various *N*, and *L*.

7 Dynamic Load Balancing

Data items can be moved from one computing unit to another during the period (B+C+D) (see Fig. 1) without affecting the correctness of the algorithm. Moving data around during this period to increase the utilization of the units in the future is called dynamic load balancing. Dynamic load balancing corrects the initial setup of the load, which may not give the best utilization of the computing units, especially if the data set consists of sparse vectors with variable numbers of non-zero elements in them. For dense vectors, the prediction of calculation time based on the number of data vectors is straightforward; a linear scale-up based on the test result from a small number of vectors will work well. But for sparse vectors, the amount of data to be loaded on to each machine becomes more difficult to estimate, due to the variation of the actual vector lengths even though the cost of partitioning and collecting SS on each subset of data is still deterministic. Only the first or maybe a second re-balancing would be needed. The necessity of a load re-balancing can be determined based on the utilization of the units calculated from the previous iteration.

8 Conclusion

We restructured the mathematics of a class of clustering algorithms to reveal a straightforward parallel implementation based on communicating only a small amount of data—SS—yielding highly efficient speed-up and scale-up for very large data sets. The ideas presented in this paper apply to iterative parameter estimation algorithms where the size of the SS is small relative to the data size. By the time this paper is accepted, we have conducted many experiments. One of them used 128 loosely connected processors. The experimental results are published in KDD 2000 workshop on distributed and parallel computing in Boston [7].

References

[1] Apostolakis, J., Coddington, P., and Marinari, E. "New SIMD algorithms for cluster labeling on parallel computers," *Intl J. of Modern Physics C*, 4(4):749-763, Aug. 1993.

[2] Baillie, C.F.,Coddington, P.D."Cluster identification algorithms for spin models," *Concurrency:Practice and Experience*,3(2):129-144, April 1991. Caltech Report C3P-855.

[3] Bradley, P., Fayyad, U. M., and Reina, C.A., "Scaling EM Clustering to Large Databases," MS Technical Report, 1998.

[4] Bradley, P., Fayyad, U. M., C.A., "Refining Initial Points for KM Clustering", MS Technical Report MSR-TR-98-36, May 1998.

[5] Dempster, A. P., Laird, N.M., and Rubin, D.B., "Maximum Likelyhood from Incomplete Data via the EM Algorithm", J. of the Royal Stat. Soc., Series B, 39(1):1-38, 1977.

[6] T. J. Fountain , "Parallel Computing : Principles and Practice," Cambridge Univ Pr 1994.

[7] Forman, G., Zhang, B.,"Linear speedup for a parallel non-approximate recasting of center-based clustering algorithms, including K-Means, K-Harmonic Means, and EM", To appear in DPKD 2000, Boston.

[8] Gersho & Gray, "Vector Quantization and Signal Compression", KAP, 1992

[9] Vipin Kumar, "Introduction to Parallel Computing," Addison-Wesley, 1994.

[10] Sanpawat Kantabutra and Alva L. Couch, "Parallel K-means Clustering Algorithm on NOWs," NECTEC Technical journal (www.nectec.or.th)

[11] MacQueen, J., "Some methods for classification and analysis of multivariate observations," Pp. 281-297, Proceedings of the 5th Berkeley symposium on mathematical statistics and probability, Vol. 1. University of California Press, Berkeley. xvii + 666 p, 1967.

[12] McKenzie, P. and Alder, M., "Initializing the EM Algorithm for Use in Gaussian Mixture Modeling", The Univ. of W. Australia, Center for Information Processing Systems, Manuscript.

[13] McLachlan, G., Krishnan, T., "The EM algorithm and extensions",John Wiley &Sons.

[14] A Commercial Recommender System http://www.netperceptions.com

[15] P. Resnick and H. Varian, "Recommender Systems," in CACM March 1997.

[16] Snyder, L., "A Programmer's Guide to ZPL (Scientific and Engineering Computation Series)", MIT Press; ISBN: 0262692171, 1999.

[17] Selim, S.Z. and Ismail, M.A., "K-Means-Type Algorithms: A Generalized Convergence Theorem and Characterization of Local Optimality", IEEE Trans. On PAMI-6, #1, 1984.

[18] Zhang, B., Hsu, M., Dayal, U., "K-Harmonic Means – A Data Clustering Algorithm", Hewllet-Packard Research Laboratory Technical Report HPL-1999-124.

[19] Zhang, T., Ramakrishnan, R., and Livny, M., "BIRCH: an efficient data clustering method for very large databases", *ACM SIGMOD Record*, Vol. 25, No. 2 (June 1996), Pages 103-114 in: SIGMOD '96.

Predictive Performance
of Weighted Relative Accuracy

Ljupčo Todorovski[1], Peter Flach[2], and Nada Lavrač[1]

[1] Department of Intelligent Systems, Jožef Stefan Institute
Jamova 39, 1000 Ljubljana, Slovenia
{Ljupco.Todorovski, Nada.Lavrac}@ijs.si
[2] Department of Computer Science, University of Bristol
The Merchant Venturers Building, Woodland Road, Bristol, UK
Peter.Flach@cs.bris.ac.uk

Abstract. Weighted relative accuracy was proposed in [4] as an alternative to classification accuracy typically used in inductive rule learners. Weighted relative accuracy takes into account the improvement of the accuracy relative to the default rule (i.e., the rule stating that the same class should be assigned to all examples), and also explicitly incorporates the generality of a rule (i.e., the number of examples covered). In order to measure the predictive performance of weighted relative accuracy, we implemented it in the rule induction algorithm CN2. Our main results are that weighted relative accuracy dramatically reduces the size of the rule sets induced with CN2 (on average by a factor 9 on the 23 datasets we used), at the expense of only a small average drop in classification accuracy.

1 Introduction

In a recent study of different rule evaluation measures in inductive machine learning [4], a new measure, named weighted relative accuracy (*WRAcc*), was proposed. This measure can be understood from different perspectives. On one hand, *WRAcc* takes into account the improvement of the accuracy relative to the default rule (i.e., the rule stating that the same class should be assigned to all examples), and also explicitly incorporates the generality of a rule (i.e., the number of examples covered). Secondly, it can be seen as a single measure trading off several accuracy-like measures such as precision and recall in information retrieval, or sensitivity and specificity as used in medical applications. Finally, we showed that weighted relative accuracy is equivalent to the novelty measure used in a descriptive induction framework.

In this paper we further investigate the first perspective, i.e., whether weighted relative accuracy is a viable alternative to standard accuracy in rule learning. We experimentally tested the hypothesis that *WRAcc* directs the learner to fewer and more general rules (because it explicitly trades off generality and accuracy) without sacrificing too much on accuracy (because it will concentrate on the classes with high default accuracy). To this end, we implemented weighted

D.A. Zighed, J. Komorowski, and J. Żytkow (Eds.): PKDD 2000, LNAI 1910, pp. 255–264, 2000.
© Springer-Verlag Berlin Heidelberg 2000

relative accuracy measure in the rule induction algorithm CN2 [2]. The original version of CN2 uses classification accuracy as a rule evaluation measure. In order to compare the predictive performance of accuracy and weighted relative accuracy, we performed experiments in twenty-three classification data sets. The results show a dramatic decrease in the number of induced rules, with only a relatively small drop in classification accuracy.

The paper is organised as follows. In Section 2 the weighted relative accuracy measure is defined. CN2 rule induction algorithm along with the changes made to incorporate weighted relative accuracy is presented in Section 3. The results of an experimental comparison of CN2 and CN2 with weighted relative accuracy on twenty-three data sets are presented and discussed in Section 4. The final Section 5, summarises the experimental results and proposes possible directions for further work.

2   Weighted Relative Accuracy

Although weighted relative accuracy can also be meaningfully applied in a descriptive induction framework, in this paper we restrict attention to classification-oriented predictive induction. Classification rules are of the form $H \leftarrow B$, where B is a condition and H is a class assignment to instances satisfying the condition. In propositional predictive rules, B is (typically) a conjunction of attribute-value pairs, and H is a class assignment. In first-order learning, frequently referred to as *inductive logic programming*, predictive rules are Prolog clauses, where H is a single positive literal and B is a conjunction of positive and/or negative literals. The difference with propositional predictive rules is that first-order rules contain variables that are shared between literals and between H and B. In this paper we restrict attention to propositional rules.

We use the following notation. $n(B)$ stands for the number of instances covered by a rule $H \leftarrow B$, $n(H)$ stands for the number of examples of class H, and $n(HB)$ stands for the number of correctly classified examples (true positives). We use $p(HB)$ etc. for the corresponding probabilities. We then have that rule accuracy can be expressed as $Acc(H \leftarrow B) = p(H|B) = \frac{p(HB)}{p(B)}$. Weighted relative accuracy is defined as follows.

Definition 1 (Weighted relative accuracy).

$$WRAcc(H \leftarrow B) = p(B)(p(H|B) - p(H)).$$

Weighted relative accuracy consists of two components: relative accuracy $p(H|B) - p(H)$ and generality $p(B)$. Relative accuracy is the accuracy gain relative to the fixed rule $H \leftarrow true$. The latter rule predicts all instances to satisfy H; a rule is only interesting if it improves upon this 'default' accuracy. Another way of viewing relative accuracy is that it measures the utility of connecting body B with a given head H. However, it is easy to obtain high relative accuracy with highly specific rules, i.e., rules with low generality $p(B)$. To this end, generality is used as a 'weight', so that weighted relative accuracy trades off generality and relative accuracy.

3   Rule Induction Algorithm CN2

CN2 [2,1] is an algorithm for inducing propositional classification rules. The list of induced classification rules can be ordered or unordered. Ordered rules are interpreted in a straight-forward manner: when classifying a new example, the rules are sequentially tried and the first rule that covers the example is used for prediction. In unordered case, all the rules are tried and predictions of those that cover the example are collected. A voting mechanism is used to obtain the final prediction. CN2 consists of two main procedures: search procedure that performs beam search in order to find a single rule and control procedure that repeatedly execute the search.

The search procedure performs beam search using classification accuracy of the rule as a heuristic function. The accuracy of the propositional classification rule if $cond$ then c is equal to the conditional probability of class c, given that the condition $cond$ is satisfied:

$$Acc(\text{if } cond \text{ then } c) = p(c|cond).$$

We replaced the accuracy measure with the weighted relative accuracy (see Definition 1):

$$WRAcc(\text{if } cond \text{ then } c) = p(cond)(p(c|cond) - p(c)), \tag{1}$$

where $p(cond)$ is the generality of the rule and $p(c)$ is the prior probability of class c. The second term in Equation 1 is the accuracy gain relative to the "default" rule if $true$ then c. Weighted relative accuracy measure trades off this relative accuracy gain with generality of the rule $p(cond)$.

Different probability estimates, like Laplace [1] or m-estimate [3], can be used in CN2 for calculating the conditional probability $p(c|cond)$. Each of them can be also used for calculating the same term in the formula for weighted relative accuracy.

Additionally, CN2 can apply significance test to the induced rule. The rule is considered to be significant, if it locates regularity unlikely to have occurred by chance. To test significance, CN2 uses likelihood ratio statistic [2] that measures the difference between the class probability distribution in the set of examples covered by the rule and the class probability distribution in the set of all training examples. Empirical evaluation in [1] shows that applying a significance test reduces the number of induced rules and also slightly reduces the predictive accuracy.

Two different control procedures are used in CN2: one for inducing ordered list of rules and the other for unordered case. When inducing an ordered list of rules, the search procedure looks for the best rule, according to the heuristic measure, in the current set of training examples. The rule predicts the most frequent class in the set of examples, covered by the induced rule. Before starting another search procedure, all examples covered by the induced rule are removed. The control procedure invokes new search, until all the examples are covered (removed).

In unordered case, the control procedure is iterated, inducing rules for each class in turn. For each induced rule, only covered examples belonging to that class are removed, instead of removing all covered examples, like in ordered case. The negative training examples (i.e., examples that belong to other classes) remain and positives are removed in order to prevent CN2 finding the same rule again. The distribution of covered training examples among classes is attached to each rule. When classifying a new example, all rules are tried and those covering the example are collected. If a clash occurs (several rules with different class predictions cover the example), the class distributions attached to the rules are summed to find the most probable class.

Our modification of CN2 affects the heuristic function only. Modified CN2 uses weighted relative accuracy as search heuristic. Everything else remain the same: in particular, the quality of learned rule sets is still evaluated by means of accuracy on a test set. Our experiments, described in the next section, demonstrate that this simple change results in a dramatic decrease of the number of induced rules, at the expense of (on average) a small drop in accuracy.

4   Experiments

We performed experiments on a collection of twenty-one domains from the UCI Repository of Machine Learning Databases and Domain Theories [6] and two data sets originating from mutagenesis domain [5]. These domains have been widely used in other comparative studies. The domains properties (number of examples, number of discrete and continuous attributes and class distribution) are given in Table 1.

Performance of the rule inducing algorithms were measured using 10-fold stratified cross validation. Cross validation is repeated 10 times using a different random reordering of the examples in the data set. The same set of re-orderings were used for all experiments. The average and standard deviation (over the ten cross validations) of the classification error on unseen examples are reported.

Two series of experiments with both versions of CN2 algorithm (unordered and ordered one) were performed. In both cases, we compared CN2 using accuracy measure (CN2-acc) and CN2 with significance test applied to the induced rules (CN2-acc-99) with CN2 using weighted relative accuracy as a search heuristic (CN2-wracc). The experimental results are presented and discussed in the following two subsections.

4.1   Unordered Rules

The classification errors of unordered rules induced with CN2-acc, CN2-acc-99 and CN2-wracc are presented in Table 2. The statistical significance of the differences is tested using paired t-tests with significance level of 99%: +/− to the left of a figure in CN2-acc or CN2-acc-99 columns means that CN2-wracc is significantly better/worse than CN2-acc or CN2-acc-99, respectively.

Table 1. Properties of the experimental data sets

Data set	#exs	#da	#ca	Class distribution
australian	690	8	6	56:44
balance	625	0	4	08:46:46
breast-w	699	9	0	66:34
bridges-td	102	4	3	15:85
car	1728	6	0	22:04:70:04
chess	3196	36	0	48:52
diabetes	768	0	8	65:35
echo	131	1	5	67:33
german	1000	13	7	70:30
glass	214	0	9	33:36:08:06:04:14
heart	270	6	7	56:44
hepatitis	155	13	6	21:79
hypothyroid	3163	18	7	05:95
image	2310	0	19	14:14:14:14:14:14:14
ionosphere	351	0	34	36:64
iris	150	0	4	33:33:33
mutagen	188	57	2	34:66
mutagen-f	188	57	0	34:66
soya	683	35	0	02:13:06:03:03:13:06:03:02:02:03:03:13:01:03:13:03:03:03
tic-tac-toe	958	9	0	35:65
vote	435	16	0	61:39
waveform	5000	0	21	33:33:34
wine	178	0	13	33:40:27

The overall predictive accuracy of CN2-wracc is smaller than the one of CN2-acc, the difference being almost 5%. CN2-wracc is significantly better in four domains, but also significantly worse in eleven out of twenty-three domains. Experiments with CN2-acc-99 confirms the results of previous empirical study in [1]: applying a significance test reduce the predictive accuracy of CN2. However, the overall predictive accuracy of CN2-wracc is still 3% smaller than the one of CN2-acc-99.

One of the bad scenarios for weighted relative accuracy is to have skewed class probability distribution. Namely, when using the *WRAcc* measure, the bigger the set of covered examples, the better is the rule. It is easier to achieve big coverage for the rules predicting the majority class, than for the rules predicting the minority classes. There is yet another handicap of *WRAcc* for rules predicting minority classes: these rules can have poor accuracy, which is still good relative to the prior probability of the minority class. Therefore, the rules (induced using CN2-wracc) predicting the minority classes tend to have very "impure" class probability distribution of the covered examples (i.e., they tend to be inaccurate). The problem with minority classes is also obvious from the empirical results. There are six domains with at least one class having prior probability below

Table 2. Classification errors (in %) of unordered rules induced with CN2-acc, CN2-acc-99 and CN2-wracc

Data set	CN2-acc	CN2-acc-99	CN2-wracc
australian	+ 17.41 ±0.50	+ 17.91 ±0.85	14.78 ±0.33
balance	- 19.95 ±0.91	- 22.83 ±1.49	28.43 ±1.50
breast-w	- 6.43 ±0.42	- 7.17 ±0.73	10.11 ±0.64
bridges-td	- 14.68 ±1.11	- 16.02 ±1.46	20.03 ±2.48
car	- 5.01 ±0.57	- 7.82 ±0.56	30.05 ±0.22
chess	- 1.58 ±0.25	- 3.25 ±0.20	5.91 ±0.01
diabetes	25.89 ±0.52	26.91 ±0.90	26.22 ±0.49
echo	33.46 ±1.86	- 31.91 ±1.94	35.24 ±2.78
german	- 26.59 ±0.59	28.34 ±0.93	28.69 ±0.91
glass	- 32.91 ±1.73	35.19 ±2.40	35.22 ±1.39
heart	22.46 ±1.40	+ 25.64 ±1.94	23.09 ±1.20
hepatitis	19.59 ±1.16	20.23 ±1.80	18.14 ±1.19
hypothyroid	1.42 ±0.25	1.53 ±0.23	1.39 ±0.10
image	+ 14.05 ±0.46	+ 14.62 ±0.52	9.32 ±0.51
ionosphere	9.29 ±0.92	10.26 ±0.88	9.90 ±0.20
iris	+ 7.48 ±0.82	+ 9.54 ±1.86	5.68 ±0.72
mutagen	- 17.13 ±1.69	18.85 ±0.86	19.03 ±0.66
mutagen-f	- 15.22 ±1.70	+ 25.35 ±1.55	20.86 ±1.78
soya	- 8.13 ±0.36	- 10.71 ±0.50	46.29 ±0.67
tic-tac-toe	- 1.49 ±0.17	- 1.98 ±0.19	26.40 ±0.65
vote	4.74 ±0.40	+ 5.79 ±0.56	4.38 ±0.01
waveform	+ 30.81 ±0.31	+ 30.39 ±0.48	27.17 ±0.35
wine	6.95 ±1.51	7.00 ±1.93	6.97 ±0.61
Average	**14.90 ±0.31**	**16.49 ±1.08**	**19.71 ±0.84**

20%: balance, bridges-td, car, glass, hypothyroid and soya. CN2-wracc performs significantly worse than CN2-acc in five and significantly worse than CN2-acc-99 in four of them. CN2-wracc performs slightly (and insignificantly) better only in hypothyroid domain. If these six domains are discarded and the averages are calculated again, the overall predictive accuracy of CN2-wracc is only 1.5% smaller than the overall accuracy of CN2-acc and 0.5% better than the overall accuracy of CN2-acc-99.

On the other hand, there are seven domains with almost uniform class distributions: australian, chess, heart, image, iris, waveform and wine. CN2-wracc significantly outperforms CN2-acc in four and CN2-acc-99 in five domains out of seven. The differences in the heart and wine domains are small and insignificant, but CN2-wracc performs significantly worse in the chess domain.

The reason for being worse in the chess domain can be in the search strategy used in CN2. When searching for the condition of the rule, CN2 starts with empty condition, specialising it with conjunctively adding the literals of the form $(A_d = v)$ for a discrete attribute A_d and $(v_1 \leq A_c \leq v_2)$ for a continuous attribute A_c. Several best conditions (the number of them provided by the beam size) are kept in the beam. Intuitively, the number of examples covered drops

Table 3. Number of unordered rules induced with CN2-acc, CN2-acc-99 and CN2-wracc

Data set	CN2-acc	CN2-acc-99	CN2-wracc
australian	35.51 ±0.72	13.28 ±0.63	2.00 ±0.00
balance	106.27 ±1.08	40.85 ±2.21	6.85 ±0.15
breast-w	37.89 ±0.74	14.30 ±0.23	5.66 ±0.12
bridges-td	15.35 ±0.45	2.73 ±0.22	3.05 ±0.08
car	120.13 ±0.69	95.05 ±0.77	6.69 ±0.13
chess	29.47 ±0.32	16.61 ±0.21	5.00 ±0.00
diabetes	46.18 ±1.43	16.39 ±0.75	2.95 ±0.11
echo	19.48 ±0.31	5.51 ±0.28	3.60 ±0.18
german	83.14 ±1.22	21.52 ±0.75	3.96 ±0.07
glass	22.40 ±0.48	15.19 ±0.40	7.45 ±0.16
heart	21.90 ±0.46	8.48 ±0.52	3.01 ±0.10
hepatitis	17.48 ±0.46	4.44 ±0.64	2.73 ±0.37
hypothyroid	23.94 ±1.20	10.65 ±0.72	2.00 ±0.00
image	39.22 ±0.72	33.21 ±0.70	7.03 ±0.05
ionosphere	17.32 ±0.19	8.71 ±0.23	3.00 ±0.00
iris	6.75 ±0.22	3.82 ±0.06	3.00 ±0.00
mutagen	16.78 ±0.35	5.28 ±0.16	3.70 ±0.12
mutagen-f	25.12 ±0.30	7.53 ±0.19	3.78 ±0.12
soya	37.70 ±0.25	35.27 ±0.23	18.20 ±0.00
tic-tac-toe	28.33 ±0.77	22.08 ±0.43	7.23 ±0.12
vote	18.41 ±0.32	7.70 ±0.15	2.00 ±0.00
waveform	208.66 ±3.35	84.39 ±2.82	3.09 ±0.09
wine	8.28 ±0.18	5.89 ±0.09	3.81 ±0.09
Average	**42.86** ±0.70	**20.82** ±0.58	**4.77** ±0.09

quicker when adding a condition involving discrete attribute when compared to a condition involving continuous attribute. Therefore, conditions involving continuous attributes can be expected to be more "WRAcc friendly".

This intuition can be empirically confirmed in seven domains consisting of discrete attributes only: breast-w, car, chess, mutagen-f, soya, tic-tac-toe and vote. In six (five) of them CN2-acc (CN2-acc-99) performs significantly better than CN2-wracc. Consider also the mutagen and mutagen-f data sets: the only difference between them is that mutagen data set has two extra continuous attributes (logP and LUMO). CN2-acc achieves better accuracy for both data sets, the difference being bigger for the one without continuous attributes (mutagen-f).

Consider now the number of rules induced by CN2-acc, CN2-acc-99 and CN2-wracc in Table 3. The unordered rule lists induced with CN2-wracc are, on average, nine times smaller than the rule lists induced with CN2-acc. Removing insignificant rules has already been used for reducing the rule lists size in CN2. However, we can see from Table 3 that CN2-wracc outperforms CN2-acc-99 in terms of rule list size: rule lists induced with CN2-wracc are, on average, four times smaller. Note also that the reduction of number of rules does not cause

Table 4. Classification errors (in %) of ordered rules induced with CN2-acc, CN2-acc-99 and CN2-wracc

Data set	CN2-acc	CN2-acc-99	CN2-wracc
australian	+ 18.93 ±0.93	+ 18.29 ±1.20	15.12 ±0.42
balance	- 17.48 ±1.30	- 22.11 ±1.47	27.68 ±0.80
breast-w	- 5.46 ±0.51	- 5.41 ±0.49	6.75 ±0.93
bridges-td	- 19.14 ±3.32	- 16.30 ±1.38	23.01 ±2.33
car	- 3.59 ±0.52	- 4.24 ±0.54	18.73 ±0.30
chess	- 0.72 ±0.11	- 1.14 ±0.17	5.72 ±0.02
diabetes	+ 29.40 ±1.17	27.70 ±1.06	27.35 ±1.25
echo	38.58 ±3.40	39.57 ±3.72	35.58 ±4.13
german	29.55 ±1.48	- 29.45 ±0.79	30.39 ±0.24
glass	- 30.94 ±2.45	- 34.52 ±1.72	37.89 ±1.36
heart	24.05 ±1.60	+ 25.79 ±1.97	23.61 ±1.50
hepatitis	22.07 ±1.71	21.45 ±2.57	23.23 ±1.23
hypothyroid	- 1.50 ±0.18	- 1.65 ±0.20	2.94 ±0.15
image	- 3.67 ±0.25	- 4.18 ±0.27	9.78 ±0.16
ionosphere	12.71 ±1.45	+ 13.13 ±1.43	12.03 ±0.65
iris	6.41 ±0.64	- 5.14 ±1.09	6.21 ±0.71
mutagen	19.84 ±2.24	19.97 ±1.47	20.37 ±1.94
mutagen-f	- 18.05 ±1.92	+ 28.30 ±2.66	22.29 ±2.37
soya	- 9.38 ±0.58	- 11.52 ±0.55	49.28 ±0.21
tic-tac-toe	- 2.70 ±0.71	- 4.31 ±1.30	30.35 ±0.83
vote	5.38 ±0.79	+ 6.48 ±0.69	4.74 ±0.25
waveform	- 22.02 ±0.58	- 22.35 ±0.63	24.06 ±0.35
wine	+ 6.37 ±1.01	5.59 ±0.79	4.90 ±1.33
Average	**15.13 ±1.25**	**16.03 ±1.22**	**20.09 ±1.02**

significant increase of rule length in terms of number of conditions in rule. The average length of rules induced with CN2-acc is 2.99 and with CN2-wracc is 3.51.

4.2   Ordered Rules

The classification errors of ordered rules induced with CN2-acc, CN2-acc-99 and CN2-wracc are presented in Table 4. The experimental results show that weighted relative accuracy is better suited towards the unordered version of the CN2 algorithm, than to the ordered version. The reason is that in each iteration of the ordered algorithm, the number of uncovered examples drops, so the coverage term in Equation 1 gets smaller. The small value of the coverage term diminishes the difference between "good" and "bad" rules. This is not the case with the covering algorithm used in unordered version of CN2, where negative covered examples are not removed.

The experimental results regarding accuracy of ordered rules more or less follow the pattern already seen in the results for unordered ones. The two notable exceptions are image and waveform domains, where the significant accuracy improvement (for unordered rules) obtained with *WRAcc* changed to significant accuracy deterioration.

Table 5. Number of ordered rules induced with CN2-acc, CN2-acc-99 and CN2-wracc

Data set	CN2-acc	CN2-acc-99	CN2-wracc
australian	34.99 ±0.84	17.43 ±1.18	2.72 ±0.11
balance	60.43 ±0.98	27.95 ±0.92	5.63 ±0.23
breast-w	30.23 ±0.46	12.87 ±0.42	11.32 ±0.19
bridges-td	9.81 ±0.35	2.55 ±0.33	2.87 ±0.07
car	51.61 ±0.89	37.58 ±0.57	9.18 ±0.38
chess	28.16 ±0.51	22.27 ±0.34	2.00 ±0.00
diabetes	49.48 ±0.56	19.65 ±1.01	4.22 ±0.24
echo	15.80 ±0.39	4.99 ±0.54	3.66 ±0.25
german	74.22 ±0.67	21.64 ±1.51	4.65 ±0.31
glass	16.23 ±0.27	10.36 ±0.28	4.00 ±0.00
heart	16.25 ±0.40	10.43 ±0.58	3.29 ±0.07
hepatitis	11.54 ±0.39	5.05 ±0.37	3.04 ±0.13
hypothyroid	20.28 ±0.61	12.92 ±0.75	2.00 ±0.00
image	29.05 ±0.37	23.89 ±0.52	8.00 ±0.00
ionosphere	10.03 ±0.32	8.74 ±0.25	2.90 ±0.09
iris	5.80 ±0.13	2.20 ±0.09	3.01 ±0.09
mutagen	11.78 ±0.39	6.02 ±0.34	3.03 ±0.13
mutagen-f	21.94 ±0.36	6.84 ±0.39	3.74 ±0.11
soya	28.19 ±0.40	21.58 ±0.16	6.00 ±0.00
tic-tac-toe	36.73 ±2.80	32.86 ±2.09	5.23 ±0.20
vote	17.89 ±0.19	7.43 ±0.23	5.17 ±0.11
waveform	175.92 ±1.32	168.41 ±3.02	6.16 ±0.13
wine	3.97 ±0.11	3.20 ±0.11	2.96 ±0.10
Average	**33.06** ±0.60	**21.17** ±0.70	**4.56** ±0.13

The comparison of the size of the ordered rule list induced with CN2-acc, CN2-acc-99 and CN2-wracc is presented in Table 5. Ordered rule lists induced with CN2-wracc are, on average, seven times smaller than the ones induced with CN2-acc and four times smaller than the ones induced with CN-acc-99.

5 Summary

In order to measure the predictive performance of a a new rule evaluation measure, weighted relative accuracy (*WRAcc*), we implemented it in the rule induction algorithm CN2. The empirical evaluation of CN2-wracc shows that *WRAcc* clearly outperforms ordinary accuracy measure used in CN2, when the size of the induced rule sets is considered. The rule sets induced with CN2-wracc are, on average, about nine times smaller in unordered case and seven times smaller in ordered case. The factor of nine (or seven) is actually huge, and can be regarded as a step towards higher comprehensibility of the rule lists induced with CN2.

Significance tests [1] of the rules have already been used in CN2 to reduce the size of the induced rule lists. However, the experiments show that the rule lists induced using *WRAcc* are on average four times smaller than the ones induced using significance test in CN2.

The price to be paid for the reduction of the rule list size is an overall drop of the predictive classification accuracy of the induced rules by 5%. When compared to the predictive accuracy of rule lists induced using significance test, the drop is about 3.5%. The drop of the accuracy is mostly due to the domains with skewed class distributions, having classes with small number of examples. In seven experimental domains with almost uniform class distributions CN2-wracc significantly outperforms CN2 also in terms of predictive accuracy.

The predictive performance of weighted relative accuracy can be further evaluated by implementing it in other rule induction algorithms. In order to confirm the improvement of the comprehensibility of the rules induced with CN2-wracc, an expert evaluation of the rules should be obtained by a human domain experts. This would be also a step towards evaluating the descriptive performance of the *WRAcc* measure.

Acknowledgements

The work reported was supported in part by the Slovenian Ministry of Science and Technology, the Joint project with Central/Eastern Europe funded by the British Royal Society and by the EU-funded project Data Mining and Decision Support for Business Competitiveness: A European Virtual Enterprise (IST-1999-11495).

References

1. Clark, P. and Boswell, R. (1991) Rule induction with CN2: Some recent improvements. In *Proceedings of the Fifth European Working Session on Learning*: 151–163. Springer-Werlag.
2. Clark, P. and Niblett, T. (1989) The CN2 induction algorithm. *Machine Learning Journal*, 3(4): 261–283.
3. Džeroski, S., Cestnik, B. and Petrovski, I. (1993) Using the m-estimate in rule induction. Journal of Computing and Information Technology, 1(1):37 – 46.
4. Lavrač, N., Flach, P. and Zupan, B. (1999) Rule Evaluation Measures: A Unifying View. In *Proceedings of the Ninth International Workshop on Inductive Logic Programming, volume 1634 of Lecture Notes in Artificial Intelligence*: 74–185. Springer-Verlag.
5. Muggleton, S., Srinivasan, A., King R. and Sternberg, M. (1998) Biochemical knowledge discovery using Inductive Logic Programming. In Motoda, H. (editor) *Proceedings of the first Conference on Discovery Science*. Springer-Verlag.
6. Murphy, P. M. and Aha, D. W. (1994) *UCI repository of machine learning databases* [http://www.ics.uci.edu/~mlearn/MLRepository.html]. Irvine, CA: University of California, Department of Information and Computer Science.

Quality Scheme Assessment in the Clustering Process

M. Halkidi, M. Vazirgiannis, and Y. Batistakis

Dept. of Informatics, Athens University of Economics & Business
Patision 76, 10434, Athens, Greece (Hellas)
{mhalk, mvazirg, yannis}@aueb.gr

Abstract. Clustering is mostly an unsupervised procedure and most of the clustering algorithms depend on assumptions and initial guesses in order to define the subgroups presented in a data set. As a consequence, in most applications the final clusters require some sort of evaluation. The evaluation procedure has to tackle difficult problems, which can be qualitatively expressed as: i. quality of clusters, ii. the degree with which a clustering scheme fits a specific data set, iii. the optimal number of clusters in a partitioning. In this paper we present a scheme for finding the optimal partitioning of a data set during the clustering process regardless of the clustering algorithm used. More specifically, we present an approach for evaluation of clustering schemes (partitions) so as to find the best number of clusters, which occurs in a specific data set. A clustering algorithm produces different partitions for different values of the input parameters. The proposed approach selects the best clustering scheme (i.e., the scheme with the most compact and well-separated clusters), according to a quality index we define. We verified our approach using two popular clustering algorithms on synthetic and real data sets in order to evaluate its reliability. Moreover, we study the influence of different clustering parameters to the proposed quality index.

1 Introduction

Data Mining is a step in the KDD process that is mainly concerned with methodologies for knowledge extraction from large data repositories. There are many data mining algorithms that accomplish a limited set of tasks and produce a particular enumeration of patterns over data sets. These main tasks, according to well established data mining methods [5], are: the definition/extraction of clusters, the classification of database values into the categories defined, and the extraction of association rules or other knowledge artifacts.

More specifically regarding clustering, the outcome (from now on referred to as *clustering scheme*) is a set of clusters characterized by their center and their limits (i.e., the bounds within which the values of the cluster objects range). In the vast majority of KDD approaches, the clustering algorithms partition a data set in a number of groups based on some parameters, such as the desired number of clusters, the minimum number of objects in a cluster, the diameter of a cluster etc. For instance, assume the data set presented in Fig. 1a. It is obvious that we can discover three clusters in the given data set. However, if we consider a clustering algorithm (e.g. K-Means) with certain parameter values (in the case of K-means the number of clusters) so as to partition our data set in four clusters, the result of clustering process would be the clustering scheme presented in Fig. 1b. In our example the clustering algorithm (K-Means) found the best four clusters in which our data set could be

D.A. Żighed, J. Komorowski, and J. Zytkow (Eds.): PKDD 2000, LNAI 1910, pp. 265–276, 2000.
© Springer-Verlag Berlin Heidelberg 2000

partitioned. We define the term "optimal" clustering scheme as the outcome of running a clustering algorithm (i.e., a partitioning) that best fits (i.e., resembles) the real partitions of the data set. It is obvious from Fig. 1b that the clustering scheme presented in it does not fit well the data set. The optimal clustering for our data set will be a scheme with three clusters. As a consequence, if the clustering algorithm parameters are assigned an improper value, the clustering method may result in a partitioning scheme that is not optimal for the specific data set leading to wrong decisions. The problems of deciding the number of clusters better fitting a data set as well as the evaluation of the clustering results has been subject of several research efforts [14]. Although various validity indices have been introduced, in practice most of the clustering methods do not use any of them. Furthermore, formal methods in database and data mining applications for finding the best partitioning of a data set, are very few [12], [17].

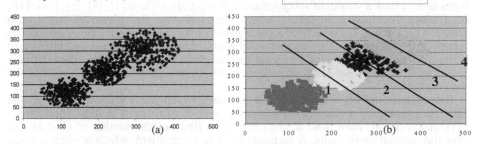

Fig. 1. a. A data set that consists of three clusters, b. The results from the application of K-means when we ask four clusters

In this paper, we present an approach for clustering scheme evaluation based on relative criteria. It aims at evaluating the schemes defined by a specific clustering algorithm, assuming different input parameter values. These schemes are evaluated using a clustering scheme quality index, which we define. Our goal it is not to propose a new clustering algorithm or to evaluate a variety of clustering algorithms, but to produce the clustering scheme with the most compact and well-separated clusters.

For the remainder of this paper, our approach can be applied and lead to good results, under the assumption that the data set can be partitioned to groups. This will prevent a misleading interpretation of the structure of a data set, when we apply a clustering algorithm. We also consider only the case of compact clusters, that is the clusters of our data set are non-ring shaped (Fig. 1). This is the most common structure of clusters presented in business databases. Moreover there are few clustering algorithms that discover successfully arbitrary shaped clusters.

The paper is organized as follows. Section two surveys related work in this area. Section three describes the proposed quality index for clustering scheme evaluation along with a qualitative and experimental evaluation of our approach. In section four, we present the algorithm for finding the best clustering scheme and we present the results of an evaluation study for our approach. We conclude in section five by summarizing and providing further research directions.

2 Related Work

Clustering is one of the main tasks in the data mining process for discovering groups and identifying interesting distributions and patterns in the underlying data. The fundamental clustering problem is to partition a given data set into groups (clusters), such that the data points in a cluster are more similar to each other than points in different clusters [8]. In the clustering process, there are no predefined classes and no examples that would show what kind of desirable relations should be valid among the data [1]. This is what distinguishes clustering from classification. Classification is a procedure of assigning a data item to a predefined set of categories [6].

There is a multitude of clustering methods available in literature, which can be broadly classified into the following types [13]: i) *Partitional clustering*, ii) *Hierarchical clustering*, iii) *Density-Based clustering, iv) Grid-based*. For each of the types there exists a wealth of subtypes and different algorithms [1], [4], [9], [14], [18] for finding the clusters. In general terms, the clustering algorithms are based on a criterion for judging the quality of a given partitioning. More specifically, they take as input some parameters (e.g. number of clusters, density of clusters) and attempt to define the best partitioning of a data set for the given parameters. Thus, they define a partitioning of a data set based on certain assumptions and *not* the "best" one that fits the data set.

Since clustering algorithms discover clusters, which are not known a-priori, the final partitioning of a data set requires some sort of evaluation in most applications [11]. Another important issue in clustering is to find out the number of clusters that give the optimal partitioning. A particularly difficult problem, which is often ignored in clustering algorithms is "how many clusters are there in the data set?". Though this is an important question that causes much discussion, the formal methods for finding the optimal number of clusters in a data set are few [12].

Previously described requirements for the evaluation of clustering results is well known in the research community and a number of efforts have been made especially in the area of pattern recognition [14]. However, the issue of cluster validity is rather under-addressed in the area of databases and data mining applications, even though recognized as important. In general terms, there are three approaches to investigate cluster validity [14]. The first is based on *external criteria*. This implies that we evaluate the results of a clustering algorithm based on a pre-specified structure, which is imposed on a data set and reflects our intuition about the clustering structure of the data set. The second approach is based on *internal criteria*. We may evaluate the results of a clustering algorithm in terms of quantities that involve the vectors of the data set themselves (e.g proximity matrix). The third approach of clustering validity is based on *relative criteria*. Here the basic idea is the evaluation of a clustering structure by comparing it with other clustering schemes, resulting by the same algorithm but with different parameter values. The two first approaches are based on statistical tests and their major drawback is their high computational cost. Moreover, the indices related to these approaches aim at measuring the degree to which a data set confirms an a-priori specified scheme.

Our work puts emphasis to the third approach of clustering evaluation, based on relative criteria. In this approach, there are two criteria proposed for clustering evaluation and selection of an optimal clustering scheme [1]: i) *Compactness*, the members of each cluster should be as close to each other as possible, ii) *Separation*, the clusters themselves to be widely spaced.

A number of cluster validity indices are described in literature using the above criteria. A cluster validity index for crisp clustering proposed in [3], attempts to identify "compact and separated clusters". Other validity indices for crisp clustering have been proposed in [2] and [10]. The implementation of most of these measures is very computationally expensive, especially when the number of clusters and number of objects in the data set grows very large [17]. For fuzzy clustering, Bezdek proposed the partition coefficient (1974) and the classification entropy (1984). The limitations of these measures are: i) their monotonous dependency on the number of clusters and ii) the lack of direct connection to the geometry of the data [2]. Other fuzzy validity measures are proposed in [7], [11], [17]. We should mention that the evaluation of proposed measures and the analysis of their reliability are limited.

Another approach for finding the best number of cluster of a data set proposed in [12]. It introduces a practical clustering algorithm based on Monte Carlo cross-validation. This approach differs significantly from the one we propose. While we evaluate clustering schemes based on widely recognized quality criteria of clustering, the evaluation approach proposed in [12] is based on density functions considered for the data set. Thus, it uses concepts related to probabilistic models in order to estimate the number of clusters, better fitting a data set, while we use concepts directly related to the data. Our approach is based on the inter-cluster and intra-clusters distances.

In the following section we introduce the details of our approach concerning the evaluation of the clustering schemes defined in data mining process.

3 Our Approach of Clustering Scheme Evaluation

The problem can be stated as follows: Given a data set of n objects containing non-categorical data, we aim at the definition of a *clustering scheme,* representing the best partitioning of the specific data set based on a well defined quality index. In the following sections we elaborate on our approach.

3.1 Quality of Clustering Schemes

The objective of the clustering methods is to provide optimal partitions of a data set. In general, they should search for clusters whose members are close to each other (or have a high degree of similarity) and well separated. Another problem we face in clustering is to decide the optimal number of clusters that fits better a data set. The majority of clustering algorithms produce a partitioning based on the input parameters (e.g. number of clusters, clusters density etc) that finally lead to a finite number of clusters. Thus, the application of an algorithm assuming different input parameters values results in different partitions of a particular data set, which are not easily comparable. A solution to this problem is to run the algorithm repetitively with different input parameter values and compare the results against a well-defined validity index. The best partitioning of our data set will maximize (or minimize) the index. In this paper we aim at the evaluation of clustering schemes produced by applying repeatedly the same algorithm with different input parameters, and at the selection of the best clustering scheme. This process is based on an index that is better described by the term *clustering scheme quality index.*

A reliable quality index must take into account both the compactness and the separation of clusters in a partitioning [11]. Thus, the reliability of an index is based on the two fundamental criteria of clustering quality and more specifically to the degree an index combines both criteria. Many validation indices are proposed in the literature. However, the evaluation of the proposed indices and the analysis of their reliability are rather under-addressed in the literature [12] and especially in the area of databases and data mining.

In the sequel, we describe a clustering scheme quality index that can be used in crisp clustering and we analyse its reliability.

3.2 Quality Index for Clustering Schemes

In this section, we define a quality index for evaluating clustering schemes based on concepts and the relative validity indices proposed in the literature.

Definitions*.
   Variance of data set. The variance of a data set X, is called •(•). The value of the p_{th} dimension is defined as follows,

$$\sigma_x^p = \frac{1}{n} \sum_{k=1}^{n} \left(x_k^p - \overline{x}^p \right)^2 \tag{1}$$

where $\overline{x}^p$ is the p_{th} dimension of $\overline{X} = \frac{1}{n} \sum_{k=1}^{n} x_k, \forall x_k \in X$ \hfill (2)

Variance of cluster i. The variance of cluster i is called $\sigma(v_i)$ and its p_{th} dimension is given by

$$\sigma_{v_i}^p = \sum_{k=1}^{n_i} \left(x_k^p - v_i^{\,p} \right)^2 \Big/ n_i \tag{3}$$

Quality index definition. The quality index of our approach is based on the validity index defined for FCM in [11]. We use the same concepts for validation, but transform them into equivalent ones for crisp clustering. Following, we give the fundamental definition for this index.

Average scattering for clusters. The average scattering for clusters is defined as

$$Scat(c) = \frac{1}{c} \sum_{i=1}^{c} \left\| \sigma(v_i) \right\| \Big/ \left\| \sigma(X) \right\| \tag{4}$$

Total separation between clusters. The definition of total scattering (separation) between clusters is given by equation (5)

$$Dis(c) = \frac{D_{max}}{D_{min}} \sum_{k=1}^{c} \left(\sum_{z=1}^{c} \left\| v_k - v_z \right\| \right)^{-1} \tag{5}$$

where D_{max} = max($\|v_i - v_j\|$) $\forall i, j \in \{1, 2,3,...,c\}$ is the maximum distance between cluster centers. The D_{min} = min($\|v_i - v_j\|$) $\forall i, j \in \{1, 2,3,...,c\}$ is the minimum distance between cluster centers.
Now, we can define a quality index based on equations (4) and (5), as follows

$$SD(c) = a\,Scat(c) + Dis(c) \tag{6}$$

where α is a weighting factor equal to $Dis(c_{max})$ where c_{max} is the maximum number of input clusters. The first term (i.e., $Scat(c)$ defined by (4)) indicates the average compactness of clusters (i.e., intra-cluster distance). A small value for this term indicates a compact cluster and as the scattering within clusters increases (i.e., they become less compact) the value of $Scat(c)$ also increases. The second term $Dis(c)$ indicates the total separation between the c clusters (i.e., an indication of inter-cluster distance). Contrary to the first term the second one, $Dis(c)$, is influenced by the geometry of the clusters centers and increase with the number of clusters. It is obvious

---

* n=number of clusters, n_i = number of objects in cluster i, c = number of clusters, v_1 = Center of cluster i, x_k^p= p_{th} dimension of x_k object, $||x_k|| = (x_k^T x_k)^{1/2}$

for previous discussion that the two terms of *SD* are of the different range, thus a weighting factor is needed in order to incorporate both terms in a balanced way. The number of clusters, c, that minimizes the above index can be considered as an optimal value for the number of clusters present in the data set.

3.3 Evaluation Study of Clustering Scheme Quality Index

In this section we present the evaluation of the quality index both qualitatively and experimentally.

Qualitative evaluation. Considering the SD index definition (eq. (6)), we can observe the following:

1. It uses cluster separation as an indication of the average scattering between clusters, minimizing thus the possibility to select a clustering scheme with significant differences in cluster distances. Also, we should mention that the total separation between clusters, *Dis(c)*, is influenced by the distribution of the cluster centers in space. As a consequence the quality index *SD* takes into account the geometry of the centers, which also influences the values of *SD*.

2. The maximum number of clusters influences the index *SD*. It is clear, from the index definition, that its values are equally affected by *Scat* (eq. (4)) and *Dis* (eq. (5)) due to the weighting factor definition, $\alpha = Dis(c_{max})$. Thus, different values of c_{max} (i.e., weighting factor) affect proportionally *SD* values. The influence of the weighting factor is an issue for further study as mentioned in [11]. However, to the best of our knowledge in recent literature there is no well-established study to analyze the effect of the weighting factor. In the sequel, we present the results of a study that we performed in order to evaluate the influence of the factor *a* to the *SD* values.

Experimental evaluation. In this study we have used real data sets to verify the results of the preceding theoretical study. More specifically, we use the Iris data set [19]. This is a biometric data set of measurements referring to flowers. The schema of this data set is R={sepal_length, sepal_width, petal_length, petal_width}. Also, we implemented and used for our study, the partitioning clustering algorithm K-Means [1] in order to define the clustering schemes. However, we should mention that any

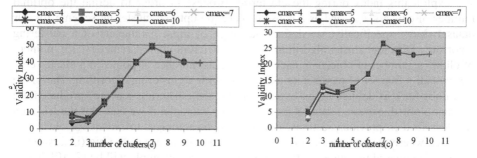

Fig. 2a: Quality index *SD* as a function of the number of clusters in :a) a two-dimensional data set (petal_length,petal_width), b) a four-dimensional data set (sepal_length, petal_length, petal_width, sepal_width).

other clustering algorithm could be used. Our focus to the K-Means algorithm for our study is justified by the following: i) It is widely used in a variety of applications, ii) Our purpose is not to evaluate a variety of clustering algorithms but to illustrate a procedure for finding the optimal partitioning of an array of clustering schemes.

The results of our experiments can be summarized as follows:
- *The index SD evaluates clustering schemes considering scattering within and between clusters.* Fig. 2a, b show the values of *SD* as a function of the number of clusters, considering two and four-dimensional data sets respectively. From these figures it is clear that the number of clusters does not influence the SD index. So it does not decrease monotonously as the number of clusters increases. Indeed the *SD* index behaves in a homogeneous and expected way almost independently of the c_{max}, c_{min} values for a particular data set. The study of *SD* in different data sets and with a variety of attributes shows that the values of *SD* are influenced by the geometry of the centers of clusters. The different shape of graphs in Fig. 2a, b is due to the difference in the geometry of the clusters' centers.
- *The index SD is slightly influenced by the maximum value of clusters' number.* As Figures 2a, b depict, the distributions of the index values for different values of c_{max} is similar. It is also noteworthy that there is no significant influence of c_{max} value on the value of SD, corresponding to the optimal clustering scheme (minimum value of SD). Hence SD results in the optimal clustering scheme almost irrespectively of the maximum number of clusters. We can only observe a slight decrease to the minimum value of the SD index when the value c_{max} becomes very small. For instance (Fig.2a) the best number of clusters is three when the c_{max} ranges from 10 to 5 while it is two when $c_{max} = 4$.

Our experiments verify that aforementioned results are valid in data sets with higher dimensionality.

4 Finding the "Best" Clustering Scheme

4.1 An Algorithm for Clustering Scheme Evaluation

In previous sections we have defined a quality index for clustering scheme evaluation. We exploit this index during the clustering process in order to define the optimal number of clusters, c_{opt}, and as a consequence, the optimal clustering scheme for our dataset. More specifically, we define the range of input parameters of a clustering algorithm. Different cluster algorithms require different input parameters (e.g. number of clusters, number of objects in each cluster, least number of points in a cluster etc). Let us call *p* this set of parameters associated with a clustering algorithm. The range of values for *p* is defined by an expert, so that the clustering schemes produced are compatible with expected data set partitions. For example, if we consider the Iris data set described in previous sections, it would be meaningless to search for clustering schemes where the number of clusters is too small (i.e., one) or too big (i.e., twenty). A suitable range of the number of clusters for the given example will be [2, 10]. Then, a clustering algorithm is performed for each value *p* and the results of clustering are evaluated using the *SD* quality index (Sect. 3.2). The number of clusters at which *SD* reaches its minimum value and also a significant local change (i.e., decrease) in its value occurs, can be considered as the best one. The basic steps of the algorithm that defines the optimal clustering scheme using *SD* as quality index is described as follows:

```
1.  Define  the  value  ranges  of  the  clustering  algorithm  input
    parameters. Let pmax and pmin are the parameters, resulting in
    schemes with the max and min number of clusters respectively.
2.  Initialize p ← pmax ;
3.  Run the clustering algorithm using p. The output is a clustering
    scheme of c clusters for the data set.
    if (p=pmax)
        a ←Disc(cmax);
        indexValue←SD(c);
        dist_between_indexValues=0;
        Copt←c }
    else  if (SD(c)<indexValue)
    if (dist_between_indexValues/abs(SD(c) -SD(c-1)))<3
        Copt←c;
        indexValue ← SD(c);
        dist_between_indexValues= SD(c) -SD(c-1);
5.  p ← p-1,
    if (p=pmax-1)
        stop
    else    goto 3.
```

The previously described procedure aims at choosing the optimal clustering scheme for a specific data set, which can be partitioned into compact clusters. It is based on: i) the iterative application of a clustering algorithm assuming different parameter values as input, and ii) the calculation of the SD quality index (see Sect. 3.2) for each clustering scheme that discovered by an algorithm. Let $[c_{min}, c_{max}]$ the range of the number of clusters, c. The clustering algorithm is repeated for successive values of c starting from c_{max} and decreasing to c_{min}. The choice of the optimal cluster number is the one for which: i) SD reaches a local minimum (there number of clusters c), ii) the slope of the linear segment (c_{i-1}, c_i) is greater than the inclination of any other linear segment (c_k, c_{k+1}), $k \in [c_{min}, c_{max}]$. We have considered for our algorithm implementation that the best quality index value corresponds to the point at which the index reaches its minimum value and at the same time the change in its value is not less than 1/3 of the previous significant change in the index value. For instance, as Fig. 4b depicts the value of SD is smaller for two than for three clusters. However, the change of the index value between three and two clusters is not significant comparing it with the previous significant change between four and three clusters (it is less than 1/3 of the index value change occurring between four and three). Thus, we conclude that the best number of clusters for Iris data set is three.

4.2 Time Complexity

It is obvious from above description that the time complexity for the definition of the optimal clustering scheme depends on the complexity of clustering algorithm we consider $(O(f(n,c,d)))$ and on the time complexity for the quality index calculation. The complexity for computing the quality index *SD*, is: $O(ndc + c^2 d)$ where d is the number of attributes (data set dimension), c is the number of clusters, n is the number of database tuples. According to our approach, the clustering algorithm and the computation of the quality index are performed for each value of the parameter p in $[p_{min}, p_{max}]$. Thus, the overall complexity for finding the optimal number of clusters is $O((p_{max}-p_{min}+1)(f(c,n,d) + ncd + c^2 d))$.

For example, in the case we use K-Means (time complexity is $O(tcnd)$) as a clustering algorithm, the overall complexity for finding the optimal number of clusters is $O((c_{max}-c_{min}+1)(tncd+c^2d))$. The graphs in Fig. 3a, b show the results of an experimental study referring to the execution time of our approach using K-Means as the clustering algorithm.

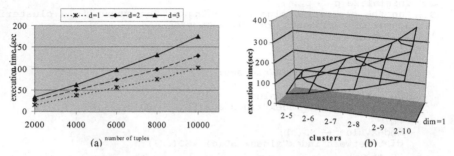

Fig. 3. a. Execution time as function of number of tuples, **b.**Execution time in seconds as function of the number of tuples

4.3 Evaluation Study of Our Approach - Experimental Results

Based on the algorithm described in Section 4.1, we implemented a clustering system for evaluating the clustering schemes and selecting the scheme that fits a data set based on a specific clustering algorithm. It is a system, which is implemented in Java and uses ODBC to connect to a database. At the current stage of development, we have implemented and applied the K-Means algorithm in order to define the clustering schemes. Nevertheless, our approach can also be applied by considering any other clustering algorithm on our data set. This claim was verified by using a hierarchical algorithm (single link algorithm)[14] on our data set. Using our system, we experimented with synthetic and real data sets so as to evaluate our approach as regards the selection of the best clustering scheme for a data set that a specific clustering algorithm defines.

Data Sets description. We evaluate the proposed approach for finding the optimal clustering scheme using two data sets: a real one based on Iris Data Set (see Sect. 3..3) and a synthetic one. A good partitioning of the data set in three groups of flowers can be obtained if we use only two features and more specifically the attributes petal_length, and petal_width of the Iris Data Set [11]. The second data set is a synthetic one and includes 10000 two- dimensional data points (Fig 5a). We produce the data set using normal distribution.

Discussion on experimental results. The goal of this experiment is to evaluate our approach regarding the selection of the optimal clustering scheme that a specific clustering algorithm can define. More specifically, we apply two-dimensional clustering to Iris Data Set considering the attributes petal_length, and petal_width. The clustering schemes are discovered using the K-means algorithm while its input parameters (number of clusters) take values between 2 and 10. Applying the steps of the algorithm (see Sect. 4.1) our system results in a clustering scheme of three clusters. We can verify this result based on Fig. 2a, which presents the quality index values as a function of the number of clusters. We observe there, that the clustering scheme with three clusters corresponds to the minimum value of quality index and also is the point at which a significant local decrease in its value occurs. Thus, according to our system three clusters is the optimal c (Fig. 4a) for our data set, which is also the number of clusters proposed by experts.

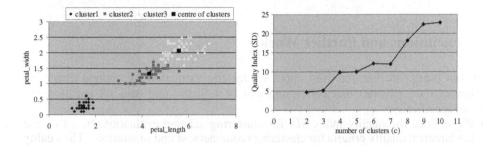

Fig 4. a .The best clustering scheme of Iris Data set, **b.** The graph of SD versus the number of clusters that corresponds to a clustering scheme of Iris Data set defined by applying an hierarchical clustering algorithm.

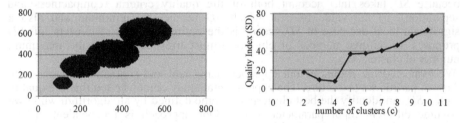

Fig 5. a.A data set that contains four clusters. **b.** The graph of SD versus the number of clusters (using K-Means).

We carried out a similar study using the synthetic data set described above. It is clear from Fig 5a that our data set consists of four clusters. This was verified by our system as the clustering algorithm (K-Means) parameter values (i.e., number of clusters) range from 2 to 10. According to the graph presented in Fig. 5b the number of clusters in which SD reaches its minimum value and also a significant decrease of its value occurs, is four.

In the sequel, we demonstrate that our approach is independent of the clustering algorithm used.

We consider the Iris Data Set (Fig. 4a) and we apply the single link algorithm [14] in order to discover clustering schemes. The algorithm will produce various schemes based on the range of values for the number of clusters that we define (e.g. [2, 10]). The result is a tree (called dendrogram) where each level corresponds to a specific number of clusters. Then, we evaluate each of the clustering schemes based on the *SD* quality index. The graph of *SD* versus the number of clusters (c) is presented in Fig. 4b. According to this graph, the optimal clustering scheme contains three clusters as in the case of K-Means. The above proves that our approach can be easily applied to any data set with compact clusters irrespective of cluster algorithm. We have to stress that our approach does not evaluate schemes discovered by different algorithms. It aims at selecting the optimal clustering scheme among the schemes that a specific clustering algorithm defines considering different input parameter values.

5 Conclusions and Further Work

In this paper, we presented an approach for the extraction of optimal clustering schemes (i.e., schemes that correspond and best fit the real data set partitioning). More specifically:

- We defined the quality index SD, for clustering scheme evaluation, based on the fundamental quality criteria for clustering (compactness and separation). The quality index is a variant of the indices defined for the FCM algorithm in [11], adapted to crisp clustering algorithms. Its definition is based on intra-cluster and inter-cluster distances.
- We evaluated the behavior of SD considering different values for the clustering algorithm input parameters (i.e., number of clusters, max number of clusters). The results of the experimental study are summarized as follows: i) the SD index reaches a local minimum as the number of clusters ranges in [c_{min}, c_{max}]. This happens because SD takes into account both of the quality criteria (compactness and separation), ii) the value of c at which SD reaches its minimum and also a significant local decrease of SD occurs, is the optimal number of clusters, iii) SD proposes an optimal number of clusters almost irrespectively of the maximum number of clusters, c_{max}.
- We evaluated the results of our approach using real and synthetic data sets with known partitioning structure. The experimental study demonstrated that the quality index SD can identify the best clustering scheme defined by an algorithm while we assume different input parameter values. Our approach is independent on the clustering algorithm used to partition the data set.

Further work will be concentrated in the following issues:

- *Comparison of the clustering scheme quality index* proposed in this paper with a number of known validity indices. We will choose and implement some known validity indices presented in relevant literature. Then, we will carry out an evaluation study considering specific data sets to compare the performance of our approach regarding the selection of the best clustering scheme with the other indices.
- *Revealing and handling of the uncertainty in data mining process.* The system presented in this paper is part of a larger scale effort that aims at extracting useful knowledge and handling uncertainty in all stages of the KDD process. This effort has resulted in a classification scheme that handles uncertainty and supports decision-making [15]. We aim at enhancing our clustering evaluation scheme in order to deal with uncertainty. Then the results of the clustering process will directly be used as a classification scheme described in [15][16].

Acknowledgements

We would like to thank C. Amanatidis for his help in the experimental study and N. Berdenis for his comments on the paper style.

References

1. Michael J. A. Berry, Gordon Linoff. Data Mining Techniques For marketing, Sales and Customer Support. John Willey & Sons, Inc, 1996.
2. Rajesh N. Dave. "Validating fuzzy partitions obtained through c-shells clustering", *Pattern Recognition Letters*, Vol .17, pp613-623, 1996
3. J. C. Dunn. "Well separated clusters and optimal fuzzy partitions", *J. Cybern.* Vol.4, pp. 95-104, 1974.
4. Martin Ester, Hans-Peter Kriegel, Jorg Sander, Xiaowei Xu. "A Density-Based Algorithm for Discovering Clusters in Large Spatial Databases with Noise", *Proceedings of 2nd Int. Conf. On Knowledge Discovery and Data Mining*, Portland, OR, pp. 226-231, 1996.
5. Usama Fayyad, Ramasamy Uthurusamy. "Data Mining and Knowledge Discovery in Databases", *Communications of the ACM.* Vol.39, No11, November 1996.
6. Usama M. Fayyad, Gregory Piatesky-Shapiro, Padhraic Smuth and Ramasamy Uthurusamy. Advances in Knowledge Discovery and Data Mining. AAAI Press 1996
7. Gath, B. Geva. "Unsupervised Optimal Fuzzy Clustering". *IEEE Transactions on Pattern Analysis and Machine Intelligence*, Vol 11, No7, July 1989.
8. Sudipto Guha, Rajeev Rastogi, Kyueseok Shim. "CURE: An Efficient Clustering Algorithm for Large Databases", *Published in the Proceedings of the ACM SIGMOD Conference*, 1998.
9. Alexander Hinneburg, Daniel Keim. "An Efficient Approach to Clustering in Large Multimedia Databases with Noise". *Proceeding of KDD '98*, 1998.
10. Zhexue Huang. "A Fast Clustering Algorithm to Cluster very Large Categorical Data Sets in Data Mining", *DMKD*, 1997
11. Ramze Rezaee, B.P.F. Lelieveldt, J.H.C Reiber. "A new cluster validity index for the fuzzy c-mean", *Pattern Recognition Letters*, 19, pp237-246, 1998.
12. Padhraic Smyth. "Clustering using Monte Carlo Cross-Validation". KDD 1996, 126-133.
13. C. Sheikholeslami, S. Chatterjee, A. Zhang. "WaveCluster: A-MultiResolution Clustering Approach for Very Large Spatial Database". *Proceedings of 24th VLDB Conference, New York, USA*, 1998.
14. S. Theodoridis, K. Koutroubas. *Pattern recognition,* Academic Press, 1999
15. M. Vazirgiannis, "A classification and relationship extraction scheme for relational databases based on fuzzy logic", in the proceedings of the *Pacific-Asian Knowledge Discovery & Data Mining '98 Conference*, Melbourne, Australia.
16. M. Vazirgiannis, M. Halkidi. "Uncertainty handling in the datamining process with fuzzy logic", in the proceedings of the IEEE-FUZZ conference, San Antonio, Texas, May, 2000.
17. Xunali Lisa Xie, Genardo Beni. "A Validity measure for Fuzzy Clustering", *IEEE Transactions on Pattern Analysis and machine Intelligence,* Vol13, No4, August 1991.
18. A.K Jain, M.N. Murty, P.J. Flyn. "Data Clustering: A Review", ACM Computing Surveys, Vol. 31, No3, September 1999.
19. Fisher,R.A. Machine readable .names file for MLC++ library. July, 1988

Algorithm for Matching Sets of Time Series

Iztok Savnik[1], Georg Lausen[1], Hans-Peter Kahle[2],
Heinrich Spiecker[2], and Sebastian Hein[2]

[1] Freiburg University
Institute for Computer Science
D-79085 Freiburg im Breisgau, Germany
{savnik,lausen}@informatik.uni-freiburg.de
[2] Freiburg University
Institute for Forest Growth
D-79085 Freiburg im Breisgau, Germany
{kahle,spiecker,hein}@uni-freiburg.de

Abstract. Time series are time-stamped sequences of values which represent a parameter of the observed processes in subsequent time points. Given a set of time series describing a set of similar processes, the model of the behavior of processes is constructed as a range of classification trees which describe the characteristics of each particular time point in series. An algorithm for matching a sequence of values with the model is used for searching common patterns in the sets of time series, and for predicting the starting time points of undated time series. The algorithm was developed and analyzed in the frame of the study of tree-ring time series. The implementation and the empirical analysis of the algorithm on the tree-ring time series are presented.

Keywords: Data mining, knowledge discovery, time-series, matching, and pattern recognition.

1 Introduction

A set of processes can be observed by studying the values of a particular parameter of processes measured in the subsequent time points. The sequence of values of the parameter for a particular process is called a *time-series*. The behavior of the processes recorded in time series can be revealed by investigating the characteristic features of the time series.

A considerable research effort has been directed recently to the development of methods for matching characteristic patterns in time series databases [2,7, 1,8,5]. The methods vary in the representation techniques for time series, the algorithms for measuring similarity between the time series, and in the search mechanisms used for mining the patterns. The problem which is related to matching subsequences in time series databases is the discovery of common patterns from a set of time series [13,9]. This problem has to our knowledge received less attention by KDD community, although it is of importance in practice for identifying the common characteristics of similar processes in applications such

D.A. Zighed, J. Komorowski, and J. Żytkow (Eds.): PKDD 2000, LNAI 1910, pp. 277–288, 2000.
© Springer-Verlag Berlin Heidelberg 2000

as monitoring and diagnosing complex systems, stock market data analysis, and medical diagnosis.

In this paper we address the problem of finding the common characteristics of the time series which describe a set of similar processes. Our main contribution is the proposal of a new algorithm for matching sequences of values with a set of time series. The problem is seen as a machine learning problem [10]. Given a description of a domain in the form of a set of training instances (time series), a theory is constructed which describes the characteristic properties of the domain. The representation of the theory is based on the "local" properties of time points — the time points are described by the characteristics of values which occurred close to the described time point. The algorithm for matching a sequence of values with the sets of training time series is realized as the procedure for the characterization of new instances by means of the computed theory. We show in the paper that the algorithm can be employed for mining the common patterns in a set of time series, and for determining the matching time points (dating) of the undated sequence of values.

The rest of the paper is organized as follows. The following section presents the matching algorithm. First, the problem of matching a sequence of values with the set of time series is defined formally. Section 2.1 presents the procedure for the construction of domain theory from the sets of time series. Further, the algorithm for matching sequences of values against the theory is detailed in Section 2.2. The experiments with the matching algorithm on tree-ring time series are described in Section 3. The results of the experiments show that the algorithm can be effectively used for dating tree-ring time series, a method which is important for revealing the past climatic and other environmental conditions, and for identifying the characteristic patterns in the behavior of a set of trees. Section 4 gives an overview of the related work and its relations to the presented work. Finally, the concluding remarks are given in Section 5.

2   Algorithm

A set of time series representing the values of a parameter of similar processes is called a *domain*. The method for matching a sequence of values with a set of time series is divided into two parts. First, the theory of domain is constructed, and, second, the matching algorithm is used to determine the time points in the time period of domain where an input sequence of values matches the domain with a given precision. Let us now define the terminology used in the paper and formally define the problem.

A time series $s = (v_1, \ldots, v_n)$ is a sequence of real numbers v_i which describe a property of entities from the domain in the subsequent time points. The time interval between the subsequent events is constant, that is, $t(v_{i+1}) = t(v_i) + \delta t$ where δt depends on the particular application domain. The application domain is described by a set of time series $\mathcal{D} = \{s_1, \ldots, s_m\}$. Each time series $s_i \in \mathcal{D}$ and $s_i = (v_{i1}, \ldots, v_{in})$ has n values with time points $t(v_{ij}) = t_j$.

The problem of matching sequences of values with the set of time series is defined as follows. Let $\mathcal{D}$ be a domain as defined above, and let $s_u = (u_1, \ldots, u_r)$ be a sequence of values where the time points $t(u_i)$ are not known. The matching procedure must find the time point $t_j \in [t_1..t_n]$ such that the sequence s_u matches with the domain $\mathcal{D}$ starting at the time point t_j. Let us now detail the construction of the theory $\mathcal{T}$ describing the features of the domain, and the algorithm for matching a sequence of values s_u with $\mathcal{D}$ by the use of theory $\mathcal{T}$.

2.1   Construction of Domain Theory

The application domain $\mathcal{D} = \{s_1, \ldots, s_m\}$ can be seen through the sequence of time points $t_1, \ldots, t_n$ for which the values of each time series s_i are specified. Given a time point $t_j \in [t_1..t_n]$, there exist one value $s_i[t_j]$ (or, v_{ij}) for each time series $s_i \in \mathcal{D}$ at time point t_j. The set of values of time series in a particular time point is denoted as $\mathcal{D}[t_j]$. Formally, $\mathcal{D}[t_j] = \{s_i[t_j]| \ i \in [1..m] \wedge s_i \in \mathcal{D}\}$. Further, the values of time series from $\mathcal{D}$ in subsequent time points $t_j, \ldots, t_{j+w}$ is the projection of $\mathcal{D}$ on the given sequence of time points resulting the subsequences of time series from $\mathcal{D}$. The projection of $\mathcal{D}$ on the time points $t_j \ldots t_{j+w}$ is defined as $\mathcal{D}[t_j..t_{j+w}] = \{(s_i[t_j], \ldots, s_i[t_{j+w}])| \ i \in [1..m] \wedge s_i \in \mathcal{D}\}$. The projection $\mathcal{D}[t_j..t_{j+w}]$ is called a *window* of $\mathcal{D}$ and w is the size of window.

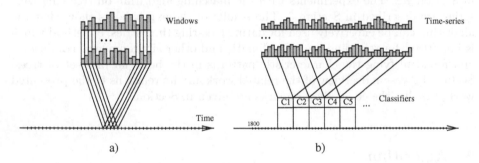

Fig. 1. Rules: a) learning cases, b) construction of classifiers.

The main idea used for the construction of the domain theory is to describe the characteristics of each particular time point in the domain by the properties of the values which are local with regards to the described time points. In other words, a time point t_j can be described by the characteristics of values which appeared some time before t_j, the values which appeared at t_j and some time after t_j. The values that are close to a time point t_j include the values from the window $\mathcal{D}[t_{j-\lfloor w/2 \rfloor}..t_{j+\lfloor w/2 \rfloor}]$ where w is the size of window. The windows of time series and the corresponding central time points are illustrated in Figure 1a.

Each window projected from the domain $\mathcal{D}$ is extended by adding to each subsequence from the window the time point of the central value of subsequence — such windows will be called *dated windows* $\mathcal{D}^*$. The dated windows are used as the basis for the construction of the classifiers for the time points of central values. More particularly, a range of dated windows is used as the input dataset of a program for the construction of classifiers. Formally, a *dataset* is defined as $\bigcup_{j \in [b..(b+d)]} \mathcal{D}^*[t_{j-\lfloor w/2 \rfloor}..t_{j+\lfloor w/2 \rfloor}]$ where w is the size of window, d is the number of windows in the range, and b is the index of the central time point of first window. The construction of the range of dated windows is illustrated by Figure 1a.

The time series from $\mathcal{D}$ can now be split into the overlapping partitions which are used for the definition of datasets. Splitting of the set of time series from the domain into partitions is illustrated by Figure 1b. Each dataset is used as the input of the program for the construction of a classifier. Note that the classifiers are in Figure 1b denoted by labels C_i. Each particular classifier describes the properties of a range of time points. The complete set of time points of the domain $\mathcal{D}$ is split into non-overlapping subsets such that each of them is covered by one of the classifiers.

2.2   Matching Algorithm

Let $\mathcal{D} = (s_1, \ldots, s_m)$ be a domain such that the time series s_i are defined for the time points $t_1, \ldots, t_n$. Further, let $C_1, \ldots, C_e$ be the set of classifiers constructed with windows of length w. The number of classifiers in the set is $e = \lfloor n/c \rfloor$ where c is the number of predicted classes by one classifier. Finally, let $s_u = (u_1, \ldots, u_r)$ be a sequence of values. The task of the matching algorithm is to find the time point $t_j \in [t_1..t_n]$ of the first value u_1 of the sequence s_u such that the sequence s_u matches with $\mathcal{D}$ starting at t_j. The precision of matching is specified by the set of parameters which will be presented shortly.

The matching algorithm extracts all possible windows of length w from the input sequence s_u. Each of the extracted windows is classified by each of the available classifiers. If the accuracy of the prediction is better than the predefined constant T_g then the prediction is called a *guess*. After a guess is obtained, it is verified using all other windows from s_u. The number of predictions which match the initial guess and the collective probability of matching is computed. A prediction which confirms with the initial guess will be in the sequel called a *hit*. In the case that the number of hits is higher than the predefined constant T_h, and at the same time, the collective probability of the match is higher than the predefined constant T_m, then we say that the sequence s_u matches the domain $\mathcal{D}$ in the predicted time point. The algorithm which implements the above sketched procedure is presented as follows.

Algorithm 1.
```
Input: Sequence of values s_u = (u_1, ..., u_r) and a range of
       classifiers R = (C_1, ..., C_e).
Output: Predictions (g, p_g) for the starting time points of s_u.
Method:
1. foreach ( window o ⊂ s_u ) do
2.     foreach ( classifier C_i ∈ R ) do
3.         (g, p_g) = predict(o, C_i);
4.             if ( p_g > T_g ) then
```

```
5.              (h, p_h) = compute_hits(R, s_u, g);
6.              if ( h > T_h ∧ p_h > T_m ) then
7.                  print_match(g, p_g);
8.          fi;
9. od;
```

We will now comment the above algorithm. The first **foreach** loop at line 1 iterates through all windows o extracted from s_u. Each of the windows o is classified by each of the classifiers from R using the function $\texttt{predict}(o, C_i)$ in line 3. The result of a prediction is a pair (g, p_g) where g is predicted time point of the first value of s_u and p_g is the probability that g is correct. In the case that p_g is greater than the threshold T_g (line 4) than g is a guess and the number of confirmations of the guess (hits) is computed in the line 5. This is done by classifying all the remaining windows of s_u. The procedure for the computation of the confirmations of guess is detailed by Algorithm 2. The threshold T_h defines the required number of confirmations of guess g and the threshold T_m defines the required collective probability for matching.

Algorithm 2.
Input: A set of classifiers $R = (C_1, \ldots, C_e)$, a sequence of values
$\quad\quad s_u = (u_1, \ldots, u_r)$, and a guess g.
Output: Number of hits h and collective probability p_h.
Method:

```
1.  function compute_hits(R, s_u, g);
2.  begin
3.      hits = 0; sump = 0;
4.      foreach ( window o ⊂ s_u ) do
5.          (p, p_p) = classify(o, R, g);
6.          if ( g = p ) then
7.              hits = hits + 1;
8.              sump = sump + p_p;
9.          fi;
10.     od;
11.     return( hits, sump/(r - w + 1) );
12. end;
```

The algorithm 2 verifies the guess g using all possible windows from s_u (line 4). In line 5 each window is classified using the function $\texttt{classify}(o, R, g)$. The classifiers which are used for the computation of the prediction are selected with respect to the guess g — we suppose that g represents the correct matching point and fix the position of s_u within the time interval of the domain $\mathcal{D}$ with respect to the time point of the guess g. The result of the function $\texttt{classify}$ is the time point p of the first value of s_u and the probability p_p that p is correct. Finally, the number of hits and the sum of matching probabilities are augmented in lines 6-9 if prediction p agrees with the guess g. The collective probability returned by the function is the average matching probability, that is, $\texttt{sump}$ divided with the number of classifications $(r - w + 1)$.

3   Analysis

The algorithm for matching sets of time series has been developed and analyzed in the frame of the study of tree-ring time series. The algorithm was implemented using the C5.0 knowledge discovery tools [15] and the `Perl` programming language. In this section we present the application domain, the implementation of matching algorithm, and the analysis of matching algorithm on the tree-ring time series. Two applications of the matching algorithm are analyzed: dating of sequences, and pattern discovery. Let us first present the tree-ring application domain, and the procedure for the acquisition of tree-ring data.

3.1   Tree-Ring Domain

The tree-ring time series are the sequences of values which represent the annual increments of the tree trunks measured on cross sections obtained after the tree is cut. The cross sections are taken from a stem height of 1.3 m. The annual radial increments are measured along eight equidistant measurement radii with a computer assisted image analysis system. The series used are the quadratic means of the eight measurements.

The tree-ring time series used in the experiments presented in the following sections are obtained from 50 European beech (Fagus sylvatica) sample trees from two geographical areas in Germany. The initial tree-ring measurement series have been detrended in order to eliminate individual tree specific variations in the time series. The result of this standardization procedure are normalized time series which are stationary in their mean and variance.

Tree rings reflect various influences of the environment on the growth of trees such as, for example, the climatic changes, the characteristics of the ground, or the events such as floods. A field of science which uses tree rings to analyze temporal and spatial patterns of processes is *Dendrochronology* [4]. The most widely used method for the study of tree-rings in Dendrochronology is called cross-dating [11]. The method has been effectively used to reconstruct the past climate for more thousands years.

3.2   Construction of Domain Theory

The domain which includes 50 tree-ring time series was split into two parts: the training time series (80%) and the test time series (20%). The analysis presented in this section is done on the set of training time series. The domain theory was constructed from the training time series as presented in Section 2.1. The test data was used for the experiments which are presented in the following two sections.

The decision trees were constructed using the C5.0 data mining tool. The rules extracted from the decision trees are used as the domain theory. The classifiers were built using the boosting technique [15]. Further, the cross-validation method was used for the estimation of the classification accuracy of the domain

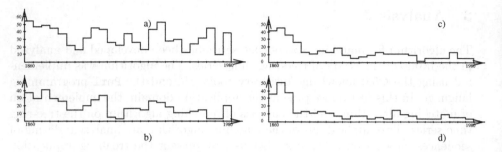

Fig. 2. Error rates of classifiers (in percents) constructed using the windows of size: a) 5, b) 10, c) 20, and d) 30.

theory. The results of the experiments with the construction of the domain theories are presented in Figure 2. In most cases 1-4 rules were computed for each of the time points. A typical example of a set of rules describing a time point 1900 is as follows.

$irp7 > 3.086$, $irp9 > 2.693$, $irp10 > 3.071$, $irp14 > 3.526$ -> class 1900 [0.965]
$irm10 > 2.861$, $irp10 <= 3.071$, $irp13 <= 3.267$ -> class 1900 [0.916]
$irm5 <= 2.835$, $irm3 <= 2.979$, $irp10 > 3.071$ -> class 1900 [0.748]

The attributes in rules are named as follows. An attribute named $irmX$ represents the value of an instance in a time point which occurred X time points before the central time point. Similarly, an attribute named $irpX$ represents the value of an instance in a time point which occurred X time points after the central time point. The classification accuracy of a rule is specified in the square brackets at the end of rule. A rule is typically based on 2 to 4 attributes.

3.3   Dating

The problem of dating sequences of values is defined as follows. Given a set of time series and an undated sequence of values, the time point t of the first value of undated sequence has to be found such that the undated sequence matches the set of time series at time point t with a given precision.

The experiments with dating were done by using 40 time series for the construction of the theories, and 10 time series as the basis for the generation of test sequences. The length of time series from the domain is from 100 to 250. For the construction of the theories we used 110 subsequent time points (years) which were defined for all time series. Further, four different theories were built for the experiments, each of them with a different size of the window used for the construction of the classifiers — the window sizes used were 5, 11, 21, and 31. The test sequences of lengths 21, 31, 51, and 71 were extracted from 10 test time series. The starting points of test sequences where chosen at random. In the experiments we dated each test sequence from the above described four groups of 10 sequences using different theories. For each run of the algorithm we collected the actual and predicted year of first value in the sequence, the

Table 1. Evaluation of matching algorithm for tree-ring dating

SeqL	WinL	NumM	AvgH	MinH	MaxH	AvgGP	AvgMP
21	5	9	75.82	52.94	94.12	0.88	0.64
31	5	7	73	59.26	92.59	0.88	0.58
51	5	8	71.53	53.19	91.49	0.89	0.56
71	5	8	72.39	55.22	89.55	0.87	0.57
21	11	5	87.27	72.72	100	0.94	0.77
31	11	8	82.14	61.90	100	0.92	0.64
51	11	10	78.78	53.66	97.56	0.92	0.61
71	11	10	79.34	59.02	98.36	0.86	0.62
31	21	6	97	90.91	100	0.90	0.81
51	21	9	87.45	64.52	100	0.92	0.68
71	21	10	84.51	58.82	100	0.93	0.67
51	31	9	93.14	76.19	100	0.90	0.71
71	31	10	90.97	65.85	100	0.92	0.70

number of hits which confirmed the match, the probability of the first guess and the collective probability of the match. Table 1 presents the results compressed in a single line for each group of ten test sequences of the same length. For each group we present: the length of sequences (SeqL); the size of window used for the construction of theory (WinL); the number of correct matches (NumM); the average, minimum and maximum number of hits (AvgH, MinH, and MaxH) expressed in percentages; the average probability of a guess (AvgGP) that leaded to matching; and the average collective probability of a correct match (AvgMP).

The parameters of Algorithm 1 which define matching of a sequence with the theory were as follows. The required percent of classifiers which agreed on the prediction T_h were 50%, the required probability of the initial guess T_g was 80%, and the required collective probability of the match T_m was 40%. Note that the thresholds T_h and T_m were set relatively low in order to allow the matchings based on the partial agreement between the subsequence and the theory; the values of AvgH in Table 1 denote the average proportion of the agreement by matchings. Finally, the prediction supported with the largest number of hits is chosen among the results of Algorithm 1.

The percentage of correctly dated test sequences in the complete set of experiments is 83.8%. However, the available set of time series describe the trees from 2 different geographical areas which may be one of the reasons for 16.2% incorrectly dated cases. This claim is supported by the fact that the incorrectly dated subsequences are, in many cases, extracted from the distinctive subset of 10 test time series.

3.4   Pattern Recognition

In this section we present the application of matching algorithm for pattern recognition. The patterns are represented as short sequences of values which capture the characteristics behavior of the observed parameter. The patterns which are used in the experiments are drawn in Figure 3. While the sequences are very specific representations of patterns, the rules which describe the properties of the time points are general — there are only a few rules which describe a particular time point. For this reason, the matching algorithm can find general patterns which are supported by a subset of time series from the domain.

The patterns which are used in the experiments on the tree-ring data are of length 5, 7 and 9 (years). We consider that in this domain the selected length of patterns (sequences) can capture the characteristic episodes in the growth of trees which may be the effects of the conditions in the environment. The first line of patterns represents temporarily defeated growth of the trees. The second and the third lines represent simple steps. Finally, the last line presents pattern where the growth rate of trees increases temporarily and than again decreases in the second part of the interval to the initial rate.

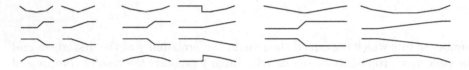

Fig. 3. Examples of patterns.

The theory which is constructed using the window of size 5 is used in all experiments. Because of the low classification accuracy of the complete theories constructed with the window of length 5 (Section 3.2), we used considerable high threshold for the collective probability of match (70%) as well as the high threshold for the required number of hits for a match (90%). Notice that the high threshold for the collective probability of match implies high accuracy of rules triggered for each particular time point in the pattern.

Let us now describe the results of the experiments presented in Table 2. The columns of the table represent: the pattern number (PatN), the length of the pattern (PatL), the number of matchings of the pattern with the theory (NumM), the average probability of guesses which leaded to matching (AvgGP), the average collective probability of the match (AvgMP), and the list of time points (TPts) where the pattern appeared. The pattern number stated in Table 2 represents the ordering number of patterns from Figure 3. The ordering of patterns is from left to right and then from the first to the fourth line (pattern number 1 is in the upper left corner). Finally, note that the number of hits obtained in matchings is not presented in the table since all possible hits were achieved for each matching.

The algorithm found from 3 to 8 instances of each pattern of length 5 in the domain. We can observe from the values of the column TPts that in some cases the neighboring patterns (horizontally in Figure 3) appear on the same time point which means that the occurrence of the pattern is more general. An

Table 2. Evaluation of matching algorithm for pattern recognition

PatN	PatL	NumM	AvgGP	AvgMP	TPts
1	5	8	0.89	0.89	1880,1888,1896,1912,1918,1929,1937,1963
2	5	7	0.86	0.86	1880,1888,1896,1929,1937,1948,1956
3	7	2	0.97	0.87	1896,1929
4	7	3	0.92	0.78	1900,1911,1921
6	9	2	0.94	0.81	1901,1912
7	5	5	0.91	0.91	1881,1889,1905,1913,1930
8	5	3	0.87	0.87	1889,1905,1913
9	7	3	0.88	0.85	1881,1930,1939
11	9	2	0.91	0.78	1898,1931
13	5	7	0.89	0.89	1892,1900,1911,1921,1928,1936,1953
14	5	4	0.89	0.89	1900,1928,1945,1955
15	7	5	0.93	0.78	1892,1900,1921,1936,1962
19	5	5	0.83	0.83	1884,1890,1899,1903,1926
20	5	4	0.85	0.85	1873,1884,1916,1926
21	7	2	0.93	0.79	1916,1932
22	7	3	0.92	0.86	1926,1931,1940

example when the similar patterns of the length 5, 7 and 9 are matched is the step represented by patterns 7, 9 and 11. The steps of length 5 and 7 appear in the year 1930 while the step of length 9 appears in the year 1931. The shift of one year in the central points of patterns appears because the steps are on the right side of the central time point in the patterns 7 and 9 while the step is on the left side of the center in the pattern 11. The patterns of length 7 and 9 are less frequent than the patterns of length 5. This reflects the nature of the processes — stable patterns in in the growth rate of trees are expected to appear only in the period of few years.

The study of the occurrences of the patterns in the data showed the correspondence between the collective probability of the match, and the support of the pattern in the data. The higher the collective probability is more the pattern is supported by the domain. Still, the support of the patterns in the data varies. The computation of more accurate estimation for the support of patterns in data would require the information about the support in data for each particular rule triggered for the time points of patterns.

4   Related Work

We distinguish between two types of methods for matching subsequences of time series. Firstly, a considerable research interest has been directed recently to the development of methods for matching subsequences in time series databases. These methods are defined to identify subsequences in particular time series. Secondly, methods for discovering characteristic subsequences of sets of time series have been recently proposed. These methods deal with the common features of sets of time series and not with the characteristics of the individual time series.

Let us first overview the methods for matching subsequences in time series databases and their relations to the proposed algorithm. As pointed out in [8], the methods differ in the representation of the time series, the comparison of time series, and in the search mechanisms. The representations of the time series used by the methods are: the normalized series [2], the representation based on the feature space obtained by Discrete Fourier Transformation [7], the piecewise linear models [8], and the piecewise linear models augmented with weights [9]. The methods for the identification of similar time series are: a simple distance function [2], the probabilistic method for the computation of the similarity measure [8], Euclidean distance between the time series [7], and the longest common subsequence of two series [6]. The search methods which were recently proposed for matching subsequences in time series databases are based on: the dynamic programming technique [2], the sequential scan for pattern identification [8], and the index data structures based on feature representation of time series [7,8].

Let us now present some of the approaches to the discovery of the common characteristic of the set of time series. The most widely known approach to the identification of the common subsequences of the set of time series is the algorithm for finding the longest common subsequences [13]. This algorithm has been used recently for the computing the similarity of sequences [6,12]. The main disadvantage of the algorithm is its exponential space and time complexity. Finally, a method for the identification of distinctive time series from the set of time series by using clustering has been introduced recently in [12].

5   Conclusions

An algorithm for matching sequences with the set of time series was presented in the paper. The algorithm can be employed for locating the starting time points of the undated sequences of values, and for locating the pattern templates which are common to the set of sequences. The algorithm was successfully used for the problems of dating and pattern recognition in tree-ring time series. We consider that the matching algorithm is general enough to be effectively used for other domains including stock market data, sensor data in engineering environments, and medical measurements.

Acknowledgments

We are thankful to Luc De Raedt and Stefan Kramer for their helpful comments and suggestions.

References

1. Agrawal, R., Lin, K.I., Sawhney, H.S., Shim, K., 'Fast Similarity Search in the Presence of Noise, Scaling, and Translation in Time-Series Databases', In *Proc. of VLDB Conference*, pp.490-501, 1995.
2. Berndt, D.J., Clifford, J., 'Finding Patterns in Time Series: A Dynamic Programming Approach', In *Advances in Knowledge Discovery and Data Mining*, U.M. Fayyad, Shapiro, G.P., Smith, P., and Uthurusamy, R., Eds., AAAI Press / The MIT Press, 1996.
3. Bollobas, B., Das, G., Gunopulos, D., Mannila, H., 'Time-Series Similarity Problems and Well-Separated Geometric Sets', *Proc. of 13th Annual ACM Symposium on Computational Geometry*, To appear, 1998.
4. Cook, E.R., Kairiukstis, L.A., *Methods of Dendrochronology: Applications in the Environmental Sciences*, Kluwer Academic Publishers, 1990.
5. Das, G., Gunopulos, D., Mannila, H., 'Finding similar time series', In *Proc. of PKDD'97*, LNCS, 1997.
6. Das, G., Gunopulos, D., Mannila, H., 'Rule discovery from time series', In *Proc. of Conf. on Knowledge Discovery and Data Mining*, AAAI Press, 1998.
7. Faloutsos, C., Ranganathan, M., Manolopoulos, Y., 'Fast Subsequence Matching in Time-Series Databases', In *Proc. of SIGMOD Conference*, 1994.
8. Keogh, E., Smyth, P., 'A Probabilistic Approach to Fast Pattern Matching in Time Series Databases', In *Proc. of Conf. on Knowledge Discovery and Data Mining*, AAAI Press, 1997.
9. Keogh, E.J., Pazzani, M.J, 'An enhanced representation of time series which allows fast and accurate classification, clustering and relevance feedback', In *Proc. of Fourth Conf. on Knowledge Discovery and Data Mining*, 1998.
10. Mitchell, T.M., *Machine Learning*, McGraw-Hill Series in Computer Science, 1997.
11. Monserud, R.A., 'Time-series analyses of tree-ring chronologies', *Forest Science 32*, pp.349-372, 1986.
12. Oates, T., 'Identifying Distinctive Subsequences in Multivariate Time Series by Clustering', In *Proc. of Conf. on Knowledge Discovery and Data Mining*, 1999.
13. Sankoff, D., Kruskall, J.B., *Time Wraps, String Edits and Macromolecules: The Theory and Practice of Sequence Comparison*, Addison-Wesley, 1983.
14. Savnik, I., Lausen, G., Kahle, H.P., Spiecker, H., Hein, S., 'Algorithm for Matching Sets of Time Series', Technical Report 134, Institut für Informatik, Universität Freiburg, Mar 2000.
15. Quinlan, J.R., *C4.5: Programs for Machine Learning*, Morgan Kauffman, 1993.

MSTS: A System for Mining Sets of Time Series

Georg Lausen, Iztok Savnik, and Aldar Dougarjapov

University of Freiburg, Institute for Computer Science
Georges-Köhler-Allee, 79110 Freiburg, Germany
{lausen,savnik,dougarja}@informatik.uni-freiburg.de

Abstract. A system to support the mining task of sets of time series is presented. A model of a set of time series is constructed by a series of classifiers each defining certain consecutive time points based on the characteristics of particular time points in the series. Matching a previously unknown series with respect to a model is discussed. The architecture of the *MSTS*-System (*M*ining of *S*ets of *T*ime *S*eries) is described. As a distinctive feature the system is implemented as a database application: time series and the models, i.e. series of classifiers, are database objects. As a consequence of this integration, advanced functionality as the manipulation of models and various forms of meta learning can be easily build on top of MSTS.

1 Introduction

Data in form of time series represents the behaviour of some time-dependent processes and can be used to exhibit the characteristics of the observed processes. In the literature there exists much work treating various aspects of time series. Matching characteristic patterns in time series databases is discussed in [2, 8, 1, 12, 5, 14]. These methods vary in the representation techniques for time series, the algorithms for measuring similarity between the time series, and in the search mechanisms used for mining patterns. Methods for the discovery of rules [6, 11] and frequent episodes [15] in time series have been proposed, as well. Rules and episodes describe the interrelations among the local shapes of time series.

The discovery of common patterns from a *set* of time series has a long tradition in statistics and one of the most widely used techniques is cross-correlation [3]. In the machine-learning framework the problem has previously been studied only by [13], to the best of our knowledge. Similar to cross-correlation, [13] base their approach on time series as a whole and classify them according to a distance measure. We take a different approach and classify concrete time points based on their local properties, i.e. properties of time points within a certain window on the whole time series. In contrast to the more mean-value oriented known technique, the resulting pointwise classification of a time series gives direct handles to explanations of the behaviour of the time series. This approach originated from a practical project to date tree trunks according to the yearly radial increase [17]; in the current paper we present a simplified and more efficient version of the algorithm and discuss its usuage as part of a data mining system.

D.A. Zighed, J. Komorowski, and J. Zytkow (Eds.): PKDD 2000, LNAI 1910, pp. 289–298, 2000.
© Springer-Verlag Berlin Heidelberg 2000

Our current contributions are as follows. First we address the problem of finding the common characteristics of the time series which describe a set of similar processes. Given a description of a domain in the form of a set of training instances (time series), a model is constructed which describes the characteristic properties of the domain. The representation of the model is based on the "local" properties of time points — the time points are described by the characteristics of values which occurred close to the described time point. To build the model a sequence of classifiers is derived (computed by C5.0 [18]), which are represented by corresponding sets of rules. An algorithm for matching a sequence of values with the sets of training time series is introduced and analyzed. This task is also called *dating* the sequence.

One important direction during the last years is the development of data mining suites to support the mining tasks by integrating different mining techniques into a common framework and supporting interoperability with a database management system (e.g. [4]). Orthogonal to this stream of work we have implemented a system with the aim to provide support not for mining in general, but for the specific task of mining for applications interested in mining sets of time series. As one example of such applications we mention experimental results concerning the analysis of tree-ring time series which represent the yearly radial increases of tree trunks [17]. The architecture and implementation of *MSTS-System* (*M*ining of *S*ets of *T*ime *S*eries) is described. As a distinctive feature, MSTS is a database application: time series and the models, i.e. series of classifiers, are database objects. MSTS has been implemented under Windows NT using C++ as programming language and the Oracle 8 database environment.

Finally, we outline the benefits of our database-centered approach. Having the domain data and the domain models in a uniform database framework not only makes it easy for the miner to choose under several precomputed classifiers for the mining task, but also provides a powerful framework to manipulate models and to perform meta learning.

The structure of the paper is as follows. Section 2 presents the formal framework for mining and section 3 an algorithm for matching. Section 4 contains the architecture of MSTS and in particular presents an ER-schema specifying the content of the database. Moreover, real world experimental results underlining the usefulness of the approach are presented. In section 5 we conclude the paper by discussing advanced functionality of MSTS.

2   Formal Framework

The following is based on [17]. A time series $s = (v_1, \ldots, v_n)$ is a sequence of real numbers v_i representing the values of a time–dependent parameter of a process. The interval between subsequent time points is assumed to be constant. A set of time series is called a *domain* $\mathcal{D} = \{s_1, \ldots, s_m\}$. Each time series $s_i \in \mathcal{D}$, $s_i = (v_{i1}, \ldots, v_{in})$ has n values with time points $t(v_{ij}) = t_j$. Let $s_u = (u_1, \ldots, u_r)$ be a sequence of values where the time points $t(u_i)$ are not known. The time series of the domain are the training cases to build a model of the domain and

s_u is a test case which has to be *dated*, i.e., whose pattern has to be matched against the model in order to predict the start time point $t(u_1)$.

The values of a time series $s \in \mathcal{D}$ in subsequent time points $t_j, \ldots, t_{j+k}$, $k \geq 0$, $j \geq 1$, $j + k \leq n$ is the projection of s on the given sequence of time points $s[t_j..t_{j+k}] = (s[t_j], \ldots, s[t_{j+k}])$. Let $w \geq 1$ and w odd, then the projection $s[t_j..t_{j+w-1}]$ is called a *window* of s and w is the size of the window; $j+w-1 \leq n$.

The main idea later used for the construction of a model is to describe the characteristics of each particular time point in the domain by the properties of the values which are local with regards to the described time points. In other words, a time point t_j can be described by the characteristics of values which appeared some time before t_j, the values which appeared at t_j and some time after t_j. To this end, in the sequel, we consider sets of windows of the form $\mathcal{D}[t_{j-\lfloor w/2 \rfloor}..t_{j+\lfloor w/2 \rfloor}] = \{s_i[t_{j-\lfloor w/2 \rfloor}..t_{j+\lfloor w/2 \rfloor}] | i \in [1..m] \wedge s_i \in \mathcal{D}\}$, where w is the size of the windows, $j > \lfloor w/2 \rfloor$ and t_j the *central (time) point* of the window. Observe that any two windows of the same time series with central points t_l, t_{l+1} will overlap in $w - 1$ time points.

Let $s_i[t_{j-\lfloor w/2 \rfloor}..t_{j+\lfloor w/2 \rfloor}]$ be a window. A window associated with its central point is called a *dated* window.[1] If $\mathcal{D}[t_{j-\lfloor w/2 \rfloor}..t_{j+\lfloor w/2 \rfloor}]$ is a given set of windows, then $\mathcal{D}^*[t_{j-\lfloor w/2 \rfloor}..t_{j+\lfloor w/2 \rfloor}]$ is the corresponding set of dated windows. A range of sets of dated windows is used as the input dataset for the construction of classifiers. Formally, a *dataset* S is defined as $S = \bigcup_{j \in [b..(b+d-1)]} \mathcal{D}^*[t_{j-\lfloor w/2 \rfloor}..t_{j+\lfloor w/2 \rfloor}]$ where each window is dated, $w \geq 1$ is the window size, $d \geq 1$ is the size of the dataset given by the number of different central points, and $b \geq \lfloor w/2 \rfloor + 1$ is the index of the first central point. Each window in a dataset is used for building a classifier as follows. The central point of the window denotes the class and all values constituting the window are the observed parameter values at the respective time points describing the corresponding class.

To simplify notation, throughout the paper we will assume $\alpha \cdot d = n - w + 1$ for some natural number α. The time series from $\mathcal{D}$ can now be split into a sequence of α datasets $S_1, \ldots, S_\alpha$, which are used for the definition of α classifiers $C_1, \ldots, C_\alpha$, where each classifier C_i, $1 \leq i \leq \alpha$ is to predict time points out of the interval $[\lfloor w/2 \rfloor + 1 + (i - 1) \cdot d, \lfloor w/2 \rfloor + i \cdot d]$. Note that any two datasets S_i, S_{i+1} overlap in $\lfloor w/2 \rfloor$ time points, however any two corresponding classifiers C_i, C_{i+1} do not overlap with respect to their predicted time points, $1 \leq i < \alpha$. Computing a classifier given a dataset, in principle, is a standard mining task; however, as we have to compute $O(n)$ classifiers, the implementation of this task needs some futher discussion which is presented in Section 4.2.

3   Matching algorithm

Let $\mathcal{D} = (s_1, \ldots, s_m)$ be a domain such that the time series s_i are defined for the time points $t_1, \ldots, t_n$. Further, let $C_1, \ldots, C_\alpha$ be the set of classifiers

---

[1] Note that a window is a mere sequence of real numbers, while a *dated* window has a reference to a time point.

Algorithm Match
```
Input: A sequence s_u = (u_1,..., u_r) with corresponding
       sequence of windows (U_1,..., U_{r-w+1}) and
       a sequence of classifiers (C_1,..., C_α).
Output: Time point t_i ∈ [t_1 ... t_n], which has been predicted
       with the maximal collected probability.
Method:
1. foreach (t_i ∈ [t_1 ... t_n] ) ColProb[t_i] = 0;
2. foreach ( window U_i ∈ (U_1,..., U_{r-w+1}) )
3.     foreach ( classifier C_j ∈ (C_1,..., C_α) ) do
4.         (t, prob) = predictCentralPoint(U_i, C_j);
5.         firstPoint = t - ⌊w/2⌋ - i + 1;
6.         ColProb[firstPoint] = ColProb[firstPoint] + prob;
7.     od;
8. foreach (t_i ∈ [t_1 ... t_n] ) ColProb[t_i] = ColProb[t_i]/(r - w + 1);
9. return( t_i ∈ [t_1 ... t_n] with maximal collected probability);
10.end;
```

Fig. 1. Matching algorithm.

constructed from α datasets of size d containing windows of size w. Finally, let $s_u = (u_1, \ldots, u_r)$, $r \geq w$, be a sequence of values which is to be dated. The task of matching then is to find the time point $t_u \in [t_1..t_n]$ of the first value u_1 of the sequence s_u such that the sequence s_u matches with $\mathcal{D}$ starting at t_u with a maximal precision. To this end all possible windows of length w are extracted from the input sequence s_u. Let $U = (U_1, \ldots, U_{r-w+1})$ be the sequence of possible windows of size w for s_u.

The matching algorithm we shall present simplifies and optimizes the approach proposed in [17]. While the algorithm in [17] tries to find a good guess for a prediction based on a threshold parameter stating a lower bound for the required probability of the guess, we directly classify each window with each classifier and store the results in a way, such that it can be guaranteed, that the time point with maximal collected probability will always be found. For length of the training sequences n and length of the test sequence m, we compute $O(n \cdot m)$ predictions. In contrast, the previous algorithm overall computes $O(\gamma \cdot n \cdot m^2)$ predictions, where $\frac{1}{m} \leq \gamma \leq 1$ depending on the threshold. Small thresholds imply large γ, i.e. a threshold of 0 implies $\gamma = 1$. This means a worst case time complexity of $O(n \cdot m^2)$ of the previous algorithm. Moreover, for a threshold greater 0 it is not guaranteed whether the time point with maximal precision will be found. This may happen because as precision the average of probabilities is computed, such that a high probability in one situation does not necessarily imply a high average.

The matching algorithm (cf. figure 1) classifies each of the windows in U with each available classifier (line 4). As the classifier predicts the central point of each window, the predicted central point is transformed into the corresponding

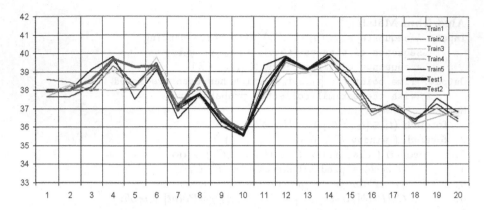

Fig. 2. Graph of a set of 5 time series and two test sequences.

predicted initial time point of s_u (line 5). ColProb then sums up the probabilities a certain time point is predicted to be the initial time point of s_u (line 6). Finally, the algorithm computes for each time point t_i the respective *collected* probability, which is computed as the average of the probabilities with respect to the number of classifications, which could predict t_i. This number is given by the number of windows in U, i.e. $(r-w+1)$. To see this assume the contrary, i.e. one time point is predicted more often than the number of windows in s_u. However this is not possible, because different classifiers predict disjoint intervalls. The final output of the algorithm then is the time point t_i, for which the collected probability is maximal; collected probabilities are our measure for precision.

To demonstrate the steps of the algorithm we will investigate two test sequence. Test sequences s_u^1 and s_u^2 and a given set of 5 training sequences are shown in figure 2. While the shape of s_u^1 was chosen such that a start point of 7 could be expected, s_u^2 was constructed by first trying to capture the shape of the series between 1 and 10, but then changing the contour of the series at time point 5 and 8 significantly.

Assume a window size $w = 5$ and a size of the dataset $d = 4$. Applying C5.0 to compute the classifiers and to predict the start points of the test sequences allows us to trace the computation for the matching tasks. For each window w and each classifier C (by writing $Ci..j$ we express that the classifier can predict as *central* point of a window a time point between i and j) the predicted first point of the considered test case s_u and the respective probability is given. Because of the underlying very small set of training examples the probabilities shown in the tables cannot be interpreted in a strong statistical sense. Results of a real practical experiment are reported in section 4.3.

Matching of s_u^1:

window	C3..6	prob	C7..10	prob	C11..14	prob	C15..18	prob
w_1	4	0.86	7	0.86	11	0.86	13	0.86
w_2	1	0.86	7	0.86	10	0.86	15	0.86
w_3	0	0.86	6	0.86	7	0.86	13	0.86
w_4	-1	0.86	2	0.86	7	0.86	10	0.86

Matching of s_u^2:

window	C3..6	prob	C7..10	prob	C11..14	prob	C15..18	prob
w_1	2	0.86	5	0.86	11	0.86	13	0.86
w_2	1	0.86	4	0.86	11	0.86	12	0.86
w_3	1	0.75	3	0.43	9	0.86	11	0.86
w_4	-1	0.86	2	0.43	9	0.86	11	0.86
w_5	-1	0.75	1	0.43	7	0.86	9	0.86
w_6	-1	0.86	1	0.86	6	0.86	9	0.43

We summarize by mentioning the best, the second and third best matching starting point:

	s_u^1	ColProb	s_u^2	ColProb
Bestmatch	**7**	0.86	**11**	0.57
next best	10	0.43	9	0.5
next best	13	0.43	1	0.48

As expected, the start point of s_u^1 is predicted by 7; for s_u^2 the algorithm predicts start point 11 which, in fact, begins a shape of the training series which is similar to the test series s_u^2. However, here the situation is not as supported by the collected probabilities as in the previous case. The collected probabilities show, that start points 1, 9, 11 are more or less equally supported.

4   MSTS: Mining Sets of Time Series

We will now present the architecture of the *MSTS*–System (*M*ining of *S*ets of *T*ime *S*eries); a detailed description can be found in [7]. In contrast to general data mining suites MSTS aims at supporting the specific task of mining sets of time series and therefore can be considered as complementing a mining suite. The mining algorithms of MSTS are based on the techniques described in section 2, 3. For building classifiers C5.0 [18] is used. The training cases for the classifiers are given by the datasets derived from a given domain. The attribute values of each case in a window of the dataset are the values of the observed parameter at a respective time points. The classes to be trained are the respective central points.

Working with MSTS supports modelling of sets of time series (called *sequences*) and matching undated sequences against a model. The sequences, resulting models and matchings are database objects. Further database objects are sessions. Sessions are the logical unit a user is working with MSTS. For a domain given by a set of sequences, different models may exist in the database. This allows to explore several alternatives when matching unknown series. Different models may arise because of varying window sizes and sizes of the datasets. Grouping modelling and matching tasks into sessions allows the user to keep track of several explorations of the same, or overlapping domains.

The information content of the database underlying MSTS is specified by the ER–schema shown in figure 3. Objects[2] of type `sequence`, `model` and `session`

---

[2] We talk about *objects* instead of *entities*.

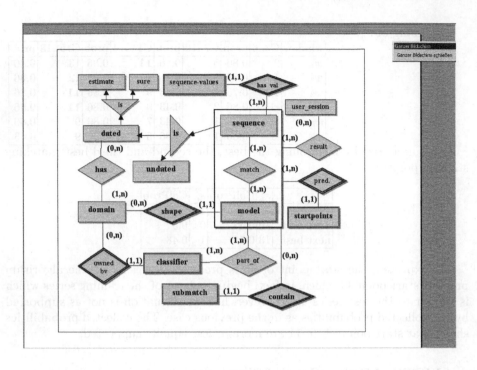

Fig. 3. ER–Schema of the database underlying MSTS; attributes are omitted.

can be created, deleted or viewed; objects of type `match` can only be viewed because they are computed by the system. Objects of type `sequence` are further classified in `dated` and `undated`, where in the first case we further distinguish `sure` and `estimated`. Estimated sequences are those whose dating has been performed by MSTS. An object of type `domain` then is given by a set of dated sequences. A model is defined by a set of classifiers and related to exactly one domain. This is represented by object types `model` and `classifier` and relationship types `shape` and `part_of`. A classifier may be part of more than one model, however these must be shapes of the same domain; therefore a classifier is related by `owned by` with exactly one domain. Matches are represented by relationship type `match`; each match is further characterized by a set of submatches, which means the elements of the table relating windows of the test sequence and the classifiers of the used model. This gives rise to object type `submatch` and relationship type `contain`. The remaining part of the ER–schema is self-explanatory. Because of limited space attributes of the types are not discussed; we only mention that `classifier` has two attributes to store the rules computed by C5.0 in binary format, such that they can be reused for prediction, and ascii format, such that they can be shown to the user.

MSTS is implemented as a database application written in C++ and connected to a Oracle 8 database server by ODBC. The coupling with the database is based on cache memory which is carefully maintained by MSTS. With respect to the sequences there is a ternary relation SEQUENCE_VALUES stored in the database, which relates a certain sequence by its ID to a certain time point giving the respective value of the parameter of interest.

When a range of classifiers has to be computed the corresponding datasets are extracted from the database as follows. The database receives an SQL-expression which extracts from SEQUENCE_VALUES the relevant subset of rows which are needed to compute the cases (windows) for constructing one dataset. This subset is sorted by time and with respect to a certain time point by sequence ID. Using only one database cursor the input file for C5.0 now can be easily derived in C++. As consequtive datasets overlap, when switching from one dataset to the next only the missing rows are extracted from the database. Note that because of sorting first by time and within one time point by sequence ID, the overlapping already sorted part can be reused. As only one database cursor is used and each row from the database is involved in exactly one sort operation, the work to build the input of C5.0 is minimized.

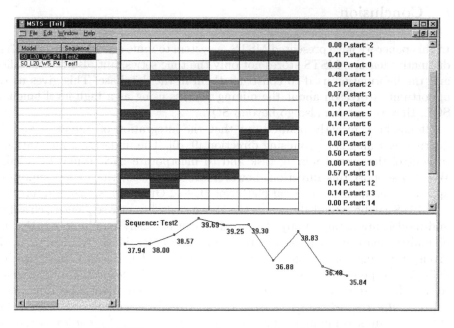

Fig. 4. MSTS screendump demonstrating the mining task.

MSTS has been successfully applied for the analysis of tree-ring time series which represent the yearly radial increases of tree trunks. The results presented in [17] are also valid for MSTS and will be mentioned shortly.[3] The domain was

---

[3] The detailed analysis reported in [17] is based on a prototypical implementation in Perl.

build out of 50 tree-ring time series of length 100 to 250. A typical example of a set of rules describing a time point 1900 is as follows (window size 21):

```
plus7 > 3.086, plus9 > 2.693, plus10 > 3.071, plus14 > 3.526 -> class 1900
[0.965]
minus10 > 2.861, plus10 <= 3.071, plus13 <= 3.267 -> class 1900 [0.916]
minus5 <= 2.835, minus3 <= 2.979, plus10 > 3.071 -> class 1900 [0.748]
```

Here plus$i$, respectively minus$i$, refers to the time point 1900+i, respectively, 1900-i. The probabilities of the rules are given in brackets. A first analysis based on inspecting the rules shows, that time point 1910 seems of some importance for predicting time point 1900. This observation could be the starting point of a deeper investigation. As an example for matching consider the following sequence: $\smile\diagup$. For a model of the domain with a window of size 5, the following is the result of matching: Number of matches = 8; Average probability = 0.89; Starting time points: 1880,1888,1896,1912,1918,1929,1937,1963.

5   Conclusion

In this paper we have presented MSTS, a system to mine sets of time series. As a distinctive feature, MSTS stores not only the time series, but also the classifiers and the information used to construct them in the database. This gives us the opportunity to reason about the mining process using as a high-level language SQL. However we can also go beyond SQL.

It has been recently pointed out that the integration of inductive and deductive reasoning is a promising direction [9, 10]. The underlying motivation is to control the overall mining task and to incorporate background knowledge. We also support this direction, however are convinced, that a less ambitious approach based on commercial platforms (C++, Oracle 8) is also of interest.

MSTS can be extended towards the above goals, because sequences, classifiers and models are database objects and thus can be easily manipulated. Because of limited space we only sketch three possible extensions. Consider a model given by a range of classifiers over time points $[t_i..t_j]$. Because each classifier defines time points by rules, it is easy to compute the *projection* of the model to $[t_i'..t_j']$, where $t_i \leq t_i' \leq t_j' \leq t_j$. Similarly, if there are models on time points $[t_{i1}..t_{j1}]$, $[t_{i2}..t_{j2}]$, where $t_{j1} = t_{i2}$ and the size of the windows are equal, then the *union* can be computed. The next extension is a form of *meta learning* called *combining* [16]. Here the predictions of the *base* classifiers performed on their validation sets form the basis for the meta-learner's training set. In our scenario the role of a classifier is taken by a model. One interesting aspect here then is the possibility to predict based on different window sizes at the same time. This allows to combine models based on small windows appropriate to reflect short term effects in the time series with models based on large window sizes appropriate to reflect long term effects. It is easy to see that the ER-schema of MSTS shown in Figure 3 provides the necessary information: a match is a relationship between a sequence and a model, which is further described by the determined startpoints. For one sequence there may exist several matches which then may form the meta-learner's training set.

6   Acknowledgement

We thank Heinrich Spieker, Hans-Peter Kahle and Sebastian Hein for many insightful discussions and the opportunity to use real world data, and thank Luc de Raedt and Stefan Kramer for valuable comments on our work.

References

1. Agrawal, R., Lin, K.I., Sawhney, H.S., Shim, K., 'Fast Similarity Search in the Presence of Noise, Scaling, and Translation in Time-Series Databases', In *Proc. VLDB*, pp.490-501, 1995.
2. Berndt, D.J., Clifford, J., 'Finding Patterns in Time Series: A Dynamic Programming Approach', In *Advances in KDD*, MIT 1996.
3. Box, G.E.P., Jenkins, G.M., 'Time Series Analysis - Forecasting and Control', San Francisco: Holden-Day, 1970.
4. http://www.spss.com/clementine/
5. Das, G., Gunopulos, D., Mannila, H., 'Finding similar time series', In *Proc. PKDD'97*, LNCS, 1997.
6. Das, G., Gunopulos, D., Mannila, H., 'Rule discovery from time series', In *Proc. KDD*, AAAI Press, 1998.
7. Dougarjapov, A., 'MSTS: Mining Sets of Time Series', *Master Thesis (in preparation)*, Institut für Informatik, Universität Freiburg, 2000.
8. Faloutsos, C., Ranganathan, M., Manolopoulos, Y., 'Fast Subsequence Matching in Time-Series Databases', In *Proc. SIGMOD*, 1994.
9. Giannotti, F., Manco, G., Nanni, M., Pedreschi, D., Turini, F., 'Integration of Deduction and Induction for Mining Supermarket Sales Data', In *Proc. SEBD'99*, 1999.
10. Giannotti, F., Pedreschi, D., 'Knowledge Discovery and Data Mining', In Tutorial Slides, EDBT, 2000.
11. Guralnik, V., Wijesekera, D., Srivastava, J., 'Pattern Directed Mining of Sequence Data', In *Proc. KDD*, AAAI Press, 1998.
12. Keogh, E., Smyth, P., 'A Probabilistic Approach to Fast Pattern Matching in Time Series Databases', In *Proc. KDD*, AAAI Press, 1997.
13. Keogh, E.J., Pazzani, M.J, 'An enhanced representation of time series which allows fast and accurate classification, clustering and relevance feedback', In *Proc. KDD*, 1998.
14. Keogh, E.J., Pazzani, M.J, 'An Indexing Scheme for Fast Similarity Search in Large Time Series Databases', In *Proc. SSDBM*, 1999.
15. Mannila, H., Toivonen, H., Verkamo, A.I., 'Discovery of frequent episodes in event sequences', *Data Mining and Knowledge Discovery Journal*, Vol.1, No.3, pp. 259-289, Nov 1997.
16. Prodromidis, A.L., Chan, P.K., Stolfo, S.J., 'Meta-Learning in Distributed Data Mining Systems: Issues and Approaches', To appear in *Advances in Distributed Data Mining*, AAAI Press, 1999.
17. Savnik, I., Lausen, G., Kahle, H.-P., Spieker, H., Hein, S., 'Algorithm for Matching Sets of Time Series', in *Proc. PKDD'00*, LNCS, 2000.
18. Quinlan, J.R., *C4.5: Programs for Machine Learning*, Morgan Kauffman, 1993.

Learning First Order Logic Time Series Classifiers: Rules and Boosting *

Juan J. Rodríguez[1], Carlos J. Alonso[1], and Henrik Boström[2]

[1] Grupo de Sistemas Inteligentes, Departamento de Informática
Universidad de Valladolid, Spain
{juanjo,calonso}@infor.uva.es
[2] Department of Computer and System Sciences
Stockholm University/KTH, Sweden
henke@dsv.su.se

Abstract. A method for learning multivariate time series classifiers by inductive logic programming is presented. Two types of background predicate that are suited for this task are introduced: interval based predicates, such as always, and distance based, such as the euclidean distance. Special purpose techniques are presented that allow these predicates to be handled efficiently when performing top-down induction. Furthermore, by employing boosting, the accuracy of the resulting classifiers can be improved significantly. Experiments on several different datasets show that the proposed method is highly competitive with previous approaches.

1 Introduction

Multivariate time series classification is useful in domains such as biomedical signals [13], continuous systems diagnosis [2] and data mining in temporal databases [6]. This problem can be tackled by extracting features of the series through some kind of preprocessing, and using some conventional machine learning method. However, this approach has several drawbacks [12]: the preprocessing techniques are usually *ad hoc* and domain specific, there are several heuristics applicable to temporal domains that are difficult to capture by a preprocess and the descriptions obtained using these features can be hard to understand. The design of specific machine learning methods for the induction of time series classifiers allows the construction of more comprehensible classifiers in a more efficient way.

When learning multivariate time series classifiers, the input consists of a set of training examples and associated class labels, where each example consists of one or more time series. The series are attributes of the examples, and are often referred to as variables, since they vary over time.

The method for learning time series classifiers that we propose in this work is based on *inductive logic programming* (ILP) and utilises two types of background predicate: i) interval based predicates, such as always(Example, Variable, Region,

---

* This work has been supported by the Spanish CYCIT project TAP 99-0344.

D.A. Zighed, J. Komorowski, and J. Żytkow (Eds.): PKDD 2000, LNAI 1910, pp. 299–308, 2000.
© Springer-Verlag Berlin Heidelberg 2000

Beginning, End), where Region is an interval in the domain of the variable, and ii) distance based predicates, such as euclidean_le(Example, Reference, Variable, Value), which considers the euclidean distance between two series.

The proposed system is neither a generic ILP system, nor is it an extension of one of the ILP systems available. In our experience, it is not enough to simply give predicates of the above types as background knowledge to an existing system, but it is also necessary to incorporate knowledge about how to process them efficiently. Hence, specific methods have been developed that allow for an efficient search for hypotheses that include these types of literals.

Moreover, we have also incorporated boosting, which is a method for generating ensembles of classifiers that has been demonstrated to improve accuracy significantly. In our case, the individual classifiers are composed only of rules with one literal in the body. When tested on several previously proposed datasets, the new method achieves better results than all previously published results on these datasets.

The rest of the paper is organized as follows. Section 2 describes the special purpose background predicates for time series classifiers. The proposed method is presented in section 3, including techniques for efficiently handling the special purpose predicates. Section 4 presents experimental results when using the new method. In section 5, it is demonstrated that these results can be improved substantially by employing boosting. Finally, we give some concluding remarks in section 6.

2 Temporal Predicates

2.1 Interval Predicates

The interval predicates make use of *regions*, which are intervals in the domain of the variable. The regions used are independent of the time, as is usual practice when working with series. The interval predicates are the following:

- always(Example, Variable, Region, Beginning, End). It is true, for the Example, if the Variable is always in this Region in the interval between Beginning and End.
- sometime(Example, Variable, Region, Beginning, End).
- true_percentage(Example, Variable, Region, Beginning, End, Percentage). It is true, for the Example, if the percentage of the time between Beginning and End where the variable is in Region is greater or equal to Percentage.

Once that it has been decided to work with temporal intervals, the use of the predicates always and sometime seems natural. Since the former predicate might be too demanding and the latter too flexible, a third one has been introduced, true_percentage.

2.2   Distance Predicates

Several machine learning methods, such as instance-based learning, are based on the use of similarity functions to measure distances between examples. The framework of ILP allows us to include several such definitions in the background knowledge. In our case, we use predicates on the following form:

<*distance*>_le(Example, Reference, Variable, Value)

which is true if the <*distance*>, for one Variable of the examples, between the Example considered and another Reference example is less or equal (_le) than Value. The predicate euclidean_le uses the *euclidean* distance. It is defined, for two univariate series s and t as: $\sqrt{\sum_{i=1}^{n}(s_i - t_i)^2}$. Its execution time is $O(n)$.

   Dynamic Time Warping (DTW) aligns a time series to another reference series in a way such that a distance function is minimized, using a dynamic programming algorithm [6]. If the two series have n points, the execution time is $O(n^2)$. The predicate dtw_le uses the minimized value obtained from the DTW as a similarity function between the two series.

3   Top-Down Induction of Time Series Classifiers

The proposed technique follows the scheme of the top-down methods in ILP [5]. The particularities arise in the selection of literals. The selection mechanisms suggested in this section could be incorporated in other ILP systems as well by using the generic method for numerical reasoning described in [18].

Obtaining Regions. In some cases, the definitions of the regions for the interval literals can be obtained from an expert. Otherwise, they can be obtained with a discretization preprocess, which obtains r disjoint, consecutive intervals. The regions considered are these r intervals (equality tests) and others formed by the union of the intervals $1 \ldots i$ (less or equal tests), as suggested by [8].

Selection of Interval Literals. Given an overly general clause that covers both positive and negative examples, the best literal to add to the body must be selected, according to some criterion. Then it is necessary to search over the space of literals. The possible number of intervals, if each series has n points, is $(n^2 - n)/2$. With the objective of reducing the search space, not all the intervals are explored. Only those that are of size power of 2 are considered. The number of these intervals is of $\sum_{i=1}^{k}(n - 2^{i-1}) = kn - 2^k - 1$ where $k = \lfloor \lg n \rfloor$.

   If p is the number of predicates considered, and v the total number of regions in the different variables, the possible number of atoms is $pvn \lg n$. In the case of predicates with additional arguments (true_percentage), it is also necessary to consider how many values are possible for them, but its number is a constant. Using a dynamic programming algorithm, the information that needs to be obtained from a window of size $2i$ is computed from two consecutive intervals of size i, with a time of $O(e)$, where e is the number of examples. Hence, the selection of the best literal requires a time of $O(epvn \lg n)$.

```
class( Example, cylinder ) :- % 213, 426
    not true_percentage( Example, x, 1_4, 22, 86, 50 ), % 193, 18
    not true_percentage( Example, x, 3, 78, 110, 30 ), % 182, 0
    !.
class( Example, bell ) :- % 213, 244
    true_percentage( Example, x, 1_4, 3, 35, 95 ), % 213, 1
    not always( Example, x, 1_4, 39, 103 ), % 213, 0
    !.
class( Example, funnel ) :- % 213, 31
    dtw_le( Example, f_35, x, 1.241447 ), % 200, 1
    not euclid_le( Example, c_162, x, 8.629407 ), % 200, 0
    !.
class( Example, cylinder ) :- % 31, 13
    not true_percentage( Example, x, 3, 9, 73, 15 ), % 31, 0
    !.
class( Example, funnel ). % 13, 0
```

Fig. 1. Rules example. The rules are ordered, organized in a decision list. They were obtained from the dataset *CBF* (Sect. 4). At the right of each literal, the number of positive and negative examples covered by the (partial) rule.

Selection of Distance Literals. Calculating the distances between all the examples requires computing $O(e^2)$ distances, which might be too costly. Furthermore, considering all the examples as reference for selecting a literal can also be too slow. Instead, in each iteration, r reference examples are randomly selected from the non-covered examples (it is possible to use only positive reference examples or positive and negative). If $d(n)$ is the time necessary for calculating the distance between two series with n points (n for the euclidean distance, n^2 for DTW), the best literal for r reference example can be calculated in $O(rv(e\,d(n) + e\lg e))$. The term $e\lg e$ is the time necessary for ordering the distances to the reference example and selecting the best value according to the criterion. Since the same example can be selected in several iterations, the calculated distances are saved.

Multiclass Problems. When there are more than two classes, it is necessary to learn a theory for each class. The question is how to apply these theories to a new example. We have employed two approaches for dealing with multiclass problems. The first one is the use of ordered rules, also named decision lists [14]. In this case, the first rule that covers the example assigns its label to it. The learning process consists of generating a rule for the class with most uncovered examples and iterate until all the examples are covered. Figure 1 shows an example of such a decision list. The second approach [19] is to learn different theories independently for each class, apply all the rules to the new example and if there is a conflict, solve it by considering the distribution of training examples covered by the rules.

4   Experimental Validation

Datasets for classification of time series are not easy to find [12]. For this reason we have used four artificial datasets and only one "real world" dataset:

- **Cylinder, Bell and Funnel (CBF).** This is an artificial problem [12], in which there are there are 3 classes: cylinder, bell and funnel. Figure 2.a shows two examples of each class. There are 266 examples of each class and each series has 128 points.
- **Control Charts.** In this dataset there are 6 classes of control charts [1]. Figure 2.b shows some examples of three of the classes. The data used were obtained from the UCI KDD Archive [4]. The number of examples is 600, with 60 points in each series.
- **Waveform.** This dataset was introduced by [9]. We used the version from the UCI ML Repository [7]. The number of examples is 900 and the number of points in each series is 21.
- **Wave + Noise.** This dataset was generated in the same way as the previous one, but 19 random points are added at the end of each example, with mean 0 and variance 1. Again, we used the dataset from the UCI ML Repository.
- **Auslan.** Auslan is the Australian sign language, the language of the Australian deaf community. Instances of the signs were collected using an instrumented glove [12]. Each example is composed by 8 series: x, y and z position, wrist roll, thumb, fore, middle and ring finger bend. There are 10 classes and 20 examples of each class. The number of points in each example is variable and currently the system does not support variable length series, so they were resampled to 20 points.

The results for each dataset and setting were obtained using five five-fold stratified cross-validations. The percentages considered for the predicate true_percentage were 5, 15, 30, 50, 70, 85 and 95. For similarity literals, at most 10 positive and 10 negative reference examples were considered when selecting literals.

Results are shown in Table 1. Euclidean is the best for the *Wave* and *Wave + Noise* using decision lists or unordered rules, and for the *Auslan* dataset with decision lists. The performance of the euclidean distance is a bit surprising considering its simplicity. Nevertheless its results are the worst for *CBF* and *Control Charts*. In these cases DTW works much better. This situation is reasonable, because the *CBF* and *Control* datasets are designed specifically to test time series classifers, and hence involve situations like shifts, compressions and expansions. These situations are not present in the *Wave* datasets, and hence euclidean works better than DTW.

The use of always and sometimes gives the best results in the *Control* dataset. In the case of decision lists, the first position is shared with the use of the predicate true_percentage. The advantages of using it over using always and sometimes are only clear for the *CBF* dataset. This is probably due to the fact that the different situations that characterize the examples have a beginning and end for the *CBF* dataset but for the *Control* dataset there are no returns to the

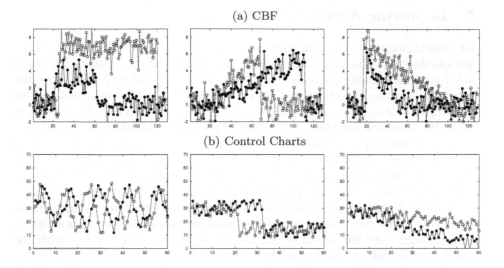

(a) CBF

(b) Control Charts

Fig. 2. Some examples of the datasets. Two examples of the same class are shown in each graph.

		EUC	DTW	AST	TRP	ALL
Decision Lists	CBF	6.67 2.39	2.16 1.22	4.51 1.38	2.76 1.25	**1.65** 1.10
	Control	4.63 1.38	3.23 1.71	**3.07** 1.64	**3.07** 1.66	3.20 1.55
	Wave	**19.64** 2.82	24.38 3.65	22.29 2.85	21.44 2.96	19.95 2.72
	Wave+Noise	**21.07** 3.04	26.78 2.78	24.80 3.35	24.62 2.44	22.31 3.17
	Auslan	**15.00** 4.21	21.60 6.33	21.40 7.00	20.10 4.97	16.60 5.49
Unordered Rules	CBF	6.87 2.16	2.11 1.02	4.31 2.03	2.88 1.28	**1.10** 0.89
	Control	6.03 2.20	4.10 1.89	**2.97** 1.23	4.13 1.98	3.60 1.64
	Wave	**19.47** 2.73	25.49 3.75	23.04 3.18	22.64 3.09	20.65 3.10
	Wave+Noise	**21.64** 2.16	28.84 2.51	24.67 2.53	25.58 3.39	22.91 2.54
	Auslan	24.60 7.49	24.20 5.94	22.80 6.55	23.80 6.13	**21.50** 7.74

Table 1. Experimental Results. The predicates considered are EUClidean, DTW, Always / SomeTime, TRue Percentage and ALL together. For each combination, the averaged error (in %) and the standard deviation of the 25 executions are shown. In boldface, the best results.

normal situation. The results obtained using all the predicates are the best for the *Auslan* dataset with unordered rules and for the *CBF* dataset. In the rest of the cases, the results are close to the best.

With respect to the use of decision lists or unordered rules, there is no clear winner regarding the error, although for the *Auslan* dataset decision lists work much better than unordered rules. Probably the inclusion of pruning methods could alter this situation.

5 Improving Accuracy by Boosting

At present, an active research topic is the use of *ensembles* of classifiers. They are obtained by generating and combining base classifiers, constructed using other machine learning methods. The target of these ensembles is to increase the accuracy with respect to the base classifiers. One of the most popular methods for creating ensembles is boosting [17], a family of methods, of which ADABOOST is the most prominent member. They work by assigning a weight to each example. In each iteration a base classifier is constructed, according to the distribution of weights. Afterwards, the weights are readjusted according to the result of the example in the base classifier. The final result is obtained by weighted votes of the base classifiers. Boosting inductive logic programming is described in [15].

Inspired by the good results of works using ensembles of very simple classifiers [17], *stumps*, we have studied base classifiers consisting only of one literal. The criterion used for selecting the best literal is to select the one with the smallest error, relative to the weights.

Multiclass Problems. There are several methods of extending AdaBoost to the multiclass case [17]. We have used ADABOOST.OC [16] since it can be used with any weak learner which can handle binary labeled data. It does not require that the weak learner can handle multilabeled data with high accuracy. The key idea is, in each round of the boosting algorithm, to select a subset of the set of labels, and train the binary weak learner with the examples labeled positive or negative depending if the original label of the example is or is not in the subset. In our concrete case, the rule inducer searches for a rule with the head:

   class(Example, [class$_1$, ... class$_k$])

This predicate means that the Example is of one of the classes in the list. Figure 3 shows an ensemble of these classifiers.

The classification of a new example is obtained from a weighted vote of the results of the weak classifiers. For each rule, if its antecedent is true the weights of all the labels in the list are incremented by the weight of the rule, if it is false the weights of the labels out of the list are incremented. Finally, the label that has been given the highest weight is assigned to the example.

5.1 Results

The results obtained with the boosting process are summarized in Table 2. For all the datasets, the benefits of using boosting are substantial. Moreover, the evolution of the error with the number of iterations is rather good. Until the limit of the number of iterations considered, the increment of the number of iterations do not cause more overfitting.

- **Cylinder, bell and funnel.** The best previously published result, according to our knowledge, with this dataset is an error of 1.9 [12], using 10 fold cross

```
class( Example, [ decreasing, downward, increasing ] ) :-    % 240, 240
    not true_percentage( Example, x, 5, 43, 59, 5 ).    % 197, 26
    % 0.456818
class( Example, [ decreasing, upward, normal ] ) :-    % 240, 240
    true_percentage( Example, x, 1_4, 14, 22, 70 ).    % 229, 84
    % 0.422232
class( Example, [ cyclic, decreasing, downward ] ) :-    % 240, 240
    sometime( Example, x, 2, 22, 54 ).    % 240, 0
    % 0.700627
class( Example, [ cyclic, decreasing, normal ] ) :-    % 240, 240
    dtw_le( Example, cyclic_54, x, 59.898940 ).    % 159, 0
    % 0.536895
. . .
```

Fig. 3. Initial fragment of an ensemble of classifiers, obtained with ADABOOST.OC, for the dataset *control charts* (Sect. 4). The weights of each individual classifier are below each classifier.

Iterations	5	10	20	30	40	50
CBF	2.41 1.21	1.50 0.85	0.75 0.60	0.60 0.53	0.58 0.65	**0.50** 0.60
Control	22.30 2.58	8.70 5.38	2.27 1.62	1.80 1.22	1.60 1.25	**1.47** 1.19
Wave	19.58 2.55	18.36 2.79	15.64 2.46	15.49 2.08	**15.00** 2.32	15.04 2.36
Wave + Noise	23.87 3.13	20.56 3.24	18.29 2.73	17.60 2.41	17.18 2.36	**16.78** 2.11
Auslan	61.50 7.14	37.90 7.90	16.70 5.72	11.50 4.39	8.40 3.88	7.60 3.27
+100 iter.		3.60 2.51	3.80 2.71	3.20 2.34	3.10 2.20	2.90 1.87
+150 iter.		3.00 1.77	3.10 2.20	2.50 1.91	2.80 2.20	2.40 2.22
+200 iter.		2.40 2.22	2.30 2.16	**1.90** 2.20	2.20 1.95	2.00 1.77

Table 2. Results obtained with boosting, for several numbers of iterations, using all the background predicates. For the *Auslan* dataset, the table includes results with more iterations (110, 120 ... 150; 160 ... 200 and 210 ... 250). In some cases, it is possible to improve these results using not all the predicates (see discussion for each dataset).

validation. From iteration no. 10, the results shown in table 2 are better than this result, and from iteration no. 20 the results are smaller than 1.0. Moreover, the results using rules without boosting (Table 1) are also better than 1.9 (1.65 and 1.10).

- **Control charts.** The only previous result we are aware of regarding this dataset is for similarity queries [1], and not for supervised classification. To check if this dataset was trivial, we tested it with C4.5, over the raw data, and obtained an average error of 8.6 (also using five five-fold cross validation). It should also be noted that obtained error is reduced to 0.82 using 100 iterations and only dtw_le.
- **Waveform.** The error of a Bayes optimal classifier on this dataset is approximately 14 [9]. The best previous result we are aware of for this dataset is an error of 15.21 [11]. That result was obtained using boosting, with decision

trees as base classifiers, which are much more complex than our base classifiers (clauses with one literal in the body). The obtained error is reduced to 14.42 using 100 iterations and euclid_le.

- **Wave + Noise.** Again, the error of an optimal Bayes classifier on this dataset is 14. This dataset was tested with bagging, boosting and variants over decision trees [3]. Although their results are given in graphs, their best error is apparently approximately 17.5. The obtained error is reduced to 14.69 using 100 iterations and euclid_le.
- **Auslan.** This is the dataset with the highest number of classes (10). In order to distinguish between this large number of classes, it turned out that 50 one-body-literal clauses were too few, so we incremented the number of iterations for this dataset, until 250. The best result previously published is an error of 2.50 [12], which is greater than the results obtained after 200 iterations, as shown in Table 2.

6 Conclusions and Future Work

A multivariate time series classification system has been developed, using ILP techniques. Two types of background predicate were considered: those based on intervals and regions and those based on similarity functions. The use of similarity literals allows a smooth integration of Instance Based Learning and ILP, leading to that similarity functions can be defined by the user and incorporated as background knowledge and that these functions appear explicitly in the rules instead of being an external element to the learned theory. Special purpose techniques have been presented that allow these predicates to be handled efficiently when performing top-down induction. Furthermore, we have demonstrated that by employing boosting, the accuracy of the resulting time series classifiers can be improved significantly. Experiments on several different datasets show that the proposed method is highly competitive with previous approaches. On all data sets, the proposed method achieves better than all previously reported results.

Boosting generic learners, which are not designed for time series, such as decision trees, can produce good results for time series classification. Nevertheless, the incorporation of temporal predicates, in the background knowledge, improves the performance of boosting, especially with respect to the size of the classifiers.

Another method to improve the accuracy of a classifier is the use of pruning techniques. Pruning also enhances the comprehensibility of classifiers, while boosting worsens it. Nevertheless, pruning and boosting are not incompatible, e.g., when boosting rules it is possible to prune them [10]. Hence, it would be convenient to incorporate pruning techniques in the rule induction system, and also to apply boosting over pruned rules.

Acknowledgements. To the maintainers of the ML [7] and KDD [4] UCI Repositories. To Mohammed Waleed Kadous, David Aha and Robert J. Alcock for, respectively, donating the *Auslan*, *Wave* and *Control Charts* datasets.

References

[1] R.J. Alcock and Y. Manolopoulos. Time-series similarity queries employing a feature-based approach. In 7th Hellenic Conference on Informatics, Ioannina, Greece, 1999.

[2] C.J. Alonso and J.J. Rodríguez. A graphical rule language for continuous dynamic systems. In Computational Intelligence for Modelling, Control and Automation, CIMCA-99. IOS Press, 1999.

[3] E. Bauer and R. Kohavi. An empirical comparison of voting classification algorithms: Bagging, boosting and variants. Machine Learning, 36(1/2), 1999.

[4] S.D. Bay. The UCI KDD archive, 1999. http://kdd.ics.uci.edu/.

[5] F. Bergadano and D. Gunetti. Inductive Logic Programming: from machine learning to software engineering. The MIT Press, 1995.

[6] D.J. Berndt and J. Clifford. Finding patterns in time series: a dynamic programming approach. In U.M. Fayyad, G. Piatetsky-Shapiro, P. Smyth, and R. Uthurusamy, editors, Advances in Knowledge Discovery and Data Mining. AAAI Press / MIT Press, 1996.

[7] C.L. Blake and C.J. Merz. UCI repository of machine learning databases, 1998. http://www.ics.uci.edu/~mlearn/MLRepository.html.

[8] H. Blockeel and L. De Raedt. Lookahead and discretization in ILP. In 7th International Workshop on Inductive Logic Programming, ILP'97. Springer, 1997.

[9] L. Breiman, J.H. Friedman, A. Olshen, and C.J. Stone. Classification and Regression Trees. Chapman & Hall, New York, 1993.

[10] W.W. Cohen and Y. Singer. A simple, fast, and effective rule learner. In 16th National Conference on Artificial Intelligence, AAAI-99, 1999.

[11] T.G. Dietterich. An experimental comparison of three methods for constructing ensembles of decision trees: bagging, boosting, and randomization. Machine Learning, 1999.

[12] M.W. Kadous. Learning comprehensible descriptions of multivariate time series. In 16th International Conference of Machine Learning (ICML-99). Morgan Kaufmann, 1999.

[13] M. Kubat, I. Koprinska, and G. Pfurtscheller. Learning to classify biomedical signals. In R.S. Michalski, I. Bratko, and M. Kubat, editors, Machine Learning and Data Mining. John Wiley & Sons, 1998.

[14] R.J. Mooney and M.E. Califf. Induction of first-order decision lists: results on learning the past tense of english verbs. JAIR, 3, 1995.

[15] J.R. Quinlan. Boosting first-order learning. In S. Arikawa and A. Sharma, editors, Algorithmic Learning Theory, 7th International Workshop, ALT'96. Springer, 1996.

[16] R.E. Schapire. Using output codes to boost multiclass learning problems. In 14th International Conference on Machine Learning (ICML-97), 1997.

[17] R.E. Schapire. A brief introduction to boosting. In 16th International Joint Conference on Artificial Intelligence (IJCAI-99). Morgan Kaufmann, 1999.

[18] A. Srinivasan and R.C. Camacho. Numerical reasoning with an ILP system capable of lazy evaluation and customized search. Journal of Logic Programming, 40(2–3), 1999.

[19] W. Van Laer, L. De Raedt, and S. Dzeroski. On multi-class problems and discretization in inductive logic programming. In 10th International Symposium on Methodologies for Intelligent Systems (ISMIS97). Springer, 1997.

Learning Right Sized Belief Networks by Means of a Hybrid Methodology*

Silvia Acid and Luis M. De Campos

Departamento de Ciencias de la Computación e I.A.
E.T.S.I. Informática, Universidad de Granada
18071 - Granada, SPAIN
{acid,lci}@decsai.ugr.es

Abstract. Previous algoritms for the construction of belief networks structures from data are mainly based either on independence criteria or on scoring metrics. The aim of this paper is to present a hybrid methodology that is a combination of these two approaches, which benefits from characteristics of each one, and to introduce an operative algoritm based on this methodology. We dedicate a special attention to the problem of getting the 'right' size of the belief network induced from data, i.e. finding a trade-off between network complexity and accuracy. We propose several approaches to tackle this matter. Results of the evaluation of the algorithm on the well-known Alarm network are also presented.

1 Introduction

Graphical models such as belief networks BN [13] have become very attractive tools because of their ability to represent knowledge with uncertainty and to efficiently perform reasoning tasks. The knowledge that they manage is in the form of dependence and independence relationships, two basic notions in the human reasoning. Both relationships are coded by means of (1) a qualitative component of the model, i.e., the directed acyclic graph (dag) and (2) a collection of numerical parameters, usually conditional probabilities which measure the strength of the dependencies displayed in the model.

In recent years, many BN learning algorithms from databases have been developed, coming from different approaches and different principles. Generally, they can be grouped in two main approaches: methods based on conditional independence tests [5,15], and methods based on a scoring metric [8,10,11,12].

The algorithm we are going to describe in this paper do not fall clearly in any of these two categories but it utilizes a hybrid methodology: it uses a specific metric and a search procedure (so, it belongs to the group of methods based on scoring metrics), although it also explicitly makes use of the conditional independences embodied in the topology of the network to elaborate the scoring metric (hence it has also strong similarities with the algorithms based on independence

---

* This work has been supported by the Spanish Comisión Interministerial de Ciencia y Tecnología (CICYT) under Project n. TIC96-0781.

D.A. Zighed, J. Komorowski, and J. Żytkow (Eds.): PKDD 2000, LNAI 1910, pp. 309–315, 2000.
© Springer-Verlag Berlin Heidelberg 2000

tests). The only precedents we know about hybrid learning algorithms are the works in [9,16,14]. The rest of the paper is organized as follows. In Section 2 we outline our methodology and we propose an algorithm, BENEDICT-*dsep* for recovering the graph with the restriction that the ordering between the variables is given. Section 3 discusses different approaches to get right sized models from the learning process. In Section 4, we carry out a comparative study of the results obtained by the algorithm using the well-known Alarm network. Finally, Section 5 contains the concluding remarks and some proposals for future works.

2   Learning Belief Networks with BENEDICT-*dsep*

The algorithm we are going to describe, BENEDICT-*dsep*, works under the assumption that the total ordering of the variables is known. This assumption, although somewhat restrictive, is quite frequent for learning algorithms. The algorithm is part of a family of algorithms [3,4] that share a common methodology [1] for learning belief networks, which we have called BENEDICT. The name is an acronym of *BE*lief *NE*tworks *DI*scovery using *C*ut-set *T*echniques, and is motivated by the use of d-separating sets or cut-sets to define the metric.

Let us briefly describe the BENEDICT methodology. The basic idea is to measure the discrepancies between the conditional independencies (d-separation statements) represented in any given candidate network G and the ones displayed by the database D. The lesser these discrepancies are, the better the network fits the data. The aggregation of all these (local) discrepencies will result in a measure $g(G, D)$ of global discrepancy between the network and the database.

To measure the discrepancy of each one of the independencies in the graphical model and the numerical model (the database), BENEDICT uses the Kullback-Leibler cross entropy:

$$Dep(X, Y|Z) = \sum_{x,y,z} P(x, y, z) \log \frac{P(x, y|z)}{P(x|z)P(y|z)},$$

where x, y, z denote instantiations of the sets of variables X, Y and Z respectively, and P is a probability estimated from the database.

As the number and complexity of the d-separation statements in a dag G may grow exponentially with the number of nodes, we cannot use all the d-separations displayed by G, but we have to focus on some selected subset of 'representative' d-separation statements and ignore the remainders. Given any candidate network G, BENEDICT calculates, from the database, the conditional dependence degrees of any two non-adjacent single variables, x_i and x_j (assuming that $x_j <_\theta x_i$) given the set of minimum size, $S_G(x_i, x_j)$, that d-separates x_i and x_j in G, $Dep(x_i, x_j|S_G(x_i, x_j))$. Of course, finding this set takes some additional effort, but it is compensated by a decreasing computing time of the corresponding dependence degree and more reliable results. The method BENEDICT uses for efficiently finding the sets $S_G(x_i, x_j)$ is described in detail in [2][1].

In order to give a score to a specific network structure G given a database D, BENEDICT uses the aggregation (the sum) of the local discrepancies, as

---

[1] In [7], approximate minimum cut-sets are used instead of exact ones.

measure of global discrepancy $g(G, D)$ (which has to be minimized). Finally, the type of search method used by BENEDICT is a simple greedy search that allows to insert into the structure the candidate arc that produces a greater improvement of the score (removal of arcs is not permitted).

Let us describe more especifically the algorithm BENEDICT-*dsep*. Initially, it starts out from a completely disconnected network. The current set L of candidate arcs to be introduced in the current network is composed by all the arcs coherent with the ordering. The scoring of this 'empty' graph G_0 is then calculated, as $g(G_0, D) = \sum_{i=2}^{n} \sum_{x_j < \iota x_i} Dep(x_i, x_j | \emptyset)$. Next, the algorithm looks for the arc whose addition to the graph results in a greater decrease of $g(., D)$, thus obtaining a new graph, G_1, containing only one arc. The process continues in this way, adding at each step, k, the single arc, say $x_j \to x_i$, which verifies

$$g(G_{k-1} \cup (x_j \to x_i), D) = \min_{\substack{x_h \to x_l \in L \\ x_h \to x_l \notin G_{k-1}}} g(G_{k-1} \cup (x_h \to x_l), D).$$

In this way arcs are added in a stepwise process until a stopping rule is satisfied.

3 Stopping Rules

In the description of the algorithm BENEDICT-*dsep* we have not specified the way in which we stop the learning process, i.e., how to decide whether the size of the graph is adequate and then do not add more arcs to the current structure. The extent to which the resultant network is a good description of the domain strongly depends on this question.

For example, if we let the graph grow until no more arcs can be included (thus obtaining a complete graph), this network has a zero global discrepancy with every database, since it does not explicitly represent any independence relationship. However, this network is far to be descriptive (although it may represent any independence relationship, that remains hidden in the conditional probability tables). In general, if we stop the learning process too late the resultant networks are very complex to be understood, estimated and used for inference purposes (moreover, an overfitting phenomenon can appear). On the other hand, if we stop too early the growing process, in order to get a simpler structure, we may lose the chance to improve it (i.e., to get a better score). So, the network displays more independences than those they are really supported by the data, thus resulting in a bad approximation of the distribution underlying the data. Therefore, we have to focus on finding appropiate stopping rules, that is, on finding a criterion for determining a good trade-off between a good description of the mechanisms which generated the data, and the simplicity of the model.

The first stopping rule we tried consisted in setting a threshold and deciding not to include an arc if the global discrepancy with the data was beneath it. A second alternative, also using thresholds, was not to include an arc if no significant improvement is achieved by adding the best arc to the current graph.

After having studied the behaviour of our algorithm using the two aforementioned stopping rules [3] (and using also a combination of both rules), by means of several databases, we get clear that these rules produced in general unsatisfactory results. The problem has two aspects: First, the threshold values are critic to the density of the learned networks. Second, it is quite difficult to determine automatically, from the data distribution, the appropriate values of these thresholds. Thus, we decided to tackle the problem of finding the optimum size of the learned network from other perspectives:

- Using independence tests to remove arcs from the set of candidate arcs, and thus avoiding to include them in the structure regardless of the value of the discrepancy reached by the current network.
- Pruning instead of stopping. Grow a network in a not very restrictive manner and then prune it by revisiting each arc (dependency) established in the building process, in order to check whether it was introduced prematurely. This is inspired by the common practise of using methods for pruning decision trees, which often produce excelent results.

3.1   Independence Tests

The basic idea for using independence tests is to consider as candidate, at every step of the learning process, only the arcs between pairs of (non-adjacent) variables which have not been found independent. In the case that the test establishes an independence relationship, then the arc linking the independent variables is deleted from the set of candidate arcs L. Therefore, the learning process will stop naturally when the set of candidate arcs becomes empty (the arcs are removed from this set either because they are inserted into the current graph or because their extreme nodes are found to be independent by the test).

In order to define the kind of test, we will take advantage of the fact that the statistic $2 * N * Dep(x_i, x_j | Z)$ is approximately χ^2-distributed with $\|Z\|(\|x_i\| - 1)(\|x_j\| - 1)$ degrees of freedom, where N is the number of samples in D, and $\|.\|$ represents the number of different states in the corresponding set of variables.

Let us see how to use the independence tests in our algorithms: At each step, once the best candidate arc $x_j \to x_i$ has been selected, it is included in the current structure G (and removed from the set of candidate arcs); then the remaining candidate arcs in L are reexamined in order to see if some may be removed because their extreme nodes are found independent by the test. The newly introduced arc changes the connectivity of the network, and can turn into independent variables that were not yet determined as independent. All these independence tests are based on values that have been already calculated, so they entail almost null additional computational cost.

3.2   The Pruning Process

The first step is to grow the network by letting the building process to work with a non very restrictive stopping criterion. In the resultant structure G, some right arcs will be introduced and also some additional (incorrect) ones. Then a review of the arcs included in the structure may reveal that some of them are superfluous and thus can be removed from the network.

An arc in G may be considered superfluous if the nodes it connects are conditionally independent (i.e., the dependence degree between the pair is very low). Perhaps this fact was not discovered before because, due to the greedy nature of the process, the d-separating set used previously to try to separate these nodes was not correct. However, the information obtained in subsequent steps (new arcs have been introduced) changes the connectivity of the network and therefore may also change the relationships between the variables. So, to verify the possible independence relationship between the linked variables, the arc is removed momentarily and the *Dep* value is calculated to be used into an independence test. If the independency is true, the arc is removed definitively, otherwise it is put back in the network.

The ordering to applying these tests is precisely the ordering in which the arcs were introduced in the structure: in this way we can reexamine the established dependences in the light of the information obtained after any arc $x_j \rightarrow x_i$ is introduced (so, the d-separating set that, at the moment in which $x_j \rightarrow x_i$ was inserted, was not able to make independent x_i and x_j, could change, thus being able to render x_i and x_j independent).

4   Experimental Results

In this section, we are going to present the results obtained by our algorithm using different databases in the learning process. All the experiments were carried out on a sun4m sparc station at 100Mhz, to reconstruct the so-called Alarm belief network. The Alarm network contains 37 variables and 46 arcs. The input data commonly used are subsets of the Alarm database, which contains 20,000 cases that were stochastically generated using the Alarm network, specifically, we use the first 500, 1000, 2000 and 3000 cases for our experiments.

The information we will show about each experiment with our algorithm is the following: *Time.-* The time, measured in minutes and seconds, spent in learning the Alarm network. $g(G, D)$.- The value of the measure of global discrepancy with respect to the database D it comes from. *Ham.-* The Hamming distance, number of different arcs (missing or added) in the learned network with respect to the original model. *N.Arc.-* Number of arcs in the learned network. *Kullback.-* The Kullback distance (cross-entropy) between the probability distribution, P, associated to the database and the probability distribution associated to the learned network, P_G, to measure how closely the probability distribution learned approximates the empirical frecuency distribution. Actually, we have calculated a decreasing monotonic linear transformation of the Kullback distance, because this one has exponential complexity and the transformation can be computed in a very efficient form [12]. The interpretation of our transformation of the Kullback distance is: the higher this parameter is the better is the network.

Our first experiment concentrates around the stopping rules, aiming to determine the quality of the different rules proposed: using thresholds, independence tests, and pruning. In the first case we use a conjunction of thresholds (low discrepancy and no significant improvement). Both thresholds were set 'ad hoc' as a fraction of the initial discrepancy calculated on the empty graph. We use the

χ^2 test for the conditional independence tests and the pruning process with a fixed confidence level of 0.99. For this experiment we used the database containing 3000 cases, and the results are displayed in table 1. From these results it is

Table 1. Comparing stopping rules: Results using BENEDICT-*dsep*.

	Time	$g(G, D)$	Kullback	Ham.	N.Arc
Thresholds	20:58	1.3754	8.2329	17	31
Indep.	7:10	0.6995	9.2213	10	50
Indep.+ pruning	7:12	0.6818	9.2036	4	44

clear that the use of independence tests improves remarkably the performance of the algorithm with respect to the use of thresholds, as much in the quality of the results as in the efficiency. They present lower discrepancy values and also higher values of the objective Kullback measure. On the other hand, we can observe that the times spent are markedly lower, than using thresholds. This gives us an idea of the savings in evaluating candidate networks obtained by removing some arcs from the candidate set L. We can also see that final process of pruning, using almost null additional time, simplifies the final structures, although slightly decreases the Kullback distances; anyway, the results are comparable. In the light of these results, in the following experiments we always will use the combination of independence tests and pruning as the stopping rule.

In the second experiment our interest is focused in evaluating the robustness of the algorithm. The results of this experiment are displayed in table 2. Several

Table 2. Using different sample sizes: Results with BENEDICT-*dsep*.

Sample Size	Time	$g(G, D)$	Kullback	Ham.	N.Arc
500	2:22	2.7586	8.9759	11	43
1000	3:49	2.4367	9.0927	9	45
2000	5:06	0.9605	9.1345	5	45
3000	7:12	0.6818	9.2036	4	44

conclusions may be drawn from this experiment. First, it seems that the learned networks approach the original one as the sample size increases, as shown by the Hamming distances and the Kullback measures. The time employed is almost linear with respect to the sample size. Specifically, for the bigger training set, the mistaken arcs are 3 missing arcs, where two of them are not strongly supported by the data. The third missing arc makes to introduce an additional arc.

5   Concluding Remarks

We have proposed a hybrid methodology for learning baycsian belief networks from databases, and an algorithm based on it. Our algorithm uses conditional independence relationships of order as low as possible, thus gaining in efficiency and reliability. The algorithm recovers reasonably well complex belief network structures from data in polynomial time, under the assumption that the ordering among the variables is known. Several possible extensions of this work, could be:

- Avoiding the requirement of an ordering on the nodes. One approach is to leave the search procedure to also cope with the directions of the arcs to be added. Another option is to perform a two-steps process: the first one is to run an algorithm for learning a good ordering [6], the second one would be carried out by our algorithm.
- Modification of the pure greedy search used, other search techniques such as branch and bound or simulated annealing, could be used in combination with our scoring metric. We could modify the search process by using arc addition as well as arc removal (or even arc reversal).

References

1. S. Acid. *Métodos de aprendizaje de redes de creencia. Aplicación a la clasificación.* PhD thesis, Universidad de Granada. Spain, 1999 (in Spanish).
2. S. Acid, L.M. de Campos. An algorithm for finding minimum d-separating sets in belief networks. in: *Proceedings of the 12th Conf. on Uncertainty in Artificial Intelligence*, 1996, 3–10.
3. S. Acid, L.M. de Campos. Benedict: An algorithm for learning probabilistic belief networks. in: *Proceedings of the 6th IPMU Conf.*, 1996, 979–984.
4. S. Acid, L.M. de Campos. A Hybrid Methodology for Learning Belief Networks: BENEDICT. *Int. Journal of Approximate Reasoning* Submitted.
5. L.M. de Campos, J.F. Huete. A new approach for learning belief networks using independence criteria. *Int. Journal of Approximate Reasoning* 24 (2000) 11–37.
6. L.M. de Campos, J.F. Huete. Approximating causal orderings for Bayesian networks using genetic algorithms and simulated annealing. in: *Proceedings of the IPMU-2000 Conf.*, to appear.
7. J. Cheng, D.A. Bell, W. Liu. An algorithm for Bayesian belief network construction from data. in: *Proceedings of the Seventh Int. Workshop on Artificial Intelligence and Statistics*, 1997, 83–90.
8. G.F. Cooper, E. Herskovits. A bayesian method for the induction of probabilistic networks from data. *Machine Learning* 9 (4) (1992) 309–348.
9. D. Dash, M.J. Druzdzel. a hybrid anytime algorithm for the construction of causal models from sparse data. in: *Proceedings of the 15th Conf. on Uncertainty in Artificial Intelligence*, 1999.
10. D.Heckerman, D.Geiger, D.M. Chickering. Learning bayesian networks: The combination of knowledge and statistical data. *Machine Learning* 20 (1995) 197–243.
11. E. Herskovits, G.F. Cooper. Kutató: An entropy-driven system for the construction of probabilistic expert systems from databases. in: P. Bonissone (Ed.), *Proceedings of the 6th Conf. on Uncertainty in Artificial Intelligence*, Cambridge, 1990, 54–62.
12. W. Lam, F. Bacchus. Learning bayesian belief networks. an approach based on the mdl principle. *Computational Intelligence* 10(4) (1994) 269–293.
13. J. Pearl. *Probabilistic Reasoning in Intelligent Systems: Networks of Plausible Inference.* Morgan Kaufmann, San Mateo, 1988.
14. M. Singh, M. Valtorta. Construction of Bayesian network structures from data: A brief survey and an efficient algorithm. *International Journal of Approximate Reasoning* 12 (1995) 111–131.
15. P. Spirtes, C. Glymour, R. Scheines. *Causation, Prediction and Search.* Lecture Notes in Statistics 81. Springer Verlag, New York, 1993.
16. P. Spirtes, T. Richardson, C. Meek. Learning Bayesian networks with discrete variables from data. in: *Proceedings of the First Int. Conf. on Knowledge Discovery and Data Mining*, 1995, 294–299.

Algorithms for Mining Share Frequent Itemsets Containing Infrequent Subsets

Brock Barber and Howard J. Hamilton

Department of Computer Science, University of Regina, Regina, SK., Canada S4S 0A2

Abstract. The share measure for itemsets provides useful information about numerical values associated with transaction items, that the support measure cannot. Finding share frequent itemsets is difficult because share frequency is not downward closed when it is defined in terms of the itemset as a whole. The Item Add-back and Combine All Counted algorithms do not rely on downward closure and thus, are able to find share frequent itemsets that have infrequent subsets. These heuristic algorithms predict which itemsets should be counted in the current pass using information available at no additional processing cost.

1 Introduction

Association rules identify items that occur together and those that are likely to occur, given that particular items have been selected (called itemsets). The discovery of association rules is a two-step process [1]: (1) discover all frequent itemsets meeting user-specified frequency criteria, and (2) generate association rules from the frequent itemsets. The second task is easier than the first [11]. Here, we study the first step in the context of itemset share, a measure of itemset importance [4].

Itemset share is the fraction of some numerical value, such as total quantity of items sold or total profit, contributed by items when they occur in an itemset. Unlike support [1], share can be applied to the non-binary numerical data associated with items in a transaction, allowing for a more insightful analysis of the impact of itemsets in terms of stock, cost or profit. In practice, itemset ranking by support and share can be significantly different [4].

Support frequency (*frequency$_{sup}$*) is downward closed, since all subsets of a frequent$_{sup}$ itemset are also frequent$_{sup}$ [3]. This property allows efficient algorithms to find all frequent$_{sup}$ itemsets while traversing only a part of the itemset lattice, e.g. [3, 6, 11]. Share frequency is also downward closed if we require each item in a frequent itemset to be frequent when it occurs in the itemset [4]. However, since share considers non-binary values, the share of an itemset can be greater than the share of its subsets. If the frequency requirement is based on the total share of the itemset, frequent itemsets might contain infrequent subsets. Thus, some frequent itemsets cannot be found using the downward closed share frequency definition. We describe heuristic algorithms to discover share frequent itemsets that do not rely on downward closure.

D.A. Zighed, J. Komorowski, and J. Zytkow (Eds.): PKDD 2000, LNAI 1910, pp. 316-324, 2000.
© Springer-Verlag Berlin Heidelberg 2000

2 Review of the Support Measure

Itemset methodology is summarized as follows. [2]. Let $I = \{I_1, I_2, ..., I_m\}$ be a set of literals, called *items*. Let $D = \{T_1, T_2, ..., T_n\}$ be a set of n transactions, where for each transaction $T \in D$, $T \subseteq I$. A set of items $X \subseteq I$ is called an *itemset*. Transaction T *contains* X if $X \subseteq T$. Each itemset X is associated with a set of transactions $T_X = \{T \in D \mid T \supseteq X\}$, the transactions containing X. The *support* s of itemset X equals $|T_X|/|D|$.

Support is illustrated using the transaction database shown in Table 1. The TID column gives the transaction identifier values. Values under each item name are quantity of item sold. To calculate support, a non-zero quantity is treated as a 1. Table 2 shows the support for each possible itemset.

Table 1. Example Transaction Database

TID	Item A	Item B	Item C	Item D
T1	1	0	1	14
T2	0	0	6	0
T3	1	0	2	4
T4	0	0	4	0
T5	0	0	3	1
T6	0	0	1	13
T7	0	0	8	0
T8	4	0	0	7
T9	0	1	1	10
T10	0	0	0	18

Table 2. Itemset Support

Item-set	s	Item-set	s
A	0.30	BC	0.10
B	0.10	BD	0.10
C	0.80	CD	0.50
D	0.70	ABC	0.00
AB	0.00	ABD	0.00
AC	0.20	ACD	0.20
AD	0.30	BCD	0.10
		ABCD	0.00

Transaction data often contains information such as quantity sold or unit profit, that support cannot consider. For example, support for items C and D is 0.8 and 0.7 respectively. However, total quantity sold for C and D is 26 and 67, respectively, so D is sold more frequently than C. Itemsets BC and BD have support of 0.10, indicating equal frequency. However, the quantity of items sold in BC and BD is 2 and 11, respectively. If items B, C and D return a net profit of $1.00, $100.00 and $0.10, then itemsets BC and BD return a net profit of $101.00 and $1.10. Yet support does not consider itemset BC to be more important than itemset BD. Support fails as a measure of relative importance of the itemsets in these instances. For target marketing, measures should consider both the frequency of an item contributing to a predictive rule and the value of the items in the prediction [7]. Support allows for neither, so measures based on specific numbers of items, such as percentage of gross sales, costs or net profit, cannot be calculated, and business payoff cannot be maximized.

3 Review of the Share Measure

A *measure attribute* (MA) is a numerical attribute associated with each item in each transaction. The *transaction measure value* of item I_p in transaction T_q, $tmv(I_p, T_q)$, is the value of a measure attribute associated with I_p in T_q. The *global measure value* of item I_p, $MV(I_p)$, is the sum of the *tmv*'s of I_p in all transactions in which I_p appears:

$$MV(I_p) = \sum_{T_q \in T_{Ip}} tmv(I_p, T_q) . \tag{1}$$

The *total measure value* (*MV*) is the sum of the global measure values for all items in *I* in every transaction in *D*, given as

$$MV = \sum_{p=1}^{m} MV(I_p) . \tag{2}$$

If x_i is the ith item of itemset *X*, the *item local measure value* of x_i in *X*, $lmv(x_i, X)$, is the sum of the transaction measure values of x_i in all transactions containing *X*, given by

$$lmv(x_i, X) = \sum_{T_q \in T_X} tmv(x_i, T_q) . \tag{3}$$

The *itemset local measure value* of *X*, $lmv(X)$, is the sum of the local measure values of each of the *k* items in *X* in all transactions containing *X*, given by

$$lmv(X) = \sum_{1}^{k} lmv(x_i, X) . \tag{4}$$

The *item share* of x_i in *X*, $SH(x_i, X)$, is the ratio of the local measure value of x_i in *X* to the total measure value, as given by

$$SH(x_i, X) = lmv(x_i, X)/MV . \tag{5}$$

The *itemset share* of *X*, $SH(X)$, is the ratio of the local measure value of *X* to the total measure value, as calculated by

$$SH(X) = lmv(X)/MV . \tag{6}$$

Table 3 gives $lmv(X)$ and $SH(X)$ for itemsets in the sample database and $lmv(x_i, X)$ and $SH(x_i, X)$ for items in these itemsets. A dash means an item is not in an itemset.

Consider again itemsets C, D, BC and BD. Itemset C, ranked higher than itemset D by support despite having a lower quantity sold, is ranked lower by share, with $SH(C)$ = 0.26 and $SH(D)$ = 0.67. Itemsets BD and BC are ranked the same by support, although the quantity of items sold in BC is less. Using share, itemset BD is ranked higher than itemset BC, with $SH(BD)$ = 0.11 and $SH(BC)$ = 0.02.

Share can be incorporated into many algorithms developed for support [5]. Approaches have been proposed for extending support to quantitative measures, e.g. [10]. We feel share is simpler and more flexible. The problem of finding frequent$_{sup}$ itemsets with $lmv(X) \geq minvalue$ has been described [8]. No solution was presented.

Table 3. Itemset Share Summary for Sample Database

Itemset	Item A		Item B		Item C		Item D		Itemset X	
	Lmv	SH	lmv	SH	lmv	SH	lmv	SH	lmv	SH
A	6	0.06	-	-	-	-	-	-	6	0.06
B	-	-	1	0.01	-	-	-	-	1	0.01
C	-	-	-	-	26	0.26	-	-	26	0.26
D	-	-	-	-	-	-	67	0.67	67	0.67
AB	0	0.00	0	0.00	-	-	-	-	0	0.00
AC	2	0.02	-	-	3	0.03	-	-	5	0.05
AD	6	0.06	-	-	-	-	25	0.25	31	0.31
BC	-	-	1	0.01	1	0.01	-	-	2	0.02
BD	-	-	1	0.01	-	-	10	0.10	11	0.11
CD	-	-	-	-	8	0.08	42	0.42	50	0.50
ABC	0	0.00	0	0.00	0	0.00	-	-	0	0.00
ABD	0	0.00	0	0.00	-	-	0	0.00	0	0.00
ACD	2	0.02	-	-	3	0.03	18	0.18	23	0.23
BCD	-	-	1	0.01	1	0.01	10	0.10	12	0.12
ABCD	0	0.00	0	0.00	0	0.00	0	0.00	0	0.00

4 Share Frequent Itemsets

Property P is *downward closed* with respect to the lattice of all itemsets if, for each itemset with property P, all of its subsets have property P [9]. Share frequency as originally defined is downward closed. Itemset X is *downward closed share frequent* (*DC-frequent*) if $\forall x_i \in X$, $SH(x_i,X) \geq minshare$, a user defined minimum share [4].

Theorem 1: DC-frequency is downward closed with respect to the lattice of all itemsets. **Proof**: To show DC-frequency is downward closed, we must show that if X is DC-frequent, then for all $X_j \subseteq X$, X_j must be DC-frequent. Suppose X is DC-frequent. By definition, for all $x_i \in X$, $SH(x_i,X) \geq minshare$. Since $X_j \subseteq X$, $lmv(x_i,X_j) \geq lmv(x_i,X)$ and $SH(x_i,X_j) = lmv(x_i,X_j)/MV \geq SH(x_i,X) = lmv(x_i,X)/MV$. Since for all $x_i \in X$, $SH(x_i,X) \geq minshare$, then for all $x_i \in X_j$, $SH(x_i,X_j) \geq minshare$. Therefore, by definition, X_j is DC-frequent. •

To find frequent itemsets with infrequent subsets, itemset X is defined to be share frequent, or simply frequent, if $SH(X) \geq minshare$. This definition removes the property of downward closure. Adding an item x_i to itemset X to create itemset Y, adds a restriction to the measure values of the items in X. Values associated with the items in X contribute to $lmv(Y)$, only when they occur with x_i. Their contribution towards $lmv(Y)$ must be less than or equal to their contribution to $lmv(X)$. However, $lmv(x_i,Y)$ is added to $lmv(Y)$, counteracting the effect of the additional restriction. Thus, $lmv(Y)$ may be less than, equal to, or greater than $lmv(X)$, depending on the relative effect of the restriction and the addition of the measure value for another item and, it is possible to have an itemset with share $\geq minshare$, whose subsets have share $< minshare$.

Theorem 2: Share frequency is not downward closed with respect to the lattice of all itemsets. **Proof**: Proof by counterexample is sufficient. Consider itemset ACD in

Table 3. Assume *minshare* = 0.20. *SH*(ACD) = 0.23 and ACD is frequent. *SH*(A) = 0.06 and A is not frequent. A is a subset of ACD so share frequency based on the share of the itemset as a whole is not downward closed. •

5 Description of Algorithms

Figure 1 shows an algorithm space consisting of six algorithms. We use an exhaustive algorithm as a starting point and specialize it using different pruning and candidate itemset generation techniques to create the algorithms. The type of pruning added is shown on the edges between nodes.

The first pass through the data collects information about all 1-itemsets. Summary information is compiled, including *MV* and TC_T, the total number of transactions. C_k is the set of candidate itemsets for the kth pass. C_2 is generated using information about the 1-itemsets and information about the candidate 2-itemsets is collected in pass 2. The process of building C_k using itemsets in C_{k-1} stops when no candidate itemsets are added to C_k. After the kth pass, the local measure value and transaction count is available for each counted k-itemset.

Candidate itemset generation and itemset pruning are done in procedure **GenerateCandidateItemsets**. A discussion of this procedure suffices to describe algorithm differences. We use two early methods for generating candidates [3], [6]. *Combination generation* is generation of k-itemsets by combining itemsets in C_{k-1}, differing only in their last item. Unless otherwise noted, our algorithms use this type of

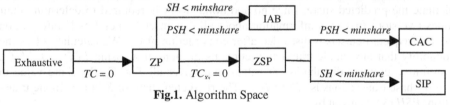

Fig.1. Algorithm Space

generation. *Item add-back generation* is generation of k-itemsets by adding to each X_i ∈ C_{k-1}, any item found in the first pass not contained in X_i. Generation of the next potential candidate itemset, X_{pc}, is represented by an iterator procedure **GenerateNextItemset**. The first call to the procedure returns the first generated itemset, and repeated calls cycle through all possible generated itemsets. When no more itemsets can be generated, the procedure returns false. We investigate two types of pruning. *Pregeneration pruning* prunes itemsets from C_{k-1} using information obtained during the k-1 pass, before any k-itemsets are generated. In *generation pruning*, potential candidate k-itemsets are generated and then pruned as required before they are added to C_k. Pregeneration (generation) pruning is done in procedure **PreGenPrune (PruneGeneratedItemset)**. Procedure **GenerateCandidateItemsets** is written as:

```
1        PreGenPrune(C_{k-1})
2        while X_pc := GenerateNextItemset() do
3            if PruneGeneratedItemset(X_pc) = false then Add X_pc to C_k
```

In the exhaustive algorithm, all possible k-itemsets are added to C_k. If m items are found in pass 1, 2^m itemsets are counted. All frequent itemsets are found.

The *Zero Pruning Algorithm* (ZP) is created from the exhaustive algorithm by adding *zero pruning*, pre-generation pruning of any itemset $X_i \in C_{k-1}$ for which $TC_{Xi} = 0$. This prevents the generation of k-itemsets from $(k-1)$-itemsets not in the data. The number of itemsets counted cannot exceed 2^m and all frequent itemsets will be found.

Even with zero pruning, X_{pc} can contain a $(k-1)$-subset, X_s, with $TC_{Xs} = 0$. The *Zero Subset Pruning Algorithm* (ZSP) adds *subset pruning*, generation pruning of any X_{pc} with $X_s \notin C_{k-1}$. The procedure **PruneGeneratedItemset** is written as:

   3.1 **foreach** $x_i \in X_{pc}$
   3.2   **foreach** $x_j \in X_{pc}$ where $i \neq j$
   3.3   add x_j to X_s
   3.4   **if** $X_s \notin C_{k-1}$ **then return true**
   3.5 **return false**

The number of itemsets counted by ZSP cannot exceed the number of itemsets counted by ZP and all frequent itemsets will be found.

The *Share Infrequency Pruning Algorithm* (SIP) is created from ZSP by adding *share infrequency pruning*, pre-generation pruning of any itemset $X_i \in C_{k-1}$ whose actual share $SH(X_i) <$ *minshare*. SIP behaves like Apriori [3], building candidate k-itemsets using only frequent itemsets from the previous pass.

The *Combine All Counted Algorithm* (CAC) is created from ZSP by adding heuristic methods to calculate the *predicted share* of X_{pc}, $PSH(X_{pc})$, and generation pruning any X_{pc} whose $PSH <$ *minshare*. For each subset X_s, there is a corresponding item x_i that is a member of X_{pc} but not a member of X_s. We use information about X_s and x_i to calculate the predicted share, since no additional work is required to determine their values (we store first pass information about all 1-itemsets). For $k > 1$, information about infrequent itemsets is discarded after construction of C_k. We calculate $P(X)$, the probability that any single transaction contains an itemset X, using $P(X) = TC_X/TC_T$. Assuming a uniform distribution of actual share over all $T \in T_X$, the share value in each of these transactions is $SH(X)/TC_X$. The predicted share of X in any single transaction, $PSH_1(X)$, is given by:

$$PSH_1(X) = P(X)*(SH(X)/TC_X) + (1 - P(X))*0 = SH(X)/TC_T . \quad (7)$$

To calculate the predicted share when an itemset, item pair occurs together, equation 8 is used if $TC_{xi} < TC_{Xs}$, equation 9 is used if $TC_{Xs} < TC_{xi}$ and the average of equations 8 and 9 is used if $TC_{xi} = TC_{Xs}$.

$$PSH = SH(x_i) + PSH_1(X_s)*TC_{xi} . \quad (8)$$

$$PSH = SH(X_s) + PSH_1(x_i)*TC_{Xs} . \quad (9)$$

The average PSH of all X_s, x_i pairs is compared to *minshare*. This value is returned by the function **GetPredictedShare** and **PruneGeneratedItemset** becomes:

   3.1 $PSH(X_{pc}) := 0$, *SubsetCount* $:= 0$
   3.2 **foreach** $x_i \in X_{pc}$
   3.3   **foreach** $x_j \in X_{pc}$ where $i \neq j$
   3.4   add x_j to X_s
   3.5   **if** $X_s \notin C_{k-1}$ **then return true**

3.6 $PSH(X_{pc}) := PSH(X_{pc}) +$ **GetPredictedShare**(x_i, X_s)
3.7 $SubsetCount := SubsetCount + 1$
3.8   **if** $PSH(X_{pc})/SubsetCount < minshare$ **then return true**
3.9   **return false**

In the *Item Add-back Algorithm* (IAB), no item with a non-zero measure value is completely discarded. Starting from the ZP algorithm, infrequency pruning is added. New itemsets are generated using item add-back generation. In the kth pass, each single item found in the first pass is added to each frequent itemset from the $(k-1)$ pass. We again use predictive pruning as described for CAC, except the predicted share value is the average *PSH* of the $(k-1)$-subset, x_i pairs available. Subset pruning is not used. The algorithm for **PruneGeneratedItemsets** differs from that for CAC only in Line 3.5, where the return true becomes a continue statement.

We now provide an example. Figure 2 gives the itemset lattice for the data in Table 1. Each node is labeled with the itemset name. Below the name are $lmv(X)$ and TC_X values, separated by a forward slash. *MV* is equal to 100 and *minshare* is assumed to be equal to 0.20. Frequent itemsets are shaded in Figure 2. For all algorithms, the first pass identifies 1-itemsets C and D as frequent itemsets.

In ZP and ZSP, 2-itemset AB is zero pruned since $TC_{AB} = 0$. Supersets of AB, (ABC, ABD and ABCD), are not generated or counted. ZSP does not subset prune any itemsets since C_{k-1} contains all subsets of the generated itemsets. All frequent itemsets are found.

In SIP, items A and B are infrequency pruned. Itemset CD is generated from the frequent items in C_1, counted in pass 2 and found to be frequent. SIP terminates because no 3-itemsets can be generated from a single 2-itemset. Frequent itemsets AD and ACD are missed, since item A was infrequency pruned after the first pass and cannot exist in any larger itemsets.

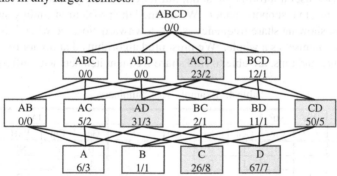

Fig. 2. Itemset Lattice

In CAC, all counted 1-itemsets are used to generate candidate 2-itemsets. Generated 2-itemsets are pruned based on predicted share value. Consider itemset AD. $SH(A) = 0.06$, $TC_A = 3$ and $SH(D) = 0.67$, $TC_D = 7$. Since $TC_A < TC_D$, we determine $PSH_1(D) = SH(D)/TC_T = 0.67/10 = 0.067$. Now $PSH(AD) = SH(A) + PSH_1(D)*TC_A = 0.06 + 0.067(3) = 0.26$. $PSH(AD) > minshare$, so AD is added to C_2. Only AD and CD meet the condition $PSH \geq minshare$, so only they are counted in pass 2. CAC terminates after pass 2 because AD and CD cannot be used to generate a 3-itemset with all subsets in C_{k-1}. CAC counted one more 2-itemset than the SIP and it was frequent. CAC missed frequent 3-itemset ACD because one of its subsets, itemset AC, was not counted in the second pass.

IAB generates all possible 2-itemsets except AB. AB is not generated since both A and B are infrequent. As for CAC, itemsets AD and CD are predicted to be frequent and counted in pass 2. After pass 2, itemsets ABD, ACD and BCD are generated by adding single items to frequent itemsets AD and CD. To determine PSH(ACD), all available 2-itemset, item pairs are examined. Only 2-itemsets AD and CD exist in C_{k-1} so C plus AD and A plus CD are examined. PSH(ACD) = 0.30 > *minshare*, so ACD is added to C_k. The predicted share of ABD and BCD do not meet the minimum share requirement. In pass 3, ACD is counted and is found to be frequent. Itemset ABCD is generated after pass three but PSH(ABCD) < *minshare*, so IAB terminates. IAB counts one 3-itemset not counted by either SIP or CAC, and it was frequent.

Table 4 summarizes the performance of the algorithms for the sample data set. A "1" in a column labeled Gen (Cnt) indicates an itemset that was generated (counted). A value of NP in a *PSH* column indicates that no prediction was made because the itemset was not generated. Rows containing frequent itemsets are shaded. The trade off between the work an algorithm does and its effectiveness is evident. ZP and ZSP found all frequent itemsets, but counted most itemsets in the lattice. SIP performed very little work but missed two frequent itemsets.

6 Conclusions

Share can provide useful information about numerical values typically associated with transaction items, that support cannot. We defined the problem of finding share frequent itemsets, showing share frequency is not downward closured when it is defined in terms of the itemset as a whole. We presented algorithms that do not rely downward closure and thus, are able to find share frequent itemsets with infrequent

Table 4. Example Task Summary

Itemset	ZP Cnt	ZSP Cnt	SIP Cnt	CAC Gen	CAC Cnt	CAC PSH	CAC SH	IAB Gen	IAB Cnt	IAB PSH	IAB SH
AB	1	1	0	1	0	0.02	0.00	0	0	NP	0.00
AC	1	1	0	1	0	0.14	0.05	1	0	0.14	0.05
AD	1	1	0	1	1	0.26	0.31	1	1	0.26	0.31
BC	1	1	0	1	0	0.04	0.02	1	0	0.04	0.02
BD	1	1	0	1	0	0.08	0.11	1	0	0.08	0.11
CD	1	1	1	1	1	0.85	0.50	1	1	0.85	0.50
ABC	0	0	0	0	0	NP	0.00	0	0	NP	0.00
ABD	0	0	0	0	0	NP	0.00	1	0	0.03	0.00
ACD	1	1	0	0	0	NP	0.23	1	1	0.30	0.23
BCD	1	1	0	0	0	NP	0.12	1	0	0.05	0.12
ABCD	0	0	0	0	0	NP	0.00	1	0	0.03	0.00
Sum	8	8	1	6	2			9	3		

subsets. Using heuristic methods, we generate candidate itemsets by supplementing the information contained in the set of frequent itemsets from a previous pass, with other information that is available at no additional processing cost. These algorithms count only those generated itemsets predicted be frequent.

References

1. Agrawal A., Imielinksi T., Swami A. 1993. Mining Association Rules between Sets of Items in Large Databases. *Proceedings of the ACM SIGMOD International Conference on the Management of Data*. Washington D.C., pp.207-216.
2. Agrawal A., Mannila H., Srikant R., Toivonen H. and Verkamo, A.I. 1996. Fast Discovery of Association Rules. *Advances in Knowledge Discovery and Data Mining*, Menlo Park, California, pp.307-328.
3. Agrawal A., Srikant R. 1994. Fast Algorithms for Mining Association Rules. *Proceedings of the 20th International Conference on Very Large Databases*. Santiago, Chile, pp.487-499.
4. Carter C.L., Hamilton H.J., Cercone N. 1997. Share Based Measures for Itemsets. *Proceedings of the First European Conference on the Principles of Data Mining and Knowledge Discovery*. Trondheim, Norway, pp.14-24.
5. Hilderman R.J., Carter C., Hamilton H.J., Cercone N. 1998. Mining Association Rules from Market Basket Data using Share Measures and Characterized Itemsets. *International Journal of Artificial Intelligence Tools*. 7(2):189-220.
6. Mannila H., Toivonen H., Verkamo A.I. 1994. Efficient Algorithms for Discovering Association Rules. *Proceedings of the 1994 AAAI Workshop on Knowledge Discovery in Databases*. Seattle, Washington, pp.144-155.
7. Masland B., Piatetsky-Shapiro G. 1995. A Comparison of Approaches for Maximizing Business Payoff of Predictive Models. *Proceedings of the Second International Conference on Knowledge Discovery and Data Mining*. Portland, Oregon, pp.195-201.
8. Ng R.T., Lakshmanan L.V.S., Han J., Pang A. 1998. Exploratory Mining and Pruning Optimizations of Constrained Association Rules. *Proceedings ACM SIGMOD International Conference on Management of Data*. Seattle, Washington, pp.13-24.
9. Silverstein C., Brin S., Motwani R. 1998. Beyond Market Baskets: Generalizing Association Rules to Dependence Rules. *Data Mining and Knowledge Discovery*, 2(1):39-68.
10. Srikant R., Agrawal R. 1996. Mining Quantitative Association Rules in Large Relational Tables. *Proceedings of the ACM SIGMOD Conference on the Management of Data*. Montreal, Canada, pp.1-12.
11. Zaki M.J., Parthasarathy, M., Ogihara M., Li W. 1997. New Algorithms for Fast Discovery of Association Rules. *Proceedings of the Third International Conference on Knowledge Discovery & Data Mining*. Newport Beach, California, pp.283-286.

Discovering Task Neighbourhoods through Landmark Learning Performances

Hilan Bensusan and Christophe Giraud-Carrier

Department of Computer Science, University of Bristol
Bristol BS8 1UB, United Kingdom
{hilanb,cgc}@cs.bris.ac.uk

Abstract. Arguably, model selection is one of the major obstacles, and a key once solved, to the widespread use of machine learning/data mining technology in business. Landmarking is a novel and promising meta-learning approach to model selection. It uses accuracy estimates from simple and efficient learners to describe tasks and subsequently construct meta-classifiers that predict which one of a set of more elaborate learning algorithms is appropriate for a given problem. Experiments show that landmarking compares favourably with the traditional statistical approach to meta-learning.

1 Introduction

The menu of data mining/machine learning techniques offers an increasingly large variety of choices. And, as with food, no one selection is universally satisfactory [15]. Instead, for every learning task, an appropriate learning algorithm must be chosen. An important goal of machine learning is to find principled ways to make this choice, which has become known as the model selection problem. Meta-learning endeavours to use machine learning itself (in a kind of recursive fashion) to the problem of model selection.

Meta-learning can be done by considering similar tasks in a reference class of tasks [1, 6, 10, 11], by inducing rules about algorithm selection [7, 11] or simply by training a meta-classifier [2, 3, 9]. In any case, tasks need to be described so that relations between tasks and learning algorithms can be automatically found. A number of strategies for task description have been used, based on statistics, information-theoretic considerations and properties of induced decision trees [2, 3, 7, 10, 11, 16]. Recently, we have proposed a novel approach, called landmarking [12, 4]. We believe this approach to be simpler, more intuitive and more effective.

Consider the class of tasks on which a learning algorithm performs well under a reasonable measure of performance (e.g., predictive accuracy) and call this class the *area of expertise* of the learner. The basic idea of landmarking is that the performance of a learner on a task uncovers information about the nature of that task. Hence, a task can be described by the collection of areas of expertise to which it belongs. We call a *landmark learner*, or simply a *landmarker*, a learning mechanism whose performance is used to describe a task. Exploring

D.A. Zighed, J. Komorowski, and J. Zytkow (Eds.): PKDD 2000, LNAI 1910, pp. 325–330, 2000.
© Springer-Verlag Berlin Heidelberg 2000

the meta-learning potential of landmarking amounts to investigating how well a landmark learner's performance hints at the location of learning tasks in the expertise space (i.e., the space of all areas of expertise). By thus locating tasks, landmarking discovers experimentally neighbouring areas of expertise. It therefore finds neighbourhoods of learning strategies and, ultimately, draws a map of the complete expertise space.

The paper is organised as follows. In section 2 we present our chosen set of landmarkers. Section 3 reports empirical results which complement those reported in [12, 4]. Finally, section 4 concludes the paper.

2   A Set of Landmarkers

In principle, every learner's performance can signpost the location of a problem with respect to other learner's expertise. In practice, however, we want landmarkers that have low time complexity.[1]

Clearly, the landmarkers may change according to the *learner pool*, i.e., the set of target learners from which one must be selected. At present, our learner pools are subsets of {C5.0tree, C5.0rule, C5.0boost, MLC++'s Naive Bayes, MLC++'s IBL, Clementine's MLP, Clementine's RBF, Ripper, Linear Discriminant, Ltree}. Given such pools, we construct the following set of landmarkers. In addition to being computationally efficient, each landmarker is intended to inform about the learning task, e.g., linear separability (see [4] for details).

1. *Decision node:* A single decision node consisting of the root of a C5.0 decision tree, where the attribute to branch on is chosen so as to maximise the information gain ratio [13].
2. *Worst node:* Same as above except that the attribute to branch on is chosen so as to minimise, rather than maximise, the information gain ratio.
3. *Randomly chosen node:* Same as above except that the attribute to branch on is chosen at random.
4. *Naive Bayes:* The standard Naive Bayes classifier [8].
5. *1-Nearest Neighbour:* The standard 1-NN classifier, based on the closest training example [8].
6. *Elite 1-Nearest Neighbour:* Same as above except that the nearest neighbour is computed based on a subset of all attributes. This elite subset is composed of the most informative attributes if the information gain ratio difference between them is small. Otherwise, the elite subset is a singleton and the learner acts like a decision node learner.
7. *Linear Discriminant:* A linear approximation of the target function [14].

For each task, the error rate of a landmarker is given by the average error over the whole instance space of the learner trained on 10 different training sets of

---

[1] In general, the cost of running the chosen landmarkers should be smaller than the cost of running the more sophisticated learners; otherwise, it would be better to run all the learners to find out which one performs best.

Meta-learner	Landmarking	Information-based	Combined
Default Class	0.460	0.460	0.460
C5.0boost	0.248	0.360	0.295
C5.0rules	0.239	0.333	0.301
C5.0tree	0.242	0.342	0.314
MLP	0.301	0.317	0.320
RBFN	0.289	0.323	0.304
LD	0.335	0.311	0.301
Ltree	0.270	0.317	0.286
IB	0.329	0.366	0.342
NB	0.429	0.407	0.363
Ripper	0.292	0.314	0.295
Average	0.298	0.339	0.312

Table 1. Meta-learning performance for different task descriptions

equal size. Each task is then described by a vector of 7 meta-attributes (i.e., the error rates of the 7 landmarkers) and labelled by the learner (from the learner pool), which gives the highest predictive accuracy based on stratified 10-fold cross-validation. We use the special label "tie" when the difference in performance between the best and worst learners is small (here, less than 10%).[2] Once a (meta-)dataset has thus been constructed, it can be (meta-)learned by any supervised learning technique.

3 Experiments with Landmarking

In the first experiment, we compare landmarking with the more traditional statistical/information theoretic approach to task description (e.g., see [9, 11]). We consider the following 6 widely-used meta-attributes: class entropy, average entropy of the attributes, mutual information, joint entropy, equivalent number of attributes and signal-to-noise ratio. For this comparative experiment, the learner pool includes all 10 learners. We use 320 artificially generated Boolean datasets with 5 to 12 attributes, where the classifications are partly rule-based and partly random. The 10 learning algorithms in the learner pool are also used as meta-learners. Error rates are based on stratified 10-fold cross-validation and the results are given in Table 1. The default class is "tie".

Table 1 shows that landmarking compares favourably with the information-based task description approach. In all but 2 cases, landmarking produces lower error rates than the information-based approach. The table also suggests that adding the information-based features to describe task impairs landmarking performance. These results are further confirmed by the learning curves of the three approaches (see Fig. 1).

---

[2] When performance differences are small, meta-learning is not only harder but also less useful since there is no great advantage of the best learner over the others.

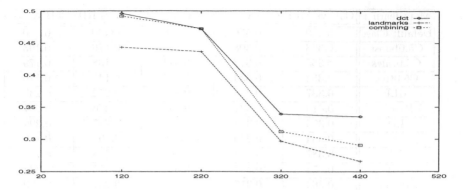

Fig. 1. Learning curves for the landmarking approach, the information-based approach and the combined approach

Note that in the above experiment three learners, namely Linear discriminant, Naive Bayes and Nearest Neighbor, are both in the set of landmarkers and in the learner pool. This can be seen as making the landmarking enterprise too easy. Notice, however, that the meta-learners are not informed of this identity, nor is it easy to uncover it since the landmarking meta-attributes values are only error rates. In fact, the meta-learner has to proceed as in the general case, finding correlations between the meta-attributes and the best learners. Experiments reported in [4], for example, suggest that the presence of Naive Bayes among the landmarkers does not significantly impact performance.

In the second experiment, we consider the performance of landmarking when the learner pool is smaller. Of course, if the learner pool is sufficiently small, it is more efficient to run all the learners in the pool to select the most suitable one. However, this experiment is intended to shed some light on the ability of landmarking to describe tasks and discriminate between areas of expertise. We restrict our attention to pairs of learners including C5.0boost (C5B), MLP, Ripper (RIP) and Naive Bayes (N-B). The results, reported in Table 2, show that landmarking with most meta-learners improves on the default class.

In the third experiment, we investigate whether knowledge induced from artificial tasks is valid when applied to more "real-world" applications. To this end, we train C4.5 (as a meta-learner) on the meta-dataset arising from our 320 Boolean problems and test the induced hypothesis on 16 benchmark datasets from UCI [5] and Daimler-Chrysler. Table 3 reports the error rate differences between the best and the worst learners. The figures preceded by ν are the error figures of cases where landmarking chooses the best learner. Hence, these are figures of errors that landmarking avoids. When landmarking selects "tie", we pessimistically make the assumption that it would select the worst learner. The results show that, in most cases, landmarking's recommendations are either right or not very damaging to the overall performance. This is to say that it selects the worst learner when the worst learner is not greatly worse than the best one.

Meta-learner	C5B vs MLP	C5B vs RIP	MLP vs RIP	N-B vs RIP	C5B vs N-B
Default Class	0.450	0.550	0.510	0.490	0.550
C5.0boost	0.328	0.308	0.366	0.176	0.193
C5.0rules	0.355	0.317	0.411	0.138	0.179
C5.0tree	0.364	0.334	0.405	0.147	0.176
MLP	0.387	0.358	0.472	0.252	0.235
RBFN	0.410	0.331	0.481	0.276	0.261
LD	0.375	0.364	0.425	0.267	0.290
Ltree	0.384	0.249	0.405	0.199	0.188
IB	0.387	0.308	0.405	0.278	0.264
NB	0.554	0.387	0.627	0.375	0.334
Ripper	0.364	0.317	0.446	0.202	0.199
Average	0.391	0.327	0.444	0.231	0.232

Table 2. Landmarking performance when selecting between pairs of learners

4   Conclusions

Landmarking is a strategy to describe tasks through the performance of simple
and efficient learners. Tasks are therefore described by their position in the
expertise space. The expertise space is itself a suggestive way of thinking of
learners and tasks. Landmarking relies on the position of a problem in that
space to represent it: "tell me what can learn you and I will tell you what
you are". Empirical results suggest that it pays off to run bare-bone, landmark
learners on a number of tasks and learn how their performance relates to that
of other, more fleshed-out learners. These results indicate that the information
about a task given by the performance of a portfolio of simple learners is useful
for meta-learning. Furthermore, they also suggest that it is safe to apply to
real-world tasks, knowledge learned from artificial ones.

Acknowledgements

This work is supported by an ESPRIT Framework IV LTR Grant (Nr. 26.357).
We thank the anonymous referees for useful comments.

References

1. D.W. Aha. Generalizing from case studies: A case study. In *Proceedings of 9th
   ICML*, pages 1–10, 1992.
2. H. Bensusan. God doesn't always shave with Occam's Razor – learning when and
   how to prune. In *Proceedings of 10th ECML*, pages 119–124, 1998.
3. H. Bensusan. *Automatic bias learning: an inquiry into the inductive basis of induction.* PhD thesis, School of Cognitive and Computing Sciences, University of
   Sussex, 1999.

Task	C5B vs MLP	C5B vs RIP	MLP vs RIP	N-B vs RIP	C5B vs N-B
mushrooms	ν 0.050	0.000	ν 0.032	ν 0.003	0.003
abalone	ν 0.500	0.033	0.088	0.047	0.015
crx	0.056	ν 0.027	0.008	0.084	ν 0.110
acetylation	ν 0.091	ν 0.357	ν 0.267	ν 0.333	ν 0.024
titanic	0.007	0.211	0.219	ν 0.216	0.004
waveform	0.040	0.176	ν 0.136	ν 0.204	0.028
yeast	0.009	0.276	ν 0.428	ν 0.422	0.004
car	0.011	ν 0.083	0.094	ν 0.021	ν 0.104
chess	ν 0.112	ν 0.380	0.269	0.195	ν 0.186
led7	ν 0.014	ν 0.042	0.028	0.038	0.004
led24	0.013	ν 0.001	0.012	0.025	0.026
tic	0.009	ν 0.005	ν 0.015	ν 0.280	ν 0.274
MONK1	ν 0.156	ν 0.256	0.000	0.218	0.245
MONK2	ν 0.141	ν 0.141	0.000	ν 0.164	ν 0.023
satimage	ν 0.431	ν 0.046	0.016	ν 0.068	ν 0.114
quisclas	0.023	ν 0.003	0.026	ν 0.258	ν 0.261

Table 3. Error rate differences between best and worst learners

4. H. Bensusan and C. Giraud-Carrier. Casa batló is in passeig de gràcia or land-marking the expertise space. In *Proceedings of ECML-2000 Workshop on Meta-Learning: Building Automatic Advice Strategies for Model Selection and Method Combination*, pages 29–46, 2000.
5. C.L. Blake and C.J. Merz. UCI repository of machine learning databases, 1998. http://www.ics.uci.edu/~mlearn/MLRepository.html.
6. P. Brazdil and C. Soares. A comparison of ranking methods for classification algorithm selection. In *Proceedings of 11th ECML*, 2000.
7. C. E. Brodley. Recursive automatic bias selection for classifier construction. *Machine Learning*, 20:63–94, 1995.
8. R. O. Duda and P. E. Hart. *Pattern classification and scene analysis*. Wiley, 1973.
9. R. Engels and C. Theusinger. Using a data metric for offering preprocessing advice in data mining applications. In *Proceedings of 13th ECAI*, pages 430–434, 1998.
10. G. Lindner and R. Studer. AST: Support for algorithm selection with a CBR approach. In *Proceedings of 3rd PKDD*, pages 418–423, 1999. (LNAI 1704).
11. D. Michie, D. J. Spiegelhalter, and C. C. Taylor, editors. *Machine Learning, Neural and Statistical Classification*. Ellis Horwood, 1994.
12. B. Pfahringer, H. Bensusan, and C. Giraud-Carrier. Meta-learning by landmarking various learning algorithms. In *Proceedings of 17th ICML*, pages 743–750, 2000.
13. J. R. Quinlan. *C4.5: Programs for Machine Learning*. Morgan Kaufmann, 1993.
14. B. D. Ripley. *Pattern Recognition and Neural Networks*. Cambridge University Press, 1995.
15. C. Schaffer. A conservation law for generalization performance. In *Proceedings of 11th ICML*, pages 259–265, 1994.
16. G. Widmer. On-line metalearning in changing context: MetaL(b) and metaL(ib). In *Proceedings of 3rd International Workshop on Multistrategy Learning*, 1997.

Induction of Multivariate Decision Trees by Using Dipolar Criteria

Leon Bobrowski[1,2] and Marek Krętowski[1]

[1] Institute of Computer Science, Technical University of Białystok, Poland
[2] Institute of Biocybernetics and Biomedical Engineering, PAS, Poland
`mkret@ii.pb.bialystok.pl`

Abstract. A new approach to the induction of multivariate decision trees is proposed. A linear decision function (hyper-plane) is used at each non-terminal node of a binary tree for splitting the data. The search strategy is based on the dipolar criterion functions and exploits the basis exchange algorithm as an optimization procedure. The feature selection is used to eliminate redundant and noisy features at each node. To avoid the problem of over-fitting the tree is pruned back after the growing phase. The results of experiments on some real-life datasets are presented and compared with obtained by state-of-art decision trees.

1 Introduction

Decision trees are hierarchical, sequential classification structures that recursively partition the feature space. The tree contains terminal nodes (leaves) and internal (non-terminal) nodes. A terminal node generates no descendants, but is designated by a class label, while an internal node contains a split, which tests the value of an expression of the attributes. Each distinct outcome of the test generates one child node, hence all non-terminal nodes have two or more child nodes. An object is classified through traversing a path from the root node until a terminal node. At each non-terminal node on the path, the associated test is used to decide which child node the feature vector is passed to. At the terminal node the class label is used to classify the input vector.

Decision trees are extensively investigated especially in statistics, machine learning and pattern recognition (see [13] for a very good multi-disciplinary survey). Most of the research effort focuses on the univariate tests for symbolic or numeric (continuous-valued) attributes (e.g. C4.5 [14], CHAID [10]). An univariate test compares a value of a single attribute to a constant, so this is equivalent to partitioning the set of observations with an axis-parallel hyper-plane. A multivariate decision tree may use as the split an expression, which exploits more than one feature (see Fig. 1). A special case of the multivariate tree that we are particular interested in is an oblique decision tree ([12]). A test in such type of the tree uses a linear combination of the attributes.

Several methods for generating multivariate trees have been introduced so far. One of the first trials was done in CART (*Classification And Regression Trees*) [5]. The system is able to search for a linear combination of the continuous-valued attributes instead of using only a single attribute. The CART system has the

D.A. Zighed, J. Komorowski, and J. Żytkow (Eds.): PKDD 2000, LNAI 1910, pp. 331–336, 2000.
© Springer-Verlag Berlin Heidelberg 2000

strong preference for univariate tests and chooses multivariate one very rare. In LMDT (*Linear Machine Decision Trees*) [6] each test is constructed by training a linear machine and eliminating variables in a controlled manner. Murthy et al. [12] introduce OC1 (*Oblique Classifier 1*), the algorithm that combines deterministic and randomized procedures to search for a good tree. The method was applied to classify a set of patients with breast cancer and showed excellent accuracy. Another tree was proposed by Chai et al., [7]. BTGA (*Binary Tree-Genetic Algorithm*) uses the maximum impurity reduction as the optimality criterion of linear decision function at each non-terminal node. The modified BTGA was used to the pap smear classification and demonstrated the high sensitivity along with the lowest possible false alarm rate. Recently, *Ltree* (*Linear tree*) was presented, which combines a decision tree with a linear discrimination by means of constructive induction [9]. At each node a new instance space is defined by insertion of new attributes that are projections over the hyper-planes given by a linear discrimination function and new attributes are propagated downward.

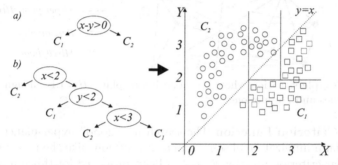

Fig. 1. An example of the simple two-class problem and its possible solutions by various decision trees: (a) multivariate and (b) univariate. If an associated test is true then we choose the left outcome, otherwise the right one.

2   Dipolar Criteria

We will take into account the N-dimensional feature vectors $\mathbf{x} = [x_1, ..., x_N]^T$ ($x_i \in \{0,1\}$ or $x_i \in R^1$) which can belong to one of the K classes ω_k ($k = 1, ..., K$). The learning set $C_k = \{\mathbf{x}^j(k)\}(j = M_{k-1} + 1, ..., M_{k-1} + m_k)$ contains m_k feature vectors $\mathbf{x}^j(k)$ from the class ω_k, where M_{k-1} is the number of objects $\mathbf{x}^j$ in the first $(k-1)$ sets C_k.

The feature space could be divided into two regions by the following hyperplane $H(\mathbf{w}, \theta) = \{\mathbf{x} : \langle \mathbf{w}, \mathbf{x} \rangle = \theta\}$, where $\mathbf{w} = [w_1, ..., w_N]$ ($\mathbf{w} \in R^N$) is the weight vector, θ is the threshold and $\langle \mathbf{w}, \mathbf{x} \rangle$ is the inner product. In the special case the weight $\mathbf{w}$ could be reduced to the unit vector $\mathbf{e}^i = [0, ..., e_i = 1, ..., 0]$. The hyper-plane $H(\mathbf{e}^i, \theta)$ is parallel to all but the i-th axis of the feature space. The learning sets are linearly separable if each of these sets C_k could be separated by some hyper-plane from the sum of all remaining sets C_i.

For simplicity of notion we introduce augmented feature space, where $\mathbf{y} = [1, x_1, ..., x_N]^T$ is the augmented feature vector, $\mathbf{v} = [-\theta, w_1, ..., w_N]$ is the weight vector and $H(\mathbf{v}) = \{\mathbf{y} : \langle \mathbf{v}, \mathbf{y} \rangle = 0\}$ is the hyper-plane.

Dipoles We recall some basic definition, for more detailed description please refer to [3] and [4]. A dipole is a pair $(\mathbf{x}^j(k), \mathbf{x}^{j'}(k'))$ of the feature vectors $\mathbf{x}^j(k)$ from the learning sets C_k, where $0 \leq j < j' \leq m$, and m is the number of the vectors in all learning sets. We call the dipole $(\mathbf{x}^j(k), \mathbf{x}^{j'}(k'))$ *mixed* if and only if the objects constituting it belong to the different learning sets C_k ($k \neq k'$). Similarly, a pair of the vectors from the same class ω_k constitutes the *pure* dipole.

Hyper-plane $H(\mathbf{v})$ splits the dipole $(\mathbf{y}^i, \mathbf{y}^j)$ if and only if: $\langle \mathbf{v}, \mathbf{y}^i \rangle \cdot \langle \mathbf{v}, \mathbf{y}^j \rangle < 0$. It means that the input vectors $\mathbf{y}^i$ and $\mathbf{y}^j$ are situated on opposite sides of the dividing hyper-plane. The necessary and sufficient condition for preserving the separability of the learning sets by the hyper-plane is the division of all mixed dipoles by the hyper-plane. The tree induction is connected with the repeated search for such a hyperplane $H(\mathbf{v})$ which "divides a possible *high* number of mixed dipoles and a possible *low* number of pure ones". These requests are related to demand for a low error rate of the resulting tree.

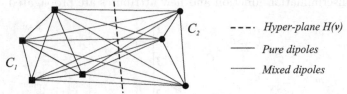

Fig. 2. A simple situation where only one hyper-plane $H(\mathbf{v})$ is enough to divide all mixed dipoles and no pure one.

Dipolar Criterion Function The search for "good" hyper-planes $H(\mathbf{v})$ could be based on minimization of some criterion functions like the perceptron one [8]. The dipolar criterion function $\Psi(\mathbf{v})$ has been proposed for this purpose [4]. Lets first define a penalty function, which is associated with each input vector $\mathbf{y}^j(k)$:

$$\varphi_j^+(\mathbf{v}) = \begin{cases} \delta^j - \langle \mathbf{v}, \mathbf{y}^j \rangle & \text{if } \langle \mathbf{v}, \mathbf{y}^j \rangle < \delta^j \\ 0 & \text{if } \langle \mathbf{v}, \mathbf{y}^j \rangle \geq \delta^j \end{cases} \tag{1}$$

and

$$\varphi_j^-(\mathbf{v}) = \begin{cases} \delta^j + \langle \mathbf{v}, \mathbf{y}^j \rangle & \text{if } \langle \mathbf{v}, \mathbf{y}^j \rangle < -\delta^j \\ 0 & \text{if } \langle \mathbf{v}, \mathbf{y}^j \rangle \geq -\delta^j \end{cases} \tag{2}$$

where $\delta^j(\delta^j \geq 0)$ is the parameter defining the margin. If $\delta^j = 0$, the penalty function is related to the error correction algorithm used in the perceptron [8]. We usually assume, that the margin is equal to 1. We can force to move $H(\mathbf{v})$ to obtain situation when $\mathbf{y}^j(k)$ is on a positive (if we use $\varphi_j^+(\mathbf{v})$) or a negative ($\varphi_j^-(\mathbf{v})$) side of $H(\mathbf{v})$. With each dipole we associate two functions, depending on our goal. Thus for a mixed dipole $(\mathbf{y}^i, \mathbf{y}^j)$ we use two function with opposite signs, because we are interested in cutting all dipoles of such type:

$$\varphi_{ij}^m(\mathbf{v}) = \varphi_i^+(\mathbf{v}) + \varphi_j^-(\mathbf{v}) \qquad (\text{or } \varphi_{ij}^m(\mathbf{v}) = \varphi_j^+(\mathbf{v}) + \varphi_i^-(\mathbf{v})) \tag{3}$$

To avoid dividing a pure dipole both end-points of it should be situated on the same side of the hyper-plane $H(\mathbf{v})$, so the penalty function is defined as follows:

$$\varphi_{ij}^p(\mathbf{v}) = \varphi_i^+(\mathbf{v}) + \varphi_j^+(\mathbf{v}) \qquad (\text{or } \varphi_{ij}^p(\mathbf{v}) = \varphi_j^-(\mathbf{v}) + \varphi_i^-(\mathbf{v})) \tag{4}$$

The choice of the form of $\varphi_{ij}^m(\mathbf{v})$ ($\varphi_{ij}^p(\mathbf{v})$) is devoted to an orientation of a dipole and for the simplicity reason we do not take into account in this paper. More detailed explanation of the orientation one can find in [4]. The dipolar criterion function $\Psi(v)$ could be represented in the following manner:

$$\Psi(\mathbf{v}) = \sum_{(i,j)\in I_p} \alpha_{ij} \cdot \varphi_{ij}^p(\mathbf{v}) + \sum_{(i,j)\in I_m} \alpha_{ij} \cdot \varphi_{ij}^m(\mathbf{v}), \tag{5}$$

where α_{ij} determines a relative importance (price) of dipole $(\mathbf{y}^i, \mathbf{y}^j)$ and $I_p(I_m)$ is the set of the pure (mixed) dipoles.

It is worth to emphasize that the mechanism of dipole's prices is very flexible. In all experiments presented in the paper we used a very simple strategy - we differentiate only prices for mixed (1.0) and pure dipoles (0.01). We work on a little bit more complicated strategy which exploits the length of dipoles.

Because $\Psi(\mathbf{v})$ is the sum of the convex and piece-wise linear penalty functions $\varphi_j^+(\mathbf{v})$ and $\varphi_j^-(\mathbf{v})$, $\Psi(\mathbf{v})$ is also a function of this type. The basis exchange algorithms, similar to linear programming methods, have been developed as an efficient tool for minimization of such a function [2]. In fact the minimization has carried out by combining the basis exchange algorithms with a search for the adequate orientations of dipoles [4].

3   Multivariate Tree Induction

Most of the existing tree induction systems proceed in a greedy, top-down fashion. At each non-terminal node, starting from the root, the best split is learned by using a criterion of optimality. The learned decision function divides the training subset into two (or more) subsets generating child nodes. The process is repeated at each newly created child node until a stopping condition is satisfied and the node is declared as a terminal node.

We are taking into consideration the binary decision tree. At the i-th non-terminal node the set S_i of the input vectors from the training data, which reaches the node is divided into two subsets S_i^L and S_i^R by the hyper-plane $H(\mathbf{v}_i)$ (or in a case of an univariate tree by $H(\mathbf{e})$):

$$S_i^L = \{\mathbf{y} : \mathbf{y} \in S_i \text{ and } \langle \mathbf{v}_i, \mathbf{y} \rangle \geq 0\} \tag{6}$$

$$S_i^R = \{\mathbf{y} : \mathbf{y} \in S_i \text{ and } \langle \mathbf{v}_i, \mathbf{y} \rangle < 0\} \tag{7}$$

All feature vectors from the set S_i^L constitute the left child of the i-th node and S_i^R the right one. To find the optimal dividing hyper-plane $H(\mathbf{v}_i)$ the dipolar criterion function $\Psi(\mathbf{v}_i)$ is used, which is constituted exclusively on dipoles created from the feature vectors belonging to S_i.

Feature Selection At each internal node of the tree we are interested in obtaining the simplest possible test, hence we try to exclude noisy (or/and irrelevant) features. This way we increase understandability of the model and what is even more important, we can improve performance. There exist many feature selection algorithms (for a good review, in the context of multivariate trees see [6]). Our goal is not the comparison of various methods, so we have to choose one.

We use Heuristic Sequential Search proposed in [6]. It is a combination of Sequential Backward Elimination (SBE) and Sequential Forward Selection (SFS) and it works as follows: firstly finds the best test based on all features and similarly the best test using only one feature. Depending which one is better SFS or SBE search is chosen. The only problem which we have to solve is: how to compare various subsets of features. We prefer the subset (and the hyper-plane build on it), which divides more mixed dipoles (and less pure ones, when a tie is observed). During the feature selection process one cannot forget the problem of "underfitting" the training data. The number of objects used to find a test should be several times greater than the number of attributes used [8] (in all experiments: "x5"). Such a situation occurs often near the leaves of the tree.

Avoiding Over-fitting It is well known fact, that the data over-fitting can decrease the classifier's performance in a significant way, especially in a noisy domain. There exist two the most common approaches to avoid this problem and to determine the correct tree size: to stop the tree growing before the perfect classification of the training data or to post-prune the tree after the full partitioning of the instance space ([5]). Although the post-pruning is definitely superior, but usually both techniques are combined and in our system we decided to follow this idea. We use a very simple stopping rule - when the number of cases in a node is lower than N_{stop} then the node is declared as terminal ($N_{stop} = 5$ in all experiments). We have chosen also one of the simplest pruning methods: *reduced-error pruning* ([14]). It exploits a separate part of the training data (in all experiments 30%) as a validation set used during the pruning phase to determine the error rates of the sub-tree.

4   Experimental Results

We have tested our approach on several datasets taken from UCI repository ([1]). The left part of Table 1 shows the classification error rates obtained by our method and C4.5, LMDT (results are taken from [11]). To estimate the error rate of an algorithm we use the testing set (when provided) or the ten-fold stratified cross-validation procedure. In the right part of Table 1 we try to present the complexity of obtained models. We include the number of nodes in the whole tree, but it is not the best measure of the complexity, especially if we want to compare univariate and multivariate trees. And even if we compare the multivariate trees we need to be careful (e.g. in the multi-class domain LMDT in each node includes more than one hyper-plane). In case of trees generated using our method we also provide the number of features used in all tests.

5   Conclusions

In the paper we presented how to induce decision trees based on dipolar criteria. The preliminary experimental results indicate that both classification accuracy and complexity are comparable with the results obtained by other systems. Furthermore many places for the possible improvements still exist. Especially the pruning strategy is very simple and we hope that applying more sophisticated method will help to improve the results. The feature selection at each non-terminal node is the most time consuming part of the induction process. At the moment we are working on integrating the feature selection with searching for the best hyper-plane, which will significantly reduce the learning time. One of the future research directions could be devoted to incorporating the variable misclassification cost into the decision tree induction. We think that it could be easily done only by a modification of dipole's prices.

Table 1. The error rates (left part) and the number of nodes in the whole tree (right part; the number of features used in all test is given in brackets) of compared systems

Dataset	Our method	C4.5	LMDT	Our method	C4.5	LMDT
breast-w	0.038	0.042	0.035	3 (7)	11	3
heart	0.191	0.196	0.163	3 (10)	23	3
housing	0.248	0.221	0.251	23 (71)	36	11
pima	0.233	0.242	0.249	13 (22)	18	11
smoking	0.292	0.305	0.350	69 (139)	1	61
vehicle	0.243	0.277	0.215	25 (122)	65	13
waveform	0.213	0.261	0.176	11 (65)	54	4

Acknowledgements: The presented work was partially supported by the grant 8 T11E 029 17 from the State Committee for Scientific Research (KBN) in Poland and by the grant W/II/1/2000 from Technical University of Białystok.

References

1. Blake, C., Merz, C.: UCI Repository of machine learning databases, available on-line: http://www.ics.uci.edu/ mlearn/MLRepository.html (1998)
2. Bobrowski, L.: Design of piecewise linear classifiers from formal neurons by some basis exchange technique. Pattern Recognition **24**(9) (1991) 862–870
3. Bobrowski, L.: Piecewise-linear classifiers, formal neurons and separability of the learning sets, In: Proc. of 13th Int. Conf. on Pattern Recognition (1996) 224–228
4. Bobrowski, L.: Data mining procedures related to the dipolar criterion function, In: Applied Stochastic Models and Data Analysis (1999) 43–50.
5. Breiman, L., Friedman, J., Olshen, R., Stone C.: Classification and Regression Trees. Wadsworth Int. Group (1984)
6. Brodley, C., Utgoff, P.: Multivariate decision trees. Mach. Learning 19 (1995) 45–77
7. Chai, B., Huang, T., Zhuang, X., Zhao, Y., Sklansky, J., Piecewise-linear classifiers using binary tree structure and genetic algorithm. Pattern Recognition **29**(11) (1996) 1905–1917
8. Duda, O.R., Heart, P.E.: Pattern Classification and Scene Analysis. J. Wiley (1973)
9. Gama, J., Brazdil, P.: Linear tree. Inteligent Data Analysis **3**(1) (1999) 1–22
10. Kass, G. V.: An exploratory technique for investigating large quantities of categorical data, Applied Statistics **29**(2) (1980) 119-127
11. Lim, T., Loh, W., Shih, Y.: A comparison of prediction accuracy, complexity, and training time of thirty-three old and new classification algorithms. Mach. Learning **40**(3) (2000) 203–228
12. Murthy, S., Kasif, S., Salzberg, S.: A system for induction of oblique decision trees. Journal of Artificial Intelligence Research 2 (1994) 1–33
13. Murthy, S.: Automatic construction of decision trees from data: A multi-disciplinary survey. Data Mining and Knowledge Discovery 2 (1998) 345–389
14. Quinlan, J.: C4.5: Programs for Machine Learning. Morgan Kaufmann (1993)

Inductive Logic Programming in Clementine

Sam Brewer[1] and Tom Khabaza[2]

Advanced Data Mining Group, SPSS (UK) Ltd
1st Floor, St. Andrew's House, West Street
Woking, Surrey GU21 1EB, UK
[1] sbrewer@spss.com ' [2] tomk@spss.com

Abstract. This paper describes the integration of ILP with Clementine. Background on ILP and Clementine is provided, with a description of Clementine's target users. The benefits of ILP to data mining are outlined, and ILP is compared with pre-existing data mining algorithms. Issues of integration between ILP and Clementine are discussed. The implementation is then described, showing how the key issues are addressed, and describing in brief the Clementine mechanisms used to integrate ILP.

1. Introduction

The Clementine data mining system offers a range of components for the requirements of a typical data mining project. These components can be combined in various ways, ranging from the integration of models with visualisation tools, through various combinations of modelling methods, to meta-modelling, the technique of generating models to describe and analyse other models.

Clementine also has a facility called the CEMI (Clementine External Module Interface) which allows executable programs, external to Clementine, to be integrated into the software. This allows the Clementine user to create a customised the system and incorporate techniques not otherwise available.

ILP (Inductive Logic Programming) is a modelling technique which has many valuable features for data mining. ILP generates rules from multiple tables and from tables with multi-row examples, and can directly incorporate background knowledge into the modelling process; this makes it ideal for applications involving complex data, such as that used to describe molecules or electronic circuits.

The incorporation of ILP into Clementine increases the range of tools available and the range of data mining projects accessible to Clementine users.

2. Background

2.1 ILP

ILP is a form of modelling algorithm based on logic programming. Logic programs are rule-based programs written in first-order logic; inductive logic programming (ILP) integrates rule induction with logic programming.

A typical rule from a logic program might be:

```
will_buy(CustID, insurance) :-
    has_bought(CustID, Prod2), price(Prod2, Price),
    Price > 5000.
```

ILP would allow the induction of such a rule from data were terms such
`has_bought(CustID, Prod), will_buy(CustID, Prod)`, and
`price(Prod, Price)` represent tables of data.
There are many varieties of ILP engine, for example:

D.A. Zighed, J. Komorowski, and J. Zytkow (Eds.): PKDD 2000, LNAI 1910, pp. 337–344, 2000.
© Springer-Verlag Berlin Heidelberg 2000

- Tilde   builds decision trees to predict the value of a target attribute;
- Warmr   produces association rules;
- Progol   produces a form of predictive rule not commonly encountered in data mining, but sometimes referred to as "compression rules".

2.2 Clementine

Clementine is a powerful and popular data mining system. Its range of techniques, from data preparation and visualisation tools through to modelling and reporting, lends itself to a wide variety of data mining applications. Many industrial sectors benefit from the use of Clementine, including finance, retail, telecoms, manufacturing, engineering and science. Clementine appeals to users from across the spectrum of data mining expertise, ranging from non-technical business analysts to data mining and machine learning experts.

Clementine's "click-and-go" user interface provides a framework within which these two extremes of the user community can be accommodated. Each tool is configured with the most widely applicable default values in order to minimise knowledge required to operate them, whilst also offering options to alter these settings according to knowledge and expertise.

For example, consider a simple modelling stream (see figure 1) containing a source node, a type node and a neural net training node. The most basic set of operations involves setting a file to be read in through the source node, using the type node to select which field is to be modeled and executing the stream. (Note that the neural net training node is used in its default state; no modelling knowledge is required.) A more expert user has the option of editing the neural net training node and changing some aspects of its configuration.

Figure 1: A Simple Clementine Modelling Stream Diagram

3. ILP Benefits to Data Mining

The strengths of Inductive Logic Programming make it suitable for applications outside the scope of current algorithms. Previous modelling techniques handle data from a single table, where one row represents one example. ILP can build models from tables containing multi-row examples and from multiple tables. Current algorithms may implicitly use extra information derived through exploration and transformation of the data; ILP can enrich the modelling process by explicitly incorporating background knowledge. This section shows that ILP has a broad potential within data mining, facilitating the data mining process for many applications not easily covered by current modelling methods.

3.1 ILP Can Extract Information From Multiple Tables

Prior to ILP, most modelling techniques handle data from a single table. If the data is stored over multiple tables, or a table consists of multi-row examples, there is a pre-processing overhead in converting the data into a single-table, single-row-example format. Apart from the processing overhead and user time involved, re-shaping the table structure may lose potentially valuable information. ILP is able to handle both multi-row examples and multiple tables; re-structuring for these purposes is not required so the potential loss of information through pre-processing is reduced.

3.2 ILP Makes Explicit Use of Background Knowledge

It is commonplace in data mining for the data miner to derive new attributes describing significant relationships within the data; this takes place during the exploration and pre-processing phases of the data mining project. The new attributes are included as inputs to modelling, and their inclusion provides a form of "background knowledge" to the modelling process. However, these derived attributes are treated in the same way as the original attributes in the dataset; the data has been manipulated to have an influence on the modelling process.

By comparison, ILP makes use of "explicit" background knowledge; this information is incorporated directly into the modelling process. It takes the form of tables of data, and of rule-based knowledge expressed as logic programs, describing links and relationships amongst the tables,.

This means that "background" information, which might previously have been discarded because of the difficulty of incorporating it into the modelling process, can be included directly under ILP.

The capability of using explicit rule-based background knowledge also means that rules discovered by ILP can be used as background knowledge later in the data mining process.

3.3 ILP Allows Rapid Prototyping of Custom Algorithms

The use of logic programs for background knowledge also enables "rapid prototyping" of novel modelling algorithms, because the type of modelling provided by ILP is partly determined by the background knowledge. This will be a very valuable feature in situations where the patterns in the data are beyond the scope of available algorithms and modelling paradigms, because it allows the data miner to try "custom algorithms" on an ad-hoc basis for quick assessment of their suitability for a problem. The bespoke algorithm would be a combination of the ILP induction algorithm and the logic program expressed in the background knowledge; this method is made possible because the ILP engines are programmable learning systems.

4. ILP Integration with Clementine

The integration of Inductive Logic Programming with Clementine takes place through the Clementine mechanism "CEMI" mechanism (Clementine External Module Interface). This tool provides an protocol to enable external programs to be "plugged in" to Clementine. A text-based specification file is used to update Clementine with the information necessary for communicating with the external executable. The ILP engines are integrated into Clementine using this device.

4.1 Guidelines for Integration

It is important that the integration of any new technique fit smoothly into the Clementine framework.

4.1.1 Integration with the Clementine stream

Clementine is a "visual" data mining tool and any integration must be consistent with the visual programming ethos. An ILP engine will be integrated as a modelling node, and used as other modelling nodes within Clementine streams. The overall design of any modelling node in a stream is to be used as follows:

1. Connect data to a modelling node;
2. Select the target field to be modelled;
3. Insert any other necessary information;
4. Press execute;
5. A model is generated.

This is the initial framework for an ILP node. It will reside in the Models palette with other algorithms, and will be placed in a stream with a simple "click-and-drop".

There are some subtle problems for ILP. Algorithms in a stream read data from one node further upstream; the ILP node must also follow this pattern, which is in conflict with its ability to handle data from multiple sources. This problem is resolved by allowing only one explicit connection to the ILP node; further data sources are added in the settings of the node.

4.1.2 The ILP Node

The second consideration is how the user interacts within the node itself. Part of the Clementine philosophy is that any tool should be usable by a sizeable subset of the Clementine user base; at one end of the spectrum, this involves users with neither machine learning nor database experience. This means that the default use of a node must involve as few settings as possible, so that a minimal amount of technical knowledge is needed. However, below this simple level must be options for customised settings.

As with other modelling nodes in Clementine, it is necessary for the target field to be specified. This is a familiar thing for the Clementine user to have to do, so including this as a compulsory setting within the node is within our guidelines.

As only one table will be explicitly sent to the ILP node, all other tables must be listed in the settings of the node. Since there is a facility for specifying data files, logic program files may also be included. Background knowledge is included in the node as user-defined files; there is also a library containing both general-purpose and domain specific files of background knowledge.

Other settings are available within the node, but it is not compulsory for the user to alter these. There may, for example, be expert settings to tune the behaviour of the ILP engine. Any settings required by the expert user but not available in the interface can be included in a file as background knowledge.

4.2 Implementation of Integration

The physical integration of the ILP node is achieved through the Clementine External Module Interface. Information such as node type (process, terminal, model) is defined here, as is the edit dialogue. All of the information contained in the ILP node is passed through the CEMI as command line arguments to the executable, which is run externally to Clementine. Output resulting from the execution of the ILP engine will be sent to a file; the location of this file is specified within the CEMI, allowing Clementine to locate and display the results via a model browser. For example, included in the command line arguments are the name of the target predicate, the pathname of the main Clementine table, all other file pathnames and any other settings.

4.2.1 Integration of Clementine Data with the ILP Engines

In order for the ILP engine to process the data tables, they must be converted into Prolog. In logic terms, a table is simply a list of facts describing a predicate's extension. Files containing Prolog background knowledge are left unchanged.

4.2.2 Configuring ILP

An ILP engine requires configuration for each learning task: this specifies what data and background knowledge is available and how it should be used. This is information of a highly technical nature, so Clementine's ILP node generates this automatically.

The generated configuration information contains "mode declarations" and type descriptions (all expressed in Prolog). Mode declarations are used by ILP engines to guide the modelling (rule-induction) process. A "head" mode declaration states the target to be modeled, and "body" mode declarations describe other attributes and relationships in the data, and the way these will be used in the rules to be constructed.

Mode declarations can be used in a variety of ways which have a direct effect on the outcome of the modelling process, and only a subset of these are supported by Clementine's ILP node. It is therefore important for the expert user to be able to override this automatic generation and apply alternative mode declarations as part of the background knowledge. Clementine also provides an intermediate level of automation, where the user can choose whether to generate mode declarations which describe whole records or individual fields.

Type definitions are also generated automatically, using the heuristic that fields of the same name are assumed to be of the same type. This (theoretically baseless) assumption greatly simplifies the user interface to ILP.

The configuration contains type descriptions, mode declarations and attribute predicates (depending upon user options).

5. Overview of Clementine ILP Nodes

Three ILP engines will be incorporated into Clementine: Tilde, Warmr and Progol. These engines are very different and demonstrate that ILP is not an algorithm, but a family of algorithms.

Tilde (Top-down Induction of Logical DEcision Trees) is based on the C4.5 algorithm; it uses the decision tree method to generate clauses which describe the set of examples given. The rules produced are similar in structure and are therefore quite familiar to the Clementine user.

Warmr is based on the Apriori association algorithm, and builds such rules from multiple tables. In comparison to Tilde which performs "predictive" learning, the output from Warmr is more "descriptive".

Progol is not based on any immediately recognisable algorithms; it produces "compression rules" by generalising from Prolog examples, which are effectively records from a table of data, which have been translated into Prolog facts. Even simple runs of the Progol node have shown that it finds rules which C5.0 needs additional (exploratory-derived) attributes to find.

5.1 The Progol Node

The Progol node edit dialogue is typical of the nodes for integrated ILP engines; this is shown in figure 2.

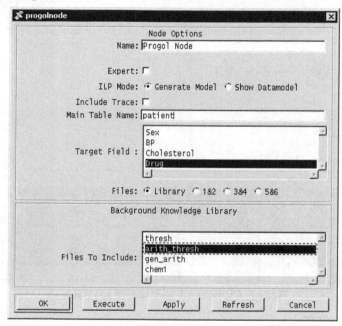

Figure 2: Progol Node Edit Dialogue

5.2 Options

Name   This is the name given to the generated model.

Expert   This is switched off by default; it reveals an extra dialogue containing some expert settings.

ILP Mode   This gives the option of generating a Progol model (the default) or generating a datamodel; this is the automatically generated file containing mode declarations and types. It may be useful for the user to view this before executing an ILP model.

Main Table Name
                      This is procedure name given to the prolog facts within the table containing the target attribute.

Target Field   This is where the user specifies the target attribute. A list of fields in the main table will appear when the node is connected; the user only has to highlight one of them.

The last pane defaults to the background library list. This can be changed to show another pane which provides a "set file" button; here the other data files and background knowledge files can be listed.

5.3 A Simple Modelling Stream Using the Integrated Progol Node

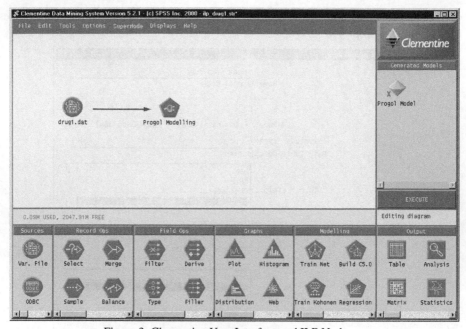

Figure 3: Clementine User Interface and ILP Node

Model Execution

After the stream has been executed, the rules generated from the ILP node form a model, which appears as a new "gem" icon on the "Generated Models" palette. When this new node is included in stream execution, data passing through the node is analysed according to the rules and a new field is created, containing predicted values. If the generated model contains rules with different predicted values which match the same record, current versions use a "first hit" approach to conflict resolution.

ILP Made Easy!

The following is a list of "click-steps" which the user has undergone in figure 3 to generate an ILP model; this demonstrates that we have achieved the design goals of incorporating ILP as a modelling algorithm in a Clementine stream (section 4.1.1.)

1. Connect data to a Progol node;
2. Select the target field to be modelled;
3. List any other files to be included;
4. Press execute;
5. A model is generated.

6. Summary

We have seen that ILP has capabilities beyond pre-exising algorithms, and that it is beneficial to have it integrated into a data mining system. ILP will enable Clementine to deal with complex data without incurring large pre-processing overheads.

ILP's ability to model data from multiple tables and with multi-record examples means that during the modelling phase, complex data can retain its structure, which may provide information. Background knowledge can also be incorporated into the modelling process, without the need for information to be added to the data itself through transformation.

This facility could be used in a more "foreground" context; the purpose of background knowledge could to alter the algorithm. This would lend ILP to rapid prototyping of bespoke algorithms.

The integration of ILP into Clementine presents some complex implementation issues, its representation in the user interface being the most obvious, as the stream construction used for Clementine nodes does not provide a perfect match with ILP. There are also complex low-level issues involved in integration. To cater for the non-technical user, the ILP node must be simplified so that various ILP components such as the mode declarations and type definitions are generated automatically. This approach provides a subset of ILP for the non-technical user. However, Clementine's integration of ILP also caters for the knowledgeable user, by providing advanced options. The range of possible applications is increased with these facilities.

The benefits of combining Clementine with ILP are twofold; firstly it will provide a set of techniques which considerably extend the capability of the software. The pre-processing of complex data and the risk of losing potentially valuable data involved, will be removed. This will facilitate data mining for such applications. Secondly, it will bring ILP to a wider and more commercial audience.

7 Acknowledgements

This work was supported by ESPRIT project number 28623 "ALADIN", which is a collaboration between Perot Systems Netherlands, British Telecom, the University of York, the Katholieke Universiteit Leuven and SPSS / Integral Solutions Ltd.

8 References

[1]   H. Blockeel & L. De Raedt, Tilde and Warmr User Manual, Version 2.0, Katholieke Universiteit Leuven, April 1999.
[2]   S. Brewer & T. Khabaza. Guidelines for ILP in Data Mining, Version 1.5, ALADIN Project Internal Report, November 1999.
[3]   L. Dehaspe, WARMR The Frequent Query Discovery Engine User's Guide 2.1, October 1998.
[4]   T. Khabaza, Note on the Integration of Inductive Logic Programming with the Clementine Data Mining System, ALADIN Project Internal Report, June 1998.
[5]   S. Muggleton & J. Firth, CProgol4.4: Theory and Use, University of York, June 1998.
[6]   Integral Solutions Ltd, "Introduction to the External Module Interface", Clementine Reference Manual, version 5, September 1998.
[7]   S. Muggleton, "Inverse Entailment and Progol", New Generation Computing, 13:245-286, 1995.
[8]   H. Blockeel and L. DeRaedt, "Top-down Induction of first order Logical Decision Trees", Artificial Intelligence 101 (1-2), 1998.
[9]   L. Dehaspe and H. Toivonen, Frequent query discovery: a unifying ILP approach to association rule mining. Technical Report CW-258, Department of Computer Science, Katholieke Universiteit Leuven, March 1998. http://www.cs.kuleuven.ac.be/publicaties/-rapporten/CW1998.html

A Genetic Algorithm-Based Solution for the Problem of Small Disjuncts

Deborah R. Carvalho[1, 2] and Alex A. Freitas[1]

[1] Pontificia Universidade Catolica do Parana (PUCPR)
Postgraduate program in applied computer science
R. Imaculada Conceicao, 1155. Curitiba – PR 805215-901. Brazil
Tel/Fax:(55) (41) 330-1669
http://www.ppgia.pucpr.br/~alex   alex@ppgia.pucpr.br
[2] Universidade Tuiuti do Parana (UTP)
Computer Science Dept.
Av. Comendador Franco, 186. Curitiba – PR 80215-090. Brazil
Tel/Fax: (55) (41) 263-3424   deborah@utp.br

Abstract. In essence, small disjuncts are rules covering a small number of examples. Hence, these rules are usually error-prone, which contributes to a decrease in predictive accuracy. The problem is particularly serious because, although each small disjuncts covers few examples, the set of small disjuncts can cover a large number of examples. This paper proposes a solution to the problem of discovering accurate small-disjunct rules based on genetic algorithms. The basic idea of our method is to use a hybrid decision tree / genetic algorithm approach for classification. More precisely, examples belonging to large disjuncts are classified by rules produced by a decision-tree algorithm, while examples belonging to small disjuncts are classified by a new genetic algorithm, particularly designed for discovering small-disjunct rules.

1 Introduction

In the context of the well-known classification task of data mining, the discovered knowledge is often expressed as a set of IF-THEN prediction rules. Typically the discovered rules are in disjunctive normal form, where each rule represents a disjunct and each rule condition represents a conjunct. A small disjunct can be defined as a rule which covers a small number of training examples [7].

In general rule induction algorithms have a bias that favors the discovery of large disjuncts, rather than small disjuncts. This preference is due to the belief that it is better to capture generalizations rather than specializations in the training set, since the latter are unlikely to be valid in the test set.

Hence, at first glance small disjuncts should not be included in the discovered rule set, since they tend to be error prone. However, a deeper study of the issue of small

D.A. Zighed, J. Komorowski, and J. Zytkow (Eds.): PKDD 2000, LNAI 1910, pp. 345-352, 2000.
© Springer-Verlag Berlin Heidelberg 2000

disjuncts reveals that in fact they can be necessary and even interesting by themselves in the context of data mining, for the following reasons:

(a) Although each disjunct covers a small number of examples, the set of all small disjuncts can cover a large number of examples. For instance [3] reports a real-world application where small disjuncts cover roughly 50% of the training examples. Therefore, if the rule induction algorithm ignores small disjuncts and discovers only large disjuncts, classification accuracy will be significantly degraded.

(b) Some small disjuncts cover examples that represent rare cases in the application domain, which constitutes an interesting concept to be discovered. Actually, bearing in mind that one of the goals of data mining is to discover previously-*unknown* rules, small-disjunct rules tend to be more interesting than large-disjunct rules, since the latter are more likely to be previously-known by the user [11].

In this paper we propose a hybrid decision tree/genetic algorithm method for rule discovery that copes with the problem of small disjuncts. The basic idea is that examples belonging to large disjuncts are classified by rules produced by a decision-tree algorithm, while examples belonging to small disjuncts (whose classification is more difficult) are classified by rules produced by a new genetic algorithm.

2 A Hybrid Decision-Tree/Genetic-Algorithm Method

We propose a hybrid method for rule discovery that combines decision trees and genetic algorithms. Decision-tree algorithms have a bias towards generality that is well suited for large disjuncts, but not for small disjuncts. On the other hand, genetic algorithms are robust algorithms which tend to cope well with attribute interactions [4], [10]. Hence, they can be more easily tailored for coping with small disjuncts, which are associated with large degrees of attribute interaction [13], [9].

The proposed method discovers rules in two training phases. In the first phase we run the C4.5 decision-tree algorithm [12]. Then, the induced decision tree with d leaves is transformed into a rule set with d rules (or disjuncts). Each of these rules is considered either as a small disjunct or as a "large" (non-small) disjunct, depending on whether or not its coverage (the number of examples covered by the rule) is smaller than or equal to a given threshold.

The second phase consists of using a genetic algorithm to discover rules covering examples belonging to small disjuncts. We have developed a new genetic algorithm (GA) for this phase. In our GA, each individual represents a small-disjunct rule.

Each run of our GA discovers a single rule (the best individual of the last generation) predicting a given class for examples belonging to a given small disjunct. We need to run our GA $d * c$ times, where d is the number of small disjuncts and c is the number of classes to be predicted. For a given small disjunct, the i-th run of the GA, $i = 1,...,c$, discovers a rule predicting the i-th class.

The genome of an individual consists of a conjunction of conditions composing the antecedent (IF part) of the rule. Each condition is an attribute-value pair - see below. The consequent (THEN part) of the rule, which specifies the predicted class, is not represented in the genome. Rather, it is fixed for a given GA run, so that all individuals have the same rule consequent during all that run.

The rule antecedent contains a variable number of rule conditions. In our GA the minimum number of conditions is always 2. The maximum number of conditions, n, depends on the small disjunct, as follows.

To represent a variable-length rule antecedent (phenotype) we use a fixed-length genome. For a given GA run, the genome of an individual consists of n genes, where $n = m - k$, m is the total number of predictor attributes in the dataset and k is the number of ancestor nodes of the decision tree leaf node identifying the small disjunct in question. Hence, the genome of a GA individual contains only the attributes that were *not* used to label any ancestor of the leaf node defining that small disjunct.

The overall structure of the genome of an individual is illustrated in Figure 1. Each gene represents a rule condition (phenotype) of the form $A_i \, Op_i \, V_{ij}$, where the subscript i identifies the rule condition, $i = 1,...,n$; A_i is the i-th attribute; V_{ij} is the j-th value of the domain of A_i; and Op_i is a logical/relational operator compatible with attribute A_i. Each gene consists of four elements, as follows:
(a) identification of a given predictor attribute, A_i, $i = 1,...,n$.
(b) identification of a logical/relational operator Op_i. For categorical (nominal) attributes, Op_i is "*in*". For continuous (real-valued) attributes, Op_i is either "$\leq$" or "$>$".
(c) identification of a set of attribute values $\{V_{ij},..., V_{ik}\}$, if the attribute A_i is categorical, or a single attribute value V_{ij}, if the attribute A_i is continuous.
(d) a flag, called the active bit B_i, which takes on 1 or 0 to indicate whether or not, respectively, the i-th condition is present in the rule antecedent (phenotype).

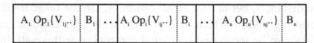

Figure 1: Structure of the genome of an individual.

To evaluate the quality of an individual our GA uses the fitness function:
$$\text{Fitness} = (TP / (TP + FN)) * (TN / (FP + TN)) , \qquad (1)$$
where TP, FN, TN and FP – standing for number of true positives, false negatives, true negatives and false positives – are well-known variables often used to evaluate the performance of classification rules – see e.g. [6].

We use tournament selection, with tournament size of 2 [8]. We also use standard one-point crossover with crossover probability of 80%, and mutation probability of 1%. Furthermore, we use elitism with an elitist factor of 1 - i.e. the best individual of each generation is passed unaltered into the next generation.

In addition to the above standard genetic operators, we have also developed a new operator especially designed for simplifying candidate rules. The basic idea of this operator, called rule-pruning operator, is to remove several conditions from a rule to make it shorter. This operator is applied to every individual of the population, right after the individual is formed.

Unlike the usually simple operators of GA, our rule-pruning operator is an elaborate procedure based on information theory [2]. Hence, it can be regarded as a way of incorporating a classification-related heuristic into a GA for rule discovery. The heuristic in question is to favor the removal of rule conditions with low information gain, while keeping the rule conditions with high information gain – see [1] for details.

Once all the $d * c$ runs of the GA are completed, examples in the test set are classified. For each test example, we push the example down the decision tree until it reaches a leaf node. If that node is a large disjunct, the example is classified by the decision tree algorithm. Otherwise we try to classify the example by using one of the c rules discovered by the GA for the corresponding small disjunct. If there is no small-disjunct rule covering the test example it is classified by a default rule. We have experimented with two strategies for defining the default rule:

• a global default rule that predicts the majority class among all small-disjunct.

• a local default rule that predicts the majority class among the examples belonging to the current small disjunct.

3 Computational Results

We have evaluated our GA on two public domain data sets from the UCI data set repository (http://www.ics.uci.edu/~mlearn/MLRepository.html).

One of the them is the Adult data set (USA census). This dataset contains 48842 examples, 14 attributes (6 are continuous and 8 are categorical), and two classes. In our experiments we have used the predefined division of the data set into a training and a test set, with the former having 32561 examples and the latter having 16281 examples. The examples that had some missing value were removed from the data set. As a result, the number of examples was slightly reduced to 30162 and 15060 examples in the training and test set, respectively.

The other data set used in our experiments was the Wave data set. This data set contains 5000 instances, 21 attributes with values between 0 and 6 and three classes. In this data set we have run a five-fold cross-validation procedure.

In our experiments a decision-tree leaf is considered a small disjunct if and only if the number of examples belonging to that leaf is smaller than or equal to a fixed size S. We have done experiments with four different values for the parameter S, namely $S = 3$, $S = 5$, $S = 10$ and $S = 15$.

We now report results comparing the performance of the proposed hybrid C4.5/GA with C4.5 alone [12]. We have used C4.5's default parameters. In each GA run the population has 200 individuals, and the GA is run for 50 generations.

We have evaluated two variants of our hybrid C4.5/GA method: global and local default rule (see above). The results for the Adult and Wave data sets are shown in Tables 1 and 2. All results refer to accuracy rate on the test set. The first column of these tables indicate the size threshold S used to define small disjuncts. The next three columns, labeled (a), (b), (c), report results produced by C4.5 alone. More precisely, columns (a) and (b) report the accuracy rate on the test set achieved by C4.5 separately for examples classified by large-disjunct rules and small-disjunct rules. Column (c) reports the overall accuracy rate on the test achieved by C4.5, classifying both large-

and small-disjunct examples. Note that the figures in this column are of course constant across all the rows, since its results refer to the case where all test examples are classified by C4.5 rules, regardless of the definition of small disjunct.

Table 1: Results comparing our hybrid C4.5/GA with C4.5 in the Adult data set.

Dis-junct size (S)	Accuracy rate of C4.5 only			Accuracy rate of C4.5/GA – global default rule			Accuracy rate of C4.5 / GA – local default rule		
	(a) large disjuncts	(b) small disjuncts	(c) overall	(d) large disjuncts	(e) small disjuncts	(f) overall	(g) large disjuncts	(h) small disjuncts	(i) overall
3	0.800	0.512	0.786	0.800	0.470	0.780	0.800	0.457	0.779
5	0.811	0.520	0.786	0.811	0.497	0.780	0.811	0.483	0.779
10	0.841	0.521	0.786	0.841	0.640	0.828	0.841	0.642	0.829
15	0.840	0.530	0.786	0.840	0.711	0.831	0.840	0.707	0.830

Table 2: Results comparing our hybrid C4.5/GA with C4.5 in the Wave data set.

Dis-junct size (S)	accuracy rate of C4.5 only			accuracy rate of C4.5/GA – global default rule			Accuracy rate of C4.5 / GA – local default rule		
	(a) large disjuncts	(b) small disjuncts	(c) overall	(d) large disjuncts	(e) small disjuncts	(f) overall	(g) large disjuncts	(h) small disjuncts	(i) overall
3	0.758	0.722	0.755	0.758	0.776	0.765	0.758	0.732	0.756
5	0.774	0.710	0.755	0.774	0.727	0.758	0.774	0.754	0.764
10	0.782	0.731	0.755	0.782	0.800	0.793	0.782	0.808	0.796
15	0.788	0.731	0.755	0.788	0.832	0.814	0.788	0.814	0.803

The next three columns, labeled (d), (e), (f), report results produced by our hybrid C4.5/GA method in the variant of global default rule. Note that the figures in column (d) are exactly the same as the figures in column (b), since our hybrid method also uses C4.5 rules for classifying examples belonging to large disjuncts. In any case, we included this redundant column in the Tables for the sakes of comprehensibility and completeness. Column (e) reports the accuracy rate on the test set for the small-disjunct rules discovered by the GA. Finally, column (f) reports the overall accuracy rate on the test achieved by our hybrid C4.5/GA method, classifying both large- and small-disjunct examples. The next three columns, labeled (g), (h), (i), refer to the results with the variant of local default rule. The meaning of these columns is analogous to the one explained for columns (d), (e), (f), respectively.

As can be seen in Tables 1 and 2, there is little difference of performance between the two variants of our hybrid C4.5/GA, and overall both variants achieved better predictive accuracy than C4.5 alone. More precisely, comparing both columns (e) and (h) with column (b) in each of those two tables we can note two distinct patterns of results. Consider first the case where a disjunct is considered as small if it covers ≤ 3 or ≤ 5 examples, as in the first and second rows of Tables 1 and 2. In this case the accuracy rate of the small-disjunct rules produced by the GA is slightly inferior to the performance of the small-disjunct rules produced by C4.5 in the Adult data set (Table 1), while the former is somewhat superior to the latter in the Wave data set. In any case, this small difference of performance referring to small-disjunct rules has a small impact on the overall accuracy rate, as can be seen by comparing both columns (f) and (i) with column (c) in Tables 1 and 2.

A different picture emerges when a disjunct is considered as small if it covers ≤ 10 or ≤ 15 examples, as in the third and fourth rows of Tables 1 and 2. Now the performance of the small-disjunct rules produced by the GA is much better than the performance of the small-disjunct rules produced by C4.5, in both data sets. For instance, comparing both columns (e) and (h) with the column (b) in the fourth row of Table 1, the GA-discovered small disjunct rules have an accuracy rate of 71.1% and 70.7%, whereas the C4.5-discovered rules have an accuracy rate of only 53%. This improved accuracy associated with GA-discovered small disjunct rules has a considerable impact on the overall accuracy rate, as can be seen comparing both columns (f) and (i) with column (c) in the third and fourth rows of Tables 1 and 2.

A possible explanation for these results is as follows. In the first case, where a disjunct is considered as small if it covers ≤ 3 or ≤ 5 examples, there are very few training examples available for each GA run. With so few examples the estimate of rule quality computed by the fitness function is far from perfect, and the GA does not manage to do better than C4.5. On the other hand, in the second case, where a disjunct is considered as small if it covers ≤ 10 or ≤ 15 examples, the number of training examples available for the GA is significantly higher - although still relatively low. Now the estimate of rule quality computed by the fitness function is significantly better. As a result, the GA's robustness and ability to cope well with attribute interaction lead to the discovery of small-disjunct rules considerably more accurate than the corresponding rules discovered by C4.5.

Although the above results are good, they do not prove by themselves that the small-disjunct rules discovered by the GA are considerably superior to the small-disjunct rules discovered by C4.5. After all, recall that the test examples belonging to small disjuncts can be classified either by a GA-discovered rule or by the default rule. This raises the question of which of these two kinds of rule is really responsible for the good results reported above.

To answer this question we measured separately the relative frequency of use of each of the two kinds of rule, namely GA-discovered rules and default rule, in the classification of test examples belonging to small disjuncts. We found that GA-discovered rules are used much more often to classify test examples belonging to small disjuncts than the default rule. More precisely, depending on the definition of small disjunct (the value of the parameter S) used, the relative frequency of use of GA-discovered rules varies between 68.7% and 95% for the Adult data set and from 84.1% to 90.2% in the Wave data set. Hence, one can be confident that the small disjunct rules discovered by the GA are doing a good job of classifying most of the test examples belonging to small disjuncts. In any case, to get further evidence we also measured separately the predictive accuracy of GA-discovered rules and default rule in the classification of test examples belonging to small disjuncts in the case of global-default rule. Overall, as expected, the GA-discovered rules have a higher predictive accuracy than the default rule. To summarize, as expected most of the credit for the good performance in the classification of small disjuncts is to be assigned to the GA-discovered rules, rather than to the default rules. (Note that there is no need to get this kind of evidence in the case of the local-default rules, since in this case there is no other way that the GA-tree hybrid could beat the tree alone.)

Turning to computational efficiency issues, each run of the GA is relatively fast, since it uses a training set with just a few examples. However, recall that in order to discover all small disjunct rules we need to run the GA $c * d$ times, where c is the number of classes and d is the number of small disjuncts. The total processing time taken by the $c * d$ GA runs varies with the number of small disjuncts, which depends on both the data set and on the definition of small disjunct (the value of S). In our experiments, the processing time taken by all $c * d$ runs of the GA was about one hour for the largest data set, Adult, and the largest number of small disjuncts, associated with S = 15. The experiments were performed on a 64-Mb Pentium II. One hour seems to us a reasonable processing time and a small price to pay for the considerable increase in the predictive accuracy of the discovered rules.

Finally, if necessary the processing time taken by all the $c * d$ GA runs can be considerably reduced by using parallel processing techniques [5]. Actually, our method greatly facilitates the exploitation of parallelism in the discovery of small disjunct rules, since each GA run is completely independent from the others and it needs to have access only to a small data set, which surely can be kept in the local memory of a simple processor node.

4 Conclusions and Future Research

The problem of how to discover good small-disjunct rules is very difficult, since these rules are error-prone due to the very nature of small disjuncts. Ideally, a data mining system should discover good small-disjunct rules without sacrificing the goodness of discovered large-disjunct rules.

Our proposed solution to this problem was a hybrid decision-tree/GA method, where examples belonging to large disjuncts are classified by rules produced by a decision-tree algorithm and examples belonging to small disjuncts are classified by rules produced by a genetic algorithm. In order to realize this hybrid method we have used the well-known C4.5 decision-tree algorithm and developed a new genetic algorithm tailored for the discovery of small-disjunct rules.

The proposed hybrid method was evaluated in two data sets. We found that the performance of our new GA and corresponding hybrid C4.5/GA method depends significantly on the definition of small disjunct. The results show that: (a) there is no significant difference in the accuracy rate of the rules discovered by C4.5 alone and the rules discovered by our C4.5/GA method when a disjunct is considered as small if it covers ≤ 3 or ≤ 5 examples; (b) the accuracy rate of the rules discovered by our C4.5/GA method is considerably higher than the one of the rules discovered by C4.5 alone when a disjunct is considered as small if it covers ≤ 10 or ≤ 15 examples.

A disadvantage of our hybrid C4.5/GA method is that it is much more computationally expensive than the use of C4.5 alone. More precisely, in a training set with about 30000 examples our hybrid method takes on the order of one hour, while C4.5 alone takes on the order of a few seconds. However the extra processing time is not too excessive, and it seems a small price to pay for the considerable increase in the predictive accuracy of the discovered rules.

An important direction for future research is to evaluate the performance of the proposed hybrid C4.5/GA method for different kinds of definition of small disjunct, e.g. relative size of the disjunct (rather than absolute size, as considered in this paper). Another research direction would be to compare the results of the proposed C4.5/GA method against rules discovered by the GA only, although in this case some aspects of the design of the GA would have to be modified.

References

1. CARVALHO, D.R. and FREITAS, A.A. A hybrid decision tree/genetic algorithm for coping with the problem of small disjuncts in data mining. *To appear in Proc. 2000 Genetic and Evolutionary Computation Conf. (GECCO-2000).* Las Vegas, NV, USA. July 2000.
2. COVER, T.M., THOMAS, J.A. (1991) *Elements of Information Theory.* John Wiley&Sons.
3. DANYLUK, A., P. and PROVOST, F.,J. (1993). Small Disjuncts in Action: Learning to Diagnose Errors in the Local Loop of the Telephone Network, *Proc. 10th International Conference Machine Learning,* 81-88.
4. FREITAS, A.A. (2000) Evolutionary Algorithms. Chapter of forthcoming *Handbook of Data Mining and Knowledge Discovery.* Oxford University Press, 2000.
5. FREITAS, A.A. and LAVINGTON, S.H. (1998) *Mining Very Large Databases with Parallel Processing.* Kluwer.
6. HAND, D.J.(1997) *Construction and Assessment of Classification Rules.* John Wiley& Sons
7. HOLTE, R.C.; ACKER, L.E. and PORTER, B.W. (1989). Concept Learning and the Problem of Small Disjuncts, *Proc. IJCAI – 89,* 813-818.
8. MICHALEWICZ, Z. (1996) *Genetic Algorithms + Data Structures = Evolution Programs.* 3rd Ed. Springer-Verlag.
9. NAZAR, K. and BRAMER, M.A. (1999) Estimating concept difficulty with cross entropy. In: Bramer, M.A. (Ed.) *Knowledge Discovery and Data Mining,* 3-31. London: IEE.
10. NODA, E.; LOPES, H.S.; FREITAS, A.A. (1999) Discovering interesting prediction rules with a genetic algorithm. *Proc. Congress on Evolutionary Comput. (CEC-99),* 1322-1329
11. PROVOST, F. and ARONIS, J.M. (1996). Scaling up inductive learning with massive parallelism. *Machine Learning 23(1),* Apr. 1996, 33-46.
12. QUINLAN, J.R. (1993). *C4.5: Programs for Machine Learning,* Morgan Kaufmann.
13. RENDELL, L. and SESHU, R. (1990) Learning hard concepts through constructive induction: framework and rationale. *Computational Intelligence 6,* 247-270.

Clustering Large, Multi-level Data Sets: An Approach Based on Kohonen Self Organizing Maps

Antonio Ciampi[1] and Yves Lechevallier[2]

[1] Department of Epidemiology & Biostatistics,
McGill University, Montreal, P.Q., Canada
and IARC, 150 Cours Albert-Thomas, Lyon
ciampi@iarc.fr
[2] INRIA-Rocquencourt,
78153 Le Chesnay CEDEX, France
Yves.Lechevallier@inria.fr

Abstract. Standard clustering methods do not handle truly large data sets and fail to take into account multi-level data structures. This work outlines an approach to clustering that integrates the Kohonen Self Organizing Map (SOM) with other clustering methods. Moreover, in order to take into account multi-level structures, a statistical model is proposed, in which a mixture of distributions may have mixing coefficients depending on higher-level variables. Thus, in a first step, the SOM provides a substantial data reduction, whereby a variety of ascending and divisive clustering algorithms become accessible. As a second step, statistical modelling provides both a direct means to treat multi-level structures and a framework for model-based clustering. The interplay of these two steps is illustrated on an example of nutritional data from a multi-center study on nutrition and cancer, known as EPIC.

1 Introduction

Appropriate use of a clustering algorithm is often a useful first step in extracting knowledge from a data base. Clustering, in fact, leads to a *classification*, *i.e.* the identification of homogeneous and distinct subgroups in data [9] and [2], where the definition of *homogeneous* and *distinct* depends on the particular algorithm used : this is indeed a simple structure, which, in the absence of *a priori* knowledge about the multidimensional shape of the data, may be a reasonable starting point towards the discovery of richer, more complex structures.

In spite of the great wealth of clustering algorithms, the rapid accumulation of large data bases of increasing complexity poses a number of new problems that traditional algorithms are not equipped to address. One important feature of modern data collection is the ever increasing size of a typical data base: it is not so unusual to work with data bases containing from a few thousands to a few millions of individuals and hundreds or thousands of variables. Now, most clustering algorithms of the traditional type are severely limited as to the

D.A. Zighed, J. Komorowski, and J. Zytkow (Eds.): PKDD 2000, LNAI 1910, pp. 353–358, 2000.
© Springer-Verlag Berlin Heidelberg 2000

number of individuals they can confortably handle (from a few hundred to a few thousands). Another related feature is the multi-level nature of the data: typically a data base may be obtained from a multi-country, multi-centre study, so that individuals are nested into centres which are nested into countries. This is an example of an elementary, known structure in the data which should not be ignored when attempting to discover new, unknown structures.

This work arises from the participation of one of its authors to the EPIC project. EPIC is a multi-centre prospective cohort study designed to investigate the effect of dietary, metabolic and other life-style factors on the risk of cancer. The study started in 1990 and includes now 23 centres from 10 European countries. By now, dietary data are available on almost 500,000 subjects. Here we initiate a new methodological development towards the discovery of dietary patterns in the EPIC data base. We look for general dietary patterns, but taking into account, at the same time, geographical and socio-economic variation due to country and centres.

Also, we limit ourselves to an analysis of data from a 24-hour recall questionnaire concerning intake of sixteen food-groups. Thus, we will only discuss clustering for 2-level systems: subjects (first level) and centre (second level), in our case.

The approach we propose is based on two key ideas :

1) A preliminary data reduction using a Kohonen Self Organizing Map (SOM) is performed. As a result, the individual measurements are replaced by the means of the individual measurements over a relatively small number of *micro-regimens* corresponding to Kohonen neurons. The micro-regimens can now be treated as new *cases* and the means of the original variables over micro-regimens as new *variables*. This *reduced* data set is now small enough to be treated by classical clustering algorithms. A further advantage of the Kohonen reduction is that the vector of means over the micro-regimens can safely be treated as multivariate normal, owing to the central limit theorem. This is a key property, in particular because it permits the definition of an appropriate dissimilarity measure between micro-regimens.

2) The multilevel feature of the problem is treated by a statistical model wich assumes a mixture of distributions, each distribution representing, in our example, a *regimen* or *dietary pattern*. Although more complex dependencies can be modeled, here we will assume that the centres only affect the mixing coefficients, and not the parameters of the distributions. Thus we look for general dietary patterns assuming that centers differ from each other only in the distribution of the local population across the general dietary patterns.

While the idea of a preliminary Kohonen reduction followed by the application of a classical clustering algorithm is not entirely new [12], [1] and [14], this work differs from previous attempts in several respects the most important of which are :

a) the Kohonen chart is trained by an the initialization based on principal component analysis;

b) the choice of clustering algorithm is guided by the multilevel aspect of the problem at hand;

c) the clustering algorithm is based on a statistical model.

Thus this work continues the author's research program which aims to develop data analytic strategies integrating KDDM and classical data analysis methods [6] and [7].

2   Data Reduction by Kohonen SOM's

We consider p measurements performed on n subjects grouped in C classes, $\{G_c, c = 1, \ldots, C\}$. We denote these measurements by $(G^{(i)}, x^{(i)}), i = 1, \ldots, n$, where for the i^{-th} subject $G^{(i)}$ denotes the class (the centre, in our example), and $x^{(i)}$ the p-vector of measurements (the 16 food-group intake variables); or, in matrix form, $\mathbf{D} = [\mathbf{G}|\mathbf{X}]$. In this section we describe the first step of the proposed approach, which consists in reducing the $n \times p$ matrix $\mathbf{X}$ to a $m \times p$ matrix, $m \ll n$.

2.1   Kohonen SOM's and PCA Initialization

We recall that in a Kohonen SOM the neurons of the rectangular sheet are associated to a grid of prototypes in the p-dimensional space which represents the row-vectors of the data matrix: the sheet is supposed to represent the grid with a minimum distortion, so that a SOM can be seen as a non-linear version of classical data reduction techniques such as a Principal Component Analysis (PCA). In order to specify a SOM, one needs to specify initial values of the sheet's connection weights and of the position of the prototypes. Then, the data points are repeatedly sent through the SOM, each passage causing an update of both the connection weights and the position of the prototypes, $i.e$ an alteration of both the sheet in 2-dimensional space and the grid in p-dimensional space. Normally, this process converges, in that the changes at each passage become negligible.

In the original approach, initial weights were chosen at random; however, as the efficacy of the algorithms crucially depends on the initialization, much effort has been devoted to improving this first step. The distinguishing feature of our construction consists in designing the sheet with the help of the results of PCA performed on $\mathbf{X}$. It is advantageous to choose the dimensions of the grid, a is the number of neurons on the first axis and b for the second axis where $m = ab$, such that :

$$\frac{a}{b} = \frac{\sqrt{\lambda_1}}{\sqrt{\lambda_2}}$$

where λ_1 and λ_2 denote the first and second eigenvalues of the PCA. Also, the initial connection weights and position of the prototypes are obtained from the two first eigenvectors of the PCA. The details are described in [8], where it is also shown that PCA initialization presents substantial practical advantages over several alternative approaches.

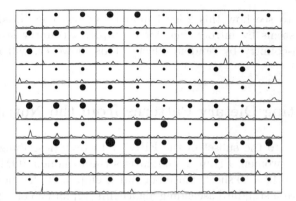

Fig. 1. Kohonen map

2.2   Binning of the Original Data Matrix Using a Kohonen Map

As a result of the training process, the SOM associates to each subject a unique neuron-prototype pair, which we shall refer to as *micro-regimen*. Each micro-regimen , $B_r, r = 1, \ldots, m$, can be considered as a *bin* in which similar individuals are grouped. We shall denote by n_r the number of subjects in B_r and by $n_{r,c}$ the number of subjects in $B_r \cap G_c$. Let also $\bar{x}_r$ and $\bar{x}_r^{(c)}$ denote the vectors of the means of $x^{(i)}$ taken over B_r and over $B_r \cap G_c$ respectively. Figure 1 gives a graphical representation of the bins [10] : in each bind the dot is proportional to bin size and the graph is a profile of the input variables.

Already at this stage, an exploratory analysis of the two-way table $\{n_{r,c}; r = 1, \ldots, m, \ c = 1, \ldots, C\}$, would be instructive: *e.g.* Correspondence Analysis (CA) of the table, ordering its rows and columns according to the factor scores and eventually clustering rows and columns, is likely to shed some light on the relationship between centers and micro-regimens.

3   Clustering Multi-level Systems

Since the number of bins $(m \ll n)$ may be chosen to be relatively small, a panoply of ascending approaches becomes accessible. Moreover, several dissimilarity-based divisive approaches are available. Among these, some conceptual clustering algorithms [5] [4] seem particularly promising as the one used in this work, see next section.

Suppose now that a clustering algorithm has been deployed. As a result, the m micro- regimens are grouped to produce $k \ll m$ macro-regimens. Furthermore, the 2-way table $\{m_{i,c}; i = 1, \ldots, k, \ c = 1, \ldots, C\}$ obtained by crossing centers with macro-regimens can be analyzed by CA as outlined in the previous section. Finally, proportions of subjects following different regimens in each centre would usefully summarize local characteristics, while a description of the clusters would give insight on the nature of the macro-regimens found in the general population.

It is easy to see how this 2-level model can be generalized to three- and multi-level systems by introducing, for example, a country level and treating centers-within-country by random effects. This, however, will not be pursued here.

4   Extracting Dietary Patterns from the Nutritional Data

We return now to the subset of the EPIC data base describing dietary habits of 4,852 French women. Figure 1 summarises the Kohonen SOM analysis of the data based on a 10×10 sheet. Since one bin is empty, 99 distinct regimens were identified. Both a standard ascending algorithm [12] and a conceptual clustering algorithm [5] applied to the micro-regimens, suggest 4, 6 or 9 classes or dietary patterns. The results of the 6-class analysis are summarised in Figure 4, which shows the first factorial plane of the CA representing the relationship between centres and dietary pattern; Figure 4, which shows the Zoom Star graphs [13] of the eight most discriminating variables describing dietary patterns; and Table 4 which gives a rough estimate of the proportions of subjects following the six dietary patterns, overall and by centre.

An example of interpretation is as follows: regimen 1 is characterized by high consumption of meat and vegetables; regimen 2 by high soups and low vegetable consumption; regimen 3 by high fish and low meat consumption (respectively 13% and 3% of the total weight of food intake); regimen 4 by high meat and low fish consumption; regimen 5 by high alcohol and meat consumption; and regimen 6 by high consumption of dairy products, eggs and vegetables and low consumption of fish, alcoholic beverage and legumes. Also, the Nord-Pas- de-Calais region is positively associated to regimen 2 and 5 and negatively to regimen 1; similarly, there is a positive association of Bretagne-Pays-de-Loire with regimen 3 and a negative association with regimen 1; and finally, Rhone-Alpes is positively associated to regimen 1.

Aknowledgments :

We wish to thank Dr. F. Clavel for having shared the French data with us, and Dr. E. Riboli for general assistance in becoming familiar with EPIC.

Table 1. Proportion of the 6 regimens: overall and by centre

Regimens	Overall	Alsace -Lorraine	Aquitaine	Bretagne-Pays -de-Loire	Ile-de -France	Languedoc -Roussillon	Nord -Pas -de-Calais	Rhone -Alpes
regim 1	0,56	0,58	0,59	0,49	0,58	0,54	0,46	0,61
regim 2	0,19	0,18	0,18	0,20	0,14	0,21	0,28	0,18
regim 3	0,08	0,08	0,07	0,12	0,09	0,08	0,08	0,06
regim 4	0,03	0,02	0,04	0,03	0,03	0,04	0,02	0,03
regim 5	0,10	0,10	0,08	0,11	0,10	0,08	0,13	0,09
regim 6	0,04	0,04	0,04	0,05	0,05	0,04	0,03	0,04

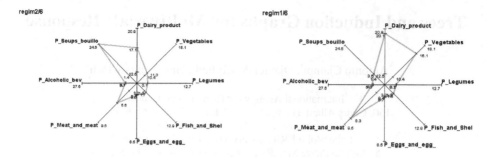

Fig. 2. Two regimens by Zoom Stars

References

1. Ambroise, C., Seże, G., Badran, F., Thiria, S.: Hierarchical clustering of Self-Organizing Maps for cloud classification. *Neurocomputing, 30,* (2000) 47–52.
2. Bock, H. H.: Classification and clustering : Problems for the future. In: E. Diday et al. (eds.): *New Approaches in Classification and Data Analysis.* Springer, Heidelberg (1993), 3–24.
3. Bock, H. H.: Clustering and neural networks. In: A. Rizzi et al. (Eds.): *Advances in Data Science and Classification.* Springer, Heidelberg (1998), 265–278.
4. Bock, H. H., Diday, E. (Eds.): *Analysis of Symbolic Data, Exploratory methods for extracting statistical information from complex data.* Studies in Classification, Data Analysis and Knowledge Organization, Springer, Heidelberg (1999).
5. Chavent, M.: A monothetic clustering algorithm. *Pattern Recognition Letters, 19,* (1998) 989–996.
6. Ciampi, A., Lechevallier, Y.: Designing neural networks from statistical models : A new approach to data exploration. Proceedings of the 1st International Conference on Knowledge Discovery and Data Mining. AAAI press, Menlo Park (1995) 45–50.
7. Ciampi, A., Lechevallier, Y.: Statistical Models as Building Blocks of Neural Networks. *Communications in Statistics, 26(4),* (1997) 991-1009.
8. Elemento, O.: Apport de l'analyse en composantes principales pour l'initialisation et la validation de cartes de Kohonen. *Journées de la Société Francophone de Classification,* Nancy (1999).
9. Gordon, A. D.: *Classification : Methods for the Exploratory Analysis of Multivariate Data.* Chapman & Hall, London (1981).
10. Hébrail, G., Debregeas, A.: Interactive interpretation of Kohonen maps applied to curves. Proceedings of the 4th International Conference on Knowledge Discovery and Data Mining. AAAI press, Menlo Park (1998) 179–183.
11. Kohonen, T.: *Self-Organizing Maps.* Springer, New York (1997).
12. Murthag, F.: Interpreting the Kohonen self-organizing feature map using contiguity-constrained clustering. *Patterns Recognition Letters, 16,* (1995) 399–408.
13. Noirhomme-Fraiture, M., Rouard, M.: Representation of Sub-Populations and Correlation with Zoom Star. *Proceedings of NNTS'98,* Sorrento (1998).
14. Thiria, S., Lechevallier, Y., Gascuel, O., Canu, S.: *Statistique et méthodes neuronales.* Dunod, Paris, (1997).

Trees and Induction Graphs for Multivariate Response

Antonio Ciampi[1], Djamel A. Zighed[2], and Jérémy Clech[1]

[1]International Agency for Research on Cancer
150, Cours Albert Thomas – 69372 Lyon Cedex 08 – France
{ciampi, clech}@iarc.fr
[2]Laboratoire ERIC – Université Lumière – Lyon2
5, Avenue Pierre Mendès-France – 69676 Bron – France
zighed@eric.univ-lyon2.fr

Abstract. We show that induction graphs can be generalized to treat more general prediction problems than those usually treated: prediction of a class variable or of a one dimensional continuous variable. We treat here the case in which the prediction concerns a multivariate continuous response. The approach used, called here GENIND1, is a combination of previous work by two of the authors (RECPAM and SIPINA). We show also that in the GENIND1 framework, clustering (unsupervised learning) as well as prediction (supervised learning) can be treated. The approach is applied to nutritional data.

1 Introduction

Induction graphs, a generalization of trees, are a powerful tool in Knowledge Discovery and Data Mining [3,8]. They can be seen as a structured extraction of predictive rules from a data set, and therefore they provide an appealing summary of the information contained therein.

In this work we initiate a new development which extends the application of both trees and induction graphs to more complex prediction problems than those usually treated by current techniques. In particular, we consider the problem of predicting a multivariate response, characterised by a random vector of which we wish to predict the mean and the variance-covariance matrix. The problem, such as formulated, has an obvious direct interest. Moreover, as we shall see, the generality of the approach will also allow us to develop clustering algorithms of the conceptual clustering type. In particular we will describe two such algorithms inspired, respectively, by the Lance and Williams algorithm [5] (see also [2,4]), and Gower's predictive classification as described in Gordon [6].

The roots of this development are in the authors previous contributions, the tree-growing algorithm RECPAM (RECursive Partition and Amalgamation) [3] and the induction graph construction method SIPINA [8]. RECPAM actually extracts from data an induction graph, but using a constructive approach which is less general than SIPINA : while in RECPAM AMalgamation of children nodes is done only at the end of the tree-growing process, in SIPINA partitioning and merging is alternated in the construction algorithm. On the other hand, RECPAM was originally conceived to predict multidimensional parameters and has been applied to handle outcome

D.A. Zighed, J. Komorowski, and J. Zytkow (Eds.): PKDD 2000, LNAI 1910, pp. 359-366, 2000.
© Springer-Verlag Berlin Heidelberg 2000

information of complex structure, such as censored survival times, count responses and multivariate normal responses [3]. This work further develops some of the ideas in [4]. We shall call the proposed approach GENIND (GENeral INDuction).

2   GENIND, a General Algorithm for Induction Graph Construction

We describe here a first version of a general induction graph construction algorithm which is essentially a restatement of RECPAM [3]. In the future we plan to develop a SIPINA-like approach as well, and so we give the proposed family of algorithms a new name, GENIND. The RECPAM-like version presented here, will be called GENIND1, the SIPINA-like approach will be called GENIND2, and so on.

The proposed algorithm, GENIND1, consists of three steps, all conditional on a given data matrix D. We suppose that the columns of D, representing, as usual, variables, are partitioned a priori into predictor and outcome variables: $D=[Z|Y]$ where Z and Y are N×n and N×m matrices, and N is the number of individuals.

STEP 1 or tree-growing, consists, like CART [1], in building recursively a binary tree by maximising the Information Gain (IG) at each node until the leaves would not be larger than a minimal size fixed by the user. Here, IG is defined as the reduction of the deviance function as discussed below. Thus, at he end of this step, a tree structure, composed by nodes and leaves, is obtained: in GENIND language, it is the large tree.

STEP 2 or pruning, operates on this large tree. The pruning sequence is obtained by minimizing the Information Loss (IL) at each step, where IL is the increase in minimized deviance from the current subtree to the smaller subtree obtained by removing a question or, more generally, a branch. Thus, this operation produces a smaller tree, called the honest tree. Pruned branches of the large tree, will be referred to as virtual branches and are drawn in dotted lines on our tree diagrams, see Figure 1. The honest tree leaves can be seen as virtual roots of virtual trees with virtual nodes and leaves.

STEP 3 or amalgamation, operates on partitions (the first one is composed by the honest tree leaves) and produces an induction graph, GENIND1 Graph. It is proper to RECPAM and is similar to STEP 2. It recursively constructs a superpartition from a (super)partition by amalgamating two sets of the given (super)partitions and each update being obtained from the current partition by merging the two sets resulting in minimum IL. The purpose of this step is to simplify the prediction and further increase generalizability.

More details of the steps are described in [3]. These three steps can be seen as a suboptimal but 'greedy' construction of a predictive model best fitting the data. The novelty in our approach is twofold. Firstly, the merging step, which leads from a tree to an induction graph, is new. Secondly, and perhaps most importantly, the approach presented here can naturally be applied to more complex response structures than the simple ones found in classification and (univariate) regression. This generality is embodied in the definition of deviance: we will show in the next two sections 3 and 4 how a proper definition of the deviance function permits the development of genuinely new algorithms.

The result of GENIND1 when applied to a data matrix D is both a GENIND1-*predictive structure* and a GENIND1-*predictor*. The *predictive structure* is the functional form of the dependence of the *parameter* $\theta = (\theta_1, \theta_2, ..., \theta_p)$ on the predictor variables z, see (1a). In general, in order to specify it, we need a function which associates to each individual of the population, a value for each of the components of θ. The parameters are features of the process generating the outcome variable vector y: for example if we can specify a statistical model for y, θ may represent the parameter of the associated probability distribution.

In order to define this *predictive structure*, let the vector of the predictors be partitioned as: $(z|x)$, where the z's are called *tree predictors* or t-predictors. We will denote by I all 'dummy' variables attached to GENIND1 elements. Thus a particular I indicates, for every individual of the population, whether or not he belongs to the GENIND1 element associated to it. More explicitely, I_g, $g = 1,..., G$, will denote the 'dummy' variables of the GENIND1 classes, I_l, $l = 1,..., L$, those of the leaves of the tree, and $I_{l(v)}$, $v = 1,...,V$, those of the virtual leaves. Clearly, the I_g's and the I_l's can be expressed as sums of the $I_{l(v)}$'s. In the equations below we use the slightly imprecise notation '$v \in l$' and '$v \in g$' to denote an index running, respectively, over the virtual leaves belonging to leaf l and to GENIND1-class g. We now can write the predictive structures associated to the honest tree and to the GENIND1-classification as follows:

Honest Tree:

$$\theta_k(z; x^{(r(k))}, x^{(lf(k))}, x^{(v(k))}) = x^{(r(k))} \cdot \beta^{(k)} + \sum_{l=1}^{L} (x^{(lf(k))} \cdot \gamma_l^{(k)}) I_l(z) + \sum_{l=1 \atop v \in l}^{L} (x^{(v(k))} \cdot \alpha_{l(v)}^{(k)}) I_{l(v)}(z) \qquad (1a)$$

GENIND1-classification:

$$\theta_k(z; x^{(r(k))}, x^{(lf(k))}, x^{(v(k))}) = x^{(r(k))} \cdot \beta^{(k)} + \sum_{l=1}^{G} (x^{(lf(k))} \cdot \gamma_l^{(k)}) I_l(z) + \sum_{l=1 \atop v \in g}^{L} (x^{(v(k))} \cdot \alpha_{l(v)}^{(k)}) I_{l(v)}(z) \qquad (1b)$$

for $k = 1,...,p$. Notice that for each component of θ there are three subsets of x, denoted x(r(k)), x(lf(k)) and x(v(k)). These are the root-, leaf- and virtual-leaf-predictors for component k, abbreviated as r-, lf- and v-predictors, to be specified by the user. Notice also that if the user specify that there are no v-variables, then the third term in the above equation is not present, and similarly for r-variables.

The simplest specification, which could be the default, is that there is only one lf-variable and that this is the constant term (*i.e.* the other two sets of variables are empty). Then the above equations become simply:

Honest Tree: *GENIND1-classification:*

$$\theta_k(z) = \sum_{l=1}^{L} \gamma_l^{(k)} I_l(z) \qquad (2a) \qquad\qquad \theta_k(z) = \sum_{g=1}^{G} \gamma_g^{(k)} I_g(z) \qquad (2b)$$

for $k = 1,...,p$. When the (θ1, θ2,...,θp) are defined as class probabilities, then the honest tree equation represents a predictive structure identical to that given by CART, AID, ID3, etc. for classification; similarly, the GENIND1-classification equation reduces to the RECPAM predictive structure for multinomial probabilities.

By GENIND1-predictor we mean the GENIND1-predictive structure, with values of the parameters which appear in the structure specified by minimization of the deviance function. We will use 'hats' in order to distinguish the predictive structure from the predictor, *e.g.* the predictor corresponding to the predictive structure (2b) will be denoted by:

$$\hat{\theta}_k(z) = \sum_{g=1}^{G} \hat{\gamma}_g^{(k)} I_g(z)$$

(3)

3 GENIND1 for Multivariate Outcome

Suppose we want to discover the relationship between a vector of multivariate outcomes y and a vector of covariates (predictors) z. Suppose also that we may assume that z only affects the mean vector and the variance-covariance matrix of y. Then: $\theta = (\mu 1,...,\mu p, \Sigma) = (\mu, \Sigma)$. The goal is to discover a predictive structure of the GENIND1 type which is an adequate representation of the relationship between z and y. Therefore our approach explicitly takes care of the associations among components of y, in contrast with the approach consisting in growing distinct decision trees for each component of y. To do so, we dispose of a data matrix D=[Z|Y]. For simplicity we are restricting ourselves to the case in which there are no x-variables other than the constant, i.e. for each component of θ one and only one of the three sets of r-, lf-, and v- variables is non-empty and this may only contain the constant term. We also limit ourselves to the situation in which the specification of where the constant term belongs is the same for all components of μ and for all components of Σ (though it may be different for μ and for Σ). A more general structure will be described elsewhere.

3.1 Deviance Function and Information Content

First, suppose that the trivial partition (root node) is an adequate representation of the data. Then for any given (μ, Σ) a natural definition for the deviance function is the sum of the Malahanobis distances of each vector from the mean:

$$dev(\boldsymbol{D}, \theta) = \sum_{i=1}^{N} (y^{(i)} - \mu)^{\mathrm{T}} \Sigma^{-1} (y^{(i)} - \mu)$$

If we assume multivariate normality, which is not necessary in this context, an even more natural definition of deviance is twice the negative log-likelihood:

$$dev(\boldsymbol{D}, \theta) = p \log |\Sigma| + \sum_{i=1}^{N} (y^{(i)} - \mu)^{\mathrm{T}} \Sigma^{-1} (y^{(i)} - \mu)$$

As described in the previous section, the algorithm repeatedly computes differences of *minimized deviances*. For instance the IG of a tree T with respect to the trivial tree T_0 given the data matrix $\boldsymbol{D}$ is defined, with obvious meaning of symbols, as:

$$IG(T : T_0 \mid D) = dev(D, \hat{\theta}_{T_0}) - dev(D, \hat{\theta}_T)$$

3.2 GENIND1-Predictor

Now, depending on the specific assumptions we want to introduce, different predictive structures may be specified. We will consider the following three cases:

1. The variance-covariance matrix is assumed known. (In practice the variance-covariance matrix is never known, but assuming it constant, it can be estimated once and for all as the sample variance-covariance matrix at the root). Then the only component of interest is the vector of the means. The predictor associated to the GENIND1-classification is simply:

$$\hat{\mu}(z) = \sum\nolimits_{g=1}^{G} \hat{\mu}_g I_g \tag{4}$$

Deviance minimization is trivial and the values of the parameters in (4) are given by the sample means at the GENIND1-classes of the components of y, with obvious meaning of symbols:

$$\hat{\mu}_g(z) = \frac{1}{N_g} \sum_{i \in g} y^{(i)}$$

2. The variance-covariance matrix is assumed to be unknown but constant. Then in the language of this work, the μ-component has constant term as lf-variable, while for the Σ-component has constant term as r-variable. This corresponds to the predictor:

$$\text{(5a)} \qquad \qquad \text{(5b)}$$

$$\hat{\mu}(z) = \sum_{g=1}^{G} \hat{\mu}_g I_g(z) \qquad \hat{\Sigma}(z) = \hat{\Sigma} = \frac{1}{N} \sum_{i=1}^{N} (y^{(i)} - \hat{\mu}(z^{(i)}))^T (y^{(i)} - \hat{\mu}(z^{(i)}))$$

3. The variance-covariance is assumed to vary the same way as the mean vector. Then, the μ-component has the same expression than (5a), whereas the Σ-component has the constant term as lf-variable, which gives:

$$\text{(6)}$$

$$\hat{\Sigma}(z) = \sum_{g=1}^{G} \hat{\Sigma}_g I_g(z) \qquad \text{with} \qquad \hat{\Sigma}_g = \frac{1}{N_g} \sum_{i \in g} (y^{(i)} - \hat{\mu}_g(z^{(i)}))^T (y^{(i)} - \hat{\mu}_g(z^{(i)}))$$

4 Conceptual Clustering Algorithm Based on GENIND

In the previous section, we have shown how to learn, from a data matrix D =[Z|Y], a prediction rule for y given z . This implies an a priori distinction between z and y, a natural one when y can be regarded as outcome. Although this situation, or supervised learning, has an obvious intrinsic interest (see example in the next section), GENIND can also serve as basis for an unsupervised learning approach.

Suppose then that no clear distinction can be made among variables, so *D=[X]* is considered as a set of measurements of the vector *x*. We are interested in discovering from *D* a structure of homogeneous and distinct classes of individuals. We also require that these classes be defined by simple statements involving some of the components of *x*, in other words we are interested in developing *conceptual clustering* algorithms. We propose here a GENIND based algorithm, called 'Factor Supervision', but of course, some other is possible like the Gower's predictive clustering approach (see also Gordon [6]).

Our approach is based on an earlier proposal [2]. It consists in transforming the unsupervised learning problem into a supervised one, with supervision provided by the first few factors extracted from [X]. Specifically, if y = (y1,y2,...,yp) denotes a vector having as components the first p principal components of X, let us consider the augmented matrix D′ =[X|Y]. Then the GENIND approach can be applied to the augmented matrix to discover clusters. Discovering such clusters means discovering clusters in a subspace where most of the 'interesting' dispersion of the original variables takes place, yet these clusters are described in terms of the original variables.

5 An Example from Nutritional Epidemiology

To illustrate the flexibility of GENIND1 both as a method for constructing predictors and as a method of conceptual clustering, we report briefly on a preliminary analysis of a data set which is a subset from a much larger epidemiological study called EPIC. It is no more than an illustration; a full report will appear elsewhere. EPIC is a multi-centre prospective cohort study designed to investigate the effect of dietary, metabolic and other life-style factors on the risk of cancer [7]. The study started in 1990 and includes now 23 centres from 10 European countries. By now, dietary data are available on almost 500,000 subjects. Here we will consider only data from a subsample of 1,201 women from the centre 'Ile-de-France'. Also, we limit ourselves to an analysis of data from a 24-hour recall questionnaire concerning intake of 16 food-groups and four energy-producing nutrients : carbohydrates, lipids, proteins and alcohol. The food-group variables, are expressed in grams of intake, and the nutrient variables are measured in Kcal.

5.1 Predicting nutrieNts from Food-Group Variables

In a first analysis, we constructed a GENIND1 predictor for nutrients using the food-group variables as predictor variables. The induction graph in Figure 1, actually a tree, was obtained by the approach of section 3, with a likelihood-based deviance and assuming that both mean vector and variance-covariance matrix vary across the leaves, see equations (5a) and (6). We obtained a large tree with 5 leaves, which reduced to an honest tree of 3 leaves after pruning; no amalgamation was possible. Two food-group variables define the tree structure : Alcohol and Meat. For leaf 1 we have the following energy consumption pattern (standard deviation in parenthesis): 201.101 (76.142) Kcal for carbohydrates, 77.047 (33.864) Kcal for lipids, 75.139 (25.261) Kcal for proteins and 0.006 (0.130) Kcal for alcohol. For leaf 2 the pattern is: 208.681 (84.166) Kcal for carbohydrates, 84.995 (38.957) Kcal for lipids, 73.494 (23.879) Kcal for proteins and 17.681 (13.701) Kcal for alcohol. Finally for leaf 3 we have 201.044 (81.174) Kcal for carbohydrates, 104.155 (43.092) Kcal for lipids, 99.647 (30.547) Kcal for proteins and 23.885 (21.198) Kcal for alcohol.

The verifications of our assumptions are in progress, but we think that if this GENIND graph is very small, it is probably du to some similar diets in the centre 'Ile-de-France'.

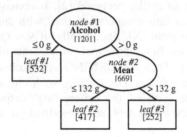

Fig. 1. GENIND1 graph for predictive clustering.

5.2 Clustering Based on Food-Group Variables: Factor Supervision

The graph in Figure 2 was obtained by applying the 'factor supervision' approach of section 4. A preliminary principal component analysis of the sixteen food-group variables yielded five principal components which explain 80% of the dispersion. These components were used as 'response variables' in a GENIND1 construction. A 7-leaves large tree was pruned to a 3-leaves honest tree which, after the AMalgamation step, yielded two GENIND1 classes. One class includes 371 subjects characterized by no consumption of 'Soups and bouillon' and 'Alcohol'. The other class includes 830 subjects which consume at least one of 'Alcohol' and 'Soups and bouillon'.

Acknowledgements

We thank Dr. F. Clavel for having made the data available and Dr. E. Riboli for general assistance in becoming familiar with the EPIC project.

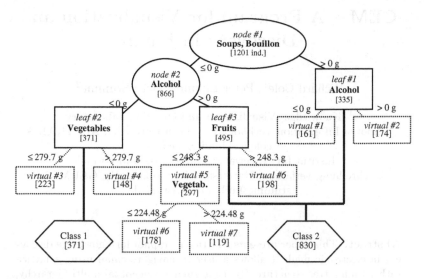

Fig. 2. GENIND1 graph for factor supervision.

References

1. Breiman, L., Friedman, J.H., Olshen, R.A., Stone, C.J.: Classification And Regression Trees. The Wadsworth Statistics/Probability Series (1984).
2. Chavent, M., Guinot, C., Lechevallier, Y., Tenenhaus, M.: Méthodes divises de classification et segmentation non supervisée: recherche d'une typologie de la peau humaine saine. Rev. Statistique Appliquée, XLVII (1999) 87-99.
3. Ciampi, A.: Constructing prediction trees from data: the RECPAM approach. Proceedings of the Prague '91 Summer School of Computational Aspects of Model Choice, Physica-Verlag, Heidelberg, (1992) 105-152.
4. Ciampi, A.: Classification and discrimination: the RECPAM approach. COMPSTAT 94, Physica Verlag, Heidelberg, (1994) 129-147.
5. Lance, G.N., Williams, W.T.: A general theory of classificatory sorting strategies. Computer Journal, (1967) 9: 373-380.
6. Gordon, A.D.: Classification. CRC Press 2nd Edition (1999).
7. Riboli E, Kaaks R.: The EPIC Project: rationale and study design. European Prospective Investigation into Cancer and Nutrition. Int J Epidemiol. (1997) 26 Suppl : S6-14.
8. Zighed, D.A., Rakotomala, R.: Graphes d'induction - Apprentissage et Data Mining. HERMES (2000).

CEM – A Program for Visualization and Discovery in Email

Richard Cole[1], Peter Eklund[1], Gerd Stumme[2]

[1] Knowledge, Visualization and Ordering Laboratory
School of Information Technology, Griffith University, AUSTRALIA
r.cole,p.eklund@gu.edu.au
[2] Institut für Angewandte Informatik und Formale
Beschreibungsverfahren, Universität Karlsruhe (TH), GERMANY
stumme@aifb.uni-karlsruhe.de

Abstract. This paper presents a lattice metaphor for knowledge discovery in electronic mail. It allows a user to navigate email using a lattice rather than a tree structure. By using such a conceptual multi-hierarchy, the content and shape of the lattice can be varied to accommodate queries against the email collection. This paper presents the underlying mathematical structures, and a number of examples of the lattice and multi-hierarchy working with a prototypical email collection.

1 Introduction

Email management systems usually store email as a tree: in analogue to the file management system. This has an advantage in that trees are easily explained to novice users as a direct mapping from the physical structure of the file system. The disadvantage is that, when the email is stored, the user preempts the way he will later retrieve the email. The tree structure forces a decision about the criteria considered as primary and secondary indexes for the email. For instance, when storing email regarding the organization of a conference, one needs to decide whether to organise the email as `Komorowski/pkdd2000/program_committee` where "Komorowski" is a primary index alternatively `conferences/pkdd/pkdd2000/` `organization`. This problem is common when a user cooperates with overlapping subjects with different topics and multiple views.

This paper profiles a *Conceptual Email Manager* called CEM. This follows earlier work [3, 2]. CEM is a lattice-based email retrieval and storage program that aids knowledge discovery by using flexible views over email. It uses a concept lattice-based data structure for storing emails rather than a tree. This permits clients to retrieve emails along different paths. In the example above, the client need not decide which path to store the email. When retrieving the email later, he can consider any combination of the catchwords in the two paths. Email retrieval is totally independent of the physical organization of the file system.

Related approaches to the above problem include the idea of a *virtual folder* that was introduced in a program called View Mail (VM)[6]. A virtual folder is a collection of email documents retrieved in response to a query. The virtual folder

D.A. Zighed, J. Komorowski, and J. Zytkow (Eds.): PKDD 2000, LNAI 1910, pp. 367–374, 2000.
© Springer-Verlag Berlin Heidelberg 2000

concept has more recently been popularised by a number of open-source projects, e. g. [8]. Our system differs from those projects in the understanding of the underlying structure – via formal concept analysis – and in its implementation.

Concept lattices are defined in the mathematical theory of Formal Concept Analysis [11]. A concept lattice derives from a binary relation assigning attributes to objects. In our application, the objects are all emails stored by the system, and the attributes catchwords like 'conferences', 'Komorowski', or 'organization'. We assume the reader to be familiar with the basic notions of Formal Concept Analysis, and otherwise refer to [5]. In the next section, we describe the mathematical structures of the CEM. Requirements for their maintenance are discussed in Section 3. We also describe how they are fulfilled by our implementation.

2   Mathematical Structure

In this section we describe the system on a structural level. We distinguish three fundamental structures: (i) a *formal context* assigning to each email a set of catchwords; (ii) a *hierarchy* on catchwords to define more general catchwords; (iii) creating *conceptual scales* as a graphical interface for email retrieval.

In CEM, we use a *formal context* (G, M, I) for storing email and assigning catchwords. G is a set containing all emails, the set M contains all catchwords. For the moment, we consider M to be unstructured (we will subsequently introduce a hierarchy on it.) The relation I indicates emails assigned to catchwords. In the previous example, the client might assign the catchwords 'Komorowski', 'pkdd2000, 'program_committee', 'conferences', 'pkdd, and 'organization' to a new email. The incidence relation is generated in a semi-automatic process: (i) an regular-expression string-search algorithm recognizes words within an email suggesting relations between email attributes, (ii) the client may accept the string-search result or otherwise modify it, and (iii) the client may attach his own attributes to the email.

Instead of a tree of disjoint folders and sub-folders, we consider the concept lattice $\mathfrak{B}(G, M, I)$ as navigation space. The formal concepts replace the folders. In particular, this means emails may appear in different concepts. The most general concept contains all email. The deeper the client moves into the hierarchy, the more specific the concepts, and the fewer emails they contain.

To support the semi-automatic assignment of catchwords to the emails, we provide the set M of catchwords with a partial order $\leq$. For this *subsumption hierarchy* we assume that a *compatibility condition* holds: $\forall g \in G$, $m, n \in M$: $(g, m) \in I$, $m \leq n \Rightarrow (g, n) \in I$ ($\ddagger$) i.e., the assignment of catchwords respects the transitivity of the partial order. Hence, when assigning catchwords to emails, it is sufficient to assign only the most specific catchwords. More general catchwords are automatically added.

For instance, the user may want to say that 'pkdd' is more specific than 'conferences', and 'pkdd2000' more specific than 'pkdd (i. e., 'pkdd2000'≤'pkdd'≤ 'conferences'). Emails about this paper are assigned by the client to 'pkdd2000' only (as well as any additional catchwords like 'cole', 'eklund' and 'stumme'). When the email-client retrieves this email, no pathname is recalled. Instead, the emails appear under the more general catchword 'conferences'. If 'conferences' provides too large a list of email, the client can refine the search by choosing a sub-term like 'pkdd', or adding a new catchword, for instance 'cole'.

Note that even though we impose no specific structure on the subsumption hierarchy $(M, \leq)$ it naturally splits three ways. One relates the contents of the emails. A second to the sender or receiver of the email and the third describes the emailing process (in-bound or out-bound). An example of a hierarchy is given in Fig. 1 which is a forest (i. e., a union of trees), but the resulting concept lattice – used as the search space – is by no means a forest. The partially order set is displayed both in the style of a folding editor and as a connected graph.

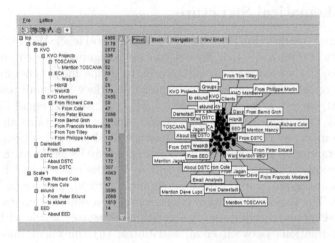

Fig. 1. Partially ordered set of catchwords: as a folding editor and connected graph.

Consider the concept generated by the conjunction of the two catchwords 'PKDD 2000' and 'conference organization'. It will have at least two incomparable super-concepts, one generated by the catchword 'PKDD 2000' and another generated by 'conference organization'. In general, the resulting concept lattice is embedded as a join-semilattice in the lattice of all order ideals (i. e., all subsets $X \subseteq M$ s. t. $x \in X$ and $x \leq y$ imply $y \in X$) of $(M, \leq)$.

Conceptual scaling deals with many-valued attributes. Often attributes are not one-valued but allow a range of values. This is modelled by a *many-valued context*. A many-valued context is roughly equivalent to a relation in a relational database with one field being a primary key. As one-valued contexts are special cases of many-valued contexts, conceptual scaling can also be applied to one-valued contexts to reduce the complexity of the visualization (readers who are interested in the exact definition of many-valued contexts and the use of conceptual scaling in general are referred to [5]). Applied to one-valued contexts, conceptual scales determine the concept lattice that arises from one vertical 'slice' of a large context:

Definition 1. A *conceptual scale* for a subset $B \subseteq M$ of attributes is a (one-valued) formal context $\mathbb{S}_B := (G_B, B, \ni)$ with $G_B \subseteq \mathfrak{P}(B)$. The scale is called *consistent* wrt $\mathbb{K} := (G, M, I)$ if $\{g\}' \cap B \in G_B$ for each $g \in G$. For a consistent scale $\mathbb{S}_B$, the context $\mathbb{S}_B(\mathbb{K}) := (G, B, I \cap (G \times B))$ is called its *realized scale*.

Conceptual scales group together related attributes. They are determined by the user, and the realized scales derived from them when a diagram is requested. CEM stores all scales defined by the client in previous sessions. The client assigns to each scale a unique name. This is modelled by a mapping $(\mathcal{S})$.

Definition 2. Let $\mathcal{S}$ be a set whose elements are called *scale names*. The mapping $\alpha \colon \mathcal{S} \to \mathfrak{P}(\mathcal{M})$ defines for each scale name $s \in \mathcal{S}$ a scale $\mathbb{S}_s := \mathbb{S}_{\alpha(s)}$.

For instance, the user may introduce a new scale that classifies emails according to being related to a conference by adding a new element 'Conference' to $\mathcal{S}$ and by defining $\alpha(\text{Conference}) := \{\text{CKP '96}, \text{AA 55}, \text{KLI '98}, \text{Wissen '99}, \text{PKDD 2000}\}$.

 Observe that $\mathcal{S}$ and M need not be disjoint. This allows the following construction deducing conceptual scales directly from the subsumption hierarchy: Let $\mathcal{S} := \{m \in M | \exists n \in M \colon n < m\}$, and define, for $s \in \mathcal{S}$, $\alpha(s) := \{m \in M | m \prec s\}$ (with $x \prec y$ if and only if $x < y$ and there is no z s.t. $x < z < y$). This means all catchwords $m \in M$, neither minimal nor maximal in the hierarchy, are considered as the name of scale $\mathbb{S}_m$ and as a catchword of another scale $\mathbb{S}_n$ (where $m \prec n$). This last construction, presented in [9], defines a hierarchy of conceptual scales for a library information system [7].

3   Requirements of the CEM

We now discuss the requirements of the CEM based on the Formal Concept Analysis. We explain how our implementation responds to these requirements. Requirements are divided along the same lines as the mathematical structures defined in Section 2; (i) assist the client edit and browse a catchword hierarchy; (ii) help visualize and modify the scale function α; (iii) allow the client to manage the assignment of catchwords to emails; (iv) assist the client search the space of emails for individual emails and conceptual groupings.

 In addition to the requirements stated above, a good email system needs to be able send, receive and display emails: processing various email formats and interacting with popular protocols which are well understood and implemented in existing email programs so not discussed further.

 The catchword hierarchy is a partially ordered set $(M, \leq)$ where each element of M is a catchword. The requirements for editing and browsing the catchword hierarchy are: (i) graphically display the structure of the $(M, \leq)$. The ordering relation must be evident to the client; (ii) make accessible to the client a series of direct manipulations to alter the ordering relation. It should be possible to create any partial order to a reasonable size limit.

 The user must be able to visualize the scale function, α, explained in Section 2. The program must allow an overlap between the set of scale labels, $\mathcal{S}$, and the set of catchwords M. Section 2 introduced the formal context (G, M, I).

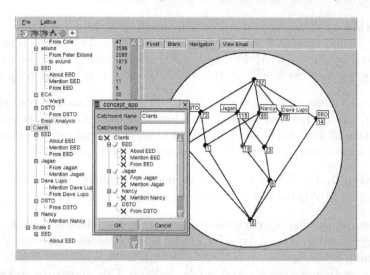

Fig. 2. Scale, catchword and concept lattice. The central dialogue box shows how α can be edited.

This formal context associates email with catchwords via the relation I. Also introduced was the notion of the *compatability condition*,($\ddagger$).

The program should store the formal context (G, M, I) and ensure the compatability condition is always satisfied. It is inevitable that the program will have to sometimes modify the formal context in order to satisfy the compatability condition after a change is made to the catchword hierarchy. The program must support two mechanisms for the association of catchwords to emails. Firstly, a mechanism where emails are automatically associated with catchwords based on the email content. Secondly, the client the should be able to view and modify the association of catchwords with emails. The program must allow the navigation of the conceptual space of the emails by drawing line diagrams of concept lattices derived from conceptual scales [5]. This is shown in Fig. 2. These line diagrams should extend to locally scaled nested line diagrams[9] shown in Fig. 3. The program must allow retrieval and display of emails forming the extension of concepts displayed in the line diagrams.

4   Implementation of CEM

This section divides the description of implementation of the CEM into a similar structure to that presented in Section 3. The user is presented with a view of the hierarchy, $(M, \leq)$ as a tree widget[1], shown in Fig. 2. The catchword hierarchy, being a partially ordered set, is a more general structure than a tree. Although the example given in Fig. 1 is a forest, no limitation is placed by the program on the structure of the partial order other than as a partial order.

The following is a definition of a tree derived from the catchword hierarchy for the purpose defining the contents and structure of the tree widget. Let $(M, \leq)$

---

[1] A widget is a graphical user interface component with a well defined behaviour usually mimicking some physical object, for example a button.

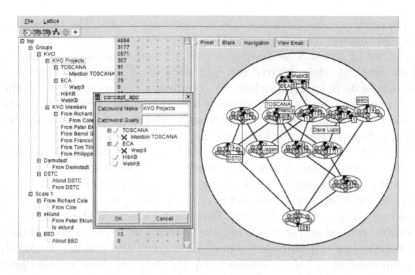

Fig. 3. Scale, catchword and nested-line diagram.

be a partially ordered set and denote the set of all sequences of elements from M by $< M >$. Then the tree derived from the catchword hierarchy is comprised by $(T, \text{parent}, \text{label})$, where $T \subseteq < M >$ is a set of tree nodes, $<>$ the empty sequence is the root of the tree, $\text{parent} : T/ <> \to T$ is a function giving the parent node of each node (except the root node), and $\text{label} : T \to M$ assigns a catchword to each tree node.

$$T = \{< m_1, \ldots, m_n > \in < M > \mid m_i \preceq m_{i+1} \text{ and } m_n \in \text{top}(M)\}$$

$$\text{parent}(< m_1, \ldots, m_n >) := < m_1, \ldots, m_{n-1} >$$
$$\text{parent}(< m_1 >) \qquad\qquad := <>$$
$$\text{label}(< m_1, \ldots, m_n >) \quad := m_1$$

Each tree node is identified by a path from a catchword to the top of the catchword hierarchy. The tree representation has the disadvantage that elements from the partial order occur multiple times in the tree and the tree can become large. If however the user keeps the number of elements with multiple parents in the partial order to a small number, the tree is manageable.

The program provides four operations for modifying the hierarchy: insert & remove catchword and insert & remove ordering. More complex operations provided to the client, moving an item in the taxonomy, are resolved internally to a sequences of these basic operations. In this section we denote the order filter of m as $\uparrow m := \{x \in M \mid m \le x\}$, the order ideal of m as $\downarrow m := \{x \in M \mid x \le m\}$, and the upper cover of m as $\succ_m := \{x \in M \mid x \succ m\}$.

The operation of insert catchword simply adds a new catchword to M, and leaves the $\le$ relation unchanged. The remove catchword operation takes a single parameter $a \in M$ for which the lower cover is empty, and simply removes a from M and $(\uparrow a) \times \{a\}$ from the ordering relation.

The operation of insert ordering takes two parameters $a, b \in M$ and inserts into the relation $\le$, the set $(\uparrow a) \times (\downarrow b)$. The operation of remove ordering takes two parameters $a, b \in M$ where a is an upper cover of b. The remove ordering operation removes from $\leq$ the set $((\uparrow a/ \uparrow (\succ_b /a)) \times (\downarrow b))$.

The set of scales S, according to the mathematization in Section 2 is not disjoint from M, thus the tree representation of M already presents a view of a portion of S. In order to reduce the complexity of the graphical interface, we make S equal to M, i.e. all catchwords are scale labels, and all scale labels are catchwords. The function α maps each catchword m to a set of catchwords. The program displays this set of catchwords, when requested by the user, using a dialog (see Fig. 2 – centre). The dialog box contains all catchwords in the down-set of m an icon (either a tick, or a cross) to indicate membership in the set of catchwords given by $alpha(m)$. Clicking the icon changes $\alpha(m)$'s membership.

By only displaying the down-set of m in the dialog box, the program restricts the definition of α to $\alpha(m) \subseteq\downarrow m$. This has an effect on the "remove ordering operation" defined on $(M, \leq)$. When the ordering of $a \leq b$ is removed the image of α function for attributes in $\uparrow a$ must be checked and possibly modified.

Each member of $(M, \leq)$ is associated with a query term, in this application is a set of *section/word pairs*. That is: Let H be the set of sections found in the email documents, W the set of words found in email documents, then a function query: $M \to \mathfrak{P}(H \times W)$ attaches to each attribute a set of section/word pairs.

Let G be a set of email. An inverted file index stores a relation $R_1 \subseteq G \times (H \times W)$ between documents and section/word pairs. $(g, (h, w)) \in R_1$ indicates that document g has word w in section h.
A relation $R_2 \subseteq G \times M$ is derived from the relation R_1 and the function query via: $(g, m) \in R_2$ iff $(g, (h, w)) \in R_1$ for some $(h, w) \in \text{query}(m)$. A relation R_3 stores user judgements saying that an email should have an attribute m. A relation R_4 respecting the compatibility condition ($\ddagger$) is then derived from the relations R_2 and R_3 via: $(g, m) \in R_4$ iff there exists $m_1 \leq m$ with $(g, m_1) \in R_2 \cup R_3$.

Inserting the ordering $b \leq a$ into $\leq$ requires the insertion of set $(\uparrow a/ \uparrow b) \times \{g \in G \mid (g, b) \in R_4\}$ into R_4. Such an insertion into an inverted file index is $O(nm)$ where n is the average number of entries in the inverted index in the shaded region, and m is the number of elements in the shaded region. The real complexity of this operation is best determined via experimentation with a large document sets and a large user defined hierarchy [1]. Similarly the removal of the ordering $b \leq a$ from $\leq$ will require a re-computation of the inverted file entries for elements in $\uparrow a$.

When new emails, G_b, are presented to CEM, the relation R_1 is updated by inserting new pairs, R_{1b}, into the relation. The modification of R_1 into $R_1 \cup R_{1b}$ causes an insertion of pairs R_{2b} into R_2 according to query(m) and then subsequently an insertion of new pairs R_{4b} into R_4.

$$R_{1b} \subseteq G_b \times (H \times W)$$
$$R_{2b} = \{(g, m) \mid \exists (h, w) \in \text{query}(m) \text{ and } (g, (h, w)) \in R_{1b}\}$$
$$R_{4b} = \{(g, m) \mid \exists m_1 \leq m \text{ with } (g, m_1) \in R_{2b}\}$$

When the user makes a judgement that an indexed email should be associated with an attribute, m, then an update must be made to R_3, which will in turn cause updates to all attributes in the order filter of m to be updated in R_4. In the case that a client retracts a judgement, saying that an email is no longer be associate with an attribute, m, requires a possible update to each attribute, n, in the order filter of m.

When the user requests that the concept lattice derived from the scale with name $s \in S$ be drawn, the program computes $\mathbb{S}_{\alpha(S)}$ from Definition 1 via the algorithm reported in [1]. In the case that the user requests a diagram combining two scales with names labels s and t, then the scale $\mathbb{S}_{B \cup C}$ with $B = \alpha(s)$ and $C = \alpha(t)$ is calculated by the program and its concept lattice $\mathfrak{B}(\mathbb{S}_{B \cup C})$ is drawn as a projection into the lattice product $\mathfrak{B}(\mathbb{S}_B) \times \mathfrak{B}(\mathbb{S}_C)$.

5   Conclusion

This paper gives a mathematical description of the algebraic structures that can be used to create a a lattice-based view of electronic mail. The claim is that this structure, its implementation and operation, aid the process of knowledge discovery in large collections of email. By using such a conceptual multi-hierarchy, the content and shape of the lattice view is varied. An efficient implementation of the index promotes client iteration.

References

1. R. Cole, P. Eklund: Scalability in Formal Concept Analysis: A Case Study using Medical Texts. *Computational Intelligence*, Vol. 15, No. 1, pp. 11-27, 1999.
2. R. Cole, P. Eklund: Analyzing an Email Collection using Formal Concept Analysis. *Proc. of the European Conf. on Knowledge and Data Discovery*, pp. 309-315, LNAI 1704, Springer, Prague, 1999.
3. R. Cole, P. W. Eklund, D. Walker: Using Conceptual Scaling in Formal Concept Analysis for Knowledge and Data Discovery in Medical Texts, *Proceedings of the Second Pacific Asian Conference on Knowledge Discovery and Data Mining*, pp. 378-379, World Scientific, 1998.
4. A. Fall: Dynamic taxonomical encoding using sparce terms. *4th Int. Conf. on Conceptual Structures*. Lecture Notes in Artificial Intelligence 1115, 1996,
5. B. Ganter, R. Wille: *Formal Concept Analysis: Mathematical Foundations*. Springer, Heidelberg 1999 (Translation of: Formale Begriffsanalyse: Mathematische Grundlagen. Springer, Heidelberg 1996)
6. K. Jones: View Mail Users Manual. *http://www.wonderworks.com/vm*. 1999
7. T. Rock, R. Wille: Ein TOSCANA-System zur Literatursuche. In: G. Stumme and R. Wille (eds.): *Begriffliche Wissensverarbeitung: Methoden und Anwendungen*. Springer, Berlin-Heidelberg 2000
8. W. Schuller: *http://gmail.linuxpower.org/*. 1999
9. G. Stumme: Hierarchies of Conceptual Scales. *Proc. Workshop on Knowledge Acquisition, Modeling and Management*. Banff, 16.–22. October 1999
10. F. Vogt, R. Wille: TOSCANA — A graphical tool for analyzing and exploring data. In: R. Tamassia, I. G. Tollis (eds.): *Graph Drawing '94*, Lecture Notes in Computer Sciences 894, Springer, Heidelberg 1995, 226–233
11. R. Wille: Restructuring lattice theory: an approach based on hierarchies of concepts. In: I. Rival (ed.): *Ordered sets*. Reidel, Dordrecht–Boston 1982, 445–470

Image Access and Data Mining: An Approach

Chabane Djeraba

IRIN, Ecole Polythechnique de l'Université de Nantes,
2 rue de la Houssinière, BP 92208 - 44322 Nantes Cedex 3, France
djeraba@irin.univ-nantes.fr

Abstract. In this paper, we propose an approach that discovers automatically visual relations in order to make more powerful the image access. The visual relationships are discovered automatically from images. They are statistical rules in the form of a → b which means: if the visual feature "a" is true in an image then the visual feature "b" is true in the same image with a precision value. The rules concern symbols that are extracted from image numerical features. The transformation of image numerical features into image symbolic features needs a visual feature book in which each book feature is the gravity center of similar features. The approach presents the clustering algorithm that creates the feature book.

1. Introduction

Image access associated to data mining may be seen as a new way of thinking retrieval of images and it opens up to a lot of new applications which have not been possible, previously. The new possibilities given by image access and data mining lies in the ability to perform "semantic queries-by-example", meaning that we can present an image of an object, color, pattern, texture, etc., and fetch the images in the image collections that most resemble the example of the query. For image collections the new possibilities lie in the ability to access efficiently and directly selected images of the database.

Our paper proposes a new way to extract automatically the hidden user semantics of images, based on basic content descriptions. Discovering hidden relations among basic features contributes to extract semantic descriptions useful to make accurate the image accesses. In our case, the relationship discovery are held into two important steps : symbolic clustering based on the new concept of visual feature book and relevant relationships discovery. The feature book is created on the basis of a new algorithm of global/local clustering and classification, powerful image descriptors and suitable similarity measures.

The rules extracted are composed of feature book items. In this paper, we consider textures and colors. The rules are qualified by conditional probability and implication intensity measures.

D.A. Zighed, J. Komorowski, and J. Zytkow (Eds.): PKDD 2000, LNAI 1910, pp. 375-380, 2000.
© Springer-Verlag Berlin Heidelberg 2000

We organize the paper as follow : in section 2, we describe how the knowledge discovery is useful to content-based image retrieval. In section 3, we present how the relationships between image descriptors are extracted. In section 4, we describe some experiment results.

2. Semantic Queries-by-Example?

The central question is : how to extract and represent the content in order to obtain accurate image access ?

To obtain accurate image access, we consider semantic representations that include image class hierarchy (images of flowers, panorama, etc.) characterized by knowledge. «semantic query by examples» specifies a query that means «find images that are similar to those specified». The query may be composed of several images. Several images accurate the quality of retrieval. For example, Several images of a «waterfall» accurate the description of the waterfall. This property makes possible the refinement of retrieval based on the feed backs (results of previous queries).

In the retrieval task, features (colors, textures) of the query specification are matched with the knowledge associated to classes (ex. natural, people, industries, etc.). The suited classes are « Natural », then the matching process focus the search on the sub-classes of Natural : « Flowers », « Mountain », « Water », « Snow », etc. The knowledge associated to flowers and waterfalls are verified, so the matching process focuses the search on the « Flower » and « Water » classes. « Flowers » and « Water » classes are leaves, so the matching process compares the features of the examples with features of the image database to determine which images are similar to the example features. The matching task is based on computing the distance between target and source image regions. When mixing several features, such as colors and textures, the resulting distance is equal to the Sum taking into account the ponderation values of the considered features. The resulting images are sorted, the shortest distance corresponds to the most similar images.

An important advantage of the semantic queries by examples is the efficiency of the content-based retrieval. When the user gives examples of image to formulate his query, and asks "find images similar to the examples", the system will not match the source image with all the images in the database. It will match the source image features with only the target image features of suited classes. If the knowledge associated to a class is globally verified, then the considered class is the suited one. Then, the system will focus the search on the sub-classes of the current one. In the target classes that contain few instances, the search is limited to sequential accesses. Another advantage is the richness of descriptions contained in the results of queries since the system presents both similar images and their classes.

The semantic queries by example needs an advanced architecture. The advanced architecture supports the knowledge in the form of simple rules. Simple rules characterize each semantic class (flowers, natural, mountain, etc.), and are extracted automatically. The rules describe relationships between visual features (colors and textures of images). Each set of rules associated to a class summarizes image contents of the class. Rules contribute in the discrimination of each class, so they represent knowledge shared by the classes. When images are inserted in the database, it is classified "automatically" in the class hierarchy. At the end of the classification process, the image is inserted in a specific class. In this case, the distance between the image and the knowledge associated to the class is the shortest one, compared to the distance between the image and the other classes. Otherwise, the instantiation relationship between the image and the class, will not be considered.

This architecture avoids efficient retrievals and browsing through classes. For example, the user may ask "find images similar to the source image but only in People classes" or "find all images that illustrate the bird class with such colors and such shapes".

3. Discovery Hidden Relations

Based on image content description, the knowledge are discovered. The discovered knowledge characterizes visual properties shared by images of the same semantic classes (Birds, Animals, Aerospace, Cliffs, etc.).

The discovery is held into two steps : symbolic clustering and relationship discoveries and validation.

```
symbolic clustering
relationship discoveries and validation
```

In the first step, numerical descriptions of images are transformed into symbolic form. The similar features are clustered together in the same symbolic features. Clustering simplifies, significantly, the extraction process. For example, an image may be composed of region1 and region2. Region1 is characterized by light red color, and region2 by water color and water texture.

Light red color is not described by a simple string, but by a color histogram [Dje 00]. Even if the region colors of different images of the same class are similar (i.e. light red), the histograms (numerical representation of color) associated with them are not generally identical.

In the second step, the knowledge discovery engine determines automatically common features between the considered images in rule form. These rules are relationships in the form of Premise => Conclusion with a certain accuracy. These rules are called statistical as they accept counter-examples.

```
(texture, water) => (color, heavy_light) (P.C. 100 %, I.I
96.08 %), (texture, waterfall) => (color, white) (P.C.
100 %, 87.4327 %), (texture, texture_bird) => (color,
color_bird) (100 %, 40,45 %)
```

We implemented a technique that clusters numerical representation of color, texture [Dje 00], by using data quantization of colors and textures, we use also the term of feature book creation. The color and texture clustering algorithms are similar, the difference is situated in the distance used.

3.1 Principle of the Algorithm

The algorithm is a classification approach based on the following observation. The scalar quantification of Lloyd developed in 1957 is valid for our vectors (color histogram, Fourier coefficients [Dje 97]), four rate distribution and for a large variety of distortion criteria. It generalizes the algorithm by modifying the feature book iteratively. This generalization is known by k-means [Lin 80]. The objective of the algorithm is to create a feature book, based on automatic classifications themselves based on a learning set. The learning set is composed of feature vectors of unknown probability density. Two steps should be distinguished :

- A first step of classification that clusters each vector of the learning set around the initial feature book that is the most similar. The objective is to create the most representative partition of the vector space.
- A second step of optimization that permits the correct adaptation in a class of the feature book vector. The gravity center of the class created in the previous step is computed.

The algorithm is reiterated in the new feature book in order to obtain a new partition. The algorithm converges to stable position by evolving at each iteration the distortion criteria. Each application of the iteration of the algorithm should reduce the mean distortion. The choice of the initial feature book will influence the local minimum that the algorithm will achieve, the global minimum corresponds to the initial feature book. The creation of the initial feature book is inspired of the splitting technique [Gra 84].

The splitting method decomposes a feature book Y_k into two different feature books $Y_{k-\varepsilon}$ and $Y_{k+\varepsilon}$, where ε is a random vector of weak energy, and its distortion depends of the distortion of the splited vector. The algorithm is then applied on the new feature book in order to optimize the reproduction vectors.

3.2 Algorithm

Based on the learning set of length equal to T, the algorithm finds a feature book of colors and textures of length equal to L, that are the most representative colors and textures of image databases.

Global Clustering

```
FeatureBook  Y_f  =  SymbolicClustering (visual  feature  =
VisualFeature,  learning set = LearningSet, Y_0, T, L)
```

Local clustering

```
FeatureBook  Y_f  =  Clustering(visual feature  =  VF,  learn-
ing set = LS, Y_0, Y_f, T, L, E)
```

The experimental results showed that the distortion values decrease quickly compared to splitting evolution. After the quick decreasing, the distortion values decrease very slowly. Conversely, The entropy increase quickly compared to splitting evolution, and then, it increases very slowly.

3.3 Relationship Discoveries and Validation

Based on the feature book, the discovery engine is triggered to discover the shared knowledge in the form of rules, and this constitutes the second step of the algorithm.

Accuracy is very important in order to estimate the quality of the rules induced. The user should indicate the threshold above which rules discovered will be kept (relevant rules). In fact, the weak rules are rules that are not representative of the shared knowledge. In order to estimate the accuracy of rules, we implement two statistical measures : conditional probability and implication intensity. The conditional probability formula of the rule $a \Rightarrow b$ makes it possible to answer the following question: ''what are the chances of proposition b being true when proposition a is true ? Conditional probability allows the system to determine the discriminating characteristics of considered images. Furthermore, we completed it by the intensity of implication [Gra 82]. For example, implication intensity requires a certain number of examples or counter-examples. When the doubt area is reached, the intensity value increases or decreases rapidly contrary to the conditional probability that is linear. In fact, implication intensity simulates human behavior better than other statistical measures and particularly conditional probability. Moreover, implication intensity increases with the considered population sample representativity. The considered sample must be large enough in order to draw relevant conclusions. Finally, implication intensity takes into consideration the sizes of sets and consequently their influence.

4. Conclusion

To demonstrate the efficiency of the semantic content-based queries, the results of the semantic content-based queries are compared with the results of queries that do not use semantic content-based queries. Since it is not possible to retrieve all relevant images, our experiment evaluates only the first 50 ranked images.

Judging on the results, it is obvious that the use of knowledge leads to improvements in both precision and recall over majority queries tested. The average improvements of advanced content-based queries over classic content-based queries are 23% for precision and 17 % for recall. Precision and recall are better for semantic-based queries than for queries that use only visual features such as color and textures.

References

[Dje 00] Djeraba C., Bouet M., Henri B., Khenchaf A. « Visual and Textual content based indexing and retrieval », to appear in International Journal on Digital Libraries, Springer-Verlag 2000.

[Gra 82] Gras Régis, THE EISCAT CORRELATOR, EISCAT technical note, Kiiruna 1982, EISCAT Report 82/34, 1982.

[Gra 84] Gray R. M. « Vector Quantization », IEEE ASSP Mag., pages 4-29, April 1984.

[Gup 97]Amarnath Gupta, Ramesh Jain «Visual Information Retrieval», A communication of the ACM, May 1997/Vol. 40, N°5.

[Haf 95] Hafner J., al. «Efficient Color Histogram Indexing for Quadratic Distance Functions». In IEEE Transaction on Pattern analysis and Machine Intelligence, July 1995.

[Jai 98] Ramesh Jain: Content-based Multimedia Information Management. ICDE 1998: 252-253

[Lin 80] Linde Y., Buzo A., Gray R. M. « An algorithm for Vector Quantizer Design », IEEE Trans. On Comm., Vol. COM-28, N° 1, pages 84-95, January, 1980.

[Moo 51]Moores C. N. «Datacoding applied to mechanical organization of knowledge» AM. Doc. 2 (1951), 20-32.

[Rag 89] Raghavan, V., Jung, G., and Bollman, P., "A Critical Investigation of Recall and Precision as Measures",   ACM Transactions on Information Systems 7(3), page 205-229

[Rij 79] C. J. Keith van Rijsbergen «Information retrieval», Second edition, London: Butterworths, 1979

[Sal 68] Salton Gerard «Automatic Information Organization and Retrieval», McGraw Hill Book Co, New York, 1968, Chapter 4.

[Zah 72] C. T. Zahn, R. Z. Roskies, « Fourier descriptors for plane closed curves », IEEE Trans. On Computers, 1972.

Decision Tree Toolkit: A Component-Based Library of Decision Tree Algorithms

Nikos Drossos, Athanasios Papagelis, and Dimitris Kalles

Computer Technology Institute, Patras, Greece
{drosnick,papagel,kalles}@cti.gr

Abstract. This paper reports on the development of a library of decision tree algorithms in Java. The basic model of a decision tree algorithm is presented and then used to justify the design choices and system architecture issues. The library has been designed for flexibility and adaptability. Its basic goal was an open system that could easily embody parts of different conventional as well as new algorithms, without the need of knowing the inner organization of the system in detail. The system has an integrated interface (ClassExplorer), which is used for controlling and combining components that comprise decision trees. The ClassExplorer can create objects "on the fly", from classes unknown during compilation time. Conclusions and considerations about extensions towards a more visual system are also described.

1. Introduction

Decision Trees -one of the major Machine Learning (ML) paradigms- have numerous advantages over other concept learners. First, they require relatively small computational power to create a model of the underlying hypothesis and this model requires a small amount of memory to be represented; this means that they can be efficiently used over relatively big amounts of information. Second, they offer a comprehensible way to organise acquired knowledge, in contrast with other learners (like Neural Networks or NaïveBayes); thus, they provide insight on the problem rather than simple classifications/predictions. Third, the classification accuracy of the underlying hypothesis obtained by decision trees is competitive with that of most concept learners. All these justify why decision trees are so popular among researchers and many ML-oriented business applications.

Due to their success and their heuristic nature, many researchers have concentrated on the problem of improving decision tree learners. Those efforts have resulted in dozens of different methods for (mainly): pre-processing data, selecting splitting attributes, pre- and post-pruning the tree, as well many different methodologies for estimating the superiority of one algorithm over the other (k-fold cross-validation, leave-one-out). However, no algorithm clearly outperforms all others in all cases.

Traditionally, every researcher wanting to try out new ideas would have to create a new learner from scratch, even though most of the tasks (like reading data and creating an internal representation) were indifferent to the researcher and irrelevant with the very heart of the new algorithm. Even when portions of previous programs were used there was still the great burden of code adaptation.

D.A. Zighed, J. Komorowski, and J. Zytkow (Eds.): PKDD 2000, LNAI 1910, pp. 381–387, 2000.
© Springer-Verlag Berlin Heidelberg 2000

The above, combined with the imperative need for smaller production times, suggest the use of reusable components that can be efficiently altered to match individual needs. This paper presents such a fully object-oriented components library that has been built with the researcher in mind [1]. This library can substantially reduce the time of deriving a new decision tree algorithm by providing building blocks of the "no-need-to-be-changed" portions of the algorithm while at the same time offering an established base where algorithms can be tested and compared to each other. Furthermore, this library is accompanied with a general tool (the ClassExplorer) where someone can easily create objects (in an interactive way) and combine them to produce a desired behaviour without having to write-compile an entire class or program.

Libraries of components are not a new concept and that holds true for Machine Learning too. The two most known ML libraries are MLC++ [2] and WEKA [3]. Both contain common induction algorithms, such as ID3 [4], C4.5 [5], k-Nearest Neighbors [6], NaïveBayes [7] and Decision Tables [8] written under a single framework. Moreover, they contain wrappers to wrap around algorithms. These include: feature selection, discretisation filters, bagging/combining classifiers, and more. Finally, they contain common accuracy estimation methods.

This work will not replace those established and global machine-learning tools. This library has a more limited scope: it focuses on a specific problem, the creation of binary decision trees. Reducing the scope provides a more solid framework for the specific problem. Furthermore, it reduces the complexity of the environment and the needed time for someone to familiarise himself with it. Moreover, the organization of the library is **component** and **not algorithm** oriented. This means, an easy way to interchange "building blocks" between different decision trees implementation, and thus, better chance to reveal the underlying causes of diversity in performance. Finally, the added value of the ClassExplorer is something that is missing from the other libraries and its utility has been proven extremely beneficial in practice. Hence, although MLC++ and WEKA are very good when one needs to have an organised base of **existing** ML algorithms they fall behind when one concentrates on the decision tree researcher/user.

The rest of this paper is organised in four sections. In the next section we present the basic characteristics of virtually any decision tree algorithm. Next, we elaborate on how to accommodate a mapping of such an algorithm to a flexible, components-based, object-oriented framework. We, then, present ClassExplorer and we conclude with intriguing topics of future research.

2. Decision Tree Building

The first step in building a decision tree learner is the acquisition of training data from some external source. Next, one has to pre-process (or filter) those data in order to bring them in an appropriate form for the concept builder. That step may include the discretisation of continuous features or the treatment of missing values. In a broader view, it may also contain the exclusion of some attributes, or the reordering of the instances inside the training instance-set.

Subsequently, the derived instance-set is used to build the decision tree. The basic framework of the top down induction of decision trees is rather simple. One has to divide the instances at hand into two distinct sub-sets using a test value. This procedure continues for every newly created sub-set until a termination (or pre-pruning) criterion is fulfilled. The last optional step is that of pruning the obtained decision tree.

Finally, having a model of the training data one can use it to classify/predict unknown instances. Figure 1 illustrates the basic steps during decision tree building.

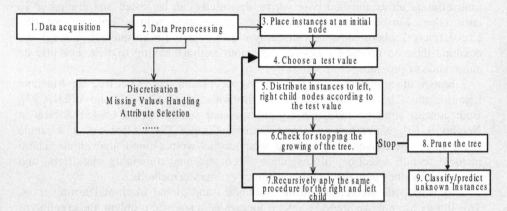

Figure 1. Basic steps during the decision tree creation/use.

Although the proposed scheme assumes a batch mode learner, it can easily be extended to incorporate incremental-mode builders.

3. Library Design

A basic requirement for the implementation of the library was the suitability of the development environment. A components library is best (if not only) suited under an object-oriented language. That is the case since inheritance and polymorphism, basic characteristics of object-oriented languages, are basic characteristics of a components library too. Another basic requirement was ease-of-use, which means the availability of integrated environment tools to boost productivity as well as capability of producing user-friendly graphical interfaces.A further desired characteristic of the development environment was that it should be a well-known and established tool with natural support for libraries and well suited to emerging technologies.

With these considerations we chose Java as the coding language. Java is fully object-oriented   (in contrast with C++ where more conservative programming approaches can be used). Furthermore, Java is known to reduce production time and its "no pointers" approach makes programs less error-prone. Moreover, it has been designed with class libraries in mind, (since even the core of the language incorporates extended libraries of classes), which means that there is an extended build-in functionality regarding the organization of the libraries and their mutual collaboration. Finally, Java offers the ability of modified or customized versions of this software that could run over the Internet.

Having determined the programming environment, we designed the library to accurately fit both the object-oriented approach and the aforementioned basic decision tree algorithm. This means that:

- We identified the basic, relatively-stable parts during the decision tree lifecycle.
- We transformed them to primitive components (abstract classes) with specific minimal functionality.
- We used this minimal functionality to interconnect components into a central decision tree framework.

For example, the minimal needed functionality of a primitive component for the pre-processing step is one function that takes as input a set of instances and returns a modified set of instances. It is up to the user to build on this basic functionality and construct complicated preprocessing classes. There is no need for other components to know what happens inside a pre-processing class; they just use its result. In other words, the interconnection framework establishes a minimum level of collaboration and agreement between the components of the library.

A simple enumeration of the basic building blocks into the abstract classes framework and the role of each one on the building of decision trees is presented in Table 1.

Table 1: Basic Components of a Decision Tree algorithm.

Component	Role on decision trees framework
Parsing	Acquisition of data from various sources
Pre-Processing	Discretization, missing values handling etc.
Splitting	Select test-values to be used during the expansion of the decision tree
Distribution	Statistics about values' distribution among problem classes
Pre-Pruning	Stop the expansion of the decision tree
Post-Pruning	Prune the tree after it has been build
Tree Build	Batch mode, incremental mode builders
Classify	Classify instances using a decision tree

Although a full description of every structure and component of the library is beyond the scope of this paper, a simple example is necessary to point out some of the system characteristics. Suppose that the shapes at the left part of Figure 2 represent the basic abstract classes of the library and that arrows between them represent interconnections between those classes. For example, if shape 1 represents a *TreeBuild* class and shape 4 represents a *Splitting* class then there must be an interconnection between them since a tree builder has to decide about splitting values during the process of tree building. The same way, shapes 2 and 3 may represent a basic *Classifier* class (that uses the output of the tree builder) and a *Distribution* class (that is used by the *Splitting* Class) accordingly.

Now, suppose that the right part of Figure 2 represent classes that inherit from the abstract classes of the library. For example, the two produced shapes next to shape 1 may represent an incremental or batch mode builder. The same way, the two produced shapes from shape 4 may represent classes that use the information gain or the chi-square criterion to choose best splitting values. Since the basic interconnections between classes are hard coded inside the basic abstract classes, it is easy to use tree-builders and splitting-value-selectors at any combination. This architecture offers the researchers an easy way to focus on the points of interest during the decision tree's lifecycle. Furthermore, it gives them the ability to test their own extensions with a variety of already built components and a selection of datasets ([9] is a possible source).

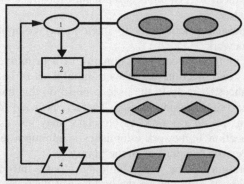

Figure 2. An example of library's architecture.

Of course, the library is accompanied with a general set of classes that represent data structures (like *Instance* or *Attribute*), containers for those structures to easily create data collections, as well as an extended set of methods to manipulate them. Furthermore, there is a general decision-tree wrapper, which helps combining the decision tree components to a decision tree entity. This wrapper contains methods to set the desired tree builder class, splitting class, pruning class, classification method etc., as well as methods to access the representative objects of those classes. Since this wrapper is aware of the components that comprise the decision tree entity, it internally handles the needed component interconnections. The library is also equipped with a number of tools (like timers or various decision tree printers) that provide additional functionality.

In its current state, the library is not directed to the implementation of standard decision tree algorithms (like C4.5) but rather to the implementation of the components that comprise those algorithms. It is up to the researcher to use already built components to simulate the behaviour of a desired algorithm. However, it is relative easy to create classes that resemble the behaviour of major current algorithms and developing a repertoire of them is a prominent topic in our agenda.

4. Class Explorer

A basic problem we confronted at the very early steps of the library's design was an easy way to accomplish basic object testing and algorithms' combinations. Although those problems could be tackled by writing and compiling new Java programs (that would act as wrappers), we found this approach rather cumbersome and thus unsuitable for research and development use.

This problem motivated the conception of the ClassExplorer, an interface tool for controlling and combining algorithms for decision trees. ClassExplorer is based on a simple interpreted programming language, which resembles Java and offers object construction "on the fly" from classes unknown during compilation time. This way we attain independence from the current state of the library while at the same time providing a flexible general tool for object manipulation. An interesting side effect of the independent nature of ClassExplorer is that, although it was developed as an accompanying tool for the library, it can be used in a variety of tasks where interactive object construction and manipulation is required.

Some of ClassExplorer's capabilities are:

- It can instantiate different objects of previously unknown classes and map different variable names to different objects.
- It offers the capability of using/combining the methods of any instantiated object.
- It can make use of non-object primitive data types as *int* or *boolean*.
- It can organize actions into scripts and thus, provide additional functionality.
- It offers a set of special instructions that, for example, manipulate the instantiated objects or change the working directory of the system, or set paths where the system should search for classes.

On the other hand, the basic limitations (or, more accurately, as of yet missing features) of ClassExplorer are:

- There is no build-in program flow control.
- There is no support for mathematical operations.

One has to distinguish between the tasks that can be accomplished using the ClassExplorer and those that have to be done using pure Java. For example, ClassExplorer cannot create a new component of the library. On the other hand, it is naturally suited to tasks where one wants to conduct experiments with different combinations of components or wants to "take a look" at intermediate results of some algorithm's component. Furthermore, the capability of organizing commands to scripts makes it possible to organize test scenarios (like *k*-fold cross-validation[1]) over one or more datasets.

5. Conclusions and Future Directions

Due to space restrictions we cannot present a full example of the library's usage. The interested user can look at *htpp://www.cti.gr/RD3/Eng/Work/DTT.html* for the library, examples and further documentation.

We anticipate that researchers and specialist business consultants could be prime users of the toolkit (they will be able to use the interconnected framework to build over new or already implemented ideas without having to worry about essential but uninteresting parts of a decision tree learner). We also expect that students will greatly benefit from using the toolkit, as it constitutes an integrated environment for practicing machine learning and software engineering skills. The ClassExplorer in particular, should prove very helpful for all users, as its interactive nature provides insight into the components behavior.

It is relatively easy to wrap some of the library components and ClassExplorer so as to construct a user-friendly interface for non-specialist business users. This interface would construct commands for the ClassExplorer, which in turn would create and manipulate the required library objects. This is a three-layer architecture as illustrated in Figure 3, where four distinct (but coupled) steps implement a decision tree development lifecycle. This would enhance the proposed toolkit with a

---

[1] The Class Explorer already contains several scripts for the most common tasks like k-fold cross validations

functionality that has underlined the commercial success of existing software, like Clementine [10].

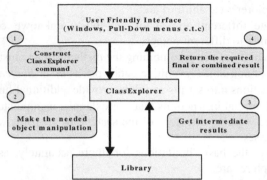

Figure 3. A 3 three-layer architecture for a business application over the library.

There is also a number of scheduled improvements that will provide added functionality and ease-of-use. We intend to incorporate a number of visual components to the library (with an interactive decision tree visualizer being the top priority). Furthermore, we intend to produce a graphical version of ClassExplorer, where every created object and its methods will be represented as visual components. The user will be able to drag-and-drop such components, visually interconnect them and simulate their combined behaviour. Finally, we intend to produce a client/server version of the library for use over Internet.

Whenever decision trees are the preferred means for knowledge representation the proposed library offers a serious alternative to other established machine-learning libraries. The reduced problem scope provides a solid and easy to familiarize-with framework; furthermore, the ClassExplorer, as a productivity tool, offers an interactive mediator between the user and the library, and component authors are assisted by the coding language (Java). Finally, the already built infrastructure provides the core in which (already scheduled) user-friendly access or networking capabilities can be seamlessly integrated.

References

[1]   Papagelis, A., Drossos, N. (1999). *A Decision Trees Components Library*. Diploma Thesis at Computer Engineering and Informatics Department, University of Patras, Greece (in Greek).

[2]   Kohavi, R., John, G., Long, R., Manley, D., Pfleger, K. (1994). MLC++: A Machine Learning Library in C++. *Proceedings of TAI'94*.

[3]   Witten, I., Frank, E. (2000). *Data Mining: Practical Machine Learning Tools and Techniques with Java Implementations*. Morgan Kaufmann Publishers, San Mateo, CA.

[4]   Quinlan, J. R. (1986). Induction of decision trees. *Machine Learning* Vol. 1, No.1, pp. 81-106.

[5]   Quinlan, J. R. (1993). *C4.5: Programs for Machine Learning*. Morgan Kaufmann , San Mateo, CA.

[6]   Aha, D., Kibler, D. (1991). Instance-based learning algorithms, *Machine Learning*, 6:37-66.

[7]   George, H.J., Langley, P. (1995). Estimating Continuous Distributions in Bayesian Classifiers. *Proceedings of the Eleventh Conference on Uncertainty in Artificial Intelligence*, pp. 338-345.

[8]   Kohavi, R. (1995). The Power of Decision Tables. *Proceeding of European Conference on ML*.

[9]   Blake, C., Keogh, E., Merz, J. (2000). *UCI Repository of machine learning databases*. Irvine, University of California, Dep. of Information and Computer Science.

[10] Clementine. http://www.spss.com/clementine/

Determination of Screening Descriptors for Chemical Reaction Databases

Laurent Dury*, Laurence Leherte, and Daniel P. Vercauteren

Laboratoire de Physico-Chimie Informatique,
Facultés Universitaires Notre-Dame de la Paix
Rue de Bruxelles, 61; B-5000 Namur (Belgium)
tel: +32-81-734534, fax: +32-81-724530
firstname.lastname@fundp.ac.be

Abstract. The development of chemical reaction databases has become crucially important for many chemical synthesis laboratories. However the size of these databases has dramatically increased, leading consequently to perfect more and more powerful search engines. In this sense, the speed and the efficiency of screening processes of the chemical information are essential criteria. Looking forward for powerful algorithms dedicated to information retrieval in chemical reaction databases, we have thus developed several new graph descriptors to find efficient indexation and classification criteria of chemical reactions.

1 Introduction

During the last two decades, organic chemists started to use computer programs in organic synthesis, especially "Computer Aided Organic Synthesis" (CAOS) programs which were developed to discover strategies for the preparation of target molecules. The use of such an approach, based on Corey's original concept of retrosynthetic analysis, was first demonstrated in 1969 by Corey and Wipke [1].

While keyword searching has been available in Chemical Abstracts (CAS) and elsewhere since the early 1970s, and (sub)structure searching in DARC and CAS ONLINE has been available since the beginning of the 1980s, "real" reaction databases, *i.e.*, containing structure diagrams, were implemented in the mid-1980s only (*cfr.*, for example, Reaction Access System (REACCS), MDL, 1982; Synthesis Library (SINLIB), 1982; Organic Reaction Access by Computer (ORAC), 1983).

Since 1990, a number of new reaction databases appeared such as CASREACT, ChemInform RX (CIRX), FIZ Chemie, ChemReact, and Beilstein Cross-Fire, and more importantly, the access to various reaction database servers was significantly improved. The current reaction databases incorporate all the elements and facilities of other types of chemical databases, such as text (for bibliographic data, reaction classifications) and/or numeric values.

---

* FRIA Ph.D. Fellow

D.A. Zighed, J. Komorowski, and J. Zytkow (Eds.): PKDD 2000, LNAI 1910, pp. 388–394, 2000.
© Springer-Verlag Berlin Heidelberg 2000

As the size of all these databases, in terms of molecular structures as well as of reactions, is increasing drastically every year, the efficiency of searching algorithms implemented within these databases is thus primordial. Optimization of these algorithms usually involves the implementation of a searching process in two separate steps. First, the screening step which determines, with the help of a set of descriptors based on the user's query, a subset of potential candidates. In a second time, a procedure usually called ABAS (Atom by Atom Search) [2] compares all components of the subset with the user's query. The ABAS part is the most time consuming step. The speed of the search is indeed inversely proportional to the size of the candidate subset given by the initial screening; this step thus controls the efficiency of the search. Therefore, the choice of the screening descriptors is very important; the adequacy of that choice determines the performance of the database search.

There are two main families of chemical descriptors which usually depend on the user's request. First, if a reaction is searched on the basis of one its molecular component, a reactant or a product, the descriptor is expressed in terms of molecular structure or substructure features. These descriptors are called molecular descriptors. Conversely, if all constituants are known, the descriptor is based on the reaction pattern, $i.e.$, the smallest set of atoms and bonds which allows to describe the transformation occurring between the products and the reactants. These descriptors are called reaction descriptors.

In order to optimize a searching algorithm, for each component of the databases, all the descriptor values need to be calculated and stored in an annex database of pointers which refer to the searched reaction database.

In this short paper, we first describe some basics about the oriented-object programming use to represent and analyse the chemical reactions. Then, we present our ideas on the development of a powerful reaction searching algorithm by describing several molecular descriptors such as reduced graph representations of molecules and reaction descriptors based on the breaking and the formation of chemical bonds.

2   Oriented-Object Programmation

In our program, an oriented-object programming language is used to manipulate an organic reaction represented using three levels of objects. As the molecular structures involved in a reaction are constituted of atoms and bonds, the first level contains this information. Each atom and bond objects, named TAtom and TBond, respectively, are next grouped in a second level which describes the molecular structure and which is called TMol. Finally, the set of molecule objects associated with the reactant(s) and product(s) are grouped in a third level object symbolizing the organic reaction (TRxn). For example, as shown in Figure 1, related to the chemical reaction $C_3H_6O + C_2H_6O_2 \longrightarrow C_5H_{10}O_2$ ($+ H_2O$), the first level of description contains the atoms $(C, C, O, ...)$ and bonds (C=O,C-C, ...). The second level of description contains the molecules $(C_3H_6O, C_5H_{10}O_2, ...)$ and the last level gathers the information about the reaction, *i.e.*, $C_3H_6O + C_2H_6O_2 \longrightarrow C_5H_{10}O_2 (+ H_2O)$. The information contained in the different objects is either predetermined and stored in the organic reaction database (we will come back to database preprocessing later) or derived from an elementary chemical analysis algorithm which allows to extract the different objects associated with a molecular (sub)structure drawn on the computer screen.

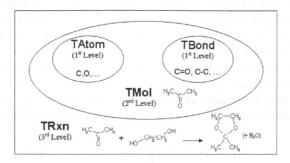

Fig. 1. Presentation of the three level of description for the chemical reaction $C_3H_6O + C_2H_6O_2 \longrightarrow C_5H_{10}O_2 (+H_2O)$.

3   Screening Descriptors

As mentioned before, the selection of the screening descriptors is of prime importance. Too strong a discrimination between the individual components of a database need to be avoided, otherwise the number of components in each subset is too small, and the number of descriptors that are needed to partition the database is close to the number of individual components in this database. The computer management of all these descriptors thus becomes heavy and the efficiency of the search algorithm strongly decreases. In conclusion, a good screening descriptor must have a discrimination power neither too small nor too large for the database in use.

In order to optimize the screening step, we propose the application of a succession of different searching criteria. The subset of potential candidates is then determined as the intersection depending on all the previously obtained sets. This approach requires that the selected descriptors are lineary independent, in other words, the subsets of potential candidate generated with these descriptors are as distinct as possible.

The discrimination power of a screening descriptor is dependent on the database in use. For example, a descriptor based on the chemical bond type only will not be interesting when working with a database mainly composed of alkane type molecules, *i.e.*, molecules whose atoms are connected through a unique kind of chemical bonds, single bonds (*i.e.*, $C-O, C-C, C-H, ...$). The screening descriptors must thus ideally be based on the principal characteristics present among the chemical structures of the database.

3.1   Molecular Descriptors

For many computer chemical applications, and particularly for the storage and retrieval in large structural databases, the structure of a molecule can be conveniently represented in terms of a graph. A graph, denoted by G, is a dimensionless mathematical object representing a set of points, called *vertices*, and the way they are connected. The connections are called *edges*. Formally, neither vertices nor edges have a physical meaning. The physical (or chemical) meaning depends only on the problem the graph is representing. In chemistry applications, when a graph is used to depict a molecular structure, *e.g.*, a planar developed formula, it is called a *molecular graph*. A comparaison between a planar developed formula and a molecular graph, applied to the reserpine structure, is presented in Figure 2. The most basic descriptors of molecular graphs are based on the features of the vertices and the edges (number of vertices, egde multiplicity, ...).

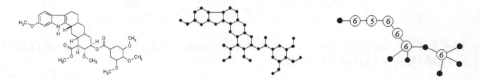

Fig. 2. Planar molecular structure, G, and $RG^{(cycle)}$ of the reserpine molecule. The open circles of $RG^{(cycle)}$ symbolize the vertices containing cyclical information, and the black circles, the vertices containing the non-cyclical information. Labels in the open circles correspond to the size of the reduced rings.

Several different reduced representations of G can be generated. Reduction involves the merging of certain features of a chemical structure into the nodes of a reduced graph. Its purpose is to bring a homeomorphic mapping from the structure onto a simpler graph, resulting, in general, in a smaller number of nodes than in the original graph [3]. As our aim is to obtain a good molecular structure descriptor for the efficient and reliable screening step of structures and substructures present in standard organic reaction databases, and as cyclic structures are widely present in organic chemistry, an obvious first idea is to work with reduced graphs (RG) based on the *cyclic* information, $RG^{(cycle)}$, contained in the molecular structures. We did not find any completely satisfying ring set definition corresponding to our needs in the literature; we have thus chosen to develop a new set of rings as well as the corresponding reduced graph [4]. This new set does not intend to compete with others in terms of strict graph theory considerations or implementation efficiency. We tried to reach a balance between the loss of information (by keeping only certain cycles), and the ease of substructure retrieval. In order to construct this set, we start with the computation of the Smallest Cycles at Edges, by searching the smallest cycles for each edge. In a second stage, the internal rings are included to the set of rings. The obtained ring set is then called the Set of Smallest Cycles at Edges. The reduced graph based on this last set, the $RG^{(cycle)}$, is the junction of two subgraphs, the Graph of Smallest Cycles at Edges, $GSCE$, and the Graph of Acyclic Subtree,

GAS. As an example, the $RG^{(cycle)}$ of the molecular structure of reserpine is also depicted in Figure 2.

3.2   Reaction Descriptors

To describe an organic reaction, it is important to find the reaction pattern. For this operation, the elementary principles of connectivity matrix analysis is used. In our case, this matrix contains in its $[i,j]$ cells the multiplicity of the bond joining the ith and j^{th} atoms of the studied molecular structure, i.e., 1 for a single bond, 2 for a double bond, ... As shown in Figure 3, the difference between the final state of the reaction, described in our algorithm by a simple connectivity matrix of the reaction product(s) (E), and the initial reactant, described with the same mathematical tool (B), gives us a new matrix called the reaction matrix (R), where $R = E - B$ [1]. This method is similar to the Dugundji-Ugi approach [5]. The final reaction matrix (R') is then obtained by the suppression of the rows and columns whose all elements equal zero. The resulting R' matrix contains all the transformations occurring during the reaction.

$$C_5H_8 + C_2H_4 \longrightarrow C_7H_{12}$$

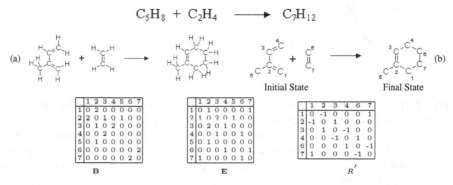

Fig. 3. Reactant connectivity matrix (B), product connectivity matrix (E), and final reaction matrix (R') of a Diels-Alder reaction. (a) Full planar representation of the molecular structure, (b) same as above without the hydrogen atoms.

With this R' matrix, we can determine many features of the reaction, as the number and the nature of the broken and formed bonds, the nature of the chemical functions implied in the transformation, and many other informations. All these informations can be used as reaction descriptors. We originally developed new descriptors based on the eigenvalues of the R' matrix in order to provide a more mathematical representation of a chemical reaction.

4   Discussion and Conclusions

One of our research aims in Computer Aided Organic Synthesis are to built efficient searching algorithms for organic reaction databases. So far, we have

---

[1] Hydrogen atoms are often omitted in description of organic reactions. They can be easily recovered from the hybridization state of the connected carbon atoms.

adopted a two-step process containing a screening step which involves the computation of a subset of potential candidates, and an ABAS step, used to find the queried reaction. In order to optimize our algorithm, new screening descriptors have been developed in function of the user's question. We have thus computed several molecular descriptors, such as reduced graphs based on cyclical information, and on the carbon/heteroatom distinction, and several reaction descriptors, based on the information contained in the reaction pattern, such as eigenvalues of the R' matrix, the number and nature of the broken and formed bonds, the nature of the chemical functions implied in the transformation. In order to analyse

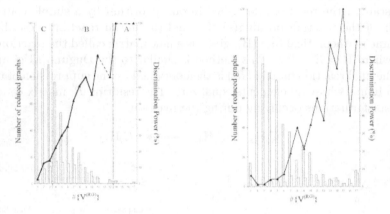

Fig. 4. Histogram showing the action of the reduced graphs (RG) based on the carbon-/heteroatom distinction (left), and based on cyclical information (right), over a test population of 540 molecular structures. Abscissa values are the number of vertices per RG, ordinate values, the number of RG. Gray columns represent the total number of graphs, white columns, only the different graphs. Black triangles symbolize the value of the discrimination power.

the discrimination power of the different screening criteria, our strategy was applied to a test database of 540 reactions. According to the results of an histogram analysis, which reports the discrimination power in function of the number of vertices in the reduced graph ($\sharp\{V^{(RG)}\}$) (Fig. 4), three different kinds of behavior can be observed. The discrimination power can be large, sufficient, or too soft. For the last kind (zone C in Fig. 4 left), the reduced graphs under study are clearly not efficient for the structures they represent. Another reduced criteria is thus necessary. The most important interest of the use of a second (or several) type of RG is that the structures that were appearing in the A, B, or C regions with one kind of reduction may appear in other regions with another kind of reduction. Indeed, as shown for example in Figure 5, a structure that appeared with the RG based on cyclical information in area C, $i.e.$, the least discriminating one, may appear in area B, with a higher discrimination power, when another criterion of reduction, the carbon/heteroatom criterion, is used. In our strategy, all the structural and reaction descriptors of the database are computed during a preprocessing step and each histogram is calculated. At each introduction or

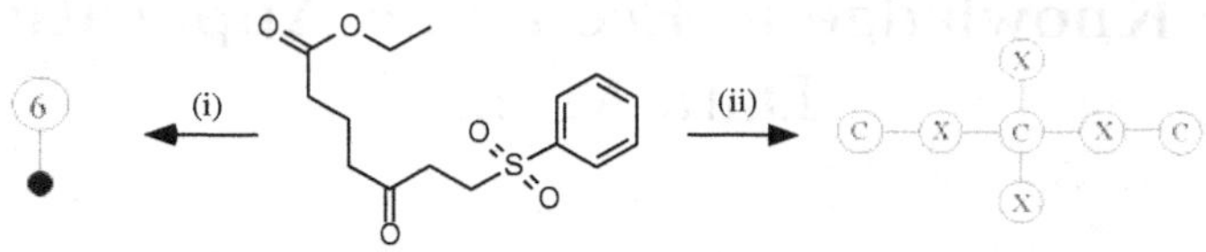

Fig. 5. Comparaison between the *RG* based on cyclical information (i) and the *RG* based on the carbon/heteroatom distinction (ii) for the structure of ethyl 4-acetyl-6-phenyl-sulfonyl-hexylate. Labels of the open circles of the $RG^{(cycle)}$ correspond to the size of the reduced rings. Black circles of the $RG^{(cycle)}$ represent non-cyclical information. Letters in open circles of the *RG* based on carbon/heteroatom reduction symbolize the vertex containing the carbon atom (C label), and the vertex containing heteroatoms (X label).

deletion of an additional chemical reaction, the database changes and thus the histograms may change too. Therefore, the histograms must be updated every time the database changes, but not, of course, every time the user is querying the database.

In conclusion, the building of the histograms associated with each screening descriptor allows to determine the best set of screening criteria, for a given user question and for a given database, and then an optimum screening stage may be established. Indeed, the discrimination power of a criteria is dependent on the database that is used. The use of molecular and reaction screening descriptors is thus a necessary step to allow a fast and efficient search for chemical reaction.

5   Acknowledgments

L.D. thanks Prof. A. Krief, Dr G.M. Downs, and T. Latour for useful discussions, and the "Fonds pour la formation à la Recherche dans l'Industrie et dans l'Agriculture" (FRIA) for his PhD fellowship.

References

1. Corey, E.J., Wipke W.T.: Computer-Assisted Design of Complex Organic Syntheses. Science **166** (1969) 179–192.
2. Bartmann, A., Maier, H., Walkowiak, D., Roth, B., Hicks M.G.: Substructure Searching on Very Large Files by Using Multiple Storage Techniques. J. Chem. Inf. Comput. Sci. **33** (1993) 539–541.
3. Gillet, V.J., Downs, G.M., Ling A., Lynch, M.F., Venkataram, P., Wood, J.V., Dethlefsen W.: Computer Storage and Retrieval of Generic Chemical Structures in Patents. 8. Reduced Chemical Graphs and Their Applications in Generic Chemical Structure Retrieval. J. Chem. Inf. Comput. Sci. **27** (1987) 126–137.
4. Dury, L., Latour, T., Leherte L., Barberis, F., Vercauteren, D.P.: A New Graph Descriptor for Molecules Containing Cycles. Application as Screening Criterion for Searching Molecular Structures within Large Databases of Organic Compounds. J. Chem. Inf. Comput. Sci., submitted for publication
5. Hippe, Z.: Artificial Intelligence in Chemistry, Structure Elucidation and Simulation of Organic Reactions. *Studies in Physical and Theoretical Chemistry;* Elsevier, PWN Polish Scientific Publishers (1991), vol. 73, 153–183.

Prior Knowledge in Economic Applications of Data Mining

A.J. Feelders

Tilburg University
Faculty of Economics
Department of Information Management
PO Box 90153
5000 LE Tilburg, The Netherlands
A.J.Feelders@kub.nl

Abstract. A common form of prior knowledge in economic modelling concerns the monotonicity of relations between the dependent and explanatory variables. Monotonicity may also be an important requirement with a view toward explaining and justifying decisions based on such models. We explore the use of monotonicity constraints in classification tree algorithms. We present an application of monotonic classification trees to a problem in house pricing. In this preliminary study we found that the monotonic trees were only slightly worse in classification performance, but were much simpler than their non-monotonic counterparts.

1 Introduction

The estimation of economic relationships from empirical data is studied in the field of econometrics. In the model specification stage of econometric modelling the relevant explanatory variables and the functional form of the relationship with the dependent variable are derived from economic theory. Then the relevant data are collected and the model is estimated and tested. Applied econometric work does not conform to this textbook approach however, but is often characterized *specification searches* ([Lea78]).

Data mining is often associated with the situation where little prior knowledge is available and an extensive search over possible models is performed. Of course one has to have some prior beliefs, for how else does one for example decide which explanatory variables to include in the model? But often the algorithm is able to select the relevant variables from a large collection of variables and furthermore flexible families of functions are used. Even though data mining is often applied to domains where little theory is available, in some cases useful prior knowledge is available, and one would like the mining algorithm to make use of it one way or the other.

One type of prior knowledge that is often available in economic applications concerns the sign of a relation between the dependent and explanatory variables. Economic theory would state that people tend to buy less of a product if its price increases (ceteris paribus), so price elasticity of demand should be negative. The

D.A. Zighed, J. Komorowski, and J. Żytkow (Eds.): PKDD 2000, LNAI 1910, pp. 395–400, 2000.
© Springer-Verlag Berlin Heidelberg 2000

strength of this relationship and the precise functional form are however not always dictated by economic theory. The usual assumption that such relationships are linear are mostly imposed for mathematical convenience.

This paper is organized as follows. In section 2 we focuss the discussion on prior knowledge for applications in economics. One of the most common forms of prior knowledge in economics and other application domains concerns *monotonicity* of relations. This subject is explored further in section 3, where we discuss monotonicity in classification tree algorithms. In section 4 we present an application of monotonic classification trees to a problem in house pricing. Finally, in section 5 we draw a number of conclusions from this study.

2   Prior Knowledge in Economic Applications

Regression analysis is by far the most widely used technique in econometrics. This is quite natural since economic models are often expressed as (systems of) equations where one economic quantity is determined or explained by one or more other quantities.

The a priori domain knowledge is primarily used in the *model specification* phase of the analysis. Such a priori knowledge is supposed to be derived largely from *economic theory*. Model specification consists of the following elements:

1. Choice of dependent and explanatory variables.
2. Specification of the functional form of the relation between dependent and explanatory variables.
3. Restrictions on parameter values.
4. Specification of the stochastic process.

In applied econometrics usually alternative specifications are tried, and the specification, estimation and testing steps are iterated a number of times (see [Lea78] for an excellent exposition of different types of *specification-searches* used in applied work). As a historical note, the search for an adequate specification based on preliminary results has sometimes been called "data mining" within the econometrics community [Lea78,Lov83]. In principle, there is nothing wrong with this approach, its combination however with classic testing procedures that do not take into account the amount of search performed have given "data mining" a negative connotation.

We shall give an example from empirical demand theory to illustrate how different types of domain knowledge may be used in the model specification phase. Empirical demand theory asserts that *ceteris paribus* an individual's purchases of some commodity depend on his income, the price of the commodity and the price of other commodities. This is reflected in a simple demand equation [Lea78]:

$$\log D_i^o = a + b \log P_i^o + c \log Y_i + d \log P_i^g + \varepsilon_i$$

where D^o denotes the purchases of oranges, P^o the price of oranges, Y denotes income, P^g denotes the price of grapefruit, and index i stands for different households. The log-linear specification is chosen primarily for convenience. It allows us to interpret the estimated coefficients as elasticities, e.g. the estimate of b is interpreted as the price elasticity of demand, and the estimate of c as the income elasticity of demand. A priori we would expect that $b < 0$ (if price increases, *ceteris paribus* demand decreases). Likewise we would expect $c > 0$ and $d > 0$ (since grapefruits are a substitute for oranges).

According to economic theory there should be absence of money illusion, i.e. if income and all prices are multiplied by the same constant, demand will not change. If we believe in the "absence of money illusion" we can add the restriction that the demand equation should be homogeneous of degree zero. For the log-linear specification this leads to the constraint $b + c + d = 0$.

3 Domain Knowledge in Trees

Tree-based algorithms such as CART[BFOS84] and C4.5[Qui93] are very popular in data mining. It is therefore not surprising that many variations on these basic algorithms have been constructed to allow for the inclusion of different types of domain knowledge such as the cost of measuring different attributes and misclassification costs. Another common form of domain knowledge concerns monotonicity of the allocation rule.

Let us formulate the notion of monotone classification more precisely. Let $(x_1, x_2, \ldots, x_p)$ denote a vector of linearly ordered features. Furthermore, let $\mathcal{X} = \mathcal{X}_1 \times \mathcal{X}_2 \times \ldots \times \mathcal{X}_p$ be the feature space, with partial ordering $\geq$, and let $\mathcal{C}$ be a set of classes with linear ordering $\geq$. An allocation rule is a function $r : \mathcal{X} \to \mathcal{C}$ which assigns a class from $\mathcal{C}$ to every point in the feature space. Let $r(\mathbf{x}) = i$ denote that an entity with feature values $\mathbf{x}$ is assigned to the ith class.

An allocation rule is monotone if

$$\mathbf{x}_1 \geq \mathbf{x}_2 \Rightarrow r(\mathbf{x}_1) \geq r(\mathbf{x}_2),$$

for all $\mathbf{x}_1, \mathbf{x}_2 \in \mathcal{X}$.

A classification tree partitions the feature space $\mathcal{X}$ into a number of hyper-rectangles (corresponding to the leaf nodes of the tree) and elements in the same hyperrectangle are all assigned to the same class. As is shown in [Pot99], a classification tree is non-monotonic if and only if there exist leaf nodes t_1, t_2 such that

$$r(t_1) > r(t_2) \text{ and } \min(t_1) \leq \max(t_2),$$

where $\min(t)$ and $\max(t)$ denote the minimum and maximum element of t respectively. A dataset $(\mathbf{x}_n, c_n)$, is called monotone if

$$\mathbf{x}_i \geq \mathbf{x}_j \Rightarrow c_i \geq c_j,$$

for all $i, j = 1, \ldots, n$.

Potharst [Pot99] provides a thorough study for the case that the training data may be assumed to be monotone. This requirement however makes the algorithms presented of limited use for data mining. For example, in loan evaluation the dataset used to learn the allocation rule would typically consist of loans accepted in the past together with the outcome of the loan (say, defaulted or not). It is very unlikely that this dataset would be monotone.

A more pragmatic approach is taken by Ben-David [BD95], who proposes a splitting rule that includes a non-monotonicity index in addition to the usual impurity measure. This non-monotonicity index gives equal weight to each pair of non-monotonic leaf nodes.

A possible improvement of this index would be to give a pair t_1, t_2 of non-monotonic leaf nodes weight $p(t_1) \times p(t_2)$, where $p(t)$ denotes the proportion of cases in leaf t. The idea behind this is that when two low-probability leaves are non-monotonic with respect to each other, this violates the monotonicity of the tree to a lesser extent than two high-probability leaves. The reader should note that $p(t_1) \times p(t_2)$ is an upperbound for the degree of non-monotonicity between node t_1 and t_2 because not all their elements have to be non-monotonic with respect to each other.

The use of a non-monotonicity index in determining the best split has certain drawbacks however. Monotonicity is a *global* property, i.e. it involves a relation between different leaf nodes of a tree. If the degree of monotonicity is measured for each possible split during tree construction, the *order* in which nodes are expanded becomes important. For example, a depth-first search strategy will generally lead to a different tree then a breadth-first search. Also, a non-monotonic tree may become monotone after additional splits. Therefore we consider an alternative, computationally more intensive, approach in this study. Rather than *enforcing* monotonicity during tree construction, we generate many different trees and *check* if they are monotonic. The collection of trees may be obtained by drawing bootstrap samples from the training data, or making different random partitions of the data in a training and test set. This approach allows the use of a standard tree algorithm except that the minimum and maximum elements of the nodes have to be recorded during tree construction, in order to be able to check whether the final tree is monotone. This approach has the additional advantage that one can estimate to what extent the assumption of monotonicity is correct. In the next section we apply this idea to an economic data set concerning house prices.

4   Den Bosch Housing Data

In this section we discuss the application of monotonic classification trees to the prediction of the asking price of a house in the city of Den Bosch (a medium sized Dutch city with approximately 120,000 inhabitants). The basic principle of a hedonic price model is that the consumption good is regarded as a bundle of characteristics for which a valuation exists ([HR78]). The price of the good is determined by a combination of these valuations:

$$P = P(x_1, \ldots, x_p)$$

In the case at hand the variables $x_1, x_2, \ldots, x_p$ are characteristics of the house. The explanatory variables have been selected on the basis of interviews with experts of local house brokers, and advertisements offering real estate in local magazines. The most important variables are listed in table 1.

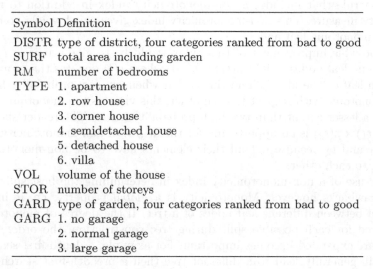

Symbol	Definition
DISTR	type of district, four categories ranked from bad to good
SURF	total area including garden
RM	number of bedrooms
TYPE	1. apartment
	2. row house
	3. corner house
	4. semidetached house
	5. detached house
	6. villa
VOL	volume of the house
STOR	number of storeys
GARD	type of garden, four categories ranked from bad to good
GARG	1. no garage
	2. normal garage
	3. large garage

Table 1. Definition of model variables

It is a relatively small data set with only 119 observations. Of all 7021 distinct pairs of observations, 2217 are comparable, of which 78 are non-monotonic. For the purpose of this study we have discretized the dependent variable (asking price) into the classes "below median" (fl. 347,500) and "above median" (the current version of the algorithm can only handle binary classification). After this discretization of the dependent variable only 9 pairs of observations are non-monotonic.

The tree algorithm used is in many respects similar to the CART program as described in [BFOS84]. The program only makes binary splits and use the gini-index as splitting criterion. Furthermore it uses cost-complexity pruning [BFOS84] to generate a nested sequence of trees from which the best one is selected on the basis of test set performance. During tree construction, the algorithm records the minimum and maximum element for each node. These are used to check whether a tree is monotone. The algorithm has been written in the Splus language [VR97].

In order to determine the effect of application of the monotonicity constraint we repeated the following experiment 100 times. The dataset was randomly partitioned (within classes) into a training set (60 observations) and test set (59 observations). The training set was used to construct a sequence of trees using cost-complexity pruning. From this sequence the best tree was selected on the basis of error rate on the test set (in case of a tie, the smallest tree was chosen). Finally, it was checked whether the tree was monotone and if not, the upperbound for the degree of monotonicity (as described in section 3) was computed.

Out of the 100 trees thus constructed, 61 turned out to be monotone and 39 not. The average misclassification rate of the monotonic trees was 14.9%, against 13.3% for the non-monotonic trees. Thus, the monotonic trees had a slightly worse classification performance. A two-sample t-test of the null hypothesis that monotonic and non-monotonic trees have the same classification error yielded a p-value of 0.0615 agains a two-sided alternative. The average degree of non-monotonicity of the non-monotonic trees was about 1.7%, which is quite low, the more if we take into consideration that this is an upper bound. Another interesting comparison is between the average sizes of the trees. On average, the monotonic trees had about 3.13 leaf nodes, against 7.92 for the non-monotonic trees. Thus, the monotonic trees are considerably smaller and therefore easier to understand at the cost of only a slightly worse classification performance.

5   Conclusion

Monotonicity of relations is a common form of domain knowledge in economics as well as other application domains. Furthermore, monotonicity of the final model may be an important requirement for explaining and justifying model outcomes. We have investigated the use of monotonicity constraints in classification tree algorithms.

In preliminary experiments on house pricing data, we have found that the predictive performance of monotonic trees was comparable to, but slightly worse than, the performance of the non-monotonic trees. On the other hand, the monotonic trees were much simpler and therefore more insightful and easier to explain. This provides interesting prospects for applications where monotonicity is an absolute requirement, such as in many selection decision models.

References

[BD95]   A. Ben-David. Monotonicity maintenance in information-theoretic machine learning algorithms. *Machine Learning*, 19:29–43, 1995.

[BFOS84]   L. Breiman, J.H. Friedman, R.A. Olshen, and C.T. Stone. *Classification and Regression Trees*. Wadsworth, Belmont, California, 1984.

[HR78]   O. Harrison and D. Rubinfeld. Hedonic prices and the demand for clean air. *Journal of Environmental Economics and Management*, 53:81–102, 1978.

[Lea78]   E. Leamer. *Specification Searches: Ad Hoc Inference with Nonexperimental Data*. Wiley, 1978.

[Lov83]   Michael C. Lovell. Data mining. *The Review of Economics and Statistics*, 65(1):1–12, 1983.

[Pot99]   R. Potharst. *Classification using decision trees and neural nets*. PhD thesis, Erasmus Universiteit Rotterdam, 1999. SIKS Dissertation Series No. 99-2.

[Qui93]   J.R. Quinlan. *C4.5 Programs for Machine Learning*. Morgan Kaufmann, San Mateo, California, 1993.

[VR97]   W.N. Venables and B.D. Ripley. *Modern Applied Statistics with S-PLUS (second edition)*. Springer, New York, 1997.

Temporal Machine Learning for Switching Control

Pierre Geurts and Louis Wehenkel

University of Liège, Department of Electrical and Computer Engineering
Institut Montefiore, Sart-Tilman B28, B4000 Liège, Belgium
{geurts,lwh}@montefiore.ulg.ac.be

Abstract. In this paper, a temporal machine learning method is presented which is able to automatically construct rules allowing to detect as soon as possible an event using past and present measurements made on a complex system. This method can take as inputs dynamic scenarios directly described by temporal variables and provides easily readable results in the form of detection trees. The application of this method is discussed in the context of switching control. Switching (or discrete event) control of continuous systems consists in changing the structure of a system in such a way as to control its behavior. Given a particular discrete control switch, detection trees are applied to the induction of rules which decide based on the available measurements whether or not to operate a switch. Two practical applications are discussed in the context of electrical power systems emergency control.

1 Introduction

Supervised learning methods, while being very mature, often are only able to handle static input or output values, either numerical or categorical. Nevertheless, many problems whose complexity makes interesting alternative approaches like automatic learning, would need to take into account temporal data. An obvious suboptimal solution is to transform the dynamic problem in order to match the static constraints of traditional learning algorithm but it does not seem to us very appropriate. Besides this simple approach, some recent research on learning classification models with temporal data are described for example in [3],[4] and [5].

In this paper, we present a temporal machine learning method able to automatically construct rules allowing to detect (predict) as soon as possible an event using past and present measurements made on a complex system. This method is able to handle directly temporal attributes and provides interpretable results. We then propose to use this method for the tuning of rules responsible for the triggering of discrete control event in complex systems. Two practical applications are discussed in the context of electrical power systems emergency control. The first one aims at avoiding loss of synchronism in the Western part of the North-American interconnection, while the goal of the second application is the detection of voltage collapse in the South-Eastern part of the French power system.

D.A. Zighed, J. Komorowski, and J. Żytkow (Eds.): PKDD 2000, LNAI 1910, pp. 401–408, 2000.
© Springer-Verlag Berlin Heidelberg 2000

The paper is organized as follows. In Section 2, we define the temporal detection problem. Section 3 is devoted to a description of the learning method we proposed to solve it. In Section 4, we discuss the two practical applications.

2   Temporal Detection Problem

When dealing with temporal data, we are given a universe U of objects representing dynamic system trajectories (or episodes, scenarios). These objects are described by a certain number of temporal candidate attributes which are functions of object and time, thus defined on $U \times [0, +\infty[^1$. We denote by $a(o, t)$ the value of the attribute a at time t for the object o, by $\mathbf{a}(o, .)$ the attribute vector $(a^1(o, .), \ldots, a^m(o, .))$ and by $\mathbf{a}_{[0,t]}(o, .)$ its restriction to the interval $[0, t]$.

In a temporal detection problem, we also assume that each scenario o is classified into one of two possible classes $c(o) \in \{+, -\}$. We then call **detection rule** a function of past and present attribute value :

$$d(\mathbf{a}_{[0,t]}(o, .)) \mapsto \{+, -\} \tag{1}$$

which classifies a scenario at each time t and has the following monotonicity property :

$$d(\mathbf{a}_{[0,t]}(o, .)) = + \Leftrightarrow \forall t' \geq t : d(\mathbf{a}_{[0,t']}(o, .)) = +. \tag{2}$$

Thus, we assume that if an object is classified into class $+$ at some time, it will remain so for all later times (monotonicity). We will call the detection time of an object by a detection rule the first time it is classified into class $+$ and we will say that an object is detected by the rule if such a time exists.

The machine learning problem can now be formulated as follows : given a random sample LS of objects whose attributes values are observed for some finite period of time $[0, t_f(o)]$, the objective is to automatically derive a detection rule which would perform as well as possible in detecting objects with class $+$ from U. Clearly a good detection rule is a rule which will detect only those objects which actually belong to class $+$, and among good rules, the better ones are those having the smallest detection times.

3   Detection Rules Model: Detection Trees

We choose to represent a detection rule by a tree whose arcs have been labeled by elementary tests. Figure 1 shows a simple detection tree and the detection rule it represents. Each test is a logical functional of the attributes :

$$T(\mathbf{a}_{[0 \ldots t]}(o, .)) \mapsto \{\mathrm{T}, \mathrm{F}\}. \tag{3}$$

which, like detection rules, exhibits a monotonicity property :

$$T(\mathbf{a}_{[0 \ldots t]}(o, .)) = \mathrm{T} \Leftrightarrow \forall t' \geq t : T(\mathbf{a}_{[0 \ldots t']}(o, .)) = \mathrm{T}. \tag{4}$$

---

[1] without loss of generality we assume start time of scenario being always 0

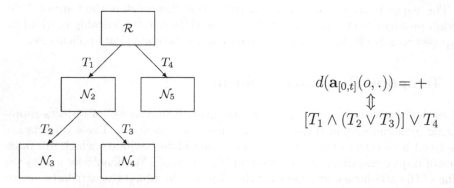

Fig. 1. An example detection tree and the corresponding detection rule

Note that the tests are not mutually exclusive; hence, using such a temporal tree, and given a value of t, a scenario is propagated through the tree along all paths starting at the root $\mathcal{R}$ (i.e. the top-node), until a condition T_i along the path is false or a terminal node is reached. If the scenario reaches at least one terminal node, it is classified into class $+$, otherwise it is classified into class $-$ at time t. Notice that the monotonicity property of tests implies the monotonicity of the tree detection.

3.1   Induction Algorithm [2]

As in TDIDT (Top Down Induction of Decision Trees [1]), temporal tree induction is based on a quality measure and is composed of two search stages : tree growing and tree pruning. Prior to tree induction, the overall learning set LS is first decomposed in two disjoint subsets : the growing sample (GS) used to evaluate the quality of candidate trees during growing, and the pruning sample (PS) used to evaluate trees during pruning.

Quality measure. In order to assess the quality of a temporal tree from a sample of objects S, we propose the following evaluation function :

$$Q(\mathcal{T}, S) = \frac{(1-\alpha)[(1-\beta)\sum_{o\in\mathcal{T}^+} f(\frac{t_{\mathcal{T}}(o)}{t_f(o)}) + \beta N_{\mathcal{T},+}] + \alpha N_{\mathcal{T},-}}{(1-\alpha)N_+ + \alpha N_-} \in [0,1], \quad (5)$$

where :

- N_+ (resp. N_-) denotes the number of samples of class $+$ (resp. $-$) and $N_{\mathcal{T},+}$ (resp. $N_{\mathcal{T},-}$) the number of them classified $+$ (resp. $-$) by the tree $\mathcal{T}$,
- $\mathcal{T}^+$ denotes the subset of S of objects of class $+$ correctly detected by the tree,
- f(.) is a monotonically decreasing function defined on $[0,1]$ such that $f(0) = 1$ and $f(1) = 0$ (in our simulation $f(x) = 1 - x$),
- α and β are two user defined parameters in $[0,1]$,
- $t_{\mathcal{T}}(o)$ is the detection time of o by the rule modeled by the tree,

– $t_f(o)$ denotes the maximal time after which observation of o stops. This time is fixed a priori for each object, and allows one to specify a time after which detection is not considered anymore useful.

This quality measure takes into account two tradeoffs : On one hand, the tradeoff between the selectivity of the rule and the degree of anticipation of the detection and on the other hand the tradeoff between non-detections and false-alarms. The first tradeoff is mainly regulated by β, the second one by α. Indeed, when β is set to 1, detection time does not affect the quality. When α is set to 0 (resp. 1), only correct detections of positive objects via $N_{\mathcal{T},+}$ (resp. correct non-detections of negative objects via $N_{\mathcal{T},-}$) are taken into account in the quality criteria. These tradeoffs are not independent : the more anticipative we want the rule to be, the more false-alarms the system will suffer in general. This relationship appears also in our quality measure as the two parameters actually influence both tradeoffs.

Growing. A greedy approach is used. The algorithm uses a stack of "open" nodes initialized to the root node corresponding to the trivial tree (which classifies all objects into class +). It then proceeds by considering the node at the top of the stack for expansion. At each expansion step, it pushes the created successor on the top of the stack, thus operating in a depth first fashion. For example, the test of the tree at figure 1 would have been added in this order : T_1, T_2, T_3 and T_4. When considering the expansion of a node, it scans the set of candidate tests and evaluates (on the set GS) the increment of detection quality which would result from the addition of each one of these tests individually. If no test can increase quality at this node, then it is removed from the stack and the algorithm proceeds with the next node on the top of the stack. Otherwise, the best test is effectively added to the node, a new successor node is created and pushed on the top of the stack while the node selected for expansion remains on the stack.

Pruning. The tree growing method generally tends to produce overly complex trees, which need to be pruned to avoid overfitting the information contained in the learning set.

The proposed pruning method is similar in principle to the standard tree pruning methods used in TDIDT. It consists in generating a nested sequence of shrinking trees $\mathcal{T}_0, \dots \mathcal{T}_K$ starting with the full tree and ending with the trivial one, composed of a single node. At each step, the method computes the incremental quality which would be obtained by pruning any one of the terminal nodes of the current tree, and removes the one (together with its test) which maximizes this quantity. It uses an independent set of objects (i.e. different from the learning set) in order to provide honest estimates of the tree qualities, and hence to detect the irrelevant parts of the initial tree. The tree of maximal quality is selected in the sequence.

Candidate tests. The choice of candidate tests depends on the application problem and also on computation efficiency criteria. Indeed, the brute force screening of all candidate tests at each expansion step implies the computation of the quality increment many times and is the most expensive step of the algorithm.

For example, in order to handle numerical attributes in our application, we consider tests in the form :

$$T(o,t) = T \Leftrightarrow \exists t' \leq t \mid \forall \tau \in [t' - \Delta t \ldots t'] : a^i(o,\tau) < (>)v_{th}, \qquad (6)$$

thus based only on one attribute. For the sake of efficiency, the user defines for each candidate attribute only a limited number of candidate values for Δt. Then the method uses a brute force search screening for each candidate attribute and Δt value all locally relevant candidate thresholds, computing their incremental quality.

4   Application to Switching Control of Power Systems

Switching (or discrete event) control of continuous systems consists of changing the structure of a system in such a way as to control its behavior. In any large scale system (such as power systems) there are generally a large number of possible discrete controls, and an even larger number of possible control laws for each of them. Given a particular discrete control (say on-off status of a switch), the question is to define a rule which decides based on the available measurements whether or not to operate the switch. Typically, measurements are incomplete and noisy to some extent which results in uncertainties. Thus the question is then to process past and present measurement in order to make as reliable decisions as possible. In this section, we will describe two practical applications of detection tree to the induction of control switch triggering rules in the context of the emergency control of power systems.

4.1   Electric Power System Emergency Control

Electric power systems are highly critical infrastructures, responsible for the delivery of electricity over large geographical areas. (For example, the Western European interconnection deserves electric energy to about 300 million customers.) In order to avoid large scale blackouts, these systems are equipped with a large number of control devices operating in different geographical areas and in different time frames. The design of these emergency control systems is difficult due to the highly non-linear, complex and uncertain character of power systems. Because it is not possible to collect data from the field to study the system behavior in highly disturbed conditions, the design is based on simulations. It is generally decomposed into successive steps :

- identification of the main problem causes (type of disturbances and instabilities which are more likely to cause a blackout);
- identification of possible remedies, in terms of control actions (emergency control has to be fast and reliable, and often uses discrete switching control);
- design of control triggering rules (choice of appropriate measurements and decision rules, which should be anticipative and reliable at the same time).

Monte-Carlo simulations together with automatic learning and data mining may be used in order to assist decision making in all these steps [7]. In what

Table 1. Results on test set (600 scenarios). P_e = error rate (%), $\#fa$ = number of false alarms (among 485 stable cases), $\#nd$ = number of non detections (among 115 instable cases), $T_\mathcal{T}^{avg}$ = average detection time of instable cases (in sec.).

β	α	$P_e(\%)$	$\#fa$	$\#nd$	$t_\mathcal{T}^{avg}$
1.0	0.3	0.17	1	0	3.441
0.5	0.3	1.17	7	0	3.056
0.0	0.3	6.5	39	0	2.239
0.2	0.3	6.83	41	0	2.232
0.2	0.1	11.67	70	0	1.967

follows, we will show how temporal machine learning of detection rules is suited for the design of control triggering rules. We will consider two different emergency control problems corresponding respectively to detection of loss of synchronism in the Western part of the North-American interconnection, and detection of voltage collapse in the South-Eastern part of the French power system.

4.2   Loss of Synchronism

In normal conditions, all the generators in a power system rotate at synchronous speed (corresponding to 50Hz in Europe, 60Hz in North-America). When a large disturbance appears (e.g. a short-circuit due to lightning), some generators start accelerating with respect to others, and, if the disturbance is very large, may loose synchronism. In this context, emergency control aims at anticipating loss of synchronism as quickly as possible so as to open some breakers to avoid cascading outages of lines and generators, which would lead to blackout. Because the phenomenon is extremely fast (a few hundred milliseconds), detection and control must be fully automatic.

The detection tree method was applied to a database composed of 1600 simulations of the North-American power system (about 25% are unstable). The objective is to design a relaying rule used to open the Pacific AC Intertie (a set of lines transmitting energy from Montana and Oregon to California). Opening these lines should reduce the impact of the instability on the overall system behavior [6], and avoid blackout.

Table 1 shows the results (obtained using a LS of 600, a PS of 400 and a TS of 600 scenarios, selected randomly in the database) corresponding to different values of parameters α and β. One can observe how different parameters lead to different types of rule characteristics : smaller values of β (as well as smaller values of α) lead to more anticipative detection (smaller value of average detection time $t_\mathcal{T}^{avg}$) at the price of a higher number of false alarms. The comparison of these results with those obtained using non-temporal machine learning [6], show that our rules are more anticipative and simpler, while being of comparable accuracy.

In all trials, 3 different attributes were used as inputs, corresponding to local measurements in the substation where the breakers are located. The amount of data processed by the detection rule learning algorithm was about 10^6 values,

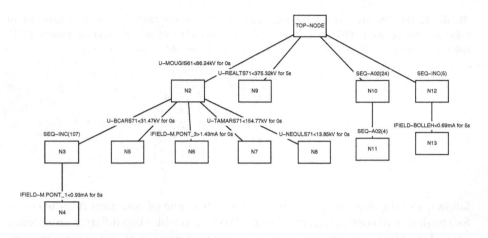

Fig. 2. A detection tree for the voltage collapse problem. Arcs are labeled with tests. SEQ-XX(n) means that the test is true when at least one out of n events (not described) of type XX occurs. SEQ-INC denotes the sequence of incidents affecting the system and SEQ-A02 denotes the triggering of lines by the installed protections. U-X denotes voltage at node X and IFIELD-Y denotes the excitation current of machine Y.

corresponding to a CPU time of 10 minutes for finalizing tree growing, pruning and testing on a 300 MHz Ultra-Sparc processor.

4.3   Voltage Collapse

In the recent years, voltage collapse has been one of the most studied problems in the area of power system emergency control, in particular due to the fact that it has been the cause of several large scale blackouts throughout the world. Usually, this phenomenon is slower than loss of synchronism (in the order of minutes), but still too fast to allow human decision making. Typical control actions consist of disconnecting temporarily some customers so as to avoid the system collapse. Of course, such control actions should be triggered only when absolutely necessary but early enough to be effective, hence the need to design appropriate decision rules which are both anticipative and selective.

We will illustrate the detection tree method on a database from a large scale study carried out on the system of Electricité de France [8]. It is composed of 1400 simulations, characterized all in all by about 800 temporal attributes (total of 2GB of raw data). In our simulations, we used only 42 of these attributes pre-selected as being potentially relevant for the problem under consideration (29 numerical time-series, and 12 sequences of events). Still, the amount of data is quite important (12.10^6 values all in all).

Figure 2 shows one particular detection tree thus obtained. To build the tree, the method has automatically selected a reduced set of attributes, together with appropriate thresholds and delays. If compared to presently used detection rules, this rule happens to be much more anticipative and accurate. Notice that the delays selected by the machine learning method are here in the order of a few seconds (compared to 50 to 100 ms in the loss of synchronism problem), which shows that the method is indeed able to identify the appropriate dynamic features of the problem.

Here, the overall CPU time required to build the rule was about 3 hours on a 300 MHz Ultra-Sparc processor, which in this particular application is negligible with respect to the time required to generate the database by simulations (3 months of CPU using a cluster of 10 high-end workstations [8]).

5   Conclusions

In this paper, we have discussed the development of an automatic induction method of detection rules and its application was proposed to the inference of triggering rules of control switch. This general framework was applied to the emergency control of electric power systems in the case of two different kinds of system failures, loss of synchronism and voltage collapse.

In the future, we will consider extensions of our detection tree method along two axes : hypothesis space and search strategy. The semantics of detection trees should be extended to allow the discovery of more complex temporal features. At its actual stage (and for efficiency criteria), the method does not permit to represent temporal constraints on the verification of tests, as for example in rules like "if T_1 becomes true and T_2 becomes true strictly after T_1 then ...". This capability has already been added to our system, but tends to be ineffective in terms of accuracy improvement on the studied problems. Expansion of the hypothesis space also will need to call into question the greedy search strategy which may not be appropriate anymore.

References

1. L. Breiman, J.H. Friedman, R.A. Olsen, and C.J. Stone. *Classification and Regression Trees.* Wadsworth International (California), 1984.
2. P. Geurts and L. Wehenkel. Early prediction of electric power system blackouts by temporal machine learning. In *Proc. of ICML-AAAI'98 Workshop on "AI Approaches to Times-series Analysis"*, Madison (Wisconsin), 1998.
3. M. W. Kadous. Learning comprehensible descriptions of multivariate time series. In *Proceedings of the Sixteenth International Conference on Machine Learning, ICML'99*, pages 454–463, Bled, Slovenia, 1999.
4. S. Manganaris. *Supervised classification with temporal data.* PhD thesis, Vanderbilt University, 1997.
5. V. Petridis and A. Kehagias. *Predictive modular neural network : applications to time series.* Kluwer Academic Publishers, 1998.
6. S. Rovnyak, C. Taylor, and Y. Sheng. Decision trees using apparent resistance to detect impending loss of synchronism. *To appear in IEEE Transactions on Power Delivery*, 2000.
7. L. Wehenkel. *Automatic learning techniques in power systems.* Kluwer Academic, Boston, 1998.
8. L. Wehenkel, C. Lebrevelec, M. Trotignon, and J. Batut. Probabilistic design of power-system special stability controls. *Control Engineering Practice*, 7(2):183–194, 1999.

Improving Dissimilarity Functions with Domain Knowledge, applications with IKBS system

David Grosser, Jean Diatta, and Noël Conruyt

IREMIA, Université de la Réunion
15, avenue René Cassin – BP 7151
97715 Saint-Denis Messag. Cedex 9, France
{grosser, jdiatta, conruyt}@univ-reunion.fr

Abstract. Some of the fundamental and theoretical issues in *Knowledge Discovery in Database* (KDD) rely on knowledge representation and the use of prior and domain knowledge to extract useful information from data. In many data exploration algorithms, dissimilarity functions do not use domain knowledge for the cases comparison. The *Iterative Knowledge Base System* (IKBS) has been designed to improve generalization accuracy of exploration algorithms through the use of structural properties of domain models. A general mathematical framework for utilizing structural properties of the domain model encompassing the definition of a *Dissimilarity Function for Structured Descriptions* is proposed. Applications are conducted with the help of IKBS on a set of databases from the UCI machine learning repository and on structured domain definition data.

Keywords. KDD, Domain Knowledge, Dissimilarity Functions, Generalization Accuracy

1 Taking advantage of Domain Knowledge in KDD

Representation issues, search complexity, use of prior and Domain Knowledge, and statistical inference are some of the core problems in KDD that are still open and require attention [6]. In Data Mining, developing methods and applications for representing knowledge about data is still a serious challenge.

In many fields of real world applications, we can capture a given aspect of the *domain knowledge* by associating attributes of the problem structure with objects linked by composition and/or specialization relationships. We can also structure the *domain definition* of nominal attributes by a hierarchy of values. These techniques enable the algorithms to take into account mutual dependencies between attributes and to compare case properties with more accuracy. For instance, in biosystematics, the scientific discipline that investigates biodiversity, the descriptions of specimens are often highly structured (composite objects, taxonomic attributes), highly noisy (erroneous or unknown data), and highly polymorphous (variable or imprecise data). To take into account this complexity, we need to define a *domain knowledge* that includes information about objects relationships, attribute types and other semantics aspects: the scope of all values,

D.A. Zighed, J. Komorowski, and J. Zytkow (Eds.): PKDD 2000, LNAI 1910, pp. 409–415, 2000.
© Springer-Verlag Berlin Heidelberg 2000

and the meaning of special values (defaults, exceptions). A *domain model* is defined by the association of a domain knowledge and reference data. It represents a given context for the discovery process concerning the *application domain*. The initial domain model is gradually enriched in the course of knowledge discovery to perfect a *domain theory* (see [7] for definitions). Thus, the *Iterative Knowledge Base System* (IKBS) [3] was developed to manage evolving and shared domain models in an object oriented formalism. It enables users to interactively incorporate objects and relations into the domain knowledge (alse called *descriptive model*) to instantiate it with a case base and to conduct supervised and unsupervised classification tasks. This paper will focus on the way to improve accuracy of data exploration algorithms with the use of Domain Knowledge. Section 2 presents a general mathematical framework for utilizing structural properties of the domain model encompassing the definition of a *Dissimilarity Function for Structured Descriptions*. In section 3, applications are conducted with the help of IKBS on a set of databases from the UCI machine learning repository [2] and on structured domain definition data dealing with corals and marine sponges systematics. As example, We show how nearest neighbor classifiers can be improved by the use of structural properties.

2   Dissimilarity Function for Structured Descriptions

There are many learning systems that depend upon a good distance function to be successful. Dissimilarity functions are used in many fields besides machine learning, including statistics, pattern recognition and in the symbolic data-analysis area. A common problem with these methods is that they adopt a syntactical and mathematical viewpoint of the dissimilarity measure that does not take into account background knowledge, and relationships between objects. In such traditional methods, attributes are independent of one another. The following sections propose a mathematical framework for defining new dissimilarity functions which use order relations between domain entities.

2.1   Structured descriptions

We define a *structured model* as a nonempty n-element object *partially ordered set* X where each object is characterized by a finite set of attributes (Fig. 1). The set of attributes of x will be denoted by A_x. A *structured description* (also called a *case*) is an instance i of a subset P of X, where for all object $x \in P$, each attribute a in A_x is assigned a value $a(i)$. The objects of P will be said *present* on i whereas those of $X \setminus P$ will be said *absent* on i.

2.2   Global description

The global description of an individual is based on (1) the order structure of X and (2) the presence/absence of objects on this individual.

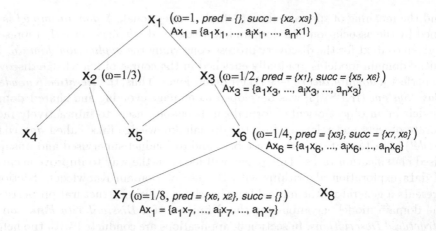

Fig. 1. Example of a structured model with filiation index ω associated to each object $x \in P$. The list of predecessors and successors is associated to each element.

To take into account the order structure of X, we will consider the following *filiation index function* ω associated to X, $\omega : X \to \mathbb{R}$ defined by

$$\omega(x) = \begin{cases} 1 & \text{if } x \text{ is maximal} \\ \underset{y \in Pred(x)}{min} \frac{\omega(y)}{|Suc(y)|} & \text{else} \end{cases}$$

To take into account the presence/absence of objects, we also consider situations in which no information is available about the presence/absence of some objets on some individuals. Such objects will be said *unknown* on the corresponding individuals. If an object x is unknown on an individual i, $p_i(x)$ will denote the probability for x to be present on i. If the objects of X are listed in a fixed linear ordering, the global description of an individual i may be identified with the n-vector $\left(\omega(x)\chi_i(x)\right)_{x \in X}$ where χ_i is defined on X by

$$\chi_i(x) = \begin{cases} 1 & \text{if } x \text{ is present on } i \\ 0 & \text{if } x \text{ is absent on } i \\ p_i(x) & \text{if } x \text{ is unknown on } i \end{cases}$$

2.3 Dissimilarity measure

The dissimilarity measure we propose in this paper is the Minkowski transform of a 2-vector. The components of this vector are the normalized *global* dissimilarity D_G and the normalized *local* dissimilarity D_L (1).

$$D(i,j) = \left(\left(\mu\, D_G(i,j) \right)^r + \left(\nu\, D_L(i,j) \right)^r \right)^{\frac{1}{r}}, r \geq 1 \qquad (1)$$

μ and ν are normalization coefficients. Following applications are conducted with $r = 1$. On unstructured databases, the component D_G is always null. In that case, the expression (1) is reduced to the local component ($\mu = 0$ and $\nu = 1$).

The local dissimilarity can be for instance the Euclidean metric, or one of those proposed by [8],i.e.:

- Heterogeneous Value Difference Metric (HVDM),
- Discretized Value Difference Metric (DVDM),
- Windowed Value Difference Metric (WVDM),
- Local dissimilarity on Heterogeneous Value (DGR) [1]

Any metric can be used in this general equation. Following application will show how the use of global dissimilarity factor can improve generalization accuracy of data exploration algorithms. The proposed *global dissimilarities* (1) consists of the Minkowski transforms on the n-dimensional vector space of global descriptions:

$$D_G(i,j) = \left(\sum_{x \in X} \omega(x)^r \left| \chi_i(x) - \chi_j(x) \right|^r \right)^{\frac{1}{r}}, r \geq 1. \qquad (2)$$

Possible extensions of various indices on presence/absence signs can be derived from this expression, which takes into account the order structure of X as well as the possible unknown objects.

3   Applications with IKBS

The *Iterative Knowledge Base System* [3] is a software that manages evolving and shared knowledge bases. Domain models and data are represented in an object oriented formalism and can be built and transformed through graphical representations. These representations are generalization or composition graphs or trees where nodes are objects of the domain and links are relationships between objects (Fig. 2 shows an example).

With IKBS, end-users can define structured domain models with different kinds of relationships between objects: composition or specialization dependencies. Another way to acquire a domain model consists of importing external databases or data tables. We thus obtain an unstructured domain model that is automatically generated complete with attributes domain definition and a case base linked to it. IKBS provides tools to interactively define structured descriptive models, hierarchical attributes, and special features such as default or exception values. Unstructured data definition can be transformed to add composition and/or specialization relationships as shown in Fig. 3.

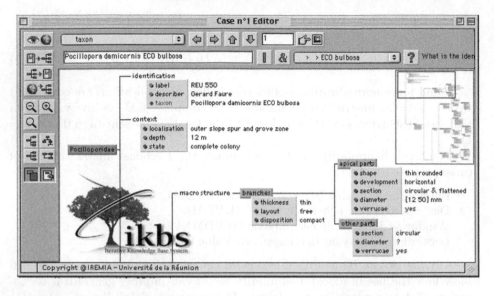

Fig. 2. Part of a structured representation pertaining to *Pocilloporidae* family (corals) in IKBS

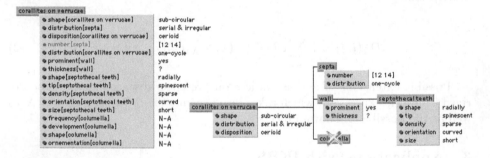

Fig. 3. The *corallites on verrucae* object transformed into a structured representation

3.1   generalization accuracy of dissimilarity functions

We present results of generalization accuracy of some dissimilarity functions on 17 databases from the *UCI Machine Learning Repository* and 3 knowledge bases from *IKBS projects* in marine biology (Fig. 4). 2 database models have only pure numeric data (Image segmentation and Vehicle) or pure symbolic data (Audiology, Monks). Others are defined by mixed features. 4 databases provide additional information on attributes: Bridges, *Pocilloporidae* and *Siderastreidae* coral families, and *Hyalonema* marine sponges. Some attributes are structured by order relationships (ordinal attributes) and organized by objects. These four structured databases were destructurized (transformed into data tables) in order to highlight the augmentation of Generalization Accuracy between unstructured and structured versions. We compare the dissimilarity functions previously mentioned. The 4 dissimilarity measures and a nearest neighbor classifier [5] (with k = 1) were programmed into IKBS. Each function was tested on 20 (+ 4 structured) datasets using 10-fold cross validation. The average generalization accuracy over all 10 trials is reported for each test in (Fig. 4). The highest accuracy achieved for each dataset is shown in bold. This application shows that DGR dissimilarity on average yields improved generalization accuracy on a collection of 24 databases. More important, it shows that using background knowledge and in particular, structures of the domain knowledge, can improve generalization accuracy with regard to any local dissimilarity.

4   Conclusion and future work

It has been shown that no learning algorithm can generalize more accurately than another when called upon to deal with all possible problems [4], unless information about the problem other than the training data is available. It follows then, that no dissimilarity function can be an improvement over another because it possesses a higher probability of accurate generalization. Its accuracy is a factor of its match with the kinds of problems that are likely to occur. Our global dissimilarity function was designed for complex data structures and is particularly well suited for data pertaining to the biological domains. Moreover, in some cases when considering tree-structures, we can obtain better performances in time of execution because attributes pertaining to absent objects are not considered. For the time being, an original version of an inductive algorithm that utilizes background knowledge has been programmed into IKBS [3] and we plan to adapt other algorithms drawn from the area of Case-Based Reasoning.

References

1. Diatta J., Grosser D., and Ralambondrainy H. A general dissimilarity measure for complex data. INF 01, IREMIA, University of Reunion Island, july 1999.
2. Merz and Murphy. Uci repository of machine learning databases. *Department of Information and Computer Science*, 1996.
3. Conruyt N. and Grosser D. Managing complex knowledge in natural sciences. *LNCS 1650, Springer Verlag*, pages 401–414, 1999.
4. Schaffer and Cullen. A conservation law for generalization performance. *In Proceedings of ML'94*, 1994.
5. Cover T. and Hart P. Nearest neighbor pattern classification. *Institute of Electrical and Electronics Enginneers Transactions on Information Theory, Vol.13, No.1*, pages 21–27, 1967.
6. Fayyad U.M., Piatetsky-Shapiro G., Padhraic Smyth, and Ramasamy Uthurusamy, editors. *From Data Mining to Knowledge Discovery: Current Challenges and Future Directions*. Advances in Knowledge Discovery and Data Mining, AAAI Press / MIT Press, 1996.
7. Klösgen W. and Zytkow J.M. *Knowledge Discovery in Databases Terminology*. Advances in Knowledge Discovery and Data Mining, AAAI Press, 1996.
8. Randall W.D. and Martinez T.R. Improved heterogeneous distance functions. *Journal of Artificial Intelligence Research, 6*, pages 1–34, 1997.

Databases	Dissimilarity	functions		
	Euclid	HVDM	WVDM	DGR
Unstructured databases				
Annealing	94.99%	94.61%	95.87%	**98.87%**
Audiology	60.50%	**77.50%**	**77.50%**	76.00%
Audiology test	41.67%	78.33%	78.33%	**88.46%**
Bridges	58.64%	59.64%	56.64%	**60.19%**
* Corals (*Pocilloporidae*)	51.12%	59.6%	59.6%	**61.06%**
* Corals (*Siderastreidae*)	72.80%	85.16%	85.40%	**86.80%**
Echocardiogram	94.82%	94.82%	**100.00%**	82.58%
Flag	48.95%	55.82%	58.74%	46.39%
Hepatitis	77.50%	76.67%	79.88%	78.71%
Images segmentation	92.86%	92.86%	93.33%	**98.10%**
LED+17 noise	42.90%	**60.70%**	**60.70%**	**60.70%**
Monks-1	77.58%	68.09%	68.09%	**79.83%**
Monk2-2	59.04%	**97.50%**	**97.50%**	96.50%
Monk2-3	87.26%	**100.00%**	**100.00%**	**100.00%**
Mushroom	**100.00%**	**100.00%**	**100.00%**	**100.00%**
Soybean (large)	87.26%	90.88%	**92.18%**	89.58%
Soybean (small)	**100.00%**	**100.00%**	**100.00%**	**100.00%**
* Sponges (*Hyalonema*)	49.21%	55.12%	55.12%	**56.8%**
Vehicle	70.93%	70.93%	65.37%	**79.02%**
Zoo	97.78%	**98.89%**	**98.89%**	98.11%
Structured databases				
Bridges	60.20%	56.24%	58.88%	**62.74%**
* Corals (*Pocilloporidae*)	53.48%	60.86%	60.86%	**63.50%**
* Corals (*Siderastreidae*)	77.30%	88.20%	88.20%	**90.00%**
* Sponges (*Hyalonema*)	51.20%	**58.00%**	**58.00%**	56.80%
Average	71.64%	79.31%	79.57%	**80.43%**

Fig. 4. % Generalization Accuracy with different dissimilarity functions, on unstructured and structured databases from UCI Machine Learning Repository and IKBS projects (*). Structured databases are utilized in unstructured and structured versions to show the interest to use global dissimilarity to improve generalization accuracy.

Mining Weighted Association Rules for Fuzzy Quantitative Items

Attila Gyenesei

Turku Centre for Computer Science (TUCS)
University of Turku, Department of Computer Science,
Lemminkaisenkatu 14, FIN-20520 Turku, Finland
gyenesei@cs.utu.fi

Abstract. During the last ten years, data mining, also known as knowledge discovery in databases, has established its position as a prominent and important research area. Mining association rules is one of the important research problems in data mining. Many algorithms have been proposed to find association rules in large databases containing both categorical and quantitative attributes. We generalize this to the case where part of attributes are given weights to reflect their importance to the user.

In this paper, we introduce the problem of mining weighted quantitative association rules based on fuzzy approach. Using the fuzzy set concept, the discovered rules are more understandable to a human.

We propose two different definitions of weighted support: with and without normalization. In the normalized case, a subset of a frequent itemset may not be frequent, and we cannot generate candidate k-itemsets simply from the frequent $(k-1)$-itemsets. We tackle this problem by using the concept of *z-potential frequent subset* for each candidate itemset.

We give an algorithm for mining such quantitative association rules. Finally, we describe the results of using this approach on a real-life dataset.

1 Introduction

The goal of data mining is to extract higher level information from an abundance of raw data. Mining association rules is one of the important research problems in data mining [1]. The problem of mining boolean association rules was first introduced in [2], and later broadened in [3], for the case of databases consisting of categorical attributes alone. Categorical association rules are rules where the events X and Y, on both sides of the rule, are instances of given categorical items. In this case, we wish to find all rules with confidence and support above user-defined thresholds (minconf and minsup). Several efficient algorithms for mining categorical association rules have been published (see [3], [4], [5] for just a few examples).

A variation of categorical association rules was recently introduced in [6]. Their new definition is based on the notion of weighted items to represent the importance of individual items.

D.A. Zighed, J. Komorowski, and J. Żytkow (Eds.): PKDD 2000, LNAI 1910, pp. 416–423, 2000.
© Springer-Verlag Berlin Heidelberg 2000

The problem of mining quantitative association rules was introduced and an algorithm proposed in [7]. The algorithm finds the association rules by partitioning the attribute domain, combining adjacent partitions, and then transforming the problem into binary one. An example of a rule according to this definition would be: "10% of married people between age 50 and 70 have at least 2 cars".

In [8], we showed a method to handle quantitative attributes using a fuzzy approach. We assigned each quantitative attribute several fuzzy sets which characterize it. Fuzzy sets provide a smooth transition between a member and nonmember of a set. The fuzzy association rule is also easily understandable to a human because of the linguistic terms associated with the fuzzy sets. Using the fuzzy set concept, the above example could be rephrased e.g. "10% of married old people have several cars".

In this paper, we introduce a new definition of the notion of weighted itemsets based on fuzzy set theory. In a marketing business, a manager may want to mine the association rules with more emphasis on some itemsets in mind, and less emphasis on other itemsets. For example, some itemsets may be more interesting for the company than others. This results in a generalized version of the quantitative association rule mining problem, which we call weighted quantitative association rule mining.

The paper is organized as follows. In the next section, we will present the definition of mining quantitative association rules using a fuzzy approach. Then we will introduce the problem of weighted quantitative association rules in Section 3. In Section 4, we give a new algorithm for this problem. In Section 5 the experimental results are reported, followed by a brief conclusion in Section 6.

2   Fuzzy Association Rules

In [8], an algorithm for mining quantitative association rules using a fuzzy approach was proposed. We summarize its definitions in what follows.

Let $I = \{i_1, i_2, \ldots, i_m\}$ be the complete item set where each i_j $(1 \leq j \leq m)$ denotes a categorical or quantitative (fuzzy) attribute. Suppose $f(i_j)$ represents the maximum number of categories (if i_j is categorical) or the maximum number of fuzzy sets (if i_j is fuzzy), and $d_{i_j}(l, v)$ represents the membership degree of v in the lth category or fuzzy set of i_j. If i_j is categorical, $d_{i_j}(l, v) = 0$ or $d_{i_j}(l, v) = 1$. If i_j is fuzzy, $0 \leq d_{i_j}(l, v) \leq 1$.

Let $t = \{t.i_1, t.i_2, \ldots, t.i_m\}$ be a transaction, where $t.i_j$, $(1 \leq j \leq m)$ represents a value of the jth attribute and can be mapped to

$$\big\{ \big(l, d_{i_j}\,(l, t.i_j)\big) \big\ | \ \text{for all } l, 1 \leq l \leq f(i_j)\big\}.$$

Given a database $D = \{t_1, t_2, \ldots, t_n\}$ with attributes I and the fuzzy sets associated with attributes in I, we want to find out some interesting, potentially useful regularities.

Definition 1. *A **fuzzy association rule** is of the form*

If $X = \{x_1, x_2, \ldots, x_p\}$ is $A = \{a_1, a_2, \ldots, a_p\}$ then $Y = \{y_1, y_2, \ldots, y_q\}$ is
$B = \{b_1, b_2, \ldots, b_q\}$, *where X, Y are itemsets and*
$a_i \in \{$*fuzzy sets related to attribute x_i*$\}$, $b_j \in \{$*fuzzy sets related to attribute y_j*$\}$

X and Y are ordered subsets of I and they are disjoint i.e. they share no common attributes. A and B contain the fuzzy sets associated with the corresponding attributes in X and Y. As in the binary association rule, "X is A" is called the antecedent of the rule while "Y is B" is called the consequent of the rule.

If a rule is interesting, it should have enough support and a high confidence value. We define the terms support and confidence as in [8]: the fuzzy support value is calculated by first summing all votes of each record with respect to the specified itemset, then dividing it by the total number of records. Each record contributes a vote which falls in [0, 1]. Therefore, a fuzzy support value reflects not only the number of records supporting the itemset, but also their degree of support.

Definition 2. *The **fuzzy support value** of itemset $\langle X, A \rangle$ in transaction set D is*

$$FS_{\langle X, A \rangle} = \frac{\sum_{t_i \in D} \Pi_{x_j \in X} d_{x_j}(a_j, t_i.x_j)}{|D|}$$

Definition 3. *An itemset $\langle X, A \rangle$ is called a **frequent itemset** if its fuzzy support value is greater than or equal to the minimum support threshold.*

We use the discovered frequent itemsets to generate all possible rules. If the union of antecedent $\langle X, A \rangle$ and consequent $\langle Y, B \rangle$ has enough support and the rule has high confidence, this rule will be considered as interesting.

When we obtain a frequent itemset $\langle Z, C \rangle$, we want to generate fuzzy association rules of the form, "If X is A then Y is B", where $X \subset Z$, $Y = Z - X$, $A \subset C$ and $B = C - A$. Having the frequent itemset, we know its support as well as the fact that all of its subsets will be also frequent.

Definition 4. *The **fuzzy confidence value** of a rule is as follows:*

$$FC_{\langle\langle X, A \rangle, \langle Y, B \rangle\rangle} = \frac{FS_{\langle Z, C \rangle}}{FS_{\langle X, A \rangle}} = \frac{\sum_{t_i \in D} \Pi_{z_j \in Z} d_{z_j}(c_j, t_i.z_j)}{\sum_{t_i \in D} \Pi_{x_j \in X} d_{x_j}(a_j, t_i.x_j)},$$

where $Z = X \cup Y$, $C = A \cup B$.

3   Weighted Quantitative Association Rules

Let itemset $\langle X, A \rangle$ be a pair, where X is the set of attributes x_j and A is the set of fuzzy sets a_j, $j = 1, \ldots, p$. We assign a weight $w_{(x,a)}$ for each itemset $\langle X, A \rangle$, with $0 \leq w_{(x,a)} \leq 1$, to show the importance of the item.

Generalizing Definition 2, we can define the weighted fuzzy support for the weighted itemset as follows:

Definition 5. *The **weighted fuzzy support** of an itemset* $\langle X, A \rangle$ *is*

$$WFS_{\langle X,A,w\rangle} = \left(\Pi_{x_j \in X} w_{(x_j,a_j)}\right) \cdot FS_{\langle X,A\rangle}$$
$$= \frac{\sum_{t_i \in D} \Pi_{x_j \in X} w_{(x_j,a_j)} d_{x_j}(a_j, t_i.x_j)}{|D|}$$

Notice the difference from [6] where the sum of weights is used, instead of the product. There are still other ways to define the combined weight. For example, the minimum of item weights would also make sense, and result in a simple algorithm. Product and minimum share the important property that if one item has zero weight, then the whole set has zero weight.

Similar to [8], a support threshold and a confidence threshold will be assigned to measure the strength of the association rules.

Definition 6. *An itemset* $\langle X, A \rangle$ *is called a **frequent itemset** if the weighted fuzzy support of such itemset is greater than or equal to the (user defined) minimum support threshold.*

Definition 7. *The **weighted fuzzy confidence** of the rule "If X is A then Y is B" as follows:*

$$WFC_{\langle\langle X,A,w\rangle,\langle Y,B,w\rangle\rangle} = \frac{WFS_{\langle Z,C,w\rangle}}{WFS_{\langle X,A,w\rangle}}$$

where $Z = X \cup Y$, $C = A \cup B$.

Definition 8. *An fuzzy association rule "If X is A then Y is B" is called an **interesting rule** if* $X \cup Y$ *is a frequent itemset and the confidence, defined in definition 7, is greater than or equal to a (user defined) minimum confidence threshold.*

The frequent itemsets in the weighted approach have the important property, that all its subsets are also frequent. Thus, we can apply the traditional bottom-up algorithm, tailored to fuzzy sets in [8].

4   Normalized Weighted Quantitative Association Rules

There is one possible problem with our definition. Even if each item has a large weight, the total weight may be very small, when the number of items in an itemset is large.

In this section, we deal with the mining of weighted association rules for which the weight of an itemset is normalized by the size of the itemset.

Definition 9. *The **normalized weighted fuzzy support** of itemset* $\langle X, A \rangle$ *is given by*

$$NWFS_{\langle X,A,w\rangle} = \left(\Pi_{x_j \in X} w_{(x_j,a_j)}\right)^{1/k} \cdot FS_{\langle X,A,w\rangle}$$

where $k = $ *size of the itemset* $\langle X, A \rangle$.

Notice that we actually use the geometric mean of item weights as the combined weight. This is a direct analogy to [6], where the arithmetic mean is applied in normalization.

Definition 10. *A k-itemset* $\langle X, A \rangle$ *is called a* **frequent itemset** *if the normalized weighted fuzzy support of such an itemset is greater than or equal to the minimum support threshold, or*

$$NWFS_{\langle X,A,w \rangle} \geq minsup$$

It is not necessarily true for all subsets of a frequent itemset to be frequent. In [8], we generated frequent itemsets with increasing sizes. However, since the subset of a frequent itemset may not be frequent, we cannot generate candidate k-itemsets simply from the frequent $(k-1)$-itemsets. All k-itemsets, which may contribute to be subsets of future frequent itemsets, will be kept in candidate generation process.

We can tackle this problem by using a new *z-potential frequent subset* for each candidate itemset.

Definition 11. *A k-itemset* $\langle X, A \rangle$ *is called a* **z-potential frequent subset** *if*

$$\left(\Pi_{x_j \in X} w_{(x_j,a_j)} \cdot \Pi_{y_j \in Y} w_{(y_j,b_j)} \right)^{1/z} \cdot FS_{\langle X,A,w \rangle} \geq minsup$$

where z is a number between k and the maximum possible size of the frequent itemset, and for each $\langle Y, B \rangle$, $Y \neq X$, *is the remaining itemset with maximum weights.*

4.1 Algorithm for Mining Normalized Weighted Association Rules

A trivial algorithm would be to solve first the non-weighted problem (weights=1), and then prune the rules that do not satisfy the weighted support and confidence. However, we can take advantage of weights to prune non-frequent itemsets earlier, and thereby cut down the number of trials.

An algorithm for mining normalized weighted quantitative association rules has the following inputs an outputs.

Inputs: A database D, two threshold values *minsup* and *minconf*.

Output: A list of interesting rules.

Notations:

D	the database
D_T	the transformed database
w	the itemset weights
F_k	set of *frequent k-itemsets* (have k items)
C_k	set of *candidate k-itemsets* (have k items)
I	complete item set
minsup	support threshold
minconf	confidence threshold

Main Algorithm $(minsup, minconf, D)$
1      $I = Search(D);$
2      $(C_1, D_T, w) = Transform(D, I);$
3      $k = 1;$
4      $(C_k, F_k) = Checking(C_k, D_T, minsup);$
5   while $(|C_k| \neq \emptyset)$ do
6   begin
7          $inc(k);$
8   if $k == 2$ then
9              $C_k = Join1(C_{k-1})$
10   else $C_k = Join2(C_{k-1});$
11         $C_k = Prune(C_k);$
12         $(C_k, F_k) = Checking(C_k, D_T, minsup);$
13         $F = F \cup F_k;$
14   end
15     $Rules(F, minconf);$

The subroutines are outlined as follows:

1. $Search(D)$: The subroutine accepts the database, finds out and returns the complete item set $I = \{i_1, i_2, \ldots i_m\}$.
2. $Transform(D, I)$: This step generates both a new transformed (fuzzy) database D_T from the original database by user specified fuzzy sets and weights for each fuzzy set. At the same time, the *candidate 1-itemsets* C_1 will be generated from the transformed database. If a 1-itemset is *frequent* or *z-potential frequent subset* then it will be kept in C_1, else it will be pruned.
3. $Checking(C_k, D_T, minsup)$: In this subroutine, the transformed (fuzzy) database is scanned and the weighted fuzzy support of candidates in C_k is counted. If its weighted fuzzy support is larger than or equal to $minsup$, we put it into the frequent itemsets F_k.
4. $Join1(C_{k-1})$: The *candidate 2-itemsets* will be generated from C_1 as in [8].
5. $Join2(C_{k-1})$: This Join step generates C_k from C_{k-1}, similar to [2].
6. $Prune(C_k)$: During the prune step, the itemset will be pruned in either of the following cases:
   - A subset of the candidate itemset in C_k does not exist in C_{k-1}.
   - The itemset cannot be a *z-potential frequent subset* of any frequent itemset.
7. $Rules(F)$: Find the rules from the *frequent itemsets* F as in [2].

5   Experimental Results

We assessed the effectiveness of our approach by experimenting with a real-life dataset. The data had 5 quantitative attributes: monthly-income, credit-limit, current-balance, year-to-date balance, and year-to-date interest.

The experiments will be done applying only the normalized version of the algorithm (as discussed in Section 4), due to reasons explained above. We use the above five quantitative attributes where three fuzzy sets are defined for each of them.

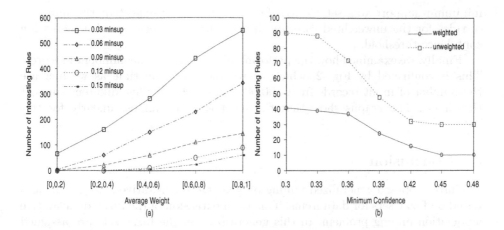

Fig. 1. Left : Random Weight Intervals. Right : Number of Interesting Rules.

Fig. 1(a) shows the increase of the number of rules as a function of average weight, for five different support thresholds. The minimum confidence was set to 0.25. We used five intervals, from which random weights were generated. The increase of the number of rules is close to linear with respect to the average weight. In the following experiments we used a random weight between 0 and 1 for each fuzzy set.

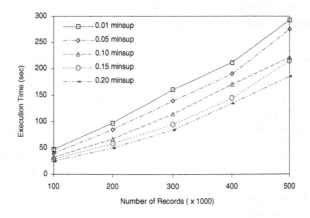

Fig. 2. Scale-up: Number of Records

Fig. 1(b) shows the number of generated rules as a function of minimum confidence threshold, for both weighted and unweighted case (as in [8]). The minimum support was set to 0.1. The results are as expected: the numbers of rules for the unweighted case are larger, but both decrease with increasing confidence threshold.

Finally, we examined how the performance varies with the number of records. This is confirmed by Fig. 2, which shows the execution time as we increase the number of input records from 100,000 to 500,000, for five different support thresholds. The graphs show that the method scales almost linearly for this dataset.

6   Conclusion

We have proposed generalized mining of weighted quantitative association rules based on fuzzy sets for data items. This is an extension of the fuzzy quantitative association mining problem. In this generalization, the fuzzy sets are assigned weights to reflect their importance to the user. The fuzzy association rule is easily understandable to a human because of the linguistic terms associated with the fuzzy sets.

We proposed two different definitions of weighted support: without normalization, and with normalization. In the normalized case, a subset of a frequent itemset may not be frequent, and we cannot generate candidate k-itemsets simply from the frequent $(k-1)$-itemsets. We tackled this problem by using the concept of *z-potential frequent subset* for each candidate itemset. We proposed a new algorithm for mining weighted quantitative association rules. The algorithm is applicable to both normalized and unnormalized cases but we prefer the former, as explained above. Therefore, the performance evaluation has been done only for the normalized version.

The results show that by associating weight to fuzzy sets, non-interesting rules can be pruned early and execution time is reduced.

References

1. Piatestky-Shapiro, G., Frawley, W.J.: Knowledge Discovery in Databases. AAAI Press/The MIT Press, Menlo Park, California (1991)
2. Agrawal, R., Imielinski, T., Swami, A.: Mining association rules between sets of items in large databases. Proceedings of ACM SIGMOD (1993) 207–216
3. Agrawal, R., Srikant, R.: Fast algorithms for mining association rules in large databases. Proceedings of the 20th VLDB Conference (1994) 487–499
4. Mannila, H., Toivonen, H., Verkamo, A.I.: Efficient Algorithms for discovering association rules. KDD-94: AAAI Workshop on KDD (1994) 181–192.
5. Toivonen, H.: Sampling large databases for association rules. Proceedings of the 20th VLDB Conference (1996)
6. Cai, C.H., Fu, Ada W.C., Cheng, C.H., Kwong, W.W.: Mining Association Rules with Weighted Items. Proc. of IEEE International DEAS (1998) 68–77.
7. Srikant, R., Agrawal, R.: Mining quantitative association rules in large relation tables. Proceedings of ACM SIGMOD (1996) 1–12.
8. Gyenesei, A.: A Fuzzy Approach for Mining Quantitative Association Rules. Turku Centre for Computer Science, Technical Report No. 336. (2000)

Centroid-Based Document Classification: Analysis and Experimental Results*

Eui-Hong (Sam) Han and George Karypis

University of Minnesota, Department of Computer Science / Army HPC Research Center
Minneapolis, MN 55455
{han,karypis}@cs.umn.edu

Abstract. In this paper we present a simple linear-time centroid-based document classification algorithm, that despite its simplicity and robust performance, has not been extensively studied and analyzed. Our experiments show that this centroid-based classifier consistently and substantially outperforms other algorithms such as Naive Bayesian, k-nearest-neighbors, and C4.5, on a wide range of datasets. Our analysis shows that the similarity measure used by the centroid-based scheme allows it to classify a new document based on how closely its behavior matches the behavior of the documents belonging to different classes. This matching allows it to dynamically adjust for classes with different densities and accounts for dependencies between the terms in the different classes.

1 Introduction

We have seen a tremendous growth in the volume of online text documents available on the Internet, digital libraries, news sources, and company-wide intranets. It has been forecasted that these documents (with other unstructured data) will become the predominant data type stored online. Automatic text categorization [20,16,12,4,8], which is the task of assigning text documents to pre-specified classes (topics or themes) of documents, is an important task that can help people to find information on these huge resources. Text categorization presents unique challenges due to the large number of attributes present in the data set, large number of training samples, attribute dependency, and multi-modality of categories. This has led to the development of a variety of text categorization algorithms [8,9,1,20] that address these challenges to varying degrees.

In this paper we present a simple centroid-based document classification algorithm. In this algorithm, a centroid vector is computed to represent the documents of each class, and a new document is assigned to the class that corresponds to its most similar centroid vector, as measured by the cosine function. Extensive experiments presented in Section 3 show that this centroid-based classifier consistently and substantially outperforms other algorithms such as Naive Bayesian [12], k-nearest-neighbors [20], and C4.5 [15], on

* This work was supported by NSF CCR-9972519, by Army Research Office contract DA/DAAG55-98-1-0441, by the DOE ASCI program, and by Army High Performance Computing Research Center contract number DAAH04-95-C-0008. Access to computing facilities was provided by AHPCRC, Minnesota Supercomputer Institute. Related papers are available via WWW at URL: http://www.cs.umn.edu/~karypis

D.A. Zighed, J. Komorowski, and J. Zytkow (Eds.): PKDD 2000, LNAI 1910, pp. 424–431, 2000.
© Springer-Verlag Berlin Heidelberg 2000

a wide range of datasets. Our analysis shows that the similarity measure used by the centroid-based scheme allows it to classify a new document based on how closely its behavior matches the behavior of the documents belonging to different classes. This matching allows it to dynamically adjust for classes with different densities and accounts for dependencies between the terms in the different classes. We believe that this feature is the reason why it consistently outperforms other classifiers that cannot take into account these density differences and dependencies.

The reminder of the paper is organized as follows. Section 3 experimentally evaluates this algorithm on a variety of data sets. Section 4 analyzes the classification model of the centroid-based classifier and compares it against those used by other algorithms. Finally, Section 5 provides directions for future research.

2   Centroid-Based Document Classifier

In the centroid-based classification algorithm, the documents are represented using the vector-space model [16]. In this model, each document d is considered to be a vector in the term-space. In its simplest form, each document is represented by the *term-frequency* (TF) vector $d_{tf} = (tf_1, tf_2, \ldots, tf_n)$, where tf_i is the frequency of the ith term in the document. A widely used refinement to this model is to weight each term based on its *inverse document frequency* (IDF) in the document collection. This is commonly done [16] by multiplying the frequency of each term i by $\log(N/df_i)$, where N is the total number of documents in the collection, and df_i is the number of documents that contain the ith term (i.e., document frequency). This leads to the *tf-idf* representation of the document, i.e., $d_{tfidf} = (tf_1 \log(N/df_1), tf_2 \log(N/df_2), \ldots, tf_n \log(N/df_n))$. Finally, in order to account for documents of different lengths, the length of each document vector is normalized so that it is of unit length, i.e., $\|d_{tfidf}\|_2 = 1$. In the rest of the paper, we will assume that the vector representation d of each document d has been weighted using *tf-idf* and it has been normalized so that it is of unit length.

In the vector-space model, the similarity between two documents d_i and d_j is commonly measured using the cosine function [16], given by

$$\cos(d_i, d_j) = \frac{d_i \cdot d_j}{\|d_i\|_2 * \|d_j\|_2},$$ (1)

where "·" denotes the dot-product of the two vectors. Since the document vectors are of unit length, the above formula simplifies to $\cos(d_i, d_j) = d_i \cdot d_j$.

Given a set S of documents and their corresponding vector representations, we define the **centroid** vector C to be

$$C = \frac{1}{|S|} \sum_{d \in S} d,$$ (2)

which is nothing more than the vector obtained by averaging the weights of the various terms present in the documents of S. We will refer to the S as the **supporting set** for the centroid C. Analogously to documents, the similarity between between a document and a centroid vector is computed using the cosine measure as follows:

$$\cos(d, C) = \frac{d \cdot C}{\|d\|_2 * \|C\|_2} = \frac{d \cdot C}{\|C\|_2}.$$ (3)

Note that even though the document vectors are of length one, the centroid vectors will not necessarily be of unit length.

The idea behind the centroid-based classification algorithm is extremely simple. For each set of documents belonging to the same class, we compute their centroid vectors. If there are k classes in the training set, this leads to k centroid vectors $\{C_1, C_2, \ldots, C_k\}$, where each C_i is the centroid for the ith class. The class of a new document x is determined as follows. First we use the document-frequencies of the various terms computed from the training set to compute the *tf-idf* weighted vector-space representation of x, and scale it so x is of unit length. Then, we compute the similarity between x to all k centroids using the cosine measure. Finally, based on these similarities, we assign x to the class corresponding to the most similar centroid. That is, the class of x is given by

$$\arg \max_{j=1,\ldots,k} (\cos(x, C_j)). \tag{4}$$

The computational complexity of the learning phase of this centroid-based classifier is linear on the number of documents and the number of terms in the training set. The computation of the vector-space representation of the documents can be easily computed by performing at most three passes through the training set. Similarly, all k centroids can be computed in a single pass through the training set, as each centroid is computed by averaging the documents of the corresponding class. Moreover, the amount of time required to classify a new document x is at most $O(km)$, where m is the number of terms present in x. Thus, the overall computational complexity of this algorithm is very low, and is identical to fast document classifiers such as Naive Bayesian.

3   Experimental Results

We evaluated the performance of the centroid-based classifier by comparing against the naive Bayesian, C4.5, and k-nearest-neighbor classifiers on a variety of document collections. We obtained the naive Bayesian results using the Rainbow [13] with the multinomial event model [12]. The C4.5 results were obtained using a locally modified version of the C4.5 algorithm capable of handling sparse data sets. Finally, the k-nearest-neighbor results were obtained by using the *tf-idf* vector-space representation of the documents (identical to that used by the centroid-based classification algorithm), and using the number of neighbors $k = 10$.

3.1   Document Collections

We have evaluted the classification algorithms on data sets from West Group [5], TREC-5 [18], TREC-6 [18], and TREC-7 [18] collections, Reuters-21578 text categorization test collection Distribution 1.0 [11], OHSUMED collection [7], and the WebACE project (WAP) [2]. The detailed characteristics of the various document collections used in our experiments are available in [6] [1]. Note that for all data sets, we used a stop-list to remove common words, and the words were stemmed using Porter's suffix-stripping algorithm

---

[1] These data sets are available from *http://www.cs.umn.edu/~han/data/tmdata.tar.gz*.

[14]. Furthermore, we selected documents such that each document has only one class (or label). In other words, given a set of classes, we collected documents that have only one class from the set.

3.2   Classification Performance

The classification accuracy of the various algorithms on the different data sets in our experimental testbed are shown in Table 1. These results correspond to the average classification accuracies of 10 experiments. In each experiment 80% of the documents were randomly selected as the training set, and the remaining 20% as the test set. The first three rows of this table show the results for the naive Bayesian, C4.5, and k-nearest neighbor schemes, whereas the last row shows the results achieved by the centroid-based classification algorithm (denoted as "Cntr" in the table). For each one of the data sets, we used a boldface font to highlight the algorithm that achieved the highest classification accuracy.

Table 1. The classification accuracy achieved by the different classification algorithms.

	west1	west2	west3	oh0	oh5	oh10	oh15	re0	re1	tr11	tr12	tr21
NB	86.7	76.5	75.1	89.1	87.1	81.2	84.0	**81.1**	**80.5**	85.3	79.8	59.6
C4.5	85.5	75.3	73.5	82.8	79.6	73.1	75.2	75.8	77.9	78.2	79.2	81.3
kNN	82.9	77.2	76.1	84.4	85.6	77.5	81.7	77.9	78.9	85.3	85.7	89.2
Cntr	**87.5**	**79.0**	**81.6**	**89.3**	**88.2**	**85.3**	**87.4**	79.8	80.4	**88.2**	**90.3**	**91.6**

	tr23	tr31	tr41	tr45	la1	la2	la12	fbis	wap	ohscal	new3
NB	69.3	94.1	94.5	84.7	**87.6**	**89.9**	**89.2**	77.9	80.6	74.6	74.4
C4.5	**90.7**	93.3	89.6	91.3	75.2	77.3	79.4	73.6	68.1	71.5	73.5
kNN	81.7	93.9	93.5	91.1	82.7	84.1	85.2	78.0	75.1	62.5	67.9
Cntr	85.2	**94.9**	**95.7**	**92.9**	87.4	88.4	89.1	**80.1**	**81.3**	**75.4**	**79.7**

Looking at the results of Table 1, we can see that naive Bayesian outperforms the other schemes in five out of the 23 data sets, C4.5 does better in one, the centroid-based scheme does better in 17, whereas the k-nearest-neighbor algorithm never outperforms the other schemes.

A more accurate comparison of the different schemes can be obtained by looking at what extent the performance of a particular scheme is statistically different from that of another scheme. We used the resampled paired t test [6] to compare the accuracy results obtained by the different classifiers. The statistical significance results using the resampled paired t test are summarized in Table 2, in which for each pair of classification algorithms, it shows the number of data sets that one performs statistically better, worse, or similarly than the other. Looking at this table, we can see that the centroid-based scheme compared to naive Bayesian, does better in ten data sets, worse in one data set, and they are statistically similar in twelve data sets. Similarly, compared to kNN, it does better in twenty, and it is statistically similar in three data sets. Finally, compared to

C4.5, the centroid-based scheme does better in eighteen, worse in one, and statistically similar in four data sets.

Table 2. Statistical comparison of different classification algorithms using the resampled paired t test. The entries in the table show the number of data sets that the classifier in the row performs better, worse or similarly than the classifier in the column.

	NB	kNN	C4.5
Cntr	10/1/12	20/0/3	18/1/4
NB		12/4/7	15/3/5
kNN			13/3/7

From these results, we can see that the simple centroid-based classification algorithm outperforms all remaining schemes, with naive Bayesian being second, k-nearest-neighbor being third, and C4.5 being the last. Note that the better performance of NB and kNN over decision tree classification algorithms such as C4.5 agrees to results reported in [3,20] using precision and recall of binary classification.

Recently, Support Vector Machines (SVM) has been shown to be very effective in text classification [8]. We were not able to directly compare the centroid-based scheme with the SVM, because the SVM code used in [8] was written for binary classification only. We plan to perform comparison studies between SVM and the centroid-based scheme by performing binary classification in the future.

4   Analysis

4.1   Classification Model

The surprisingly good performance of the centroid-based classification scheme suggests that it employs a sound underlying classification model. The goal of this section is to understand this classification model and compare it against those used by other schemes.

In order to understand this model we need to understand the formula used to determine the similarity between a document x, and the centroid vector C of a particular class (Equation 3), as this computation is essential in determining the class of x (Equation 4). From Equation 3, we see that the similarity (i.e., cosine) between x and C is the ratio of the dot-product between x and C divided by the length of C. If S is the set of documents used to create C, then from Equation 2, we have that:

$$x \cdot C = x \cdot \left(\frac{1}{|S|} \sum_{d \in S} d \right) = \frac{1}{|S|} \sum_{d \in S} x \cdot d = \frac{1}{|S|} \sum_{d \in S} \cos(x, d).$$

That is, the dot-product is the average similarity (as measured by the cosine function) between the new document x and all other documents in the set. The meaning of the length of the centroid vector can also be easily understood using the fact that $\|C\|_2 = \sqrt{C \cdot C}$. Then, from Equation 2 we have that:

$$\|C\|_2 = \sqrt{\left(\frac{1}{|S|} \sum_{d \in S} d\right) \cdot \left(\frac{1}{|S|} \sum_{d \in S} d\right)} = \sqrt{\frac{1}{|S|^2} \sum_{d_i \in S} \sum_{d_j \in S} \cos(d_i, d_j)}.$$

Hence, the length of the centroid vector is the square-root of the average pairwise similarity between the documents that support the centroid.

In summary, the classification model used by the centroid-based document assigns a test document to the class whose documents better match the behavior of the test document, as measured by average document similarities. It computes the average similarity between the test document and all the other documents in that class, and then it amplifies that similarity, based on how similar to each other are the documents of that class. If the average pairwise similarity between the documents of the class is small (i.e., the class is *loose*), then that amplification is higher, whereas if the average pairwise similarity is high (i.e., the class is *tight*), then this amplification is smaller. More detailed ananlysis can be found in [6].

4.2 Comparison with Other Classifiers

One of the advantages of the centroid-based scheme is that it summarizes the characteristics of each class, in the form of the centroid vector. A similar summarization is also performed by naive Bayesian, in the form of the per-class term-probability distribution functions.

The advantage of the summarization performed by the centroid vectors is that it combines multiple prevalent features together, even if these features are not simultaneously present in a single document. That is, if we look at the prominent dimensions of the centroid vector (i.e., highest weight terms), these will correspond to terms that appear frequently in the documents of the class, but not necessarily all in the same set of documents. This is particularly important for high dimensional data sets for which the coverage of any individual feature is often quite low. Moreover, in the case of documents, this summarization has the additional benefit of addressing issues related to synonyms, as commonly used synonyms will be represented in the centroid vector (see [6] for some of the centroid vectors of data sets used in the experiments). For these reasons, the centroid-based classification algorithm (as well as naive Bayesian) tend to perform better than the C4.5 and the k-nearest neighbor classification algorithms.

The better performance of the centroid-based scheme over the naive Bayesian classifier is due to the method used to compute the similarity between a test document and a class. In the case of naive Bayesian, this is done using Bayes rule, assuming that when conditioned on each class, the occurrence of the different terms is independent. However, this is far from being true in real document collections [10]. The existence of positive and negative dependence between terms of a particular class causes naive Bayesian to compute a distorted estimate of the probability that a particular document belongs to that class.

On the other hand, the similarity function used by the centroid-based scheme does account for term dependence within each class. From the discussion in Section 4, we know that the similarity of a new document x to a particular class is computed as the ratio of two quantities. The first is the average similarity of x to all the documents in the class, and the second is the square-root of the average similarity of the documents within the class. To a large extent, the first quantity is very similar, in character, to the probability estimate used by the naive Bayesian algorithm, and it suffers from similar over- and under-estimation problems in the case of term dependence.

However, the second quantity of the similarity function, (i.e., the square-root of the average similarity of the documents within the class) does account for term dependency. This average similarity depends on the degree at which terms co-occur in the different documents. In general, if the average similarity between the documents of a class is high, then the documents have a high degree of term co-occurrence (since the similarity between a pair of documents computed by the cosine function, is high when the documents have similar set of terms). On the other hand, as the average similarity between the documents decreases, the degree of term co-occurrence also decreases. Since this average internal similarity is used to amplify the similarity between a test document and the class, this amplification is minimal when there is a large degree of positive dependence among the terms in the class, and increases as the positive dependence decreases. Consequently, this amplification acts as a correction parameter to account for the over- and under-estimation of the similarity that is computed by the first quantity in the document-to-centroid similarity function. We believe that this feature of the centroid-based classification scheme is the reason that it outperforms the naive Bayesian classifier in the experiments shown in Section 3.

5   Discussion and Concluding Remarks

In this paper we focused on a simple linear-time centroid-based document classification algorithm. Our experimental evaluation has shown that the centroid-based classifier consistently and substantially outperforms other classifiers on a wide range of data sets. We have shown that the power of this classifier is due to the function that it uses to compute the similarity between a test document and the centroid vector of the class. This similarity function can account for both the term similarity between the test document and the documents in the class, as well as for the dependencies between the terms present in these documents.

There are many ways to further improve the performance of this centroid-based classification algorithm. First, in its current form it is not well suited to handle multi-modal classes. However, support for multi-modality can be easily incorporated by using a clustering algorithm to partition the documents of each class into multiple subsets, each potentially corresponding to a different mode, or using similar techniques to those used by the generalized instance set classifier [9]. Second, the classification performance can be further improved by using techniques that adjust the importance of the different features in a supervised setting. A variety of such techniques have been developed in the context of k-nearest-neighbor classification [19], some of which can be extended to the centroid-based classifier [17].

References

1. L. Baker and A. McCallum. Distributional clustering of words for text classification. In *SIGIR-98*, 1998.
2. D. Boley, M. Gini, R. Gross, E.H. Han, K. Hastings, G. Karypis, V. Kumar, B. Mobasher, and J. Moore. Document categorization and query generation on the world wide web using WebACE. *AI Review (accepted for publication)*, 1999.
3. W.W. Cohen. Fast effective rule induction. In *Proc. of the Twelfth International Conference on Machine Learning*, 1995.
4. W.W. Cohen and H. Hirsh. Joins that generalize: Text classification using WHIRL. In *Proc. of the Fourth Int'l Conference on Knowledge Discovery and Data Mining*, 1998.
5. T. Curran and P. Thompson. Automatic categorization of statute documents. In *Proc. of the 8th ASIS SIG/CR Classification Research Workshop*, Tucson, Arizona, 1997.
6. E.H. Han and G. Karypis. Centroid-based document classification algorithms: Analysis & experimental results. Technical Report TR-00-017, Department of Computer Science, University of Minnesota, Minneapolis, 2000. Available on the WWW at URL *http://www.cs.umn.edu/~karypis*.
7. W. Hersh, C. Buckley, T.J. Leone, and D. Hickam. OHSUMED: An interactive retrieval evaluation and new large test collection for research. In *SIGIR-94*, pages 192–201, 1994.
8. T. Joachims. Text categorization with support vector machines: Learning with many relevant features. In *Proc. of the European Conference on Machine Learning*, 1998.
9. Wai Lam and Chao Yang Ho. Using a generalized instance set for automatic text categorization. In *SIGIR-98*, 1998.
10. D. Lewis. Naive (bayes) at forty: The independence assumption in information retrieval. In *Tenth European Conference on Machine Learning*, 1998.
11. D. D. Lewis. Reuters-21578 text categorization test collection distribution 1.0. *http://www.research.att.com/~lewis*, 1999.
12. A. McCallum and K. Nigam. A comparison of event models for naive bayes text classification. In *AAAI-98 Workshop on Learning for Text Categorization*, 1998.
13. Andrew Kachites McCallum. Bow: A toolkit for statistical language modeling, text retrieval, classification and clustering. http://www.cs.cmu.edu/ mccallum/bow, 1996.
14. M. F. Porter. An algorithm for suffix stripping. *Program*, 14(3):130–137, 1980.
15. J. Ross Quinlan. *C4.5: Programs for Machine Learning*. Morgan Kaufmann, San Mateo, CA, 1993.
16. G. Salton. *Automatic Text Processing: The Transformation, Analysis, and Retrieval of Information by Computer*. Addison-Wesley, 1989.
17. S. Shankar and G. Karypis. A feature weight adjustment algorithm for document classification. In *SIGKDD'00 Workshop on Text Mining*, Boston, MA, 2000.
18. TREC. Text REtrieval conference. *http://trec.nist.gov*.
19. D. Wettschereck, D.W. Aha, and T. Mohri. A review and empirical evaluation of feature-weighting methods for a class of lazy learning algorithms. *AI Review*, 11, 1997.
20. Y. Yang and X. Liu. A re-examination of text categorization methods. In *SIGIR-99*, 1999.

Applying Objective Interestingness Measures in Data Mining Systems

Robert J. Hilderman and Howard J. Hamilton

Department of Computer Science
University of Regina
Regina, Saskatchewan, Canada S4S 0A2
{hilder,hamilton}@cs.uregina.ca

Abstract. One of the most important steps in any knowledge discovery task is the interpretation and evaluation of discovered patterns. To address this problem, various techniques, such as the chi-square test for independence, have been suggested to reduce the number of patterns presented to the user and to focus attention on those that are truly statistically significant. However, when mining a large database, the number of patterns discovered can remain large even after adjusting significance thresholds to eliminate spurious patterns. What is needed, then, is an effective measure to further assist in the interpretation and evaluation step that ranks the interestingness of the remaining patterns prior to presenting them to the user. In this paper, we describe a two-step process for ranking the interestingness of discovered patterns that utilizes the chi-square test for independence in the first step and objective measures of interestingness in the second step. We show how this two-step process can be applied to ranking characterized/generalized association rules and data cubes.

1 Introduction

Techniques for finding association rules have been widely reported in the literature, commonly within the context of discovering buying patterns from retail sales transactions [1]. In the *share-confidence framework* [6], an *association rule* is an implication of the form $X \Rightarrow Y$, where X and Y are items (or sets of items), and the implication holds with *confidence* c and *share* s, if the number of items contained in X comprise $c\%$ of the number of items contained in $X \cup Y$, and the number of items contained in $X \cup Y$ comprises $s\%$ of the total number of items in the database. A *characterized itemset* is an itemset in which the corresponding transactions have been partitioned into classes based upon attributes which describe specific characteristics of the itemset [6]. A *generalized itemset* is one where the values of one or more characteristic attributes are generalized according to a taxonomic hierarchy [13].

As a result of the widespread adoption of on-line analytical processing, the study of data cubes is also receiving attention in the literature [4]. A *data cube*, also known as a summary table, is a redundant, multidimensional projection of a relation. Data cubes describe materialized aggregate views that group the

D.A. Zighed, J. Komorowski, and J. Zytkow (Eds.): PKDD 2000, LNAI 1910, pp. 432–439, 2000.
© Springer-Verlag Berlin Heidelberg 2000

unconditioned data in the original database along various dimensions according to SQL groupby operators associated with measure attributes. Dimensions in data cubes can also be generalized according to taxonomic hierarchies [4].

The number of patterns generated by a knowledge discovery task can exceed the capacity of a user to analyze them efficiently and effectively, and this is a widely recognized problem. In response to this problem, various techniques/metrics have been proposed to prune those patterns that are most likely to be considered uninteresting or not useful. Many successful techniques utilize the chi-square test for independence [12] which is based upon the differences between the expected number of occurrences for a discovered attribute value combination and the observed number. The primary assumption is that these differences will be less for independent attributes. A chi-square value that has a low probability of occurring strictly by chance leads to a rejection of the hypothesis that the attributes are *independent*, and attributes that are not independent are considered to be *associated*. Other statistics, known as *measures of association* [3], can be used to determine the relative strength of discovered patterns.

Although the chi-square test and measures of association can be used to reduce the number of patterns that must be considered by a user, when mining a large database, the number of patterns discovered can remain large even after adjusting significance thresholds to eliminate spurious patterns. What is needed, then, is an effective measure to assist in the interpretation and evaluation step that ranks the interestingness of the remaining patterns prior to presenting them to the user.

In this paper, we describe a two-step process for ranking the interestingness of discovered patterns that utilizes the chi-square test for independence in the first step and objective measures of interestingness in the second step. We show how this two-step process can be applied to ranking characterized/generalized association rules and data cubes. We introduced this use of diversity measures for ranking discovered patterns in [7, 8]. We also identified five diversity measures, known as the *PHMI set* (i.e., principled heuristic measures of interestingness), that satisfy five principles of interestingness proposed in [9].

2   Background

The five diversity measures in the PHMI set are shown in Figure 1. These diversity measures consider the frequency or probability distribution of the values in some numeric measure attribute to assign a single real-valued index that represents the interestingness of a discovered pattern relative to other discovered patterns. Let m be the total number of values in the numeric measure attribute. Let n_i be the i-th value. Let $N = \sum_{i=1}^{m} n_i$ be the sum of the n_i's. Let p be the actual probability distribution of the n_i's. Let $p_i = n_i/N$ be the actual probability for t_i. Let q be a uniform probability distribution of the values. Let $\bar{q} = 1/m$ be the probability for t_i, for all $i = 1, 2, \ldots, m$ according to the uniform distribution q. For a thorough discussion of the PHMI set, see [7, 8].

$$I_{Variance} = \frac{\sum_{i=1}^{m}(p_i - \bar{q})^2}{m-1}$$

$$I_{Simpson} = \sum_{i=1}^{m} p_i^2$$

$$I_{Shannon} = -\sum_{i=1}^{m} p_i \log_2 p_i$$

$$I_{Total} = m * I_{Shannon}$$

$$I_{McIntosh} = \frac{N - \sqrt{\sum_{i=1}^{m} n_i^2}}{N - \sqrt{N}}$$

Figure 1. The PHMI set of diversity measures

3   Applications

In this section, we present two applications for objective measures of interestingness in data mining systems: ranking (1) characterized/generalized association rules and (2) data cubes. Due to space limitations, we do not describe the chisquare test for independence, as this topic is covered in other work [12], and we do not use it in the derived examples that follow. Instead, use of the the chi-square test for pruning results is demonstrated in the experimental results of the next section. Also, we do not dwell on techniques for generating characterized/generalized association rules and data cubes, as these techniques have also been covered in other work [1, 4, 7]. Here, we assume that these techniques are understood, at an intuitive level at least, and restrict our presentation to the use of diversity measures for ranking discovered patterns.

Input is provided by a sales database consisting of the *Transact* and *Cust* tables, shown in Tables 1 and 2, respectively. In the *Transact* table, the *TID* column describes the transaction identifier, the *LI* column describes a unique line item identifier within the corresponding transaction identifier, the *CID* column describes the identifier of the customer who initiated the transaction, the *Loc* column describes the location where the transaction was processed, the *ItemNo* column describes the item sold in the corresponding line item, the *Qty* column describes the quantity of the corresponding item that has been sold. In the *Cust* table, the *CID* column is the customer identifier, the *Loc* column describes the location where the customer lives, and the *Name* column describes the name corresponding to the customer identifier. The *Transact* and *Cust* tables can be joined on the *Transact.CID* and *Cust.CID* columns. The *Cust.Loc* column shown in the *Transact* table is a result of such a join, and is shown for reader convenience in the presentation that follows. The values in the *Transact.Loc* and *Cust.Loc* columns can be generalized according to the DGG shown in Figure 2.

3.1   Characterized/Generalized Association Rules

Using the characterized itemset generation algorithm, *CItemset* [5, 6], from the share-confidence framework, a minimum share threshold of 30%, a mini-

Table 1. The *Transact* table

TID	LI	CID	Loc	ItemNo	Qty	Cust.Loc
1	1	4	2	A	2	3
1	2	4	2	B	1	3
1	3	4	2	C	4	3
2	1	1	3	A	2	4
2	2	1	3	B	3	4
2	3	1	3	C	1	4
2	4	1	3	D	2	4
3	1	2	1	A	3	2
3	2	2	1	C	5	2
3	3	2	1	E	1	2
4	1	3	4	D	5	1
5	1	1	2	B	2	4
5	2	1	2	C	1	4
6	1	4	1	B	7	3
6	2	4	1	C	3	3
6	3	4	1	E	2	3
7	1	3	3	A	5	1
7	2	3	3	C	8	1
8	1	2	4	D	6	2
8	2	2	4	E	3	2
9	1	3	2	A	2	1
9	2	3	2	B	4	1
9	3	3	2	C	1	1
10	1	1	3	C	5	4
11	1	4	1	B	4	3
11	2	4	1	C	6	3
11	3	4	1	D	2	3
11	4	4	1	E	7	3
12	1	2	4	A	3	2
12	2	2	4	C	8	2
12	3	2	4	D	1	2
13	1	3	3	E	2	1

Table 2. The *Cust* Table

CID	Loc	Name
1	4	Smith
2	2	Jones
3	1	White
4	3	Black

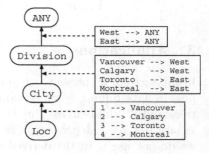

Figure 2. The *Loc* DGG

mum confidence threshold of 50%, the multi-attribute generalization algorithm, *All_Gen* [10, 11], and the SQL statement SELECT TID, Cust.Loc, Qty FROM Transact, Cust WHERE Transact.CID = Cust.CID, two of the many possible association rules and the corresponding summaries that can be generated according to the DGG in Figure 2, are shown in Table 3. In Table 3, the *Rule* $(x \Rightarrow y)$ column describes the discovered association rule, the *Share* and *Conf.* columns describe the *global share* and *count confidence* of the corresponding association rule, respectively, as described in [5, 6], the *Node* column describes the level to which the values in the *Cust.Loc* column have been generalized according to the DGG in Figure 2, the *TIDs* column describes the transactions from the *Transact* table aggregated in each row as a result of generalizing the values in the *Cust.Loc* column (TIDs are not actually saved in practice), the *Cust.Loc* column describes the characteristic attribute, the *Qty(x)* and *Qty(y)* columns describe the *local item count*, as described in [5, 6], for the antecedent and consequent, respectively, of the corresponding association rule, the *Qty* $(x \cup y)$ column describes the sum of *Qty(x)* and *Qty(y)*, and the *Count* column describes the number of transactions aggregated in each row.

Table 3 shows that association rule $C \Rightarrow A$ has share and confidence of 39.6% and 61.4%, respectively. Share is calculated as the sum of the quantity of all items in itemset $\{C, A\}$ divided by the quantity of all items in the *Transact* table (i.e., 44/111 = 39.6%). Confidence is calculated as the quantity of item C in itemset $\{C, A\}$ divided by the sum of the quantity of all items in itemset $\{C, A\}$ (i.e., 27/44 = 61.4%). The values of *Qty(C)* and *Qty(A)*, corresponding to

Table 3. Characterized/generalized association rules generated

Rule $(x \Rightarrow y)$	Share (%)	Conf. (%)	Node	TIDs	Cust.Loc	Qty(x)	Qty(y)	Qty $(x \cup y)$	Count
$C \Rightarrow A$	39.6	61.4	City	3, 12	Calgary	13	6	19	2
				7, 9	Vancouver	9	7	16	2
				1	Toronto	4	2	6	1
				2	Montreal	1	2	3	1
			Division	3, 7, 9, 12	West	22	13	35	4
				1, 2	East	5	4	9	2
$B \Rightarrow C$	33.3	56.8	City	2, 6, 11	Montreal	14	10	24	3
				5, 9	Calgary	6	2	8	2
				1	Toronto	1	4	5	1
			Division	1, 2, 6, 11	East	15	14	29	4
				5, 9	West	6	2	8	2

transactions 3 and 12, are 13 and 6, respectively, and calculated as the quantity of items C and A sold in the two transactions.

The values in the Qty $(x \cup y)$ and *Count* columns, called *vectors*, describe distributions of the quantity of items and the number of transactions, respectively, in the corresponding itemset, and these distributions can be used by the measures in the PHMI set to determine the relative interestingness of the corresponding association rule. For example, in Table 3, the vectors in the Qty $(x \cup y)$ column of the *City* and *Division* summaries for association rule $C \Rightarrow A$ are $(19, 16, 6, 3)$ and $(35, 9)$, respectively, and for association rule $B \Rightarrow C$ are $(24, 8, 5)$ and $(29, 8)$, respectively. The interestingness for these four summaries, according to $I_{Variance}$ (due to space limitations, we only discuss the results obtained for $I_{Variance}$) is 0.008815, 0.174587, 0.076211, and 0.161066, respectively. Thus, the rank order of the four association rules (from most to least interesting) is $C \Rightarrow A$ (*Division*), $B \Rightarrow C$ (*Division*), $B \Rightarrow C$ (*City*), and $C \Rightarrow A$ (*City*).

3.2   Data Cubes

Using the multi-attribute generalization algorithm, *All_Gen* [10, 11], and the SQL statement CREATE VIEW ItemsByLoc (Transact.Loc, Item, Cust.Loc, TotalQty) AS SELECT Transact.Loc, Item, Cust.Loc, SUM (Qty)) AS TotalQty FROM Transact, Cust WHERE Transact.CID = Cust.CID GROUPBY Item, Transact.Loc, Cust.Loc, four of the eight data cubes that can be generated according to the DGG in Figure 2, are shown in Figure 3. Figure 3 actually describes four data cubes because each cell contains two values; the top value is the quantity of items aggregated from the transactions, and the bottom value is the number of transactions aggregated. The *Item* attribute is on the vertical dimension, *Transact.Loc* on the horizontal, and *Cust.Loc* on the diagonal. The *City* and *Division* labels describe the level to which the values in both the *Transact.Loc* and *Cust.Loc* dimensions have been generalized according to the DGG in Figure 2. The other four possible data cubes (not shown) are obtained by generalizing only one of the *Transact.Loc* and *Cust.Loc* dimensions, respectively, in each cube.

Within the context of data cubes, objective interestingness measures can be

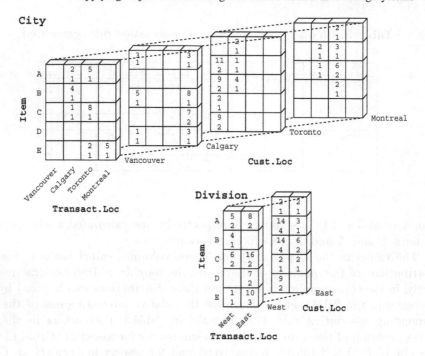

Figure 3. Data cubes generated

applied in three ways: (1) to the whole data cube, (2) to slices, and (3) to rows and columns (rows and columns example not shown).

Whole Data Cube. The *Qty* vectors for the *City* and *Division* data cubes are $(11, 9, 9, 8, 8, 7, 6, 5, 5, 4, 4, 3, 3, 3, 3, 2, 2, 2, 2, 2, 2, 2, 1, 1, 1, 1)$ and $(16, 14, 14, 10, 9, 8, 7, 6, 6, 5, 4, 3, 2, 2, 2, 2, 1)$, respectively. The interestingness of the *City* and *Division* data cubes, according to $I_{Variance}$, is 0.000761 and 0.001807, respectively. Thus, the rank order of the two data cubes is *Division* and *City*.

Slices. The *Qty* vectors for the *Vancouver*, *Calgary*, *Toronto*, and *Montreal* slices in the *Cust.Loc* dimension of the *City* data cube are $(8, 5, 5, 4, 2, 2, 1)$, $(8, 7, 5, 3, 3, 3, 1)$, $(11, 9, 9, 4, 2, 2, 1)$, and $(6, 3, 2, 2, 2, 1)$, respectively. The interestingness of the four slices, according to $I_{Variance}$, is 0.007969, 0.006931, 0.011740, and 0.011879, respectively. Thus, the rank order of the four slices is *Montreal*, *Toronto*, *Vancouver*, and *Calgary*.

4   Experimental Results

A series of experiments were run using *DGG-Interest*, an extension to *DB-Discover*, a research data mining tool developed at the University of Regina [2].

DGG-Interest evaluates the summaries generated by *DB-Discover* using the proposed two-step process (again, we present only the $I_{Variance}$ results).

Input data was supplied by the NSERC Research Awards and Customer databases [7, 10, 11]. Summary results for six representative discovery tasks, where two to four attributes have been selected for discovery, are shown in Table 4. In Table 4, the *Task* column describes a unique discovery task identifier, the *Attributes* column describes the number of attributes selected, the *Generated* column describes the number of summaries generated by the corresponding discovery task, the *Pruned* (*%Pruned*) column describes the number (percentage) of summaries in which no significant association between attributes was found in the first step, and the *Associated* (*%Associated*) column describes the number (percentage) of summaries in which a significant association was found in the first step, and which are available for ranking in the second step. For example, in *N-2*, an NSERC discovery task, two attributes were selected, 22 summaries were generated, 14 (63.6%) were pruned, and a significant association was discovered between attributes in the remaining eight (36.3%), which were available for ranking in the second step.

Table 4. Summary results for seven representative discovery tasks

Task	Attributes	Generated	Pruned	%Pruned	Associated	%Associated
N-2	2	22	14	63.6	8	36.3
N-3	3	70	43	61.4	27	38.6
N-4	4	186	143	76.9	43	23.1
C-2	2	340	325	95.6	15	4.4
C-3	3	3468	3288	94.8	180	5.2
C-4	4	27744	26163	94.3	1581	5.7

Detailed results for the *N-2* discovery task are shown in Table 5. In Table 5, the *Summary* column describes a unique summary identifier, the *Tuples* column describes the number of tuples in each summary, the *Attributes* column describes the number of attributes containing more than one unique domain value, the χ^2-*Status* column describes the result of the chi-square test, the χ^2-*Value* column describes the calculated chi-square value, the *DF* column describes the degrees of freedom for the chi-square test, the $I_{Variance}$ and *Rank* columns describe the calculated interestingness and rank determined by $I_{Variance}$ after pruning those containing no significant associations. In the chi-square calculation, any zeroes occurring in the work contingency table associated with each summary are considered to be structural zeroes.

References

1. T. Brijs, G. Swinnen, K. Vanhoof, and G. Wets. Using association rules for product assortment decisions: A case study. In *Proceedings of the Fifth ACM SIGKDD International Conference on Knowledge Discovery and Data Mining (KDD'99)*, pages 254–260, San Diego, California, August 1999.

Table 5. Detailed results for the N-2 discovery task

Summary	Tuples	Attributes	χ^2-Status	χ^2-Value	DF	$I_{Variance}$	Rank
1	5	2	Not Associated	0.343263	2	0.079693	-
2	11	2	Associated	11.343847	5	0.013534	1.0
3	10	2	Not Applicable	-	-	0.041606	-
4	21	2	Associated	33.401486	18	0.011547	4.0
5	50	2	Associated	77.999924	45	0.002078	6.0
6	3	1	Not Applicable	-	-	0.128641	-
7	6	1	Not Applicable	-	-	0.018374	-
8	4	1	Not Applicable	-	-	0.208346	-
9	4	2	Not Applicable	-	-	0.208346	-
10	9	2	Not Associated	7.003831	6	0.050770	-
11	21	2	Not Associated	23.186095	15	0.008896	-
12	10	1	Not Applicable	-	-	0.041606	-
13	5	1	Not Applicable	-	-	0.024569	-
14	9	1	Not Applicable	-	-	0.017788	-
15	2	1	Not Applicable	-	-	0.377595	-
16	2	2	Not Applicable	-	-	0.377595	-
17	9	2	Not Associated	6.731139	4	0.018715	-
18	16	2	Associated	12.431117	8	0.010611	2.0
19	40	2	Associated	63.910313	36	0.002986	5.0
20	67	2	Associated	97.799623	72	0.001582	7.0
21	17	2	Associated	18.221498	12	0.012575	3.0
22	30	2	Not Associated	24.920839	24	0.006470	-

2. C.L. Carter and H.J. Hamilton. Efficient attribute-oriented algorithms for knowledge discovery from large databases. *IEEE Transactions on Knowledge and Data Engineering*, 10(2):193–208, March/April 1998.
3. L.A. Goodman and W.H. Kruskal. *Measures of Association for Cross Classifications*. Springer-Verlag, 1979.
4. J. Han, W. Ging, and Y. Yin. Mining segment-wise periodic patterns in time-related databases. In *Proceedings of the Fourth International Conference on Knowledge Discovery and Data Mining (KDD'98)*, pages 214–218, New York, New York, August 1998.
5. R.J. Hilderman, C.L. Carter, H.J. Hamilton, and N. Cercone. Mining association rules from market basket data using share measures and characterized itemsets. *International Journal on Artificial Intelligence Tools*, 7(2):189–220, June 1998.
6. R.J. Hilderman, C.L. Carter, H.J. Hamilton, and N. Cercone. Mining market basket data using share measures and characterized itemsets. In X. Wu, R. Kotagiri, and K. Korb, editors, *Proceedings of the Second Pacific-Asia Conference on Knowledge Discovery and Data Mining (PAKDD'98)*, pages 159–173, Melbourne, Australia, April 1998.
7. R.J. Hilderman and H.J. Hamilton. Heuristic measures of interestingness. In J. Zytkow and J. Rauch, editors, *Proceedings of the Third European Conference on the Principles of Data Mining and Knowledge Discovery (PKDD'99)*, pages 232–241, Prague, Czech Republic, September 1999.
8. R.J. Hilderman and H.J. Hamilton. Heuristics for ranking the interestingness of discovered knowledge. In N. Zhong and L. Zhou, editors, *Proceedings of the Third Pacific-Asia Conference on Knowledge Discovery and Data Mining (PAKDD'99)*, pages 204–209, Beijing, China, April 1999.
9. R.J. Hilderman and H.J. Hamilton. Principles for mining summaries: Theorems and proofs. Technical Report CS 00-01, Department of Computer Science, University of Regina, February 2000. Online at http://www.cs.uregina.ca/research/Techreport/0001.ps.
10. R.J. Hilderman, H.J. Hamilton, and N. Cercone. Data mining in large databases using domain generalization graphs. *Journal of Intelligent Information Systems*, 13(3):195–234, November 1999.
11. R.J. Hilderman, H.J. Hamilton, R.J. Kowalchuk, and N. Cercone. Parallel knowledge discovery using domain generalization graphs. In J. Komorowski and J. Zytkow, editors, *Proceedings of the First European Conference on the Principles of Data Mining and Knowledge Discovery (PKDD'97)*, pages 25–35, Trondheim, Norway, June 1997.
12. B. Liu, W Hsu, and Y. Ma. Pruning and summarizing the discovered associations. In *Proceedings of the Fifth ACM SIGKDD International Conference on Knowledge Discovery and Data Mining (KDD'99)*, pages 125–134, San Diego, California, August 1999.
13. R. Srikant and R. Agrawal. Mining generalized association rules. In *Proceedings of the 21th International Conference on Very Large Databases (VLDB'95)*, pages 407–419, Zurich, Switzerland, September 1995.

Observational Logic Integrates Data Mining Based on Statistics and Neural Networks

Martin Holeňa

Institute of Computer Science, Academy of Sciences of the Czech Republic
Pod vodárenskou věží 2, CZ-182 00 Praha 8, Czech Republic,
martin@cs.cas.cz, http://www.cs.cas.cz/~martin

Abstract. In data mining, artificial neural networks have become one of important competitors of traditional statistical methods. They increase the potential of discovering useful knowledge in data, but only if the differences between both kinds of methods are well understood. Therefore, integrative frameworks are urgently needed. In this paper, a framework based on the calculus of observational logic is presented. Basic concepts of that framework are outlined, and it is explained how generalized quantifiers can be defined in an observational calculus to capture data mining with statistical and ANN-based methods.

1 Introduction

In the area of data mining, traditional statistical methods are often competed by methods relying on more recent approaches, including artificial neural networks, [3, 5, 15]. Those methods increase the potential of knowledge discovery and can bring us closer to the aim of extracting all the useful knowledge contained in the data. However, a prerequisite is that specific features of each kind of methods (e.g., the ability of statistical methods to discover structures which cannot be attributed to random influences and noise, or the universal approximation capabilities of neural networks), the differences between them with respect to the applicability conditions and to the meaning of the obtained results are well understood. Therefore, integrative frameworks are urgently needed.

In this paper, an integrative framework based on the calculus of of observational logic is presented. It has been elaborated already in the seventies, as a theoretical foundation of the method Guha, one of the first methods for automated knowledge discovery [7]. Originally, observational calculus has been proposed as a means for a unified treatment of different kinds of statistical methods. Recent research indicates that the framework can be successfully extended also behind statistical methods of knowledge discovery [8, 12, 14]. Basic concepts pertaining to the calculus are recalled in the following section, whereas the remaining two sections explain how generalized quantifiers can be defined in the calculus to capture statistical hypotheses testing and the extraction of rules from data with artificial neural networks. The framework is illustrated on small fragments from two real-world applications.

D.A. Zighed, J. Komorowski, and J. Zytkow (Eds.): PKDD 2000, LNAI 1910, pp. 440–445, 2000.
© Springer-Verlag Berlin Heidelberg 2000

2   Observational Logic and Its Importance for Data Mining

Monadic observational predicate calculus is a collection of *unary predicates* with a single object variable x, and of *generalized quantifiers*. A unary predicate, in general, states a range of values or even one specific value of a variable capturing some single property of an individual object (e.g., a patient), such as in
`disease duration > 10 years,`
`sex = male.`
Consequently, an open formula built from unary predicates by means of logical connectives corresponds to some combined property (combination of single properties), of an individual object, e.g.,
`sex = male ∧ disease duration > 10 years ∧ grand mal seizures.`
On the other hand, closed formulae state properties characterizing in some way the whole set of objects (e.g., patients) underlying the considered data. For example, $(\forall x)$ `age > 15`, meaning that we deal with data about adult patients. In the case of binary quantifiers or other quantifiers of a higher arity, the closed formula $(Qx)(\varphi_1, \ldots, \varphi_m)$, built from an m-ary generalized quantifier Q and open formulae $\varphi_1, \ldots, \varphi_m$, in general states some relationship between properties corresponding to $\varphi_1, \ldots, \varphi_m$. Examples will be given in the next section.

As *models* of a monadic observational predicate calculus, *data matrices* are used, which for the considered objects record the values of variables capturing the considered properties. More precisely, each column records the values of one variable. The interpretation in the model of a predicate concerning that variable is a $\{0, 1\}$-valued column vector with 1s on positions corresponding to those rows of the data matrix for which the values of the variable fulfil the predicate. Similarly, the interpretation of an open formula is the column vector that results from combining the interpretations of its constituent predicates in the boolean algebra of $\{0, 1\}$-valued vectors. For example, the interpretation of
`sex = male ∧ disease duration > 10 years ∧ grand mal seizures,`
is the vector with ones on positions corresponding to those rows of the data matrix for which the three constituent predicates are simultaneously valid, and with zeros on all remaining positions. Finally, the interpretation of the closed formula $(Qx)(\varphi_1, \ldots, \varphi_m)$ is obtained by applying some $\{0, 1\}$-valued function $\mathrm{Tf}_{(Qx)(\varphi_1, \ldots, \varphi_m)}$, called the *truth function* of $(Qx)(\varphi_1, \ldots, \varphi_m)$ to the data matrix serving as the model. The value of the truth function can, in general, depend on the whole data matrix, more precisely on those its columns that are needed to obtain the interpretations of $\varphi_1, \ldots, \varphi_m$. However, in many important cases that dependence reduces to a dependence on the interpretation of $\varphi_1, \ldots, \varphi_m$. In such a case, the truth function is a function on the set of all dichotomous matrices with m columns, is denoted simply Tf_Q and called the truth function of the quantifier Q. For example, the truth functions of the classical unary *existential quantifier* $\exists$ and *universal quantifier* $\forall$ are defined $\mathrm{Tf}_\exists(M) = 1$ iff $\sum M_i > 0$ and $\mathrm{Tf}_\forall(M) = 1$ iff $M_1 = M_2 = \cdots = 1$, respectively.

If the observational calculus includes at least one of the quantifiers $\forall$ and $\exists$, all tautologies derivable in the classical boolean predicate logic can be derived also here. In addition, a number of other tautologies are derivable also for various classes of generalized quantifiers of a higher arity [7]. Using those tautologies, the validity of some relationships can be deduced from the validity of other relationships, without having to evaluate the formula capturing the deduced relationship. This can substantially increase the efficiency of data mining.

3   Generalized Quantifiers for Statistical Hypotheses Testing

Let the data matrix serving as a model of a monadic observational calculus can be viewed as a realization of a random sample from some probability distribution P.

Through appropriate constraints on P, it is possible to capture probabilistic relationships between the properties corresponding to the columns of that data matrix. Assuming that P is a priori known to belong to some set $\mathcal{D}$ of probability distributions, a *statistical test* assesses the validity of the constraint $H_0 : P \in \mathcal{D}_0$ for a given $\mathcal{D}_0 \subset \mathcal{D}$ against the alternative $H_1 : P \in \mathcal{D}_1 = \mathcal{D} \setminus \mathcal{D}_0$. The test is performed using some random variable t, the test statistic, and some borel set C_α, the critical region of the test on a significance level $\alpha \in (0,1)$, which are connected to H_0 through the condition $(\forall P \in \mathcal{D}_0) P(t \in C_\alpha) \leq \alpha$. In addition, the test often makes use of the fact that if the considered model is a realization of a random sample, then also the matrix M of interpretations of open formulae $\varphi_1, \ldots, \varphi_m$ is a realization of a random sample, more precisely a realization of a random sample from a multinomial distribution. That fact allows to choose some test statistic t that is a function of that multinomial random sample, and to define a generalized quantifier $\sim$ corresponding to the constraint H_1 by means of a truth function depending only on M,

$$\mathrm{Tf}_\sim(M) = 1 \text{ iff } t_M \in C_\alpha, \tag{1}$$

where t_M is the realization of t provided the realization of the random sample from P yields the matrix of interpretations M.

The existing implementations of Guha include about a dozen quantifiers based on statistical hypotheses testing [7,16]. All of them are binary, the most fundamental being the *lower critical implication* $\rightarrow_\alpha^p$, $p \in (0,1)$, which corresponds to the binomial test (with a significance level $\alpha \in (0,1)$), the *Fisher quantifier* $\sim_\alpha^F$, corresponding to the one-sided Fisher exact test, and the χ^2 *quantifier* $\sim_\alpha^{\chi^2}$, corresponding to the χ^2 test. For example in [21], the following relationships have been discovered using the Fisher quantifier:
sex = male $\wedge$ disease duration > 10 years $\wedge$
                    $\wedge$ only grand mal seizures $\sim^F$ memory quotient > 90,
disease duration > 10 years $\wedge$ only grand mal seizures $\wedge$
                    $\wedge$ course of the disease = good $\sim^F$ memory quotient > 90.

A second example comes from an application of Guha to the area of river ecology [13]. Table 1 shows antecedents of formulae found to hold in the collected data and expressible as $(\bigwedge_{i \in \mathcal{I}} E_i \sim_{10\%}^F RD) \wedge (\bigwedge_{i \in \mathcal{I}} E_i \rightarrow_{10\%}^{\frac{2}{3}} RD)$, where $\mathcal{I}$ is some subset of ecological factors, E_i for $i \in \mathcal{I}$ are predicates stating ranges of values of the respective ecological factors, and RD is a predicate stating the occurrence of the species "Robackia demeierei". More precisely, RD states that the number of individuals of "Robackia demeierei" per sample is at least $\frac{1}{10} n_{\max}$, where $n_{\max}$ denotes the maximal number of individuals of this species per sample encountered in the data.

4   Quantifiers for ANN-Based Rule Extraction

The extraction of knowledge from data by means of artificial neural networks has received much attention not only in connection with data mining, but also in connection with pattern recognition [1,2,17–19]. Actually, already the mapping learned by a neural network incorporates knowledge about the implications that certain values of the input variables have for the values of the output variables. Usually, however, the ANN-based knowledge extraction aims at the more comprehensible representation of those implications as *rules* [4,10,18,20].

A comprehensive survey of ANN-based rule extraction methods has been given in [1] (see also [19]), where a classification scheme for those methods has been proposed as well. In the context of that classification scheme, two remarks

Table 1. Antecedents of found formulae $(\bigwedge_{i \in \mathcal{I}} \mathrm{E}_i \sim^F_{10\%} \mathrm{RD}) \wedge (\bigwedge_{i \in \mathcal{I}} \mathrm{E}_i \rightarrow^{\frac{2}{3}}_{10\%} \mathrm{RD})$

attained significance:	$\sim^F_{10\%}$	$\rightarrow^{\frac{2}{3}}_{10\%}$
90%il of grain diameter < 3mm ∧ flow velocity > 0.2m/s	0.008%	3.6%
mean grain diameter < 1.5 mm ∧ flow velocity > 0.2m/s	0.008%	3.6%
glowable proportion < 5% ∧		
90%il of grain diameter < 3mm ∧ flow velocity > 0.2m/s	0.33%	5.5%
glowable proportion < 5% ∧		
mean grain diameter < 1.5mm ∧ flow velocity > 0.2m/s	0.33%	5.5%

can be made to the possibility to capture the extraction of rules using a generalized quantifier of an observational calculus:

- As far as the primary dimension of the proposed scheme is concerned, the *expressive power* of the rules, the fact that the truth function of a generalized quantifier is $\{0, 1\}$-valued implies that only *boolean rules* can be covered. In connection with the other major class of rules, fuzzy rules, ongoing research into fuzzy generalized quantifier should be mentioned [6, 9, 11, 12].
- With respect to the second dimension, the *translucency*, the extracted rules should concern only network inputs and outputs. The reason is that hidden neurons are not explicitly assigned to properties of real objects, thus they do not correspond to open formulae of an observational calculus.

Even with the choices with respect to those two dimensions fixed, there are innumerable possibilities how to design a rule extraction method. Here, a particular method described in [14] is considered. Due to the limited extent of this contribution, the method cannot be recalled in detail, instead only its basic principles are listed:

1. As the neural network architecture, a *multilayer perceptron (MLP) with one hidden layer*, m input neurons and n output neurons is used.

2. To each hidden and output neuron, a *piecewise-linear sigmoidal activation function* is assigned.

3. Any hyperrectangle in $\Re^n$ (in particular, any set of output value tuples for which a given conjunction of unary predicates holds) is the map, via F, of some finite set of polyhedra in $\Re^n$.

4. A replaceability relation $\approx^{\mu_P}_{\varepsilon}$ is defined between the set $\mathcal{P}_m$ of all polyhedra in $\Re^n$ and the set $\mathcal{H}_m$ of all hyperrectangles in $\Re^n$, being determined by as system of monotone measures $\mu = (\mu_P)_{P \in \mathcal{P}_m}$, and by a tolerance $\varepsilon > 0$. Practically important monotone measures are the empirical distribution of the data $\hat{\mu}$, and its weighted version $\hat{\mu}_P$, which is for $P \in \mathcal{P}_m$ with $\hat{\mu}(P) \neq 0$ defined $(\forall S \subset \Re^m)\hat{\mu}_P(S) = \frac{\hat{\mu}(S)}{\hat{\mu}(P)}$. For $P \in \mathcal{P}_m$ with $\hat{\mu}(P) = 0$, the measure $\hat{\mu}_P$ can be defined arbitrarily, depending on whether we want to admit replaceability of such polyhedra.

This method allows to state the fact that particular values of some input variables imply particular values of some output variables, using the formula

$$\bigwedge_{k \in \mathcal{I}} \pi_k \rightarrow^{\mu}_{\varepsilon} \bigwedge_{k \in \mathcal{O}} \pi_k , \qquad (2)$$

(more pedantically, $(\rightarrow^{\mu}_{\varepsilon} x)(\bigwedge_{k \in \mathcal{I}} \pi_k, \bigwedge_{k \in \mathcal{O}} \pi_k)$), where $\mathcal{I} \subset \{1, \ldots, m\}$, $\mathcal{O} \subset \{1, \ldots, n\}$, for each $k \in \mathcal{I}$, or $k \in \mathcal{O}$, π_k is a unary predicate stating some

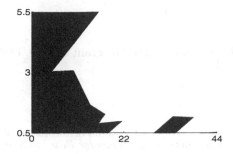
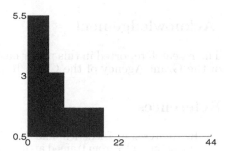

Fig. 1. Example of found polyhedra   **Fig. 2.** Example of found hyperrectangles

interval I_k of values of the k-th input variable, or the k-th output variable, respectively, and $\rightarrow_\varepsilon^\mu$ is a binary quantifier corresponding tro the replaceability relation $\approx_\varepsilon^{\mu_P}$. The reader is defined to [14] for the definition of its truth function.

For comparison with Table 1, Figure 1 depicts a 2-dimensional projection of the union of polyhedra that a particular coordinate F_k of the mapping F learned by a MLP maps to a particular interval I_k. In Figure 1, the considered F_k counts the number of individuals of "Robackia demeierei" per sample, and $I_k = \langle \frac{1}{10} n_{\max}, +\infty)$. The depicted union of polyhedra is projected to the coordinates corresponding to the ecological factors "glowable proportion" and "90 %il of grain diameter". Figure 2 shows a projection, to the same coordinates, of those hyperrectangles that replace any polyhedron P from the union depicted in Figure 1 according to the replaceability relation $\approx_{10\%}^{\hat\mu_P}$, where $\hat\mu_P$ is the weighted empirical distribution. The projection in Figure 2 reflects the following rules that can be extracted from the data using (2) with $\mu = (\hat\mu_P)_{P \in \mathcal{P}_m}$:

glowable proportion < 5.39 % $\rightarrow_{10\%}^\mu$ RD,
glowable proportion < 18.69 % $\wedge$ mean grain diameter < 1.17 mm
                    $\wedge$ 90 %il of grain diameter < 1.7 mm $\rightarrow_{10\%}^\mu$ RD,
glowable proportion < 9.04 % $\wedge$ mean grain diameter < 1.62 mm
                    $\wedge$ 90 %il of grain diameter < 3.06 mm $\rightarrow_{10\%}^\mu$ RD.

Needless to say, it is now very easy to compare such extracted rules with result obtained using statistical hypotheses testing. This impressively illustrates the usefulness of the employed integrative framework.

5   Conclusion

This paper tried to show that observational logic can be used as a framework for integrating traditional statistical methods for knowledge discovery in data with methods based on artificial neural networks. This framework provides a common theoretical view of both kinds of methods while preserving their specific advantages. Due to the different underlying paradigms and different initial assumptions, artificial neural networks do not necessarily yield the same results as statistical methods. Hence, a coincidence between relationships discovered in the data by both kinds of methods increases the chance that those relationships pertain to the reality behind the data, to the phenomena that generated them. On the other hand, if a relationship found by means of some quantifier can not be confirmed using other quantifiers, including a quantifier based on the other paradigm, then such as relationship is a natural starting point for further, deeper investigations.

Acknowledgement

The research reported in this paper has been supported by the grant 201/00/1489 of the Grant Agency of the Czech Republic.

References

1. R. Andrews, J. Diederich, and A.B. Tickle. Survey and critique of techniques for extracting rules from trained artificical neural networks. *Knowledge Based Systems*, 8:378–389, 1995.
2. C.M. Bishop. *Neural Networks for Pattern Recognition*. Clarendon-Press, Oxford, 1995.
3. K. Cios, W. Pedrycz, and R. Swiniwarski. *Data Mining Methods for Knowledge Discovery*. Kluwer Academic Publishers, Dordrecht, 1998.
4. W. Duch, R. Adamczak, and K. Grabczewski. Extraction of logical rules from neural networks. *Neural Processing Letters*, 7:211–219, 1998.
5. Y. Frayman and L. Wang. Data mining using dynamically constructed recurrent fuzzy neural networks. In *Research and Development in Knowledge Discovery and Data Mining*, pages 122–131. Springer-Verlag, Berlin, 1998.
6. P. Hájek. *Metamathematics of Fuzzy Logic*. Kluwer Academic Publishers, Dordrecht, 1998.
7. P. Hájek and T. Havránek. *Mechanizing Hypothesis Formation*. Springer-Verlag, Berlin, 1978.
8. P. Hájek and M. Holeňa. Formal logics of discovery and hypothesis formation by machine. In *Discovery Science*, pages 291–302. Springer-Verlag, Tokyo, 1998.
9. P. Hájek and L. Kohout. Fuzzy implications and generalized quantifiers. *International Journal of Uncertainty, Fuzziness and Knowledge-Based Systems*, 4:225–233, 1996.
10. M.J. Healy and T.P. Caudell. Acquiring rule sets as a product of learning in a logical neural architecture. *IEEE Transactions on Neural Networks*, 8:461–474, 1997.
11. M. Holeňa. Exploratory data processing using a fuzzy generalization of the Guha approach. In *Fuzzy Logic*, pages 213–229. John Wiley and Sons, New York, 1996.
12. M. Holeňa. Fuzzy hypotheses for Guha implications. *Fuzzy Sets and Systems*, 98:101–125, 1998.
13. M. Holeňa. Traditional and modern artificial intelligence explores ecological data. In *STeP 2000: Millenium of Artificial Intelligence*, 2000.
14. M. Holeňa. Statistical, logic-based, and neural networks based methods for mining rules from data. In *Multisensor and Sensor Data Fusion*. NATO Science Series Publishers, to appear.
15. H. Lu, R. Setiono, and H. Liu. Effective data mining using neural networks. *IEEE Transactions on Knowledge and Data Engineering*, 8:957–961, 1996.
16. J. Rauch. Logical calculi for knowledge discovery in databases. In *Principles of Data Mining and Knowledge Discovery*, pages 47–57. Springer-Verlag, Berlin, 1997.
17. B.D. Ripley. *Pattern Recognition and Neural Networks*. Cambridge University Press, 1996.
18. R. Setiono. Extracting rules from neural networks by pruning and hidden unit splitting. *Neural Computation*, 9:205–225, 1997.
19. A.B. Tickle, R. Andrews, M. Golea, and J. Diederich. The truth is there: Directions and challenges in extracting rules from trained artificial neural networks. *IEEE Transactions on Neural Networks*, 9:1058–1068, 1998.
20. G.G. Towell and J.W. Shavlik. Extracting refined rules from knowledge-based neural networks. *Machine Learning*, 13:71–101, 1993.
21. J. Zvárová, J. Preiss, and A. Sochorová. Analysis of data about epileptic patients using Guha method. *International Journal of Medical Informatics*, 45:59–64, 1997.

Supporting Discovery in Medicine by Association Rule Mining of Bibliographic Databases

Dimitar Hristovski[1], Sašo Džeroski[2], Borut Peterlin[3], and Anamarija Rožić-Hristovski[4]

[1]IBMI, Medical Faculty; Vrazov trg 2/2, 1105 Ljubljana, Slovenia
dimitar.hristovski@mf.uni-lj.si
[2]Institute Jozef Stefan; Jamova 39, 1000 Ljubljana, Slovenia
saso.dzeroski@ijs.si
[3]Department of Human Genetics, Clinical Center Ljubljana; Zaloska, 1000 Ljubljana, Slovenia
borut.peterlin@guest.arnes.si
[4]Central Medical Library, Medical Faculty; Vrazov trg 2/2, 1105 Ljubljana, Slovenia
anamarija@ibmi.mf.uni-lj.si

Abstract. The paper presents an interactive discovery support system for the field of medicine. The intended users of the system are medical researchers. The goal of the system is: for a given starting concept of interest, discover new, potentially meaningful relations with other concepts that have not been published in the medical literature before. We performed two types of preliminary evaluation of the system: 1) by a medical doctor and 2) by automatic means. The preliminary evaluation showed that our approach for supporting discovery in medicine is promising, but also that some further work is needed, especially on limiting the number of potential discoveries the system generates.

Introduction

The two main questions addressed by this paper are: 1. Is it possible to discover new, potentially meaningful relations between medical concepts by searching and analysing the documents from a bibliographic database such as Medline? and 2. To what degree can be the discovery process automated? As an attempt to deal with these issues we developed an interactive discovery support system based on association rule mining of the Medline bibliographic database. Its intended use is as a generator for research ideas that should be then investigated by traditional medical methods.

The idea of discovering new relations from a bibliographic database was introduced by Swanson [1,2] who managed to make seven medical discoveries that have been published in relevant medical journals [1,3].

Our system is based on Swanson's ideas, but there are however, several notable differences between our approach and theirs. Instead of using title words as a representation of the Medline documents' meaning, we use the MeSH descriptors. The MeSH descriptors are assigned to documents by human experts during the document indexing stage. We believe that the MeSH descriptors represent more precisely what a particular document is about. We use association rules as a measure of relationship between medical concepts while Swanson uses word frequencies. We have built a

D.A. Zighed, J. Komorowski, and J. Żytkow (Eds.): PKDD 2000, LNAI 1910, pp. 446-451, 2000.
© Springer-Verlag Berlin Heidelberg 2000

large association rule base by pre-calculating and storing the association rules in a database management system. This allows us to build a truly interactive discovery support system with fast response.

Background

Swanson's work. Swanson's discovery support process is based on the concepts of complementary literatures and noninteractive literatures [1,2]. If one set of articles (XY) reports an interesting relation between concepts X and Y, and a different set of articles (YZ) reports a relation between Y and Z, but nothing has been published concerning a possible link between X and Z, then XY and YZ are called complementary literatures. Generally, XY and YZ are complementary if a potentially new relation can be inferred by considering them together that can not be inferred from either of them separately. For example, X might be a disease, Y a physiological function associated with X and Z a substance or drug which induces or regulates the physiological function Y. If the readers and authors of one literature are not acquainted with the other, and vice versa, as might often be the case with two different specialties, then the two literatures are noninteractive. By combining the concepts of complementary and noninteractive literatures, Swanson developed the concept of undiscovered public knowledge meaning that although the literatures XY and YZ represent publicly available knowledge, the potentially new relation between X and Z remains undiscovered and is a valuable source of new discoveries.

In the beginning, Swanson performed the discovery process manually by searching the Medline database [2]. Later he added software support for some of the stages of the process. His current system is called ARROWSMITH and is described in detail in [3].

Medline. The Medline database is a product of the US National Library of Medicine (NML). Because of its coverage and free accessibility, Medline is the most important bibliographic database in the field of biomedicine. Each citation is associated with a set of MeSH (Medical Subject Headings) terms that describe the content of the item [4].

Medical Subject Headings (MeSH). MeSH comprises NLM's controlled vocabulary and thesaurus used for indexing articles and for searching MeSH-indexed databases, including Medline [4].

Unified Medical Language System (UMLS). The Unified Medical Language System (UMLS) project that NLM began in 1986 was undertaken with the goal of providing a mechanism for linking diverse medical vocabularies as well as sources of information. UMLS consists of three components: the Metathesaurus, Specialist Lexicon and Semantic Network [4,5].

System Description

Goal and Basic Premises. The system we developed is an interactive discovery support system for the field of medicine and is supposed to be used as a generator of new, potentially meaningful relations between a starting known concept of interest and other concepts.

The Medline database is used heavily by medical researchers. Traditionally it is used to check what is new in the literature on a particular topic of interest or to check if a medical discovery has already been published. In contrast to the traditional use of Medline, our system actively helps in the discovery process by generating potentially new discoveries and research ideas by analyzing the Medline database.

We used the major MeSH descriptors assigned to a Medline record as a representation of the contents of the article the record is about. Some of the MeSH descriptors are designated as major (preceded by an asterisk in the Medline record). Major descriptors are those that are the main topic of the article.

We used association rules [6] between pairs of medical concepts as a method to determine which concepts are related to a given starting concept. In our system an association rule of the form

$X \rightarrow Y$ *(confidence, support)*

means that in *confidence* percent of articles containing X, Y is present and that there are *support* number such articles. In other words, we take concept co-occurrence as an indication of a relation between concepts. If X is a disease, for example, then some possible relations might be: *has-symptom*, *is-caused-by*, *is-treated-with-drug* and so on. We do not try to find out the kind of relation. This can not be done by using the MeSH descriptors assigned to an article because there is no information about the relation between the descriptors.

Algorithm. We calculated all the associations between the major MeSH descriptors. We did this regardless of the confidence and support values and for two Medline time segments: 1990-1995 and 1996-1999. The calculated associations are stored in a database management system: there are currently more than 11.000.000 associations in the rule base. The calculation of the association rules was much simplified by the use of the data contained in the UMLS.

The large association rule base is a foundation upon which the algorithm for discovering new relations between concepts proceeds as described in Table 1. The user of the system should then evaluate the proposed (X, Z) pairs and select among them those that deserve further investigation.

Table 1. The algorithm for discovering new relations between medical concepts.

1. Let X be a given starting concept of interest.
2. Find all concepts Y such that there is an association rule X -> Y.
3. Find all concepts Z such that there is an association rule Y -> Z.
4. Eliminate those Z for which an association X -> Z already exists.
5. The remaining Z concepts are candidates for a new relation between X and Z.

Because in Medline each concept can be associated with many other concepts, the possible number of X -> Z combinations can be extremely large. In order to deal with this combinatorial problem, the algorithm incorporates *filtering (limiting)* and *ordering* capabilities. The default filtering that can not be relaxed is that only the associations between major MeSH headings are considered by the system. The related concepts can be limited by the semantic type to which they belong. Each MeSH descriptor belongs to one or more semantic types. For example, if the starting concept X is a disease (semantic type *disease or syndrome*) then the user can request that Y concepts are of semantic type *pathologic function* and that Z concepts are of semantic type *pharmacologic substance*. The last possibility for limiting the number of related concepts is by setting thresholds on the support and confidence measures of the association rules in steps 2. and 3. of the algorithm. In fact, all of the filtering options can be interactively set alone or several of them in combination.

The goal of the ordering is to present best candidates first to make human review as easy as possible. Currently the default ordering is by the decreasing association rule confidence, but it is also possible to order by support or semantic type.

Implementation. Figure 1. shows the user interface of our discovery support system. The user starts a discovery session by searching for a starting concept X, which is usually from his own research area. The concepts Y related to the starting concept are found by pressing the *Find Related* button and are presented in the *Related Concepts1* frame. Before finding related concepts the user can specify limits and the order of the related concepts. Similarly the Z concepts related to the Y concepts are found and shown in the *RELATED CONCEPTS2* frame. The frame *RELATED CONCEPTS2* contains an important additional field designated as *"Discovery?"*. This value of this field is YES if a relation (association) between the starting concept X and the current concept Z does not exist in the appropriate Medline segment and *NO* if such a relation exists.

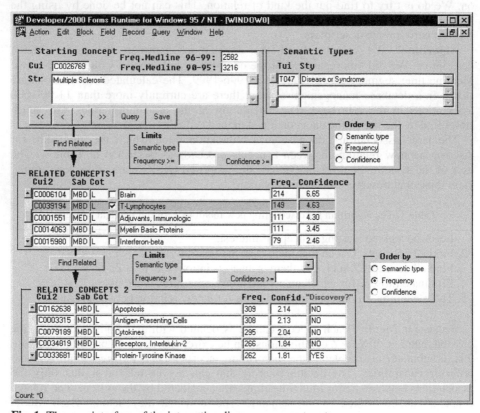

Fig. 1. The user interface of the interactive discovery support system.

Evaluation

The ultimate proof of the system is to (help) discover medical discoveries that can be published in relevant medical journals. However, we have only managed to do a preliminary evaluation so far.

Medical Doctor. The main goal of the evaluation done by the medical doctor was to check if the Y concepts, which the system finds to be related to the starting concept X, have medical sense. He selected as a starting concept *multiple sclerosis (MS)* which could be defined as a demyelinating disease of the central nervous system of putative autoimmune origin.

The first 5 concepts related (associated) with MS ordered by decreasing support of the association are: 1. *MRI magnetic resonance imaging* (diagnostics); 2. *Brain* (anatomical structure – organ involved); 3. *Interferon* (treatment).; 4. *T-lymphocytes* (pathogenesis); 5. *Myelin basic protein (MBP)*(pathogenesis).

Among the first 20 concepts there are 6 related to pathogenesis, 4 to treatment, 3 to diagnostic methods, 3 to symptoms, 2 to target organs-anatomical structures and 2 are related to general disease categories.

It could be estimated that the concepts found as related by the system are associated with the current main focus of medical endeavors in the field of MS which is still oriented to treatment and therefore to better understanding of pathogenesis.

Automatic. The goal of the automatic evaluation was to see how many of the potential discoveries made by the system at some point of time become realised at a later time. For us, a potential discovery is a relationship between two concepts proposed by our system, but not present in Medline at some point of time. We consider the potential discovery realised if the two concepts later appear together in a document in the Medline database. In other words, the goal of the evaluation was to see how good our system was in predicting what discoveries would be made in the future.

We approached this goal by first dividing the Medline database and the corresponding association rules into two segments according to the publication date of the documents stored: the older segment is from 1990 to 1995 and the newer segment is from 1996 to 1999. We then analysed *Multiple sclerosis (MS)*, the same starting concept as in the medical doctor's evaluation.

MS appears in 2582 documents in the older segment. It is related to 1610 distinct concepts. When analysing the old segment, the system proposed 15617 concepts as potential discoveries. MS is related to 662 new concepts in the new segment that it was not related to in the old segment. Our system successfully predicted 95.5% (632 out of 662) realised discoveries in the new segment. However, only 4% (632 out of 15617) of the proposed potential discoveries got realised. It should be stressed that MS was not related to 15965 out of 17575 distinct concepts appearing in the older segment. The system proposed 97.8% of the concepts MS was not yet related to as potential discoveries. The conclusion is that without using limits on the strength of relationship the system is very successful at predicting future discoveries, but proposes far too many potential discoveries. Then we repeated the evaluation with two values for thresholds on the support level of the association rules. In one case the threshold was set to the average support of the associations between one concept and the others (AVGS) and in the other case it was set to 2*AVGS. Only associations with support greater or equal to the threshold were taken into account. The number of proposed potential discoveries dropped from 15617 without thresholds to 6848 for AVGS and to 3151 for the 2*AVGS threshold. The percent of successfully predicted realised discoveries dropped from 95.5% (632 of 662) without thresholds to 78.7% (521 of 662) for AVGS and to 55.2% (366 of 662) for 2*AVGS. However, the ratio of realised to proposed potential discoveries improved from 4% (632 out of 15617) without thresholds to 7.6% (521 of 6848) for AVGS and to 11.6% (366 of 3151) for 2*AVGS. We conclude from this that with the use of proper thresholds the usability of the system is much better because a smaller number of better potential discoveries are generated.

Discussion and Further Work

The paper presented an interactive discovery support system for the field of medicine. For a given starting medical concept it discovers new, potentially meaningful relations with other concepts that have not been published in the medical literature before. The proposed relations should be evaluated and verified by a qualified medical professional.

As a measure of the relation between concepts we use association rules calculated from the Medline bibliographic database. As part of the preliminary evaluation of the system, a medical doctor confirmed that most of the relations between concepts found by association rules are meaningful. In the other part of the evaluation, which was done by automatic means, the system proved to be successful at predicting future discoveries. However, this came at the expense of generating a large number of potential discoveries that have to be judged and verified by the user of the system.

We have several ideas for improving the system. One of them is to develop a Web version of the system that will increase the number of users. More users means better chance for real medical discoveries. The preliminary evaluation showed that properly set thresholds are crucial for successful use of the system. Thus, we plan to work on setting good default values for the thresholds that can be changed by the user if necessary. Another important way to improve the system is to include additional information sources, such as molecular biology sequence databases.

References

1. Swanson, D.R.: Fish oil, Raynaud's syndrome, and undiscovered public knowledge. Perspect Biol Med. 1986 Autumn;30(1):7-18.
2. Swanson, D.R.: Online search for logically-related noninteractive medical literatures: a systematic trial-and-error strategy. J Am Soc Inf Sci. 1989 Sep;40(5):356-8.
3. Swanson, D.R., Smalheiser, N.R.: An interactive system for finding complementary literatures: a stimulus to scientific discovery. Artif. Intell. 91 (1997) 183-203.
4. U.S. National Library of Medicine. http://www.nlm.nih.gov/<30.04.2000>
5. Humphreys, B.L., Lindberg, D.A.B., Schoolman, H.M., Barnett, G.O.: The Unified Medical Language System: an informatics research collaboration. JAMIA 1998;5(1):1-11.
6. Agrawal, R. et al: Fast discovery of association rules. In U. Fayyad et al, editors, Advances in Knowledge Discovery and Data Mining. MIT Press, Cambridge, MA. (1996)

Collective Principal Component Analysis from Distributed, Heterogeneous Data

Hillol Kargupta, Weiyun Huang, Krishnamoorthy Sivakumar, Byung-Hoon Park, and Shuren Wang

School of Electrical Engineering and Computer Science
Washington State University
Pullman, WA 99164-2752, USA
{hillol, whuang1, siva, bhpark, swang}@eecs.wsu.edu

Abstract. Principal component analysis (PCA) is a statistical technique to identify the dependency structure of multivariate stochastic observations. PCA is frequently used in data mining applications. This paper considers PCA in the context of the emerging network-based computing environments. It offers a technique to perform PCA from distributed and heterogeneous data sets with relatively small communication overhead. The technique is evaluated against different data sets, including a data set for a web mining application. This approach is likely to facilitate the development of distributed clustering, associative link analysis, and other heterogeneous data mining applications that frequently use PCA.

1 Introduction

Representation plays an important role in data analysis and modeling. Principal component analysis (PCA) offers a technique to construct an adaptive representation of the data that exploits its underlying distribution. PCA detects the dependencies among the features defining the data and constructs a number of basis vectors (often fewer than the number of features) that can be used to represent the data in a compact and efficient manner. PCA is frequently used in data mining applications [1,2].

PCA involves computation of the dominant eigenvectors of the covariance matrix generated by the data. There exist several well-known efficient techniques for performing PCA [3,4]. Most of the established techniques for PCA require the data set to be centrally located. However, the emergence of network-based computation has offered a new challenge to the traditional practice of PCA. This has introduced a new important dimension to the PCA problem — distributed sources of data. Performing PCA from distributed data tables is of growing importance and it is still an open question. To the best of our knowledge, there does not exist any technique that can perform distributed PCA from heterogeneous data sites (i.e. sites with different tables storing data for different features) with minimal data communication.

This paper offers one solution to this problem—the Collective PCA. It contributes a new technique to the growing body of the Distributed Knowledge Discovery (DKD) [5,6,7] literature.

D.A. Zighed, J. Komorowski, and J. Żytkow (Eds.): PKDD 2000, LNAI 1910, pp. 452–457, 2000.
© Springer-Verlag Berlin Heidelberg 2000

Section 2 presents the Collective PCA (CPCA) algorithm. Section 3 presents the experimental results. Section 4 concludes this paper.

2   Collective PCA

Let us first consider the CPCA algorithm from an abstract algorithmic perspective. Let X be the entire relational data set and it is stored in different tables at different sites. In this paper we shall consider heterogeneous data sites [7] where different sites observe different sets of features. In particular, each site has data regarding a particular subset of the n features. For the sake of simplicity we shall assume that each site has only one table. Let us assume that there are s sites and X can be partitioned as $X = [X_1, X_2, \ldots, X_s]$, where X_i is a $m \times n_i$ submatrix of X that is available at site i. A PCA of this distributed data in a centralized fashion would involve moving data to one central site and calculating the eigenvalues/eigenvectors of the covariance matrix $X'X$ of the global data matrix X. The amount of data to be moved is $O(mn)$, where m is the number of data samples (rows of the global data table X) and n is the total number of features (columns of the global data table X). For typically DDM-applications this amount of data communication is either prohibitive because of the limited bandwidth or impractical because of logistics and/or security related reasons.

In the CPCA approach, we perform the computations for PCA (to the extent possible) locally, thereby minimizing the amount of data communication and the computation at the central site. First we perform a local PCA on the data partition X_i at site i. Let A_i be the $n_i \times k_i$ matrix whose columns are the k_i eigenvectors corresponding to the k_i largest eigenvalues of $X_i'X_i$. Let

$$Y_i = X_i A_i \tag{1}$$

be the principal components computed at site i. The choice of k_i will depend on the eigenvalues of the local covariance $X_i'X_i$ and would be dictated by our error tolerance. It is a tradeoff between dimensionality reduction and accuracy.

These principal components can be used to obtain a reasonable approximation to the original data as follows:

$$\hat{X}_i = Y_i A_i' \tag{2}$$

The relative mean-squared error (RMSE) between X_i and $\hat{X}_i$ can be expressed in terms of the eigenvalues of the covariance matrix (Σ_x) as follows:

$$\frac{E[(X_i - \hat{X}_i)(X_i - \hat{X}_i)']}{E[X_i X_i']} = \frac{\sum_{j=k+1}^n \lambda_j}{\sum_{j=1}^n \lambda_j}. \tag{3}$$

If the eigenvalues of Σ_x are "spread-out" in the sense that $\lambda_{\max}/\lambda_{\min}$ is large, the data X maybe represented (with a small RMSE) by a small number of principal components.

In all our experiments, we chose k_i to yield a RMSE of 0.1 (see eq. (3)). Since the number of rows in large tables (m) is very large, we select a subset of

c samples (rows) (where $c \ll m$), with uniform probability. With some abuse of notation, we will denote by Y_i the $c \times k_i$ matrix consisting of the principal components and only c selected samples. The individual Y_i and A_i from each site are then transmitted to a central site. At the central site, a new $c \times k$ data matrix Y is formed by putting together the data from the individual sites $Y = [Y_1, Y_2, \ldots, Y_s]$, where $k = \sum_{i=1}^{s} k_i$. The data communication involved here is $O(ck)$ for the Y_i's and $O(\sum_i n_i k_i)$ for the A_i's (compare with $O(mn)$ for the centralized PCA case, where $c \ll m$ and $k = \sum_{i=1}^{s} k_i \ll n$). Typically, the number of samples c that are selected is much larger than the number of features n_i at site i. Therefore, the overhead involved in transmitting A_i would be negligible compared to the overall data transmission involved.

At the central site, in principle, we need to reconstruct the original data X from the Y_i's and A_i's (see eq. (2)) and perform a PCA based on this reconstructed data. It follows from eq. (1) that Y is a linear transformation of X with $Y = XA$, where A is a block diagonal matrix with matrices $A_1, \ldots, A_s$ on the diagonal. Since the PCA is invariant to linear transformations [8], we can work with the Y data instead of X. The primary advantage is that we can exploit the dimensionality reduction already achieved at each of the local sites. Let v_i, $i = 1, 2, \ldots, p$ be the eigenvectors, corresponding to the p largest eigenvalues of the covariance matrix of Y. Then $w_i = Av_i$ are the required eigenvectors of the covariance $X'X$ of the original data X. The following section presents the experimental results.

3   Experiments with the CPCA

This section presents results of applying the CPCA to two experimental test suites. The first experimental suite tests the scalability of CPCA to large data sets. The second suite considers the application of CPCA in a real-life application scenario.

3.1   Experiment Suite: I

This section tests the overall performance including the scalability of the CPCA. We used the Quest Synthetic Data Generation Code [1] to generate a data set with 60,000 observations and 200 features. Although this code generates labeled data, we did not use the class-label for the unsupervised CPCA approach.

We partitioned the data set into two subsets, each containing 100 features. We select the set of PCs that gives a RMSE of 0.1. Figure 1(Left) shows two curves, each representing the variation of the angle (in radian) between a CPCA-generated dominant principal component and the corresponding principal direction generated from the centralized data. Note that the angle between two vectors represents the distance between them and therefore it is a good measure of the accuracy of the estimated PCs.

---

[1] $http : //www.almaden.ibm.com/cs/quest/syndata.html$

Table 1. Size of data set, number of sites, and number of selected local principal components.

Data	No. of rows	No. of features	No. of sites	Total number of selected local PCs
Quest	60000	200	2	161
Quest	60000	200	10	164

Figure 1(Right) shows similar performance when the data set is distributed among ten sites. Table 1 shows the size of the global data set and the total number of selected local principal components from all the sites.

We see from the figures that the CPCA achieves good compression. Suppose our global data set is m by n, and is distributed among s sites, such that each site contains n_i features. Number of selected local principal component is k_i, and the total number of selected local principal components is $k = \sum_{i=1}^{s} k_i$. Usually k is quite small compared to n. Let us denote the number of sampled rows by c; for our experiments we chose a value of c such that $c < 0.2 * m$. The total communication cost is $O(c * k + \sum_{i=1}^{s} n_i * k_i)$, which is very small compared to $O(m * n)$, the cost to move the whole data set to one site.

3.2   Experiment Suite: II

In this section we illustrate the technique in the context of a practical application scenario. PCA is frequently used for high-dimensional text-analysis applications. Therefore text analysis should be an ideal candidate for applying the CPCA. However, in a distributed environment, applications may become more interesting and challenging when relevant data sets involve text, numeric, and other non-numeric features. In the following we describe one such case.

Consider the financial news stories regularly posted on the Internet. These news stories are often very useful for investors, portfolio managers, and others.

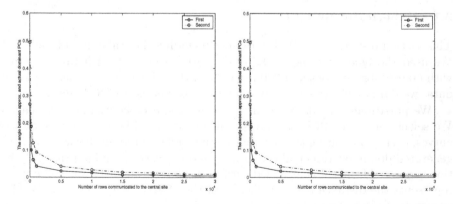

Fig. 1. (Left) Performance of CPCA for the Quest Synthetic data with two data sites. (Right) Performance of CPCA for the Quest Synthetic data with ten data sites.

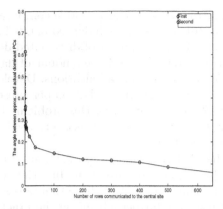

Data site	No. of feature	No. of selected local PCs
Profile Information	73	2
Sector data	15	4
News	793	110

Fig. 2. (Left)Performance of CPCA for the Company Profile data with 3 data sites. (Right) Dimension compression in case of the Company Profile data.

Typically many of these stories are associated with some companies. Announcements regarding new products, quarterly revenue, legal battles, mergers, and partnerships of companies often dominate such business news. On the other hand, there exist many sites in the Internet that offer valuable information about the background and current financial profile about almost all major companies. For example Yahoo finance, CNN finance are only a few among many excellent sources for such information. The web sites of the company itself also provides quite useful information such as new developments, products, and others.

The experiment performed in this section considers a real-life applications using three data tables, collected using the information available over the Internet. One table corresponds to news corpora. Each document in this corpora corresponds to a key. For our application we used a unique code associated with every company name as the key. A second table is constructed that stores the financial and background feature values of a company, again indexed by its company symbol. A third data table is constructed using the data about the sector the company belongs to. This table is also indexed by the company name. We considered a data set involving 1,027 companies and the three tables are located at three different sites. The CPCA technique is applied on this three data sets. The RMSE threshold is set to 0.1. Table 2(Right) shows the number of features at each site and corresponding number of chosen PCs. As we see, the CPCA technique offers a big compression factor. Figure 2(Left) shows two curves, each representing the variation of the angle between a CPCA-generated dominant principal component and the corresponding principal direction generated by the centralized approach. The following section concludes this paper.

4   Conclusions

Distributed data analysis is playing an increasingly important role in knowledge discovery from multiple databases distributed over a network. The growing popularity of environments with "thin" computing devices (e.g. wearable computers, palmtops, and laptops) connected to powerful servers through limited bandwidth (possibly wireless) networks is aiding the rapid development of distributed data mining techniques that do not require communication of large amount of data. PCA is frequently used for high dimensional data mining applications. Development of distributed PCA algorithm is therefore important. This paper presented the collective PCA technique which offers one solution to this problem. The experimental results demonstrated that the CPCA can be effectively used for analyzing high dimensional data of large quantity.

We are in the process of applying the CPCA technique for distributed clustering of heterogeneous data. We are also working towards bounding the error introduced by the approximations used by the CPCA. We believe that this simple but effective technique will be quite useful in further advancing the practice of the DKD technology.

Acknowledgement

The authors thank American Cancer Society for supporting part of this research.

References

1. Boley, D.: Principal direction divisive partitioning. Data Mining and Knowledge Discovery **2** (1998) 325–344
2. Faloutsos, C., Korn, F., Labrinidis, A., Kotidis, Y., Kaplunovich, A., Perkovic, D.: Quantifiable data mining using principal component analysis. Technical report (1997) Institute for Systems Research, University of Maryland technical Report TR 97-25.
3. Golub, G.H., Loan, C.F.V.: Matrix Computations. The Johns Hopkins University Press, Baltimore (1989)
4. Anderson, E., Bai, Z., Bischof, C., Blackford, S., Demmel, J., Dongarra, J., Croz, J.D., Greenbaum, A., Hammarling, S., McKenney, A., Sorensen, D.: LAPACK Users' Guide. Third edn. Society for Industrial & Applied Mathematics (1999)
5. Chan, P.K., Stolfo, S.J.: Sharing learned models among remote database partitions by local meta-learning. In Simoudis, E., Han, J., Fayyad, U., eds.: The Second International Conference on Knowledge Discovery and Data Mining, AAAI Press (1996) 2–7
6. Grossman, R., Bailey, S., Kasif, S., Mon, D., Ramu, A., Malhi, B.: The preliminary design of papyrus: A system for high performance, distributed data mining over clusters, meta-clusters and super-clusters. Fourth International Conference of Knowledge Discovery and Data Mining, New York, New York, Pages 37–43 (1998)
7. Kargupta, H., Park, B., Hershbereger, D., Johnson, E.: Collective data mining: A new perspective toward distributed data mining. To be published in the Advances in Distributed and Parallel Knowledge Discovery, Eds: Hillol Kargupta and Philip Chan, AAAI/MIT Press (1999)
8. Jackson, J.E.: A User's Guide to Principal Components. John Wiley (1991)

Hierarchical Document Clustering
Based on Tolerance Rough Set Model

Saori Kawasaki, Ngoc Binh Nguyen, and Tu Bao Ho

Japan Advanced Institute of Science and Technology
Tatsunokuchi, Ishikawa, 923-1292 JAPAN
{skawasa,binh,bao}@jaist.ac.jp

Abstract. Clustering is a powerful tool for knowledge discovery in text collections. The quality of document clustering depends not only on clustering algorithms but also on document representation models. We develop a hierarchical document clustering algorithm based on a tolerance rough set model (TRSM) for representing documents, which offers a way of considering semantics relatedness between documents. The results of validation and evaluation of this method suggest that this clustering algorithm can be well adapted to text mining.

1 Introduction

We introduce a document representation model, namely *tolerance rough set model* (TRSM). Rough set theory, a mathematical tool to deal with vagueness and uncertainty introduced by Pawlak [3], has been successful in many applications, in particular in data mining [4]. TRSM employs a tolerance relation instead of an equivalence relation in the original rough set model (see also [5]). We develop a TRSM hierarchical clustering algorithm for documents that exploits semantics relatedness between documents. The algorithm consists of two phases of making representation of each document using TRSM (section 2) and grouping documents by an agglomerative clustering with their approximations (section 3). We report our experiments on five test collections in section 4.

2 A Tolerance Rough Set Model for Documents

Denote the set of M full text documents by $\mathcal{D}$, and the set of N terms from $\mathcal{D}$ by $\mathcal{T}$. Each full text document $d_j \in \mathcal{D}$ is mapped into a list of terms t_i weighted by their importance in the document, as $d_j = (t_{1j}, w_{1j}; t_{2j}, w_{2j}; \ldots; t_{rj}, w_{rj})$ with $w_{ij} \in [0, 1]$. Then the TRSM is used for enriching the document representation in terms of semantics relatedness by creating tolerance classes of terms in $\mathcal{T}$ and approximations of subsets of documents. Terms are viewed better using overlapping classes, which can be generated by *tolerance relations* R (with reflexive and symmetric properties) in an universe U instead of an equivalence relation (with

D.A. Zighed, J. Komorowski, and J. Żytkow (Eds.): PKDD 2000, LNAI 1910, pp. 458–463, 2000.
© Springer-Verlag Berlin Heidelberg 2000

reflexive, symmetric and transitive properties) used in the original rough set model. For formulating tolerance classes of index terms of documents, we employ the co-occurrence of index terms in all documents from $\mathcal{D}$. Denote by $f_{d_j}(t_i)$ the number of occurrences of term t_i in d_j, and by $f_{\mathcal{D}}(t_i)$ the number of documents in $\mathcal{D}$ that term t_i occurs in, and by $f_{\mathcal{D}}(t_i, t_j)$ the number of documents in $\mathcal{D}$ in which two index terms t_i and t_j co-occur. We define an uncertainty function I depending on a threshold θ as

$$I_\theta(t_i) = \{t_j \mid f_{\mathcal{D}}(t_i, t_j) \geq \theta\} \cup \{t_i\} \tag{1}$$

It is clear that the function I_θ defined above satisfies the condition of $t_i \in I_\theta(t_i)$ and $t_j \in I_\theta(t_i)$ iff $t_i \in I_\theta(t_j)$ for any $t_i, t_j \in \mathcal{T}$, and so I_θ is both reflexive and symmetric. This function corresponds to a tolerance relation $\mathcal{I} \subseteq \mathcal{T} \times \mathcal{T}$ that $t_i \mathcal{I} t_j$ iff $t_j \in I_\theta(t_i)$, and $I_\theta(t_i)$ is the tolerance class of index term t_i. A vague inclusion function ν, which determines how much X is included in Y, is defined as $\nu(X, Y) = |X \cap Y|/|X|$. Using this the membership function μ for $t_i \in \mathcal{T}, X \subseteq \mathcal{T}$ can be defined as $\mu(t_i, X) = \nu(I_\theta(t_i), X) = |I_\theta(t_i) \cap X|/|I_\theta(t_i)|$. With these definitions we can define a tolerance space as $\mathcal{R} = (\mathcal{T}, I, \nu, P)$ in which the *lower approximation* $\mathcal{L}(\mathcal{R}, X)$ and the *upper approximation* $\mathcal{U}(\mathcal{R}, X)$ in $\mathcal{R}$ of any subset $X \subseteq \mathcal{T}$ can be defined as

$$\mathcal{L}(\mathcal{R}, X) = \{t_i \in \mathcal{T} \mid \nu(I_\theta(t_i), X) = 1\} \tag{2}$$

$$\mathcal{U}(\mathcal{R}, X) = \{t_i \in \mathcal{T} \mid \nu(I_\theta(t_i), X) > 0\} \tag{3}$$

Index terms in each document are weighted as follows.

$$w_{ij} = \begin{cases} (1 + \log(f_{d_j}(t_i))) \times \log \frac{M}{f_{\mathcal{D}}(t_i)} & \text{if } t_i \in d_j, \\ \min_{t_h \in d_j} w_{hj} \times \frac{\log(M/f_{\mathcal{D}}(t_i))}{1+\log(M/f_{\mathcal{D}}(t_i))} & \text{if } t_i \in \mathcal{U}(\mathcal{R}, d_j) \setminus d_j \\ 0 & \text{if } t_i \notin \mathcal{U}(\mathcal{R}, d_j) \end{cases} \tag{4}$$

The vector length normalization is then applied to document representation.

3   TRSM-Based Hierarchical Clustering Algorithm

Figure 1 describes the general TRSM-based hierarchical clustering algorithm that is an extension of the hierarchical agglomerative clustering algorithm. The main point here is at each merging step where upper approximations of documents are used in finding two closest clusters to merge by group-average link. It allows to use cluster representatives to calculate the similarity between clusters instead of averaging similarities of all document pairs included in clusters with average complexity $O(N^2)$. This algorithm constructs a *polythetic* representative R_k for each cluster $C_k, k = 1, \ldots, K$. The following rules form the cluster representatives: (i) Initially, $R_k = \phi$, (ii) For all $d_j \in C_k$ and for all $t_i \in d_j$, if $f_{C_k}(t_i)/|C_k| > \sigma$ then $R_k = R_k \cup \{t_i\}$, (iii) If $d_j \in C_k$ and $d_j \cap R_k = \phi$ then $R_k = R_k \cup \text{argmax}_{t_i \in d_j} w_{ij}$. The weights of terms t_i in R_k are first averaged by of weights of this terms in all documents belonging to C_k, that means

$w_{ik} = (\sum_{d_j \in C_k} w_{ij})/|\{d_j : t_i \in d_j\}|$, then normalized by the length of the representative R_k. Three common coefficients of Dice, Jaccard and Cosine [1] are implemented to calculate the similarity between pairs of documents d_{j_1} and d_{j_2}. Two main advantages of using upper approximations are: (i) To reduce the

1.	Given: a collection of M documents $\mathcal{D} = \{d_1, d_2, \ldots, d_M\}$
2.	a similarity measure $sim : \mathcal{P}(\mathcal{D}) \times \mathcal{P}(\mathcal{D}) \to R$
3.	**for** $j = 1$ **to** M **do**
4.	$C_j = \{d_j\}$ **end**
5.	$H = \{C_1, C_2, \ldots, C_M\}$
6.	$i = M + 1$
7.	**while** $|H| > 1$
8.	$(C_{n_1}, C_{n_2}) = \text{argmax}_{(C_u, C_v) \in H \times H} sim(\mathcal{U}(\mathcal{R}, C_u), \mathcal{U}(\mathcal{R}, C_v))$
9.	$C_i = C_{n_1} \cup C_{n_2}$
10.	$H = (H \setminus \{C_{n_1}, C_{n_2}\}) \cup \{C_i\}$
11.	$i = i + 1$

Fig. 1. TRSM-based hierarchical agglomerative clustering algorithm

number of zero-valued coefficients, and (ii) The upper approximations formed by tolerance classes make it possible to relate documents that may have few (even no) terms in common with the user's topic of interest or the query.

4   Validation and Evaluation

The first columns of Table 1 summarizes test collections used in our experiments. The clustering quality for each test collection depends on parameter θ in TRSM and on σ in clustering algorithm. Note that the higher value of θ the large upper approximation and the smaller lower approximation of a set X. In Table 3 we can see how retrieval effectiveness relates to different values of θ. To avoid biased experiments when comparing algorithms we take default values $\theta = 15$,

Table 1. Results of clustering tendency

Col	Subject	nb doc	nb terms	nb queries	nb rel. doc.	% average of relevant doc.						
						0	1	2	3	4	5	average
JSAI	Art. Int.	802	1813	20	32	19.9	19.8	18.5	18.5	11.8	11.5	2.2
CACM	Comp. Sci.	3200	6520	64	15	50.3	22.5	12.8	7.9	4.2	2.3	1.0
CISI	Library	1460	4414	76	40	45.4	25.8	15.0	7.5	4.3	1.9	1.1
CRAN	Aeronautics	1400	3182	225	8	33.4	32.7	19.2	9.0	4.6	1.0	1.2
MED	Medicine	1033	4841	30	23	10.4	18.7	18.6	21.6	19.6	11.1	2.5

Table 2. Synthesized results about the stability

	Percentage of changed data						
	1%	2%	3%	4%	5%	10%	15%
$\theta = 2$	2.84	5.62	7.20	5.66	5.48	11.26	14.41
$\theta = 3$	3.55	4.64	4.51	6.33	7.93	12.06	15.85
$\theta = 4$	0.97	2.65	2.74	4.22	5.62	8.02	13.78

and when comparing algorithms, default values $\theta = 15$, and $\sigma = 0.1$ for all five test collections.

4.1   Validation of Clustering Tendency

We employ the *nearest neighbor test* by considering, for each relevant document of a query, how many of its n nearest neighbors are also relevant; and by averaging over all relevant documents for all queries in a test collection in order to obtain single indicators. Table 1 reports the experimental results synthesized from those done on five test collections. Columns from 7 to 12 stand for the percentage average of the relevant documents in a collection that had 0, 1, 2, 3, 4, and 5 relevant documents in their sets of 5 nearest neighbors. The last column shows the average number of relevant documents among 5 nearest neighbors of each relevant document. The result on tendency suggests that the TRSM clustering method is appropriate for the retrieval purpose.

4.2   Validation of Clustering Stability

Table 2 shows the experimental results of clustering stability for JSAI test collection with different values of s from 210 experiments with $s\% = 1\%$, 2%, 3%, 4%, 5%, 10% and 15%. For each value 2, 3, and 4 of θ, the experiments are done 10 times each for a reduced database of size $(100 - s)\%$ of $\mathcal{D}$. We randomly removed a specified amount of $s\%$ documents from the collection, then re-determined the new tolerance space for the reduced database to perform the TRSM clustering algorithm and evaluate the change of clusters due to the change of the database. Note that a little change of data implies a possible little change of hierarchy (about at the same percentage as for $\theta = 4$). The experiments for other test collections have nearly the same results. It suggests that the TRSM hierarchical clustering are stable.

4.3   Evaluation of Cluster-Based Retrieval Effectiveness

The experiments evaluate effectiveness of the TRSM cluster-based retrieval by comparing it with full retrieval by using the common measures of *precision* and *recall*. Table 3 shows precision and recall of the TRSM-based full retrieval and the VSM-based full retrieval (Vector Space Model) where the TRSM-based retrieval is done with values 30, 25, 20, 15, 10, 8, 6, 4, and 2 of θ. We see that

Table 3. Precision and recall of full retrieval

θ	JSAI		CACM		CISI		CRAN		MED	
	P	R	P	R	P	R	P	R	P	R
30	0.934	0.560	0.146	0.231	0.147	0.192	0.265	0.306	0.416	0.426
25	0.934	0.560	0.158	0.242	0.151	0.194	0.266	0.310	0.416	0.426
20	0.934	0.560	0.159	0.243	0.150	0.194	0.268	0.311	0.416	0.426
15	**0.934**	**0.560**	**0.160**	**0.241**	**0.155**	**0.204**	**0.257**	**0.301**	**0.415**	**0.421**
10	0.934	0.560	0.141	0.221	0.142	0.178	0.255	0.302	0.414	0.387
8	0.934	0.560	0.151	0.254	0.138	0.172	0.242	0.291	0.393	0.386
6	0.945	0.550	0.141	0.223	0.146	0.178	0.233	0.271	0.376	0.365
4	0.904	0.509	0.137	0.182	0.152	0.145	0.223	0.241	0.356	0.383
2	0.803	0.522	0.111	0.097	0.125	0.057	0.247	0.210	0.360	0.193
VSM	**0.934**	**0.560**	**0.147**	**0.232**	**0.139**	**0.184**	**0.258**	**0.295**	**0.429**	**0.444**

Table 4. Precision and recall of the TRSM cluster-based and full retrieval

Col.	1.2% (0.18)		1.8% (0.16)		2.9% (0.14)		8.0% (0.11)		16.9% (0.09)		full search	
	P	R	P	R	P	R	P	R	P	R	P	R
JSAI	0.950	0.472	0.948	0.485	0.949	0.502	0.939	0.541	0.938	0.559	0.934	0.560
CACM	0.048	0.037	0.096	0.068	0.100	0.084	0.116	0.194	0.105	0.262	0.160	0.241
CISI	0.181	0.043	0.180	0.061	0.180	0.089	0.130	0.183	0.112	0.261	0.155	0.204
CRAN	0.121	0.127	0.140	0.149	0.139	0.173	0.139	0.214	0.112	0.245	0.257	0.301
MED	0.477	0.288	0.530	0.324	0.565	0.375	0.518	0.460	0.422	0.531	0.415	0.421

Table 5. Precision and recall of the TRSM and VSM cluster-based retrieval

Col.	2.9% of $\mathcal{D}$ ($\gamma = 0.14$)				8.0% of $\mathcal{D}$ ($\gamma = 0.11$)				16.9% of $\mathcal{D}$ ($\gamma = 0.09$)			
	TRSM		VSM		TRSM		VSM		TRSM		VSM	
	P	R	P	R	P	R	P	R	P	R	P	R
JSAI	0.949	0.502	0.947	0.501	0.939	0.541	0.947	0.518	0.938	0.559	0.939	0.549
CACM	0.100	0.084	0.075	0.479	0.116	0.194	0.075	0.479	0.105	0.262	0.075	0.479
CISI	0.180	0.089	0.099	0.366	0.130	0.183	0.099	0.366	0.112	0.261	0.099	0.366
CRAN	0.139	0.173	0.066	0.519	0.139	0.214	0.066	0.519	0.112	0.245	0.066	0.519
MED	0.565	0.375	0.520	0.430	0.518	0.460	0.458	0.521	0.422	0.531	0.375	0.585

precision and recall for JSAI are high, and they are higher and stable for the other collections with $\theta \geq 15$. With these values of θ, the TRSM-based retrieval effectiveness is comparable or somehow higher than that of VSM. Table 4 reports the average of precision and recall for all queries in experiments for TRSM cluster-based retrieval in a subset $\mathcal{D}'$ of $\mathcal{D}$ that contains clusters closed to the query. We experiment with various proportion (%) of $|\mathcal{D}'|$ to $|\mathcal{D}|$, and full retrieval in whole $\mathcal{D}$ (accordingly, values of γ). In several cases (JSAI, CISI, and MED) just searching a small part of $\mathcal{D}$, says 1.2% or 1.8%, TRSM cluster-based search reaches precision higher than that of full search. Also, the TRSM cluster-based search achieved recall higher than that of full retrieval on most collections when $|\mathcal{D}'|$ is about 17% of $|\mathcal{D}|$. Table 5 reports the effectiveness of TRSM cluster-based

Table 6. Performance Measurements of the TRSM Cluster-based Retrieval

Col.	Size (MB)	Nb of Queries	TRSM Time	Clustering Time	Full Search (sec)	Cluster Search (sec)	Memory (MB)
JSAI	0.1	20	2.4s	14.9s	0.8	0.1	8
CACM	2.2	64	22m2.2s	26m46.8s	13.3	1.2	201
CISI	2.2	76	13m16.8s	4m49.8s	40.1	3.4	84
CRAN	1.6	225	23m9.9s	3m6.9s	20.5	1.8	71
MED	1.1	30	40.1s	1m30.8s	2.5	0.3	25

retrieval (TRSM) versus VSM cluster-based retrieval (VSM) when $|\mathcal{D}'|$ is 2.9%, 8.0%, and 16.9% of $|\mathcal{D}|$. It shows that TRSM cluster-based retrieval often achieves precision higher than that of VSM cluster-based retrieval thought its recall is somehow lower.

4.4   Evaluation of TRSM Hierarchical Clustering Efficiency

A direct implementation of procedures in the first phase requires the time complexity of $O(M + N^2)$, but we implemented them by applying the quick-sort algorithm of $O(N \log N)$ to make the indexing files, then created the TRSM related files for the term co-occurrence, tolerance classes, upper and lower approximations files in the time of $O(M + N)$. We can note that the search for clusters requires in average $\log M$, then the search will be done with a subset of documents in the clusters. However, the time complexity of the clustering is of $O(M^2 + N)$, and the space is of $O(M^2 + N)$ because of using an $M \times M$-matrix to store the similarities/distances between clusters in the hierarchy. All the experiments reported in this paper were performed on a conventional workstation GP7000S Model 45 (Fujitsu, 250 MHz Ultra SPARC-II, 512 MB). Table 6 summaries the time for generating the TRSM files, clustering, full search, cluster-based search, and the required memory size for each collection.

References

1. Fakes, W. B. and Baeza-Yates, *Information Retrieval. Data Structures and Algorithms* (eds.), Prentice Hall, 1992.
2. Ho, T. B. and Funakoshi K., "Information retrieval using rough sets", *Journal of Japanese Society for Artificial Intelligence*, Vol. 13, N. 3, 1998, 424–433.
3. Pawlak, Z., *Rough sets: Theoretical Aspects of Reasoning about Data*, Kluwer Academic Publishers, 1991.
4. Polkowski, L. and Skowron, A., *Rough Sets in Knowledge Discovery 2. Applications, Case Studies and Software Systems* (eds.), Physica-Verlag, 1998.
5. Skowron, A. and Stepaniuk, J., "Generalized approximation spaces", *The 3rd International Workshop on Rough Sets and Soft Computing*, 1994, 156–163.

Application of Data-Mining and Knowledge Discovery in Automotive Data Engineering

Jörg Keller[1], Valerij Bauer[1], and Wojciech Kwedlo[2]

[1] DaimlerChrysler AG, Machine Learning, FT3/KL, P.O.-Box 2360, D-89013 Ulm Germany
{joerg.keller, valerij.bauer}@.daimlerchrysler.com
[2] Technical University of Bialystok, Wiejska 45a, 15-351 Bialystok, Poland
wkwedlo@ii.pb.bialystok.pl

Abstract. In this paper the authors present a powerful and efficient alternative to Neural Networks (NN) by application of Knowledge Discovery and Data-Mining (KDD) methods for real world data in vehicle design, particularly for automotive Data Engineering (DE) mechanisms and processes. Typical tasks in automotive engineering are dependency analysis, classification of concepts and prediction of characteristic design parameters. From the point of view of a design engineer the main drawback of a NN-based approach is a lack of clear interpretation of the results. For classical, statistical tasks an application of an instance-based method, e.g. K-Nearest-Neighbors (KNN), represents an appropriate alternative for the engineer. By application of rule-based methods the authors demonstrate an alternate in conceptual design, which, in contrast to NN, allows to interpret the results and proof or enhance designers knowledge. The approach of this paper is based on a novel application of an Evolutionary Decision Rule Learner with Multivariate Discretization (EDRL-MD) for classification, and of M6 for regression learning.

Keywords: Automotive Data Engineering, Evolutionary Decision Rule Learning, Knowledge Discovery, Data-Mining

Introduction

In automotive engineering designers have the task to develop vehicle concepts, which are limited by governmental restrictions, e.g. exhaust emissions and fuel consumption, and in regard to the customers requests, e.g. power, torque, acceleration or maximum speed. Development cycles of vehicles decrease constantly. More product niches are served by the manufactures. Therefore new vehicle concepts have to be developed rapidly.

In motivation, to overcome the time consumption by Computational Aided Design (CAD), traditionally the application of NN was a favourite AI technique over the last decade. But the results of NN can not be visualised, it is more like a black-box for the design engineer, who wants to proof the results for plausibility. Rule-based systems would offer a well-performing, alternative. Derived design rules are transparent and can be duplicated by the design engineer. Due to data quality, non plausible results could be detected. For objective, machine learning (ML) methods will be used

D.A. Zighed, J. Komorowski, and J. Zytkow (Eds.): PKDD 2000, LNAI 1910, pp. 464-469, 2000.
© Springer-Verlag Berlin Heidelberg 2000

inautomotive data engineering for Knowledge Discovery and Data mining (KDD). Design rules and knowledge could be extracted from the voluminous databases in automotive industry. In this paper we describe the ML method employment in automotive Data Engineering. In our approach we demonstrate, with M6 for regression learning, and a novel application of an Evolutionary Decision Rule Learner with Multivariate Discretization (EDRL-MD) for classification, a powerful and efficient alternative to NN for concept design in CAD.

In section 2 we will illustrate the used domain data (real world data), a vehicle data warehouse. Section 3 describes our applied methods of regression and classification in automotive Data Engineering. In section 4 experiments and the empirical results are demonstrated. In section 5 we will finish with our conclusions.

Automotive Data Engineering

ML methods and tools in a global product and tool developing process have to be evaluated for data engineering. Data fusion has to be done for an automated data engineering data consistency has to be proved constantly. Furthermore data has to be free of redundancy. Rezende et al. [3] describe how to generate a unified database interface for multiple heterogeneous databases for automotive application, which has been used by the authors as a data base for data-mining.

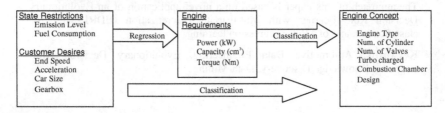

Fig. 1. Typical employment of classification and regression (prediction) in automotive data engineering

The data used for our analysis was taken from DaimlerChrysler internal databases. In our approach, ML for automotive data engineering, we have to employ with several layers of data: cylinder head, engine, drive assembly, car body, additional aggregates and or units. The engine, as subset, contains geometrical, mechanical or fluid data. Emission and fuel consumption are dependent on, e.g. turbulence generation in the combustion chambers, mixing of fuel injection and air. For our approach on ML in data engineering we selected core experiments, exemplary, to examine the subjects:

- vehicle-concept classification

- regression learning on exhaust emission and fuel consumption

A table was selected with more than 1036 records and 5 attributes, fuel consumption, exhaust emission, power, torque, max speed, vehicle weight, valve concept (no. of valves per cylinder) and gearbox type.

Employed Methods for Data Engineering

3.1   Employed Methods for Regression

In the table below the variables we took for our regression task are shown. Because of very high positive correlation between engine power engine torque of about $r \cong 0.99$, we agree on prediction only of engine power.

Table 1. Input and Output Variables for Regression

Input Variables	Var. Type	Output Variables	Var. Type
Car Weight	Numeric	Engine Power	Numeric
Car Speed	Numeric	Engine Capacity	Numeric
Acceleration	Numeric		
Gearbox Type	Symbolic		

3.1.1   NN

For our core experiments the training method *Quick* was selected in CLEMENTINE, SPSS/ISL 1998 [1], because of the higher accuracy. Prevent overtraining button was switched on and the training was stopped on default. 50% of the data set were chosen for training model and 50% for testing. The accuracy of the model was tested with cross-validation method by dividing the data set in 10 sub sets and using nine for training and one for testing for every sub set. The authors applied a NN capacity & power architecture for the experiments:

- Input layer: 4 neurones
- Hidden layer: 4 neurones
- Output layer: 2 neurones
- Predicted accuracy: 96.36%

3.1.2   KNN

The KNN Model for prediction of engine power and capacity were built on 3 Nearest Neighbours using the Euclidean metric for calculating distances. Six examples were removed from Data set to test them for prediction and compare with real Data. Results are shown in the Table 6.

3.1.3   M6

In our approach we used for regression modelling the M6 method of Quinlan [4]. By making regression with M6 only one output variable is possible to define at once. It was necessary to build one model for engine power and one for engine capacity prediction.

3.2   Employed Methods for Classification

For our classification experiments we used EDRL-MD system [2]. EDRL-MD learns decision rules using an *evolutionary algorithm*. The main novelty of this system consists in *multivariate discretization* i.e. the simultaneous discretization of all continuous-valued attributes.

The implementation of EDRL-MD used in our experiments employs a different fitness function than the original version described in [2]. The fitness of the ruleset RS_c the class c from the other classes is defined as:

$$f(RS_c) = \frac{\Pr(RS_c)}{Compl(RS_c)}$$

(1)

where $\Pr(RS_c)$ is the probability of a correct classification of an example given by:

$$\Pr(RS_c) = \frac{p+N-n}{P+N}$$

(2)

where the number of the positive and negative examples covered by the ruleset is denoted by p and n respectively; the total number of positive and negative examples in the learning set is denoted by P and N. $Compl(RS_c)$ is the complexity of the ruleset given by:

$$Compl(RS_c) = \alpha \log(L+1) + 1$$

(3)

where α is a user supplied parameter and L is the total number of elementary conditions (selectors) in the learning set. In our experiments we used $\alpha = 0.0005$.

Core Experiments and Test Cases

The test cases have been done for the parameters: engine power and capacity.

4.1   Regression in Automotive Data Engineering

Analysing the regression results, we noticed, that the highest deviation (Absolute Error>400 cm^3) by prediction of Engine Capacity occurred by high volume Engines (>3000 cm^3) with missing values. M6 maximum relative error for engine capacity prediction amounts 24% and maximum relative error for engine power prediction amounts 17%, those error rates are less than using NN method for regression. The correlation of relative errors for corresponding fields of NN-Model and M6-Model result in poor correlation (0.036 and –0.062).

4.2   Classification Learning

In this section the results of our classification experiment are presented. We compared the performance of EDRL-MD to that of two neural networks: multi-layer perceptron network (MLP) with one hidden layer and radial basis function network (RBFN).

Table 2. Results of KNN-Prediction for Engine Power and Capacity, REC Relative Error for Engine Capacity; REP Relative Error for Engine Power

	KNN-Cap.	KNN-Power	Capacity	Power	REC	REP
Vehicle 1	2330	100	2496	110	6,7(%)	9,1(%)
Vehicle 2	1995	79	1994	77	0,1(%)	3,0(%)
Vehicle 3	2381	102	1997	80	19,2(%)	27,1(%)
Vehicle 4	2496	110	2926	135	14,7(%)	18,5(%)
Vehicle 5	2524	107	3226	145	21,8(%)	26,4(%)
Vehicle 6	2151	92	2151	75	0,0(%)	22,7(%)

Table 3. Results of Prediction

Car Type	Capacity	M6-Capacity	Error(%)	NN-Capacity	Error(%)
Vehicle 1	2496	2571,1	3,01	2560	2,56
Vehicle 2	1994	1874,7	5,98	1803	9,58
Vehicle 3	1997	2076,5	3,98	2165	8,41
Vehicle 4	2926	3162,9	8,10	2787	4,75
Vehicle 5	3226	3489,5	8,17	3075	4,68
Vehicle 6	2151	2017,2	6,22	2180	1,35
Vehicle 1	110	108,3	1,51	112,0	1,82
Vehicle 2	77	59,2	23,13	61,0	20,78
Vehicle 3	80	79,9	0,18	80,0	0,00
Vehicle 4	135	132,6	1,79	136,0	0,74
Vehicle 5	145	156,9	8,20	146,0	0,69
Vehicle 6	75	76,2	1,63	76,0	1,33

Table 4. Classification errors and learning times in seconds for EDRL-MD, Multi-Layer-Perceptron (MLP) and Radial Basis Function Network (RBFN) .

METHOD	EDRL-MD	MLP	RBFN
ERROR RATE	25.2%	55.5%	34.5%
LEARNING TIME	313.2s	105.8s	59.9s

Both networks were trained using commercial CLEMENTINE [1] data mining system. The results of that comparison are shown on Table 4. The error rate estimated by ten-fold crossvalidation and the CPU time needed to build the classifier are presented. The learning time is the average of ten iterations of a single ten-fold crossvalidation run. All the algorithms were run on Sun Ultra-10 workstation with 300 MHz CPU. Using Mc Nemar's test we found that the difference in error rate between EDRL-MD and the other algorithms is highly (p-value<0.00001) significant.

Conclusions

This paper demonstrates that ML and DM give impact on a quick concept design description in automotive data engineering. In spite of the former used NN , rule-based methods offer for design engineers the advantage of comprehensible design rules, which are readable. The employment of the regression models for prediction of engine requirements show, that both methods are able to make sufficing prediction from engineering point of view. For classification of design concepts EDRL-MD has the significantly lower error rate than MLP or RBFN. However this improvement is achieved at the expense of increased computational complexity. Nevertheless all methods can be successfully applied for engineering purposes, to help designer in the concept phase.

Acknowledgement

Wojciech Kwedlo was supported by Deutscher Akademischer Austauschdienst (DAAD) under the grant A/99/12865.

References

1. CLEMENTINE (1998). Data-Mining Tool, *SPSS/ISL, Reference Manual Version 5*
2. Kwedlo, W; Kretowski, M. (1999). An evolutionary algorithm using multivariate discretization for decision rule induction. Principles of Data Mining and Knowledge Discovery. *3rd European Symposium PKDD'99. Springer LNCS 1704.*
3. Rezende, F.d.F; Oliveira, G.d.S.; Pereira, R.C.G.; Hermsen, U.; Keller, J. (1999). A Unified Database Interface for Multiple Heterogeneous Databases. *Proceedings of 2nd Workshop on Engineering Federated Information Systems EFIS'99, May 5-7, 1999, Kühlungsborn, Germany, eds. Conrad, S.; Hasselbring, W.; Saake, G., infix*
4. Quinlan, J.R. (1992) Learning with Continuos Classes. *Basser Department of Computer Science, University of Sydney*

Towards Knowledge Discovery from cDNA Microarray Gene Expression Data

Jan Komorowski[1], Torgeir R. Hvidsten[1], Tor-Kristian Jenssen[1],
Dyre Tjeldvoll[1], Eivind Hovig[2], Arne K. Sandvik[3], and Astrid Lægreid[3]

[1] Knowledge Systems Group, Department of Information and Computer Science,
Norwegian University of Science and Technology, 7491 Trondheim, Norway
jan.komorowski@idi.ntnu.no, http://www.idi.ntnu.no/grupper/KS-grp/
[2] Department of Tumor Biology, Institute for Cancer Research,
The Norwegian Radium Hospital, Oslo, Norway
[3] Department of Physiology and Biomedical Engineering,
Norwegian University of Science and Technology, Trondheim, Norway

Abstract. The advent of the so-called cDNA microarrays has offered
the first possibility to obtain a global understanding of biological processes in living organisms by simultaneous readouts of tens of thousands of
genes. Initial experiments suggest that genes with similar function have
similar expression patterns in microarray experiments. Until now, most
approaches to computational analysis of gene expressions have used unsupervised learning. Although in some cases unsupervised methods may
be sufficient, the complexity of the biological processes is so high that it
is unlikely that purely syntactical analyses are capable of fully exploiting the richness of the microarray data. In addition, it seems natural to
re-use the existing biological (background) knowledge.

In this paper, we present some elements of a methodology for knowledge
discovery from microarray experiments. Two source of bio-medical knowledge are used: Ashburner's gene ontology and our own literature-derived
network of gene-gene relations obtained by analysing Medline citation records. Predictive models can be induced and their classification quality
validated through the ROC/AUC analysis and applied to provide hypotheses regarding the function of unclassified genes. The methodology
has been so far tested on publicly available gene expression data and its
results evaluated by molecular biologists and medical researchers.

1 Introduction

Until recently, molecular biological research has been investigating the genetic
and biochemical mechanisms that underly inter- and intra-cellular organisation
using experimental methods that produce extremely low levels of information in
relation to the huge number of parameters that govern living systems. With the
advent of the so-called microarray technology [1], it is now possible to observe in
parallel several thousands of genes. Coupled with the large scale data generating
programs, such as the Human Genome Project, very large sources of information
about the living systems are becoming available. However, the avalanche of data

D.A. Zighed, J. Komorowski, and J. Żytkow (Eds.): PKDD 2000, LNAI 1910, pp. 470–475, 2000.
© Springer-Verlag Berlin Heidelberg 2000

is overwhelming and it is now realized that new methods to process this data are needed.

The microarrays provide a view onto cellular organisation of life through quantitative data on mRNA expression levels. These may be used to *discover* the biological function of genes [2]. It has been shown in early experiments [4] that genes of similar function tend to produce similar patterns in microarray experiments. With one exception, [3], most of the approaches uses unsupervised learning to discover the functional classes in expression data. In the unsupervised methods, a distance function is used to define similarity of gene expressions and a clustering algorithm is applied to find groups of genes. Hierarchical clustering [4] and self-organising maps [5] have been used.

The main thesis of this paper is that knowledge discovery from gene expressions requires new approaches to knowledge discovery, if the complexity barrier is to be ever broken. In our approach we move away from strictly unsupervised methods that seem to dominate the current state of art in this area, and show how to use various forms of background knowledge to discover and validate new knowledge. Individual genes are first *annotated* using available (background) knowledge such as various genomic databases (e.g. SWISSPROT), literature repositories (e.g. Medline) and ontologies (e.g. Ashburner's Gene Ontology [6]). Gene expressions are clustered according to some method, preferably in a way that lends itself to a biological interpretation. Having clusters with annotations, it is possible to validate their correctness. The annotations also effectively provide a training sample from which a model can be induced and whose quality of classification can be evaluated in some standard way (e.g. with the ROC/AUC analysis) and applied to classify unknown genes.

2   Datamining Methods

Our methodology is illustrated on a publicly available data set [7] previously analysed by Iyer et al. [8]. Iyer et al. studied the human fibroblast response to serum. This cell has a pivotal structural role in connective tissue and in important processes such as wound healing. The temporal changes in mRNA level of 8613 human genes were measured at 12 time points in the time period between 0 minutes and 24 hours after serum stimulation. A subset of 517 genes whose expression changed substantially in response to serum was selected for further analysis. An agglomerative implementation of the hierarchical clustering method was used to cluster the 517 genes into groups on the basis of the similarity of their expression profiles over the entire period of 24 hours. Ten clusters were identified containing 452 of the 517 genes.

Template-based clustering. Requiring that gene expressions be similar over the entire period of 24 hours seems to be somewhat unrealistic for the process at hand. Instead, we proposed a clustering that is defined on time sub-intervals. Since the same gene may be involved in different processes, we wanted to obtain clusters which include genes with expression profiles that show similarity in one time period but be rather different in another one. To this end we introduced a method called *template-based clustering*. Constraints defining the exact properties of a gene matching one template were prepared according to the uncertainties of the data. These templates have been used on all possible combinations of subintervals in order to assign the genes to clusters (see Fig. 1). According to domain experts, the resulting set of clusters describes well the dynamics in the data. Most importantly, the clusters display characteristic temporal changes of gene expression levels during a very large number of subintervals. This large variety of patterns of changes in gene expression reflects the expected complexity of molecular biological events during the fibroblast serum and thus, significantly simplifies the task of biological interpretation.

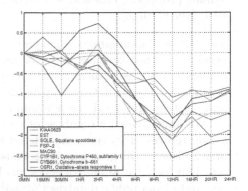

Fig. 1. A sample template-based cluster: Genes with a *decreasing* expression profile between 2 hours and 12 hours.

From clusters to decision classes. The need for a structured way of relating genes to each other based on their function or process has prompted the creation of *gene ontologies*. A gene ontology is a hierarchical structure where parent nodes give a more general description of a gene than their children. The leaf nodes give an accurate description of each individual gene. An ontology is a formalised source of high quality biological knowledge that will be important in much of the research on gene expressions. In particular, Ashburner's Gene Ontology [6] was developed using two model organisms, fruit fly (*Drosophila melanogaster*) and ordinary yeast (*Saccharomyces cerevisiae*). Surprisingly many genes in these simple organisms are homologous to human genes. This can be exploited simply by placing the human gene in the same location in the ontology as its homologous counterpart in these organisms. The information associated with a gene as a result of its location in the ontology is referred to as a gene *annotation*.

The annotations provide essentially a way of testing the quality, or rather classificatory power of a cluster before making assumptions about unknown genes located in it. We prove the point by means of the example in the earlier Fig. 1. This cluster contains eight genes of which four are named. Two of the named genes, Squalene epoxidase and Cytochrome P540, have annotation "cholesterol bio-synthesis". Another, CYB561, has annotation "electron transport" which may indicate a possible role in cholesterol synthesis. The gene OSR1, Oxidative-stress responsive-1, encodes a protein induced by oxidative stress, which is a process related to the redox processes involved in cholesterol bio-synthesis. EST, KIAA0623, FSP-2 and MAC30 encode proteins of unknown functions. To the biological domain experts, the relation between the annotated functions of the named genes is striking. The clustering of genes involved in redox processes and more specifically cholesterol synthesis, which is detected by this method may suggest that some of the genes with unknown function may also be involved in similar processes. This and several other related results are new in comparison with Iyer et al. [8].

Validation of annotations – pubgene. A large part of the biological knowledge needed to annotate genes is not organised in ontologies, but exists as text documents. We present here a method for mining gene-gene relations from such documents. These relations may be used both for validating existing annotations and clusters, and obtaining new annotations.

We have created a database of gene-gene relations by analysing Medline citation records. Two genes are related if they have been mentioned in the same article. Hence, the gene-gene links constitute a literature-derived network of genes. Our local list of human genes has been compiled by collecting and structuring data from public databases available through ftp or http. The main source of information has been the list of approved genes/symbols from the HUGO Nomenclature Committee [9,10].

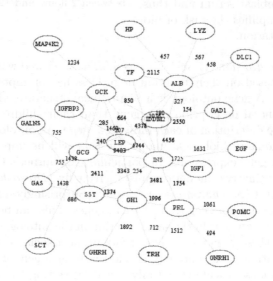

Fig. 2. A sample neighbourhood-network for the gene insulin (INS).

In the graphical display on our website [11] the neighbourhood of each gene is presented as a graph centered around the gene (see Fig. 2). Neighbours of the gene appear as nodes with edges weighted by the number of co-citations. The user can move around in the network by changing the gene in focus by clicking neighbours in the graph.

Altogether we compiled a list containing 14961 genes. Based on a subset of Medline containing all publications from the years 1992 - 1999, we found citations for 7561 (50%). Of these, 6978 had one or more co-citations (47%). The total number of articles was about 3 million, of which approximately 15% gave a hit for one or more genes.

Our work demonstrates that literature co-occurrence can be exploited to extract biologically meaningful information. Moreover, we also show that the approach can be carried out on a large scale. Details are described in [12].

Learning gene classifiers. Most processes that involve genes can be only descri-bed by a combination of several templates over a number of time sub-intervals. Consequently, we mine sets of template–interval pairs that together describe a process and obtain minimal sets of discerning pairs. Then, a set of IF-THEN rules over a training set of annotated genes is induced. This rule set holds both the descriptive knowledge about the temporal aspect of a given process and the predictive knowledge that can be used to classify genes for which no process is known (unknown genes). In order to induce a classifier, we use the rough set framework [13] for rule induction and voting for classification as implemented in the ROSETTA system [14,15]. The classification quality is validated through the ROC/AUC analysis in order to estimate the expected correctness of the classification of genes for which no process is known.

In the fibroblast data we have identified about 300 known genes, most of them having more than one annotation. These genes seem to be involved in over 20 different processes, of which 16 include sufficiently many genes (e.g. 10 genes) to induce a predictive model. Early results show that we in fact are capable of inducing a model that recognise these 16 processes. One example is the genes involved in *blood coagulation* which in a 5×3-fold cross-validation setting can be classified with an AUC value of 0.90. A thorough discussion of these results is to be found in [16].

3   Conclusions

A suite of knowledge discovery tools that support gene expression analysis, an-notation and visualisation has been created (e.g. [17,18]). We have made sig-nificant steps towards an automatic use of background knowledge in discovery from gene expressions. It should be noticed that standard methods of valida-ting clusters against background knowledge are not feasible here due to such factors as, for instance, a very large number of genes, hierarchical structure of this knowledge and the uncertainty of the genomic information.

The tools are now in use in our project on developing genomic classifiers from microarray data and background knowledge. Biologists and medical researchers find the tools particularly interesting since paradigms of biomedical background knowledge are often well captured. We have corroborated our hypothesis that knowledge-based tools are likely to gain an edge in knowledge discovery in such complex field as molecular biology.

References

1. Schena M, Shalon D, Davis R and Brown PO, Quantitative monitoring of gene expression patterns with a cDNA microarray, *Science*, 270:467–470, 1995.
2. Deboucek and Goodfellow, *Nature Genetics*, 21 (1 Suppl):48–52, 1999.
3. Brown MPS, Grundy WN, Cristianini N, Sugnet CW, Furey TS, Ares M and Haussler D, Knowledge-based analysis of microarray gene expression data by using support vector machines, *PNAS*, No. 1, Vol. 97:262–267, 1999.
4. Eisen M, Spellman P, Brown P and Botstein D, Cluster analysis and display of genome-wide expression pattern, *Proc. Natl. Acad. Sci. USA*, 95:1464–1480, 1998.
5. Kohonen T, The Self-Organizing Map, *Proceedings of the IEEE*, Vol. 78, No. 9:1464–1480, 1990.
6. The Ashburner Gene Ontology homepage, `http://genome-www.stanford.edu/GO/`.
7. The transcriptional program in the response of human fibroblasts to serum on the WEB `http://genome-www.stanford.edu/serum/`.
8. Iyer VR, Eisen MB, Ross DT, Schuler G, Moore T, Lee JCF, Trent JM, Staudt LM, Dudson Jr. J, Boguski MS, Lashkari D, Shalon D, Botstein D and Brown PO, The transcriptional program in the response of human fibroblasts to serum, *Science*, 283:83–87, 1999.
9. White JA, et al., Guidelines for human gene nomenclature, *Genomics*, 45(2):468–471, Oct 15 1997.
10. White JA, et al., The HUGO Nomenclature Committee home page `http://www.gene.ucl.ac.uk/nomenclature`.
11. Jenssen TK, The PubGene home page `http://www.idi.ntnu.no/grupper/KS-grp/microarray/pubgen/genes.cgi`.
12. Jensen T-K, Lægreid A, Komorowski J and Hovig E, *A literature network of human genes for high-throuput gene-expression analysis*, submitted for publication, June 2000.
13. Pawlak Z, Rough Sets, *International Journal of Computer and Information Sciences*, Vol. 11:341–356,1982.
14. Komorowski J, Skowron A and Øhrn A, *The Rosetta system*, to appear in *Handbook of Data Mining and Knowledge Discovery*, (W. Klösgen, J. Zytkow, Eds.), Oxford University Press, 2000.
15. Komorowski J and Øhrn A, Modelling Prognostic Power of Cardiac Tests Using Rough Sets, *Artificial Intelligence in Medicine*, Vol. 15, No. 2:167–191, 1999.
16. Hvidsten TR, Komorowski J, Lægreid A and Sandvik, *Discovery of gene functions and processes from gene expressions and ontologies*, submitted for publication, July 2000.
17. Hvidsten TR, Jenssen T-K, Komorowski J, Lægreid A, Sandvik A and Tjeldvoll D, *Template-based gene expression analysis*, in "Currents in Computational Molecular Biology – RECOMB 2000", edited by S. Miyano, R. Shamir and T. Takagi, pp. 10–11, ISBN 4-946443-61-4, Universal Academy Press, Inc, April 8-11, 2000, Tokyo, Japan.
18. Jenssen T-K, Lægreid A, Komorowski J and Hovig E, *PubGene: Discovering and visualising gene-gene relations*, in "Currents in Computational Molecular Biology – RECOMB 2000", edited by S. Miyano, R. Shamir and T. Takagi, pp. 48–49, ISBN 4-946443-61-4, Universal Academy Press, Inc, April 8-11, 2000, Tokyo, Japan.

Mining with Cover and Extension Operators

Marzena Kryszkiewicz

Institute of Computer Science, Warsaw University of Technology
Nowowiejska 15/19, 00-665 Warsaw, Poland
mkr@ii.pw.edu.pl

Abstract. Mining around association rules discovered in a large database is an important problem. In the paper, we consider the case, when a user wants to mine around the given set of association rules, but does not have access to the original database. We show how to reason with a set of rules by means of the cover and extension operators. Since the number of association rules can be huge, we introduce the concept of maximal covering rules. The algorithms for mining with the cover and extension operators are offered.

1 Introduction

The problem of discovery of association rules was introduced in [1] for sales transaction database. The association rules identify sets of items that are purchased together with other sets of items. Users are often interested in mining around the discovered set of rules. This is especially important, when the user is not allowed to access the database (e.g. for security reasons) and can deal only with a fraction of the rules that were provided by some trusted person. However, the user may be willing to induce as much knowledge as possible from the provided set of rules.

In this paper we show how to reason with a set of rules by means of the cover operator. We show how to find common knowledge and how to derive new rules as well as assess they support and confidence without accessing the database. Additionally, we introduce the notion of an extension operator that altogether with the cover operator can augment the original knowledge considerably. Since the number of association rules can be huge, we introduce the concept of maximal covering rules. The algorithms for mining with the cover and extension operators are offered as well.

2 Association Rules and Cover Operator

Let $I = \{i_1, i_2, ..., i_m\}$ be a set of distinct literals, called *items*. Any set of items will be called an *itemset*. Let D be a set of transactions, where each transaction T is a subset of I. An *association rule* is an expression $X \Rightarrow Y$, where $\varnothing \neq X, Y \subset I$ and $X \cap Y = \varnothing$. *Support* of an itemset X is denoted by $sup(X)$ and defined as the percentage (or the number) of transactions in D that contain X. *Support* of the association rule $X \Rightarrow Y$ is denoted by $sup(X \Rightarrow Y)$ and defined as $sup(X \cup Y)$. *Confidence* of $X \Rightarrow Y$ is denoted by $conf(X \Rightarrow Y)$ and defined as $sup(X \cup Y) / sup(X)$. The problem of mining association rules is to generate all rules that have sufficient support and confidence. In the sequel, the set of all association rules whose support is greater than s and

D.A. Zighed, J. Komorowski, and J. Zytkow (Eds.): PKDD 2000, LNAI 1910, pp. 476-482, 2000.
© Springer-Verlag Berlin Heidelberg 2000

confidence is not less than c will be denoted by $AR(s,c)$. If s and c are understood, then $AR(s,c)$ will be denoted by AR.

A notion of a *cover operator* was introduced in [3] for deriving a set of association rules from a given association rule without accessing a database. The *cover C* of the rule $X \Rightarrow Y$, $Y \neq \varnothing$, was defined as follows:

$$C(X \Rightarrow Y) = \{X \cup Z \Rightarrow V | Z, V \subseteq Y \text{ and } Z \cap V = \varnothing \text{ and } V \neq \varnothing\}.$$

Each rule in $C(X \Rightarrow Y)$ consists of a subset of items occurring in the rule $X \Rightarrow Y$. The antecedent of any rule r covered by $X \Rightarrow Y$ contains X and perhaps some items from Y, whereas r's consequent is a non-empty subset of the remaining items in Y.

Property 1 [3]. Let $r: (X \Rightarrow Y)$ and $r': (X \Rightarrow Y)$ be association rules.

$r' \in C(r)$ iff $X' \cup Y' \subseteq X \cup Y$ and $X' \supseteq X$ iff $X' \subseteq X \cup Y$ and $X' \supseteq X$ and $Y' \subseteq Y$.

Property 2 [3]. Let r and r' be association rules.

If $r' \in C(r)$ then $sup(r') \geq sup(r)$ and $conf(r') \geq conf(r)$.

Clearly, if $r' \in C(r)$ then $C(r') \subseteq C(r)$. The number of different rules in the cover of the association rule $X \Rightarrow Y$ is equal to $3^m - 2^m$, where $m = |Y|$ (see [3]).

3 Set-Theoretical Intersection of Covers

Investigation of the relationships among rules is a typical operation of mining around rules. In particular, having discovered rules r and r' one may wonder which association rules can be induced both from r and r', i.e. belong to $C(r) \cap C(r')$. In this section we examine the properties of set-theoretical intersection of covers of rules.

Property 3. Let $r: X \Rightarrow Y$, $r': X' \Rightarrow Y'$.

$$C(r) \cap C(r') = \begin{cases} C(s), \text{ where } s \text{ is the rule} : X \cup X' \Rightarrow Y \cap Y' \\ \quad \text{if } X \cup X' \subseteq (X \cup Y) \cap (X' \cup Y') \wedge Y \cap Y' \neq \varnothing; \\ \varnothing \quad \text{otherwise.} \end{cases}$$

Proof: Let $r: X \Rightarrow Y$, $r': X' \Rightarrow Y'$, and $r'': X'' \Rightarrow Y''$. By Property 1, $r'' \in C(r)$ iff $X'' \subseteq X \cup Y \wedge X'' \supseteq X \wedge Y'' \subseteq Y$ and $r'' \in C(r')$ iff $X'' \subseteq X' \cup Y' \wedge X'' \supseteq X' \wedge Y'' \subseteq Y'$. Hence, $r'' \in C(r) \cap C(r')$ iff $(X \cup Y) \cap (X' \cup Y') \supseteq X'' \supseteq X \cup X' \wedge Y'' \subseteq Y \cap Y'$. In addition, we note $Y \cap Y'$ must be different from $\varnothing$, otherwise r'' would have an empty consequent Y''. Thus, only rules $r'': X'' \Rightarrow Y''$, where $X'' \supseteq X \cup X'$ and $Y'' \subseteq Y \cap Y'$, belong to $C(r) \cap C(r')$ provided $X \cup X' \subseteq (X \cup Y) \cap (X' \cup Y')$ and $Y \cap Y' \neq \varnothing$ The set of such rules constitutes the cover of the rule $X \cup X' \Rightarrow Y \cap Y'$.

Example 1. Let us consider the intersection $C(ab \Rightarrow cde)$ and $C(ac \Rightarrow bde)$. By Property 3, it is equal to $C(abc \Rightarrow de) = \{abc \Rightarrow de, abc \Rightarrow d, abc \Rightarrow e, abcd \Rightarrow e, abce \Rightarrow d\}$.

Now, we generalize Property 3 for the case of n rules, where $n \geq 2$.

Property 4. Let $r_1: X_1 \Rightarrow Y_1, ..., r_n: X_n \Rightarrow Y_n$.

$$C(r_1) \cap ... \cap C(r_n) = \begin{cases} C(s), \text{ where } s \text{ is the rule}: X_1 \cup ... \cup X_n \Rightarrow Y_1 \cap ... \cap Y_n \\ \quad \text{if } X_1 \cup ... \cup X_n \subseteq (X_1 \cup Y_1) \cap ... \cap (X_n \cup Y_n) \wedge \\ \quad Y_1 \cap ... \cap Y_n \neq \emptyset; \\ \emptyset \qquad \text{otherwise.} \end{cases}$$

Property 4 tells us that the intersection of the covers of any number of rules is either the cover of one rule or is an empty set.

4 Inducing Knowledge by Means of Cover Operator

Let $R \subseteq AR$. We define *cover of the set of rules R* (denoted by $C(R)$) as follows:

$$C(R) = \cup_{r \in R} C(r).$$

The number of rules induced by the cover operator from R can be greater than the number of rules in R. In general, $C(R) \supseteq R$. By Property 2, each rule belonging to the cover of another rule has confidence and support not worse than the covering rule. So, for each new rule r in $C(R) \setminus R$, we can assess its support and confidence by choosing maximum confidence and support of the rules covering r. The assessment of support and confidence of r can be even more precise if we take into account itemsets of all known rules (the supports of which are known) and their antecedents (the supports of which can be computed from rules supports and confidences) as follows:

Let $r \in C(R) \setminus R$ and $F(R)$ be the family of itemsets of rules R and their antecedents, i.e. $F(R) = \{X \cup Y | X \Rightarrow Y \in R\} \cup \{X | X \Rightarrow Y \in R\}$. Then,

- *assessed support* $aSup(r: X \Rightarrow Y, R) = \max\{sup(f) | f \in F(R) \wedge X \cup Y \subseteq f\}$.

- *assessed confidence* $aConf(r: X \Rightarrow Y, R) = aSup(r, R) / \min\{sup(f) | f \in F(R) \wedge X \supseteq f\}$.

The real support of the rule r will not be less than $aSup(r, R)$ and the confidence will not be less than $aConf(r, R)$. Below we show an example of inducing new knowledge without accessing the database.

Example 2. Let us consider two rules discovered from some hospital database: $r_1: \{X\} \Rightarrow \{U, M\}$ (supp.=15%, conf.=60%), $r_2: \{X, U\} \Rightarrow \{O\}$ (supp.=5%, conf.=20%), where X stands for (*medical treatment* = X), U for (*result* = *Unsuccessful*), M for (*marital status* = *Married*) and O for (*age* = *Old*). Applying the cover operator to r_1 one will obtain e.g. the following rule: $r_3: \{X\} \Rightarrow \{U\}$. Knowing supports and confidences of the rules r_1 and r_2 we can derive supports of the following itemsets: $sup(\{X, U, M\}) = 15\%$, $sup(\{X\}) = 15\% / 60\% = 25\%$, $sup(\{X, U, O\}) = 5\%$, $sup(\{X, U\}) = 5\% / 20\% = 25\%$. Hence,

- $aSup(r3) = max(sup(\{X,U,M\}), sup(\{X,U,O\}), sup(\{X,U\})) = 25\%$,

- $aConf(r3) = aSup(r3) / min\{sup(\{X\})\} = 25\% / 25\% = 100\%$.

Thus r_3 has support not less than 25% and confidence not less than 100%.

The above example shows how it can be insecure to provide a user with the knowledge which seems to be unimportant. It may turn out that the cover operator produces the results which the rules provider would not like to reveal. This is common fear when sharing the knowledge with competing companies. The competing company may find out too much from originally minor knowledge.

Straightforward computation of the knowledge $C(R)$ augmented by the cover operator consists in taking union of the covers of all rules in R. Obviously, such an approach is not efficient. In particular, some rules in $C(R)$ will be generated many times, e.g. if $r,r' \in R$ and $r' \in C(r)$, then all rules induced by the cover operator from r' will be also induced from r. In the next section we introduce the notion of *maximal covering rules* and show how to use them for efficient derivation of new rules from R.

5 Maximal Covering Rules

Let $R \subseteq AR$. We define *maximal covering rules (MCR)* for R as follows:

$$MCR(R) = \{r \in R | \nexists\ r' \in R, r' \neq r \text{ and } r \in C(r')\}.$$

A maximal covering rule for the rule set R does not belong to the cover of any other rule in R.

Property 5. Let $R \subseteq AR$. Then, $C(R) = C(MCR(R))$.

The computation of the knowledge $C(R)$ induced by the cover operator from R can consist of two steps: 1) compute $R' = MCR(R)$; 2) compute $C(R')$. The redundancy of computation of association rules will be now restricted to computation of the overlapping covers of some maximal covering rules. The algorithm *FindMaxCoveringRules* is an example implementation of Step 1.

Further on, we assume that rules R are kept in such a way that for an itemset Z: 1) it is easy to find all rules created from itemsets being subsets of Z; 2) it is easy to find all rules with the antecedents being subsets of Z. This can be obtained e.g. by applying two hashing trees structures [2], one for the access of type 1 and the second one for the access of type 2. We also assume that $rules(f,R)$ is the function returning all rules in R that were created from f (i.e. $rules(f,R) = \{(r: X \Rightarrow Y) \in R | X \cup Y = f\}$).

```
Algorithm. FindMaxCoveringRules
input: set of rules R;
output: set of maximal covering rules R';
F = {itemset(r) | r∈R};
R' = ∅;
while F ≠ ∅ do begin
  f = a maximal itemset in F;
  V = all subsets of f in F;
  for each itemset v in V \ {f} do begin
    for each rule r' in rules(v,R) do
      if there is r in rules(f,R) such that r'.antecedent ⊇ r.antecedent then
        remove r' from R;
    for each rule r' in rules(f,R) do
      if there is r≠r' in rules(f,R) such that r'.antecedent ⊇ r.antecedent then
        remove r' from R;
  endfor;
  move rules(f,R) from R to R';
  F = F \ {f};
endwhile;
return R';
```

The *FindMaxCoveringRules* algorithm uses Property 1 to find maximal covering rules R' for the given rule set R.

6 Inducing Knowledge by Means of Extension Operator

The cover operator allow us to induce shorter rules from longer ones. However, under some conditions it is also possible to induce longer rules from shorter ones. Let us assume $r: X \Rightarrow Y$ and there is an itemset Z being a superset of $X \cup Y$ the support of which is the same as the support of r. Then, the rule $r': X \Rightarrow Z \setminus X$ will have the same support and confidence as r. The operator that allows us to induce such rules for r from the information on the set of rules R will be called an *extension operator* and will be denoted by $E(r,R)$. It is defined as follows:

$$E(r: X \Rightarrow Y, R) = \{X \Rightarrow (X' \cup Y') \setminus X \mid \exists r': X' \Rightarrow Y', X \cup Y \subseteq X' \cup Y' \wedge sup(r) = sup(r')\}.$$

Property 6. Let $r' \in E(r,R)$. Then:

$$sup(r') = sup(r), conf(r') = conf(r), r \in C(r'), E(r',R) \subseteq E(r,R).$$

$E(R)$ will denote the *extension of the set of rules R* and will be defined as follows:

$$E(R) = \bigcup_{r \in R} E(r,R).$$

Example 3. Let $R = \{r_1: ab \Rightarrow cde\ (s_1,c_1),\ r_2: abc \Rightarrow d\ (s_2,c_2),\ r_3: def \Rightarrow abc\ (s_3,c_3),\ r_4: dh \Rightarrow abc\ (s_4,c_4)\}$, where (s_i,c_i) are values of support and confidence of each i-th rule. Additionally, we assume that $s_1 = s_3$, $s_2 = s_4$ and $s_1 \neq s_2$. We observe that, $itemset(r_1) \subseteq itemset(r_3)$ and $itemset(r_2) \subseteq itemset(r_4)$. Hence, $r_5: ab \Rightarrow cdef\ (s_1,c_1) \in E(r_1,R)$ and $r_6: abc \Rightarrow dh\ (s_2,c_2) \in E(r_2,R)$. Let us note that $r_2 \in C(r_1)$, so $itemset(r_2) \subseteq itemset(r_1)$. Nevertheless, we could not rediscover r_1 from r_2 by the extension operator since $s_1 \neq s_2$. Actually, $E(R) = \{r_1, r_2, r_3, r_4, r_5, r_6\}$.

7 Inducing Knowledge by Means of Both Operators

Let us start with the property that shows how the rule sets augments when applying the extension and cover operators several times.

Property 7. Let $R \subseteq AR$.
- $E(E(R)) = E(R)$,
- $C(C(R)) = C(R)$,
- $MCR(MCR(R)) = MCR(R)$,
- $C(E(R)) \supseteq E(R)$,
- $E(C(R)) \supseteq C(R)$,
- $E(R) \supseteq E(MCR(R))$.

In the sequel of this section, we consider how to compute $C(E(R))$ efficiently. Since $C(E(R)) = C(MCR(E(R)))$ (by Property 5), we will concentrate on the problem of computing $MCR(E(R))$. To this end one can call the *FindMaxCoveringRules* algorithm with $E(R)$ as an argument.

Example 4. Let R be the set of rules $\{r_1, r_2, r_3, r_4\}$ from Example 3 and $E(R) = \{r_1, r_2, r_3, r_4, r_5, r_6\}$ as computed in Example 3. According to Property 3 the following pairs of rules (r_1,r_2), (r_1,r_5), (r_1,r_6), (r_2,r_5), (r_2,r_6), (r_3,r_5) and (r_5,r_6) have non-empty intersection of their covers, namely: $C(r_1) \cap C(r_2) = C(r_1) \cap C(r_6) = C(r_2) \cap C(r_5) = C(r_2) \cap C(r_6) = C(r_5) \cap C(r_6) = C(abc \Rightarrow d)$; $C(r_1) \cap C(r_5) = C(ab \Rightarrow cde)$; $C(r_3) \cap C(r_5) =$

$C(abdef{\Rightarrow}c)$. Thus, $|C(E(R))| = |C(r_1)| + |C(r_2)| - |C(r_1){\cap}C(r_2)| + |C(r_3)| + |C(r_4)| + |C(r_5)| - |C(r_1){\cap}C(r_5)| - |C(r_3){\cap}C(r_5)| + |C(r_6)| - |C(r_5){\cap}C(r_6)| = 106$ rules.

On the other hand, $MCR(E(R))$ generated by *FindMaxCoveringRules(E(R))* would be equal to $\{r_3, r_4, r_5, r_6\}$. Hence, $|C(MCR(E(R)))| = |C(r_3)| + |C(r_4)| + |C(r_5)| + |C(r_6)| - |C(r_3){\cap}C(r_5)| - |C(r_5){\cap}C(r_6)| = 106$ rules (as expected).

Let us also observe that without applying the extension operator we would derive much less rules, namely: $|C(R)| = |C(MCR(R))| = |C(\{r_1, r_3, r_4\})| = |C(r_1)| + |C(r_3)| + |C(r_4)| = 57$ rules.

Applying the *FindMaxCoveringRules* algorithm to $E(R)$ in order to compute $MCR(E(R))$ is not the best solution. Let us remind that every extended rule covers the rule from which it was generated. This means that $MCR(E(R))$ cannot contain $r{\in}R$ unless $E(r,R) = \{r\}$. Actually, all rules in $E(r,R)$ will not belong to $MCR(E(R))$ unless they are built from maximal itemsets in $\{itemset(r)| r{\in}E(r,R)\}$. This observation was applied in the *FindCovering_of_ExtendedRules* algorithm. For the algorithm, we assume that $rules(V,R)$, where V is a family of itemsets and R is a set of rules, returns all rules in R built from itemsets in V (i.e. $rules(V,R) = \{r: X{\Rightarrow}Y{\in}R| X{\cup}Y = f, f{\in}V\}$).

```
Algorithm. FindCovering_of_ExtendedRules
input: set of rules R;
output: MCR(R') = MCR(E(R));
F = {itemset(r) | r∈R};
R' = ∅;
while F ≠ ∅ do begin
  f = a maximal itemset in F;
  V = all subsets of f in F with the same support as f;
  for each group of rules in rules(V,R) with the same antecedent, say a, do
    R' = R' ∪ {a⇒f \ a};
  F = F \ V;
endwhile;
return FindMaxCoveringRules(R');
```

8 Related Work

The notion of the cover operator was used as a basis for the construction of *representative rules* [3] that constitute a least set of rules that covers all association rules. A set of *representative association rules* wrt. support s and confidence c ($RR(s,c)$) was defined as follows: $RR(s,c) = \{r{\in}AR(s,c)| \not\exists\ r'{\in}AR(s,c), r'{\neq}r$ and $r{\in}C(r')\}$. Clearly, $C(RR(s,c)) = AR(s,c)$. One can easily observe that $MCR(AR(s,c)) = RR(s,c)$. In addition, $C(AR(s,c)) = E(AR(s,c)) = AR(s,c)$, i.e. no new knowledge will be added by applying the extension and/or cover operators to $AR(s,c)$.

9 Conclusions

Both the cover and extension operator can augment the original rule base considerably. The new rules may be even more interesting than the original ones. The number of association rules can be huge, so we proposed their condensed representation called maximal covering rules. It plays the same role for the given rule set as representative rules for all association rules. In addition, it was shown that the intersection of covers of rules constitutes a cover of a rule or is an empty set.

References

1. Agrawal, R., Imielinski, T., Swami, A.: Mining Associations Rules between Sets of Items in Large Databases. In: Proc. of the ACM SIGMOD Conference on Management of Data. Washington, D.C. (1993) 207-216
2. Agrawal, R., Mannila, H., Srikant, R., Toivonen, H., Verkamo, A.I.: Fast Discovery of Association Rules. In: Fayyad, U.M., Piatetsky-Shapiro, G., Smyth, P., Uthurusamy, R. (eds.): Advances in Knowledge Discovery and Data Mining. AAAI, Menlo Park, California (1996) 307-328
3. Kryszkiewicz, M.: Representative Association Rules. In: Proc. of PAKDD '98. Melbourne, Australia. LNAI 1394. Springer-Verlag (1998) 198-209

A User-Driven Process for Mining Association Rules

Pascale Kuntz, Fabrice Guillet, Rémi Lehn, and Henri Briand

IRIN – IRESTE/Ecole polytechnique de l'université de Nantes
La Chantrerie BP 60601 44306 Nantes CEDEX 3 – France
{kuntz, guillet, lehn, briand}@irin.univ-nantes.fr

Abstract. This paper describes the components of a human-centered process for discovering association rules where the user is considered as a heuristic which drives the mining algorithms via a well-adapted interface. In this approach, inspired by experimental works on behaviors during a discovery stage, the rule extraction is dynamic : at each step, the user can focus on a subset of potentially interesting items and launch an algorithm for extracting the relevant associated rules according to statistical measures. The discovered rules are represented by a graph updated at each step, and the mining algorithm is an adaptation of the well-known A Priori algorithm where rules are computed locally. Experimental results on a real corpus built from marketing data illustrate the different steps of this process.

1 Introduction

As well-kwown by data analysts (e.g. [3]), appropriate visualization supports often give an insight into data that would be more difficult to get from looking at long listings of output, and consequently makes the user-interaction easier. During the upstream steps comprising the KDD process i.e. data mining and post-treatment where discovered knowledge is evaluated, visualization tools are targeted to several major objectives [16]. They allow to intelligibly describe basic relationships -often statistical- between data, and they contribute to the exploratory analysis by facilitating comparisons with more sophisticated models associated with hypotheses on data. Hence, numerous models previously developed in statistical analysis or machine learning have been integrated in software. The rapid development of new technologies such as virtual reality further stimulates the investigation of new paradigms of data representation (e.g. [6]). All of these tools make data and result handling easier, but most of the existing systems reported in the literature still let the decision-makers cope on their own with various diagrams. They do not explicitly take into account the cognitive capacities, in particular the domain knowledge, of each user. Unfortunately, the expression of the decision-maker's goals still remains a sensitive problem in KDD, and the current complexity of the knowledge representation systems limits their transfer into a KDD process.

D.A. Zighed, J. Komorowski, and J. Żytkow (Eds.): PKDD 2000, LNAI 1910, pp. 483–489, 2000.
© Springer-Verlag Berlin Heidelberg 2000

A pragmatic approach to get round these difficulties is to include the analyst into the discovery loop : the decision-maker is considered as a heuristic which drives the mining algorithms *via* a well-adapted interface. Such a human-centered approach has known an increasing development in Decision Aid during the last decade -from decision making in selection tasks to manufacturing and process control [17], and some models start being developed in KDD [5].

In this paper, we are concerned with discovery of association rules $X \to Y$ where X and Y are disjoint itemsets describing objects in a database. We describe a human-centered mining process which extracts rules bringing into play items that are *a priori* relevant for the user. Our approach has been inspired by experimental works on user behaviors during a discovery stage [2], [18], and by some cognitive mechanisms studied in decision models [19]. These works show on not only that, due to his short-term memory capacity, a decision maker manipulates a small amount of information at each step, but also that the processed information that leads to decisions has to be large enough for individual or social justification. Hence, in our case, we develop a dynamic model where the user can focus on a small subset of items at each step and can globally extract a significant set of association rules represented by a graphical model.

2   Rule Representation

We consider a large set of n objects $O = \{o_1, ..., o_n\}$ described by the items $I = \{i_1, ..., i_p\}$. Although our approach may be generalized to more complex data, here we restrict ourselves to binary items. In this case, each object o_i is described by a subset X_i of I : $X_i = \{i_1, i_4, i_5\}$ means that only items i_1, i_4 and i_5 are present for o_i,

Graphical Models for Implications

Amongst visualization techniques, graphs often have a privileged place, in particular for rule relationship representation. They can be used at the same time as theoretical models and as representation tools[7].

For purely logical rules, i.e. without any counter-examples, Galois lattices [4] have known a renewed interest in the last decade in Combinatorial Data Analysis for searching implications in binary data [8]. Let us recall that each node of the lattice is a pair, composed of a subset of O and a subset of I and that the set of pairs is ordered by the standard set inclusion relationship applied to O and I. This partially ordered set can be represented by a Hasse diagram but, unfortunately, determining a Galois lattice along with its Hasse diagram is a computationally difficult problem [15], and the representation become inextricable for numerous items.

We here develop a simpler model which does not take into account the Galois connection between subsets of O and I, and whose incrementality is directed by the user. A rule network is represented by a directed acyclic graph $G = (V, A)$ where each vertex of V is a potentialy interesting itemset, and each arc of A represents a significant implication between two itemsets. Formally, a vertex is a subset X of I, and an arc exists between two vertices X and X' if the rule

$X \rightarrow X' \setminus X$ is valid according to statistical measures described below. For instance, if $X = \{i_4, i_7\}$ and $X' = \{i_4, i_7, i_9\}$, an arc between X and X' is associated with the rule $[i_4 \wedge i_7] \rightarrow i_9$.

Statistical Measures of Rule Quality

Quality of discovered rules is measured here by statistical measures. Intuitively, a rule $X \rightarrow Y$ is statistically significant if it is covered by a large number of objects and refuted by few. Many coefficients have been presented in the literature to quantify these features, but from experimental comparisons [11] and [13], we retain three complementary measures :

1) The support $\rho(X \rightarrow Y)$ is the proportion of transactions which satisfy the rule. Let $g(X)$ be the maximal transaction subset covered by the itemset X. Then , $\rho(X \rightarrow Y)$ is the ratio $|g(X \cup Y)| / |O|$.

2) The conditional probability -or confidence- $\pi(X \rightarrow Y) = |g(X \cup Y)| / |g(X)|$.

3) The intensity of implication $\varphi(X \rightarrow Y)$ measures the "surprise" of having few counter-examples for the $X \rightarrow Y$ rule as compared with a random law. This measure was introduced by Gras [12] to improve the evaluation of rule confidence. The basic idea of intensity of implication is to compare the number of counter-examples $N(g(X), g(\overline{Y}))$ of the rule $X \rightarrow Y$ with the expected number $N(U, \overline{U'})$ where U and U' are two randomly selected subsets of O considered to be equals to respectively $|g(X)|$ and $|g(Y)|$. Then, the rule is surprising when the probability $P = \Pr(N(g(X), g(\overline{Y})) \leq N(U, \overline{U'}))$ is small. The intensity of implication is defined by $\varphi(X \rightarrow Y) = 1 - P$.

3   Interactive Visualization of the Model

The successive mining steps of the user are modeled by a series of graphs $G_t = (V_t, A_t)$ with the vertex set V_t and the arc set A_t defined in section 2. The graph G_t at step t is deduced from the graph G_{t-1} at step $t - 1$ by vertex and arc insertions.

Initially, the graph G_0 only contains vertices representing itemsets X of cardinality one associated with a support $|g(X)| / |O|$ fixed by the user. At each step t, the user selects a vertex and the significant associated rules are automatically generated by the algorithm described below. Then, a new graph, G_{t+1}, is drawn to update the knowledge. By moving sliders, he can also, at each t, modify the threshold values of the quality measures of the rules. This operation entails the deletion or the insertion of some arcs.

To make the interactivity easy, the different graph layouts must be intelligible. Common adopted readability criteria attempt to characterise readability by the means of general combinatorial optimisation goals such as minimizing arc crosses and minimizing the sum of the arcs' lengths to avoid long lines which can create confusion. Moreover, a drawing convention which precise vertex and arc characteristics must be specified. We consider here polyline drawings with vertices arranged in vertical layers.

Each layer is associated with a degree of
precision in the knowledge state : layers at
the left correspond to general characteristics described by few attributes whereas
layers at the right are more specific.

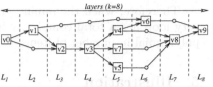

When t is fixed, this polyline drawing problem is known to be NP-complete
[9]. Different efficient algorithms have been proposed, but they do not take into
account the user interactivity. If a modification is performed, the algorithm runs
again and produces a new drawing which may be completely different from the
previous one. This requires the user to continually re-analyse the entire drawing,
and important changes between consecutive layouts highly disturb the interpretation task. Hence, our interactive mining process requires drawing methods
which preserve the "user's mental map" as much as possible. The formalization of
these cognitive constraints and their treatment remain a difficult question which
is far beyond the scope of this paper and we refer to [14] for the description of
an efficient genetic algorithm we developed for this problem.

4   Extraction Algorithm

The rule computation can be split up into two separated stages. Initially, large
itemsets, i.e. itemsets X with a support $|g(X)| \, / \, |O|$ larger than a given threshold
$minSup$ and a cardinality bounded by a constant $maxDepth$, are computed with
the same procedure as in the first stage of the well-known A $Priori$ algorithm
[1]. There, at each step t, once a vertex has been selected by the user on G_t, the
associated rules are extracted with a particular local procedure.

By selecting an itemset X, the user can trigger two different algorithms
according to his own goals :

A1 for computing new "local rules" of type $X \to \{i\}$ with item $i \in I \setminus X$,
which are more specific. This case corresponds to a "forward chaining" which
highlights conclusions that are infered from the current state.

A2 for computing more general rules of type $X \setminus \{i\} \to \{i\}$ with item $i \in X$.
This case corresponds to a "backward chaining" which highlights premises that
allow to conclude on the current state.

Let us denote by L_k the set of large itemsets of cardinality k, and by
$c_X = (X, |g(X)|)$ the description of each large itemset X. Figure 1 describes
the algorithm **A1**, triggered at step t by the user selection of item X. We suppose that $X \in L_k$. The algorithm **A1** first selects into L_{k+1}, the subset S_X of
itemsets including X. Secondly the set R_X of the more specific rules is deduced
from S_X : R_X is the set of rules $X \to Y$ so that $X \setminus Y \in S_X$.

In this case, the number of computed rules is bounded by the number of
items $|I|$ and consequently, the algorithm **A1** has a polynomial complexity in
$O(|I|)$.

The algorithm **A2** follows a similar principle : the set R'_X of the more general
rules is computed by selecting into L_{k-1} the subset S'_X of itemsets included in
X. The complexity is in $O(|I|)$.

Therefore, when the mining process is composed of T steps, the rule discovery loop has a total complexity in $O(T.|I|)$ which has the advantage of being polynomial in number of items.

5 Illustration

We present experimental results on a real database which contains the results of a marketing survey on domestic appliances. Questions were mainly focused on ownership, satisfaction, future purchase projects concerning 23 appliance categories. The database comprises 5074 answers associated with 115 boolean attributes as for instance rx which means *"the consumer whishes to replace the machine coded by x"*.

This survey aims at optimizing the marketing operations for these categories : determining what kind of equipment to propose to consumers owning another category, ... Let us notice that, in this case, no consumer category is defined *a priori* and that the search is non-supervised.

Figures **2a.**, **2b.** and **2c.** show different steps of the mining process of a user. Figure **2a.** corresponds to a common initial situation. A single item i.e. a large itemset of L_1 is selected and the associated rules are extracted by the algorithm A1. Here the item $\boxed{\texttt{h11}}$ means *"the consumer is happy with his washing machine"* and a discovered rules is for instance : $\boxed{\texttt{h11}} \rightarrow \boxed{\texttt{h01}\wedge\texttt{h11}}$ which means that *"a consumer who is satisfied by his washing machine is also satisfied with his TV set"*. All these rules have a confidence and a intensity of implication value greater or equal than 0.9.

Figure **2b.** describes the discovered graph after three successive requests concerning consumers satisfied with their washing machine: a "backward chaining" on $\boxed{\texttt{o11}\wedge\texttt{h11}}$ followed by a "forward chaining" on $\boxed{\texttt{o11}}$ and $\boxed{\texttt{o01}\wedge\texttt{o11}\wedge\texttt{o15}}$ ("ownership of TV set, video tape recorder, washing machine and refrigerator").

In order to discover more general informations concerning the notion of ownership of common appliances, the user engaged backward chaining (algorithm **A2**) from the previously discovered itemset $\boxed{\texttt{o01}\wedge\texttt{o02}\wedge\texttt{o11}\wedge\texttt{o15}}$ (Fig. **2c.**). He tries to determine wether the ownership of some of these appliances entails the ownership of the others. The graph shows that, for the chosen rule quality thresholds, the ownership of a video tape recorder and a refrigerator entails the ownership of the whole set of these usual appliances while the ownership of a TV set or a washing machine does not (see the first step in backward inference from $\boxed{\texttt{o01}\wedge\texttt{o02}\wedge\texttt{o11}\wedge\texttt{o15}}$ on Fig. **2c.**).

Integrating the user in the KDD process as a heuristic in the rule mining task not only dramatically decreases the rule search space, but also allows the user to focus on potentially meaningful knowledge. It also opens new research perspectives in improving the intelligibility of the knowledge representation. We currently investigate on two distinct directions to enhance the readibility of the rule networks. We aim at eliminating the "redundant knowledge" by concealing the rules that the user can easily infer himself and we carry on perfecting a highly-interactive rule interface.

● A1 : **More specific rules** :

$X := C_X$.itemset; $k := | X |$ // *Clicked itemset and its size*
$S := \{C_Y \in L_{k+1} \mid X \subset C_Y$.itemset$\}$ // *large $_{k+1}$ itemsets access*
$R_X := \emptyset$ // *the set of more specific rules*
For each $C_Y \in S$ // $\mid S \mid \leq \mid I_u \mid - \mid X \mid << \mid L_{k+1} \mid$
  | $Y := C_Y$.itemset
  | r.rule $:= X \to Y \setminus X$ // *a new rule ($\mid Y \setminus X \mid = 1$)*
  | r.support $:= C_Y$.support   // *its quality indices*
  | r.confidence $:= C_Y$.support $/ C_X$.support
  | $C_{Y \setminus X} := \{c \in L_1 \mid c$.itemset $= Y \setminus X\}$
  | r.intensity $:= \varphi(\mid I_u \mid, C_X$.support,
                       C_Y.support, $C_{Y \setminus X}$.support$)$ // *formula (1)*
  | $R_X := R_X \cup \{r\}$
End for each

Fig. 1. Algorithm **A**1 for local rule computation. (minConf and minInt are respectively the confidence threshold and the intensity of implication threshold.)

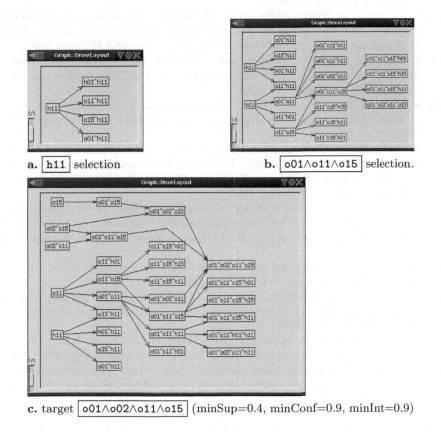

a. |h11| selection   b. |o01∧o11∧o15| selection.

c. target |o01∧o02∧o11∧o15| (minSup=0.4, minConf=0.9, minInt=0.9)

Fig. 2. Discovered rules following different vertex selection.

References

1. R. Agrawal, H. Mannila, R. Srikant, H. Toivonen, and A.-I. Verkamo. Fast discovery of association rules. In Fayyad et al. [10], pages 307–328.
2. I. Bandhari. Attribute focussing: machine-assisted knowledge discovery applied to software production process control. *Knowledge Acquisition*, 6:271–294, 1994.
3. J. Bertin. *Graphics and Graphic Information Processing*. Berlin: Springer, 1981.
4. G. Birkoff. *Lattice Theory*. A.M.S, 1967.
5. J.R. Brachman and T. Anand. The process of knowledge discovery in databases: a human-centered approach. In Fayyad et al. [10], pages 37–58.
6. C. Brunk, J. Kelly, and R. Kohavi. Mineset: An integrated system for data mining. In *Proc. of the 3rd Int. Conf on Knowledge Discovery and Data Mining*, pages 135–139. AAAI Press, 1997.
7. W. Buntine. Graphical models for discovering knowledge. In Fayyad et al. [10], pages 59–82.
8. V. Duquenne. Latticial structures in data analysis. *Theoretical Computer Science*, 217(2):407–436, 1999.
9. P. Eades and N. Wormald. Edge crossings in drawings of bipartite graphs. *Algorithmica*, 11:379–403, 1994.
10. U.M. Fayyad, G. Piatetsky-Sapiro, and P. Smyth. From data mining to knowledge discovery. In U.M. Fayyad, G. Piatetsky-Sapiro, and P. Smyth, editors, *Advances in Knowledge Discovery and Data Mining*, pages 2–34. AAAI Press, 1996.
11. L. Fleury, H. Briand, J. Philippé, and C. Djeraba. Rule evaluation for knowledge discovery in databases. In *Proc. of the 6th Int. Conf. on Database and Expert System App.*, pages 405–414, 1995.
12. R. Gras, H. Briand, and P. Peter. Structuration sets with implication intensity. In *Proc. of the Int. Conf. on Ordinal and Symbolic Data Analysis*, pages 147–156, 1996.
13. S. Guillaume, F. Guillet, and J. Philipp . Improving the discovery of association rules with intensity of implication. In *Proc. of the Second European Symposium on Principles of Data Mining and Knowledge Discovery*, pages 318–327. Lect. Notes in Comp. Sc., Springer-Verlag, 1998.
14. F. Guillet, P. Kuntz, and R. Lehn. A genetic algorithm for visualizing networks of association rules. In *Proc. of the 12th Int. Conf. on Industrial and Engineering App. AI and Expert Sys.*, pages 145–154. Lect. Notes in Comp. Sc., Springer-Verlag, 1999.
15. A. Gu noche. Building the galois lattice of a binary relation (in french). *Math matiques, Informatique et Sciences Humaines*, 109:41–53, 1990.
16. D.A. Keim and H.-P. Kriegel. Visualization techniques for mining large databases: A comparison. *Trans. on Knowledge and Data Engineering, Special Issue on Data Mining*, 8(6):923–938, 1996.
17. P. Lenca, editor. *Proc. of the 10th Mini Euro Conf.: Human Centered Processes*. Brest: Ecole Nationale Sup rieure des T l communications de Bretagne, 1999.
18. J. Philippé, T. Teusan, S. Baquedano, and C. Bourcier. The implicative analysis in a knowledge extraction context for perfecting decision aid systems in behavioral analysis (in french). In *Actes des journ es sur La fouille dans les donn es par l'analyse implicative (to appear)*, 2000.
19. O. Svenson. Decision rules and information processing in decision making. In *Human decision making*. Bodafors: Bodaforlaget Doxa, 1983.

Learning from Labeled and Unlabeled Documents: A Comparative Study on Semi-Supervised Text Classification

Carsten Lanquillon

DaimlerChrysler AG
Research and Technology 3
D-89013 Ulm, Germany
carsten.lanquillon@daimlerchrysler.com

Abstract. Supervised learning algorithms usually require large amounts of training data to learn reasonably accurate classifiers. Yet, for many text classification tasks, providing labeled training documents is expensive, while unlabeled documents are readily available in large quantities. Learning from both, labeled and unlabeled documents, in a semi-supervised framework is a promising approach to reduce the need for labeled training documents. This paper compares three commonly applied text classifiers in the light of semi-supervised learning, namely a linear support vector machine, a similarity-based tfidf and a Naïve Bayes classifier. Results on a real-world text datasets show that these learners may substantially benefit from using a large amount of unlabeled documents in addition to some labeled documents.

1 Introduction

With the enormous growth of on-line information available through the World Wide Web, corporate intranets, electronic news feeds, and other sources, the problem of automatically classifying text documents into predefined categories is an important issue in many information organization and management tasks.

These classification problems can be solved by applying supervised learning algorithms to learn classifiers from labeled training examples that predict the class label of new, previously unseen documents. In order to learn reasonably accurate classifiers, we must be provided with enough labeled training examples. This commonly requires a person to read many documents and to decide on the class label to be given to each of these documents. This is a tedious and time consuming process. Thus, for complex learning tasks, providing sufficiently large sets of labeled training examples becomes prohibitive. By contrast, unlabeled documents are often readily available in large quantities. Therefore, a promising idea is to utilize unlabeled documents in addition to labeled documents when learning a classifier. We refer to learning from both, labeled and unlabeled data, as semi-supervised learning. See [6,9] for a discussion of related work.

In this paper, we empirically compare three commonly applied text learning algorithms, namely a linear support vector machine, a similarity-based tfidf and

D.A. Zighed, J. Komorowski, and J. Żytkow (Eds.): PKDD 2000, LNAI 1910, pp. 490–497, 2000.
© Springer-Verlag Berlin Heidelberg 2000

a Naïve Bayes classifier in a semi-supervised framework [5,6,9]. Classification accuracy is reported on an independent test set. Further, we investigate the effect on semi-supervised learning when varying the number of features.

The remainder of this paper is organized as follows. Section 2 gives a brief introduction to text classification and some traditional learning algorithms. In Section 3, a general framework for learning from labeled and unlabeled documents is described. Initial experimental results that support the idea of semi-supervised learning are reported in Section 4. Section 5 concludes this paper.

2 Text Classification

The task of text classification is to automatically classify documents into a predefined number of classes. Each document can be in multiple, exactly one, or no class. In the experiments presented in Section 4, the task is to assign each document to exactly one class. When using supervised learning algorithms in this particular setting, a classifier can try to represent each class simultaneously. Alternatively, each class can be treated as a separate binary classification problem where each binary problem answers the question of whether or not a document should be assigned to the corresponding class [5].

2.1 Document Representation

In information retrieval, documents are often represented as feature vectors, and a subset of all distinct words or word stems occurring in the given documents are used as features. Words that frequently occur in many documents (*stop words* such as "and", "or" etc.) or words that occur only in very few documents may be removed. Further, measures such as *information gain* can be used for feature selection [14]. Each feature is given a weight which depends on the learning algorithm at hand. This leads to an attribute-value representation of text. Possible weights are, for example, binary indicators for the presence or absence of features, plain feature counts—*term frequency (tf)*—or more sophisticated weighting schemes, such as multiplying each term frequency with the *inverted document frequency (idf)* [11]. Finally, each feature vector may be normalized to unit length to abstract from different document lengths.

2.2 Learning Algorithms

A variety of text learning algorithms has been studied and compared in the literature, for example see [1,2,13]. We focus on three widely applied classifiers.

One of the learning algorithms we apply is the multinomial Naïve Bayes classifier (NB) which uses plain term frequency as feature weights [9]. The idea of the Naïve Bayes classifier is to use the joint probabilities of words (features) and classes to estimate the probabilities of the classes given a document. A document is then assigned to the most probable class. It is computationally very efficient because of the simplifying assumptions that the feature weights are conditionally independent given the class label. Although this independence assumption is usually violated in practice, this method often yields good performance [2,14].

Further, we use a similarity-based method based on *tfidf* weights which we denote as *single prototype classifier (SPC)*. It is a variant of Rocchio's method for relevance feedback [10] applied to text classification and is also described as the *Find Similar* algorithm in [1]. The classifier models each class with exactly one prototype computed as the average (centroid) of all available training documents belonging to that class. We use a scheme for setting feature weights which is denoted as *ltc* in the Smart system. A document is assigned to the class of the prototype to which it has the largest cosine similarity.

Support vector machines (SVMs) are successfully applied to text classification problems [1,3,13]. SVMs can only solve binary classification problems. More complex classification tasks must be composed of binary decisions. We apply a SVM in its basic form, which is a hyperplane that separates examples of the two classes with maximum margin. The class label of a new document is determined based on which side of the hyperplane it is.

3   Semi-Supervised Learning

This section discusses the use of unlabeled in addition to labeled documents when learning a text classifier. Having given an argument for the use of unlabeled documents, we describe a general framework for semi-supervised learning.

3.1   Why Does Using Unlabeled Data Help?

As pointed out in [5,9], it is well known in information retrieval that words in natural language occur in strong co-occurrence patterns [12]. While some words are likely to co-occur in one document, others are not. When using unlabeled documents we can exploit information about word co-occurrences that is not accessible from the labeled documents alone. Albeit independent from the class labels, this information can help to enhance classification accuracy.

3.2   General Framework

We use a general framework for learning from labeled and unlabeled documents that can be instantiated with any learning algorithm [6]. It is a generalization of the approach proposed in [9]. The class labels of unlabeled documents are treated as missing values, and an EM-like scheme is used to alternately predict the missing class labels and build a new classifier based on both, the labeled and unlabeled documents together with the predicted pseudo class labels as described below. Table 1 gives an outline of this general framework.

Given a set of training documents D, for some subset of the documents $d_i \in D^l$ we know the true class labels, and for the rest of the documents $d_i \in D^u$, the class labels are unknown. Thus we have a disjoint partitioning of our training documents into a labeled set and an unlabeled set of documents $D = D^l \cup D^u$.

Table 1. Framework for semi-supervised learning.

- **Inputs:** Sets D^l and D^u of labeled and unlabeled documents.
- Build initial classifier, H, based only on the labeled documents, D^l.
- Loop while class memberships of unlabeled documents, U^u, change:
  - **(E-step)** Use current classifier, H, to score unlabeled documents.
  - Transform scores of unlabeled documents into class memberships, U^u.
  - **(M-step)** Re-build H based on D^l and D^u, with labels obtained from U^u.
- **Output:** Classifier, H, for predicting class labels of unseen documents.

The task is to build a classifier based on the training documents, D, for predicting the class label of new, previously unseen documents.

First, an initial classifier, H, is build based only on the labeled documents, D^l. Then the algorithm iterates the following three steps until the class memberships given to the unlabeled documents, D^u, by the current classifier, H, do not change from one iteration to the next. Corresponding to the **E-step**, the current classifier, H, is used to classify each unlabeled documents. The classifier may respond with any type of classification scores whose interpretation need not be probabilistic. In order to abstract from the classifier's response, in the next step we transform these scores into class memberships, yielding a class membership matrix, $U^u \in [0,1]^{c \times |D^u|}$, where c is the number of classes. The sum of class memberships of a document over all classes is assumed to be one. Possible transformations are, for instance, normalizing the scores to unity which yields soft memberships, or using hard memberships, e.g. setting the largest score to one and all other scores to zero. The choice of the transformation function depends on the classifier at hand such that the classifier knows how to make use of the class membership matrix, U^u. Notice that using hard memberships always allows us to use any traditional supervised learning algorithm. Now, provided with the class membership matrix, U^u, a new classifier, H, can be build from both, the labeled and unlabeled documents. This corresponds to the **M-step**. The final classifier, H, can then be used to predict the class labels of new, previously unseen examples.

3.3 Instantiations

In order to apply this algorithmic framework, the underlying classification algorithm, and the function for transforming classification scores must be specified. In the following, we instantiate the framework with the three text learning algorithms described in Section 2 to obtain semi-supervised text learners.

When using a Naïve Bayes classifier and leaving the resulting probabilistic classification scores unchanged, we end up with the algorithm given in [9]. This instantiation, which we refer to as *ssNB*, has a strong probabilistic framework and is guaranteed to converge to a local minimum [9].

Next, we will apply the single prototype classifier in combination with a transformation of classification scores into hard class memberships, yielding the *ssSPC* classifier. Initial experiments showed that using soft class memberships for this classifier does not yield reasonable results. Notice that this instantiation turns out to be a variation of the well known *hard k-means* clustering algorithm [7]. The difference is that the memberships of the labeled documents remain fixed during the clustering iterations. The ssSPC algorithm is guaranteed to converge to a local minimum after a finite number of iterations [6].

Finally, we use a linear SVM with hard class memberships in the semi-supervised framework to yield the *ssSVM* classifier. Obviously, using soft class memberships for a conventional SVM does not make sense because an example is always on exactly one side of the separating hyperplane (unless it is on that hyperplane). Since we do not have any guarantee of convergence, we specify a maximum number of iterations for the EM scheme.

4   Experimental Results

This section gives empirical evidence that combining labeled and unlabeled documents with certain text classifiers using the algorithmic framework in Table 1 can improve traditional text classifiers. Due to space limitation, experimental results are reported on only on the 20 Newsgroups dataset, which is publicly available at *http://www.cs.cmu.edu/~textlearning*. We use a modified version of the *Rainbow* system and the to run our experiments [8]. In addition, the SVM^{light} implementation is used to learn SVMs [4]. We follow the description in [9] to setup the experiments with the semi-supervised classifiers described Section 3, with the corresponding traditional supervised learners, and with a transductive linear support vector machine [5].

4.1   Dataset and Protocol

The 20 Newsgroups dataset consists of 20017 articles divided almost evenly among 20 different UseNet discussion groups [2]. The task is to classify an article into the one of the twenty newsgroups to which it was posted. When tokenizing the documents, UseNet headers are skipped, and tokens are formed from contiguous alphabetic characters. We do not apply stemming, but remove common stop words. The remaining words are used as vocabulary. We create a test set of 4000 documents and an unlabeled set of 10000 documents. Labeled training sets are formed by partitioning the remaining documents into non-overlapping sets. All sets are created with equal number of documents per class. Where applicable, up to ten trials with disjunct labeled training sets are run for each experiment. Results are reported as averages over these trials.

4.2   Results

Figure 1 shows the effect of using unlabeled documents in addition to labeled documents when learning a classifier with the different classifiers. The learning curve shows the classification accuracies of the traditional and semi-supervised

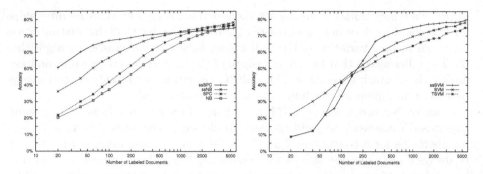

Fig. 1. Classification accuracy of the partially supervised learning framework (ss) using the single prototype classifier (SPC), the Naïve Bayes classifier (NB) and the linear support vector machine (SVM) compared to the traditional classifiers and a transductive linear support vector machine (TSVM) on the 20 Newsgroups dataset.

single prototype classifer (SPC) and Naïve Bayes classifier (NB) (left) and the traditional, semi-supervised and transductive linear SVM (right) when varying the number of labeled documents. The horizontal axes indicate the number of labeled training documents on a log scale. Notice, for instance, that 20 training documents correspond to one training document per class. The vertical axes indicate the average classification accuracy on the test sets.

Traditional SPC and SVM achieve approximately the same results. Only with many labeled training documents, the SVM slightly outperforms the SPC by about two points. The traditional NB is generally worse than SPC and SVM. The learning curves of the plain supervised learners illustrate that learning reasonably accurate classifiers requires a large amount of labeled training documents.

The semi-supervised SPC and NB perform substantially better than the traditional variants when the amount of labeled training documents is small. For instance, with 100 training examples, ssSPC achieves 67% accuracy on the test set while the traditional SPC reaches only about 42%. Thus, the classification error is reduced by about 43%. As one would expect, the more labeled documents are available, the smaller the performance increase. For large numbers of labeled documents, accuracy even slightly degrades when using unlabeled documents with ssSPC.

The results of the semi-supervised SVM are different. When only a small amount of labeled documents is available, adding unlabeled documents drastically hurts performance as compared to plain SVM. With more labeled data, however, ssSVM also starts to outperform its traditional variant. Compared to SPC and NB, learning a SVM involves a much more complex search. We thus hypothesize that the SVM requires more labeled documents in order to benefit from the unlabeled documents. Further note that using the SVM in the semi-supervised setting requires much more computing time than the other learning algorithms due to the more complex search. The transductive SVM cannot compete at all. It starts as bad as ssSVM and never gets better than plain SVM. Note, howe-

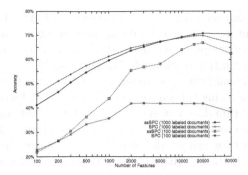

Fig. 2. Classification accuracy of traditional and semi-supervised SPC when varying the number of features for two amounts of labeled training examples.

ver, that the transductive SVM is actually designed to enhance performance on the unlabeled data used during training rather than on an independent test set. And in fact, further experiments show that TSVM outperforms SVM on the unlabeled dataset in the presence of many labeled training examples.

In Section 3 we argue that classification accuracy can be increased by exploiting co-occurrence patterns in unlabeled documents. The performance should therefore crucially depend on the number of features. Figure 2 shows the accuracies of the traditional and semi-supervised SPC for 100 and 1000 labeled training documents when varying the number of features from 100 to more than 50000 feature, which comes close to using all features. For 1000 labeled training documents, using unlabeled documents does not help a lot building the classifier. And this does not depend on the number of features. The respective curves are very similar. Things are quite different when using only 100 labeled documents. Here we have a rather large performance gain. Except for the decrease towards the end, using more feature generally yields a larger performance increase. When using too few features, we do not get any help from the unlabeled documents.

5 Conclusions and Future Work

This paper compared three commonly applied text classifiers in the light of semi-supervised learning with labeled and unlabeled documents, which is an important issue when hand-labeling documents is expensive but unlabeled documents are readily available in large quantities.

Empirical results on a real-world text classification task show that the single prototype classifier and the Naïve Bayes classifier can benefit substantially when incorporating unlabeled documents with some labeled documents into the learning process. We also see some improvement in classification accuracy when using support vector machines. However, the results are not as consistent as those for the single prototype and the Naïve Bayes classifiers. The support vector machines seem to require more labeled documents in order to benefit from a large set of unlabeled documents. Note that adding unlabeled documents to a larger number of labeled training documents even hurts classification accuracy in some cases. Future work will focus on preventing the unlabeled documents from degrading performance. An interesting approach is to introduce a weight to adjust the contribution of unlabeled documents [9].

Further experiments with a semi-supervised k-nearest neighbor rule show no improvement at all. This is because there is no proper learning process. Consequently, this algorithm does not generalize and it cannot make use of additional co-occurrence patterns inherent to unlabeled documents. Note, that this supports our argument for using unlabeled documents given in Section 3.

An important finding is that in the semi-supervised setting the feature set should not be reduced too much since additional features may contain valuable information even though cross-validation on the labeled training data would suggest using less features. Note that the calculation of many common feature selection measures depend on the class labels. For unlabeled documents, however, this label is missing. Therefore, an interesting research issue is to use feature selection methods that are especially developed for clustering tasks.

References

1. S. Dumais, J. Platt, D. Heckerman, and M. Sahami. Inductive learning algorithms and representation for text categorization. In *Proceedings of the Seventh International Conference on Information and Knowledge Management*, 1998.
2. T. Joachims. A probabilistic analysis of the Rocchio algorithm with tfidf for text categorization. In *Proceedings of ICML'97*, 1997.
3. T. Joachims. Text categorization with support vector machines: Learning with many relevant features. In *Proceedings of ECML'98*. Springer Verlag, 1998.
4. T. Joachims. *Advances in Kernel Methods – Support Vector Learning*, chapter Making large-Scale SVM Learning Practical. MIT-Press, 1999.
5. T. Joachims. Transductive inference for text classification using support vector machines. In *Proceedings of ICML'99*, 1999.
6. C. Lanquillon. Partially supervised text classification: Combining labeled and unlabeled documents using an EM-like scheme. In *accepted at ECML2000*, 2000.
7. J. MacQueen. Some methods for classification and analysis of multivariate observations. In *Proceedings of the Fifth Berkeley Symposium on Mathematical Statistics and Probability*, pages 281–297, 1967.
8. A. K. McCallum. Bow: A toolkit for statistical language modeling, text retrieval, classification and clustering. http://www.cs.cmu.edu/~mccallum/bow, 1996.
9. K. Nigam, A. McCallum, S. Thrun, and T. Mitchell. Text classification from labeled and unlabeled documents using EM. *Machine Learning*, 2000. To appear.
10. J. J. Rocchio Jr. Relevance feedback in information retrieval. In *The SMART Retrieval System*. Prentice-Hall, 1971.
11. G. Salton. *Automatic Text Processing: The Transformation, Analysis, and Retrieval of Information by Computer*. Addison-Wesley, 1989.
12. C. J. van Rijsbergen. A theoretical basis for the use of co-occurrence data in information retrieval. *Journal of Documentation*, 33(2):106–119, 1977.
13. Y. Yang and X. Liu. A re-examination of text categorization methods. In *Proceeding of the ACM SIGIR Conference*, pages 42–49, 1999.
14. Y. Yang and J. O. Pedersen. A comparative study on feature selection in text categorization. In *Proceedings of ICML'97*, pages 412–420, 1997.

Schema Mining: Finding Structural Regularity among Semistructured Data

P.A Laur[1], F. Masseglia[1,2], and P. Poncelet[1]

[1] LIRMM UMR CNRS 5506, 161, Rue Ada, 34392 Montpellier Cedex 5, France
{laur,massegli,poncelet}@lirmm.fr
[2] Laboratoire PRiSM, Université de Versailles, 45 Avenue des Etats-Unis, 78035
Versailles Cedex, France

Abstract. Motivated by decision support problems, data mining has been extensively addressed in the few past years. Nevertheless, the proposed approaches mainly concern flat representation of the data and to the best of our knowledge, not much effort has been spent on mining interesting patterns from such structures. In this paper we address the problem of mining structural association of semistructured data, or in other words the discovery of structural regularities among a large database of semistructured objects. This problem is much more complicated than the classical association rule one, since complex structures in the form of a labeled hirearchical objects partially ordered has to be taken into account.

1 Introduction

Motived by decision support, the problem of mining association rules or sequential patterns has recently received a great deal of attention [AIS93,AS94, BMUT97,FPSSU96,SON95,Toi96,SA96,MCP98]. Nevertheless, the proposed approaches mainly concern flat representation of the data and to the best of our knowledge, not much effort has been spent on mining interesting structures from such a data [WL98,WL99]. In fact, this problem is much more complicated than the classical association rule or sequential patterns, since complex structures in the form of a labeled hirearchical objects partially ordered has to be taken into account.

In this paper, we present a new algorithm, called SCM (Schema Mining) for mining structural regularities of semistructured data. The rest of this paper is organized as follows. In section 2, the problem is stated and illustrated. The algorithm SCM is described in section 3. Related work is briefly presented in section 4. Finally section 5 concludes the paper.

2 Problem Statement

In this section we give the formal definition of the structural association mining problem. First we formulate the semistructured data which widely resumes

D.A. Zighed, J. Komorowski, and J. Żytkow (Eds.): PKDD 2000, LNAI 1910, pp. 498–503, 2000.
© Springer-Verlag Berlin Heidelberg 2000

the formal description of the Object Exchange Model (OEM) defined for representing structured data [AQM+97,BDHS96,ABS00]. Second we look at the structural association mining problem in detail.

The data model that we use is based on the OEM model designed specifically for representing semistructured data. We assume that every object o is a tuple consisting of an *identifier*, a *type* and a *value*. The *identifier* uniquely identifies the object. The type is either *complex* or some identifier denoting an atomic type (like integer, string, gif-image, etc.). When type is complex then the object is called a *complex object* and value is a set (or list) of *identifiers*. Otherwise the object is an *atomic object*, and its value is an atomic value of that type. As we consider set semantics as well as list semantics, we use a circle node to represent an identifier of a set value and a squared node to represented an identifier of a list value.

We can thus view OEM data as a graph where the nodes are the objects and the labels are on the edges. In this paper we assume that there is no cycle in the OEM graph. We also require that: (i) *identifier(o)* and *value(o)* denotes the identifier and value of the object o; (ii) *object(id)* denotes the unique object with an identifier *id*; (iii) Each atomic object has no outgoing edges. (iv) If two edges connect the same pair of nodes in the same direction then they must have different label. We thus assume that we are provided with a labeling function $F_E : E \rightarrow L_E$ where L_E is the domain of edge labels. A single path in the graph is an alternating sequence of objects and labels $< o_1 l_1 o_2 ... o_{k-1} l_{k-1} o_k >$ beginning and ending with objects, in which each label is incident with the two nodes immediately preceding and following it. The number of labels from the source object to the target node in a path, k, is the length of the path. As we consider nested structures, we can consider that the length is similar to the nested level of the structure. Let P_k the set of all paths p where the length of p is k. We now consider multiple path defined as follows: a *multiple path* (or *path* for short) is a set of single paths such as the source object is the same in all the single paths. The length of the path is the maximal length of all single paths. As we are only interested in structural regularity, in the following we do not consider atomic values anymore and we use symbol $\perp$ in order to denote an atomic value in the graph.

A path p_m is a *sub-path* of a path p_n if p_m is included in p_n. We define the inclusion in the following way: A path $p_a =< o_{a_1} l_{a_1} o_{a_2} ... o_{a_{k-1}} l_{a_{k-1}} o_{a_n} >$ is included in another path $p_b =< o_{b_1} l_{b_1} o_{b_2} ... o_{b_{k-1}} l_{b_{k-1}} o_{b_m} >$ if and only if there exists integers $i_1 < i_2 < ... < i_n$ such that $o_{a_1} = o_{b_{i_1}}$, $o_{a_2} = o_{b_{i_2}}, ..., o_{a_n} = o_{b_{i_n}}$.

Example 1 The path {*category*: $\perp$, *name*: $\perp$, *address*: {*street*: $\perp$, *city*: $\perp$, *zipcode*: $\perp$}, *price*: $\perp$} is not a sub-path of {*category*: $\perp$, *name*: $\perp$, *address*: {*street*: $\perp$, *city*: $\perp$, *zipcode*: $\perp$}, *nearby*: {*price*: $\perp$, *category*: $\perp$, *name*: $\perp$}, *nearby*: {*category*: $\perp$, *name*: $\perp$, *address*: $\perp$, *address*: $\perp$, *price*: $\perp$, *zipcode*: $\perp$}} since the label *price* is not at the same level in the graph.

Let DB be a set of transactions where each transaction T consists of transaction-id and a multiple path embedded in the OEM graph and involved in the transaction. A support value ($supp(s)$) for a multiple path in the OEM graph gives its number of actual occurrences in DB. In other words, the support of a path is defined as the fraction of total data sequences that contain p. In order to decide whether a path is frequent or not, a minimum support value ($minSupp$) is specified by user, and the multiple path expression is said *frequent* if the condition $supp(s) \geq minSupp$ holds.

Given a database DB of customer transactions the problem of regularity mining, called *Schema mining*, is to find all maximal paths occurring in DB whose support is greater than a specified threshold (minimum support). Each of which represents a *frequent path*. From the problem statement, discovering structural association sequential patterns resembles closely to mining association rules [AIS93,AS94,BMUT97,FPSSU96,SON95,Toi96] or sequential patterns [AS95, SA96]. However, elements of handled association transactions have structures in the form of a labeled hirearchical objects, and a main difference is introduced with partially ordered references. In other word we have to take into account complex structures while in the association rules or sequential patterns elements are atomics, i.e. flat sets or lists of items.

3   ScM Algorithm

In this section we introduce the ScM algorithm for mining structural association of semistructured data in a large database. We split the problem into the following phases:

1. **Mapping phase:** The transaction database is sorted with transaction id as a major key and values embedded in a set-of are sorted according to the lexicographic order. In order to efficiently find structural regularity among path expressions, each path expression is mapped in the following way. If the object value is a part of a *set-of* value, then we merge an 'S' to the label (resp. a 'L' for a *list-of* value). Furthermore, in order to take into account the level of the label into the transaction, we append to the label an integer standing for the level of the label in the nested structure. When two labels occur at the same level and if the first one directly follows the second one, they are grouped together in the same set otherwise they form a new set of labels. The ordering of the *list-of* value is taken into account by creating a new set of labels. The "composite" transaction which results from the union of such sets obtained from original transaction describe a *sequence of label* and the ordering of such a sequence may be seen as the way of navigating through the path expression.

2. **Mining Phase:** The GENERAL algorithm is used to find the frequent paths in the database.

Example 2 In order to illustrate the mapping phase, let us consider the following transactions $[t_1, \{a :< e : \{f : \perp, b : \perp\}, d :< g : \perp, h : \perp >>, c : \perp\}]$. According to the previous discussion, the transaction t_1 is mapped into the following: $[t_2, (Sa_1) \ (Le_2) \ (Sf_3Sb_3) \ (Lg_3)(Lh_3)(Ld_2) \ (Sc_1)]$.

Our approach for mining structural regularities fully resumes the fundamental principles of the classical association rule problem [AIS93]. At each step k, the DB is browsed for counting the support of current candidates (procedure VERIFY_CANDIDATE). Then the frequent sequence set L_k can be built. From this set, new candidates are exhibited for being dealt at the next step (procedure CANDIDATE-GENERATION). The algorithm stops when the longest frequent paths, embedded in the DB are discovered thus the candidate generation procedure yields an empty set of new candidates. Due to lack of space we do not provide the full description of the algorithms but rather an overview.

In order to improve the efficiency of the generation as well as the organization of candidates paths we use a prefix tree structure close to the structure used in [MCP98] and improved in [MPO99]. At the kth step, the tree has a depth of k. It captures all the candidate k-path as well as all frequent paths: once candidates are counted and determined frequent they remain in their proper position in the tree and become frequent paths. Any branch, from the root to a leaf stands for a candidate path, and considering a single branch, each node at depth l ($k \geq l$) captures the lth label of the path. Furthermore along with a label, a terminal node provides the support of the path from the root to the considered leaf (included). Sets cutting is captured by using labelled edges. In order to illustrate this difference, let us consider two nodes, one being the child of the other. If the labels emboddied in the node originally occurred in the same set, the edge linking the nodes is labelled with a dashed link otherwise it is labelled with a line.

In order to improve the efficiency of retrievals when candidates paths are compared to data-sequences, we assume that each node is provided with two hashtables. The fist one is used to consider labels occurring in the same set while the other one is used for taking into set cutting.

Candidate_Generation

The candidate generation builds, step by step, the tree structure. At the beginning of step 2, the tree has a depth of 1. All nodes at depth 1 (frequent labels) are provided with children supposed to capture all frequent labels. This means that for each node, the created children are a copy of its brothers. When the kth step of the general algorithm is performed, the candidate generation operates on the tree of depth k and yields the tree of depth $k+1$. For each leaf in the tree, we must compute all its possible continuations of a single item. Exactly like at step 2, only frequent items can be valid continuations. Thus only paths captured by nodes at depth 1 are considered. Associated leaves, by construction of the tree, are brothers.

Verify_Candidate
In fact, to discovery candidates paths included in a data sequence, the data-sequence is progressively browsed beginning with its first label, then with the second, and so on. From the item, a navigation is performed through the tree until a candidate path is fully detected. This is done by comparing successively following items in the data sequence with the children of the current node in the prefix-tree structure. Finally when a leaf is reached, the examined path support the candidate and its counting value must be incremented.

The ScM algorithm is implemented using Gnu C++. Due to lack of space we do not report experiments about our approach but interested reader may refer to [Lau00].

4   Related Work

To the best of our knowledge there is few work on mining such a structural regularity in a large database [DT98]. Nevertheless, our work is very related to the problem of mining structural association of semistructured data proposed in [WL98,WL99] where a very efficient approach for mining such regularities is provided. The author propose a very efficient approach and they give some pruning strategies in order to improve the candidate generation. Nevertheless our work has some important differences. Unlike their approach we are insterested in all structures embedded in the database while they are interested in mining tree expression which are defined as a path from the root of the OEM graph to the atomic values. According to this definition of the tree expression they cannot find regularities such as *identity*: {*address*: < *street*: ⊥, *zipcode*: ⊥ >} (Cf. section 2). In fact, when parsing the database in order to find frequent tree, they are only provided with maximal tree and when only a part of the tree is frequent it is not discovered.
Discovering structural information from semistructured data has been studied in some interesting works. In this context, they are some propositions for extracting the typing of semistructured data. For instance in [NAM98] they extract the structure implicit in a semistructured schema. This approach is quite different from our since we address the repetition of a structure in a schema. Nevertheless we assume that our transaction database has been preprocessed using such a technique in order to provide an OEM graph where the raw data has been casted in terms of this structure.

5   Conclusion

In this paper we present the ScM approach for mining structural association of semistructured objects in a large database. Our approach is based on a specific generation of candidates efficiently performed using a new structure as well as a very well adapted mapping of the initial database.

References

[ABS00]   S. Abiteboul, P. Buneman, and D. Suciu. *Data on the Web*. Morgan Kaufmann, 2000.

[AIS93]   R. Agrawal, T. Imielinski, and A. Swami. Mining Association Rules between Sets of Items in Large Databases. In *SIGMOD'93*, May 1993.

[AQM⁺97]   S. Abiteboul, and al. The Lorel Query Language for Semi-Structured Data. *International Journal on Digital Libraries*, 1(1):68–88, April 1997.

[AS94]   R. Agrawal and R. Srikant. Fast Algorithms for Mining Generalized Association Rules. In *VLDB'94*, September 1994.

[AS95]   R. Agrawal and R. Srikant. Mining Sequential Patterns. In *ICDE'95*, March 1995.

[BDHS96]   P. Buneman, S. Davidson, G. Hillebrand, and D. Suciu. A Query Language and Optimization Techniques for Unstructured Data. In *SIGMOD'96*.

[BMUT97]   S. Brin, R. Motwani, J.D. Ullman, and S. Tsur. Dynamic Itemset Counting and Implication Rules for Market Basket Data. In *SIGMOD'97*, May 1997.

[DT98]   L. Dehaspe, H. Toivonen and R.D. King. Fining Frequent Substructures in Chemical Coumpounds. In *KDD'98*, August 1998.

[FPSSU96]   U.M. Fayad, G. Piatetsky-Shapiro, P. Smyth, and R. Uthurusamy, editors. *Advances in Knowledge Discovery and Data Mining*. AAAI Press, 1996.

[KS95]   D. Konopnicki and O. Shmueli. W3QS: A Query System for the World-Wide Web. In *VLDB'95*, September 1995.

[Lau00]   P.A. Laur. Recherche de regularités dans des bases de données d'objets complexes. Technical Report, LIRMM, France, June 2000.

[MPO99]   F. Masseglia, P. Poncelet and R. Cicchetti. An Efficient Algorithm for Web Usage Mining. In *Networking and Information Systems Journal*, October 1999.

[MCP98]   F. Masseglia, F. Cathala, and P. Poncelet. The PSP Approach for Mining Sequential Patterns. In *PKDD'98*, September 1998.

[NAM98]   S. Nestorov, S. Abiteboul, and R. Motwani. Extracting Schema from Semistructured Data. *SIGMOD'98*, 1998.

[SA96]   R. Srikant and R. Agrawal. Mining Sequential Patterns: Generalizations and Performance Improvements. In *EDBT'96*, September 1996.

[SON95]   A. Savasere, E. Omiecinski, and S. Navathe. An Efficient Algorithm for Mining Association Rules in Large Databases. In *VLDB'95*, pages 432–444, Zurich, Switzerland, September 1995.

[Toi96]   H. Toivonen. Sampling Large Databases for Association Rules. In *VLDB'96*, September 1996.

[WL98]   K. Wang and H.Q. Liu. Discovering Typical Structures of Documents: A Road Map Approach. In *ACM SIGIR*, August 1998.

[WL99]   K. Wang and H. Liu. Discovering Structural Association of Semistructured Data. *IEEE TKDE*, 1999.

[Wor97]   Work97. The Workshop on Management of Semistructured Data. In *www.research.att.com/s̃uciu/workshop-papers.html*, May 1997.

Improving an Association Rule Based Classifier

Bing Liu, Yiming Ma, and Ching Kian Wong

School of Computing
National University of Singapore
3 Science Drive 2, Singapore 117543
{liub, maym, wongck}@comp.nus.edu.sg

Abstract. Existing classification algorithms in machine learning mainly use heuristic search to find a subset of regularities in data for classification. In the past few years, extensive research was done in the database community on learning rules using exhaustive search under the name of association rule mining. Although the whole set of rules may not be used directly for accurate classification, effective classifiers have been built using the rules. This paper aims to improve such an exhaustive search based classification system CBA (*Classification Based on Associations*). The main strength of this system is that it is able to use the most accurate rules for classification. However, it also has weaknesses. This paper proposes two new techniques to deal with these weaknesses. This results in remarkably accurate classifiers. Experiments on a set of 34 benchmark datasets show that on average the new techniques reduce the error of CBA by 17% and is superior to CBA on 26 of the 34 datasets. They reduce the error of C4.5 by 19%, and improve performance on 29 datasets. Similar good results are also achieved against RIPPER, LB and a Naïve-Bayes classifier.

1 Introduction

Building effective classification systems is one of the central tasks of data mining. Past research has produced many techniques and systems (e.g., C4.5 [10], and RIPPER [3]). The existing techniques are, however, largely based on heuristic/greedy search. They aim to find only a *subset* of the regularities that exist in data to form a classifier.

In the past few years, the database community studied the problem of rule mining extensively under the name of association rule mining [1]. The study there is focused on using exhaustive search to find all rules in data that satisfy the user-specified minimum support (minsup) and minimum confidence (minconf) constraints.

Although the complete set of rules may not be directly used for accurate classification, effective and efficient classifiers have been built using the rules, e.g., CBA [7], LB [8] and CAEP [4]. The major strength of such systems is that they are able to use the most accurate rules for classification. This explains their good results in general. However, they also have some weaknesses, inherited from association rule mining.

- Traditional association rule mining uses only a single minsup in rule generation, which is inadequate for unbalanced class distribution (this will be clear later).
- Classification data often contains a huge number of rules, which may cause combinatorial explosion. For many datasets, the rule generator is unable to generate rules with many conditions, while such rules may be important for classification.

D.A. Zighed, J. Komorowski, and J. Zytkow (Eds.): PKDD 2000, LNAI 1910, pp. 504-509, 2000.
© Springer-Verlag Berlin Heidelberg 2000

This paper aims to improve the CBA system (*Classification Based on Associations*) by dealing directly with the above two problems. It tackles the first problem by using *multiple class minsups* in rule generation (i.e., each class is assigned a different minsup), rather than using only a single minsup as in CBA. This results in a new system called msCBA. Experiments on a set of 34 benchmark problems show that on average msCBA achieves lower error rate than CBA, C4.5 (tree and rules), and a Naïve-Bayse classifier (NB), LB and RIPPER (CAEP is not available for comparison).

The second problem is more difficult to deal with directly as it is caused by exponential growth of the number of rules. We deal with it indirectly. We try to find another classification technique that is able to help when some rules from msCBA are not accurate. The decision tree method is a clear choice because decision trees often go very deep, i.e., using many conditions. We then propose a technique to combine msCBA with the decision tree method as in C4.5. The basic idea is to use the rules of msCBA to segment the training data and then select the classifier that has the lowest error rate on each segment to classify the future cases falling into the segment. This composite method results in remarkably accurate classifiers.

2 Association Rule Mining for Classification

Association rule mining is stated as follows [1]: Let $I = \{i_1, i_2, ..., i_m\}$ be a set of items. Let D be a set of transactions (the dataset), where each transaction d (a data record) is a set of items such that $d \subseteq I$. An *association rule* is an implication of the form, $X \rightarrow Y$, where $X \subset I$, $Y \subset I$, and $X \cap Y = \varnothing$. The rule $X \rightarrow Y$ holds in the transaction set D with *confidence c* if $c\%$ of transactions in D that support X also support Y. The rule has *support s* in D if $s\%$ of the transactions in D contains $X \cup Y$.

Given a set of transactions D (the dataset), the problem of mining association rules is to discover all rules that have support and confidence greater than the user-specified minimum support (called *minsup*) and minimum confidence (called *minconf*). An efficient algorithm for mining association rules is the Apriori algorithm [1].

Mining association rules for classification: The Apriori algorithm finds association rules in a transaction data of items. A classification dataset, however, is normally in the form of a relational table. Each data record is also labeled with a class. The table data can be converted to transaction data as follows: We first discretize each continuous attribute into intervals (see e.g., [5] [6] on discretization algorithms). After discretization, we can transform each data record to a set of (*attribute, value*) pairs and a class label, which is in the transaction form. A (*attribute, value*) pair is an *item*.

For classification, we only need to generate rules of the form $X \rightarrow c_i$, where c_i is a possible class. We call such rules the *class association rules* (CARs). It is easy to modify the Apriori algorithm to generate CARs. We will not discuss it here (see [7]).

3 Classifier Building in CBA

After all rules (CARs) are found, a classifier is built using the rules. In CBA, a set of high confidence rules is selected from CARs to form a classifier (this method is also used in msCBA). The selection of rules is based on a total order defined on the rules.

Definition: Given two rules, r_i and r_j, $r_i \succ r_j$ (also called r_i precedes r_j or r_i has a higher precedence than r_j), if the confidence of r_i is greater than that of r_j, or if their confidences are the same, but the support of r_i is greater than that of r_j, or if both the confidences and supports of r_i and r_j are the same, but r_i is generated earlier than r_j;

Let R be the set of CARs, and D the training data. The basic idea of the classifier-building algorithm in CBA is to choose a set of high precedence rules in R to cover D. A CBA classifier is of the form:

$$<r_1, r_2, ..., r_n, \textit{default\_class}>. \tag{1}$$

where $r_i \in R$, $r_a \succ r_b$ if $b > a$. In classifying an unseen case, the first rule that satisfies the case classifies it. If no rule applies, the default class is used. A simple version of the algorithm for building such a classifier is given in Figure 1. [7] presents an efficient implementation of the algorithm. It makes at most two passes over the data.

R = sort(R);   /* according the precedence $\succ$ */
for each rule $r \in R$ in sequence **do**
    if there are still cases in D AND r classifies at least one case correctly **then**
        delete all training examples covered by r from D;
        add r to the classifier
    end
end
add the majority class as the default class to the classifier.

Fig. 1. A simple classifier-building algorithm

4   Improving CBA

4.1   Using Multiple Minimum Class Support

The most important parameter in association rule mining is the minsup. It controls how many rules and what kinds of rules are generated. The CBA system follows the classic association rule model and uses a single minsup in its rule generation. We argue that this is inadequate for mining of CARs because many practical classification datasets have uneven class frequency distributions. If we set the minsup value too high, we may not find sufficient rules of infrequent classes. If we set the minsup value too low, we will find many useless and over-fitting rules for frequent classes. To solve the problems, msCBA adopts the following (*multiple minimum class supports*):

minsup$_i$: For each class c_i, a different *minimum class support* is assigned. The user only gives a total minsup, denoted by *t_minsup*, which is distributed to each class:

$$\text{minsup}_i = \text{t\_minsup} \times \text{freqDistr}(c_i). \tag{2}$$

The formula gives frequent classes higher minsups and infrequent classes lower minsups. This ensures that we will generate sufficient rules for infrequent classes and will not produce too many over-fitting rules for frequent classes.

4.2   Seeking Help from Other Techniques

As we mentioned earlier, for many datasets, the rule generator is unable to generate rules with many conditions (i.e., long rules) due to combinatorial explosion. When such long rules are important for classification, our classifiers suffer. Here, we propose a combination technique. The aim is to combine msCBA with a method that is able to find long rules. Clearly, the decision tree method is a natural choice because decision trees often go very deep, i.e., using many conditions. In our implementation, we also include the Naïve-Bayes method (NB) as NB comes free from msCBA (the probabilities needed by NB are all contained in the 1-condition rules of msCBA).

The proposed combination method is based on the competition of different classifiers on different segments of the training data. The key idea is to use one classifier to segment the training data, and then choose the best classifier to classify each segment.

Let A be the classifier built by msCBA, T be the decision tree built by C4.5, and N be the Naïve-Bayse classifier. We use the rules in A to segment the data. For the set of training examples covered by a rule r_i in A, we choose the classifier that has the lowest error on the set of examples to replace r_i. That is, if r_i has the lowest error, we keep r_i. If T has the lowest error, we use T to replace r_i. If r_i is replaced by T, then in testing when a test case satisfies the conditions of r_i, it is classified by T instead of r_i. The same applies to N. The algorithm is given in Figure 2.

From line 3-6, we compute the number of errors made by r_i, T, and N on the training examples covered by each r_i. $Error_i$, $Error_{Ti}$ and $Error_{Ni}$ are initialized to 0. From line 9-11, we use T (or N) to replace r_i if T (or N) results in fewer errors on the training examples covered by r_i. $X \rightarrow$ (use T) means that in testing if a test case satisfies X (the conditions of r_i), T will be used to classify the case.

```
1    construct the three classifiers, A, T, N;
2    for each training example e do
3        find the first rule r_i in A that covers e
4        if r_i classifies e wrongly then Error_i = Error_i + 1 end
5        if T classifies e wrongly then Error_Ti = Error_Ti + 1 end
6        if N classifies e wrongly then Error_Ni = Error_Ni + 1 end
7    endfor
8    for each rule r_i (X → c_j) in R do      /*X is the set of conditions */
9        if Error_i ≤Error_Ti and Error_i ≤Error_Ni then  keep r_i
10       elseif Error_Ti ≤Error_Ni then use X → (use T) to replace r_i
11       else use X → (use N) to replace r_i
12   endfor
```

Fig. 2. The combination algorithm

5 Experiments

We now compare the classifiers built by msCBA, CBA, C4.5 (tree and rules, Release 8), RIPPER, NB, LB, and various combinations of msCBA, C4.5 and NB. The evaluations are done on 34 datasets from UCI ML Repository [9]. We also used Boosted C4.5 (the code is obtained from Zijian Zheng [11]) in our comparison. We ran all the systems using their default settings. We could not compare with existing classifier combination methods as we were unable to obtain the systems.

In all the experiments with msCBA, *minconf* is set to 50%. For *t_minsup*, from our experience, once *t_minsup* is lowered to 1-2%, the classifier built is already very accurate. In the experiment results reported below, we set *t_minsup* to 1%.

Columns 1–11 are error rates err(%); columns 12–21 are ratios **Vs msCBA+C4.5+NB**; columns 22–23 are err(%).

#	Dataset		1 msCBA	2 CBA	3 C4.5 tree	4 C4.5 rules	5 RIPPER	6 NB	7 LB	8 C4.5+NB	9 msCBA+NB	10 msCBA+C4.5	11 msCBA+C4.5+NB	12 msCBA	13 CBA	14 C4.5 tree	15 C4.5 rules	16 RIPPER	17 NB	18 LB	19 C4.5+NB	20 msCBA+NB	21 msCBA+C4.5	22 Boosted C4.5	23 msCBA+Boosted C4.5
1	anneal	CV-10	2.1	3.6	7.5	5.2	4.6	2.7	3.6	7.6	2.1	2.2	2.8	1.319	.768	.369	.533	.606	1.045	.778	.365	1.319	1.259	4.3	1.4
2	australian	CV-10	14.6	13.4	14.8	15.3	15.2	14.0	13.6	14.6	14.2	14.2	14.2	.972	1.059	.959	.927	.934	1.011	1.053	.971	1.000	1.000	15.9	13.5
3	auto	CV-10	19.9	27.2	17.6	19.9	23.8	32.1	28.1	19.0	18.5	20.9	18.5	.928	.679	1.049	.928	.776	.575	.656	.971	1.000	.882	15.1	18.5
4	breast-w	CV-10	3.7	4.2	5.6	5.0	4.0	2.4	2.7	5.6	3.1	3.0	2.4	.659	.581	.436	.488	.604	1.004	.897	.437	.775	.811	3.1	2.2
5	cleve	CV-10	17.1	16.7	21.5	21.8	21.1	17.1	17.1	16.8	16.8	17.5	17.5	1.023	1.047	.813	.802	.829	1.023	1.021	1.042	1.039	.999	20.5	18.2
6	crx	CV-10	14.6	14.1	15.0	15.1	14.6	14.6	12.9	15.1	14.5	14.2	14.3	.981	1.016	.955	.948	.978	.979	1.110	.951	.999	1.009	15.7	13.6
7	diabetes	CV-10	25.5	25.3	26.1	25.8	25.3	24.6	24.4	25.7	25.5	24.8	24.8	.973	.980	.950	.961	.980	1.008	1.016	.965	.973	1.000	25.2	24.8
8	german	CV-10	26.5	26.5	28.4	27.7	27.8	24.6	24.7	28.6	25.5	24.8	24.9	.940	.940	.877	.899	.896	1.012	1.008	.871	.976	1.003	29.4	25.2
9	glass	CV-10	26.1	27.4	30.4	31.3	35.0	29.4	30.8	28.5	26.5	29.5	29.5	1.130	1.076	.970	.942	.842	1.003	.958	1.035	1.111	1.000	24.7	24.8
10	heart	CV-10	18.1	18.5	21.8	19.2	19.6	18.1	18.2	21.1	18.1	17.4	17.0	.940	.920	.781	.886	.867	.939	.938	.807	.940	.979	20.7	20.0
11	hepatitis	CV-10	10.9	15.1	18.2	19.4	17.5	15.0	15.6	18.2	17.5	16.2	16.2	.858	1.074	.891	.836	.926	1.082	1.038	.892	.926	1.000	17.5	18.2
12	horse	CV-10	17.6	18.7	14.7	17.4	14.7	20.6	20.7	13.1	17.4	17.1	17.1	.972	.914	1.163	.983	1.163	.828	.828	1.309	.984	.979	18.7	16.0
13	hypo	CV-10	1.0	1.7	0.7	0.8	0.8	1.5	1.5	1.0	1.1	0.9	0.9	.870	.512	1.243	1.088	1.145	.565	.540	.916	.813	.926	1.1	0.9
14	ionosphere	CV-10	7.7	8.2	10.5	10.0	11.4	12.0	8.8	10.8	8.3	8.3	8.3	1.073	.787	.690	.826	.725	.690	.937	.763	1.000	1.000	6.8	8.3
15	iris	CV-10	5.3	7.1	4.7	4.7	5.3	6.0	5.3	4.7	5.3	2.7	2.7	.508	.377	.570	.580	.503	.446	.503	.575	.506	1.000	5.3	2.7
16	labor	CV-10	13.7	17.0	22.3	20.7	18.5	14.0	12.3	19.0	13.7	12.0	12.0	.876	.706	.538	.580	.663	.857	.973	.632	.878	1.000	8.3	8.3
17	led7	CV-10	28.1	27.8	30.5	27.8	30.8	26.6	26.6	26.7	26.7	27.0	26.1	.930	.940	.848	.939	.836	.981	.981	.979	.979	.967	26.2	25.1
18	lymph	CV-10	22.1	19.6	23.8	28.5	20.8	24.4	19.7	23.2	22.1	17.6	17.6	.795	.896	.738	.616	.846	.720	.893	.759	.795	1.000	18.3	16.9
19	pima	CV-10	27.1	27.6	25.8	24.5	26.3	24.5	24.7	25.4	26.1	22.7	22.0	.812	.797	.853	.898	.837	.898	.890	.866	.845	.971	27.3	24.1
20	sick	CV-10	2.8	2.7	1.1	1.5	1.9	3.9	3.0	1.1	1.9	1.9	1.9	.684	.689	1.691	1.240	.964	.473	.628	1.632	1.000	.940	1.3	1.7
21	sonar	CV-10	22.5	21.7	28.4	29.8	27.9	23.0	24.0	27.9	22.5	23.0	21.6	.960	.995	.760	.724	.775	.937	.900	.775	.960	.940	20.2	23.0
22	tic-tac-toe	CV-10	0.4	0.1	13.8	0.6	2.4	30.1	32.1	14.5	0.2	23.0	0.2	.500	2.000	.014	.333	.083	.007	.006	.014	1.000	.009	3.3	0.2
23	vehicle	CV-10	31.0	31.3	28.5	27.4	31.4	30.5	30.5	28.4	31.0	29.8	34.3	1.106	1.095	1.203	1.251	1.092	.856	1.124	1.209	1.107	1.151	24.3	29.0
24	waveform21	CV-10	20.3	20.8	22.8	21.9	20.5	19.3	17.5	23.0	20.0	17.7	17.7	.871	.858	.775	.807	.864	.914	1.009	.767	.896	.998	18.2	16.0
25	wine	CV-10	5.0	8.4	7.3	7.3	8.5	9.5	1.7	7.8	5.0	5.0	5.0	1.000	.595	.685	.686	.588	.526	1.455	.641	.829	1.000	4.0	4.5
26	zoo	CV-10	3.2	5.4	7.8	7.8	11.0	13.7	5.8	2.0	2.9	2.9	2.9	.909	.539	.373	.373	.265	.212	.500	1.455	1.000	1.000	0.0	2.9
27	Adult	test	16.7	14.4	14.6	14.1	15.6	15.8	14.2	14.0	14.0	14.9	13.9	.833	.963	.950	.984	.887	.877	.974	.992	.130	.933	16.2	16.2
28	Chess	test	2.0	1.9	0.5	1.1	1.9	12.9	7.2	0.5	2.2	0.3	0.3	.142	.149	.560	.248	.149	.022	.039	.596	.130	1.000	0.3	0.2
29	DNA	test	10.3	15.4	7.3	6.9	8.3	6.6	7.3	6.7	7.5	8.3	7.9	.771	.514	1.086	1.094	.960	1.202	1.094	1.191	1.057	.960	5.3	6.2
30	Letter	test	30.0	29.5	12.3	13.7	15.2	25.0	16.1	12.3	24.5	14.0	13.8	.459	.467	1.120	1.006	.905	.550	.856	.905	.563	.987	5.2	8.9
31	Satimage	test	14.6	15.9	14.6	14.8	15.1	18.0	13.5	15.7	14.5	14.0	14.1	.966	.890	.966	.953	.937	.786	1.048	.898	.976	1.007	10.3	12.3
32	Segment	test	6.0	6.8	4.0	6.6	7.8	6.2	5.7	4.6	6.6	3.8	3.8	.631	.559	.605	.484	.605	.628	.676	.826	.605	1.000	3.1	4.2
33	Soybean Big	test	7.5	7.5	10.5	9.6	9.2	8.9	8.8	10.5	6.6	7.0	7.0	.941	.941	.869	.731	.762	.605	.800	.667	1.067	1.000	6.1	7.0
34	Waveform40	test	24.3	24.4	29.6	30.5	28.8	21.7	21.4	29.2	23.1	23.2	23.2	.954	.951	.784	.761	.806	1.088	1.086	.795	1.007	1.000	20.9	23.1
	Average		14.9	15.5	16.0	15.6	15.8	16.9	15.3	15.5	14.2	13.5	13.3	0.96	0.83	0.82	0.81	0.78	0.79	0.90	0.86	0.91	0.99	13.1	13.0
	won-lost-tied: msCBA+C4.5+NB vs the other methods		28-5-1	26-8-0	27-7-0	29-5-0	31-3-0	24-10-0	23-11-0	26-8-0	21-7-6	14-4-16													
	* average ratios after the largest and smallest ratios are removed													0.96	0.82	0.82	0.81	0.79	0.80	0.86	0.86	0.92	0.99		

Table 1: Experiment Results

* One problem with average error ratios is that when the actual error rates are very small, ratios tend to have extreme values. Here, we recompute the average ratios of msCBA+C4.5+NB vs the other methods after the largest and smallest values are removed.

Experiment results are shown in Table 1. The error rates on the first 26 datasets are obtained from 10-fold cross-validation, while on the last 8 datasets they are obtained from the test sets. All the composite methods involving C4.5 uses C4.5 tree.

Error rate comparison: For each dataset, column 1-11 show the error rates of msCBA, CBA, C4.5 tree, C4.5rules, RIPPER, NB, LB, C4.5+NB (C4.5 tree combinedwith NB), msCBA+NB, msCBA+C4.5 and msCBA+C4.5+NB respectively. Column 12-21 show the ratios of the error rate of msCBA+C4.5+NB vs. the other methods. From the table we see that on average the error rate of msCBA is lower than every other individual method. It is also clear that over the 34 datasets, the composite methods are superior to individual methods. msCBA+C4.5+NB gives the lowest error rate on average. It reduces the error of C4.5 tree (or rules) by 18% (or 19%) on average, and its won-lost-tied record against C4.5 tree (or rules) is 27-7-0 (or 29-5-0). It reduces the error of msCBA by 14%, and its won-lost-tied record against msCBA is 28-5-1. Similar good results are also achieved against CBA, RIPPER, NB and LB.

msCBA+C4.5 and msCBA+C4.5+NB have similar performances. This confirms our intuition that msCBA's weakness is overcome by deep trees of C4.5. Column 22 and 23 show the error rates of boosted C4.5 and msCBA+boostedC4.5. We see that msCBA+C4.5+NB's results are comparable to boosted C4.5, and its won-lost-tied record against boosted C4.5 is 18-15-1. Since boosted C4.5 is regarded as one of the best classifiers, we can say that msCBA+C4.5+NB is also among the best.

6 Conclusion

This paper aims to improve an exhaustive search based classification system CBA. It first identified two weaknesses of the system, and it then proposed two new techniques to deal with the problems. The techniques produce markedly better classifiers.

References

1. Agrawal, R. & Srikant, R. 1994. Fast algorithms for mining association rules. VLDB-94.
2. Chan, P. & Stolfo, J. S. 1993. Experiments on multistrategy learning by meta-learning. Proc. Second Intl. Conf. Info. Know. Manag., 314-323.
3. Cohen, W. 1995. Fast effective rule induction. ICML-95.
4. Dong, G., Zhang, X. Wong, L. Li, J. 1999. CAEP: classification by aggregating emerging patterns. Discovery-Science-99.
5. Fayyad, U. & Irani, K. 1993. Multi-interval discretization of continuous-valued attributes for classification learning. IJCAI-93, 1022-1027.
6. Kohavi, R., John, G., Long, R., Manley, D., & Pfleger, K. 1994. MLC++: a machine-learning library in C++. Tools with artificial intelligence, 740-743.
7. Liu, B., Hsu, W. & Ma, Y. 1998. Integrating classification and association rule mining. KDD-98.
8. Meretkis, D. & Wuthrich, B. 1999. Extending naïve bayes classifiers using long itemsets. KDD-99.
9. Merz, C. J. & Murphy, P. 1996. UCI repository of machine learning database. [http://www.cs.uci.edu/~mlearn].
10. Quinlan, J. R. 1992. C4.5: program for machine learning. Morgan Kaufmann.
11. Zheng, Z. and Webb, G. 1999. Stochastic attribute selection committees with multiple boosting: Learning more accurate and more stable classifier committees. *PAKDD-99.*

Discovery of Generalized Association Rules with Multiple Minimum Supports[1]

Chung-Leung Lui and Fu-Lai Chung

Department of Computing, Hong Kong Polytechnic University,
Hunghom, Kowloon, Hong Kong.
{cscllui, cskchung}@comp.polyu.edu.hk

Abstract. Mining association rules at multiple concept levels leads to the discovery of more concrete knowledge. Nevertheless, setting an appropriate *minsup* for cross-level itemsets is still a non-trivial task. Besides, the mining process is computationally expensive and produces many redundant rules. In this work, we propose an efficient algorithm for mining generalized association rules with multiple *minsup*. The method scans appropriately $k+1$ times of the number of original transactions and once of the extended transactions where k is the largest itemset size. Encouraging simulation results were reported.

1. Introduction

Generalized association rules (GAR) [9] and multiple level association rules (MLAR) [2,3] were proposed to find rules at multiple concept levels. Unfortunately, a dilemma exists when setting *minsup* in GAR. First, if it is set too high, low-level rules may not have *minsup*. Second, if it is set too low, this will produce too many rules [7]. On the other hand, MLAR can set different *minsup* at each level, but extension to cross-level itemsets seems to be difficult. Furthermore, both methods generate many redundant rules. Pruning algorithms [3,9] were then proposed, but they always remove descendant rules, retaining rules at the most general concept that may often lead to rules corresponding to prior knowledge [4]. Efficiency is another major concern. If the largest itemset discovered is k-itemset, the number of scanning of database in GAR is roughly $k+1$ times of the number of extended transactions. MLAR scans roughly $k+1$ times of the number of original transactions for each level of taxonomy.

In this work, a new framework based on concept generalization [5-6,8] for mining generalized association rules with multiple *minsup* is proposed. It scans roughly $k+1$ times of the number of original transactions and once of the extended transactions, if the largest size of discovered itemsets is k. The *minsup* of each itemset can be different. Hence, low-level rules can have *minsup* while high-level rules are prevented from combinatorial explosion. The paper is organized into five sections. The new framework and major mining steps are described in section 2 and 3 respectively. Simulation results are reported in section 4 and the final section concludes the paper.

---

[1] Ack: supported by the H.K. Polytechnic University Research Studentship, project no. G-V722

D.A. Zighed, J. Komorowski, and J. Zytkow (Eds.): PKDD 2000, LNAI 1910, pp. 510-515, 2000.
© Springer-Verlag Berlin Heidelberg 2000

2. A New Framework for Mining Generalized Association Rules

In Fig.1, the basic framework for mining *multiple-level frequent itemsets* is shown. It is efficient by running once the Apriori algorithm on the original transactions and scanning once the extended transactions. As the support at low concept level is implicitly smaller, finding rules with *minsup* at higher level requires first generating a list of itemsets with their support < *minsup*. We call them as *less frequent itemsets* and the support parameter as *sub-minimum support*, γ. Note here that the smaller the value of γ, the larger the potential to discover hidden high-level rules, but the slower the mining process. The concept generalization process, it is divided into two parts, namely, inter-itemset and intra-itemset generalizations. Firstly, the *less frequent itemsets* are divided into two groups, i.e., the *frequent itemsets* and the itemsets with γ < *minsup*. The latter group will then undergo the inter-itemset generalization. The results (*inter-generalized itemsets*) will be merged with the *frequent itemsets* and they will then be passed to the intra-itemset generalization. After itemset generalization, all subsets of the generalized itemsets are generated and any duplication is deleted because any subset of a large itemset must be large [1]. Those generalized itemsets and its subsets are *multiple-level less frequent itemsets*. They are finally filtered by multiple *minsup* to produce the *multiple-level frequent itemsets*. In this step, we have to count the support of each candidate itemset by scanning the extended transactions only once. As proposed in [9], we can pre-compute the ancestors for each item and add those ancestors that are in *multiple-level less frequent itemsets*. Following the same idea, we propose to pre-compute the generality of each item. This helps to speed up concept generalization and multiple *minsup*. With the final *multiple-level frequent itemsets*, it is a simple step to apply rule generation algorithm [1] to produce generalized association rules.

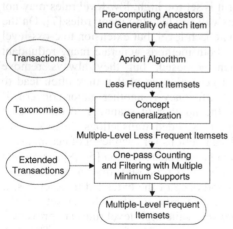

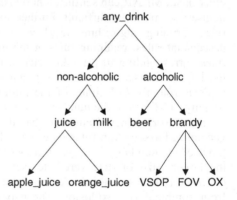

Fig. 1. Framework for mining *multiple-level frequent itemsets*

Fig. 2. Taxonomy of Drink

3. Major Mining Steps

3.1 Concept Generalization of Itemsets

Concept generalization of itemsets is a process of forming a general description of a class of objects (leaf-level itemsets) given a set of taxonomies. With reference to Fig.2, let us consider the following three itemsets t1:(*drink apple_juice*), t2:(*drink milk*), and t3:(*drink orange_juice*). Since all itemsets are characterized by the same predicate *drink*, the conceptual distance [5] between them is exclusively determined by the distance between their arguments. Intuitively, distance(*t1, t3*) < distance(*t1, t2*). Let us also consider the following generalizations $G(t_i, t_j)$ of two itemsets t_i and t_j: G(*t1, t3*):(*drink juice*). While *orange_juice* is an instance, *juice* is a generalization with 2 instances. Thus, we define the degree of generality of an argument $g(a)$ as:

$$g(a) = \frac{number\ of\ instances\ of\ "a"}{number\ of\ instances\ of\ the\ domain\ of\ "a"} \tag{1}$$

e.g., g(*juice*) = 2/7. This definition applies also to the so-called linear (quantitative) variables [8]. To sum up, Let P_i be a predicate; a_i, b_i and c_i be the arguments of P_i:

t1: $(P_1\ a_1)$ & $(P_2\ a_2)$ & ... $(P_i\ a_i)$ ... & ... $(P_n\ a_n)$
t2: $(P_1\ b_1)$ & $(P_2\ b_2)$ & ... $(P_i\ b_i)$ ... & ... $(P_n\ b_n)$
G(*t1, t2*): $(P_1\ c_1)$ & $(P_2\ c_2)$ & ... $(P_i\ c_i)$ ... & ... $(P_n\ c_n)$

Let $g(c_i)$ be the degree of generality of c_i. The estimation of the contribution of P_i to the dissimilarity D and similarity S between t1 and t2 are by the following functions:

$$D(t1, t2, P_i) = 0.5(\Sigma g(c_i))/n \tag{2}$$
$$S(t1, t2, P_i) = 1 - D(t1, t2, P_i) \tag{3}$$

We distribute the total contribution of P_i between similarity and dissimilarity in accordance with the generality degree of its arguments. The range of similarity and dissimilarity are 0.5 to 1 and 0 to 0.5 respectively, but their total must be 1. The estimation of conceptual distance between t1 and t2 (corresponding to G), as a function of D and S is f(*t1, t2, G*) = D(*t1, t2, G*)/S(*t1, t2, G*). We take the minimum of f(*t1, t2, G*) over all possible generalizations of t1 and t2 as the conceptual distance:

$$concept\_dist(t1, t2) = \min_{G(t1, t2)} \{f(t1, t2, G(t1, t2))\} \tag{4}$$

3.1.1 Inter-itemset Generalization

When same predicate appears in two itemsets whose size and number of common items are equal, they can be inter-generalized. The common items should be extracted out first, and they will be added back after generalization. For instance, *Food D* is a common item and it is extracted. *Drink X* and *drink Y* are generalized to *drink Z*. *Food D* is being added back lastly. Thus, we have t1:(*drink X*)(*food D*), t2:(*drink Y*)(*food D*) → t3:(*drink Z*)(*food D*). Unfortunately, the problem becomes complicated if there is more than one item in each itemset. For example, we have two combinations in the following case:

Combination 1:

t1:(*drink apple_juice*)(*drink beer*)

t2:(*drink orange_juice*)(*drink VSOP*)

→ t3:(*drink juice*)(*drink alcoholic*)

Combination 2:

t1:(*drink apple_juice*)(*drink beer*)

t2:(*drink VSOP*)(*drink orange_juice*)

→ t3:(*drink any_drink*)(*drink any_drink*)

In this case, only one generalization is chosen as new itemset where its conceptual distance is minimum. For example, combination 1 is chosen for inter-generalization.

Combination 1:

$D(t1, t2, drink) = 0.5*[2/7+4/7]/2 = 0.21$

$S(t1, t2, drink) = 1-D(t1, t2, drink) = 0.79$

Conceptual distance $= 0.21/0.79 = 0.27$

Combination 2:

$D(t1, t2, drink) = 0.5*[7/7+7/7]/2 = 0.5$

$S(t1, t2, drink) = 1-D(t1, t2, drink) = 0.5$

Conceptual distance $= 0.5/0.5 = 1$

3.1.2 Intra-itemset Generalization

The problem of intra-itemset generalization is much simpler. For those items with common predicate, they are generalized. For instance, t1:(*drink X*)(*drink Y*)(*food D*) → t1a:(*drink Z*)(*food D*). The intra-generalized itemset does not replace the original one, i.e., itemsets t1 & t1a co-exist. The newly generated itemsets are then undergone intra-itemset generalization recursively. The following is an example:

<Case 1> t1:(*drink beer*)(*drink FOV*)(*drink XO*) → t1a:(*drink alcoholic*)

<Case 2> t1:(*drink beer*)(*drink FOV*)(*drink XO*) → t1b:(*drink alcoholic*)(*drink XO*)

<Case 3> t1:(*drink beer*)(*drink FOV*)(*drink XO*) → t1c:(*drink FOV*) (*drink alcoholic*)

<Case 4> t1:(*drink beer*)(*drink FOV*)(*drink XO*) → t1d:(*drink beer*) (*drink brandy*)

If a generalized itemset consists of an item and its ancestor, the itemset should be deleted. Since the support for an itemset t that contains both an item x and its ancestor x' will be the same as the support for the itemset t-x' [9]. So, case 2 and 3 are deleted.

3.2 Multiple Minimum Supports

Now, we propose a method to calculate *minsup* for each itemset dynamically from its generality. Let t be an itemset, and x_i be an item in t. The generality of t is:

$$G(t) = min[g(x_1), g(x_2), ..., g(x_n)] \qquad (5)$$

where the predicate of x_i can be different. Here, we define *lower-minimum support* and *upper-minimum support* for lowest level and highest level itemsets respectively.

Definition 1: The *lower-minimum support* β is a user-defined parameter. This is the *minsup* for the lowest concept level itemsets, and $\beta > \gamma$.

Definition 2: The *upper-minimum support* α is a user-defined parameter. This is the *minsup* for the highest concept level itemsets, and $\alpha > \beta$.

Based upon α and β, the *minsup(t)* for each itemset t is computed independently as:

$$minsup(t) = \beta + \{(\alpha-\beta) * g(t)\} \qquad (6)$$

Here, *minsup(t)* varies from β to α, depending on the generality g(*t*). For example, if α = 20, β = 10, and g(*t*) = 0.2, then *minsup(t)* = 12. Subsequently, each *multiple-level less frequent itemset* can be filtered by its *minsup* dynamically.

4. Simulation Results

We simulated two sets of experiment. First, the performance of the proposed algorithm with single *minsup* against the GAR algorithm [9] was compared. Second, the effectiveness of the proposed mining algorithm with multiple *minsup* was demonstrated. The synthetic data generation program was obtained from the IBM Almaden Research Center [http://www.almaden.ibm.com/cs/quest]. Details can be found in [1,9]. The default parameters for both experiments are shown in Table 1 & 2.

Table 1. Parameters for data generation

Parameter	Value
Number of transactions	100K
Average size of the transactions	5
Number of items	1,000
Number of roots	10
Number of levels	5

Table 2. Parameters for the experiments

Parameter	Value
Minimum support	10
Minimum confidence	80
Sub-minimum support (Concept Generalization only)	0.3
r-interest (Cumulate only)	1.1

In the first simulation, the performance of the GAR (Cumulate) algorithm with redundant rules pruned [9] and the proposed method (Concept Generalization) with single *minsup* was compared and shown in Fig.3. We changed the number of transactions from 100K to 400K. The number of rules found by both algorithms shows little changes, but the proposed method scales up better than the Cumulate algorithm in execution time. In the second simulation, the performance of the proposed method with multiple *minsup* was evaluated. We first ran the algorithm with single *minsup* at 10% and 20%. Then, we fixed the β at 10% and varied the α from 10% to 20%. The results are plotted in Figs.4&5. At α = 10%, the number of itemsets/rules found is exactly the same as that of single *minsup* fixed at 10%. At α = 20%, the number of itemsets/rules is smaller than single *minsup* with 20%, because not all itemsets' *minsup* are 20%. In fact, their *minsup* vary from 10% to 20%.

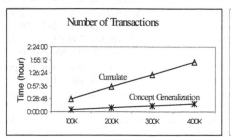
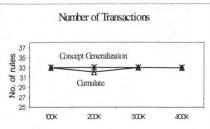

Fig. 3. Simulation Results I

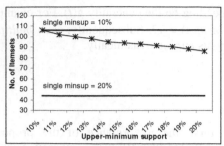

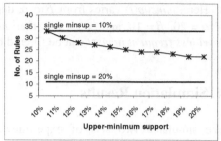

Fig. 4. Number of itemsets discovered with different *upper-minimum supports*

Fig. 5. Number of rules discovered with different *upper-minimum supports*

5. Conclusions

We have introduced a new approach to the discovery of generalized association rules with multiple minimum supports. It eliminates some of the limitations of existing generalized or multiple level association rule mining algorithms, e.g., using single *minsup* or setting it at each level manually. More importantly, the number of scanning of database required by the proposed algorithm is approximately $k+1$ times of the number of original transactions plus one pass of the extended transactions, if the largest itemset being discovered is k-itemset; while other methods have to scan $k+1$ times of the number of extended transactions or more.

Reference

1. R. Agrawal and R. Srikant, "Fast algorithms for mining association rules," *Proc. VLDB Conference*, Santiago, Chile, pp.487-499, Sept. 1994.
2. J. Han and Y. Fu, "Discovery of multiple-level association rules from large databases," *Proc. 21ˢᵗ VLDB Conference*, Zurich, Switzerland, pp.420-431, Sept. 1995.
3. J. Han and Y. Fu, "Mining multiple-level association rules in large databases", *IEEE Trans. on Knowledge and Data Engineering*, Vol.11, No.5, pp.798-805, Sept. 1999.
4. M. Klemettinen, et. al., "Finding interesting rules from large sets of discovered association rules," *Proc. 3ʳᵈ Int. Conf. on Information and Knowledge Management*, pp.401-408, 1994.
5. Y. Kodratoff, and J. G. Ganascia, "Improving the generalization step in learning," *Machine Learning: An Artificial Intelligence Approach*, vol.2, LA, pp. 215-244, 1986.
6. Y. Kodratoff, and G. Tecuci, "Learning based on conceptual distance," *IEEE Transactions on Pattern Analysis and Machine Intelligence*, Vol.10, No.6, pp.897-909, Nov. 1988.
7. B. Liu, W. Hsu, and Y. Ma, "Mining association rules with multiple minimum supports," *Proc. ACM SIGKDD Conf. on KDD*, San Diego, CA, pp. 337-341, August, 1999.
8. R. S. Michalski, and R. Stepp, "Learning by observation," *Machine Learning: An Artificial Intelligence Approach*, pp.163-190, 1983.
9. R. Srikant and R. Agrawal, "Mining generalized association rules," *Proc. of the 21ˢᵗ VLDB Conf.*, Zurich, Switzerland, pp.407-419, Sept. 1995.

Learning Dynamic Bayesian Belief Networks Using Conditional Phase-Type Distributions

Adele Marshall[1], Sally McClean[1], Mary Shapcott[1], and Peter Millard[2]

[1]School of Information and Software Engineering, University of Ulster, Jordanstown.
BT37 0QB, Northern Ireland. UK
{AH.Marshall, SI.McClean, CM.Shapcott}@ulst.ac.uk
[2] Department of Geriatric Medicine, St George's Hospital, London. SW17 0QT, UK
P.Millard@sghms.ac.uk

Abstract. In this paper, we introduce the Dynamic Bayesian Belief Network (DBBN) and show how it can be used in data mining. DBBNs generalise the concept of Bayesian Belief Networks (BBNs) to include a time dimension. We may thus represent a stochastic (or probabilistic) process along with causal information. The approach combines BBNs for modelling causal information with a latent Markov model for dealing with temporal (survival) events. It is assumed that the model includes both qualitative (causal) and quantitative (survival) variables. We introduce the idea of conditional phase-type (C-Ph) distributions to model such data. These models describe duration until an event occurs in terms of a process consisting of a sequence of phases - the states of a latent Markov model. Our approach is illustrated using data on hospital spells (the process) of geriatric patients along with personal details, admissions reasons, dependency levels and destination (the causal network).

1 Introduction

Bayesian Belief Networks (BBNs) are statistical graphical models which encapsulate causal information and provide a framework for describing and evaluating probabilities using a network of inter-related variables. BBNs have been discussed by a number of authors. [1] has described the use of such models as providing a "unified qualitative and quantitative way for representing and reasoning with probabilities and independencies". Dependency may result from some sort of causal mechanism coming into play. Such causality is often represented by means of a BBN [2]. Benefits of using such a representation include the incorporation of prior knowledge, the provision of validation and insight, and the ability to learn causal interactions [3]. Such an approach may lead to the development of probabilistic expert systems [4].

Dynamic Bayesian Belief Networks (DBBNs) generalise BBNs to include a time dimension. We may thus represent a stochastic (or probabilistic) process along with causal information [5][6]. There have been various efforts to extend the standard theory of BBNs to allow time to be modelled. [7] provide a temporal definition ofcausal dependence where a set of variables indexed by time is associated with each cause and with an effect. Other approaches have been introduced by [8-13].

D.A. Zighed, J. Komorowski, and J. Zytkow (Eds.): PKDD 2000, LNAI 1910, pp. 516-523, 2000.
© Springer-Verlag Berlin Heidelberg 2000

In statistical theory, Markov models are often used to represent stochastic processes. Structured phase-type (Ph) distributions [14] describe duration until an event occurs in terms of a process consisting of a sequence of latent phases. For example, duration of stay in hospital can be thought of as a series of transitions through phases such as: acute illness, intervention, recovery or discharge. This captures how a domain expert conceptualises the process. We can prove that any such statistical distribution may be represented arbitrarily closely by one of phase-type form [15].

2 The Model

Our aim is to provide an approach which uses probabilistic temporal reasoning to describe a process which commences with an initiating event and may terminate in a number of different ways. The initiating event may be described in terms of a BBN. We begin by presenting a motivating example, as follows.

2.1 A Motivating Example:

Consider a situation where, at time zero, an elderly person has a traumatic event, such as a stroke or a fall, and is admitted to hospital where she remains until discharge at time t. Discharge may be either due to death or discharge alive to one of a number of destinations such as "own home" or "nursing home". On admission to hospital, data are available on a number of variables which may be regarded as causal with respect to the traumatic event e.g. domestic situation, or previous state of health. The ability to predict this length of stay in hospital is important both to the geriatrician and to the hospital manager.

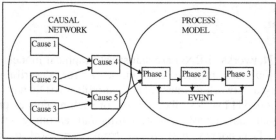

Fig. 1. The underlying representation describes a DBBN in terms of a number of interrelated causal nodes which temporally precede and predetermine (in a probabilistic sense) the effect node(s) which constitute the process. The effect node(s) here are characterised by a continuous positive random variable(s), the duration, described by a phase-type distribution

We introduce a model which combines a BBN (causal network) with a process model (Fig. 1.). The Process Model may be defined in a manner similar to that of [5][16] where we consider an event <E> which initiates a process P at time zero and <P, t> indicates that the process P is active at time t. Then prob<P, t> is the probability that the process is still active at time t. Prob<P, t> is known as the *survivor function*, denoted by F(t) and, for a continuous time representation, its derivative f(t) is the *probability density function* (p.d.f.) of the time for which the process is active. Here we define f(t) by

f(t) δt = prob(process terminates in (t, t+δt) I process is still active at t).

Some previous work has been done on data of this sort, (including both qualitative quantitative variables) mainly involving the introduction of conditional Gaussian (CG) distributions [17][18]. We here introduce the idea of Conditional Phase-type (C-Ph) distributions which are more appropriate for process data.

3 Learning DBBNs Using Conditional Phase-Type Distributions

3.1 Phase-Type Distributions

To describe the probability prob<P, t> that the process is still active at time t we will use a sub-class of phase-type distributions, the Coxian sub-class of phase-type distributions [19]. These have the latent states (or phases) ordered with the process starting in the first of these and then developing by either sequential transitions through these phases or transitions into the absorbing state (the terminating event, phase $n+1$). Such phases may then be used to describe stages of a process which terminates at some stage. For example, in the context of durations of hospital treatment, patients may be thought of as passing through (latent) stages of diagnosis, assessment, rehabilitation and long-stay care where most patients are eventually rehabilitated and discharged. We define a phase-type distribution, as follows: Let $\{X(t); t \geq 0\}$ be a (latent) Markov chain in continuous time with states $\{1, 2, ...n, n + 1\}$ with $X(0) = 1$, and

$$prob\{ X(t+\delta t) = i + 1 | X(t) = i \} = \lambda_i \delta t + o(\delta t) \tag{1}$$

and for $i = 1, 2, ...n$

$$prob\{ X(t + \delta t) = n + 1 | X(t) = i \} = \mu_i \delta t + o(\delta t). \tag{2}$$

Here states $\{1, 2, ...n\}$ are latent (transient) states of the process and state $n + 1$ is the (absorbing) state initiated by the terminating event, λ_i represents the transition from state i to state (i+1) and μ_i the transition from state i to the absorbing state $(n+1)$. [20] developed the theory of Markov chains such as those defined by (1) and (2). The *probability density function* of T is as follows;

$$f(t) = \mathbf{p} \exp\{Qt\}\mathbf{q} \tag{3}$$

where:

$$\mathbf{p} = (1\ 0\ 0\ ...0\ 0), \tag{4}$$

$$\mathbf{q} = -\mathbf{Q1} = (\mu_1\ \mu_2...\mu_n)^T. \tag{5}$$

and the matrix $\mathbf{Q}$ is:

$$\mathbf{Q} = \begin{bmatrix} -(\lambda_1 + \mu_1) & \lambda_1 & 0 & \cdots & 0 & 0 \\ 0 & -(\lambda_2 + \mu_2) & \lambda_2 & \cdots & 0 & 0 \\ \vdots & \vdots & \vdots & \cdots & \vdots & \vdots \\ 0 & 0 & 0 & \cdots & -(\lambda_{n-1} + \mu_{n-1}) & \lambda_{n-1} \\ 0 & 0 & 0 & \cdots & & -\mu_n \end{bmatrix}, \tag{6}$$

with the λ_i's and μ_i's from (1) and (2) (Fig. 2). Equation (3) gives the Coxian phase-type distribution (p.d.f.), with the λ's describing transitions through the ordered transient states $\{1, 2, ...n\}$ and the μ's transitions into the (single) absorbing state $n+1$.

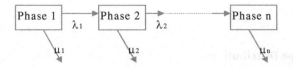

Fig. 2. An illustration of phase-type distributions where λ_i represents the transition from phase i to phase (i+1) and μ_i the transition from phase i to the absorbing phase ($n+1$)

3.2 The Dynamic Bayesian Belief Network (DBBN)

In the DBBN we have causal nodes $\mathbf{C}=\{C_1,...C_m\}$ and process nodes $\mathbf{Ph} =\{Ph_1,...Ph_n\}$ where the causal nodes comprise a BBN in the usual sense. We may therefore represent the joint distribution of $\mathbf{C}$ by:

$$P(\mathbf{C}) = \prod_i P(C_i \mid pa(C_i)) \tag{7}$$

where pa is the parent set of C_i. The distribution of the process nodes $\mathbf{Ph}$ is then given by the p.d.f.:

$$f(t \mid pa(process)) = \mathbf{p} \exp\{\mathbf{Q}t\}\mathbf{q} \tag{8}$$

where pa(process) are the causal nodes which are parents of the process and $\mathbf{p}$, $\mathbf{Q}$ and $\mathbf{q}$ vary according to the values of the causal nodes in pa(process). The process may therefore be considered locally independent of the causal network.

When the topology of the causal network is known a priori we may induce the DBBN by using maximum likelihood methods to fit parameters to phase-type distributions for each set of values in pa(process). A sequential procedure (unsupervised learning) is adopted whereby increasing numbers of n phases are tried, starting with n=1 (corresponding to the exponential distribution), until no improvement in the fit to the data can be obtained by adding a new phase. Such an approach may be implemented by using a series of likelihood ratio tests [21]. If the causal network topology is not known a priori, likelihoods for different models may be compared until the optimal topology is found.

3.3 Learning Incomplete DBBNs

It is frequently the case that DBBNs represent data which is incomplete. Duration of stay is typically *missing* since learning of processes is often incremental and for some individuals the length of stay will not be known when data collection ceases; typically the cessation of data collection occurs at the current time point. Such data are known in the statistical literature as *right censored* and are a common feature of survival data of which our process is an example. More generally, such data are referred to as being *incomplete*. For our model such incompleteness may be straightforwardly taken account of by incorporating appropriate terms into the likelihood functions. The likelihoods may then be maximised using the EM (Expectation-Maximisation) algorithm [21][22]. Use of the EM algorithm with phase-type distributions is described by [23].

4   An Application to Geriatric Medicine

The data analysed in this example relate to geriatric patients at St George's Hospital, London. Durations of hospital stay (the process) were available for patients along with personal details, admission reasons, patient dependency levels and outcome (the causal network). In total there were 4730 patients details recorded in the clinical computer system (Clinics) between 1994-1997. The significance of each individual Clinics variable on patient length of stay was tested using Chi-square tests, the most significance of these are included in Fig. 3.

These variables have been used as input into CoCo [24] a computer package capable of learning BBNs. The package uses the maximum likelihood approach to obtain the causal networks which best describe the data set. It does so by comparing the complete model with all possible candidate models. Fig. 3 is one of various candidate models that can represent the causal relations between the Clinics variables.

The BBN consists of a number of interrelated causal nodes which are grouped into blocks according to the time domain to which they relate. The causal network in Fig. 3 comprises of four such blocks. Whenever a patient is admitted to a geriatric ward, their personal details are recorded (block 1 – personal details), they then undergo an initial medical examination to investigate their condition or illness (block 2 – admission reasons). Following the medical examination, the patient is assessed for level of dependency to carry out everyday activities. The score given to patients is known as the Barthel score which is categorised into grades and used in the third block in the BBN. After spending a period of time in hospital (duration of stay) the patient will experience one of three events; discharge from hospital, transfer to another ward or hospital or death (destination). This is the terminating event, the last node in the BBN. The network is further described in [25].

Fig. 3 is an illustration of how the Clinics variables might be interrelated. In consultation with the geriatricians and the Chi-square test results of significance, we selected three of the most significant of these variables to produce a simplified model similar in form to that in Fig. 1. The nodes (gender and stroke) represent the causal network while duration of stay, the continuous variable defines the process model.

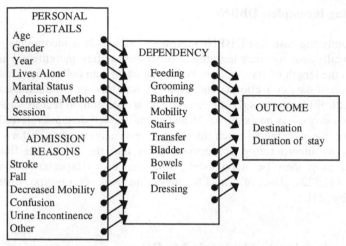

Fig. 3. An illustration of how the Clinics variables might be interrelated.

Similar data to the Clinics data set is described in [26] where a two phase distribution is fitted and reported to be a reasonable approximation to the data. We have here fitted a general phase-type distribution to the continuous variable, duration of stay. We induced the network displayed in Fig. 4.

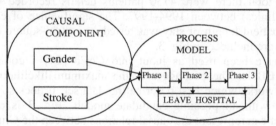

Fig. 4. DBBN for the Clinics data set where there are two causal nodes, gender and stroke, which temporally precede the process variable, 'duration of stay'

The number of phases in each of the cohorts of patient length of stay was induced by unsupervised learning. The p.d.f. of the Coxian phase-type distribution is represented by equation (7). The parameters from (1) and (2), which describe this distribution, are determined using maximum likelihood function:

$$L = \prod_{g=1}^{2} \prod_{s=1}^{2} (p_{gs}) \prod_{i}^{n_{gs}} f_{gs}(t_i) \tag{9}$$

where g represents the gender, s the stroke variable, n_{gs} the number of patients in cohort gs and f_{gs} the corresponding p.d.f. Each cohort of patients is separately fitted to a phase-type distribution. The sum of these represents the log-likelihood function (Table 1). MATLAB software [27] is used to implement the Nelder-Mead algorithm [28] to perform the likelihood ratio tests which determine the optimal number of phases in the distribution. The estimates for the μ_i and λ_i parameters of the phase-type distributions are displayed in Table 2. The optimality was also assessed using the least squares approach; an alternative to the likelihood ratio test. Both approaches yielded similar results.

Table 1. Phase-Type Distributions - displays the results of fitting phase-type distributions to the data. The * indicates the optimal number of phases for each particular patient cohort.

Patient Cohort	Log-Likelihood
Male Stroke	n=1, L=-481.87; n=2, L=-452.77*; n=3, L=-452.46
Male Non-Stroke	n=1, L=-7828.3; n=2, L=-6415.1*; n=3, L=-6414.1
Female Stroke	n=1, L=-775.96; n=2, L=-705.49*
Female Non-Stroke	n=1, L=-17251; n=2, L=-13725; n=3, L=-13705; n=4, L=-13700*

Table 2. Estimates for the C-Ph Distribution

Patient Cohort	Estimates of Parameters
Male Stroke	$\mu_1 = 0.0272$ $\mu_2 = 0.0012$ $\lambda_1 = 0.0008.$
Male Non-Stroke	$\mu_1 = 0.0043$ $\mu_2 = 0.0007$ $\lambda_1 = 0.0020.$
Female Stroke	$\mu_1 = 0.0256$ $\mu_2 = 0.0009$ $\lambda_1 = 0.0011.$
Female Non-Stroke	$\mu_1 = 0.0389$ $\mu_2 = 0.0005$ $\mu_3 = 0.0011$ $\mu_4 = 0.0001$ $\lambda_1 = 0.0013$ $\lambda_2 = 0.0009$ $\lambda_3 = 0.0003.$

5 Summary and Conclusion

There is extensive literature on graphical models and BBNs. However the extension of such models to dynamic situations, where time is an important factor, is an important research area which requires further attention. We believe that our approach has much to contribute in the development of this important research area.

In this paper our concern has been with providing a new technique for mining DBBNs. Our approach combines ideas from statistics for the analysis of survival data with extensive literature on BBNs. We have thus provided a methodology for probabilistic temporal reasoning. We have introduced a particular type of latent Markov model - the conditional Phase-type distribution - to describe local dependencies in relation to a stochastic process. Our resulting Dynamic Bayesian Belief Network is hybrid, in that we use discrete variables for the causal model and a continuous variable for the stochastic process. We use the EM algorithm to learn from data which are statistically incomplete and which include latent variables. A general approach to such problems has been introduced and illustrated by a simple example.

References

1.   Buntine, W.: A Guide to the Literature on Learning Probabilistic Networks from Data. IEEE Transactions on Knowledge and Data Engineering, Vol. 8(2) (1996) 195-210
2.   Pearl, J.: Causal Diagrams for Empirical Research. Biometrika **82** (1995) 669-710
3.   Friedman, N., Goldszmidt, M., Heckerman, D., Russell, S.: Challenge: What is the Impact of Bayesian Networks on Learning? Proc. 15th Int. Joint Conf. on AI (1997) 10-15
4.   Spiegelhalter, D., Cowell, R.G. Learning in Probabilistic Expert Systems. In Bernando, J.M., Berger, J.M., Dawid, A.P., Smith, A.F.M. (eds.) Bayesian Statistics **4**. Oxford University Press (1992) 447-465
5.   Dean, Y., Kanazawa, K.: A Model for Reasoning about Persistence and Causation. Computational Intelligence, Vol. 5(3) (1989) 142-150

6.   Russell, S., Binder, J., Koller, D., Kanazawa, K.: Local Learning in Probabilistic Networks with Hidden Variables. Proc. 14th Int. Joint Conf. on AI (1995) 1146-1152
7.   Heckerman, D., Breese, J.S.: A New Look at Causal Independence. Proc. 10th Conf. UAI (1997)
8.   Aliferis, C.F., Cooper, G.F.: A Structurally and Temporally Extended BBN Model: Definitions, Properties and Modeling Techniques. Proc 12th Conf. On UAI (1996) 28-39
9.   Arroyo-Figuerca, G., Sucar, L.E., Villavicencio, A.: Probabilistic Temporal Reasoning and its Application to Fossil Power Plant Operation. Expert Sys. with Apps. **15** (1998) 317-324
10.   Berzuini, C., Bellazzi, R., Quaglini, S.: Temporal Reasoning with Probabilities, Proc. of Workshop on UAI (1989)14-21
11.   Kjaerulff, A.: A Computational Scheme for Reasoning in Dynamic Probabilistic Networks. Proc. 8th Conf. On UAI (1992) 121-129
12.   Nicholson, A.E., Brady, J. M.: Dynamic Belief Networks for Discrete Monitoring. IEEE Trans. on Systems, Man and Cybernetics **34**(11) (1994) 1593-1610
13.   Santos E.(Jnr), Young, J.D.: Probabilistic Temporal Networks: A Unified Framework for Reasoning with Time and Uncertainty. Int. J. Approx. Reasoning, **20** (1999) 263-291
14.   Neuts, M.: Structured Stochastic Matrices of M/G/1 Type and Their Application. Marcel Dekker, NY (1989)
15.   Faddy, M.: Examples of Fitting Structured Phase-Type Distributions, Applied Stochastic Models and Data Analysis **10** (1994) 247-255
16.   Hanks, S., Madigan, D., Gavrin, J.: Probabilistic Temporal Reasoning with Endogenous Change. Proc. 11th Conf. on UAI (1995)
17.   Lauritzen, S.L., Wermuth, N.: Graphical Models For Associations Between Variables, Some of Which are Qualitative and Some Quantitative. Annals of Statistics **17** (1989) 31-57
18.   Gammerman, A., Luo, Z., Aitken, C.G.G., Brewer, M.J.: Computational Systems for Mixed Graphical Models. Adaptive Computing & Information Processing. UNICOM, (1994)
19.   Cox, D. Miller, H.D.: The Theory of Stochastic Processes, Methuen, London (1965)
20.   Faddy, M., McClean, S.: Analysing Data on Lengths of Stay of Hospital Patients Using Phase-Type Distributions. Applied Stochastic Models and Data Analysis (2000) to appear
21.   Dempster, A.P., Laird, N.M., Rubin, D.B.: Maximum Likelihood from Incomplete Data via the EM Algorithm. Journal of the Royal Statistical Society B, **39** (1977) 1-38
22.   Lauritzen, S.L.: The EM Algorithm For Graphical Association Models With Missing Data. Computational Statistics and Data Analysis **19** (1995) 191-201
23.   Aalen, O.O.: Phase Type distributions in Survival Analysis. Scand. J. Stat. **4** (1995) 447-463
24.   Badsberg, J.H.: A Guide to CoCo - An Environment for Graphical Models. Inst. of Elect. Sys., Dept. of Maths. & Computer Science, Aalborg University, Denmark (1992)
25.   Marshall, A., McClean, S., Shapcott, M., Hastie, I., Millard, P.: Developing a BBN for the Management of Geriatric Hospital Care. J. Health Care Mgt. Sc. (2000) to appear
26.   McClean, S.I., Millard, P.H.: Patterns of Length of Stay After Admission in Geriatric Medicine - An Event History Approach. The Statistician **42** (1993) 263-274
27.   MATLAB. Reference Guide, The MathsWorks Inc., Natick, Massachusetts (1992)
28.   Bunday, B.D.: Basic Optimisation Methods. Edward Arnold Publishers Ltd, London (1984)

Discovering Differences in Patients with Uveitis through Typical Testors by Class

José Fco. Martínez-Trinidad[1], Miriam Velasco-Sánchez[2], E. E. Contreras-Arevalo[2]

[1]Centro de Investigación en Computación, IPN
Juan de Dios Batiz s/n esq. Othón de Mendizabal, UPALM; C.P. 07738, México D.F.
e. mail: fmartine@cic.ipn.mx
[2]Fundación Hospital Nuestra Señora de la Luz (FHNSL)
Ezequiel Montes No. 135, Col. Tabacalera, CP 06030, México D.F.

Abstract. An analysis of clinical features on patients with uveitis attending to their localization through typical testors is presented. The main goal for the physician was to discover feature sets such that distinguish four groups of patients with uveitis. In order to solve the problem was used the Testor Theory in the Logical Combinatorial Pattern Recognition context.

1. Introduction to Clinical Problem

Uveitis is a group of diseases that affect typically the intermediate layer of the eye. This pathology represents between 3% and 5% of ophthalmological consultation. Uveitis had been subject of different classifications the most frequently is based in the location of the lesion in anterior uveitis, intermediate uveitis and posterior uveitis [1].

Uveitis can affect corneal endothelium by inflammation or retrokeratic deposits. Endothelial damage depends upon the intensity and location of the inflammation [2,3]. We studied a group of patients with active uveitis by using specular microscopy. The specular microscopy is a procedure that was initially used by David Maurice in 1968 [4]. In our study we included patients with anterior uveitis, pars planitis and toxoplasmosis.

The goal of our research was to discover feature sets that allow us distinguishing patients in the following four groups of uveitis: pars planitis, toxoplasmosis, anterior uveitis (in patients before 40 years old) and anterior uveitis (after 40 years old).

2. Typical Testors by Class

The Logical Combinatorial Pattern Recognition (LCPR) has as main characteristic that allow us handle information of the objects (their descriptions) under study in terms of a set of features or attributes [5]. These features may be of any nature, that is to say, numeric or not numeric. It allows us to model the problem according to the specialist do it in the practice, instead of force the mathematical tools to adapt to the practical problem.

D.A. Zighed, J. Komorowski, and J. Zytkow (Eds.): PKDD 2000, LNAI 1910, pp. 524–529, 2000.
© Springer-Verlag Berlin Heidelberg 2000

Let $\Omega=\{O_1,...,O_m\}$ a finite set of objects. Let $R=\{x_1,...,x_n\}$ a set of features, each feature x_i has associated a set of admissible values denoted as M_i $i=1,...,n$. A description $I(O_p)$ is defined for each object O_p $p=1,...,m.$, as a finite sequence of values of features $(x_1(O_p),...,x_n(O_p))$. Also for every variable is defined a comparison criterion for the admissible values $C_i:M_i\times M_i \rightarrow L_i$ $i=1,...,n$. It is known that the set Ω is divided in a finite number of subsets $K_1,...,K_l$ called classes. Each class K_j $j=1,...,l$ is defined by a number of objects. This information about objects and classes is in a learning matrix (LM) as the one shown in the Table 1.

The LCPR allows computing feature sets that distinguish objects in a class against the other classes through *Testors Theory* [6].

If the problem is related with Medicine and the number of the objects in each class is greater than one, it means that characterization is done on patients by means of their symptoms and signs associated with a particular disease. In this way the objects O_p in LM are the patients, features x_i $i=1,...,n$ are the symptoms and classes K_j $j=1,...,l$ are the different diseases a patient can suffer.

Table 1. Learning Matrix with m objects, n features and l classes

LM	Objects	Features		
		x_1	...	x_n
K_1	O_1	$x_1(O_1)$	...	$x_n(O_1)$
	...	...		...
	O_p	$x_1(O_p)$	...	$x_n(O_p)$
...	...	...	...	...
K_l	O_q	$x_1(O_q)$	...	$x_n(O_q)$
	...	...		...
	O_m	$x_1(O_m)$	...	$x_n(O_m)$

In this context, a set of features $\{x_{i1},...,x_{is}\}$ of LM is called a *testor by class* (*TC*) for the class K_i from LM if after deleting from LM all columns except $\{x_{i1},...,x_{is}\}$, all rows of LM corresponding to K_i are different to rows corresponding to K_j $j\neq i$. A *TC* is called *typical testor by class* (*TTC*) if none of its proper subsets is a *TC* [7]. *TTC* will be used too instead of *typical testors by class* indistinctly.

In general, we can say that finding all typical testors of a matrix is equivalent to doing an exhaustive search on all feature subsets, which are $2^{|R|}$, where $|R|$ is R's cardinal number. When the number of rows, columns or classes increases, the time necessary for this procedure could grow until it becomes impossible to be computed. As a consequence of this, many studies have been developed; in fact, this problem alone is a research line within Testor Theory framework [6].

Through computation of *TTC*, it is possible to discover feature sets that allow distinguishing patients in different classes.

Let Ψ be the set of all *TTC* for K_i $i=1,...,l$, of a given problem and $\Psi(x)$ the subset of all *TTC* that contain feature x. To each feature x we assign the value obtained by the equation 1.

$$F(x)=|\Psi(x)|/|\Psi| \qquad (1)$$

The value $F(x)$ is called informational weight based in the apparition frequency of the feature in the *TTC* and we identify it as a measure of its importance [6]. In addition, we assign to each *TTC* a weight based in the informational weight of the features as follows:

$$\sigma(t) = \frac{1}{|t|} \sum_{x_i \in t} F(x_i) \qquad (2)$$

It is clear according to this approach that between more frequent is a feature in the set of *TTC* (*the feature is more useful than other to distinguish objects belonging to different classes*) most important will be it. Consequently the *TTC* with the highest weight will be those containing features with the highest weight. Therefore, we consider that a typical testor with highest weight is more important to distinguish objects belonging to different classes than a typical testor with low weight.

3. Methodology

In our study, we include patients treated in the FHNSL from March to August in 1999. These patients satisfied the following clinical criteria: active uveitis, first inflammatory manifestation, specular microscopy, fulfillment of treatment, complete clinical file. We exclude those patients with: diabetes mellitus of any type, contact lens user, associated or previous ocular or corneal pathology, previous ocular trauma or ocular surgery, other systemic pathologies (neurological disease).

In the specular microscopy, the features shown in the table 2 were analyzed.

Table 2. Clinical features used to describe patients with uveitis

Feature	Admissible values	Type
Age (A)	10-80	Quantitative
Gender (G)	M= Male, F= Female	Qualitative Boolean
Cellular Density (CD)	500-2500	Quantitative
Pleomorphism (Pl)	absent light moderate severe	Qualitative Ordinal
Polimegathism (Po)	absent light moderate severe	Qualitative Ordinal
Tyndall (T)	- + ++ +++	Qualitative Ordinal
Flare (F)	- + ++ +++	Qualitative Ordinal
Guttatas (Gu)	0 I II III	Qualitative Ordinal

From the 38 studied patients, 5 were eliminated (because them did not satisfy clinical criteria) and 33 were considered. The patients were separated in four classes taking account the localization of the uveitis as is shown in the table 3.

Note that the classes toxoplasmosis and pars planitis have only three and four patients respectively, nevertheless the physicians consider that these descriptions of patients are representatives of these groups of patients.

As first step, we modeled jointly with the specialists, the comparison criteria for the values of the features. For qualitative features was used a comparison criterion based in the matching, see equation (3).

$$C_{x_i}\left(x_i\left(I(O_i)\right), x_i\left(I(O_j)\right)\right) = \begin{cases} 1 & \text{if } x_i\left(I(O_i)\right) \neq x_i\left(I(O_j)\right) \\ 0 & \text{otherwise} \end{cases} \qquad (3)$$

In the case of quantitative features was used a comparison criterion with threshold, that is to say, two values are different if the absolute difference between them is greater than a threshold, see equation (4).

$$C_{x_i}\left(x_i\left(I(O_i)\right), x_i\left(I(O_j)\right)\right)=\begin{cases}1 & if\ \left|x_i\left(I(O_i)\right)-x_i\left(I(O_j)\right)\right|>\varepsilon \\ 0 & otherwise\end{cases} \tag{4}$$

The specialists according to their expertise proposed the threshold for this criterion. So the threshold used for age feature was $\varepsilon=5$ and $\varepsilon=100$ for cellular density feature.

Table 3. Sample of patients with uveitis

	Num	A	G	CD	Pl	Po	T	F	Gu
Pars Planitis	P1	12	F	1500	light	moderate	-	-	I
	p2	12	M	1500	absent	absent	++	-	0
	P3	35	M	2200	absent	absent	++	-	0
	P4	13	F	2000	light	light	++	-	0
Toxoplasmosis	P5	20	M	1400	absent	absent	+	-	0
	P6	16	F	1000	severe	severe	+	-	0
	P7	30	F	1200	absent	light	+	-	0
Anterior Uveitis <=40	P8	38	M	1200	light	light	+	-	I
	P9	21	F	1400	light	light	+	++	I
	P10	22	M	1200	absent	light	-	-	0
	P11	30	M	1200	light	light	+++	-	I
	P12	38	F	1200	moderate	light	++	+++	II
	P13	24	M	800	severe	severe	++	++	I
	P14	30	F	1000	severe	severe	++	+	II
	P15	19	F	1200	severe	severe	+	-	I
	P16	40	F	1200	light	light	++	-	I
Anterior Uveitis >40	P17	52	F	600	severe	severe	+	-	III
	P18	70	F	600	light	severe	+++	++	III
	P19	49	M	600	moderate	severe	+++	++	III
	P20	66	F	700	moderate	severe	+	-	II
	P21	44	M	1200	light	light	++	-	0
	P22	53	F	900	light	moderate	+++	+++	II
	P23	49	F	1200	light	light	++	+	II
	P24	80	F	600	severe	severe	+	-	II
	P25	45	F	1200	absent	absent	++	-	0
	P26	47	F	800	severe	severe	++	-	II
	P27	41	M	1200	light	light	+++	+++	I
	P28	43	M	1400	moderate	absent	++	+++	0
	P29	42	M	1000	moderate	light	++	+	II
	P30	70	M	600	moderate	severe	+	-	II
	P31	44	F	800	moderate	moderate	++	++	I
	P32	55	F	900	severe	moderate	++	++	II
	P33	42	M	700	moderate	severe	+	++	III

It exists several algorithms to compute the typical testors [8], in this particular problem was applied the CC algorithm. The *TTC* for the four classes of patients are shown in the table 4.

4. Discussion and Conclusions

In the first column of tables 4a to 4d, we can see the sets of features that allow us to distinguish the patients of the respective class against the rest patients (*TTC*). In the fourth column appears the weight associated to the feature in the column third.

Remind that the weight associated to the features is computed using (1). On basis of this weight is computed the weight associated to the typical testor through (2), it appears in the second column. Note that if we observe the subdescriptions in a particular class, attending only to the features in the typical testor for this class, all those are different to the subdescriptions in the other classes. Therefore, through *TTC* computation it is possible discover the sets of features that allow us detect the differences between the patients under study.

Table 4. Typical testors by class for 4a) Pars Planitis; 4b) Toxoplasmosis; 4c)Anterior uveitis <=40 and 4d) Anterior uveitis > 40 classes.

Typical Testor	Weight	Feature	Weight	Typical Testor	Weight	Feature	Weight
{CD, T, F}	0.4583	A	0.75	{T,G}	0.49995	A	0.5714
{CD, Pl, Tl}	0.4167	G	0	{CD, PO,T}	0.52376	G	0
{A,T,Gu}	0.5416	CD	0.625	{CD, Pl, T}	0.52376	CD	0.7142
{A, Po}	0.4375	Pl	0.375	{A, Po, Gu}	0.42853	Pl	0.2857
{A, Pl,T}	0.5833	Po	0.125	{A, CD, Gu}	0.5713	Po	0.2857
{A, CD, Gu}	0.5416	T	0.625	{A, CD, T, F}	0.49995	T	0.5714
{A,CD,T}	0.6666	F	0.125	{A, CD, Pl}	0.5237	F	0.1428
{A, CD, Pl}	0.5833	Gu	0.25			Gu	0.4285

	4a				4b		

Typical Testor	Weight	Feature	Weight	Typical Testor	Weight	Feature	Weight
{Po,T,F,Gu}	0.7083	A	0.6666	{CD, Po,T,F,Gu}	0.72	A	0.4
{Pl,T,F,Gu}	0.6666	G	0	{CD, Pl,T,F,Gu}	0.72	G	0
{A, T,F,Gu}	0.7916	CD	0.3333	{CD,Pl,Po,F,Gu}	0.68	CD	0.8
{A,Po,F,Gu}	0.7083	Pl	0.1666	{A, F, Gu}	0.73	Pl	0.4
{A,CD,F,Gu}	0.7083	Po	0.3333	{A, CD, T, Gu}	0.70	Po	0.4
{A,CD,T,Gu}	0.6666	T	0.6666			T	0.6
		F	0.8333			F	0.8
		Gu	1			G	1

	4c				4d		

After compute the *TTC*, it is possible to evaluate the informational weight for features based in its importance to distinguish patients of a class against the rest. Remind that we consider that between more frequent is a feature in the set of *TTC*, then the feature is more useful than other to distinguish objects belonging to different classes. Consequently, it is possible to use the weight of features to associate a weight to the typical testors as is expressed in (2). This weight can be interpreted as the importance that the typical testor has to distinguish patients of a particular class against the rest. In some cases may be useful only consider the *TTC* with highest weight (See the rows in cursive in the tables 4a-4d).

The typical testors by class are important to select those sets of features that are most efficient to preserve class separability. So, these sets of features can be used to classify new patients.

As future work, in a second phase, we pretend to construct a computer system to realize differential diagnosis of intraocular inflammatory diseases. We will use the algorithms based in partial precedence as *voting algorithms, Kora-Ω* and *representative sets* [9]. These algorithms allow us in natural way consider both *TTC* and their weights, in order to classify new patients. The idea of these algorithms is to classify objects according to their description and their similitude with the objects already classified in the K_i classes. The similarity is evaluated analyzing subdescriptions of objects instead of whole descriptions. The *TTC* define the set of features that should be compared through a convenient similarity function and the weight of the testors weigh this similarity. Therefore, It would be possible to apply any of these algorithms to LM in order to classify a new object, obtaining the decision criteria for the class it belongs to.

In this paper, a methodology to discover differences between patients with uveitis was presented. In order to solve this problem *TTC* were used.

The analysis here presented constitute a first step for solve the differential diagnosis of intraocular inflammatory diseases. Besides, based on this work the construction of a computer system in order to realize differential diagnosis of intraocular inflammatory diseases was proposed as future work.

Of course, we will add feedback to the process enriching jointly with the physicians the sample of patients in LM.

Finally, the methodology proposed in this paper is of extremely useful, not only to discover differences between patients with uveitis, but also, to discover differences between patients of any diseases.

References

1. American Academy Of Ophthalmology: Intraocular Inflammation And Uveitis. *In Immunology* (Eds. Margaret Dennit), American Academy Of Ophthalmology Press. 1996, pp. 15-80.
2. Nozik R A.: Uveitis. In *Ophthalmology Clinics Of North America*. (Saunders Philadelphia) 1993, pp. 1-22, 24-50.
3. Zantos S G.: Guttate Endothelial Changes with Anterior Eye Inflammation. *Br J Ophthalmol*. 65 (1981) 101-103.
4. Maurice, D. M.: Cellular membrane activity in the corneal endothelium of the intact eye. *Experientia*. 24 (1968) 1094.
5. Ortiz-Posadas M.R., Martínez-Trinidad J.F. and Ruiz-Shulcloper J: A new approach to differential diagnosis of diseases. *International Journal of Bio-Medical Computing* 40 (1996) 179-185.
6. Lazo-Cortes M. and Ruiz-Shulcloper.: Determining the feature relevance for non-classically described objects and a new algorithm to compute typical fuzzy testors. *Pattern Recognition Letters*. 16 (1995) 1259-1265.
7. Lazo-Cortes M., Douglas de la Peña, Quintana Gómez. Testores por clase. Una aplicación al Reconocimiento de caracteres. In *Proceedings of the III Iberoamerican Workshop on Pattern Recognition*, Politécnica México, D.F. 1998, pp. 229-236.
8. Sanchez-Díaz G. and Lazo-Cortés M.: Modelos algorítmicos paralelos y distribuidos para el cálculo de testores típicos. In *Proceedings of the II Iberoamerican Workshop on Pattern Recognition*, Politécnica México, D.F. 1997, pp. 135-140.
9. Ruiz-Shulcloper J. and Lazo-Cortés M.: Mathematical Algorithms for the Supervised Classification Based on Fuzzy Partial Precedence, *Mathematical and Computer Modelling* 29 (1999) 111-119.

Web Usage Mining: How to Efficiently Manage New Transactions and New Clients

F. Masseglia[1,2], P. Poncelet[2], and M. Teisseire[2]

[1] Laboratoire PRiSM, Univ. de Versailles, 45 Avenue des Etats-Unis, 78035 Versailles Cedex, France
[2] LIRMM UMR CNRS 5506, 161 Rue Ada, 34392 Montpellier Cedex 5, France
{massegli, poncelet, teisseire}@lirmm.fr

Abstract. With the growing popularity of the World Wide Web (Web), large volumes of data such as user address or URL requested are gathered automatically by Web servers and collected in access log files. Recently, many interesting works have been published in the Web Usage Mining context. Nevertheless, the large amount of input data poses a maintenance problem. In fact, maintaining global patterns is a non-trivial task after access log file update because new data may invalidate old client behavior and creates new ones.

keywords: data mining, Web usage mining, sequential patterns, incremental mining.

1 Introduction

With the growing popularity of the World Wide Web (Web), large volumes of data such as address of users or URLs requested are gathered automatically by Web servers and collected in access log files. Recently, many interesting works have been published in the context of the Web Usage Mining and very efficient approaches have been proposed for mining user patterns [5,2,6,8,1,3].
Nevertheless, the access log file is not a static file because new updates are constantly being applied on it: new records are frequently added to record client behaviors. The issue of maintaining such user patterns becomes essential because new transactions or new clients may be updated over time. In this case, some existing user patterns would become invalid after database while some new user patterns might appear. To the best of our knowledge not much effort has been spent on maintaining such user pattern in the Web usage mining context.

In this paper we address the problem of incremental Web usage mining, i.e. the problem of maintaining user patterns over a significantly long period of time. We propose an efficient approach, called IseWum (Incremental Sequence Extraction for Web usage mining), for maintaining user patterns either when new transactions are added to the access log files or when new visitors access the Web server. In section 2, the problem is stated and illustrated. Our proposal is described in section 3. Finally section 4 concludes the paper.

D.A. Zighed, J. Komorowski, and J. Żytkow (Eds.): PKDD 2000, LNAI 1910, pp. 530–535, 2000.
© Springer-Verlag Berlin Heidelberg 2000

2   Problem Statement

Let DB be the original database and $minSupp$ the minimum support. Let db be the increment database where new transactions are added to DB. We assume that each transaction on db has been sorted on visitor-id and transaction time. $U = DB \cup db$ is the updated database containing all sequences from DB and db. Let L^{DB} be the set of frequent sequences in DB. The problem of incremental mining of sequential patterns is to find frequent sequences in U, noted L^U, with respect to the same minimum support.

An example

Ip address	Time	URL accessed
res1.newi.ac.uk	01/*Jan*/1998	`/api/java.io.BufferedWriter.html`
res1.newi.ac.uk	01/*Jan*/1998	`/api/java.util.zip.CRC32.html`
res1.newi.ac.uk	04/*Feb*/1998	`/api/java.util.zip.CRC32.html`
res1.newi.ac.uk	18/*Feb*/1998	`/atm/logiciels.html`
res1.newi.ac.uk	18/*Feb*/1998	`/relnotes/deprecatedlist.html`
acasun.eckerd.edu	11/*Jan*/1998	`/api/java.io.BufferedWriter.html`
acasun.eckerd.edu	11/*Jan*/1998	`/api/java.util.zip.CRC32.html`
acasun.eckerd.edu	16/*Jan*/1998	`/html4.0/struct/global.html`
acasun.eckerd.edu	29/*Jan*/1998	`/postgres/html-manual/query.html`
acces.francomedia.qc.ca	05/*Jan*/1998	`/api/java.io.BufferedWriter.html`
acces.francomedia.qc.ca	05/*Jan*/1998	`/api/java.util.zip.CRC32.html`
acces.francomedia.qc.ca	12/*Feb*/1998	`/postgres/html-manual/query.html`
acces.francomedia.qc.ca	16/*Feb*/1998	`/html4.0/struct/global.html`
ach3.pharma.mcgill.ca	06/*Feb*/1998	`/perl/perlre.html`
ach3.pharma.mcgill.ca	08/*Feb*/1998	`/perl/struct/perlst.html`

Fig. 1. An access-log file example

In order to illustrate the problem of incremental Web usage mining let us consider the part of the access log file given in figure 1. Let us assume that the minimum support value is 50%, thus to be considered as frequent a sequence must be observed for at least two visitors. The only frequent sequences, embedded in the access log, are the following: <(/api/java.io.BufferedWriter.html   /api/java.util.zip.CRC32.html) (/html4.0/struct/global.html)>,   and   <(/api/java.io.BufferedWriter.html /api/java.util.zip.CRC32.html) (/postgres/html-manual/query.html)>. because they could be detected for both *acasun.eckerd.edu* and *access.francomedia.qc.ca*.

Let us now consider the problem when new visitors and new transactions are appended to the original access log file after some update activities. Figure 2 describes the increment access log. We assume that the support value is the same. As a new visitor has been added to the access log file (*acahp.mg.edu*), to be considered as frequent a pattern must now be observed for at least three visitors. According to this constraint the set of user patterns of the original database is reduced to: <(/api/java.io.BufferedWriter.html /api/java.util.zip.CRC32.html) >. This pattern

Ip address	Time	URL accessed
acasun.eckerd.edu	$8/Mar/1998$	`/atm/logiciels.html`
acasun.eckerd.edu	$8/Mar/1998$	`/perl/perlre.html`
acasun.eckerd.edu	$8/Mar/1998$	`/relnotes/deprecatedlist.html`
acasun.eckerd.edu	$17/Mar/1998$	`/java-tutorial/ui/animLoop.html`
acasun.eckerd.edu	$17/Mar/1998$	`/java-tutorial/ui/BufferedDate.html`
acces.francomedia.qc.ca	$06/Mar/1998$	`/atm/logiciels.html`
acces.francomedia.qc.ca	$06/Mar/1998$	`/perl/perlre.html`
acces.francomedia.qc.ca	$12/Mar/1998$	`/java-tutorial/ui/animLoop.html`
acces.francomedia.qc.ca	$12/Mar/1998$	`/perl/struct/perlst.html`
acahp.mg.edu	$08/Mar/1998$	`/api/java.io.BufferedWriter.html`
acahp.mg.edu	$08/Mar/1998$	`/postgres/html-manual/query.html`
acahp.mg.edu	$18/Apr/1998$	`/relnotes/deprecatedlist.html`
acahp.mg.edu	$18/Apr/1998$	`/java-tutorial/ui/animLoop.html`

Fig. 2. An increment access log

is frequent since it appears in the sequence of the visitors: *res1.newi.ac.uk*, *acasun.eckerd.edu* and *acces.francomedia.qc.ca*. Nevertheless, by introducing the increment access log file, the set of frequent sequences in the updated file is: <(/api/java.io.BufferedWriter.html /api/java.util.zip.CRC32.html) (/atm/logiciels.html)>, <(/api/java.io.BufferedWriter.html) (/relnotes/deprecatedlist.html)>, <(/api/java.io.BufferedWriter.html) (/java-tutorial/ui/animLoop.html>, <(/postgres/html-manual/query.html) (/java-tutorial/ui/animLoop.html>, and <(/perl/perlre.html)>.

Let us have a closer look to the sequence <(/api/java.io.BufferedWriter.html /api/java.util.zip.CRC32.html) (/atm/logiciels.html)>. This sequence could be detected for visitor *res1.newi.ac.uk* in the original file but it is not a frequent sequence. Nevertheless, as the URL /atm/logiciels.html occurs three times in the updated file, this sequence also matches with transactions of *acasun.eckerd.edu* and *acces.francomedia.qc.ca*. Let us now consider the sequence <(/api/java.io.BufferedWriter.html) (/relnotes/deprecatedlist.html)>. This sequence becomes frequent since, with the increment, it appears in *res1.newi.ac.uk*, *acasun.eckerd.edu* and the new visitor *acahp.mg.edu*.

3   Proposal

Let us consider that k stands for the length of the longest frequent sequences in DB. We decompose the problem as follows:

1. Find all new frequent sequences of size $j \leq (k + 1)$. During this phase three kinds of frequent sequences are considered:
   - Sequences embedded in DB could become frequent since they have sufficient support with the incremental database, i.e. same sequences as in the original file appear in the increment.

- New frequent sequences embedded in *db* but we did not appear in the original database.
- Sequences of *DB* might become frequent when adding items of *db*.

2. Find all frequent sequences of size $j > (k + 1)$.

The second sub-problem can be solved in a straightforward manner since we are provided with frequent $(k + 1)$-sequences discovered in the previous phase. Applying a GSP-like [7] approach at the $(k + 1)^{th}$ step, we can generate the candidate $(k + 2)$-sequences and repeat the process until all frequent sequences are discovered. At the end of this phase all frequent sequences embedded in U are found. Hence, the problem of incremental mining of sequential patterns is reduced to the problem of finding frequent sequences of size $j \leq (k + 1)$.

To discover frequent sequences of size $j \leq (k+1)$, the ISEWUM approach contains a number of iterations. When 1-frequent sequences are found in the updated database, they are used to generate new candidates and to find previous frequent sequences in the original database occurring, according to the timewise order, before such sequences. The main concern of the next iterations is to find new j-candidates ($j \leq (k+1)$) which can be extensions of frequent sequences previoulsy found. These features combined together with the GSP-like approach when $j > (k+1)$ form the core of the ISEWUM approach and make our approach faster than re-run the mining process from scratch, i.e. for the whole updated database. The first iteration is addressed in the next part and is followed by the presentation of the remaining iterations.

First iteration First, we scan the increment *db* and we count the support of individual items. We are thus provided with the set of items occurring at least once in *db*. Next, combining this set with the set of items embedded in *DB* we determine which items of *db* are frequent in the updated database, i.e. items for which the minimum support condition hold. Finally, as we assume to be provided with the support of each frequent sequences of the original database, we can update this support if new customers are added in the increment.

We use the frequent 1-sequences in *db* to generate new candidates by joining L_1^{db} with L_1^{db}. We scan on *db* and obtain the 2-sequences embedded in *db*. This phase is quite different from the GSP approach since we do not consider the support constraint. We assume that a candidate 2-sequence is a 2-sequence if and only if it occurs at least once in *db*. The main reason is that we do not want to provide the set of all 2-sequences, but rather to obtain the set of potential extensions of items embedded in *db*. In other words, if a candidate 2-sequence does not occur in *db* it cannot of necessity be an extension of an original frequent sequence of *DB*, and then cannot give a frequent sequence for U. In the same way, if a candidate 2-sequence occurs in *db*, this sequence might be an extension of previous sequences in *DB*. Next, we scan the original database to find out if these sequences are frequent in the updated database or not.

An additional operation is performed on the items discovered frequent in *db*. The main idea of this operation is to retrieve in *DB* the frequent sub-sequences of L^{DB} preceding, according to the timewise order, items of db. So, during the scan, we also obtain the set of frequent sub-sequences preceding items of db. From this set, by appending the items of db to the frequent sub-sequences we obtain a new set of frequent sequences of size $j \leq (k+1)$.

At the end of the first scan on U, we are thus provided with a new set of frequent 2-sequences as well as a new set of frequent sequences having a length lower or equal to $k+1$.

j^{th} **Iteration (with $j \leq (k+1)$)** Let us assume that we are at the jth pass with $j \leq k+1$. In these subsequent iterations, we start by generating new candidates from the two seed sets found in the previous pass. The supports for all candidates are then obtained by scanning U and those with minimum support become frequent sequences. These two sets are then used to generate new candidates. The process iterates until all frequent sequences are discovered or $j = k+1$.

At the end of this phase, $L^{U_{k+1}}$, the set of all frequent sequences having a size lower or equal to $k+1$, is obtained from L^{DB} and maximal frequent sequences obtained during the previous iteration.

j^{th} **Iteration (with $j > (k+1)$)** Now, as all frequent sequences of size $j \leq (k+1)$ are discovered, we find the new frequent j-sequences in L^{U} where $j > k+1$. First we extract from $L^{U_{k+1}}$ frequent $(k+1)$-sequences. New candidate $(k+2)$-sequences are then generated applying a GSP like approach and the process iterates until no more candidates are generated.

Pruning out non maximal sequences, we are provided with L^{U} the set of all frequent sequences in the updated database.

4   Conclusion

In this paper we present the IsEWUM approach for incremental Web usage mining of user patterns in access log files. In order to assess the relative performance of the IsEWUM approach when new transactions or new clients were appended to the original access log file we have carried out some experiments on the access log file obtained from the Lirmm Home Page. Figure 3 compares the execution times, when varying the minimum support value, of our approach and GSP from scratch, i.e. when re-mining the new updated access log file. The original log used for this experiment contains entries corresponding to the request made during March of 1998 and the updated file contains entries made during April of 1998. As we can observe our incremental approach clearly outperforms the GSP approach. Due to space limitation, we do not provide detailled results on other experiments which could be found in [4].

Access Log

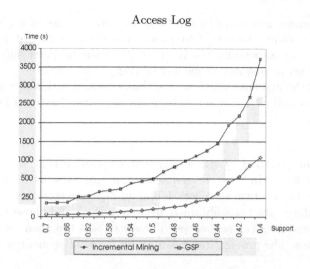

Fig. 3. Execution times for the Lirmm Home Page server

References

1. D.W. Cheung, B. Kao, and J. Lee. Discovering User Access Patterns on the World-Wide Web. In *PAKDD'97*, February 1997.
2. R. Cooley, B. Mobasher, and J. Srivastava. Web Mining: Information and Pattern Discovery on the World Wide Web. In *ICTAI'97*, November 1997.
3. F. Masseglia, P. Poncelet, and R. Cicchetti. WebTool: An Integrated Framework for Data Mining. In *DEXA'99*, August 1999.
4. F. Masseglia, P. Poncelet, and M. Teisseire. Incremental Mining of Sequential Patterns in Large Databases. Technical report, LIRMM, France, January 2000.
5. B. Mobasher, N. Jain, E. Han, and J. Srivastava. Web Mining: Pattern Discovery from World Wide Web Transactions. Technical Report TR-96-050, University of Minnesota, 1996.
6. M. Spiliopoulou and L.C. Faulstich. WUM: A Tool for Web Utilization Analysis. In *EDBT Workshop WebDB'98*, March 1998.
7. R. Srikant and R. Agrawal. Mining Sequential Patterns: Generalizations and Performance Improvements. In *EDBT'96*, September 1996.
8. O. Zaïane, M .Xin, and J. Han. Discovering Web Access Patterns and Trends by Applying OLAP and Data Mining Technology on Web Logs. In *Proceedings on Advances in Digital Libraries Conference (ADL'98)*, April 1998.

Mining Relational Databases

Frédéric Moal, Teddy Turmeaux, and Christel Vrain

LIFO, Université d'Orléans,
rue Léonard de Vinci, BP 6759,
45067 Orléans cedex 02, France
{moal,turmeaux,cv}@lifo.univ-orleans.fr

Abstract. In this paper, we propose a classification system to induce an intentional definition of a relation from examples, when background knowledge is stored in a relational database composed of several tables and views. Refinement operators have been defined to integrate in a uniform way different induction tools learning numeric and symbolic constraints. The particularity of our approach is to use integrity constraints over the database (keys and foreign keys) to explore the hypotheses space. Moreover new attributes can be introduced, relying on the aggregation operator "group by".

1 Introduction

Nowadays, most of the data sources are stored in large relational database systems in which information is expressed in a given normal form and attributes are often distributed over several tables. Current data mining systems [2] often require data to be stored in a single relation and therefore relational applications have to be flattened into a single table, losing interesting information about the structure of the database and leading to huge, intractable relation. On the other hand, Inductive Logic Programming (**ILP** [4]) is devoted to relational learning in a logical framework, and some algorithms have been successfully applied to Data Mining tasks [5]. However, characteristics of Data Mining tasks differ from usual ILP ones: the search space is more restricted (no recursive rules), and the data size is much higher.

In this paper, we present a classification system that is able to handle relational databases and to use information about the structure of the database to guide the search of good hypotheses. For efficiency reasons, we have chosen a separate and conquer strategy relying on a hill climbing heuristic, as in the system FOIL [6]. We have defined new refinement operators based on the schema of the database and on integrity constraints. Moreover, a new quality criterion is used to avoid the generation of too specific rules at the end of the process. Finally, the classical ILP expressiveness has been extended by introducing the *group-by* SQL operator.

The paper is organized as follows. Section 2 formalizes the problem in the framework of relational algebra. Section 3 is devoted to our system, its quality criterion and its refinement operators. In Section 4, we present experiments on

D.A. Zighed, J. Komorowski, and J. Żytkow (Eds.): PKDD 2000, LNAI 1910, pp. 536–541, 2000.
© Springer-Verlag Berlin Heidelberg 2000

a classical problem for relational learners (mutagenesis [5]), and we conclude on perspectives in Section 5.

2   Relational Algebra

The Data Mining task we address here is a classification task. It can be seen as the search of a "good" query over the database, the answer of which covers many positive examples and rejects many negative examples. A natural way to express queries is relational algebra [1], that can easily be translated into SQL.

2.1   Notations

The database schema is denoted by $\mathcal{D}$, with $\mathcal{D} = \{R_1(U_1), \dots, R_n(U_n)\}$ where R_i is a relation and U_i is the set of attributes of the relation R_i.

To avoid naming conflicts in expressions, attributes are referenced by their positions in the relation. For instance, in the relation $person(id, name, age)$, $person[2]$ stands for the attribute $name$. Moreover, we use uppercase letters for the schema $\mathcal{D}$ and its relations R_i, and the same lowercase letters for their corresponding instances; an instance d of the database $\mathcal{D} = (R_1, \dots, R_n)$ is a set $\{r_1, \dots, r_n\}$ where r_i is an instance, i.e. a set of tuples, of the relation R_i.

We consider the following relational algebra operators:

- $R \bowtie_{i=j} S$ denotes the join of the tuples of r and s on the ith attribute of R and the jth attribute of S;
- $\sigma_C(R)$ denotes the selection of the tuples t of r that satisfy the constraint C, a propositional formula over attributes of R;
- $\pi_i(R)$ denotes the projection of r on its ith attribute;
- $R \cup S$ denotes the union of the tuples of r and s.

A relation is defined *in intension* by a relational algebra expression, it is defined *in extension* by the list of its tuples. If E denotes a relational algebra expression over the relations of $\mathcal{D}$, and if d is an instance of the database, then $E(d)$ is the set of tuples of d satisfying E.

In this paper, we focus on two usual classes of integrity constraints: keys and foreign key links. A *key* $R_1[i]$ means that given a value for the ith attribute of the relation r_1, there exists a single tuple in r_1 with that value. A *foreign key link* links an attribute in a relation to a key attribute in another relation: $R_1[i] \rightarrow R_2[j]$ means that for each tuple with a given value for the ith attribute of the relation r_1, there exists a single tuple in the relation r_2 with this value for the jth attribute.

2.2   The Mutagenesis Problem

The aim of this problem [5] is to discriminate between molecules, depending on their mutagenic activity (*active* or *inactive*). The dataset consists of generic 2D chemical compound descriptions, described by:

- a measure of hydrophobicity of the molecule, denoted by *LogP*;
- a measure of the energy of the lowest unoccupied molecular orbital in the molecule, denoted by *Lumo*;
- a boolean attribute *I*1, identifying compounds with 3 or more benzyl rings;
- a boolean attribute *Ia*, identifying a subclass of compounds (acenthryles);
- the atoms that occur in a molecule, with their elements (carbon, zinc, ...), Quanta type and partial charge;
- the bonds between atoms, with their bond type (aromatic, ...).

This information is represented by the following relational model:

$$\mathcal{D} = \{\ Compound(DrugId, Lumo, Logp, I1, Ia)$$
$$Atom(AtomId, DrugId, Element, Quanta, Charge),$$
$$Bond(AtomId1, AtomId2, BondType)\ \}$$

with the keys $\{Compound[1], Atom[1], Bond[1,2]\}$ and the foreign key links $\{Atom[2] \rightarrow Compound[1], Bond[1] \rightarrow Atom[1], Bond[2] \rightarrow Atom[1]\}$.

2.3   Formulation of the Problem

Given a database $\mathcal{D}$ and an instance d of $\mathcal{D}$, and given a concept to learn defined *in extension* by two relations on the same attribute, a positive example relation e^+ and a negative one e^-, find an *intentional* definition E over the database $\mathcal{D}$ that is complete ($e^+ \subseteq E(d)$) and consistent ($e^- \cap E(d) = \emptyset$).

When these requirements are too strong, the aim is only to find a definition which covers many positive examples and rejects many negative examples.

As explained in Section 3 and due to our search strategy and to our definitions of operators, the relational expressions that are learned in our system are written: $\pi_a(\ldots \sigma_{C_{i_3}}(\sigma_{C_{i_2}}(\sigma_{C_{i_1}}(R_{i_1}) \bowtie R_{i_2}) \bowtie R_{i_3}) \ldots)$. A projection must be applied on the learned hypothesis to restrict the relation to the attribute defining the examples.

3   Architecture of the System

The underlying algorithm is a refinement of the classical *separate-and-conquer* algorithm: the basic idea is to search for a rule that covers a part of the positive examples, to remove these examples, and to start again with the remaining positive examples. In our system, two parameters are given defining respectively the minimum rate of covered positive examples (MinPosRate) and the maximum rate of covered negative examples (MaxNegRate). Each iteration builds a hypothesis that is the best refinement of the current one according to a quality criterion.

3.1   Quality Criterion

Algorithms based on a *separate-and-conquer* strategy lead to overly specific rules at the end of the process. To overcome this problem, we propose a new quality evaluation function, which takes into account the number of rules that have already been learned.

The quality of a refinement h' of a hypothesis h is therefore computed by the formula: $t_{pos}^n * t_{neg}^2$, where t_{pos} (resp. t_{neg}) is the ratio of the number of positive (resp. negative) examples covered (resp. rejected) by h' out of the number of positive (resp. negative) examples covered by h, and n is the number of the iteration. With such an expression, when n increases, low values for t_{pos} penalize the quality function. This solution reduces the number of rules in the final solution; on the other hand, the last generated rules are longer, since more refinements are necessary.

3.2   Refinement Operators

Our approach is a top-down one: the system starts from a general rule that covers all the examples and iteratively specializes it so that it covers less and less tuples representing negative examples. Given a hypothesis h, a refinement operator generates hypotheses h' that are more specific than h.

Classical Operators

Selection refinement $\mathcal{O}_S$: Given a hypothesis $h = \pi_a(R)$, $\mathcal{O}_S$ produces a set of hypotheses $\mathcal{O}_S(h) = \{\pi_a(\sigma_C(R))\}$, where C is a propositional formula over the attributes of R. The constraint C is obtained by using classical algorithms in discrimination tasks as for example finding a discriminant hyperplane.

Join refinement $\mathcal{O}_J$: Given a current hypothesis $h = \pi_a(R)$, where R is a composition of selections and joins on relations $R_1, \ldots, R_n$, the operator generates a set of hypotheses h' such as: $h' = \{\pi_a(R \bowtie_{i=j} S)\}$, with $S(A_1, \ldots, A_n) \in \mathcal{D}$, i references the kth attribute of a relation R_m in $R_1 \ldots, R_n$, and $R_m[k]$ and $S[j]$ are connected with a foreign key link. The use of keys and foreign key links restricts the possible joins and thus enables to prune the search space efficiently.

For instance, on the mutagenesis database, the hypothesis $\pi_2(Atom)$ is refined to $\pi_2(Atom \bowtie_{2=1} Compound)$, using the link with Compound, or to $\pi_2(Atom \bowtie_{1=1} Bond)$ or $\pi_2(Atom \bowtie_{1=2} Bond)$ using the two links between *Atom* and *Bond*.

New Operators

Composition: The application of these two operators allows to search the hypotheses space. Nevertheless, as pointed out in [7], when using a hill-climbing algorithm which chooses the best hypothesis at each step, some literals which bring no discrimination power are not introduced while they could be very useful to introduce new discriminant features.

In our system, the two operators are composed: first, the *join* refinement is applied, and for each refinement the *selection* refinement is applied. The *composition* refinement operator is then $\mathcal{O}_C = \mathcal{O}_S \circ \mathcal{O}_J$. Join refinements are evaluated according to the discriminant power of the new attributes they introduce, and the best composition is chosen.

Aggregate refinement operator. It refines a hypothesis $h = \pi_a(R)$, where R is a composition of selections and joins on relations $R_1, \ldots, R_n$ into $\pi_a(R \bowtie_{i=j} Group(S,j))$, such that $S(A_1, \ldots, A_j, \ldots, A_n) \in \mathcal{D}$, i references the kth attribute of a relation R_m in $R_1, \ldots, R_n$, and there is a foreign link $S[i] \rightarrow R_m[k]$. This definition can be easily extended to group a relation over more than one attribute. The only restriction is that the attributes used in the join appear in these grouping attributes.

4   Experiments and Discussion

The mutagenesis database consists of generic chemical compound descriptions, as shown in Section 2.2. The complete database is composed of 230 compounds which have been divided into two distinct subsets. We focus on the subset identified as "Regression friendly" to compare our results with some results of Attribute-based algorithms. This subset is composed of 125 active molecules (positive examples) and 63 inactive molecules (negative examples).

The goal is to find discriminant rules for the active molecules. All the results are obtained using a 10-fold cross-validation, where every molecule is used nine times in the training set and once in the test set.

Typically, the accuracy is the number of well classified examples divided by the total number of examples. To compute the real accuracy of our system on this problem, we adopt the following procedure: for each of the 10 cross-validation run, 1/10 of the examples composed the test set and 9/10 are used as the learning set. This learning set is then used to determine the parameters MAXNEGRATE and MINPOSRATE. for various values of the parameters, we randomly select 80% of this learning set to induce relational expressions, testing the accuracy on the 20% remainder. The parameters that lead to the best accuracy over these 20% are then used with the complete learning set. The accuracy of the learned hypothesis is computed over the initial testing set.

The predictive accuracies of various systems, extracted from [5], are presented in Table 1. Progol [3] is an Inductive Logic Programming system, and the other are attribute-based algorithms. The "default" algorithm classifies all the examples to the majority class (active).

Table 1 states that our system is competitive in terms of predictive accuracy, despite the hill-climbing strategy we use. This strategy is interesting because it tests much less hypotheses than more exhaustive search strategies and this is a very important feature to deal with large databases.

5   Conclusion

In this paper, we have proposed a new system to deal with classification tasks in the framework of relational databases. We have developed new refinement operators based on the database schema and integrity constraints.

Table 1. Predictive Accuracies (from a 10-fold cross-validation).

METHOD	ESTIMATED ACCURACY
OUR SYSTEM	0.89
PROGOL	0.88
LINEAR REGRESSION	0.89
NEURAL NETWORK	0.89
DECISION TREE	0.88
DEFAULT	0.66

The results of our experiments seem to confirm the interest of our approach, in terms of accuracy and computational cost. The low computational cost allows the user to iteratively refine its search with different parameters and biases. In our opinion, such an iteration is a necessary stage in all Data Mining tasks. However, these results must be validated with other databases.

The use of the database structure for the design of refinement operators offers some perspectives: the study of this approach in the framework of Object Databases which allows to express finer structural constraints; the extension of the tuples expressiveness with Constraints Databases[8].

Acknowledgements: This work is partially supported by "Région Centre".

References

[1] P. Atzeni and V. De Antonellis. *Relational Database Theory.* Benjamin/Cummings Publ. Comp., Redwood City, California, 1993.

[2] U. M. Fayyad, G. Piatetsky-Shapiro, P. Smyth, and R. Uthurusamy, editors. *Advances in Knowledge Discovery and Data Mining.* MIT Press, Mento Park, 1996.

[3] S. Muggleton. Inverse entailment and Progol. *New Generation Computing,* 13:245–286, 1995.

[4] S. Muggleton and L. De Raedt. Inductive logic programming: Theory and methods. *The Journal of Logic Programming,* 19 & 20:629–680, May 1994.

[5] S. Muggleton, A. Srinivasan, R. King, and M. Sternberg. Biochemical knowledge discovery using Inductive Logic Programming. In H. Motoda, editor, *Proc. of the first Conference on Discovery Science,* Berlin, 1998. Springer-Verlag.

[6] J. R. Quinlan. Learning logical definitions from relations. *Machine Learning,* 5(3):239–266, 1990.

[7] G. Silverstein and M. Pazzani. Relational cliches: constraining constructive induction during relational learning. In *Proceedings of the Sixth International Workshop on Machine Learning,* Los Altos, CA, 1989. Kaufmann.

[8] T. Turmeaux and C. Vrain. Learning in constraint databases. In *Discovery Science, Second International Conference,* volume 1721 of *LNAI,* pages 196–207, Berlin, december 1999. Springer.

Interestingness in Attribute-Oriented Induction (AOI): Multiple-Level Rule Generation

Maybin K. Muyeba and John A. Keane

Department of Computation, UMIST, Manchester, M60 1QD, UK
muyeba,jak@co.umist.ac.uk

Abstract. Attribute-Oriented Induction (*AOI*) is a data mining technique that produces simplified descriptive patterns. Classical *AOI* uses a predictive strategy to determine distinct values of an attribute but generalises attributes indiscriminately i.e. the value 'ANY' is replaced like any other value without restrictions. *AOI* only produces interesting rules by using interior concepts of attribute hierarchies. The COMPARE algorithm that integrates *predictive* and *lookahead* methods and of order complexity $O\ (np)$, where n and p are input and generalised tuples respectively, is introduced. The latter method determines distinct values of attribute clusters and greatest number of attribute values with a '*common parent*' (except parent 'ANY'). When generating rules, a rough set approach to eliminate redundant attributes is used leading to more interesting multiple-level rules with fewer 'ANY' values than classical *AOI*.

1 Introduction

Attribute-Oriented Induction (*AOI*) [1] is a data mining technique that produces descriptive patterns by generalisation. Each attribute has a predefined *concept hierarchy* often referred to as *domain knowledge*. Concept hierarchy values (specifically *interior concepts* as shown in figure 1) are used repeatedly to replace low-level attribute values and generate a *prime relation*. Attributes that do not have concept hierarchies or posses a large number of distinct values (e.g. *keys* to relations) are dropped except in the *key-preserving AOI* [2]. The number of allowed attributes and tuples in the final table is controlled by *attribute* and *rule thresholds* T_a and T_r respectively. The end result is a *final generalised relation* or *rule table* of descriptive patterns. Rules produced in this way can use preserved *keys* to perform efficient data queries. For example, a rule describing American cars with low mileage can be used to query specific properties like engine size, type of steering etc from the database. Generally, *leaf concepts* and value 'ANY' are not interesting to the user as they represent known facts. The interesting values however lie in interior concepts of an attribute's concept hierarchy according to [3]. Classical *AOI* does not consider interestingness of the generated rules as the value 'ANY' or 'don't-care' is used without restrictions.

D.A. Zighed, J. Komorowski, and J. Zytkow (Eds.): PKDD 2000, LNAI 1910, pp. 542-549, 2000.
© Springer-Verlag Berlin Heidelberg 2000

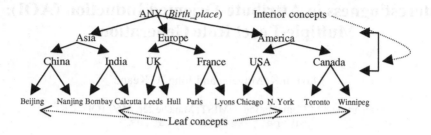

Fig. 1. Concept hierarchy for attribute '*Birth_place*'

We propose an approach to obtain interesting multiple-level rules (rules with high and low-level concepts) and regulate the degree of the rule table by generalising clusters of attribute values with a 'similar' *next* high-level concept (a '*common parent*'). A rough set approach is used to eliminate irrelevant attributes for rule generation [4].

The rest of the paper is organised as follows: section 2 presents related work; section 3 introduces concepts and terminologies; section 4 introduces the algorithm and analysis; section 5 shows an application of the approach using public domain data; section 6 presents conclusions and future work.

2   Related Work

Interestingness measures are either objective or subjective [5]. Objective approaches to foster interestingness in *AOI* can be categorised as *lookahead* and *predictive* [6]. The former are based on selecting an attribute for generalisation using heuristic measures based on the attribute's concept hierarchy structure [8]. Predictive strategies use thresholds [4] or predetermine the interestingness potential of interior concepts of an attribute's hierarchy. The rough set approach is used to determine dependencies between attributes. In, classification, information gain is used to choose an attribute to use as a root of the classification tree [9]. This results in correctly classified trees. Our approach uses distinct attribute values to choose an attribute and generalise clusters of values for the attribute to produce multiple-level rules [7].

3   Terminology and Definitions

Suppose a database has m attributes and n tuples. Assume that a concept hierarchy H_i is defined for each attribute A_i, i=2..m and a prime relation of p tuples exists.

Definition 3.1 A Concept hierarchy H_i of A_i is a *poset* (partially ordered set) $(H, \prec)$ where H_i is a finite set of concepts and $\prec$ is a partial order on H_i.

Definition 3.2. An attribute A_i is *generalisable* if there exists a concept hierarchy for the attribute (i.e. there are higher level concepts which subsume the given attribute values). Otherwise it is *nongeneralisable*.

Definition 3.3 A concept y of attribute A_i is the *nearest ancestor* of concept x if x, $y \in$ H_i with $x \prec y$, $x \neq y$, and there is no other concept $z \in H_i$ such that $x \prec z$ and $z \prec y$.

Definition 3.4 A concept y of attribute A_i is an *interior concept* of H_i if y is a non-leaf concept and y $\neq$ 'ANY'.

Definition 3.5 Two concept values a_r, $a_s \in A_i$ have a *'common parent'* $p \in H_i$ if p is the nearest ancestor of a_r and a_s.

Definition 3.6 A *key* $t_p(A_i)$ of a tuple t_p is the first numeric attribute value, i=1, $p <= n$.

Definition 3.7 A *characteristic rule* [7] is a rule whose left hand side is the query condition of the target data and the right hand side is a conjunction of generalised (attribute, value) pairs.

Definition 3.8 Two tuples t_p and t_q are *equivalent* if $t_p(A_i) = t_q(A_i)$, $i \neq 1$, i= 2,..,m.

Definition 3.9 A *merge* of two tuples t_p and t_q occurs by determining their equivalence, incrementing the count of total equivalent tuples by one and deleting one tuple.

The first task the algorithm in the next section deals with is maintaining interestingness through the *AOI* generalisation process. After reducing the complexity of the data by *AOI*, the second part which is rule generation, introduces a rough set approach. Rules are then represented as decision rules. When decision rules are generated, rules redundancies and rule inconsistencies may arise. Our proposed algorithm inherently solves this problem by the nature of the generalisation process (see application example). Definitions and detail theory of decision rules, rule redundancy and inconsistencies are given in [4].

4   Common Parent Rule Algorithm (COMPARE)

The COMmon PArent RulE (COMPARE) algorithm performs *AOI* primitives as follows:
1. Retrieve the input and generalise each tuple during input so that no tuple has leaf concepts at rule generation stage except for attributes with a concept hierarchy depth[1] of 1. This also reduces the number of tuples in memory to a small relation.
2. An *attribute threshold* T_a and an evaluation function[2] for choosing the next attribute for generalisation are used. For each chosen attribute, all attribute values with a *'common parent'* (excluding 'ANY') are grouped together and counted. The group with the highest count is generalised until thresholds are reached.
3. *Rule generation* only considers reduction of number of tuples by further generalisation of selected attribute values and controlled by a rule threshold Tr. Significant values[3] are used to eliminate less significant attributes by starting with tuples differing in one attribute value. This approach, like in [4], ensures that we remain with non-redundant attributes for rule generation.

---

[1] Depth of root node is zero and any other node is 1 more than the depth of its parent
[2] The function d_{weight}/t_{weight} is used in [4]
[3] The chi-square statistic X^2

4.1   Analysis

When an attribute A_i is chosen for generalisation using the evaluation function, attribute values with a *'common parent'* in the concept hierarchy H_i are identified and grouped together. When k similar attribute values are encountered (i.e. they have a *'common parent'*), k keys are inserted in a $(p+c)x(p+c)$ array and a count of the flags (denoting each flag as 'z') representing 'a similar value encounter' is recorded, where c allows for all 'z' flags and is small. The number of *distinct* values in each group is evaluated taking all 'z' flags into account and the group with the highest value is a candidate for further generalisation. This process is repeated with other groups until the evaluation function is satisfied using threshold T_a.

INPUT Task relevant data of N tuples, m attributes A_i, i=1,..,m, thresholds T_a, T_r
OUTPUT A set of interesting multiple-level rules
Step 1.   Input tuples, generalise each tuple
           Merge equivalent tuples and store in a prime table
Step 2. For ($i = 1$ to m)
           Cluster values of A_i with *'common parents'*
           Generalise cluster with highest distinct values
               While (count distinct $A_i > T_a$)
                   Generalise cluster with next greatest distinct values
                   Evaluate count distinct of A_i
Step 3. Compute significant values (SIG_i) for each attribute A_i
Step 4. While (threshold T_r not reached)
           For each A_i, starting with lowest SIG_i of A, $i \in (2,..,m)$
           If two tuples differ in only ONE attribute A_i
               Replace attribute values with 'ANY'
               merge equivalent tuples and evaluate threshold
Step 5. Output tuples in final table

Fig. 2. The COMPARE algorithm

The complexity of the algorithm in figure 2 is evaluated as follows: To input n tuples to a generalised prime relation is $O(np)$ [2] and finding distinct values for each attribute A_i is $O(m)$, $0<i\leq m$, for m attributes and p generalised values [6]. For h clusters of A_i where $C_{ji}=\{C_{1i},..,C_{hi}\}$, $|C_{ji}| < p/m$, $\forall j=1,..,h$, i.e. the size of clusters C_{ji} are not too different for each attribute. Inserting 'similar value encounter' in the $(p+c)x(p+c)$ array is less than $O(p^2)$ and negligible. Assuming the cardinality of each corresponding cluster is the same for each attribute (i.e. $|C_{11}|=|C_{12}|=|C_{13}|$ etc except for concept distributions), then the number of tuples is expressed as

$$\sum_{i=1}^{m}\sum_{j=1}^{h}(|C_{ji}|) = p.$$

After choosing attribute A_i, to find common parent for attribute values in j^{th} cluster C_{ji} for k concepts and generalising, the complexity is $O(k|C_{ji}|)$. For h clusters of m

attributes, the complexity is O $(kmh|C_{ji}|)$ in the worst case, where each C_{ij} is generalised for small k and m. Since $|C_{ji}|<p/m$, $hm|C_{ji}|=p$, $\forall_{j,i}$ and therefore the order complexity is O (kp) or just O (p).

The complexity of the algorithm by using attribute thresholds is O (np) + O (m) + O (p) = O (np). To merge the prime relation using least significant values for m attributes and inserting in a final table of q tuples, the complexity is O (pqm^2). Thus, if we say $p=n_i$, and $q=n_2$ as in [6], we have O $(m^2 n_i n_o)$. Thus we have order complexity O $(np)+O$ $(p)+$ O (m^2pq). For small p and q in large databases, the complexity is reduced to O (np).

4.2   Example Application

Table 1. Car Information

Cno	Model	Fuel	Disp	Weight	Cyl	Power	Turbo	Comp	Tran	Mileage
1	Ford Escort	EFI	Medium	876	6	High	Yes	High	auto	Medium
2	Dodge Shadow	EFI	Medium	1100	6	High	No	Medium	manu	Medium
3	Ford Festiva	EFI	Medium	1589	6	High	No	High	manu	Medium
4	Chevrolet Corvette	EFI	Medium	987	6	High	No	Medium	manu	Medium
5	Dodge Stealth	EFI	Medium	1096	6	High	No	High	manu	Medium
6	Ford Probe	EFI	Medium	867	6	High	no	Medium	manu	Medium
7	Ford Mustang	EFI	Medium	1197	6	High	no	High	manu	Medium
8	Dodge Daytona	EFI	Medium	798	6	High	yes	High	manu	High
9	Chrysler Le Baron	EFI	Medium	1056	4	Medium	no	Medium	manu	Medium
10	Dodge Sprite	EFI	Medium	1557	6	High	no	Medium	manu	Low
11	Honda Civic	2-BBL	Small	786	4	Low	no	High	manu	High
12	Ford Escort	2-BBL	Small	1098	4	Low	no	High	manu	Medium
13	Ford Tempo	2-BBL	Small	1187	4	Medium	no	High	auto	Medium
14	Toyota Corolla	EFI	Small	1023	4	Low	no	High	manu	High
15	Mazda 323	EFI	Medium	698	4	Medium	no	Medium	manu	High
16	Dodge Daytona	EFI	Medium	1123	4	Medium	no	Medium	manu	Medium
17	Honda Prelude	EFI	Small	1094	4	High	yes	High	manu	High
18	Toyota Paseo	2-BBL	Small	1023	4	Low	no	Medium	manu	High
19	Chevrolet Corsica	EFI	Medium	980	4	High	yes	Medium	manu	Medium
20	Chevrolet Beretta	EFI	Medium	1600	6	High	no	Medium	auto	Low
21	Chevrolet Cavalier	EFI	Medium	1002	6	High	no	Medium	auto	Medium
22	Chrysler Le Baron	EFI	Medium	1098	4	High	no	Medium	auto	Medium
23	Mazda 626	EFI	Small	1039	4	Medium	no	High	manu	High
24	Chevrolet Corsica	EFI	Small	980	4	Medium	no	High	manu	High
25	Chevrolet Lumina	EFI	Small	1000	4	Medium	no	High	manu	High

{Honda_Civic,..,Honda_Accord} ⊂ Honda
{Toyota_Tercel,...,Toyota_Camry} ⊂ Toyota
{Mazda 323,..., Mazda_626} ⊂ Mazda   {Ford,...,Chevrolet} ⊂ USA{Car}
{Ford_Escort,...,Ford_Taurus} ⊂ Ford   {Honda,...,Mazda} ⊂ Japan {Car}
{Chev_Covertte,...,Chev_Corsica} ⊂ Chevrolet {Japan (Car),..., USA(Car)} ⊂ Any (Model)
{Dodge_stealth,...,Dodge_Dynasty} ⊂ Dodge   {Light, Medium,Heavy} ⊂ Any (Weight)
{0..800} ⊂ Light
{801..1200} ⊂ Medium
{1201..1600} ⊂ Heavy

Fig. 3. A concept hierarchy for Table 1

Consider a collection of car information in Table 1 and its associated concept hierarchy shown in figure 3. From the tables, the attributes '*Fuel*', '*Disp*', '*Cyl*', '*Power*', '*Turbo*', '*Comp*', '*Tran*' and '*Mileage*' have at most 3 distinct values and their concept hierarchies are not given. If we choose to investigate characteristics of cars that determine *mileage* from the given data, then *mileage* will be a *decision attribute* or target class and all other attributes will be a conjunction of generalised attribute value pairs. This eliminates the attributes *'Fuel'*, *'Disp'*, *'Cyl'* and *'Turbo'* by analogy and expert knowledge. Another attribute *'Cno'* for car number to act as a *key* for each tuple as defined earlier is introduced.

4.3   Results

A Generalised Car Information System is first obtained as shown in table 2.

Table 2. A table generated from COMPARE algorithm (18 tuples)

Cno	Model	Weight	Power	Comp	Tran	Mileage	Count
1	USA_CAR	Medium	High	High	Auto	Medium	1
2	USA_CAR	Medium	High	Medium	Manu	Medium	4
3	USA_CAR	Heavy	High	High	Manu	Medium	1
5	USA_CAR	Medium	High	High	Manu	Medium	2
8	USA_CAR	Light	High	High	Manu	High	1
9	USA_CAR	Medium	Medium	Medium	Manu	Medium	2
10	USA_CAR	Heavy	High	Medium	Manu	Low	1
11	Honda	Light	Low	High	Manu	High	1
12	USA_CAR	Medium	Low	High	Manu	Medium	1
13	USA_CAR	Medium	Medium	High	Auto	Medium	1
14	Toyota	Medium	Low	High	Manu	High	1
15	Mazda	Light	Medium	Medium	Manu	High	1
17	Honda	Medium	High	High	Manu	High	1
18	Toyota	Medium	Low	Medium	Manu	High	1
20	USA_CAR	Heavy	High	Medium	Auto	Low	1
21	USA_CAR	Medium	High	Medium	Auto	Medium	2
23	Mazda	Medium	Medium	High	Manu	High	1
24	USA_CAR	Medium	Medium	High	Manu	High	2

Table 3. Significant values for attributes

Attribute	χ^2	Attribute	χ^2
Weight	17.54	Tran	4.53
Model	12.86	Comp	3.84
Disp	7.08	Fuel	0.63
Cyl	5.94	Turbo	0.63
Power	5.68		

Attribute and rule thresholds of 3 and 9 were used to compare the results obtained in [6]. Significant values for attributes in table 2 are shown in table 3. Applying the COMPARE algorithm generates table 4. The values 'ANY' are represented by '-'. After attribute thresholds are satisfied, rule thresholds need to be satisfied by further processing of table 4. Table 5 shows the generated table using the user-defined rule threshold T_r=9.

Table 4. A table after merging tuples differing by one attribute (12 tuples)

Cno	Model	Weight	Power	Comp	Tran	Mileage	Count
1	USA_CAR	Medium	High	-	-	Medium	8
3	USA_CAR	Heavy	High	High	Manu	Medium	1
6	USA_CAR	Meium	Medium	-	Manu	Medium	4
8	USA_CAR	-	Medium	High	Manu	High	3
10	USA_CAR	Heavy	Low	Medium	Manu	Low	1
11	Honda	Light	Low	High	Manu	High	1
13	USA_CAR	Medium	Medium	Hgh	Auto	Medium	1
14	Toyota	Medium	Low	-	Manu	High	2
15	Mazda	Light	Medium	Medium	Manu	High	1
17	Honda	Medium	High	High	Manu	High	1
20	USA_CAR	Heavy	High	Medium	Auto	Low	1
23	Mazda	Medium	Medium	High	Manu	High	1

Table 5. Final Table (8 tuples)

Cno	Model	Weight	Power	Comp	Tran	Mileage	Count
1	USA_CAR	Medium	High	-	-	Medium	9
3	USA_CAR	-	-	High	Manu	Medium	2
8	-	Light	-	High	Manu	High	2
9	USA_CAR	Medium	Medium	-	-	Medium	3
10	USA_CAR	Heavy	High	Medium	-	Low	2
14	Toyota	Medium	Low	-	Manu	High	2
15	Mazda	Light	Medium	Medium	Manu	High	4
17	-	Medium	-	High	Manu	High	4

Notice that for attribute '*Model*' in table 5, the concepts 'Toyota' and 'Mazda' are less general than 'USA_CAR' or 'JAPAN_CAR'. With a rule threshold, no rule inconsistencies arise as the number of rules would be few. Arguably, a small rule threshold means producing more general rules that may not be interesting. Rule 1 (first row of table 5) shows that about 32% of cars having 'Medium' weight, 'High' power and 'Medium' mileage are USA cars. If any of the less significant attributes like '*Tran*', '*Comp*', or '*Power*' were removed, no rule inconsistencies arise or even when all three are removed together. The rules with keys 1 and 9 are logically included in the rule with *key* 3 when '*Comp*' and '*Tran*' are generalised to 'ANY'. If attributes '*Tran*', '*Comp*', '*Model*' and '*Power*' are removed, rules with *keys* 1 and 9 are inconsistent with rules with *keys* 14 and 17 i.e. „if (weight=medium) then (mileage=Medium)" is inconsistent with „if (weight=medium) then (mileage=High)" respectively. So we keep the attribute '*Model*'. Similarly, if '*Power*', '*Model*' and '*Tran*' were removed, the rule „if (weight=medium) then (mileage=Medium)" with *keys* 1 and 9 is inconsistent with rule „if (weight=medium) then (mileage=High)" with *key* 14 unless '*Model*' or '*Power*' is kept and so on.

Table 6. Final Table (9 tuples)

Model	Weight	Power	Comp	Tran	Mileage
-	Heavy	-	Medium	-	Low
USA_CAR	Medium	High	-	-	Medium
USA_CAR	Medium	-	Medium	-	Medium
-	Medium	-	-	Auto	Medium
USA_CAR	-	Low	-	-	Medium
-	Heavy	-	High	-	Medium
-	-	Medium	High	Manu	High
Japan	-	-	-	-	High
-	Light	-	-	-	High

Using this approach, fewer or no inconsistent rules are generated unless most attributes are dropped. This is because step 2 of the algorithm clusters attribute values with *common parents* making tuples less equivalent. Analysing table 6 from [6] shows that attribute '*Model*' has no values 'Toyota' and 'Mazda' as they have been over generalised to value 'Japan'. Also, more significant attributes like 'Model' and 'Weight' (table 5) have larger numbers of 'non-ANY' attribute values than those in table 6. This comparison is also true for less significant attributes. In general, the number of 'don't-care' or 'ANY' values in table 5 (12 of them) as compared to those in table 6 (26 of them) shows how cautiously the algorithm generalises the data by delaying replacement of 'ANY'. This is advantageous in preserving interestingness of the rules.

5   Conclusion

We have presented the COMPARE algorithm of complexity O (*np*) that integrates predictive and *lookahead strategies*, by use of thresholds and *'common parent'* respectively, for rule interestingness in *AOI*. Using these two approaches and the rough set approach for removing noisy data at rule generation, interesting multiple-level rules are produced in *AOI*. Further work can be summarised as follows:

♦ Repeatedly check rule thresholds when merging. Both our approach and classical *AOI* overlook this. For example, a rule threshold of 9 may produce 8 rules instead of 9 (see tables 5,6). In table 4, merging tuples with *keys* represented by *key* pairs (1,6), (8,23) reduces the tuples from 12 to 10. The next merge is pair (10,20) and the rule threshold of 9 is satisfied. Therefore, we need not merge the next pair (11,17). In large databases, this would be important for preserving interestingness.

♦By Integrating with relevant attribute selection [6] prior to mining and the interestingness approach presented, more interesting rules may be produced.

References

1. Han, J.; Cercone, N. and Cai, Y. 1991. „*Attribute-Oriented Induction in Relational Databases*" In G. Piatetsky-Shapiro and W. J. Frawley, editors, Knowledge Discovery in Databases, pp 213-228.
2. Muyeba, K. M., and Keane, J.A 1999. „*Extending Attribute-Oriented Induction as a Key-Preserving Data Mining Method*" In Proceedings of the Third European Conference on Principles of Data Mining and Knowledge Discovery (PKDD'99), Prague, Czech. Republic, pp 448-455.
3. Fudger, D. and Hamilton, H. 1993. „*Heuristic Evaluation of Databases for Knowledge Discovery with DBLEARN*" International Workshop on Rough Sets and Knowledge Discovery, Banff, Canada, pp 29-39.
4. Shan, N.; Howard, J. H. and Cercone, N. 1996. „*GRG: Knowledge Discovery Using Information Generalisation, Information Reduction and Rule Generation*" In International Journal of Artificial Intelligence tools, 5: (1&2), pp 99–112.
5. Silberschatz, A. and Tuzhilin, A. 1995. „*On subjective measures of Interestingness in Knowledge Discovery*" In Proceedings of the First International Conference on Knowledge Discovery and Data Mining (KDD'95), Montreal, Canada, pp 275-281.
6. Barber, B. and Howard, H. J. 1997. „*A Comparison of Attribute Selection Strategies for Attribute-Oriented Generalisation*" In International Symposium on Methodologies for Intelligent Systems (ISMIS'97), Charlotte, NC, pp 106-116.
7. Yongjian, F. 1996. „*Discovery of Multiple-level Rules from Large Databases*", Ph.D. thesis, Simon Fraser University, Canada.
8. Carter, C.L.; Hamilton, H. J. and Cercone, N. 1994. „*The Software Architecture of DBLEARN*", Technical Report CS-94-04, University of Regina.
9. Quinlan, J. R. 1986. „*Induction of decision trees*", Machine Learning, Vol. 1, pp 81-106.

Discovery of Characteristic Subgraph Patterns Using Relative Indexing and the Cascade Model

Takashi Okada and Mayumi Oyama

Center for Information & Media Studies, Kwansei Gakuin University
1-1-155, Uegahara, Nishinomiya 662-8501, Japan
{okada,oyama}@kwansei.ac.jp

Abstract. Relational representation of objects using graphs reveals much information that cannot be obtained by attribute value representations alone. There are already many databases that incorporate graph expressions. We focus on syntactic trees in language sentences, and we attempt to mine characteristic subgraph patterns. The mining process employs two methods: relative indexing of graph vertices and the cascade model. The former extracts many linear subgraphs from the database. An instance is then represented by a set of items, each of which indicates whether a specific linear subgraph is contained within the graph of the instance. The cascade model is a rule induction method that uses levelwise expansion of a lattice. The basic assumption of this mining process is that characteristic subgraphs may be well represented by the concurrent appearance of linear subgraphs. The resulting rules are shown to be a good guide for obtaining valuable knowledge in linguistics.

1 Introduction

Structured objects can be represented very effectively by using graphs. Graphs can express general relationships in data that cannot be obtained by the usual attribute value expressions, and many databases therefore incorporate graph representations. For example, the structural formulae in chemistry, syntactic trees in natural language, and circuits in engineering all use graphs. We put our focus on the mining method applicable to all of these graph-structured objects.

Recently, interest in graph-structured objects has increased in the fields of machine learning and data mining; there has been work on GBI (graph based induction) [1], ILP (inductive logic programming) [2], and the association rule for graphs [3, 4]. ILP has been applied to SAR (structure activity relationships) problems in chemistry and shown useful [5]. However, these methods have not sufficiently considered all respects of universal validity, applicability to a variety of problems, and required computational resources, and there is a need for a new, efficient method.

The principal aim of this paper is to propose a mining scheme that is generally applicable to different kinds of graph-structured objects. Section 2 explains the new mining process, which consists of item generation and application of the cascade model. In Section 3, the procedure is applied to the analysis of syntactic parse trees in a corpus, and we can get reasonable results. The method has also shown its usefulness in a preliminary study on the mutagenicity of chemical compounds [6].

D.A. Zighed, J. Komorowski, and J. Zytkow (Eds.): PKDD 2000, LNAI 1910, pp. 550-557, 2000.
© Springer-Verlag Berlin Heidelberg 2000

2   Mining Methods

2.1 General Scheme

We propose a mining scheme that consists of two steps. In the first, we generate thousands of attributes from a set of instance graphs; each attribute denotes whether a specific subgraph is contained within the graph. The method of relative indexing of vertices restricts the subgraphs to linear types without branching, and provides an affordable number of attributes. Each graph can then be described as a tuple in a table with thousands of columns. Other properties of a graph can be included as an attribute of the table.

The second step is to find dependencies among attributes. There are many possible methods; we can employ a decision tree to derive classification rules for some attributes. Alternatively, the subgraph patterns of a graph could be regarded as items in a basket and the association rules method could be applied. In this paper, we employ the cascade model to derive rule expressions. We chose this model because it is able to derive characteristic, and/or classification, rules in a single unified framework, with a pruning method that can suppress combinatorial explosion of lattice size, even with high item density.

The resulting rules can act as a guide in extracting valuable knowledge from a database. The rules are expressed not by the target subgraph, but by the concurrent appearance of plural linear subgraphs that are interpreted to provide knowledge.

2.2 Relative Indexing of Graph Vertices

Our method can be explained as follows, using a syntactic tree as an example. Suppose that we wish to find characteristic patterns in the syntactic tree associated with the verb "think". An example of the tree is shown in Figure 1; it contains 8 leaf vertices and 6 non-leaf vertices.

If we extract all possible subgraphs from this kind of trees, the number of attributes will be too large for most mining methods to handle, and we therefore need to impose some restrictions on the subgraph pattern. To do so, we introduce a new scheme: the

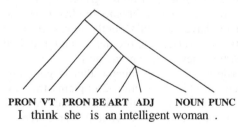

PRON VT PRON BE ART ADJ   NOUN PUNC
I think she is an intelligent woman .

Fig. 1. Sample syntactic tree.

Table 1. Leaf vertices and relative indices.

Word	Part of speech	Index
I	PRON	1.2./.1
she	PRON	1./.2.1
is	BE	1./.2.2.1
an	ART	1./.2.2.2.1
intelligent	ADJ	1./.2.2.2.2
woman	NOUN	1./.2.2.2.3
. (period)	PUNC	1.2.1./.2

relative indexing of graph vertices. This scheme assumes that a subgraph is linear and consists of two parts.

- Two meaningful vertices.
- The relationship between the two vertices.

We can fix one of the two meaningful vertices to the leaf vertex, [VT: think], as our aim is to analyze the syntactic pattern based on its usage. As the non-leaf vertices in this tree possess no valuable information except topology, we can restrict the source of the other meaningful vertex to the leaf vertices. Therefore, the attributes employed are the 7 subgraphs between "think" and 7 leaf nodes, as shown in Table 1.

The next problem is the expression of the relationship between the selected leaf vertices. As the resulting rules are depicted using these expressions, we expect the original graph structure to recover as much as possible from the attributes' expression. The syntactic tree is an ordered tree and the edges branching from a vertex can be numbered. Therefore, we assign a relative index to each leaf node, as shown in the last column of Table 1. The relative index of the word "I" is given by "1.2./.1", where "/" indicates the root vertex of the minimum subtree containing the two words, as shown in Figure 2. Starting from the position of "/", the numbers on the left (right) side, delimited by periods, indicate the sequence of edge indices to the word "think" ("I"). Here, we have assigned the edge index 1 to the leftmost edge. The resulting relative index is given by concatenating the two indices to "think" and "I".

We can define a unique relative index for any vertex. Consequently, we can recover the relative positions of the words from the index unambiguously. However, this indexing scheme may require modification, depending on the problem considered. For example, when treating chemical compounds, the edges in a graph are not ordered, and therefore we cannot give an unambiguous index between a pair of vertices.

The characteristic subgraph that is to be mined may very well be a general graph that cannot be represented as a linear graph. There is then the question as to whether a set of generated linear subgraphs can stand in for a general subgraph in the representation of a rule, and this is the core point by which to judge the current method. We anticipate that in most cases the concurrent appearance of linear subgraphs in a rule can substitute for a general subgraph. We inspect this hypothesis later in this paper.

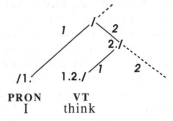

Fig. 2. Relative indexing between "think" and "I".

2.3 The Cascade Model

The cascade model was originally proposed by Okada [7]. It can be considered as an extension of association rule mining. The method creates an itemset lattice, where [attribute▪value] pairs are employed as items that constitute itemsets. Links in the lattice are selected and expressed as rules, by examining the distribution of the RHS attribute values along all the links. A sudden change of distribution along a link will bring the two terminal nodes of that link into focus. Suppose that the itemset at the upper end of a link is [A: y], and that an item [B: n] is added along the link. If a sharp increase in [C: y] is then found along this link, we can write a rule with the following expression:

$$\text{IF [B: n] added on [A: y] THEN [C: y]}$$

where the added item [B: n] is the main condition of the rule, and the items on the upper end of the link ([A: y]) are considered as the preconditions. Any number of items can be put into the RHS of a rule if its distribution shows a strong interaction with the main condition.

Subsequently, the sum of squares criterion for categorical data was introduced to improve the definition of rule strength [8]. The formulation of the model was also extended to cover the mining of classification rules and characteristic rules in a unified framework. The problem of combinatorial explosion in the number of lattice nodes was also resolved by a new pruning methodology [9]. The cascade model is implemented as DISCAS software using lisp, and it is used in this work.

3. Application to Syntactic Trees

3.1 Problem Definition and Computation

Mining from corpus data may lead to new knowledge in linguistics, which may be reflected in improvements in natural language processing. We used the Electronic Dictionary Research (EDR) English corpus, which contains 160,000 sentences, with syntactic tree data for each [10]. As an example, we extracted sentences containing the verb "think" and tried to find characteristic patterns that were associated with this word. Among the 1,001 sentences retrieved, there were 134 and 867 sentences that contained VI (intransitive verb) and VT (transitive verb), respectively.

The corpus treats all blanks between words in a sentence as a special kind of word; we omitted these blanks to simplify the trees. There is a linguistic tag on every non-leaf vertex in the tree of this corpus, but it proved too difficult to interpret these tags and we were forced to omit them. After preprocessing, the resulting tree had the structure shown in Figure 1. The details of the corpus data, including definition of parts of speech, can be accessed over the Internet [10].

Generating an attribute using the scheme in Section 2.2 provides the option of using another indexing scheme through numbering the edges from right to left. As there is no reason to prefer one indexing scheme to the other, we employed both schemes to generate relative indices. The attribute format was set to the concatenation of the index and the part of speech columns in Table 1. Using the two indexing schemes, every word except "think" generates two attribute records. The number of records created from 1001 sentences was 28010, of which 10469 were recognized as different. The verb class, VI or VT, was also added as an attribute.

The cascade model was used to mine for characteristic rules, using the parameter values (*minsup*: 0.05, *thres*: 0.05, *thr-BSS*: 0.05) [9]. The pruning condition defined by the *thres* value can eliminate most attributes from the actual computation, and in this case left only 29 attributes for construction of the lattice. That is, if the two values *y* and *n* of an attribute have a very unbalanced distribution they do not contribute to forming the characteristic subgraph patterns.

The lattice construction took 7 seconds, giving 359 nodes, using a 266MHz Pentium II computer. The first rule set gave us 5 rules, which explained about half of the total sum of squares in the problem.

3.2 Rule Interpretation Scheme

The strongest rule is the first rule of the first rule set and has the expression shown in Figure 3. The main condition of this rule indicates the existence of AUX (auxiliary verb) at the position [1-/-2-2], where hyphens are delimiters among edge indices numbered from the right. Six RHS clauses are shown in decreasing order of *BSS* values. The underlined row is included to show the information of the main condition item. The last line shows that among 1,001 sentences, 117 satisfy the main condition, and the sum of *BSS* values for all attributes is 640 along this link.

The position of AUX indicated by the main condition is illustrated as **I** in Figure 4, where dashed lines denote the possibility of edges at the indicated locations. The first line of the RHS part indicates the existence of the subgraph **II**. The percentage of subgraph **II** increases from 11.7% to 100% along this rule link, and the associated *BSS* value is 91.2. We can therefore say that the appearance of subgraph **I** is always accompanied by that of **II**. As the frequencies of these two subgraphs are the same,

```
IF  [1-/-2-2AUX: y] added on []
THEN [2./.1.1AUX: y]        11.7%->100.0%; BSS: 91.2
THEN [1-/-2-2AUX: y]        11.7%->100.0%; BSS: 91.2
THEN [1-/-2-1ADV: y]        11.6%-> 98.3%; BSS: 88.0
THEN [2./.1.2ADV: y]        11.6%-> 98.3%; BSS: 88.0
THEN [1-2-1-/-2PRON: y]     14.3%-> 84.6%; BSS: 57.9
THEN [2.1.2./.1PRON: y]     14.3%-> 84.6%; BSS: 57.9
     Cases: 1001 -> 117              Sum_BSS:640.
```

Fig. 3. A rule expression for the verb "think" by the cascade model.

they will always appear together, so the actual main condition of the rule can be expressed by subgraph **III** in Figure 4.

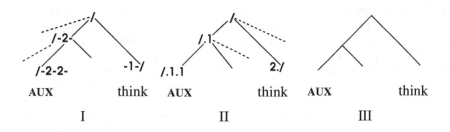

Fig. 4. Subtree expression of the main condition of a rule.

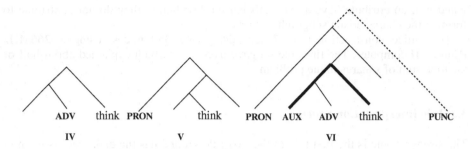

Fig. 5. Characteristic subgraph patterns found in a rule.

The lines 3-4 and 5-6 indicate the high confidence for concurrent appearance of subgraphs **IV** and **V** when the main condition is satisfied. In conclusion, the overall rule interpretation is shown by **VI** where the subgraph pattern, depicted by the solid lines, appears very frequently when the auxiliary verb is located at the position shown by the bold lines. Also indicated in **VI,** by the dotted lines, is the punctuation symbol that appears with the confidence of 61.5%.

Another example of a RHS clause with a large *BSS* value is shown below,

```
THEN [1.2./.1 PRON: n]   57.6% -> 100.0%; BSS:21.0
```

The item of this RHS clause indicates the nonexistence of the specified pattern. Two interpretations are possible for this description. One suggests the existence of words other than PRON at this location; the other leads to the nonexistence of the location itself in the tree, since either no words exist, but rather a subtree, or the existence of the location is incompatible with the main condition. We can see that the location and subgraph **VI** are contradictory in this rule, and therefore this clause does not add useful information.

3.3 Characteristic Patterns

The first rule set contained 5 characteristic rules, from which we constructed 3 characteristic patterns, **VI – VIII**, shown in Figure 6, following the procedure given in the previous section. One rule has few supporting sentences and the other only discriminates a group of sentences from those characterized by the three patterns. Therefore, we can conclude that these are the major patterns associated with the usage of "think". These patterns are exclusive to each other, and cover 56% of all sentences.

The patterns in Figure 6 are shown with an example sentence, the precondition description, the number of cases, and a *BSS* value. The sub-pattern shown by bold lines indicates the main condition, while solid lines are concurrent ones. No significant changes in the VI/VT ratio were observed in these patterns.

In fact, we can see these patterns frequently in various media. How are we to understand the absence of nouns at the locations of pronouns in these patterns? We have to be careful in the interpretation of the patterns. Actually, we can expect proper nouns and noun phrases at the same locations, but proper nouns are less frequent than pronouns and the noun phrase is not recognized in the corpus. Incorporating these kinds of items should result in more impressive patterns.

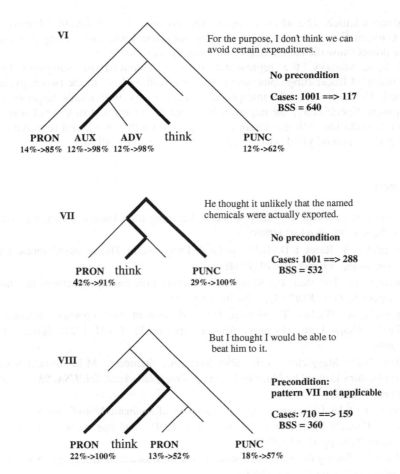

Fig. 6. Characteristic subgraph patterns for the verb "think".

The results obtained here can be regarded as a type of statistic on syntactic pattern. Extensive application of this method is expected to lead to new knowledge in the field of linguistics.

4. Concluding Remarks

Combination of the relative indexing of graph vertices and the cascade model has led to successful data mining in linguistics. A preliminary study on the mutagenicity of chemical compounds also showed the usefulness of the current method [6]. The ordered directed tree of sentence syntactic structure presents a clear contrast to the unordered undirected graph of chemical structure formulae. Both applications generate thousands of subgraph patterns as attributes, from which the efficient pruning strategy of the cascade model is able to select less than one hundred attributes to construct a lattice. The whole computation process is very efficient. Moreover, the search is exhaustive, using the given pruning parameters with the mining process. All of these points show the excellence of this mining method.

The basic strategy, the representation of a characteristic subgraph by the superposition of linear subgraphs, seems to work well, at least in the two applications examined. However, the rule interpretation process by individuals requires future development. Specifically, the negation item can be interpreted in several ways, and constant consultation with the database is required. Further work and research of this mining process should yield positive results in various applications.

References

[1]   Yoshida, K., Motoda, H.: CLIP: Concept Learning from Inference Patterns, *Artificial Intelligence*, **75**, pp.63-92 (1995).

[2]   Muggleton, S., Raedt, L.D.: Inductive Logic Programming:Theory and Methods. *J. Logic Programming*, **19**, pp.629–679 (1994).

[3]   Dehaspe, L., Toivonen, H., King, R.D.: Finding Frequent Substructures in Chemical Compounds, *Proc. KDD-98*, pp.30–36, AAAI (1998).

[4]   Inokuchi, A., Washio, T., Motoda, H.: Derivation of the Topology Structure from Massive Graph Data, *Discovery Science*, pp.330-332, LNAI 1721, Springer-Verlag (1999).

[5]   King, R.D., Muggleton, S.H., Srinivasan, A., Sternberg, M.J.: Structure-Activity Relationships Derived by Machine Learning. *Proc. Natl. Acad. Sci. USA*, **93**, pp.438–442 (1996).

[6]   Okada, T.: SAR Discovery on the Mutagenicity of Aromatic Nitro Compounds Studied by the Cascade Model, Proc. Int. Workshop KDD Challenge on Real-world Data at PAKDD-2000, pp.47-53, (2000).

[7]   Okada, T.: Finding Discrimination Rules Using the Cascade Model, *J. Jpn. Soc. Artificial Intelligence*, **15**, pp.321-330 (2000).

[8]   Okada, T.: Rule Induction in Cascade Model based on Sum of Squares Decomposition, *Principles of Data Mining and Knowledge Discovery (Proc. PKDD'99)*, pp.468-475, LNAI 1704, Springer-Verlag (1999).

[9]   Okada, T.: Efficient Detection of Local Interactions in the Cascade Model, *Knowledge Discovery and Data Mining (PAKDD 2000)*, pp.193-203, LNAI 1805, Springer-Verlag (2000).

[10]   Japan Electronic Dictionary Research Institute: EDR Electronic Dictionary, http://www.iijnet.or.jp/edr/index.html.

Transparency and Predictive Power
Explaining Complex Classification Models

Gerhard Paass and Jörg Kindermann

GMD - German National Research Center for Information Technology
53754 Sankt Augustin, Germany, mailto: paass,kindermann@gmd.de

Abstract. Complex classification models like neural networks usually have lower errors than simple models. They often have very many interdependent parameters, whose effects no longer can be understood by the user. For many applications, especially in the financial industry, it is vital to understand the reasons why a classification model arrives at a specific decision. We propose to use the full model for the classification and explain its predictive distribution by an explanation model capturing its main functionality. For a real world credit scoring application we investigate a spectrum of explanation models of different type and complexity.

1 Introduction

During the last years the availability of business process data allowed the utilization of sophisticated prediction and classification models in enterprises. A prominent example is credit scoring, where a customer is classified as solvent or insolvent based on a vector $\mathbf{x}$ of features. The unknown model parameters are trained using data on credit histories of customers. There is a plethora of models starting from simple linear models to complex neural networks or support vector machines. Many experimental evaluations (e.g. [8]) showed that more complex procedures in general yield better results in terms of classification errors. These methods usually have many interdependent parameters whose meaning no longer can be understood by the user.

There is a trade-off between the classification reliability and transparency of approaches. It is, however, possible to use a complex *full model* optimized with respect to predictive power and one or more simplified *explanation models*, which approximate the predictions of the full model as well as possible. If the difference between both predictions is small we can explain the results of the full model by evaluating the simple structure of the explanation model.

In the literature the approximation of a neural network by a rule system was investigated [1]. Craven [6] proposed to use tree models for the same task. Becker et al. [2] explored the explanation properties of naive Bayesian classifiers while Madigan et al. [9] explained Bayes probabilistic networks. In 1997 we developed a new credit scoring procedure for the DSGV, a large group of German banks [8]. This paper describes the results of a follow-on project funded by DSGV where we analyzed which type of explanation model is best suited to explain the predictions of the full model.

D.A. Zighed, J. Komorowski, and J. Zytkow (Eds.): PKDD 2000, LNAI 1910, pp. 558–565, 2000.
© Springer-Verlag Berlin Heidelberg 2000

2   Criteria for Selecting an Explanation Model

2.1   Notation

As *full model* we used a *Bayesian model* $p(y = 1|\mathbf{x}, \theta) = f(\mathbf{x}, \theta)$ for our credit scoring task. It predicts the *credit risk* for $\mathbf{x}$, i.e. the probability that a customer with feature vector $\mathbf{x}$ will not repay a credit ($y = 1$). Instead of estimating an *'optimal'* parameter θ we derived the *Bayesian posterior distribution* $p(\theta|\mathbf{y}, \mathbf{X}) = p(\mathbf{y}|\mathbf{X}, \theta)p(\theta) / \int p(\mathbf{y}|\mathbf{X}, \theta)p(\theta)d\theta$ of parameters given data $\mathbf{y}, \mathbf{X}$ and prior distribution $p(\theta)$. The posterior distribution describes the plausibility of different parameters. It is approximated by a sample $\{\theta_1, \ldots, \theta_B\}$ of parameters, which is generated by Markov Chain Monte Carlo methods. The model is based on decision trees.

For a new input $\mathbf{x}_0$ we may calculate the risk values $p(y = 1|\mathbf{x}_0, \theta_b)$ with respect to all $\theta_b \in \{\theta_1, \ldots, \theta_B\}$. These values approximate the *predictive distribution* describing the uncertainty of the credit risk. If we state the consequences of decisions in terms of a loss function, we should classify the case $\mathbf{x}_0$ such that the average expected loss is minimal. This implies that $\mathbf{x}_0$ is classified as insolvent if the mean risk is larger than some threshold π_{dec}

$$P(y = 1|\mathbf{x}_0) \approx \frac{1}{B} \sum_b p(y = 1|\mathbf{x}_0, \theta_b) > \pi_{dec} \tag{1}$$

A change in the loss function (e.g. interest rates) merely leads to a shift in π_{dec}.

In this framework an *explanation model* can be considered as a simplified model $p(y = 1|\mathbf{x}, \psi) = g(\mathbf{x}, \psi)$, which on the one hand should be easy to understand, and on the other hand should approximate the classifications of the full model as good as possible. To keep the situation simple only an optimal model $\hat{\psi}$ is derived, e.g. by maximizing the posterior density with respect to ψ.

2.2   Comprehensibility

The explanation of a classification model is based on the following factors:

- The agreement of the results with intuitive common *factual rationalizations*.
- Insight into the *internal "mechanics"* of the model.
- The *empirical validation* of results by simple statistics.

Undoubtedly an explanation by factual rationalizations is most convincing. The internal mechanics of a model is comprehensible only for models with simple structural features:

Analysis of Subpopulations. The data is partitioned into disjoint subsets according to the values of some features. Each subset is characterized by a simple "submodel" to enhance comprehension [7, p.185].

If-Then Rules. Rules are especially evident, if the conclusions correspond to the common lines of reasoning in the application.

Figure 1: Scorecard for six variables with up to three categories: low (L), medium (baseline) and high (H). The scores are the positive or negative differences to the baseline representing a_0. The scores for a single case are indicated by lines.

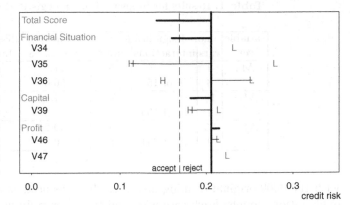

Linear Relations. The behavior of linear models is simple to understand.

Independent Variables. Even nonlinear models are easier to understand, if each feature adds an independent term to the result.

Subsets of Variables. Comparison of the results for different subsets demonstrate the contradictory effects of the subsets for the specific case.

To compare the comprehensibility of explanation models, we mainly use the *number of parameters* of the explanation model. This criterion is increased or decreased according to the *transparency* as well as the *empirical validity*. The latter quantities are highly subjective and give only a rough indication of the relative performance of procedures.

2.3 Classification Similarity and Quality

It is the task of the explanation model to characterize the distribution of predicted risks $p(y = 1|\mathbf{x}, \theta_b)$ of the full model. As the loss functions vary only to a small extent, the threshold π_{dec} is restricted to a small range. Only in this range the value of the decision risk $p(y = 1|\mathbf{x}, \theta)$ affects the decision (c.f. (1)).

Outside$_{90}$: The main criterion for classification *dissimilarity* of the full model and the explanation model. the credit risk predicted by the explanation model is *not* between the 5%-quantile and the 95%-quantile of the predictive distribution (section 2.1) of the full model.

Err$_{8.75}$: As an additional criterion we assess the *classification error* of the explanation model itself. We determine the percentage of rejected 'good' loans when 8.75% of bad loans are accepted. The Err$_{8.75}$-value for the full model was 32.0.

3 Comparing Explanation Models for Credit Scoring

To improve the fit btw. explanation and full models we created *synthetic data* by calculating the credit risk for 11000 unclassified genuine input vectors. Together

Table 1. Results for Stepwise Linear Logistic Regression.

variable set	no. of significant ...			classif. error	dissimilarity
	vars	interactions	param.	$Err_{8.75}$ %	$Outside_{90}$ %
$\mathcal{M}_6$	4		5	57.6	13.7
	3	6 of 15	10	56.6	12.9
$\mathcal{M}_{17}$	10		11	54.8	13.2
	4	11 of 136	16	51.8	13.1
$\mathcal{M}_{39}$	25		26	49.2	12.1
	9	16 of 741	26	50.8	12.5

with the 6000 original training instances this was used as training input for the explanation models. Each case was used twice for training with output 1 and 0 and sampling weights equal to the credit risk $p(y = 1|\mathbf{x}, \theta)$ and its complement.

We used three different sets of input features selected by domain experts: The set $\mathcal{M}_6$ of six most relevant variables, the set $\mathcal{M}_{17}$ containing additional medium important variables and a comprehensive set $\mathcal{M}_{39}$ [11, p. 135ff].

3.1 Linear Logistic Model

For a given feature vector $\mathbf{x} = (x_1, \ldots, x_k)$ this procedure approximates the credit risk by $g(\mathbf{x}, \theta) = \exp(\eta)/(1 + \exp(\eta))$ with $\eta = \theta_0 + \sum_{i=1}^{k} \theta_i x_i$. This is a classical credit scoring model used in many analyses [3].

Transparency. If coefficient θ_i is positive, the risk grows as attribute x_i is increased. Psychological experiments show that linear relations are far simpler to comprehend than nonlinear relations [7, p.129]. However strongly correlated features and corresponding parameter values with high covariance reduce the interpretability of parameters.

Classification Similarity. We used stepwise regression to identify the most important parameters. Even for the synthetic training set only four of the six most important variables are significant (see table 1). In comparison with later results the dissimilarity and the classification error are rather high. In summary this indicates the presence of non-linearity in the full model. Therefore linear models are not useful for explaining credit risk for this application.

3.2 Scorecards

This approach assumes that each feature x_i has an independent additive effect. We divide the domain of each feature into m subsets, each of which gets a point *score* a_{ij}. The sum of independent scores for a specific case is the total score $y = a_0 + \sum_{i=1}^{k} \sum_{j=1}^{m-1} a_{ij} x_{ij}$ where $x_{ij} = 1$ in the j-th subset of x_i and 0 elsewhere. The a_{ij} can be estimated from training data using stepwise linear regression [12, p.220] to detect significant scores. Scorecards are widely used in the financial industry [13]. A scorecard for the set $\mathcal{M}_6$ is shown in figure 1.

Table 2. Results for Scorecards Estimated by Stepwise Regression.

variable set	no. of categ.	signif. param.	classif. $Err_{8.75}$	dissimilarity $Outside_{90}$
$\mathcal{M}_6$	3	10	49.4	12.9
	5	20	47.5	12.4
$\mathcal{M}_{17}$	3	21	43.5	11.0
	5	40	41.9	10.2
$\mathcal{M}_{39}$	3	31	38.9	9.4
	5	53	37.9	8.2

Table 3. Results for Decision Trees for Different Sets of Features.

variable set	no. of rules	classif. $Err_{8.75}$	dissimilarity $Outside_{90}$
$\mathcal{M}_6$	27	52.8	11.4
	146	45.5	10.4
$\mathcal{M}_{17}$	23	51.2	10.8
	116	40.3	8.8
$\mathcal{M}_{39}$	26	42.1	9.5
	115	37.6	7.2

Transparency. The transparency of decisions and the simplicity of the 'mechanics' of scorecards is very high. Psychological research shows that the use of scores corresponds to our intuitions [7, S.189]. In summary scorecards are good explanation models.

Classification Similarity. In spite of the simple model structure scorecards have a surprisingly good classification similarity and quality (table 2). Note that there is an additional drop if we either increase the number of features or the number of categories within features. Generalized versions of scorecards [4, S.249ff] did not improve the results.

3.3 Decision Trees

Decision trees are frequently used for credit scoring [10],[4]. In contrast to linear models and scorecards, decision trees can represent any functional relation arbitrary well. They have an inherent representation bias, as they are well-suited for step-like functions.

Transparency. Decision trees have a double advantage in terms of transparency. First the internal mechanics of the method is simple as the credit risk within subsets defined by a few splits is constant. Second the credit risk itself is determined as the empirical frequency of 'bad' cases in this subset, leading to a good empirical validation. The number of rules (nodes) in the tree can be controlled. Large trees no longer offer a good explanation, as the user can only cope with a limited number of rules. An additional problem arises from repeated splits with respect to a single feature, as this is more difficult to understand.

Classification Similarity. The classification dissimilarity in table 3 is roughly in the range of figures for scorecards. In summary the scorecards have slight advantages compared to decision trees.

3.4 Explanation Models in Leaves of Decision Trees

A natural extension is the combination of approaches. In this section we combine decision trees with the following Scorecards and One- and two-dimensional nonlinear loess models as 'leaf'-models. We first developed an integrated approach

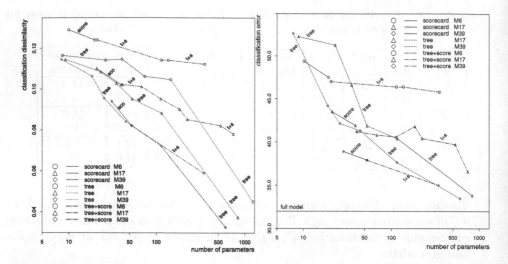

Fig. 2. Classification dissimilarity of scorecards, decision trees and scorecards in the leaves of decision trees.

Fig. 3. Classification error of scorecards, decision trees and scorecards within decision trees.

which simultaneously determined the trees as well as the scorecards with the aim to reduce the overall error. It turned out that this algorithm didn't offer any advantages. Therefore we simply generated a tree of the desired size and subsequently estimated scorecards within the leaves using stepwise regression.

Transparency. A scorecard is easier to understand than a decision tree with the same number of parameters. As scorecard merely is the sum of independent feature scores. A decision tree usually develops highly interlocked subset definitions, whose multidimensional relation is difficult to comprehend. On the other hand small decision trees are easy to understand. The combination of small trees and scorecards also is relatively transparent.

Classification Similarity. In the figures 2 and 3 the classification errors and dissimilarities of scorecards, trees and scorecards within trees are compared in relation of the number of parameters. In terms of dissimilarity it turns out that for the variables in $\mathcal{M}_6$ trees work better than scorecards and scorecards in leaves. For the variables in $\mathcal{M}_{17}$ all tree approaches have similar performance up to about 40 parameters. Finally for the set $\mathcal{M}_{39}$ all three procedures are nearly identical. Taking into account that scorecards are somewhat easier to understand than pure trees this gives scorecards and scorecards within trees an advantage.

As a second model we used loess model within the leaves of trees. Loess is a nonparametric approximation approach which is generated by the superposition of many simple local models defined in a small subset [5, p.93ff]. For each leaf subset we determined loess models with respect to one or two variables. The

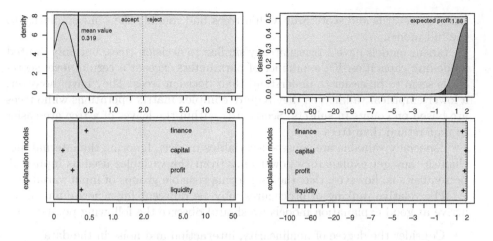

Fig. 4. Densities of predicted credit risk (left) and profit (right) for a case projected to the base population. Below are mean predictions of the subset-of variable models.

form of the curve gives a good idea of the resulting changes in credit risk if the feature is modified. However, in terms of classification error and similarity this approach performed worse than decision trees or scorecards.

3.5   Subset-of-Variables Models

The credit risk may be predicted by a model which includes only the variables of a specific group, e.g. profit. By comparing the prediction based on different variable groups we get a picture on the perhaps antagonistic effects of the groups. The model for this explanation may be a 'black box'. Therefore we may use the best model at hand, in our case the Bayesian procedure.

Figure 4 shows in its upper part the density of the predicted credit risk as well as the profit of the full model for a specific case. The classification is determined by the mean value of the credit risk (vertical line), which is below the decision threshold (dotted line), indicating that the case is classified as 'solvent'. Below are the mean predictions of four subset-of-variables models corresponding to the groups finance, capital, profit and liquidity. Finance as well as the liquidity indicate a higher credit risk, whereas the capital and profit suggest a lower risk. In that sense the height of the risk is 'caused' by finance and liquidity. The fact that the prediction is performed by an optimized model adapted to the training data gives a good empirical validation of the approach.

4   Summary

As a major result there was always some dissimilarity between the full Bayesian model and the simple explanation models for our data. Beside of this fact decision trees, scorecards and scorecards within trees had roughly the same similarity to the full model.

Linear models have a transparency similar to decision trees, but only limited modelling capacities. If the number of parameters exceeds a certain level scorecards seem to be easier to understand than decision trees. They have, however, a lower empirical validity, as they require complex matrix operations while trees are based on counts in subsets. Scorecards within the leaves of a tree are easier to understand than trees.

Especially valuable are subset-of-variables models. Ignoring the internal mechanics they get explanatory power only from the variables used as inputs. A prerequisite is, however, that the user forms sensible groups of input variables.

The relative ranking of procedures may be different for a new domain. To arrive at good explanation models we should observe the following points:

- Consider the degree of nonlinearity, interaction and noise in the data.
- Evaluate comprehensibility and classification similarity of selected models.
- Ask potential users for an assessment and use good or optimal models.
- Develop an intuitive graphical representation and an interactive interface.

References

1. R. Andrews and J. Diederich, editors. *Rules and Networks*, Brisbane, 1996. Queensland University of Technology, Neurocomputing Research Centre.
2. B. Becker, R. Kohavi, and D. Sommerfield. Visualizing the simple bayesian classifier. In *KDD Workshop on Issues in the Integration of Data mining and Data Visualization*, 1997.
3. Steven V. Campbell. Workouts: Reorganize or liquidate? prediction model gives insights for small-business loans. *Commercial Lending Review*, 11:78–85, 1996.
4. J. M. Chambers and T. J. Hastie. *Statistical Models in S*, volume Monograph 35. Wadsworth & Brooks, Pacific Grove, California, 1992.
5. W. S. Cleveland. *Visualizing Data*. Hobart Press, Summit, New Jersey, 1993.
6. M. W. Craven. *Extracting Comprehensible Models From Trained Neural Networks*. PhD thesis, University of Wisconsin - Madison, 1996.
7. R. M. Hogarth. *Judgement and Choice*. Wiley, Chichester, 2nd edition, 1987.
8. J. Kindermann and G. Paass. Model switching for Bayesian classification trees with soft splits. In J. Zytkow and M. Quafafou, editors, *Principles of Data Mining and Knowledge Discovery*, pages 148–157. Springer, 1998.
9. D. Madigan, K. Mosurski, and R.G. Almond. Explanation in belief models. *Journal of Computational and Graphical Statistics*, 6:160–181, 1997.
10. E. Rosenberg and A. Gleit. Quantitative methods in credit management: A survey. *Operations Research*, 42:589–613, 1994.
11. C. Uthoff. *Erfolgsoptimale Kreditwürdigkeitsprüfung auf der Basis von Jahresabschlüssen und Wirtschaftsauskünften mit Künstlichen Neuronalen Netzen*. M & P Verlag für Wissenschaft und Forschung, Stuttgart, 1997.
12. W. N. Venables and B. D. Ripley. *Modern Applied Statistics with S-Plus*. Springer Verlag, New York, 2nd edition, 1997.
13. Mark Zmiewski and Beverly Foster. Credit-scoring speeds small business loan processing. *Journal of Lending & Credit Risk Management*, 79(3):42–47, 1996.

Clustering Distributed Homogeneous Datasets*

Srinivasan Parthasarathy and Mitsunori Ogihara

Department of Computer Science
University of Rochester
Rochester, NY 14627–0226
{srini,ogihara}@cs.rochester.edu

Abstract. In this paper we present an elegant and effective algorithm for measuring the similarity between homogeneous datasets to enable clustering. Once similar datasets are clustered, each cluster can be independently mined to generate the appropriate rules for a given cluster. The algorithm presented is efficient in storage and scale, has the ability to adjust to time constraints, and can provide the user with likely causes of similarity or dis-similarity. The proposed similarity measure is evaluated and validated on real datasets from the Census Bureau, Reuters, and synthetic datasets from IBM.

1 Introduction

Large business organizations, like Sears, with nation-wide or international interests usually rely on a homogeneous distributed database to store their transaction data. This leads to multiple data sources with a common structure. Traditional methods to analyze such distributed databases involve combining the individual databases into a single logical entity before analyzing it. The problem with this approach is that the data contained in each individual database can have totally different characteristics, such as "The customers in the South Carolina store rarely buy winter-related products while those in a store in Maine may buy such items a lot", leading to a loss of potentially vital information. Another option, mining each database individually is unacceptable as this will likely result in too many spurious patterns (outliers) being generated, wherein it will become harder to decide which patterns are important. Our solution is to first *clusters the datasets*, and then to apply the *traditional distributed mining* approach to generate a set of rules for each resulting cluster.

The primary problem with clustering homogeneous datasets is to identify a suitable distance (similarity) metric. In Section 2 we develop our similarity measure. We then show how one can cluster distributed homogeneous database sources based on our similarity metric in a novel communication efficient manner in Section 3. We then experimentally validate our approach on real and synthetic datasets in Section 4. Finally, we conclude in Section 5.

---

* This work was supported in part by NSF grants CDA–9401142, CCR–9702466, CCR–9701911, CCR–9725021, INT–9726724, and CCR–9705594; an NIH grant RO1-AG18231; and an external research grant from Digital Equipment Corporation.

D.A. Zighed, J. Komorowski, and J. Zytkow (Eds.): PKDD 2000, LNAI 1910, pp. 566–574, 2000.
© Springer-Verlag Berlin Heidelberg 2000

2 Similarity Measure

Similarity is a central concept in data mining. Recently there has been considerable work in defining intuitive and easily computable measures of similarity between complex objects in databases [2,9,8]. The similarity between attributes can also be defined in different ways. An internal measure of similarity is defined purely in terms of the two attributes. The *diff* [11] method is an internal measure where the distance between the probability density function of two attributes is used as a measure of difference between two attributes. An external measure of similarity [5] compares the attributes in terms of how they are individually correlated with other attributes in the database. Das *et al.* [5] show that the choice of the other attributes (called the probe set), reflecting the examiner's viewpoint of relevant attributes to the two, can strongly affect the outcome. However, they do not provide any insight to automating this choice ("first guess") when no apriori knowledge about the data is available. Furthermore, while the approach itself does not limit probe elements to singleton attributes, allowing for complex (boolean) probe elements and computing the similarities across such elements can quickly lead to problems of scale in their approach. We propose an external measure of similarity for homogeneous datasets. Our similarity measure compares the datasets in terms of how they are correlated with the attributes in the database. By restricting ourselves to frequently occurring patterns (associations), as probe elements, we can leverage existing exact [3] and approximate [13] solutions for such problems to generate the probe sets. Furthermore by using associations as the initial probe set we are able to obtain a "first guess" as to the similarity between two attributes. Also using associations enables one to interactively customize [1] the probe set to inculcate examiner bias and probe for the causes of similarity and dis-similarity.

2.1 Association Mining Concepts

We first provide basic concepts for association mining that are relevant to this paper, following the work of Agrawal *et al.* [3]. Let $\mathcal{I} = \{i_1, i_2, \cdots, i_m\}$ be a set of m distinct *attributes* [1], also called *items*. A set of items is called an *itemset* where for each nonnegative integer k, an itemset with exactly k items is called a *k-itemset*. A *transaction* is a set of items that has a unique identifier *TID*. The *support* of an itemset A in database $\mathcal{D}$, denoted $\sup_{\mathcal{D}}(A)$, is the percentage of the transactions in $\mathcal{D}$ containing A as the subset. The itemsets that meet a user specified *minimum support* are referred to as *frequent* itemsets or as *associations*. We use our group's ECLAT [12] algorithm to compute the frequent itemsets (*associations*).

2.2 Similarity Metric

Let A and B respectively be the set of associations for a database $\mathcal{D}$ and that for a database $\mathcal{E}$. For an element $x \in A$ (respectively in B), let $\sup_{\mathcal{D}}(x)$ (respectively $\sup_{\mathcal{E}}(x)$) be the frequency of x in $\mathcal{D}$ (respectively in $\mathcal{E}$). Our metric is:

$$Sim(\mathcal{D}, \mathcal{E}) = \frac{\sum_{x \in A \cap B} \max\{0, 1 - \alpha|\sup_{\mathcal{D}}(x) - \sup_{\mathcal{E}}(x)|\}}{\|A \cup B\|}$$

---

[1] To handle continuous attributes we adopt a novel discretization method not described here due to lack of space.

where α is a scaling parameter. The parameter α has a default value of 1 and can be modified to reflect the significance the user attaches to variations in supports. For $\alpha = 0$ the similarity measure is identical to $\frac{\|A \cap B\|}{\|A \cup B\|}$, i.e., support variance carries no significance. Sim values are bounded and lie in [0,1]. Sim also has the property of *relative ordinalility*, wherein if $Sim(X,Y) > Sim(X,Z)$, then X is more similar to Y than it is to Z. These two properties are essential for our clustering algorithm.

2.3   Interactive Similarity Mining

An important point raised by Das *et al.* [5] is that using a different set of probes could potentially yield a different similarity measure. That property created the need for pruning/constraining the probe space to select appropriate probe sets. In our case the action of modifying the probe set corresponds to modifying the association sets of the input databases, and that is achieved either by modifying the minimum support or by restricting that associations should satisfy certain conditions (Boolean properties over attributes). Supporting such interactions is accomplished by leveraging interactive association mining [1]. In addition to the interactions supported in [1] we also support **influential attribute** identification. This interaction basically identifies the (set of) probe attribute(s) that contribute most to the similarity or dissimilarity of the two datasets.

We define the influence of an attribute a as follows. Let A' be the subset of A containing all associations that include the attribute a. Similarly we define B' with respect to B. Then the influence (INF) and relative influence (RINF) of an attribute a can be defined as follows:

$$INF(a, \mathcal{D}, \mathcal{E}) = \sum_{x \in A' \cap B'} \max\{0, 1 - \alpha | \sup_{\mathcal{D}}(x) - \sup_{\mathcal{E}}(x)|\},$$

$$RINF(a, \mathcal{D}, \mathcal{E}) = \frac{\sum_{x \in A' \cap B'} \max\{0, 1 - \alpha | \sup_{\mathcal{D}}(x) - \sup_{\mathcal{E}}(x)|\}}{\|A' \cup B'\|}$$

By definition the attribute which has the largest INF value is the largest contributor to the similarity measure. Also, the attribute which has the lowest INF value is the principle cause for dissimilarity. The user is presented with tuples of the form (INF(a,A,B), RINF(a,A,B)). The RINF values are useful to the user as it conditions the user relative to the size of $A' \cup B'$. As it turns out, these tuples are very useful for probing unexpected/interesting similarity results.

2.4   Sampling and Association Rules

The use of sampling for approximate, quick computation of associations has been studied in the literature [13]. While computing the similarity measure, sampling can be used at two levels. First, if generating the associations is expensive (for large datasets) one can sample the dataset and subsequently generate the association set from the sample, resulting in huge I/O savings. Second, if the association sets are large one can estimate the distance between them by sampling, appropriately modifying the similarity measure presented above. Sampling at this level is particularly useful in a distributed setting when the association sets, which have to be communicated to a common location, are very large.

3 Clustering Datasets

Clustering is commonly used for partitioning data [7]. The technique we adopt for clustering datasets is greedy tree clustering. We use the similarity metric defined in Section 2 in our clustering algorithm. Input to the algorithm is the number of clusters in the final result or a user specified *merge cutoff*. At the start of the clustering process each database constitutes a unique cluster. Then we repeatedly merge the pair of clusters with the highest similarity into one cluster until there are the desired number of clusters left or if merging any pair of clusters involves merging clusters that exhibit a *Sim* value that is below the minimal threshold (*merge cutoff*). As our similarity metric is based on associations, there is an issue of how to merge their association lattices when two clusters are merged. A solution would be to combine all the datasets in both clusters (treating them as one logical entity) and recompute the associations, but this would be time-consuming and involve heavy communication and I/O overheads (all the datasets will have to be re-accessed). Another solution would be to intersect the two association lattices and use the intersection as the lattice for the new cluster, but this would be very inaccurate. We take the half-way point of these two extremes.

Suppose we are merging two clusters $\mathcal{D}$ and $\mathcal{E}$, whose association sets are respectively A and B. The value of $\sup_{\mathcal{D}}(x)$ is known only for all $x \in A$ and that of $\sup_{\mathcal{E}}(x)$ is known only for all $x \in B$. The support of x in the merged cluster is estimated as

$$\frac{\sup_{\mathcal{D}}(x) \cdot \|\mathcal{D}\| + \sup_{\mathcal{E}}(x) \cdot \|\mathcal{E}\|}{\|\mathcal{D}\| + \|\mathcal{E}\|}.$$

When x does not belong to A or to B, we will approximate the unknown sup-value by a "guess" θ [2], which can be specific to the cluster as well as to the association x.

4 Experimental Analysis

In this section we experimentally evaluate our similarity metric. We first evaluate the performance and sensitivity of computing this metric using sampling, under various support thresholds in a distributed setting. We then evaluate the sensitivity of our metric to choice of α. Finally we demonstrate the efficacy of our dataset clustering technique to synthetic datasets from IBM and on a real dataset from the Census Bureau, and evaluate the results obtained.

4.1 Setup

All the experiments (association generation, similarity computation) were performed on DECStation 4100s containing four 600MHz Alpha 21164 processors, with 256MB of memory per processor. In order to model distributed market basket data we generated 12 different synthetic datasets ranging from 90MB to 110MB, which are generated adopting the procedure described in [3]. These databases mimic the transactions in a retailing environment. Table 1 shows the databases used and their properties. The number of transactions is denoted as $numT$, the average transaction size as T_l, the average maximal

---

[2] Our strawman approach to estimate the value of θ is to randomly guess a value between 0 and the minimum support, since if it does not appear in the set of associations it must have a support less than the minimum support. The second approach, which we evaluate is to estimate the support of an itemset based on the available supports of its subsets.

potentially frequent itemset size as I, the number of maximal potentially frequent item-sets as $\|L\|$, and the number of items was 1000. For each $\|L\|$, I pair (generation pool) we created 3 different (a, b and c) databases by tweaking the database generation program parameters. We refer the reader to [3] for more detail on the database generation.

Database	$numT$	T_l	I	$\|L\|$	Size (range)
$D1_{a,b,c}$	2000000	8-9	2000	4-5	85-90MB
$D2_{a,b,c}$	2000000	10-12	6000	2-4	95-100MB
$D3_{a,b,c}$	2100000	9-10	4000	3-4	100-102MB
$D4_{a,b,c}$	2250000	10-11	10000	6-8	110-120MB

Fig. 1. Database properties

We also use the Reuters-21578 collection [10] to evaluate some aspects of our work. The original data set consists of 21578 articles from the Reuters newswire in 1987. Each article has been tagged with keywords. We created a basket dataset from this collection by representing each news article as a transaction and each keyword being an item. About 1800 articles had no keywords and the average transaction length was between 2 and 3. From this basket dataset we created 12 *country datasets*. Each *country dataset* contains the transactions (country names were among the keywords) that refer to that particular country.

The Census data used in this work was derived from the County Business Patterns (State) database from the Census Bureau. Each dataset (one dataset per state, eight states) contains one transaction per county. Each transaction contains items which highlight information on subnational economic data by industry. Each industry is divided into small, medium and large scale concerns. The original data has numeric attributes (countably infinite) corresponding to number of such concerns occurring in the county which we discretized into three (high, medium, small) categories.

4.2   Sensitivity to Sampling Rate

In Section 2 we mentioned that sampling can be used at two levels to estimate the similarity efficiently in a distributed setting. If association generation proves to be expensive, one can sample the transactions to generate the associations and subsequently use these associations to estimate the similarity accurately. Alternatively, if the number of associations in the lattice are large, one can sample the associations to directly estimate the similarity. We evaluate the impact of using sampling to compute the approximate the similarity metric below.

For this experiment we breakdown the execution time of computing the similarity between two of our databases $D3_a$ and $D4_a$ under varying sampling rates. The two data-sets were located in physically separate locations. We measured the total time to generate the associations for a minimum support of 0.05% and 0.1% (Computing Associations) for both datasets (run in parallel), the time to communicate the associations from one machine (Communication Overhead) to another and the time to compute the similarity

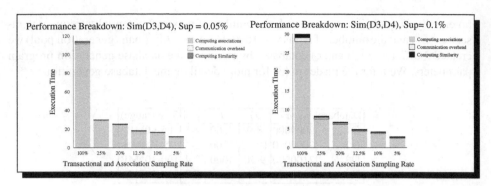

Fig. 2. Sampling Performance

metric (Computing Similarity) from these association sets. Transactional sampling influences the computing of associations while association sampling influences the latter two aspects of this experiment. Under association sampling, each processor computes a sample of its association set and sends it to the other, both then compute a part of similarity metric (in parallel). These two values are then merged appropriately, accounting for duplicates in the samples used. While both these sampling levels (transaction and association) could have different sampling rates, for expository simplicity we chose to set both at a common value. We evaluate the performance under the following sampling rates, 5%, 10%, 12.5%, 20%, and 25%. Figure 2 shows the results from this experiment.

Breaking down the performance it is clear that by using sampling at both levels the performance improves dramatically. A sampling rate of 10% yields an overall speedup of 8. From the figure it is easy to see that the dominant factor in this experiment is computing the associations. However, in a higher latency, lower bandwidth environment (current experiment was in a LAN environment), as will be the case when computing the similarity across distributed datasets interconnected via commodity networks the communication overhead will play a more dominant role.

The above experiment affirms the performance gains from association and transactional sampling. Next, we evaluate the quality of the similarity metric estimated using such approximation techniques for two minimum support values (0.05% and 0.1%). From Table 1 it is clear that using sampling for estimating the similarity metric can be very accurate (within 2% of the ideal (Sampling Rate 100%)) for all sampling rates above 5%. We have observed similar results (speedup and accuracy) for the other dataset pairs as well.

4.3 Sensitivity to Support

From Table 1 it should also be clear that the similarity metric is affected by the support. Choosing the "appropriate support" parameter for measuring similarity is an open research problem currently under investigation. The heuristic we use for choosing "appropriate supports" is that the cardinality of the resulting set of associations should lie within a user-specified interval. For the synthetic datasets we used the interval [10000,100000].

Table 1. Sampling Accuracy: Sim($D3_a$,$D4_a$

Support	SR-100%	SR-25%	SR-20%	SR-10%	SR-5%
0.05%	0.136	0.135	0.134	0.135	0.139
0.1%	0.12	0.12	0.12	0.12	0.11

4.4   Similarity Metric: Sensitivity to α

In this section we first evaluate the sensitivity of our similarity metric to the choice of α (alpha). Recall from Section 2 that the choice of α corresponds to the significance the user associates with variation in actual supports.

We evaluate the similarity between three dataset pairs for varying values of α. The results are shown in figure 3A. We wanted to evaluate the robustness to choice of α when two datasets were basically quite similar ($D3_a$,$D3_b$), and when two datasets were basically quite dissimilar ($D3_a$,$D2_a$). We also wanted to see the relative effect between dataset similarity pairs ($D3_a$,$D2_a$ vs $D3_a$,$D4_a$) as α was varied. We varied α from 0.5 to 500 (x axis: log scale). As can be seen from the graph when the datasets are similar the similarity metric is pretty robust to variation in α (up to $\alpha = 200$). When both datasets were dis-similar then they were robust to changes in α when the matches (common itemsets) also had similar support values ($D3_a$,$D4_a$). However, when the matches did not have similar support values ($D2_a$,$D3_a$) increasing α caused a relatively sharp decline in similarity values. This is highlighted by the crossover between the two graphs ($D3_a$,$D4_a$) and ($D2_a$,$D3_a$). This crossover highlights an important fact, *the choice of alpha can affect dataset clustering*. For $\alpha > 100$, dataset $D3_a$ is closer to $D4_a$, but for $\alpha < 100$, $D3_a$ is closer to $D2_a$.

4.5   Dataset Clustering

We now evaluate the efficacy of clustering homogeneous distributed datasets based on similarity. We used the synthetic datasets described earlier as a start point. We ran a simple tree-based clustering algorithm on these twelve datasets. Figure 3B shows the result. The numbers attached to the joins are the *Sim* metric with $\alpha = 1.0$. Clearly the datasets from the same origin are merged first. Given four as the desired number of clusters (or a merge cutoff of 0.2), the algorithm stops right after executing all the merges depicted by full lines, combining all the datasets from the same generation pool (as described in Section 4.1) into single clusters and leaving apart those from different pools highlighting the importance of clustering datasets before mining for rules. We next validate our approach on real datasets.

4.6   Census Dataset Evaluation

We asked our algorithm to cluster eight state datasets into four clusters. The clustering algorithm returned the clusters [IL, IA, TX], [NY, PA], [FL], and [OR,WA].

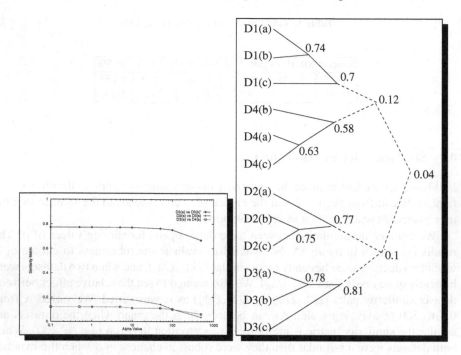

Fig. 3. A: Sensitivity to α B: Clustering Synthetic Datasets

An interesting by-play of our preprocessing step, discretization of the number of industrial concerns into three categories (high, middle and low), is that states with larger counties (area-wise), such as PA, NY and FL tend to have higher associativity (since each county has many industries) and thereby tend to have less affinity to states with lower associativity. On probing the similarity between IA, IL and TX [3] the most **influential attribute** is found to be agricultural concerns (no surprise there). This experiment validates our approach as the automatically derived clusters are sensible ones (geographically co-located). Interestingly, we found that the Census data benefits, performance-wise, from association sampling due its high associativity.

4.7   Reuters Dataset Evaluation: Attribute Similarity

Here three clusters were requested by us. The resulting clusters were [India, Pakistan, S. Korea], [USSR, Poland, Argentina], and [Japan, USA, Canada, UK, France, Brazil]. A surprising fact about the first cluster is that India and Pakistan have exactly four transactions in which they co-occur. Our **influential-attribute** algorithm identified that Pakistan and India have common trading partners (UK, France) and are involved in the trade of similar items (various oils and various grains/wheat). In the second cluster the inclusion of Argentina is not intuitive. Probing the cause for this our algorithm was able

---

[3] Cattle farming is also grouped under agricultural concerns in the Census data.

to identify [4] that although Argentina and Poland, have no co-occurring transactions, they are involved in the trade for similar items (wheat, grain, oilseed, etc.) resulting in the strong *Sim* value. Similarly the similarity between Argentina and USSR was found to be due to agricultural trade. The third cluster essentially consists of advanced countries. Most of which include Brazil as a strong trading partner resulting in that countries presence in the cluster.

5   Conclusions

In this paper we propose a method to measure the similarity among homogeneous databases and show how one can use this measure to cluster similar datasets to perform meaningful distributed data mining. An interesting feature of our algorithm is the ability to interact via informative querying to identify attributes influencing similarity. Experimental results show that our algorithm can adapt to time constraints by providing quick (speedup of 4-8) and accurate estimates (within 2%) of similarity. We evaluated our work on several datasets, synthetic and real, and show the effectiveness of our techniques.

Our similarity metric is sensitive to the support and the choice of α. As part of ongoing work we are investigating whether we can automate/guide the choice of these parameters subject to certain user/domain constraints. We are also evaluating the effectiveness of the current merging criteria, and exploring other criteria, as described in Section 3. As part of future work we will focus on evaluating and applying dataset clustering to other real world distributed data mining tasks.

References

1. C. Aggarwal and P. Yu. Online generation of association rules. In *ICDE'98*.
2. R. Agrawal, C. Faloutsos, and A. Swami. Efficient similarity search in sequence databases. In *Foundations of Data Organization and Algorithms*, 1993.
3. R. Agrawal, H. Mannila, R. Srikant, H. Toivonen, and A. Inkeri Verkamo. Fast discovery of association rules. In U. Fayyad and et al, editors, *Advances in Knowledge Discovery and Data Mining*, pages 307–328. AAAI Press, Menlo Park, CA, 1996.
4. R. Agrawal and R. Srikant. Fast algorithms for mining association rules. In *VLDB'94*.
5. G. Das *et al.* Similarity of attributes by external probes. In *KDD* 1998.
6. L. Devroye. A course in density estimation. In *Birkhauser:Boston MA*, 1987.
7. U. M. Fayyad, G. Piatetsky-Shapiro, and P. Smyth. The KDD process of rextracing useful information from volumes of data. *Communications of ACM*, 39(11):27–34, 1996.
8. R. Goldman *et al.* Proximity search in databases. In *VLDB'98*.
9. H. Jagadish, A. Mendelzon, and T. Milo. Similarity based queries. In *PODS*, 1995.
10. D. Lewis. *www.research.att.com/ lewis/reuters21578.html*, 1997.
11. R. Subramonian. Defining *diff* as a data mining primitive. In *KDD* 1998.
12. M. J. Zaki *et al.* New algorithms for fast discovery of association rules. In *KDD*, 1997.
13. M. J. Zaki *et al.* Evaluation of Sampling for Data Mining of Association Rules In *RIDE*, 1997.

---

[4] We emphasize here that the influential attribute detection algorithm was able to automatically identify the most influential attributes causing the similarity in these cases as opposed to requiring supervision [5].

Empirical Evaluation of Feature Subset Selection Based on a Real-World Data Set

Petra Perner[1] and Chid Apte[1]

[1]Institute of Computer Vision and Applied Computer Sciences
Arno-Nitzsche-Str. 45,04277 Leipzig
ibaiperenr@aol.com http://www.ibai-research.de
[2]IBM T.J. Watson Research Center, Yorktown Heights, NY 10598, USA

Abstract. Selecting the right set of features for classification is one of the most important problems in designing a good classifier. Decision tree induction algorithms such as C4.5 have incorporated in their learning phase an automatic feature selection strategy while some other statistical classification algorithm require the feature subset to be selected in a preprocessing phase. It is well know that correlated and irrelevant features may degrade the performance of the C4.5 algorithm. In our study, we evaluated the influence of feature pre-selection on the prediction accuracy of C4.5 using a real-world data set .We observed that accuracy of the C4.5 classifier can be improved with an appropriate feature pre-selection phase for the learning algorithm.

1 Introduction

Selecting the right set of features for classification is one of the most important problems in designing a good classifier. Very often we don't know a-priori what the relevant features are for a particular classification task. One popular approach to address this issue is to collect as many features as we can prior to the learning and data modeling phase. However, irrelevant or correlated features, if present, may degrade the performance of the classifier. In addition, large feature spaces can sometimes result in overly complex classification models that may not be easy to interpret.

In the emerging area of data mining applications, users of data mining tools are faced with the problem of data sets that are comprised of large numbers of features and instances. Such kinds of data sets are not easy to handle for mining. The mining process can be made easier to perform by focussing on a subset of relevant features while ignoring the other ones. In the feature subset selection problem, a learning algorithm is faced with the problem of selecting some subset of features upon which to focus its attention.

In this paper, we present our study on features subset selection and classification with C4.5 algorithm. In Section 2, we briefly describe the criteria used for feature selection and the feature selection methods. Although, C4.5 has a feature selection strategy included in its learning performance it has been observed that this strategy is not optimal. Correlated and irrelevant attributes may degrade the performance of the induced classifier. Therefore, we use feature subset selection prior to the learning

D.A. Zighed, J. Komorowski, and J. Zytkow (Eds.): PKDD 2000, LNAI 1910, pp. 575-580, 2000.
© Springer-Verlag Berlin Heidelberg 2000

phase. The CM algorithm selects features based upon their rank ordered *contextual merits* [4].The feature selection strategy used by C4.5 and the CM algorithm are reviewed in Section 2. For our experiments, we used a real data set that includes features extracted from x-ray images which describe defects in a welding seam. It is usually unclear in these applications what the right features are. Therefore, most analyses begin with as many features as one extract from the images. This process as well as the images are described in Section 3.

In Section 4, we describe our results. We show that the prediction accuracy of the C4.5 classifier will improve when provided with a pre-selected feature subset. The results show that the feature subsets created by CM algorithm and the feature subset normally extracted by C4.5 have many features in common. However , the C4.5 selects some features that are never selected by the CM algorithm. We hypothesize that irrelevant features are weeded out by the CM feature selection algorithm while they get selected by the C4.5 algorithm. A comparison of the feature ranking done by the CM algorithm with the ranking of the features done by C4.5 for the first 10 features used by C4.5 shows that there is a big difference. Finally, our experiments also indicate that model complexity does not significantly change for the better or worse when pre-selecting features with CM.

2 Feature Subset Selection Algorithms

According to the quality criteria [8] for feature selection, the model for feature selection can be distinguished into the filter model and the wrapper model [1,7]. The wrapper model attempts to identify the best feature subset for use with a particular algorithm, while the filter approach attempts to assess the merits of features from the data alone. Although the wrapper model can potentially produce the best resulting classifier, it does so by doing an exhaustive search over the entire feature space. Various search strategies have been developed in order to reduce the computation time [9] for wrapper algorithms. The filter approach on the other hand is a greedy search based approach that is computationally not as expensive. The feature selection in C4.5 may be viewed as a filter approach, while the CM algorithm may be viewed as a wrapper approach.

2.1 Feature Selection Done by Decision Tree Induction

Determining the relative importance of a feature is one of the basic tasks during decision tree generation. The most often used criteria for feature selection is information theoretic based, such as the Shannon entropy measure I for a data set. If we subdivide a data set using values of an attribute as separators, we obtain a number of subsets. For each of these subsets we can compute the information value. If the the information value of a subset n is i_n, then the new information value is given by $I_i = q_n i_n$, where q_n is the subset of data points with attribute value n. I_i will be smaller than I, and the difference $(I - I_i)$ is a measure of how well the attribute has discriminated between different classes. The attribute that maximizes this difference is selected.

The measure can also be viewed as a class separability measure. The main drawback of the entropy measure is its sensitivity to the number of attributes values

[11]. Therefore C4.5 uses the gain ratio. However, this measure suffers the drawback that it may choose attributes with very low information content of the attribute itself [2].

C4.5 [10]uses a univariate feature selection strategy. At each level of the tree building process only one attribute, the attribute with the highest values for the selection criteria, is picked out of the set of all attributes. Afterwards the sample set is split into sub-sample sets according to the values of this attribute and the whole procedure is recursively repeated until only samples from one class are in the remaining sample set or until the remaining sample set has no discrimination power anymore and the tree building process stops.

As we can see feature selection is only done at the root node over the entire decision space. After this level, the sample set is split into sub-samples and only the most important feature in the remaining sub-sample set is selected. Geometrically it means, the search for good features is only done in orthogonal decision subspaces, which might not represent the real distributions, beginning after the root node. Thus, unlike statistical feature search strategies [3] this approach is not driven by the evaluation measure for the combinatorial feature subset; it is only driven by the best single feature. This might not lead to an optimal feature subset in terms of classification accuracy.

Decision trees users and researchers have recognized the impact of applying a full set of features to a decision tree building process versus applying only a judiciously chosen subset. It is often the case that the latter produces decision trees with lower classification errors, particularly when the subset has been chosen by a domain expert. Our experiments were intended to evaluate the effect of using multivariate feature selection methods as pre-selection steps to a decision tree building process.

2.2 Contextual Merit Algorithm

For our experiment, we used the contextual merit (CM) algorithm [4]. This algorithm employs a merit function based upon weighted distances between examples which takes into account complete feature correlation's to the instance class. The motivation underlying this approach was to weight features based upon how well they discriminate instances that are close to each other in the Euclidean space and yet belong to different classes. By focusing upon these nearest instances, the context of other attributes is automatically taken into account.

To compute contextual merit, the distance d_{rs}^{k} between values z_{kr} and z_{ks} taken by feature k for examples r and s is used as a basis. For symbolic feature, the inter-example distance is 0 if $z_{kr} = z_{ks}$, and 1 otherwise. For numerical features, the inter-example distance is $\min\left(\dfrac{z_{kr} - z_{ks}}{t_k}, 1\right)$, where t_k is a threshold for feature k (usually 1/ 2 of the magnitude of range of the feature). The total distance between examples r and s is $D_{rs} = \sum_{k=1}^{N_f} d_{rs}^{k}$, and the contextual merit for a feature f is $M_f = \sum_r \sum_{s} w_{rs}^{f} d_{rs}^{f}$, where N is the total number of examples, C_r is the set of examples not in the same class as examples r, and w_{rs}^{f} is a weight function chosen so that examples that are close together are given greater influence in determining each features merit. In practice , it has been observed that $\dfrac{1}{D_{rs}^2}$ if s is one of k nearest neighbors to r, and 0 otherwise, provides robust behavior as a weight function. Additionally, using $\log_2 \# \overline{C(r)}$ as the value for k has also exhibited robust behavior. This approach to computing and ordering features by their merits has been observed to be very robust, across a wide range of examples.

3 Our Data Set

A detailed description of the data set can be found in [6]. Here we can try to briefly sketch out how the data set was created and what features were used.

The subject of investigation is the in-service inspection of welds in pipes of austenitic steel. The flaws to be looked for in the austenitic welds are longitudinal cracks due to intergranular stress corrosion cracking starting from the inner side of the tube. The radio-graphs are digitized with a spatial resolution of 70 mm and a gray level resolution of 16 bit per pixel. Afterwards they are stored and decomposed into various Regions of Interest (ROI) of 50x50 pixel size. The essential information in the ROIs is described by a set of features which are calculated from various image-processing methods.

The final data set contains 36 parameters collected for every ROI. The data set consists of altogether 1924 ROIs with 1024 extracted from regions of no disturbance, 465 from regions with cracks and 435 from regions with under-cuts.

4 Results

Table 1 illustrates the error rate for the C4.5 classifier when using all features as well as error rates for different feature subsets. The error rate was estimated using cross-validation. The improvement in accuracy is two percent for the pruned case. To interpret this improvement, we use a classification analysis conducted earlier [5], where performance actually peaked and then deteriorated as the number of features was increased. We observe similar behavior in our experiments. Classification error is at its minimum when the feature subset size is 20. This is in contrast to the feature subset size of 28 that C4.5 selects when presented with the entire feature set, with no pre-selection.

It may be argued that it is not worth doing feature subset selection before tree induction since the improvement in prediction accuracy is not so dramatic. However, the importance of an improvement, however small, clearly depends on the requirements of the application for which the classifier is being trained. We further observed (Table 4) that about 67% of the total features are used similarly by CM and C4.5, while about 33% of the features are exclusively selected by CM, and 16% are exclusively selected byC4.5. Table 3 shows that the tree does not necessarily become more compact even if a reduced set of features is used. The tree actually becomes even larger in the case with the best error rate. We therefore cannot draw any useful conclusion about feature set size and its relation to model complexity. We also observe (Table 4) that in comparing the two trees generated by C4.5 with and without CM's pre-selection, the feature used for splitting at the root node changes.

5 Conclusion

We have studied the influence of feature subset selection based on a filter and wrapper approach to the performance of C4.5. Our experiment was motivated by the fact that C4.5 uses a non-optimal feature search strategy. We used the CM algorithm for feature subset selection which measures importance of a feature based on a contextual merit function. Our results show that feature subset selection can help to improve the prediction accuracy of the induced classifier. However, it may not lead to more compact trees and the prediction accuracy may not increase dramatically.

The main advantage may be that fewer features required for classification can be important for application such as image interpretation where computational costs for extracting the features may be high and require special purpose hardware. For such domains, feature pre-selection to prune down the feature set size may be a beneficial analysis phase.

Table 1. Error Rate for Different Feature Subsets

	Test=Design		Crossvalidation	
Parameters	unpruned	pruned	unpruned	Pruned
all	0,9356	1,6112	24,961	24,545
10	1,5073	3,7942	29,4332	28,7051
15	1,4033	3,0146	26,365	26,4171
20	1,5073	2,5988	23,7649	22,7769
24	0,9356	1,7152	24,493	23,5049
28	0,9875	1,7152	25,117	24,077

Table 2. Error Rates for Different Sizes Feature Sets

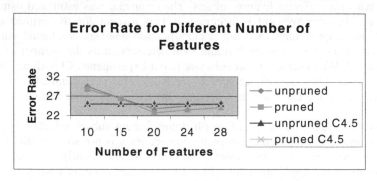

Table 3. Number of Nodes and Edges

Number of Features	10	15	20	24	28	37
Nodes	236	204	178	166	164	161
Edges	237	206	176	137	161	159

Attributes	10	15	20	24	28	C 4.5
2	0	0	1	1	1	1
3	1	1	1	1	1	1
4	1	1	1	1	1	1
5	0	0	0	0	0	1
6	0	1	1	1	1	1
7	0	0		1	1	1
8	0	1	1	1	1	0
9	1	1	1	1	1	1
10	1	1	1	1	1	0
11	0	0	0	0	0	1
12	0	0	0	1	1	1
13	0	0	0	0	0	1
14	0	0	0	0	0	1
15	0	0	0	0	1	
16	0	0	0	0	0	1
17	0	1	1	1	1	0
18	0	0	0	0	1	1
19	1	1	1	1	1	1
20	0	0	0	1	1	0
21	0	0	0	0	0	1
22	1	1	1	1	1	0
23	0	1	1	1	1	1
24	1	1	1	1	1	
25	0	0	0	0	0	1
26	0	0	0	0	0	1
27	0	0	1	1	1	
28	0	0	0	1		
29	0	1	1	1	1	
30	0	0	0	0	1	1
31	1	1	1	1	1	1
32	0	0	1	1	1	
33	0	0	0	0	1	
34	0	0	1	1	1	1
35	1	1	1	1	1	
36	1	1	1	1	1	
37	0	0	1	1	1	1
Number used	10	15	20	24	28	31

Rank	Name
1	4
2	22
3	23
4	3
5	10
6	24
7	17
8	19
9	9
10	31
11	36
12	35
13	8
14	6
15	34
16	29
17	2
18	27
19	37
20	32

Number in Tree	Name
11	9
21	3
22	27
31	3
32	31
33	24
34	34
41	37
42	31
43	3
44	10
45	31
46	34
47	2
48	35
51	None
52	None
53	8
54	6
55	None
56	None
57	None
58	None
59	9
5A	17
5B	24
5C	27
5D	None
5E	None
5F	22
5G	24

Table 4   Ranked Feature and the first 10 Features used by Decision Tree

References

[1] T.M Cover, ‚On the possible ordering on the measurement selection problem‘, IEEE Transactions, **SMC-7**(9), 657-661, (1977).

[2] R.Lopez de Mantaras, ‚A distance-based attribute selection measure for decision tree induction‘, *Machine Learning, 6*(1991), 81-92.

[3] Keinosuke Fukunaga, *Introduction to Statistical Pattern Recognition*, AcademicPress, 1990.

[4] Se June Hong, ‚Use of contextual information for feature ranking and discretization‘, *IEEE Trans. on Knowledge Discovery and Data Engineering*, (1996).

[5] G. F. Hughes, ‚On the mean accuarcy of statistical pattern recognizers‘, IEEE Transactions, **IT-14**(1), 55-63, (1968).

[6] C. Jacobsen, U.Zscherpel,and P.Perner, *A Comparison between Neural Networks and Decision Trees*, 144-158, Machine Learning and DataMining in Pattern Recognition, Springer Verlag,1999.

[7] R.Kohavi and G.H. John, *The Wrapper Approach*, 33-50, Feature Extraction Construction and Selection, Kluwer Academic Publishers, 1998.

[8]   M. Nadler and Eric P. Smith, *Pattern Recognition Engeenering*, John Wiley&Sons Inc., 1993.

[9] P.Pudil, J. Navovicova, and J.Kittler, ‚Floating search methods in feature selection‘, *Pattern Recognition Letters, 15*(1994), 1119-1125.

[10]   J.R. Quinlan, C4.5:Programs for Machine Learning, Morgan Kaufmann, 1993.

[11] A.P. White and W.Z. Lui,‘Bias in the information-based measure in decision tree induction‘, *Machine Learning, 15*(1994),321-329.

Discovery of Ambiguous Patterns in Sequences Application to Bioinformatics

Gerard Ramstein[1], Pascal Bunelle[1], and Yannick Jacques[2]

[1] IRIN, Ecole polytechnique universitaire de Nantes,
Rue Christian Pauc, La Chantrerie, BP 60602, F-44036 Nantes Cedex 3, France
gramstei@ireste.fr
[2] INSERM U.463, Institut de Biologie,
9 Quai Moncousu, F-44035 Nantes Cedex 1, France
yjacques@nantes.inserm.fr

Abstract. An important issue in data mining concerns the discovery of patterns presenting a user-specified minimum support. We generalize this problematics by introducing the concept of ambiguous event. An ambiguous event can be substituated for another without modifying the substance of the concerned pattern. For instance, in molecular biology, researchers attempt to identify conserved patterns in a family of proteins for which they know that they have evolved from a common ancestor. Such patterns are flexible in the sense that some residues may have been substituated for others during evolution. A[B C] is an example of notation of an ambiguous pattern representing the event A, followed by either the event B or C. A new scoring scheme is proposed for the computation of the frequency of ambiguous patterns, based on substitution matrices. A substitution matrix expresses the probability of the replacement of an event by another. We propose to adapt the Winepi algorithm [1] to ambiguous events. Finally, we give an application to the discovery of conserved patterns in a particular family of proteins, the cytokine receptors.

1 Introduction

Data mining techniques focus on the discovery of regularities underlying large collections of data and reveals new insight into the hidden organisation of these raw data. One important problem of data mining thus concerns the search of frequent patterns in sequential data, i.e. collections of sequences of events. We propose in this paper an extension of the Winepi algorithm [1] for the discovery of ambiguous patterns. Such patterns can be encountered in many situations. For example in insurance, a cardio-vascular history may present various forms such as arterial hypertension, infarction, phlebistic, etc. All these forms do not have the same gravity degree and therefore should be differently weighted. An interest of our method is indeed that it permits to characterize the representativeness of an event. In supermarket basket data, the purchase of a printer can be differently evaluated according to the power of the device or its characteristics (e.g laser or inck jet printer). Note that the concept of ambiguous event

D.A. Zighed, J. Komorowski, and J. Zytkow (Eds.): PKDD 2000, LNAI 1910, pp. 581–586, 2000.
© Springer-Verlag Berlin Heidelberg 2000

encompasses the notion of taxinomy, but is more flexible ; we do not impose that ambiguous events must belong to a same class.

We call pattern a sequence of events that frequently occur in the observed database. Generally a pattern is constrained to belong to a sliding window that limits the search space. A window of size w can then provide patterns of size 1 up to w. For temporal transactions, this restricts the events to be bounded within a maximal time interval. An ambiguous pattern contains at least one event that present a given degree of polymorphism. Let suppose that an event A may be substituted for another event B and that $w = 4$, then the sequences CEAD and CBFD are two possible occurrences of the same pattern CAD. A better representation of this pattern is C[A B]D ; this notation denotes the interchangeable events inside a pair of brackets.

In this paper we shortly describe the original Winepi algorithm proposed by Toivonen and Mannila [1], then we introduce the concept of ambiguous event as well as the definition of substitution costs, We next show how it can be implemented into the Winepi algorithm. Finally one presents an application to bioinformatics : the discovery of conserved patterns in a family of protein sequences, namely the cytokine receptors.

2   Data mining algorithms for sequence analysis

The AprioriAll algorithm [2] finds all the patterns having a minimal support, but presents some inconvenients such as an expensive data transformation and the absence of timing window. GSP [3] is an improvment of AprioriAll that also generalizes the concept of item by using diffcrent levels of a user-defined taxonomy. These algorithms are based on the generation of candidate patterns and the counting of these candidates to check their actual frequency. Similar techniques have been used by Mannila and Toivonen [1], [4]. We propose in this paper an adaptation of Winepi [1] to the discovery of ambiguous patterns.

3   Principle of Winepi

An event E is a pair (A, t) where $A \in \Omega$ is an event type and t is an integer. The index t may denote the temporal occurrence of the event E in a sequence s, or simply its position. An episode $\alpha = (A_1, t_1), (A_2, t_2) \ldots (A_k, t_k)$ is an event sequence ordered by the time (i.e. $t_1 \leq t_2 \leq \ldots \leq t_k$) such as $|t_k - t_1| \leq w$, where w is the length of the timing window. As we do not care of the occurrence t_i of each individual event in α, we will only define the episode by the succession of its events, i.e. $\alpha = A_1 \ldots A_k$. A pattern represents a frequent episode ; the frequency (or support) of the pattern is the proportion of sequences that contain this pattern. THe frequency of a pattern must be greater than an input parameter $minFreq$.

The basic structure of Winepi comprises two key aspects ; the generation of potentially frequent episodes, called candidate episodes, and the actual counting of these candidates. The algorithm makes several passes over the data, at each pass, it increases the size k of the episodes. Winepi starts with a size $k = 1$ and considers that each individual event belonging to an alphabet Ω is a potential pattern. Then k is increased of one unit at each pass. The algorithm terminates when there are no frequent sequences at the end of a pass or when there are no candidate episodes generated :

> compute $C_1 = \{A \in \Omega\}$;
> $k := 1$;
> while $C_k \neq \emptyset$ do
>> // sequence set pass
>> compute the pattern set $P_k = \{\alpha \in C_k | freq(\alpha) \geq minFreq\}$;
>> $k := k + 1$;
>> // candidate generation
>> compute C_k from P_{k-1};

3.1   Generation of episodes

An episode is an ordered sequence of events that is generated by the joining of two subepisodes. A subepisode α' of size $k' \leq k$ contains k' events belonging to the episode α and occurring in the same order (e.g. ABC is a subepisode of ADBEF but not ACB). The candidate generation phase is based on a joining procedure that takes two patterns from the pattern set P_{k-1} to form a new episode α of size k. A pattern p_1 joins with a pattern p_2 if the subepisode obtained by dropping the first event of p_1 is the same as the subepisode obtained by dropping the last event of p_2 : the episode $\alpha = A_1 \ldots A_k$ is a new candidate if there exist in P_{k-1} two subepisodes verifying $p_1 = A_1^1 \ldots A_{k-1}^1 1^1$, $p_2 = A_2^2 \ldots A_k^2$ such as $A_i^1 = A_i^2$ $\forall i \in [2, k-1]$.

The joining procedure provides a set of potential candidates that is next checked by a pruning procedure. This latter removes all the candidates that cannot be frequent ; the pruning is based on the fact that all the subepisodes of a frequent episode must be frequent. Therefore, if there exist at most one subepisode that does not have a minimum support, the potential episode created by the joining procedure will not be inserted in C_k.

3.2   Computation of the frequency of the episodes

During the second phase, the algorithm counts the occurrences of the elements of C_k in the sequence set. The algorithm performs this computation in a single pass and takes advantage of the sliding window by recognizing episodes through incremental updates. Winepi is based on a set $\mathcal{A}$ of state automatum that accept candidate episodes and ignore all other input. When a new event E enters in the window, an automaton a will be initialize for all the episodes begining with E and will be added to $\mathcal{A}$. When the same event E leaves the window, a is removed from the set $\mathcal{A}$. An efficient data structure stores $\mathcal{A}$ and handles the corresponding state transitions.

3.3   Ambiguous events

In Winepi, the concept of frequency (or support) of an episode α relies on a counting of the observed occurrences of α in the database. The introduction of the concept of ambiguous event necessitates to modify this notion of frequency. We propose a scoring scheme that is commonly used in different fields and especially in bioinformatics : substitution matrices [5]. A substitution matrix $\mathcal{M}$ expresses the cost $w(i,j)$ associated to the substitution of the event i by the event j. We denote $compatible(X)$ the set of events that can be substituted for X : $compatible(X) = \{Y \in \Gamma | w(X,Y) > 0\}$.

We propose to generalize the original concept of frequency that has been developed in Winepi. An exact occurrence of an episode ABC will conserve a score of 1 while ambiguous occurrences will have a lower score. Let consider the sequence $s = (s_1 s_2 \ldots sk)$ as a potential occurrence of a given episode $e = (e_1 e_2 \ldots e_k)$. We then define a scoring function $f(s, e)$ as follows :

$$f(s,e) = \frac{1}{k} \sum_{i=1}^{k} w(s_i, e_i)/w(e_i, e_i)$$

where w(a,b) are the coefficients of $\mathcal{M}$. Let consider the event alphabet $\Omega = \{A, B, C\}$ and the substitution matrix $\mathcal{M}$ given in Table 1. In this example, we

Table 1. Substitution matrix

	A	B	C
A	+4	+2	-1
B	+2	+5	-3
C	-1	-3	+7

have f(ABC,ABC)=1, f(BBC,ABC)=0.83, f(CCC,ABC)=0.05. In practise, we will reject incompatible events ; an occurrence of an event X will be considered as an admissible substitution of the event Y only if it belongs to $compatible(Y)$. Before a formal presentation of our novel algorithm, let consider on a simple example how we associate a current score to each episode. Let suppose that the window size w is 4, that the observed sequence is ABCA and that the concerned episode α is AA. The first event A enters the window and creates an new instance of an automaton a_1 associated to α, since this input corresponds to the first event of α. The partial score is $w(A, A)/w(A, A) = 1$. The next event B is considered as a substitution of A and the automaton writes that it reaches its second position with a score of $w(B, A)/w(A, A) = 0.5$. The third event C is neglicted since it does not belong to $compatible(A)$. The algorithm next considers the last event A and notices that it corresponds to the second position of α with a better score : the last score is then overwritten. As it also corresponds to a first position of α, a new instance a_2 of automaton is created as well. Finally the first event leaving the window involves the removal of the automaton a_1. This latter reaches its last state, which means that α has been detected, with a global score of $(1+1)/2 = 1$.

The Winepi algorithm is rather sophisticated and for concision sake, we will not give its detailed presentation ; it uses different data structures that optimize the management of $\mathcal{A}$, the interested reader can refer to [1] for further details. We will only stress on the major modifications that we made, concerning the ambiguous events. We have maintained the principle of a list *waits* that contains alll the episodes that are waiting for their *ith* element. For instance, the event A is waited by the episode $\gamma = AC$ on its first position and by $\beta = BA$ on its second position, if one assumes that the event B has already been met. The corresponding lists are : $A.waits[1] = \gamma$ and $A.waits[2] = \beta$. The main difference with the original version of Winepi lies in the fact that a new event Y entering in the window is now considered as a potential ambiguous event and is replaced by each event X belonging to *compatible*(Y).

The computation of the frequency is also replaced by scores that are dynamically updated for each new event X. We need to associate to a current episode α an array of partial scores $\alpha.s$. The size of this array is k and $\alpha.s[i]$ is the current score of the episode $\alpha[1] \ldots \alpha[i]$. We initialise $\alpha.s[i] = 0 \forall i \in [0..k]$ (for convenience, we add an extra score $\alpha.s[0]$ that always remains 0). We give below the algorithm that takes into account an input event Y.

```
for i := k downto 1 do
    for all X ∈ compatible(Y) do
        for all α ∈ X.waits[i] do
            α.s[i] := α.s[i − 1] + f(X, α[i]);
```

Note that by definition all episodes are always waiting for their first event. The frequency of an episode is the sum of the best scores obtained on each sequence, divided by the number of sequences.

4   Discovery of conserved patterns in protein sequences

Knowledge discovery is a promising new research area in bioinformatics [6]. We propose an application to the identification of conserved patterns in a family of proteins. Proteins are linear macromolecules that are assembled from twenty different amino acids hooked together in long chains by chemical bonds. These amino acids are coded by a standard alphabet (e.g L states for Lysine), so that the protein can be represented as a simple string of characters. Cytokines are soluble proteins that are involved in various aspect of cellular proliferation and differentiation, especially in hematopoiesis and immune reactions. Their cell surface receptors are composed of the non-covalent association of two or three transmembrane receptor subunits. These subunits display conserved structural features indicating that they have evolved from a common ancestor. The receptor family that we use comprises 31 sequences of various lengths, from 229 up to 1165 residues. The overall signature of this family is rather complex, so we will focus only on a particular region that is especially characteristic for which the following patterns have been found : `G-W-S-E-W-S-P, score=0.827` and `G-W-S-D-W-S-P score=0.821`. A refinement phase should be added to our tool to obtain the PROSITE-like pattern `G-W-S-[E D]-W-S-P`.

5   Conclusion and future works

The concept of ambiguous event permits to consider polymorphic data and our novel definition of the frequency defines the degree of transformation that is acceptable for an ambiguous pattern. We show that sequence analysis algorithms can be adapted to take into account this new paradigm ; we successfully apply our scoring scheme to the Winepi algorithm. No doubt that knowledge discovery will be a major issue in computational biology, as it has been predicted by many authors [6]. Our first results show that our algorithm is robust and efficient, but that the presentation of the results must be improved. We are currently working on the adaptation of the Minepi algorithm [4] so that it accepts ambiguous events. Minepi is an alternative method for the discovery of patterns proposed by the authors of Winepi. The Minepi algorithm is more powerful than Winepi, in the sense that it only treats the actual occurrences of the patterns in the database. This efficiency has a conterpart : Minepi is more memory expensive than Winepi. But we are confident that this algorithm may be applied to the case of protein sequences.

References

1. H. Mannila, H. Toivonen, and AI Verkamo. Discovering frequent episodes in sequences. In *First International Conference on Knowledge Discovery and Data Mining (KDD'95)*, pages 210 – 215. AAAI Press, August 1995.
2. R. Agrawal and R. Srikant. Mining sequential patterns. In *11th International Conference on Data Engineering*, March 1995.
3. Ramakrishnan Srikant and Rakesh Agrawal. Mining sequential patterns: Generalizations and performance improvements. In Peter M. G. Apers, Mokrane Bouzeghoub, and Georges Gardarin, editors, *Proc. 5th Int. Conf. Extending Database Technology, EDBT*, volume 1057 of *Lecture Notes in Computer Science, LNCS*, pages 3–17. Springer-Verlag, 25–29 March 1996.
4. Heikki Mannila and Hannu Toivonen. Discovering generalized episodes using minimal occurrences. In *2nd International Conference on Knowledge Discovery and Data Mining (KDD'96)*, pages 146 – 151. AAAI Press, August 1996.
5. S Henikoff and JG Henikoff. Amino acid substitution matrices from protein blocks. *Proc. Natl. Acad. Sci. USA*, 89:10915–10919, 1992.
6. Janice Glasgow, Igor Jurisica, and Raymond Ng. Data mining and knowledge discovery in databases. In *Pacific Symposium on Biocomputing (PSB '99)*, January 1999.

Action-Rules: How to Increase Profit of a Company

Zbigniew W. Ras[1,2] and Alicja Wieczorkowska[1,3]

[1] UNC-Charlotte, Computer Science Dept., Charlotte, NC 28223, USA
[2] Polish Academy of Sciences, Inst. of Comp. Science, Warsaw, Poland
[3] Polish-Japanese Institute of Information Tech., 02-018 Warsaw, Poland
ras@uncc.edu   or   awieczor@uncc.edu

Abstract. Decision tables classifying customers into groups of different profitability are used for mining rules classifying customers. Attributes are divided into two groups: stable and flexible. By stable attributes we mean attributes which values can not be changed by a bank (age, marital status, number of children are the examples). On the other hand attributes (like percentage rate or loan approval to buy a house in certain area) which values can be changed or influenced by a bank are called flexible. Rules are extracted from a decision table given preference to flexible attributes. This new class of rules forms a special repository of rules from which new rules called action-rules are constructed. They show what actions should be taken to improve the profitability of customers.

1 Introduction

In the banking industry, the most widespread use of data mining is in the area of customer and market modeling, risk optimization and fraud detection. In financial services, data mining is used in performing so called trend analysis, in analyzing profitability, helping in marketing campaigns, and in evaluating risk. Data mining helps in predictive modeling of markets and customers, in support of marketing. It can identify the traits of profitable customers and reveal the "hidden" traits. It helps to search for the sites that are most convenient for customers as well as trends in customer usage of products and services. Examples of specific questions include:

1. What are micromarkets for specific products?
2. What is the likelihood that a given customer will accept an offer?
3. What actions improve the profitability of customers?
4. What customers switch to other services?

Even if it is very tempting to use limited tools and seek quick answers to very specific questions, it is important to keep a broad perspective on knowledge. For instance, while the direct data make a particular offer look profitable, a more thorough analysis may reveal cannibalization of other offers and the overall decrease in profit.

Specific questions about profitability must be answered from a broader perspective of customer and market understanding. For instance, customer loyalty can be often as important as profitability. In addition to short term profitability the decision makers must also keep eye on the lifetime value of a customer. Also, a broad and detailed

D.A. Zighed, J. Komorowski, and J. Zytkow (Eds.): PKDD 2000, LNAI 1910, pp. 587–592, 2000.
© Springer-Verlag Berlin Heidelberg 2000

understanding of customers is needed to send the right offers to the right customers at the most appropriate time. Knowledge about customers can lead to ideas about future offers which will meet their needs.

2 Information Systems and Decision Tables

An information system is used for representing knowledge. Its definition, presented here, is due to Pawlak [2] .

By an information system we mean a pair $S = (U,A)$, where:

1.   U is a nonempty, finite set called the universe,
2.   A is a nonempty, finite set of attributes i.e. $a:U \rightarrow V_a$ for $a \in$ A, where V_a is called the domain of a.

Elements of U are called objects. In our paper objects are interpreted as customers. Attributes are interpreted as features, offers made by a bank, characteristic conditions etc.

In this paper we consider a special case of information systems called decision tables [2]. In any decision table together with the set of attributes a partition of that set into conditions and decisions is given. Additionally, we assume that the set of conditions is partitioned into stable conditions and flexible conditions. We consider decision tables with only one decision attribute to be seen as "profit ranking" with values being integers. This attribute classifies objects (customers) with respect to the profit for a bank. Date of Birth is an example of a stable attribute. Interest rate on any customer account is an example of a flexible attribute (dependable on bank). We adopt the following definition of a decision table:

A decision table is any information system of the form $S = (U, A_1 \cup A_2 \cup \{d\})$, where $d \notin A_1 \cup A_2$ is a distinguished attribute called decision. The elements of A_1 are called stable conditions, whereas the elements of A_2 are called flexible conditions.

The cardinality of the image $d(U) = \{$ k: $d(x)=k$ for some $x \in$ U$\}$ is called the rank of d and is denoted by $r(d)$. Let us observe that the decision d determines the partition $CLASS_S(d) = \{X_1,X_2,...,X_{r(d)}\}$ of the universe U, where $X_k = d^{-1}(\{k\})$ for $1 \leq k \leq r(d)$. $CLASS_S(d)$ is called the classification of objects in S determined by the decision d.

In this paper, as we mentioned before, objects correspond to customers. Also, we assume that customers in $d^{-1}(\{k1\})$ are more profitable for a bank than customers in $d^{-1}(\{k2\})$ for any $k2 \leq k1$. The set $d^{-1}(\{r(d)\})$ represents the most profitable customers for a bank. Clearly the goal of any bank is to increase its profit. It can be achieved by shifting some customers from the group $d^{-1}(\{k2\})$ to $d^{-1}(\{k1\})$, for any $k2 \leq k1$. Namely, through special offers made by a bank, values of flexible attributes of some customers can be changed and the same all these customers can be moved from a group of a lower profit ranking to a group of a higher profit ranking.

Careful analysis of decision tables classifying customers can help us to identify groups of customers within any $d^{-1}(\{k1\})$ who may accept attractive offers from a competitive bank. In some of these cases, our action rules will suggest what offers should be sent to these customers moving them to a lower profit ranking (but still profitable for the bank) group instead of loosing them entirely.

3 Action Rules

In this section we describe a method to construct action rules from a decision table containing both stable and flexible attributes.

Before we introduce several new definitions, assume that for any two collections of sets X, Y, we write, $X \subseteq Y$ if $(\forall x \in X)(\exists y \in Y)[\, x \subseteq y\,]$. Let $S = (U, A_1 \cup A_2 \cup \{d\})$ be a decision table and $B \subseteq A_1 \cup A_2$. We say that attribute d depends on B if $CLASS_S(B) \subseteq CLASS_S(d)$, where $CLASS_S(B)$ is a partition of U generated by B (see [2]). Assume now that attribute d depends on B where $B \subseteq A_1 \cup A_2$. The set B is called d-reduct in S if there is no proper subset C of B such that d depends on C. The concept of d-reduct in S was introduced to induce rules from S describing values of the attribute d depending on minimal subsets of $A_1 \cup A_2$. In order to induce rules in which the THEN part consists of the decision attribute d and the IF part consists of attributes belonging to $A_1 \cup A_2$, subtables $(U, B \cup \{d\})$ of S where B is a d-reduct in S should be used for rules extraction. By Dom(r) we mean all attributes listed in the IF part of a rule r. For example, if $r = [\,(a1,3)*(a2,4) \rightarrow (d,3)]$ is a rule then Dom(r) = {a1,a2}. By d(r) we denote the decision value of a rule. In our example d(r) = 3.

If r1, r2 are rules and $B \subseteq A_1 \cup A_2$ is a set of attributes, then r1/B = r2/B means that the conditional parts of rules r1, r2 restricted to attributes B are the same. For example if r1 = [(a1,3) $\rightarrow$ (d,3)], then r1/{a1} = r/{a1}.

Example 1. Assume that S = ({x1,x2,x3,x4,x5,x6,x7,x8}, {a,c} {b} {d}) be a decision table represented by Figure 1. The set {a,c} lists stable attributes, in {b} we have flexible attributes and d is a decision attribute. Also, we assume that H denotes customers of a high profit ranking and L denotes customers of a low profit ranking.

In our example r(d)=2, r(c)=3, r(a)=2,
$CLASS_S(d)$={{x1,x2,x3,x4,x5,x6},{x7,x8}}, $CLASS_S(\{b\})$={{x1,x3,x7,x8}, {x2,x4},{x5,x6}}, CLASS({a,b})={{x1,x3},{x2,x4},{x5,x6}, {x7,x8}}, $CLASS_S(\{a\})$={{x1,x2,x3,x4},{x5,x6,x7,x8}}, $CLASS_S(\{c\})$={{x1,x3},{x2,x4}, {x5,x6,x7,x8}}, CLASS({b,c})={{x1,x3},{x2,x4},{x5,x6},{x7,x8}}.

So, CLASS({a,b}) $\subseteq$ $CLASS_S(d)$ and CLASS({b,c}) $\subseteq$ $CLASS_S(d)$. It can be easily checked that both {b,c} and {a,b} are d-reducts in S.

Rules can be directly derived from d-reducts and the information system S. In our example, we get the following rules:

$(a,0) \wedge (b,S) \rightarrow (d,L)$,	$(b,S) \wedge (c,0) \rightarrow (d,L)$,
$(a,0) \wedge (b,R) \rightarrow (d,L)$,	$(b,R) \wedge (c,1) \rightarrow (d,L)$,
$(a,2) \wedge (b,P) \rightarrow (d,L)$,	$(b,P) \wedge (c,2) \rightarrow (d,L)$,
$(a,2) \wedge (b,S) \rightarrow (d,H)$,	$(b,S) \wedge (c,2) \rightarrow (d,H)$.

	a	b	c	d
x1	0	S	0	L
x2	0	R	1	L
x3	0	S	0	L
x4	0	R	1	L
x5	2	P	2	L
x6	2	P	2	L
x7	2	S	2	H
x8	2	S	2	H

Fig. 1

We use information system S to simplify them. We get :

$(a,0) \rightarrow (d,L)$, $(c,0) \rightarrow (d,L)$,

$(b,R) \rightarrow (d,L)$, $(c,1) \rightarrow (d,L)$,

$(b,P) \rightarrow (d,L)$, $(a,2) \wedge (b,S) \rightarrow (d,H)$, $(b,S) \wedge (c,2) \rightarrow (d,H)$.

Now, let us assume that $(a, v \rightarrow w)$ denotes the fact that the value of attribute a has been changed from v to w. Similarly, the term $(a, v \rightarrow w)(x)$ means that $a(x)=v$ has been changed to $a(x)=w$. Saying another words, the property (a,v) of a customer x has been changed to property (a,w).

Assume now that $S = (U, A_1 \cup A_2 \cup \{d\})$ is a decision table, where A_1 is the set of stable attributes and A_2 is the set of flexible attributes. Assume that rules r1, r2 have been extracted from S and $r1/A_1 = r2/A_1$, $d(r1)=k1$, $d(r2)=k2$ and $k1 < k2$. Also, assume that $(b1, b2,..., bp)$ is a list of all attributes in $Dom(r1) \cap Dom(r2) \cap A_2$ on which r1, r2 differ and $r1(b1)= v1$, $r1(b2)= v2,..., r1(bp)= vp$, $r2(b1)= w1$, $r2(b2)= w2,..., r2(bp)= wp$.

By (r1,r2)-action rule on $x \in U$ we mean a statement:

$[(b1, v1 \rightarrow w1) \wedge (b2, v2 \rightarrow w2) \wedge...\wedge (bp, vp \rightarrow wp)](x) \Rightarrow [(d,k1) \rightarrow (d,k2)](x)$.

If the value of the rule on x is true then the rule is valid. Otherwise it is false.

Let us denote by $U^{<r1>}$ the set of all customers in U supporting the rule r1. If (r1,r2)-action rule is valid on $x \in U^{<r1>}$ then we say that the action rule supports the new profit ranking k2 for x.

Example 2. Assume that $S = (U, A_1 \cup A_2 \cup \{d\})$ is a decision table from the Example 1, $A_2 = \{b\}$, $A_1 = \{a,c\}$. It can be checked that rules r1=[(b,P) $\rightarrow$ (d,L)], r2=[(a,2)$\wedge$(b,S) $\rightarrow$ (d,H)], r3=[(b,S)$\wedge$(c,2) $\rightarrow$ (d,H)] can be extracted from S. Clearly x5, x6 $\in U^{<r1>}$. Now, we can construct (r1,r2)-action rule executed on x:

$[(b, P \rightarrow S)](x) \Rightarrow [(d,L) \rightarrow (d,H)](x)$.

It can be checked that this action rule supports new profit ranking H for x5 and x6.

Example 3. Assume that $S = (U, A_1 \cup A_2 \cup \{d\})$ is a decision table represented by Figure 2. Assume that $A_1 = \{c, b\}$, $A_2 = \{a\}$.

	c	a	b	d
x1	2	1	1	L
x2	1	2	2	L
x3	2	2	1	H
x4	1	1	1	L

Fig. 2

Clearly $r1=[(a,1) \wedge (b,1) \rightarrow (d,L)]$, $r2=[(c,2) \wedge (a,2) \rightarrow (d,H)]$ are optimal rules which can be extracted from S. Also, $U^{<r1>} = \{x1, x4\}$. If we construct (r1,r2)-action rule: $[(a, 1 \rightarrow 2)](x) \Rightarrow [(d,L) \rightarrow (d,H)](x)$.

then it will certainly support new profit ranking H for x1 but only possibly for x4.

Algorithm to Construct Action Rules

Input: Decision table $S = (U, A_1 \cup A_2 \cup \{d\})$,

A_1 – stable attributes, A_2 – flexible attributes, $\lambda 1$, $\lambda 2$ – weights.

Output: R – set of action rules.

Step 0. $R := \varnothing$.

Step 1. Find all d-reducts $\{D_1, D_2, \ldots, D_m\}$ in S which satisfy the property
$card[D_i \cap A_1]/card[A_1 \cup A_2] \leq \lambda 1$
(reducts with a relatively small number of stable attributes)

Step 2. FOR EACH pair (D_i, D_j) of d-reducts (found in step 1) satisfying the property $card(D_i \cap D_j)/card(D_i \cup D_j) \leq \lambda 2$ DO
find set R_i of optimal rules in S using d-reduct D_i,
find set R_j of optimal rules in S using d-reduct D_j.

Step 3. FOR EACH pair of rules $(r1, r2)$ in $R_i \times R_j$ having different THEN parts DO
if $r1/A_1 = r2/A_1$, $d(r1)=k1$, $d(r2)=k2$ and $k1< k2$, then
if $(b1, b2, \ldots, bp)$ is a list of all attributes in $Dom(r1) \cap Dom(r2)$
$\cap A_2$ on which $r1, r2$ differ and
$r1(b1)= v1, r1(b2)= v2, \ldots, r1(bp)= vp$,
$r2(b1)= w1, r2(b2)= w2, \ldots, r2(bp)= wp$
then the following (r1,r2)-action rule add to R:
$if\ \ [(b1,\ v1 \rightarrow w1)\ \wedge\ (b2,\ v2 \rightarrow w2)\ \wedge \ldots \wedge\ (bp,\ vp \rightarrow wp)](x)\ then\ [(d,\ k1) \rightarrow (d,k2)](x)$

The resulting $(r1,r2)$-action rule says that if the change of values of attributes of a customer x will match the term
$[(b1,\ v1 \rightarrow w1)\ \wedge\ (b2,\ v2 \rightarrow w2)\ \wedge \ldots \wedge\ (bp,\ vp \rightarrow wp)](x)$
then the ranking profit of custumer x will change from $k1$ to $k2$.

This algorithm was initially tested on a sampling containing 20,000 tuples extracted randomly from a large banking database containing more than 10 milion customers. We used DataLogicR$^+$ for rule extraction. Both "roughness" and "rule precision threshold" has been set up to 0.85. We found many pairs of rules (r1,r2) meeting the conditions required by the above algorithm for generating (r1,r2)-action rules. For instance, we extracted the following three rules:

[A79 > 610.70] ∧ [A75 > 640.50] ∧ [A73 > 444.00] → [PROFIT = 1],

[A79 ≤ 610.70] ∧ [A75 > 640.50] ∧ [A78 ≤ 394.50] → [PROFIT = 1],

[A75 ≤ 640.50] → [PROFIT = 3].

Attributes A73, A75, A78, A79 are all flexible.

The following action rule can be generated:

if [A75, (A75 > 640.50) → (A75 ≤ 640.50)](x)

then [(PROFIT = 1) → (PROFIT = 3)].

Similar method is used to identify groups of customers who may accept attractive offers from a competitive bank. In such a case, action rules suggest what offers should be sent to all these customers moving them to lower profit ranking (but still profitable for the bank) groups instead of loosing them entirely.

4 Conclusion

Proposed algorithms identify the customers which may move from one profit ranking group to another profit ranking group if values of some flexible attributes describing them are changed. Also, the algorithms show what attribute values should be changed and what their new values should be (what type of offers should be sent by a bank to all these customers).

References

1. Chmielewski M. R., Grzymala-Busse J. W., Peterson N. W., Than S., "The Rule Induction System LERS - a Version for Personal Computers", in *Foundations of Computing and Decision Sciences*, Vol. 18, No. 3-4, 1993, Institute of Computing Science, Technical University of Poznan, Poland, 181-212.
2. Pawlak Z., "Rough Sets and Decision Tables", in *Lecture Notes in Computer Science* **208,** Springer-Verlag, 1985, 186-196.
3. Skowron A., Grzymala-Busse J., "From the Rough Set Theory to the Evidence Theory", *in ICS Research Reports* 8/91, Warsaw University of Technology, October, 1991

Aggregation and Association in Cross Tables

Gilbert Ritschard[1] and Nicolas Nicoloyannis[2]

[1] Dept of Econometrics, University of Geneva, CH-1211 Geneva 4, Switzerland
`ritschard@themes.unige.ch`
[2] Laboratoire ERIC, University of Lyon 2, C.P.11 F-69676 Bron Cedex, France
`nicolas.nicoloyannis@univ-lyon2.fr`

Abstract. The strength of association between the row and column variables in a cross table varies with the level of aggregation of each variable. In many settings like the simultaneous discretization of two variables, it is useful to determine the aggregation level that maximizes the association. This paper deals with the behavior of association measures with respect to the aggregation of rows and columns and proposes an heuristic algorithm to (quasi-)maximize the association through aggregation.

1 Introduction

This paper is concerned with the maximization of the association in a cross table. In this perspective, it deals with the effect of the aggregation of the row and column variables on association measures and proposes an aggregation algorithm to (quasi-)maximize given association criteria. Indeed, the strength of the association may vary with the level of aggregation. It is well known for instance that aggregating identically distributed rows or columns does not affect the Pearson or Likelihood ratio Chi-squares (see for example [4], p. 450) but increases generally the value of association measures as illustrated for instance by the simulations carried out in [5].

Maximizing the strength of the association is of importance in several fields. For example, when analyzing survey sample data it is common to group items to avoid categories with too few cases. For studying the association between variables, as it is often the case in social sciences, it is, then, essential to understand how the association may vary with the grouping of categories in order to select the grouping that best reflects the association. This issue was discussed, for instance, by Benzécri [1] with respect to the maximization of the Pearson's Chi-square.

A second motivation concerns discretization that is a major issue in supervised learning. In this framework, a joint aggregation of the predictor and the response variable should be more efficient than an independent discretization of each variable. Nevertheless, with the exception of the joint dichotomization process considered by Breiman et al. [2], it seems that optimal solutions for partitioning values exist in the literature only for a single variable (see for instance [8] for a survey). There is thus an obvious need for tools allowing a joint general

D.A. Zighed, J. Komorowski, and J. Żytkow (Eds.): PKDD 2000, LNAI 1910, pp. 593–598, 2000.
© Springer-Verlag Berlin Heidelberg 2000

optimal discretization of two or more variables. The results presented here for jointly partitioning the row and column values in an unfixed number of classes are a first step in this direction.

The joint optimal partition can be found by scanning all possible groupings. Since the number of these groupings increases exponentially with the number of categories, such a systematic approach is, however, generally untractable. An iterative process is thus introduced for determining the (quasi-)optimal partition by successively aggregating two categories.

The distinction between nominal and ordinal variables is of first importance in this partitioning issue. Indeed, with ordinal variables only an aggregation of adjacent categories makes sense.

It is worth mentioning that the optimization problem considered here differs from that of Benzécri [1], for which Celeux et al. [3] have proposed an algorithm based on clustering techniques. These authors consider only partitions into a fixed number of classes and maximize Pearson's Chi-square. Unlike this framework, our settings allow varying numbers of rows and columns. We have therefore to rely on normalized association measures to compare configurations.

Section 2 illustrates with an example how the aggregation of row and column values may affect the Chi-square statistics and the association measures. The formal framework and the notations are described in Section 3. Section 4 specifies the complexity of the enumeration search of the optimal solution and proposes an heuristic algorithm heuristic. Section 5 summarizes a sensitivity analysis of association measures. Finally, Section 6 proposes some concluding remarks.

2   Example and Intuitive Results

Consider the following cross table between a row variable x and a column variable y

$$M = \quad \begin{array}{c|cccc} x \backslash y & A & B & C & D \\ \hline a & 10 & 10 & 1 & 1 \\ b & 10 & 10 & 1 & 1 \\ c & 1 & 1 & 10 & 10 \\ d & 1 & 1 & 10 & 10 \end{array} \quad .$$

Intuitively, aggregating two identical columns, $\{A, B\}$ or $\{C, D\}$, should increase the association level. The same holds for a grouping of rows $\{a, b\}$ or $\{c, d\}$. On the other hand, grouping categories B and C for example reduces the contrast between column distributions and should therefore reduce the association. Let us illustrate these effects by considering the aggregated tables

$$M_y^+ = \quad \begin{array}{c|ccc} x \backslash y & A & B & \{C, D\} \\ \hline a & 10 & 10 & 2 \\ b & 10 & 10 & 2 \\ c & 1 & 1 & 20 \\ d & 1 & 1 & 20 \end{array} \qquad M_{xy}^+ = \quad \begin{array}{c|ccc} x \backslash y & A & B & \{C, D\} \\ \hline a & 10 & 10 & 2 \\ b & 10 & 10 & 2 \\ \{c, d\} & 2 & 2 & 40 \end{array}$$

and

$$M_y^- = \begin{array}{c|ccc} x\backslash y & A & \{B,C\} & D \\ \hline a & 10 & 11 & 1 \\ b & 10 & 11 & 1 \\ c & 1 & 11 & 10 \\ d & 1 & 11 & 10 \end{array} \qquad M_{xy}^- = \begin{array}{c|ccc} x\backslash y & A & \{B,C\} & D \\ \hline a & 10 & 11 & 1 \\ \{b,c\} & 11 & 22 & 11 \\ d & 1 & 11 & 10 \end{array}\,.$$

Table 1. Association measures for different groupings

	M	M_y^+	M_{xy}^+	M_y^-	M_{xy}^-
rows	4	4	3	4	3
columns	4	3	3	3	3
df	9	6	4	6	4
Pearson Chi-square	58.91	58.91	58.91	29.45	14.73
Likelihood Ratio	68.38	68.38	68.38	34.19	17.09
Tschuprow's t	0.47	0.52	0.58	0.37	0.29
Cramer's v	0.47	0.58	0.58	0.41	0.29
Goodman-Kruskal $\tau_{y\leftarrow x}$	0.22	0.40	0.40	0.13	0.07
Goodman-Kruskal $\tau_{x\leftarrow y}$	0.22	0.22	0.40	0.11	0.07
Uncertainty Coefficient $u_{y\leftarrow x}$	0.28	0.37	0.37	0.19	0.09
Uncertainty Coefficient $u_{x\leftarrow y}$	0.28	0.28	0.37	0.14	0.09
Goodman-Kruskal γ	0.68	0.77	0.80	0.63	0.57
Kendall's τ_b	0.55	0.60	0.65	0.45	0.37
Somers' $d_{y\leftarrow x}$	0.55	0.55	0.65	0.41	0.37
Somers' $d_{x\leftarrow y}$	0.55	0.65	0.65	0.49	0.37

Table 1 gives the values of a set of nominal (t, v, τ, u) and ordinal (γ, τ_b, d) association measures for cross table M and the four aggregated tables considered. The nominal τ and u and the ordinal d are directional measures. For more details on these measures see fo instance [5]. According to the distributional equivalence property, the Chi-square statistics remain the same for tables M, M_y^+ and M_{xy}^+. However, the association measures increase as expected with the grouping of similar columns and rows. The figures for the aggregated tables M_y^- and M_{xy}^- show that the aggregation of columns or rows very differently distributed reduces both the Chi-squares and the association measures.

3 Notations and Formal Framework

Let x and y be two variables with respectively r and c different states. Crossing variable x with y generates a contingency table $T_{r\times c}$ with r rows and c columns. Let $\theta_{xy} = \theta(T_{r\times c})$ denote a generic association criterion for table $T_{r\times c}$. Let P_x be a partition of the values of x, and P_y a partition of the states of y. Each couple

(P_x, P_y) defines then a contingency table $T(P_x, P_y)$. The optimization problem considered is the maximization of the association among the set of couples of partitions (P_x, P_y)

$$\max_{P_x, P_y} \theta\big(T(P_x, P_y)\big) \tag{1}$$

For ordinal variables, hence for interval or ratio variables, only partitions obtained by aggregating adjacent categories make sense. We consider then the restricted program

$$\begin{cases} \max\limits_{P_x, P_y} \theta\big(T(P_x, P_y)\big) \\ \text{u.c.} \quad P_x \in \mathcal{A}_x \text{ and } P_y \in \mathcal{A}_y \end{cases} \tag{2}$$

where $\mathcal{A}_x$ and $\mathcal{A}_y$ stand for the sets of partitions obtained by grouping adjacent values of x and y. Letting $\mathcal{P}_x$ and $\mathcal{P}_y$ be the unrestricted sets of partitions, we have for $c, r > 2$, $\mathcal{A}_x \subset \mathcal{P}_x$ and $\mathcal{A}_y \subset \mathcal{P}_y$. Finally, note that for ordinal association measures that may take negative values, maximizing the strength of the association requires to take the absolute value of the ordinal association measure as objective function $\theta\big(T(P_x, P_y)\big)$.

4   Complexity of the Optimal Solution

4.1   Complexity of the Enumerative Approach

To find the optimal solution, we have to explore all possible groupings of both the rows and the columns, i.e. the set of couples (P_x, P_y). The number of cases to be checked is given by $\#\mathcal{P}_x \#\mathcal{P}_y$, i.e. the number of row groupings times the number of column groupings.

For a nominal variable, the number of possible groupings is the number $B(c) = \#\mathcal{P}$ of partitions of its c categories. It may be computed iteratively by means of Bell formula

$$B(c) = \sum_{k=0}^{c-1} \binom{c-1}{k} B(k)$$

with $B(0) = 1$. For $c = r$, the number $B(c)B(r)$ of configurations to be explored is then for example respectively 25, 225, 2'704 and 41'209 for $c = 3, 4, 5, 6$ and exceeds $13 \cdot 10^9$ for $c = r = 10$.

For ordinal variables, hence for discretization issues, only adjacent groupings are considered. This reduces the number of cases to browse. The number $G(c) = \#\mathcal{A}$ of different groupings of c values is

$$G(c) = \sum_{k=0}^{c-1} \binom{c-1}{k} = 2^{(c-1)} \ .$$

There are thus respectively $G(c)G(r) = 16, 64, 256, 1'024$ configurations to browse for a square table with $c = r = 3, 4, 5, 6$, and more than a million for $c = r = 10$.

4.2 An Heuristic

Due to the limitation of the enumerative approach, we propose an iterative process that successively aggregates the two row or column categories that most improve the association criteria $\theta(T)$. Such an heuristic may indeed not end up with the optimal solution, but perhaps only with a quasi-optimal solution.

Formally, the configuration (P_x^k, P_y^k) obtained at step k is the solution of

$$
\begin{cases}
\max_{P_x, P_y} \ \theta\big(T(P_x, P_y)\big) \\[2mm]
\text{u.c. } P_x = P_x^{(k-1)} \text{ and } P_y \in \mathcal{P}_y^{(k-1)} \\[2mm]
\text{or} \\[2mm]
\phantom{\text{u.c. }} P_x \in \mathcal{P}_x^{(k-1)} \text{ and } P_y = P_y^{(k-1)}
\end{cases}
\tag{3}
$$

where $\mathcal{P}_x^{(k-1)}$ stands for the set of partitions on x resulting from the grouping of two classes of the partition $P_x^{(k-1)}$.

For ordinal variables, $\mathcal{P}_x^{(k-1)}$ and $\mathcal{P}_y^{(k-1)}$ should be replaced by the sets $\mathcal{A}_x^{(k-1)}$ and $\mathcal{A}_y^{(k-1)}$ of partitions resulting from the aggregation of two adjacent elements.

Starting with $T^0 = T_{r \times c}$ the table associated to the finest categories of variables x and y, the algorithm successively determines the tables $T^k, k = 1, 2, \dots$ corresponding to the partitions solution of (3). The process continues while $\theta(T^k) \geq \theta(T^{(k-1)})$ and is stopped when the best grouping of two categories leads to a reduction of the criteria.

The *quasi-optimal grouping* is the couple (P_x^k, P_y^k) solution of (3) at the step k where

$$
\theta\big(T^{(k+1)}\big) - \theta\big(T^k\big) < 0 \quad \text{and} \quad \theta\big(T^k\big) - \theta\big(T^{(k-1)}\big) \geq 0
$$

By convention, we set the value of the association criteria $\theta(T)$ to zero for any table with a single row or column. The algorithm then ends up with such a single value table, if and only if all rows (columns), are equivalently distributed.

5 Effect of Aggregating Two Categories

For the heuristic proposed, it is essential to understand how the association criteria behave in response to the aggregation of two categories. We have therefore carried out an analytical sensitivity analysis reported in [7] of which we summarize here the main results.

Chi-square statistics remain constant when the two aggregated categories are equivalently distributed and decrease otherwise.

Chi-square based measures: Tschuprow's t can increase when r or c decreases. Cramer's v may only increase when the aggregation is done on the variable with the smaller number ($\min\{r, c\}$) of categories.

Nominal PRE measures: $\tau_{y \leftarrow x}$ and $u_{y \leftarrow x}$ may only increase with an aggregation on the dependent variable.

Ordinal measures: Their absolute value may increase for an aggregation on any variable. In the case of an aggregation of two equivalently distributed categories $|\gamma|$ and $|\tau_b|$ increase while $|\tau_c|$ increases if $\min\{r, c\}$ decreases and $|d_{y \leftarrow x}|$ increases when the aggregation is done on the independent variable x and remains constant otherwise.

6   Further Developments

This paper is concerned with the issue of finding the partitions of the row and column categories that maximizes the association. The results presented, on the complexity of the solution and on the sensitivity of the association criteria toward aggregation, are only preliminary materials. A lot remains to be done, especially concerning the properties and implementation of the algorithm sketched in Section 4.2. Let us just mention two important aspects. First, we have to empirically assess the efficiency of the heuristic. We are presently building simulation designs to check how the quasi-optimal solution provided by the algorithm may differ from the true global optimal solution. From the results of Section 4.1, the comparison with the true solution is only possible for reasonably sized starting tables, say tables with six rows and six columns. Secondly, it is worth to take account of the higher reliability of large figures. The algorithm will therefore be extended by implementing Laplace's estimates of the probabilities to increase the robustness of the solution.

References

1. J.-P. Benzécri. *Analyse des données. Tome 2: Analyse des correspondances*. Dunod, Paris, 1973.
2. L. Breiman, J. H. Friedman, R. A. Olshen, and C. J. Stone. *Classification And Regression Trees*. Chapman and Hall, New York, 1993.
3. G. Celeux, E. Diday, G. Govaert, Y. Lechevallier, and H. Ralambondrainy. *Classification automatique des données*. Informatique. Dunod, Paris, 1988.
4. J. D. Jobson. *Applied Multivariate Data Analysis. Vol.II Categorical and Multivariate Methods*. Springer-Verlag, New York, 1992.
5. M. Olszak and G. Ritschard. The behaviour of nominal and ordinal partial association measures. *The Statistician*, 44(2):195–212, 1995.
6. R. Rakotomalala and D. A. Zighed. Mesures PRE dans les graphes d'induction: une approche statistique de l'arbitrage généralité-précision. In G. Ritschard, A. Berchtold, F. Duc, and D. A. Zighed, editors, *Apprentissage: des principes naturels aux méthodes artificielles*, pages 37–60. Hermes Science Publications, Paris, 1998.
7. G. Ritschard, D. A. Zighed, and N. Nicoloyannis. Optimal grouping in cross tables. Technical report, Department of Econometrics, University of Geneva, 2000.
8. D. A. Zighed, S. Rabaseda, R. Rakotomalala, and F. Feschet. Discretization methods in supervised learning. *Encyclopedia of Computer Science and Technology*, 40(25):35–50, 1999.
9. D. A. Zighed and R. Rakotomalala. *Graphes d'induction: apprentissage et data mining*. Hermes Science Publications, Paris, 2000.

An Experimental Study of Partition Quality Indices in Clustering

Céline Robardet[1], Fabien Feschet[1], and Nicolas Nicoloyannis[2]

[1] LASS - UMR 5823 - bât 101 - Université Lyon 1 - Villeurbanne France
[2] ERIC - bât L - Université Lyon 2 - Bron France

Abstract. We present a preliminary study to define a comparison protocol to evaluate different quality measures used in supervised and unsupervised clustering as objective functions. We first define an order on the set of partitions to capture the common notion of a good partition towards the knowing of the ideal one. We demonstrate the efficiency of this approach by providing several experiments.

keywords. Unsupervised clustering, partitions ordering, partition quality indices.

1 Introduction

Unsupervised clustering aims at organizing a data set by grouping objects into clusters to discover at best their relations. This clustering can conduct to overlapping groups, however it is simpler to look for partitions. The clustering should gather the couple of objects the most similar and should separate the couple the most dissimilar. Most clustering algorithms can be described through the concept of similarity and the optimization procedure. A lot of algorithms [8,2] use an objective function to operationally express a compromise between the intra-cluster proximity and the inter-cluster farness. It is also possible to use the EM algorithm [2] where two objects are closed if they can be considered as a realization of the same random variable. Another important family of methods can be pointed out under the name of conceptual clustering algorithms [6,5]. They have been originally constructed with nominal variables, but extensions to other type of data are available [7]. The particularity of those methods is that they aim to build understandable partitions. Those algorithms rely on non parametric probabilistic measures to define clusters and two objects are closed if they have the same value on *most* of the variables. Clusters are such that their probability vectors have the greatest entropy on each variable. Conversely to supervised learning where a class variable is known on a training data set, there are no references in unsupervised clustering. Thus, beside clusters construction, one might take special care to the relevance of the discovered data organization. It is necessary to check the validity of the obtained partition. In this article, we present an evaluation of some similarity measures used in supervised and unsupervised clustering as objective functions. We study their capability to discern,

D.A. Zighed, J. Komorowski, and J. Żytkow (Eds.): PKDD 2000, LNAI 1910, pp. 599–604, 2000.
© Springer-Verlag Berlin Heidelberg 2000

in an ideal case, the best partition among all partitions. We are not concern with clusters validity. We only compare several objective functions regarding their discrimination capabilities. The variability of the similarity measures on the set of partitions must be necessarily as high as possible to ensure a non random choice in the set of measure equivalent partitions as it can be found in ISAAC [11] for instance. Even if the search strategy is important in the cluster construction, similarity measures must be sufficiently discriminant.

The organisation of the paper is the following. In the section one, we present the measures we have evaluated. Then, we present our strategy to have a meaningful comparison. Then, some results are given and discussed and some concluding remarks are given.

2   Objective Functions

We choose to evaluate only objective functions which require nominal variables, and not based on any kind of metrics. Thus we only study the behavior of an objective function and not a structuring of a data space. In the following we used the functions φ_β defined on $[0;1]$ such that $\varphi_\beta(x) = \frac{2^{\beta-1}}{2^{\beta-1}-1}x(1-x)^{\beta-1}$. This permits us to introduce generalized entropy [12] in the measures.

The Category Utility Function
This function is the one used in the well known conceptual clustering algorithm COBWEB [6] and in other related systems like ITERATE [1] and called category utility. It is a trade-off between intra-class similarity and inter-class dissimilarity,

$$CU = \frac{\sum_k P(C_k) \sum_i \sum_j \left[P(A_i = V_{ij} \mid C_k)^2 - P(A_i = V_{ij})^2 \right]}{K}$$

CU rewards the clusters which most reduce the collective impurity over all variables. This function has a form closed to the GINI index used in supervised clustering. Indeed, CU is the weighted average of the GINI index: $CU = \frac{\sum_k P(C_k) \times \text{GINI}}{K}$ in order to make this index independent of the number of clusters.

Quinlan's Gain Ratio
Other objective functions used in supervised clustering can be adapted for unsupervised clustering [5]. The adapted Quinlan's Gain ratio does not depend on the number of clusters and is given by,

$$\sum_i \frac{\sum_k P(C_k) \sum_j \left[\varphi_\beta(P(A_i = V_{ij}|C_k)) - \varphi_\beta(P(A_i = V_{ij})) \right]}{-\sum_k \varphi_\beta(P(C_k))}$$

3   Comparing Measures through Partitions Ordering

Since we compare measures on the set of partitions, simply taking values on the same partitions does not lead to a meaningful comparison. Indeed, the resulting values on two different partitions can not be compared until the partitions are compared and ordered towards the clustering objectives. Thus, we consider data sets which are generated as the expression of a partitioning PI called the ideal case. Our comparisons aim at discovering if, among all the partitions, PI is measured as the best one by the different similarity indices and if those measures are sufficiently discriminating on the whole set of partitions. However, having PI, a natural choice to compare two partitions is to create a distance d such that $d(PI, P)$ permits to order the partitions P in reference to the ideal case PI. Of course, two different partitions can be at the same distance of PI. In order to have a significant measure, we use the two constituents of a partition: the objects and the vectors of variables values taken by the objects. We then build two distances, $\mu_{\mathcal{O}}$ and $\mu_{\mathcal{V}}$, such that the whole distance is constructed from both measures.

3.1 Comparing Clusters

A Distance Between Two Clusters Taken on the Objects
Marczewski and Steinhaus [10] present a distance for the comparison of two sets. Let $\mathcal{P}(X)$ be the class of all subsets of the finite set X. Then[1],

$$\forall C_k, C_{k'} \in \mathcal{P}(X), \ \mu_{\mathcal{O}}(C_k, C_{k'}) = \begin{cases} \frac{|C_k \Delta C_{k'}|}{|C_k \cup C_{k'}|} & \text{if } |C_k \cup C_{k'}| > 0 \\ 0 & \text{otherwise} \end{cases}$$

A Distance Taken on the Variables
On the same principle, we define a distance between two probabilistic vectors, $\forall C_k, C_{k'} \in \mathcal{P}(X)$,

$$\mu_{\mathcal{V}}(C_k, C_{k'}) = \frac{1}{m \, p} \sum_{i=1}^{p} \sum_{j=1}^{m} |P(A_i = V_{ij} \mid C_k) - P(A_i = V_{ij} \mid C_{k'})|$$

3.2 Distance between Partitions

An Hausdorff Like Distance
Following Karonski and Palka [9], the previous distances can be extended to the case of comparing two partitions P_1 and P_2 given by $P_i = \{C_{i1}, \dots, C_{iI_i}\}$. Given a measure μ between two sets, we can construct an Hausdorff like distance:

$$\mathcal{D}_\mu(P_1, P_2) = \tfrac{1}{2} \left[\max_{i \in I_1} \min_{j \in I_2} \mu(C_{1i}, C_{2j}) + \max_{j \in I_2} \min_{i \in I_1} \mu(C_{1i}, C_{2j}) \right]$$

In order that the distance take into account the two previous measures, we use the following distance,

$$\mathcal{D}_{\min - \max}(P_1, P_2) = \sqrt{\mathcal{D}^2_{\mu_{\mathcal{O}}}(P_1, P_2) + \mathcal{D}^2_{\mu_{\mathcal{V}}}(P_1, P_2)}$$

---

[1] with $A\Delta B = A \cup B - 2A \cap B$ the symmetric difference

With the min-max distance, we hope to obtain a good discrimination of the partitions regarding to both aspects taken in account in conceptual clustering.

An Improved Measure
Through experiments, we observed that the distance $\mathcal{D}_{\min - \max}$ has some drawbacks. This measure is not very sensitive to minor changes because of its principle of worth case. That is why we propose another distance based on the search of the matching of maximal cardinality and minimal weight inside a bipartite graph [4]. The graph $G = (X_1 \cup X_2, E)$ associated to two partitions, has the clusters of the two partitions as nodes and all edges between X_1 and X_2 exist and are weighted by either $\mu_\mathcal{O}$ or $\mu_\mathcal{V}$. Notice that we can restrict the number of edges by selecting those whose weights are sufficiently low. The value of the matching permits to order the partitions. It decreases when partitions closer to PI, in the sense of the used measure, are taken and by definition is zero when PI is taken.

4   The Results

In the following study we reduce our investigations to boolean variables. The results obtained on such variables can not be extrapolated to nominal variables without bias, but a such constraint is necessary to reduce the number of parameters. We use an artificial dataset given by a diagonal block matrix, each block containing only 1 and the sizes of the blocks vary. This also defines PI. The partitions compared to PI are either the whole set, when the combinatory permits (8 objects and 8 variables), or random selected partitions otherwise (60 objects and 15 variables). In case of a random selection, 30000 partitions are sampled. Finally, we introduce some noise in order to evaluate the noise resistance of the measures. It is generated by random permutations of 1 and 0 in the boolean matrix. To measure the influence of β, we also study different values in $\{0.5, 1, 2, 3\}$.

Comparison of the Two Indexes
On the figure 1, CU function is plotted relatively to both measures in the exhaustive case. As expected, the matching index is more discriminant than the min-max distance. This is an important result which confirm the sensitivity of the matching measure. Notice that we observed similar results for all other measures and noise levels.

The Ideal Case
Given PI, we first study the influence of β using Quinlan gain ratio (see the figure 2 on the left). The effect of the parameter β is to spread out the values of the measures thus having more variations with β different to one. Using $\beta = 0.5$ leads to the measure with smaller variations and thus to the less discrimant one. The others choice give similar behavior with only differences in the scale. Thus, we choose to keep the original version of the gain ratio for comparison purposes. However, some experiments should be extended to better precise the influences of a high value of β. Having chosen a β for Quinlan gain ratio, we can make a comparison with the CU measure (see the figure 2 on the right). The variations

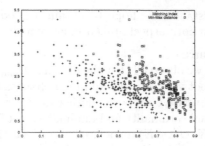

Fig. 1. Relative performance of the two orderings (x is distances, y is CU)

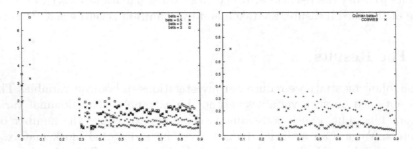

Fig. 2. (x is order, y is measure): (left) influence of β. (right) CU vs Quinlan.

of CU are very small so that nearly all partitions seem to be similar for that measure, except the extremal one. Following these preliminary results, Quinlan measure seems to be a better measure than CU even if more experiments are necessary to conclude.

The Noisy Ideal Case
In order to simulate a real case, we introduce some noise in the boolean matrix (see the figure 3). Following our experiments, Quinlan measure appears to be more noise resistant than CU. With a 5 percent noise level, it behaves like in the ideal case. When noise increases, some partitions take aberrant values (see the figure 3 (left)). However, this measure remains regular when CU becomes very perturbated (see the figure 3 (right)).

5   Conclusion

In this article, we have presented a new ordering of partitions to objectively compare the behavior of different quality measures which can be used in unsupervised learning. Through our experiment protocol, we have first established that our index has a better discrimant power than previous one of Marczewski and Steinhaus. Secondly, we shown than Quinlan gain ratio is noise resistant

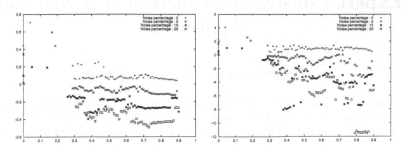

Fig. 3. Noise influence (x is order, y is measure): (left) Quinlan - (right) CU

and more discrimant than the other functions. When generalizing it with φ_β, it appeared that a bad choice was $\beta = 0.5$. Other values were not significantly different in our experiments. Due to lake of space, we have not reported here all our results but they all confirm the conclusions done in this paper. Let us also report that we also studied an adapted version of the Mantaras function [3] which behaved, in this set of experiments, like Quinlan gain ratio.

References

1. G. Biswas, J. Weinberg, and C. Li. Iterate: a conceptual clustering method for knowledge discovery in databases. Technical report, CS Departement, Vanderbilt university, Nashville, 1995.
2. G. Celeux and E. Diday et al. *Classification automatique des données*. Dunod, 1988.
3. R. López de Màntaras. A distance based attribute selection measure for decision tree induction. *Machine Learning*, 6:81–92, 1991.
4. J. Edmonds. Maximum matching and a polyhedron with 0-1 vertices. *Res. Nat. Bureau Standards*, 69B(1-2):125–130, 1965.
5. Doug Fisher. Iterative optimization and simplification of hierarchical clusterings. *Journal of Artificial Intelligence Research*, 4:147–180, 1996.
6. Douglas H. Fisher. Knowledge acquisition via incremental conceptual clustering. *Machine Learning*, 2:139–172, 1987.
7. W. Iba and P. Langley. Unsupervised learning of probabilistic concept hierarchies. Technical report, Inst. for the study of learning and expertise, Pablo Alto, 1999.
8. A. K. Jain and R. C. Dubes. *Algorithms for clustering data*. Prentice Hall, Englewood cliffs, New Jersey, 1988.
9. M. Karonski and Z. Palka. On marczewski-steinhaus type distance between hypergraphs. *Zastosowania Mat. Appli. Mathematicae*, 16:47–57, 1977.
10. E. Marczewski and H. Steinhaus. On a certain distance of sets and the corresponding distance of functions. *Colloquium Mathematicum*, 6:319–327, 1958.
11. L. Talavera and J. Béjar. Efficient construction of comprehensible hierarchical clusterings. In J. M. Żytkow and M. Quafafou, editors, *2nd PKDD*, volume 1510, pages 93–101. Springer-Verlag, 1998.
12. L. Wehenkel. On uncertainty measures used for decision tree induction. In *Info. Proc. and Manag. of Uncertainty*, pages 413–418, 1996.

Expert Constrained Clustering: A Symbolic Approach

Fabrice Rossi and Frédérick Vautrain

LISE/CEREMADE (CNRS UMR 7534), Université Paris-IX/Dauphine,
Place du Maréchal de Lattre de Tassigny, 75775 Paris Cedex 16, France
e-mails: rossi@ceremade.dauphine.fr, vautrain@ceremade.dauphine.fr

Abstract. A new constrained model is discussed as a way of incorporating efficiently *a priori* expert knowledge into a clustering problem of a given individual set.

The first innovation is the combination of fusion constraints, which request some individuals to belong to one cluster, with exclusion constraints, which separate some individuals in different clusters. This situation implies to check the existence of a solution (ie if no pair of individuals are connected by fusion and exclusion constraints).

The second novelty is that the constraints are expressed in a symbolic language that allows compact description of group of individuals according to a given interpretation.

This paper studies the coherence of such constraints at individual and symbolic levels. A mathematical framework, close to the Symbolic Data Analysis[3], is built in order to define how a symbolic description space may be interpreted on a given individual set. A partial order on symbolic descriptions (which is an usual assumption of Artificial Intelligence), allows a symbolic analysis of the constraints. Our results provide an individual but also a symbolic clustering.

1 Introduction

In order to take into account prior expert knowledge, it is quite common to implement **constraints** in classification algorithms [4]. The general goal of classification is to find a satisfying partition of the population (set of individuals). Adding constraints allows to reduce the set of acceptable partitions. In this paper we consider only fusion and exclusion constraints. A fusion constraint implies to keep specified individuals into one cluster. For instance, we will not accept partitions in which individuals with a given property (e.g., small size) are classified into separated clusters. An exclusion constraint is exactly the opposite: we ask for specified individuals to remain in distinct clusters.

It is possible to take into account fusion and exclusion constraints at the same time, which introduces a **coherency problem**: some sets of fusion and exclusion constraints cannot be satisfied together by any partition. For instance, if we ask for x to belong to y's cluster (fusion) and also for x to be in a different cluster than y (exclusion), it is obvious that no partition can satisfy both constraints. The

D.A. Zighed, J. Komorowski, and J. Zytkow (Eds.): PKDD 2000, LNAI 1910, pp. 605–612, 2000.
© Springer-Verlag Berlin Heidelberg 2000

coherency problem (solution existence) can be solved with a simple algorithm, as shown in [2]. The proposed algorithm works on a population and checks if a constraint set is coherent. When it is, the algorithm gives the smallest partition which satisfies the constraints (a partition P is smaller than a partition Q when each cluster of P is included in a cluster of Q). One of the drawbacks of this algorithm is that it works directly at the individual level, which can be slow on big populations because the global analysis is needed to compute the smallest partition: it is not simple in general to work on subsets of the population and to merge the results. Moreover, working at the population level gives results that can be difficult to analyze.

In order to improve this algorithm, we propose to work at a *symbolic level*. We will describe individual subsets thanks to a symbolic approach that allows short representation of such subsets. For instance, a subset C of individuals might be defined thanks to a conjunction of properties that all the individuals of C satisfy, e.g., individuals whose weight belongs to a specified interval.

A mapping called ext (for *extension*) interprets each symbolic description on the set of individuals. It allows to build the actual individual subset associated to a description. Each constraint is defined as a pair of symbolic descriptions (d, d'). The interpretation mapping translates each pair into a classical constraint on the population, as follows:

1. if (d, d') is a fusion constraint then, individuals which are described by d and individuals which are described by d' must belong to one cluster;
2. if (d, d') is an exclusion constraint then, individuals which are described by d and individuals which are described by d' must belong to different clusters;

In this paper, we prove that under reasonable assumptions on the description space, the coherency problem can be studied directly at the symbolic level. The main advantage of this approach it to reduce the processing cost because each description can cover a lot of individuals. Moreover, when the constraints are coherent, the proposed algorithm gives a symbolic description of the smallest partition. Another interesting point is that the interpretation mapping can be changed without modifying the coherency result, as long as it stays in a broad class of acceptable mappings.

The rest of the paper is organized as follows: we start by defining the proposed mathematical framework. Then we present three important results. First we provide a mixed approach that allows to study symbolic constraints in relation to their interpretation on a given set of individuals. Then, we give our pure symbolic coherency results.

Due to size constraints, proof are omitted and can be found in [6].

2   Mathematical framework

2.1   Description space and context definition

The considered constrained classification problem involves:

1. the **description space** (called $\mathcal{D}$) corresponds to high level (symbolic) description of individuals (or groups of individuals);
2. the **population** (called Ω) is a set of individuals that can be described thanks to the description space;
3. the **interpretation mapping** (called ext) is a map from $\mathcal{D}$ to $\mathcal{P}(\Omega)$ (the set of all subsets of Ω). It transforms a symbolic description into a set of individuals that are correctly described by this description. For a description a, $\text{ext}(a)$ is called the **extension** of a.

   The pair $\mathcal{G} = (\Omega, \text{ext})$ is called a **context** for $\mathcal{D}$.

In general, the description space $\mathcal{D}$ is fixed, but the population or the interpretation mapping can change. For instance, if $\mathcal{D}$ uses very high level description such as *"tall"*, the exact interpretation of *"tall"* might depend on the chosen semantic (e.g., if *"tall"* is applied to human beings, it might describe men higher than 1m80 at current time, or men higher than 1m70 several centuries ago). If the population is stored in a database the extension mapping can be considered as a two step process: first we translate the symbolic description into a SQL SELECT query and then we let the database compute the corresponding result which is the extension of the description.

In general, it is possible to provide an order denoted $\leq$ for $\mathcal{D}$ (see for instance [5] for examples of such symbolic orders). In our framework, we need such an order to be able to provide pure symbolic analysis (see section 4). The only technical assumption needed on the ordered set $(\mathcal{D}, \leq)$ is that any totally ordered subset of $\mathcal{D}$ must be bounded below.

The intuitive meaning of the symbolic order is linked to description precision. More precisely, $a \leq b$ means that a is less general than b. In other words, a is a particular case of b, for instance *"blue"* is less general than *"red or blue"*. The interpretation mapping **must** translate this meaning into an equivalent property on the population. Technically, we have:

Definition 1 *An **ordered description space** is a pair $(\mathcal{D}, \leq)$, where $\mathcal{D}$ is a description space and $\leq$ an order on $\mathcal{D}$. If Ω is a population and ext is a mapping from $\mathcal{D}$ to $\mathcal{P}(\Omega)$, the context $\mathcal{G} = (\Omega, \text{ext})$ is **consistent** with $(\mathcal{D}, \leq)$ if and only if*

1. *ext is an increasing mapping from $(\mathcal{D}, \leq)$ to $(\mathcal{P}(\Omega), \subset)$;*
2. *for any a and b in $\mathcal{D}$:*

$$\min(a, b) = \emptyset \Rightarrow \text{ext}(a) \cap \text{ext}(b) = \emptyset, \tag{1}$$

where $\min(a, b) = \{c \in \mathcal{D} \mid c \leq a \text{ and } c \leq b\}$.

Keeping in mind that $u \leq v$ means u is less general than v, $\min(a, b) = \emptyset$ means that no description can be more precise than a and b at the same time. Therefore, if there are some individuals in $\text{ext}(a) \cap \text{ext}(b)$, this means that there is no way to describe those individuals more precisely than with a or b. This is a symptom of *inadequacy* of the description space to the context and we assume therefore that this condition cannot happen.

Example Let's consider a population Ω a subset of COLOR $\times$ WEIGHT(g) $\times$ HEIGHT(cm), where COLOR $= \{Yellow, Blue, Red\}$, WEIGHT$(g) = [1, 100]$ and HEIGHT$(cm) = [1, 25]$. Each individual ω (a spare part) is defined as a triplet $\omega = (\omega_C, \omega_W, \omega_H)$.

Let $\mathcal{D} = \mathcal{P}^*$ (COLOR)$\times \mathcal{I}^*$ (WEIGHT) $\times$ HEIGHT a set of symbolic descriptions where $\mathcal{P}^*$ (COLOR) is the set of the subsets (parts) of COLOR minus the set $\varnothing$, $\mathcal{I}^*$ (WEIGHT) is the set of the subsets (intervals) of WEIGHT minus the set $\varnothing$ and HEIGHT$= \{Small, Tall\}$.

We denote a symbolic description $d = (d_C, d_W, d_H)$ and the symbolic order on $\mathcal{D}$ is defined as follows:

$$d \preceq d' \Leftrightarrow d_C \subseteq d'_C, d_W \subseteq d'_W, d_H = d'_H$$

An interpretation mapping (ext) can be defined for instance as follows:

$$\begin{cases} \text{ext} \left((d_C, d_W, Small) \right) = \{\omega \in \Omega \mid \omega_C \in d_C \, , \, \omega_W \in d_W \, , \, 1 \leq \omega_H < 15\} \\ \text{ext} \left(d_C, d_W, Tall \right) = \{\omega \in \Omega \mid \omega_C \in d_C \, , \, \omega_W \in d_W \, , \, 15 \leq \omega_H \leq 25\} \end{cases}$$

2.2   Constraints

Definition 2 *A **constraint set** for a description space $\mathcal{D}$ is a pair of subsets of $\mathcal{D}^2$, (F, E). The first subset F represents **fusion** constraints and the second subset E represents **exclusion** constraints.*

2.3   Constrained binary relation

Definition 3 *Let $\mathcal{D}$ be a description space and (Ω, ext) be a context for $\mathcal{D}$. Let (F, E) be a constraint set for $\mathcal{D}$. A binary relation r on Ω is **compatible** with (F, E) if and only if:*

$$\forall (a, b) \in F, \forall x \in \text{ext}(a), \; \forall y \in \text{ext}(b), \; r(x, y) \tag{2}$$

$$\forall (c, d) \in E, \forall x \in \text{ext}(c), \; \forall y \in \text{ext}(d), \; \neg r(x, y) \tag{3}$$

Our goal is to study constrained classification: in general, we will focus on equivalence binary relations.

2.4   Notations

Let R be a subset of T^2, an arbitrary set. R is the graph of a binary relation on T and we can define the following sets:

- $^t R$ is the transposed (or dual) relation defined by: $(x, y) \in {}^t R \Leftrightarrow (y, x) \in R$;
- R^s is the symmetric closure of R, defined by: $R^s = R \cup {}^t R$;
- if S is a subset of T^2, RS, the product (or composition) of R and S, is defined as follows: $RS = \{(a, b) \in T^2 \mid \exists c \in T \text{ such that } (a, c) \in R \text{ and } (c, b) \in S\}$

- R^k is the k-th power of R, defined as follows: $R^0 = I_T$ (I_T is the diagonal of T, i.e., $I_T = \{(x, x) \mid x \in T\}$) and $R^k = \{(x, y) \in T^2 \mid \exists z \in T \text{ so that } (x, z) \in R^{k-1} \text{ and } (z, y) \in R\}$, for $k > 0$. Thanks to the previous definition, this can be rewritten in $R^k = R^{k-1}R$;
- R^+ is the transitive closure of R, defined by: $R^+ = \bigcup_{k \geq 1} R^k$;
- R^* is the reflexive and transitive closure of R, defined by: $R^* = R^+ \cup I_T$;
- $\mathcal{S}(R)$ is the support of R, defined by: $\mathcal{S}(R) = \{x \in T \mid \exists y \in T \text{ so that } (x, y) \in R \text{ or } (y, x) \in R\}$, which can be simplified into $\mathcal{S}(R) = \{x \in T \mid \exists y \in T \text{ so that } (x, y) \in R^s\}$.

3   Population based analysis

The purpose of this section is to analyze the coherency problem at symbolic level, but taking into account the interpretation and the population.

3.1   Fusion constraint closure

To state our results, we need first to introduce some technical definitions:

Definition 4 *Let R and S be two binary relations on the same set T. R is said to be **stable with respect to** S if and only if for all a, b, c and d in T, we have:*

$$(a, b) \in R, \; (b, c) \in S \text{ and } (c, d) \in R \Rightarrow (a, d) \in R, \tag{4}$$

i.e., $RSR \subset R$.

In informal terms, this means that S does not introduce short-circuits in R.

Definition 5 *Let R and S be two binary relations on the same set T. We call R_S the closure of R by S, which is defined as the smallest binary relation that contains R and that is stable with respect to S. Let us define $R_S^0 = R$ and for $k > 0$, $R_S^k = R_S^{k-1}SR$. We have the following properties:*

1. $R_S = \bigcup_{k \geq 0} R_S^k$;
2. *is R is a symmetric relation on T and is S is a symmetric relation, then R_S is a symmetric relation;*
3. *is R is a transitive relation on T, then R_S is a transitive relation.*

Definition 6 *Let F be a binary relation on $\mathcal{D}$ a description space and let $\mathcal{G} = (\Omega, \text{ext})$ be a context for $\mathcal{D}$.*

We denote $F_{\mathcal{G}} = \{(a, b) \in F \mid \text{ext}(a) \neq \emptyset, \text{ext}(b) \neq \emptyset\}$ ($E_{\mathcal{G}}$ is defined in a similar way).

We denote $\widetilde{F}_{\mathcal{G}}$ the closure of $((F_{\mathcal{G}})^s)^+$ by S, where S is defined by:

$$S = \{(a, b) \in \mathcal{D} \mid \text{ext}(a) \cap \text{ext}(b) \neq \emptyset\} \tag{5}$$

$\widetilde{F}_{\mathcal{G}}$ *is a symmetric and transitive relation. Moreover, for each $a \in \mathcal{S}(F_{\mathcal{G}})$, $(a, a) \in \widetilde{F}_{\mathcal{G}}$, which means that $\widetilde{F}_{\mathcal{G}}$ is an equivalence relation on $\mathcal{S}(F_{\mathcal{G}})$.*

This closure is an useful tool for classification study: if we have $(a, b) \in F$, where F is the fusion part of a constraint set, this means that elements in $\text{ext}(a)$ and $\text{ext}(b)$ must be related by compatible binary relations. If $(c, d) \in F$, the same property is true for $\text{ext}(c)$ and $\text{ext}(d)$. Let us assume now that $\text{ext}(b) \cap \text{ext}(c) \neq \emptyset$, and let y be an element of this intersection. Let now x be an element of $\text{ext}(a)$ and z an element of $\text{ext}(d)$. If r is a compatible binary relation on Ω, we have $r(x, y)$ and $r(y, z)$. Therefore, if r is transitive (this is the case for classification, where r is an equivalence relation), we have $r(x, z)$. Therefore, r is also compatible with $F \cup \{(a, d)\}$ and we have, by construction, $(a, d) \in \widetilde{F}_{\mathcal{G}}$.

3.2   Constraint set coherency on a population

Definition 7 *Let (F, E) be a constraint set on $\mathcal{D}$ a description space and let $\mathcal{G} = (\Omega, \text{ext})$ be a context for $\mathcal{D}$. We say that (F, E) is **coherent on** $\mathcal{G}$, and we note $F \vartriangleleft_{\mathcal{G}} E$, if and only if the following conditions are satisfied:*

1. *for all $(a, b) \in E$, $\text{ext}(a) \cap \text{ext}(b) = \emptyset$;*
2. *for all $(a, b) \in F$ and $(c, d) \in E$, $\text{ext}(a) \cap \text{ext}(c) = \emptyset$ or $\text{ext}(b) \cap \text{ext}(d) = \emptyset$.*

We can now give our first result:

Theorem 1 *Let (F, E) be a constraint set on $\mathcal{D}$ a description space and let $\mathcal{G} = (\Omega, \text{ext})$ be a context for $\mathcal{D}$. The following properties are equivalent:*

1. *there is a binary equivalence relation on Ω compatible with (F, E);*
2. *$\widetilde{F}_{\mathcal{G}} \vartriangleleft_{\mathcal{G}} E$.*

Moreover, the smallest equivalence relation on Ω compatible with (F, E) can be defined as follows: $r(x, y)$ if and only if there are $(a, b) \in \widetilde{F}_{\mathcal{G}}$ with $x \in \text{ext}(a)$ and $y \in \text{ext}(b)$.

3.3   Discussion

The practical implications of theorem 1 are important. It builds an equivalence relation on $\mathcal{S}(F_{\mathcal{G}})$, a subset of the description space. This equivalence relation is used for two purposes:

1. it gives a simple criterion for the existence of a solution to the constrained clustering problem;
2. the smallest clustering is the "image" of the symbolic equivalence relation by ext.

In a database system, the advantages of this approach are obvious: rather than working on the global population, we simply have to compute intersection of extensions. As in general, computing an extension is simply a SQL SELECT query, it is straightforward to calculate the intersection query. Therefore we never need to extract the full population from the database, but only to make some queries to first build $\mathcal{S}(F_{\mathcal{G}})$ and then to check if $\widetilde{F}_{\mathcal{G}} \vartriangleleft_{\mathcal{G}} E$.

Whereas this approach gives results easier to interpret than the pure population based approach given in [2], it is still based on the population and results depend on the context $\mathcal{G}$.

4   Pure symbolic analysis

In this section, we use the symbolic order so as to provide a pure symbolic answer to the coherency problem: given a constraint set (F, E) on an ordered description space $(\mathcal{D}, \leq)$, can we prove that (F, E) is coherent enough so that for any consistent context $\mathcal{G}$, there will exist a compatible equivalence relation (i.e., a solution to the constrained clustering problem).

4.1   Description set based closure

Definition 8 *Let F be a binary relation on $(\mathcal{D}, \leq)$ an ordered description space. We call $\widetilde{F}$ the closure of $(F^s)^+$ by S, where S is defined by:*

$$S = \{(a, b) \in \mathcal{D} \mid \min(a, b) \neq \emptyset\} \tag{6}$$

$\widetilde{F}$ *is a symmetric and transitive relation. Moreover, for each $a \in \mathcal{S}(F)$, $(a, a) \in \widetilde{F}$, which implies that $\widetilde{F}$ is an equivalence relation on $\mathcal{S}(F)$.*

4.2   Constraint set coherency

Definition 9 *Let (F, E) be a constraint set on $(\mathcal{D}, \leq)$ an ordered description space. We say that (F, E) is **coherent on** $\mathcal{D}$, and we note $F \lhd E$, if and only if the following conditions are satisfied:*

1. *for all $(a, b) \in E$, $\min(a, b) = \emptyset$;*
2. *for all $(a, b) \in F$ and $(c, d) \in E$, $\min(a, c) = \emptyset$ or $\min(b, d) = \emptyset$.*

Theorem 2 *Let (F, E) be a constraint set on $(\mathcal{D}, \leq)$ an ordered description space such that any totally ordered subset of $\mathcal{D}$ is bounded below. The following properties are equivalent:*

1. *$\widetilde{F} \lhd E$*
2. *for each consistent context $\mathcal{G} = (\Omega, \text{ext})$, there exists a equivalence relation r on Ω compatible with (F, E)*

4.3   Discussion

The practical implications of theorem 2 are quite important:

1. the computation is done at a pure symbolic level, which reduces in general the cost compared to previous approaches (if the population is stored in a database, the pure symbolic analysis makes no access to this database);
2. if obtained, the coherency result applies to *any* consistent context, which avoids to assume that the studied population is exhaustive or fixed (for instance). Moreover, this allows to change the interpretation, if needed;

3. the equivalence relation built on $\mathcal{S}(F)$ does not give directly (through the interpretation mapping) the smallest compatible equivalence relation on a given context, but only one among many possible equivalence relations. Some clusters are constructed by the closure calculation, based on stability through min. On a particular context, it might happen that $\min(a, b) \neq \emptyset$ but that $\text{ext}(a) \cap \text{ext}(b) = \emptyset$. If the description space is well suited to describe the population, a possible interpretation is to say that the population is not exhaustive. For instance, if we consider *"Red or Blue"* and *"Red or Yellow"* descriptions, the description space will in general contain a *"Red"* description (this is for instance the case in the example defined in section 2.1). It might happen that the population does not contain a *"Red"* individual. At the symbolic level, such individual potentially exists, and **must** be taken into account. The equivalence relation induced by $\widetilde{F}$ take them into account and is therefore bigger than the smallest that might available. The symbolic approach shows therefore relationship that are hidden on a particular context.

5   Conclusion

In this paper, we have introduced a new approach for studying symbolic constrained classification. This approach allows to work both on a population and at a pure symbolic level. The symbolic tools give efficient algorithms (especially when the population is big) as well as easy result analysis.

The proposed symbolic approach has been implemented in the SODAS framework [1] and is currently being benchmarked against the population based approach.

References

1. Hans-Hermann Bock and Edwin Diday, editors. *Analysis of Symbolic Data. Exploratory methods for extracting statistical information from complex data.* Springer Verlag, 2000.
2. Roland De Guio, Thierry Erbeja, and Vincent Laget. A clustering approach for GT family formation problems. In *1st international conference on Engineering Design and Automation*, pages 18–21, March 1997.
3. Edwin Diday. L'analyse des données symboliques : un cadre théorique et des outils. Technical Report 9821, LISE/CEREMADE (CNRS UMR 7534), Université Paris-IX/Dauphine, Mars 1998.
4. A. D. Gordon. A survey of constrained classification. *Computational Statistics & Data Analysis*, 21:17–29, 1996.
5. Amedeo Napoli. Une introduction aux logiques de descriptions. Technical Report 3314, Projet SYCO, INRIA Lorraine, 1997.
6. Fabrice Rossi and Frédérick Vautrain. Constrained classification. Technical report, LISE/CEREMADE (CNRS UMR 7534), Université Paris-IX/Dauphine, April 2000.

An Application of Association Rules Discovery to Geographic Information Systems

Ansaf Salleb and Christel Vrain

LIFO - Université d'Orléans - BP 6759
45067 Orléans Cedex 2 - France
{salleb,cv}@lifo.univ-orleans.fr

Abstract. In this paper, we are interested in the problem of extracting spatial association rules in Geographic Information Systems (GIS). We propose an algorithm that extends existing methods to deal with spatial and non-spatial data over multiple layers. It handles hierarchical, multi-valued attributes, and produces general spatial association rules. We also present a prototype, which has been applied on a real and large geographic database in the field of mineral exploration.

1 Introduction

In this paper, we are interested in Spatial Data Mining [4,6,8], which leads to practical applications in many domains, such as geographic information systems (GIS). We present an application of mining association rules between geographic layers according to spatial and non-spatial properties of the objects.

The concept of association rules introduced by Agrawal [1] has been extended by Koperski and Han [7] to Spatial Data. For instance, the rule:

$Is(X, largetown) \wedge Intersect (X, highway) \rightarrow AdjacentTo(X, water)$ (86%)

expresses that 86 % of the large towns intersecting highways are also adjacent to water areas (rivers, lakes, ...).

They have proposed an algorithm to discover such rules. In their works, non-spatial information is organized into hierarchies, and rules including this information are learned. Nevertheless, information belonging to different levels of hierarchies cannot appear in a same rule. We have extended their framework to introduce non-spatial predicates in rules, even when attributes are not hierarchical, and we have applied it to a real application provided by BRGM [1].

The application we are interested in is explained in Section 2 and formalized in Section 3. In Section 4, we present an algorithm for extracting spatial association rules in GIS and a prototype that has been developed and applied on geographic data provided by BRGM. We conclude with a discussion and perspectives, in Section 6.

---

[1] BRGM ('Bureau des Recherches Géologiques et Minières') is a French public institution, present in all regions of France and in some 40 countries abroad. Based on the Earth Sciences, its expertise relates to mineral resources, pollution, risks, and the management of geological environment (http://www.brgm.fr).

D.A. Zighed, J. Komorowski, and J. Żytkow (Eds.): PKDD 2000, LNAI 1910, pp. 613–618, 2000.
© Springer-Verlag Berlin Heidelberg 2000

2   The Application

A GIS stores data in two distinct parts: spatial data representing an object by its geometric shape (a point, a line or a polygon) and its location on the map, and non-spatial data storing properties of the geographic objects. Moreover, a GIS uses the notion of thematic layers in order to separate data into classes of same kinds. Each thematic layer is described by a relational table where each tuple is linked to a geometric object. For instance, the GIS Andes developed by BRGM [2] is a homogeneous information system of the entire Andes Cordillera. There are different layers in the system, which can be combined in any way by the user (Figure 1). Our goal is to find association rules between a given thematic layer and others layers, according to spatial relations and non-spatial attributes.

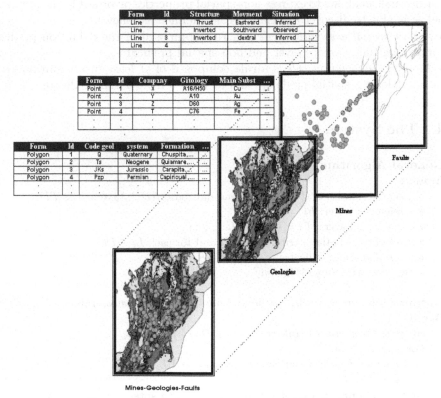

Figure 1. Three layers of an extract of the Andes

3   Problem Formalization

We define two kinds of layers: **the Reference Layer** is the thematic layer which contains the spatial objects we focus on. It is unique, and in the following, the variable x is associated to the Reference Layer. **A Descriptive Layer** contains objects that have spatial relations with those of the reference layer.

In our work, a spatial association rule is a rule like:
$$P_1, P_2, ..., P_n \rightarrow Q_1, Q_2, ..., Q_m$$
where P_i, Q_j, are **predicates**. A predicate may be:

1. a definition predicate: It defines a thematic layer, its syntax is:
Definition_Predicate(var). For instance, in the example of Section 2, *Mine(x)* is a definition predicate for the layer Mines.

2. a spatial predicate: It links the reference layer with a descriptive layer. Its general syntax is: *Spatial_Predicate(var_i, var_j)*.
For instance, *Near_to(x,z)* denotes the proximity between x and z.

3. a non-spatial predicate: It describes a layer, its general syntax is:
NonSpatial_Predicate (var, Value). In our example, *Main_Substance(x, Ag)* expresses that the mine x has *silver* as main substance.

Our formalism allows to express non-spatial properties on objects. The algorithm below is able to discover association rules defined as follows:

• A *non spatial association rule* is a rule composed of the definition predicate concerning the reference layer and non-spatial predicates.

• A *spatial association rule* is a rule composed of at least two definition predicates, at least a spatial predicate and at least a non-spatial predicate.

4   The System

General Algorithm
Inputs:

> • a spatial database SDB, a relational database RDB, taxonomies TDB.
> • a reference Layer: Rl
> • descriptive Layers: $Dl = \{L_i/i = 1, ..., n\}$
> • a set of non-spatial attributes for Rl and for each $L_i \in Dl$
> • a set of spatial predicates to compute: SP
> • $Buffer, MinSupp, MinConf$

Outputs: solid, multi-levels, spatial and non-spatial association rules.
Begin

> **Step 1: Creation of link tables:** $LDB = \bigcup_{i=1}^{n} LDB_i$
> For each $L_i \in Dl$
>> • search Spatial Relations between Rl and L_i
>> • update LDB_i
>
> **Step 2: Computation of frequent predicate sets**
> • Creation of sets of predicates
> • For each example E_j of LDB
>> • Search the predicates sets that are true for E_j and increase their support
>
> • Keep predicate sets that have sufficient support
> **Step 3: Generation of strong Spatial Association Rules**

End

Details of the Algorithm
Let us notice first that each spatial object in a given layer has a unique key or

identifier. The algorithm is based on the building of *link tables*, each one relating the *reference layer* to a *descriptive layer*. The structure of a link table is composed of the following fields: the reference layer key, non-spatial attributes of the reference layer, a spatial relation linking an object of the *reference layer* and an object of the *descriptive layer*, the *descriptive layer* key, and non-spatial attributes of the descriptive layer. An *example E_j* is a set of tuples, *belonging to different link tables*, that have the same reference layer key j [3].

The support of a rule represents the percent of examples verifying all the predicates of the rule at the same time. A rule holds with confidence $c\%$, if $c\%$ of the examples of the *Reference layer* verifying the predicates given in the condition of the rule, also satisfy the predicates of the consequence of the rule.

Stop 1 creates a link table LDB_i between the reference layer Rl and each descriptive layer of Dl, by searching for each object O_j of Rl, and for each layer of Dl, the objects O verifying at least a spatial predicate $sp(O_j, O)$. Then, frequent predicate sets are computed. For each link table LDB_i, a conjunction composed of a single predicate is created, as follows:

• Each non-spatial attribute of Rl (resp. L_i in the LDB_i) becomes a non-spatial binary predicate with variable x (resp. y_i).

• The set of constants that can appear in the second argument of the predicates are computed as follows:

— For each non-hierarchical attribute, we extract in the LDB_i its set of values.

— For each hierarchical attribute, values are obtained from its taxonomy.

• To each possible value of an attribute corresponds a predicate with this value as constant (second argument of the predicate, see Section 3).

• Each variable y_i must be linked to x by a spatial predicate (because the support concerns only the reference layer).

• The support of each predicate in the LDB_i is computed, and frequent predicates are kept in a list L.

• As in *apriori*, predicate sets of size k are generated by combination of frequent predicate sets of size (k-1), and only those with a sufficient support are kept.

Step 3 is a classical process for generating association rules from the set L.

Example: Let us suppose that we specify in the inputs the following parameters:

• Reference layer Rl = *Mines*, with non-spatial attributes: {*Gitology, Main_Subst*}. Note that Gitology is a hierarchical attribute (it takes its possibles values from the nodes of its taxonomy: $A_1, A_2, A_3, \dots$).

• Descriptive layer L_1 = *Geologies*, with attributes: {*System, Code_Geol*}.

• $Sp = \{$*Included_in, Near_to*$\}$, Buffer = 10 km, MinSupp=5% and MinConf=40%

We aim at finding relations between *Mines* deposits, represented by points, and the nearest *Geologies* (polygons).

The link table LDB_1 is constructed, as shown in Figure 2. For instance, the example E_j is composed of two tuples of *key=2* in the link table and means that the mine number 2 is included in Geology 102 and near to Geology 3.

The support table of large predicate sets is then built. Note here that the predicate sets with frequency less than 130 (which represents 5% of 2618 mines in the Rl) are filtered out (Figure 3).

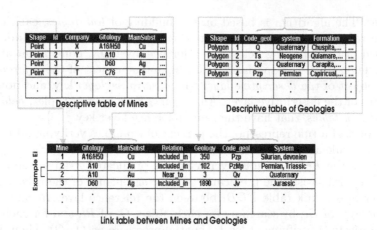

Figure 2. Link table of the example

Num	Frequent Predicate Sets	Frequency
1	Gitology (x, A)	533
2	Gitology (x, A1)	491
3	Main_substance (x, Ag)	529
4	Main_substance (x, Au)	459
5	Geologic_Code (y, Pzs), Is_inside(x,y)	384
6	System (y, Ordovician), Is_inside(x,y)	214
7	Gitology (x, A), Gitology (x, A1)	491
8	Gitology (x, A1), Main_substance (x, Ag)	180
9	Gitology (x, A1), Geologic_Code (y, Pzs), Is_inside(x,y)	177
10	Geologic_Code (y, Pzs), System (y, Ordovician), Is_inside(x,y)	209

Figure 3. An extract of the support table

Based on the support table, the algorithm generates the following rule with a support of 6.67%:

$$Mine(x) \wedge Geology(y) \wedge Code\_Geol(y,Pzs) \wedge Included\_In(x,y)$$
$$\rightarrow Gitology(x,A1) \quad (6.67\%, 46.09\%)$$

which expresses that 46.09% of mines included in geologies of code *Pzs* have as gitology *A1*. We can classify generated association rules into three kinds:

- **Statistic rules**: they give the repartition of items in a hierarchy, such as:
$$Mine(x) \wedge Gitology(x, A) \rightarrow Gitology(x, A1) \quad (92.12\%)$$

- **Control rules**: experts can also verify some known correlations:
$$Mine(x) \wedge Gitology(x, H12) \rightarrow MainSubstance(x, AU) \quad (89.32\%)$$

- **New rules**: as for instance, the following rule with confidence 43.75%:
$$Mine(x) \wedge Fault(z) \wedge Gitology(x, C5) \wedge Near\_to(x, z) \rightarrow Structure(z, Strike\_slip)$$

We have implemented the algorithm presented in this paper, as a research prototype, named ARGIS. It has been developed in Avenue©, an object oriented language available in ArcView©[2]. In ARGIS, the user can formulate the inputs by means of an interactive graphical user interface. ARGIS handles multi-valued attributes and taxonomies, and the user can choose the levels that interest him.

---

[2] A GIS developed and commercialized by ESRI

We have experimented the prototype on 3 layers of GIS Andes, a database of about 150 mega bytes (15 MB for the Mines layer, 130 MB for the Geology layer and 5 MB for the Fault layer) composed of about 23 thousands of records, each time using a reference layer and a descriptive layer. The process has discovered about 70 spatial association rules, and some have been considered as very interesting by the experts.

5   Discussion and Conclusion

The system presented in this paper is an extension of a previous work of Koperski and Han [7], and an application of mining spatial association rules in GIS. First we have added non-spatial predicates to spatial ones. Second, as suggested by Koperski in [5], we can mine rules at cross-levels of taxonomies. However, to get interesting rules at low levels of the hierarchies, the support must be low. This leads to a lot of rules because we cannot give a variable support threshold according to the level in the hierarchies as done by Koperski, whose method is guided by a scroll of all the hierarchies at the same time. In order to handle a large number of layers, and in view of the phenomenal growth of data, the efficiency of the system must be improved in the following directions: pruning useless and redundant rules, interaction with statistics, dealing with numeric data, parallelization and use of multidimensional trees.

Acknowledgments: We wish to thank the BRGM, especially the Geology and Metallogeny Research group for providing us the opportunity to experiment our algorithm on their databases. **This work is supported by BRGM and by 'Région Centre'.**

References

1. R. Agrawal and R. Srikant. Fast Algorithms for Mining Association Rules in Large Databases. In *Proc. 1994 Int. Conf. VLDB*, pages 487–499, 1994.
2. D. Cassard. Gis andes: A metallogenic gis of the andes cordillera. In *4th Int. Symp. on Andean Geodynamics*, pages 147–150. IRD Paris, October 1999.
3. L. Dehaspe and L. De Raedt. Mining association rules in multiple relations. In *Proc. of the 7th Int. Workshop on ILP*. Springer, 1997.
4. M. Ester, H.-P. Kriegel, and J. Sander. Spatial data mining: A database approach. *Lecture Notes in Computer Science*, 1262:47–66, 1997.
5. K. Koperski. *A progressive Refinement Approach To Spatial Data Mining*. PhD thesis, Computing Science, Simon Fraser University, 1999.
6. K. Koperski, J. Adhikary, and J. Han. Spatial data mining: Progress and challenges. In *SIGMOD'96 Workshop.DMKD'96, Canada*, June 1996.
7. K. Koperski and J. Han. Discovery of spatial association rules in geographic information databases. *Lecture Notes In Computer Science*, 951:47–66, 1995.
8. J. F. Roddick and M. Spiliopoulou. A bibliography of temporal, spatial and spatio-temporal data mining research. In *SIGKDD Explorations*, June 1999.

Generalized Entropy and Projection Clustering of Categorical Data

Dan A. Simovici, Dana Cristofor, and Laurentiu Cristofor

University of Massachusetts at Boston,
Department of Mathematics and Computer Science,
Boston, Massachusetts 02446, USA,
{dsim,dana,laur}@cs.umb.edu

Abstract. We generalize the notion of entropy for a set of attributes of a table and we study its applications to clustering of categorical data. This new concept allows greater flexibility in identifying sets of attributes and, in a certain case, is naturally related to the average distance between the records that are the object of clustering. An algorithm that identifies clusterable sets of attributes (using several types of entropy) is also presented as well as experimental results obtained with this algorithm.

1 Introduction

In this paper we investigate clustering on categorical data, a problem that poses challenges that are quite distinct from numerical data clustering (see [AGGR98] and [CFZ99]). This type of data can be encoded numerically; however, this introduces artificial total orderings on the domains of the attributes involved which have the potential to create artificial agglomerations that are meaningless from a practical point of view. Our approach to clustering is based on identifying projections that have sufficient concentrations of values (determined by certain types of entropies associated to attribute sets). The sets of records that are formed by grouping on these sets of attributes qualify as clusters if they do not differ too much on the remaining attributes; this requirement can be enforced by limiting the value of a certain generalization of conditional entropy. Our work extends the information-theoretical approach to clustering found in [CFZ99].

Let $\mathbb{R}$ be the set of reals. The k-dimensional simplex is the set $\mathsf{SIMPLEX}_{k-1} = \{(p_1, \ldots, p_k) \in \mathbb{R}^k \mid p_i \geq 0 \text{ and } p_1 + \cdots + p_k = 1\}$. A function $f : \mathbb{R} \longrightarrow \mathbb{R}$ is *concave on a set* $S \subseteq \mathbb{R}$ if $f(\alpha x + (1 - \alpha)y) \geq \alpha f(x) + (1 - \alpha)f(y)$ for $\alpha \in [0, 1]$ and $x, y \in S$. The function f is *sub-additive* (*supra-additive*) on S if $f(x + y) \leq f(x) + f(y)$ $(f(x+y) \geq f(x) + f(y))$ for $x, y \in S$. In [CHH99] a concave impurity measure is defined as a concave real-valued function $i : \mathsf{SIMPLEX}_{k-1} \longrightarrow \mathbb{R}$ such that $i(\alpha\mathbf{p} + (1 - \alpha)\mathbf{q}) = \alpha i(\mathbf{p}) + (1 - \alpha)i(\mathbf{q})$ implies $\mathbf{p} = \mathbf{q}$ for $\alpha \in [0, 1]$ and $\mathbf{p}, \mathbf{q} \in \mathsf{SIMPLEX}_{k-1}$ and, if $\mathbf{p} = (p_1, \ldots, p_k)$, then $i(\mathbf{p}) = 0$ if $p_i = 1$ for some i, $1 \leq i \leq k$.

The corresponding *frequency-weighted impurity measure* is the real-valued function $I : \mathbb{N}^k \longrightarrow \mathbb{R}$ given by $I(n_1, \ldots, n_k) = Ni(n_1/N, \ldots, n_k/N)$, where

D.A. Zighed, J. Komorowski, and J. Żytkow (Eds.): PKDD 2000, LNAI 1910, pp. 619–625, 2000.
© Springer-Verlag Berlin Heidelberg 2000

$N = \sum_{i=1}^{k} n_i$. Both the Gini impurity measure and the Shannon entropy can be generated using a simple one-argument concave function $f : [0,1] \longrightarrow \mathbb{R}$ such that $f(0) = f(1) = 0$. In this paper we additionally require two extra conditions which are satisfied by many functions.

Definition 1. *A concave function $f : [0,1] \longrightarrow \mathbb{R}$ is a generator if $f(0) = f(1) = 0$, f is subadditive and for $(a_1, \ldots, a_n) \in \mathsf{SIMPLEX}_{n-1}$ and $k \in [0,1]$ we have $f(ka_1) + \cdots + f(ka_n) \leq k(f(a_1) + \cdots + f(a_n)) + f(k)$.*

The monogenic impurity measure *induced by the generator f is the impurity measure generated by concave impurity measure i of the form $i(p_1, \ldots, p_k) = f(p_1) + \cdots + f(p_k)$, where $(p_1, \ldots, p_k) \in \mathsf{SIMPLEX}_{k-1}$.*

It is easy to verify that such functions as $f_{\mathrm{gini}}(p) = p - p^2$, $f_{\mathrm{sq}}(p) = \sqrt{p} - p$, $f_{\mathrm{shannon}}(p) = -p \log p$, or f_{peak} (given by $f_{\mathrm{peak}}(p) = p$ for $0 \leq p \leq 0.5$ and $f_{\mathrm{peak}}(p) = 1 - p$ for $0.5 < p \leq 1$) satisfy the conditions that define generators. Since f is concave it satisfies Jensen's inequality $f(p_1) + \cdots + f(p_k) \leq k f\left(\frac{1}{k}\right)$ for every $(p_1, \ldots, p_k) \in \mathsf{SIMPLEX}_{k-1}$ and this implies that the largest value of the sum $f(p_1) + \cdots + f(p_k)$ is achieved if and only if $p_1 = \cdots = p_k = \frac{1}{k}$. Therefore, for the monogenic impurity measure generated by the function f we have $0 \leq i(p_1, \ldots, p_k) \leq k f(\frac{1}{k})$ for $(p_1, \ldots, p_k) \in \mathsf{SIMPLEX}_{k-1}$.

For relational terminology and notations see, for example, [ST95].

2 Impurity of Sets and Partitions

We introduce the notion of impurity of a subset of a set S relative to a partition. In turn, this notion is used to define the impurity of a partition relative to another partition.

Definition 2. *Let f be a generator, S be a set and let $\mathsf{PART}(S)$ be the set of all partitions of S. The impurity of a subset L of S relative to a partition $\pi \in \mathsf{PART}(S)$ is the monogenic impurity measure induced by f:*

$$\mathsf{IMP}_{\pi}^{f}(L) = |L| \left(f\left(\frac{|L \cap B_1|}{|L|}\right) + \cdots + f\left(\frac{|L \cap B_n|}{|L|}\right) \right),$$

where $\pi = \{B_1, \ldots, B_n\}$. The specific impurity *of L relative to π is*

$$\mathsf{imp}_{\pi}^{f}(L) = \frac{\mathsf{IMP}_{\pi}^{f}(L)}{|L|} = f\left(\frac{|L \cap B_1|}{|L|}\right) + \cdots + f\left(\frac{|L \cap B_n|}{|L|}\right).$$

Jensen's inequality implies that the largest value of the impurity of a set $L \subseteq S$ relative to a partition $\pi = \{B_1, \ldots, B_n\}$ of S is $|L| \cdot n \cdot f(\frac{1}{n})$.

Theorem 1. *Let K, L be two disjoint subsets of the set S and let $\pi \in \mathsf{PART}(S)$. Then, we have $\mathsf{IMP}_{\pi}^{f}(K \cup L) \geq \mathsf{IMP}_{\pi}^{f}(K) + \mathsf{IMP}_{\pi}^{f}(L)$.*

Corollary 1. *If K, L are subsets of S such that $K \subseteq L$ and $\pi \in \mathsf{PART}(S)$, then $\mathsf{IMP}_{\pi}^{f}(K) \leq \mathsf{IMP}_{\pi}^{f}(L)$.*

Theorem 2. *Let $\pi = \{B_1, \ldots, B_n\}$ and $\sigma = \{C_1, \ldots, C_m\}$ be two partitions of a set S. If $\pi \leq \sigma$, then $\mathsf{IMP}_{\sigma}^{f}(K) \leq \mathsf{IMP}_{\pi}^{f}(K)$ for every subset K of S.*

3 f-Entropy

Let $\tau = (T, H, \rho)$ be a table. For $X \subseteq H$ define the equivalence $\equiv_X$ by $u \equiv_X v$ if $u[X] = v[X]$; denote the corresponding partition of the set ρ of tuples by π_X. Every block of π_X corresponds to a distinct value of $\mathsf{aDom}_\tau(X)$. Therefore, $|\pi_X| = |\mathsf{aDom}_\tau(X)|$. If U, V are two subsets of H such that $U \subseteq V$, then $\pi_V \le \pi_U$.

Definition 3. *Let $\tau = (T, H, \rho)$ be a table and let X, Y be sets of attributes, $X, Y \subseteq H$. The f-entropy of X is $\mathcal{H}^f(X) = \mathsf{imp}^f_{\pi_X}(\rho)$.*

The conditional f-entropy *of X on Y is $\mathcal{H}^f(X|Y) = \sum_{j=1}^m \frac{|C_j|}{|\rho|}\mathsf{imp}^f_{\pi_X}(C_j)$, where $\pi_Y = \{C_1, \dots, C_m\}$.*

In other words, the conditional entropy $\mathcal{H}^f(X|Y)$ is the average value of the specific impurity of the blocks of the partition π_Y relative to the partition π_X.

Since $\pi_\emptyset = \{\rho\}$, it follows that $\mathcal{H}^f(X) = \mathcal{H}^f(X|\emptyset)$ for every set of attributes X. Therefore, we have $\mathcal{H}^f(X) = \sum_{i=1}^n f\left(\frac{|B_i|}{|\rho|}\right)$, where $\pi_X = \{B_1, \dots, B_n\}$. For the conditional entropy we have $\mathcal{H}^f(X|Y) = \frac{1}{|\rho|}\sum_{j=1}^m |C_j| \sum_{i=1}^n f\left(\frac{|B_i \cap C_j|}{|C_j|}\right)$. The concavity of the generator f implies that the largest value of the f-entropy of a set of attributes X occurs when the blocks of the partition π_X are of equal size. Therefore, we have $\mathcal{H}^f(X) \le n f(\frac{1}{n})$, where n is the number of blocks of π_X, that is, the cardinality of the active domain of X.

Theorem 3. *Let X, X', Y, Y' be sets of attributes of the table $\tau = (T, H, \rho)$. If $X \subseteq X'$ and $Y \subseteq Y'$, then $\mathcal{H}^f(X|Y) \le \mathcal{H}^f(X'|Y)$, $\mathcal{H}^f(X|Y) \ge \mathcal{H}^f(X|Y')$ and $\mathcal{H}^f(X) \le \mathcal{H}^f(X')$ for every generator f.*

Theorem 4. *Let X, Y be two sets of attributes of a table $\tau = (T, H, \rho)$. We have $\mathcal{H}^f(X|Y) \le \mathcal{H}^f(X)$ and $\mathcal{H}^f(XY) \le \mathcal{H}^f(X|Y) + \mathcal{H}^f(Y)$.*

Corollary 2. *For every sets of attributes X, Y we have $\mathcal{H}^f(XY) \le \mathcal{H}^f(X) + \mathcal{H}^f(Y)$.*

4   Clusters for Categorical Data

Let $\tau = (T, H, \rho)$ be a table, X be a set of attributes $X \subseteq H$ and assume that $\mathsf{aDom}_\tau(X)$ consists of N elements, $\mathsf{aDom}_\tau(X) = \{x_1, \dots, x_N\}$. Each block B_i of π_X corresponds to a value x_i of the active domain of X for $1 \le i \le N$, and x_i appears under X with the frequency $\frac{|B_i|}{|\rho|}$ for $1 \le i \le N$.

Let $d_{\tau,X}$ be a distance defined on $\mathsf{aDom}_\tau(X)$. The average distance between the elements of $\mathsf{aDom}_\tau(X)$ is $E(d_\tau(X)) = \frac{\sum_{i=1}^N \sum_{j=1}^N \{d_{\tau,X}(x_i, x_j)|B_i|\cdot|B_j|\}}{|\rho|^2}$. The next result connects one of the generalizations of entropy with the average distance and serves as a foundation for the applications of information-theoretical methods to clustering.

Let $d_{\tau,X}$ be the distance on $\mathsf{aDom}_\tau(X)$ defined by: $d_{\tau,X}^{0,1}(x,x') = 0$ if $x = x'$; otherwise, $d_{\tau,X}^{0,1}(x,x') = 1$ for $x, x' \in \mathsf{aDom}_\tau(X)$.

Theorem 5. *The average distance $E(d_{\tau,X}^{0,1})$ equals $\mathcal{H}^{f_{gini}}(X)$.*

Theorem 5 suggests that by limiting the entropy of a set of attributes X we limit the average distance between the projections of the tuples of the table on X and, therefore, increase our chances to find large clusters of tuples based on their similarity on X. A projection of a table τ on a set of attributes X presents an interest for us if it satisfies two criteria:

1. The f-entropy $\mathcal{H}^f(X)$, referred to as the *internal entropy* of X is limited. This insures that the values X-projections congregate towards a small number of clusters.
2. The impurity of each of the blocks of the partition π_X relative to the partition π_{H-X} is small. This insures that the tuples created by clustering the X-projection do not diverge excessively on the remaining attributes of the table. The definition of conditional f-entropy shows that such sets of attributes can be identified by limiting the value of the conditional entropy $\mathcal{H}^f(H - X|X)$, referred to as the *external* entropy of X is limited.

Definition 4. *Let $\tau = (T, H, \rho)$ be a table. An (f, α)-clusterable set of attributes in τ is a set $X \subseteq H$ such that $\mathcal{H}^f(X) \leq \alpha$ and X is maximal with this property, that is, $X \subset Y$ implies $\mathcal{H}^f(Y) > \alpha$.*

The clusters determined by the (f, α)-clusterable set X are the groups of records that have equal values on X.

Next, we present an algorithm that can be used to identify clusterable sets of attributes. The algorithm is based on the monotonicity of the f-entropy shown in Corollary 1. Namely, if U is a set of attributes such that $\mathcal{H}^f(U) > \alpha$, then for every set V such that $U \subset V$ we have $\mathcal{H}^f(V) > \alpha$. Thus, once a set U is disqualified from being (f, α)-clusterable, no superset of U can be (f, α)-clusterable. Of course, if U, U' are (f, α)-clusterable, the set UU' is a candidate for the collection KAttrSets and we apply the apriori-gen technique from [AMS+96] for idenfifying those sets that can be included in KAttrSets. If $X \subseteq X'$, then, by Theorem 3, we have $\mathcal{H}^f(H - X|X) \geq \mathcal{H}(H - X'|X')$. Thus, to limit the external entropy, we focus on the maximal (f, α)-clusterable sets of attributes.

We use three collections of sets of attributes: Candidates, KAttrSets and MaxAttrSets. At the completion of the algorithm shown in Figure 1 the collection MaxAttrSets will contain the (f, α)-clusterable sets. The inputs of the algorithm are the generator function f and the value α representing the limit of internal entropy. There are N attributes in our dataset. The external entropies of the sets of MaxAttrSets are also computed and we sort the maximal (f, α)-clusterable sets on their external f-entropies.

The algorithm was run on a database extracted from the "agaricus-lepiota" dataset from UCI (http://www.ics.uci.edu/AI/ML/Machine-Learning.html) by

```
Candidates = empty; KAttrSets = empty; MaxAttrSets = empty;
add to Candidates all one-attribute sets;
for step = 1 to N do
  scan dataset and compute for each set S from Candidates
    the internal entropy of S;
  for each set of attributes S from Candidates do
    if H(S) <= alpha do
      add S to KAttrSets;
      add S to MaxAttrSets;
    endif;
  apriori_gen(Candidates, KAttrSets);
  if Candidates is empty
    break;
endfor;
remove non maximal sets from MaxAttrSets;
```

Fig. 1. Algorithm for Finding the (f, α)-clusterable sets

retaining the first 15 attributes of a table documenting 23 attributes for a number of 8124 types of mushrooms. We experimented using three different function generators for different values of α: f_{shannon}, f_{gini}, and f_{sq}. The results are summarized below.

α	f_{gini}		α	f_{gini}		α	f_{shannon}		f_{sq}	
	time	sets		time	sets		time	sets	time	sets
0.5	1	5	0.92	59	92	1	1	5	3	14
0.6	1	8	0.94	100	140	2	4	19	13	58
0.7	3	16	0.96	231	156	4	56	140	97	233
0.8	8	33	0.98	573	165	6	519	225	265	311
0.9	42	93				8	1541	110	527	311

Elementary computations show that if $0 \leq p \leq 0.025$, then $f_{\text{gini}}(p) \leq f_{\text{shannon}}(p) \leq f_{\text{sq}}(p)$; for $0.025 < p \leq 0.39$ we have $f_{\text{gini}}(p) \leq f_{\text{sq}}(p) \leq f_{\text{shannon}}(p)$. Finally, if $0.39 < p \leq 1$, we have $f_{\text{sq}}(p) \leq f_{\text{gini}}(p) \leq f_{\text{shannon}}(p)$. Thus, f_{sq} "penalizes" the sets of attributes whose partitions contain small blocks (containing fewer than 2.5% of the set of all tuples); such sets are not good candidates for defining clusters of records. It is easy to see that if an attribute set partitions a set of tuples in k blocks, then its maximum entropy would be $1 - \frac{1}{k}$ for the f_{gini}-entropy, $\log k$ for the f_{shannon}-entropy, and $\sqrt{k} - 1$ for the f_{sq}-entropy. Considering this, by specifying a limit for the entropy of an attribute, we can get an approximative idea of how many large blocks its partition will have.

As an example of clusters determined by an (f, α)-clusterable set of attributes, consider the attribute set *bruises gill-attachment gill-spacing* obtained when we used f_{shannon} with $\alpha = 2$, f_{gini} with $\alpha = 0.7$, and f_{sq} with $\alpha = 1$ and had a minimal external f_{sq}-entropy. The clusters formed by grouping on these attributes contain 210, 3330, 1208, 3272 and 104 records, respectively, as shown in Figure 2.

Set of attributes			Block B		
bruises	gill-attachment	gill-spacing	count	external clustering	$\mathsf{imp}^{f_{\mathrm{sq}}}_{\pi_{H-X}}(B)$
no	attached	close	210	12 × 16 tuples 18 × 1 tuple	3.554
no	free	close	3330	72 × 18 tuples 228 × 6 tuples 48 × 4 tuples 48 × 2 tuples 18 × 1 tuple	18.690
no	free	crowded	1208	216 × 4 tuples 144 × 2 tuples 56 × 1 tuple	18.899
yes	free	close	3272	144 × 12 tuples 120 × 8 tuples 72 × 4 tuples 148 × 2 tuples	19.830
yes	free	crowded	104	48 × 2 tuples 8 × 1 tuples	6.440

Fig. 2. Clusters Formed by Grouping

It is clear that the group with minimal impurity is obtained for *bruises* = 'no', *gill-attachment* = 'attached', and gill-spacing = 'closed'.

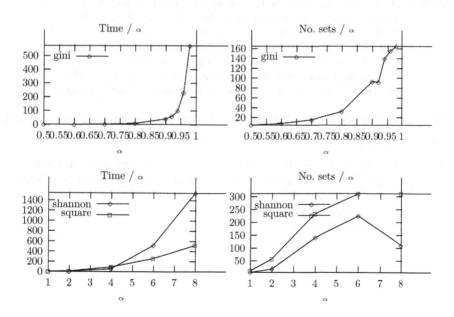

5 Open Problems

Our technique for identifying clusters could be improved by performing an initial step of eliminating all attributes that have a high f-entropy. These attributes can be considered as "noise" for the purpose of clustering, so by ignoring them we would clean the data and would make the values of external entropy more meaningful.

An investigation of classes of generator functions is necessary to determine the types of clusterable sets that are "favored" by these classes of generators.

The definition of clusters as groups of records determined by certain sets of attributes that have limited external impurities must be further refined and techniques that allow the user to choose a covering of the set of records by (f, α)-clusters must be developed.

Also, an important open question is the axiomatization of the notion of f-entropy of an attribute set. Such an axiomatization should relax known axiomatizations of the notion of entropy (see [JS99]) to accomodate the generalizations discussed in this paper.

References

[AGGR98]   R. Agrawal, J. Gehrke, D. Gunopulos, and P. Raghavan. Automatic subspace clustering for high dimensional data for data mining applications. In *Proceedings of the 1998 ACM International Conference on Management of Data*, pages 94–105, 1998.

[AMS⁺96]   R. Agrawal, H. Mannila, R. Srikant, H. Toivonen, and A. Inkeri Verkamo. Fast discovery of association rules. In U. M. Fayyad, G. Piatetsky-Shapiro, P. Smyth, and R. Uthurusamy, editors, *Advances in Knowledge Discovery and Data Mining*, pages 307–328. AAAI Press, Menlo Park, 1996.

[CFZ99]   C. H. Cheng, A. W. Fu, and Y. Zhang. Entropy-based subspace clustering for mining numerical data. In *Proceedings of the Fifth ACM SIGKDD International Conference on Knowledge Discovery and Data Mining*, pages 84–93, 1999.

[CHH99]   D. Coppersmith, S.J. Hong, and J.R.M. Hosking. Partitioning nominal attributes in decision trees. *Data Mining and Knowledge Discovery*, 3:197–217, 1999.

[JS99]   S. Jaroszewicz and D. A. Simovici. On axiomatization of conditional entropy. In *Proceedings of the 29th International Symposium for Multiple-Valued Logic*, pages 24–31, Freiburg, Germany, 1999.

[ST95]   D. A. Simovici and R. L. Tenney. *Relational Database Systems*. Academic Press, New York, 1995.

Supporting Case Acquisition and Labelling in the Context of Web Mining

Vojtěch Svátek and Martin Kavalec

Department of Information and Knowledge Engineering
and Laboratory of Intelligent Systems,
University of Economics, Prague, 13067 Praha 3, Czech Republic
{svatek|xkavm04}@vse.cz

Abstract. Case acquisition and labelling are important bottlenecks for predictive data mining. In the web context, a cascade of supporting techniques can be used, from general ones such as user interfaces, through filtering based on keyword frequency, to web-specific techniques exploiting public search engines. We show how a synergistic application of multiple techniques can be helpful in obtaining and pre-processing textual data, in particular for ILP-based web mining. The (two-fold) learning task itself consist in construction and disambiguation of categorisation rules, which are to process the results returned by web search engines.

1 Introduction

Most research efforts in the data mining community is concentrated on algorithms for *discovering* regularities in data. Comparably less attention is paid to the problems related to *acquisition* of this data, and their rendering to the form required by the learning algorithm. The characteristic feature of *supervised predictive learning* (which is topical for this paper) is the existence of a distinct goal (class) attribute, the values of which are assumed to be given, for each learning case. If the underlying reasoning task (classification) is novel in the sense that there are no historical data including the "real" class (such as the results of past loan contracts, in the credit-risk assignment task), then the actual mining process has to be preceded by a (usually) manual, dedicated process of class assignment, also denoted as *labelling*. Unfortunately, this process may present an important bottleneck in the overall data mining task. A specific form of the classification task is *document classification*, where, instead of explicit attributes, we can (and have to) deal with sequences of words and other symbols. If we refine the task even further, we can proceed to the classification of *web pages*. This leads us into an enormous space of documents, representing potential data usable for mining. However, the adequate documents are typically dispersed over many servers, and their properties are to a great extent unpredictable.

In this paper, we present several techniques that have been used to support data acquisition and labelling for a data mining task in the context of web search. In section 2 we present our view of the document categorisation problem, with

D.A. Zighed, J. Komorowski, and J. Żytkow (Eds.): PKDD 2000, LNAI 1910, pp. 626–631, 2000.
© Springer-Verlag Berlin Heidelberg 2000

respect to web documents, and in particular for the case of limited input data. In section 3, we describe the individual case acquisition and labelling techniques used as preprocessing within our web-mining tasks. In section 4, we show some preliminary results of the most complex mining task, obtained via relational (ILP) learning. Finally, we review some related work (section 5) and outline perspectives for the future (section 6).

2   Document Categorisation in the Web Context

Document categorisation is often understood as assignment of (possibly hierarchical) subject topics (such as "Computers", "Finance" or "Medicine") to documents. Subject topics are definitely useful for supporting web navigation; their utility for other tasks such as search and filtering is, however, spurious, as they are often strongly correlated with the actual user's query/profile. In the Vseved meta-search project (see [1] for more detail), we have proposed three "query-orthogonal" typologies, which can be applied on most WWW documents more-or-less independent of each other as well as of the subject topic. Most of this paper deals with *bibliographic categorisation* ("article", "bibliography", "pricelist" and the like), following the Dublin Core metadata system [7].

The input for document categorisation may be quite heterogeneous, ranging from explicit metadata or simple data such as URLs and page titles, to abstract concepts extracted from free text or images. In our work, we have concentrated on the situation when the amount of information is rather limited. This is typically the case for on-line *meta-search* systems, operating solely on a few data items they receive from the primary search engines in response to user queries, and on-line systems for support of *navigation*, which are, again, often constrained to using URLs when making judgements about the locations referenced by the page currently browsed. The importance of the URL in document categorisation lays in the insight it may provide to the directory structure at the host server, beside the information content of the server domain as a whole. Recent studies [5] show that humans can make significant deductions about the content of URLs (in particular, of "longer" ones); most of these deductions can be modelled as simple heuristic rules and performed by computer.

In the current project, we have attempted to learn a rulebase relating web document types to terms from and structure of the URL, as well as other information returned by search engines – name, size, date and textual "snippet" of the page. For the learning task, we have used a fast and straightforward *frequency analysis* of terms and symbols from URLs, completed with structure-sensitive but costly *inductive logic programming* (ILP) using all information mentioned above. We have shown in [6] that a few dozens of pure URL-based rules obtained thanks to frequency analysis can more-or-less successfully assign some generic category to approx. 70-90% of pages retrieved by search engines; 30-60% of the assignments account for bibliographic categories. Future experiments will show the impact of the newly-introduced, ILP-based disambiguation on these figures.

3   Case Acquisition and Labelling

Various techniques have been used, in turn, to eventually obtain labelled data.

Collection of Generic URLs. In order to obtain a large set of generic URLs required for the initial frequency analysis, we have submitted extremely general (to say, "empty") *queries to search engines* – e.g. in the form +domain:com in the case of AltaVista. The URLs of the "hits" returned have been then parsed to their constituent parts and to individual terms, abstract (e.g. calendar) concepts have been deduced, and, for the resulting structured collections of terms and concepts, relative frequencies have been computed, while discounting multiple occurrences of the same term from the same server (for details see [6]). The lists of terms and concepts served as a base for manual formulation of category-recognition rules.

Supporting Identification of Ambiguous Terms. In the process of manual formulation of recognition rules, *search engines* have served again, namely to verify the reliability of key terms. Queries of the sort +url:<term> or +url:<term> -host:<term> have been posed, and the hits visually inspected (only the first 20–30 per query, which should suffice to identify significant deviations from the main, expected, meaning). Some very frequent terms, such as "art" (article, but also page about art), "bio" (biography, but also page about biology), "cat" (catalogue, but also page about cats) or "pub" (list of publications, publicity page, or page about restaurants) have been then submitted for disambiguation.

Collection of Input Data for the Disambiguation Task. To obtain (still unlabelled) cases for the ILP learning (disambiguation) task, *search engines* have been used as in the previous step. This time, however, a fully automatic process has been employed: a special program has called the search engine, extracted the hits from the output pages, and parsed them into their descriptive elements: URL, title, extracted text, size and date.

Frequency-Based Case Filtering. During the visual inspection of hits corresponding to ambiguous terms, we have observed that in addition to clearly identifiable positive and negative examples of the category in question (e.g., for the URL term "art", scientific and newspaper articles vs. pages dealing with art), some "problematic" ones have been obtained. The difficulty was usually related to one of the following: unexpected, *marginal semantic* of the term (e.g. art can also stand for Arthur, artificial or artillery...), *cumulation of semantics* (e.g. articles about art), or *complete ignorance* (e.g. pages in an uncommon language, or with the visible part unrelated to the term). Obviously, such examples would not be of much use in inductive learning of the target concept, and their manual labelling would be waste of time. In order to eliminate "problematic" cases without human intervention, we have experimented with a frequency-based filter. The filter is based on the assumption that an example may be useful for learning only if it has some property that can be found in a sufficient number of other examples. In

Table 1. Effect of frequency-based filtering

	"Clear"	"Problematic"	Total
Before filtering	54 (62%)	33 (38%)	87
After filtering	50 (72%)	19 (28%)	69

text mining[1], the example must contain a sufficiently frequent term. We can thus select a set of terms (and abstracted concepts) with frequency above a certain threshold and keep only those examples containing them. We have experimented with filtering on a small set of 87 examples, of which approx. 38% have been previously identified as "problematic" in one of the senses above. By means of the frequency filter, we have eliminated all examples but those containing at least two "frequent" terms. In this way, approx. 20% of examples have been rejected, and in the resulting subset, only 28% of examples were "problematic"; at the same time, we have lost only few "clear"examples (Tab. 1). Note that the overall cutoff of the volunteers' time gained by frequency filtering may be even higher than the plain difference in the count of examples, since a large proportion of "problematic" examples were actually difficult to evaluate even by human, and, in addition to their inutility in learning, they would make the user spend more time on them than on "clear" ones.

Interactive Semantic Indexing. The actual labelling (assignment of semantic indices) was done by 2-3 volunteers, using an interactive program, which displayed the *information about each page* (i.e. re-structured output of the search engine), in turn, offered a *menu of semantic indices*[2], *recorded* the answer, and enabled to *backtrack* to previous answers and change them, if necessary. The labelling results of different people have been compared, and only the cases with identical index obtained (the degree of concordance was usually rather high, which can be attributed to prior frequency filtering) have been converted to predicate representation and submitted to the inductive learner.

4   Preliminary Results of ILP-Based Mining

The ILP task has been performed on the following predicate representation. sterm(Id, Term, Pos) indicates the occurrence of term *Term* in the snippet of example *Id*, on position *Pos*. Analogously, we use tterm for the page title, dterm for the directory part of URL, and fterm for the filename. no_fterm(Id]) and no_dterm(Id) mean that the URL doesn't contain a filename or directory part,

---

[1] It is an interesting question whether frequency-based filtering could also be used when labelling tabular data; in principle, there is no hindrance to that.

[2] Such as, for the "art" cases: *scientific/newspaper article, page about art, catalogue article (goods), article of law, other, undecidable.*

respectively. `owner(Id)` and `no_owner(Id)` indicate whether the URL contains the owner indication (`~`⟨*user*⟩ at the beginning of the path) or not. `in_srv(Id)` and `not_in_srv(Id)` specify whether the term to be disambiguated occurs in the name of server or not. Finally, `nextto(Pos1, Pos2)` holds if $Pos2 = Pos1 + 1$. It is used to express the adjacency of terms.

As inductive learner, we are currently using *Aleph*[3]; For the above mentioned example of "scientific/newspaper article" semantic index, it has returned (depending on the settings) approx. 10–15 positive and a similar number of negative rules. Some interesting positive ones were e.g.:

```
[Rule 10] [Pos cover = 9 Neg cover = 0]
pos_example(A) :- sterm(A,by,B), fterm(A,art,C), no_owner(A).
[Rule 13] [Pos cover = 8 Neg cover = 0]
pos_example(A) :- fterm(A,'_num',B), fterm(A,art,C).
```

Rule no.10 probably covers some articles placed on specialised publishing servers, since on a personal homepage the author would probably name the file according to the topic of the article rather than by "art", and also would not explicitly state the authorship using "by". The '_num' symbol in rule no.13 is the abstraction of number, thus if the filename contains art and a number, it is probably an article. For negative examples, we got e.g.:

```
[Rule 2] [Pos cover = 31 Neg cover = 0]
neg_example(A) :- sterm(A,x,B).
[Rule 4] [Pos cover = 17 Neg cover = 0]
neg_example(A) :- no_fterm(A).
```

Rule no.2 clearly covers some artwork, since online galleries often state physical dimensions, in ASCII, as *width* x *height*. Rule no.4 states the obvious fact that articles are not stored in the directory index page (URL with no filename). We can see the "truly relational" predicate `nextto` has not been needed, since "propositional" predicates sufficed to discriminate between positive and negative examples, (e.g. in Rule 2 above, the "x" symbol alone has "substituted" the sequence "number-x-number").

5   Related Work

Using data output by search engines for inductive learning has been the topic in the MetaCrawler-STC project [8]. Unlike our project, the task consisted in *clustering* the hits with respect to subject topic rather than in classification to predefined (moreover, bibliographic) categories. In this sense, our work is rather similar to the AdEater project [3], which concentrated on the *binary task* of distinguishing between banner ads (which can be understood as a sort of

---

[3] An implementation of *Progol*, available at `http://web.comlab.ox.ac.uk/ oucl/research/areas/machlearn/Aleph/aleph.html`

bibliographic category) and other graphics on web pages: an interactive, graphic tool has been used to assist users in labelling examples as positive or negative, and a decision tree has been induced over terms from URLs of images plus some additional information. Further, Mitchell's group at CMU [2] attempts to use ILP in order to recognise the page type. The input for learning is, however, fulltext analysis (including HTML), which implies the use of a more complex predicate representation. In terms of using ILP for term disambiguation, our project is also akin to some natural language disambiguation projects [4].

6 Future Work

In the paper, we have presented several techniques supporting the acquisition and labelling of cases to be input to the learning process, in the specific context of mining web search results. Some of the techniques are likely to be reused for other web-mining tasks, in particular for information extraction from the page fulltexts. Future work should also concentrate on assessing the utility of the ILP approach. It provides a comfortable way to specify background knowledge and term containment in examples; it is however unclear whether its representational power is indispensable for the tasks like the one above.

The research on this topic has been partially supported by Grant no.VS96008 of the Czech Ministry of Education, "Laboratory of Intelligent Systems".

References

1. Berka P., Sochorová M., Svátek V., Šrámek D.: The VSEved System for Intelligent WWW Metasearch. In: (Rudas I. J., Madarasz L., eds.:) INES'99 – IEEE Intl. Conf. on Intelligent Engineering Systems, Stara Lesna 1999, 317-321.
2. Craven M., DiPasquo D., Freitag D., McCallum A., Mitchell T., Nigam K., Slattery S.: Learning to Extract Symbolic Knowledge from the World Wide Web. In: Proc. of 15th AAAI, Madison, WI, 1998.
3. Kushmerick N.: Learning to remove Internet advertisements. In: 3rd Int. Conf. on Autonomous Agents, 1999.
4. Popelínský L., Pavelek T.: Mining Lemma Disambiguation Rules from Czech Corpora. In: PKDD'99 - Principles of Data Mining and Knowledge Discovery, Prague, 1999, 498-503.
5. Stanyer D., Procter R.: Human Factors and the WWW: Making sense of URLs. In: (Brewster S., Cawsey A., Cockton G., eds.:) Human-Computer Interaction – Interact'99 (Vol.2). The British Computer Society, 1999, 59-60.
6. Svátek V., Berka P.: URL as starting point for WWW document categorisation. In: (Mariani J., Harman D.:) RIAO'2000 – Content-Based Multimedia Information Access, CID, Paris, 2000, 1693-1702.
7. Weibel S., Kunze J., Lagoze C., Wolf M.: Dublin Core Metadata for Resource Discovery. IETF #2413. The Internet Society, September 1998.
8. Zamir O., Etzioni O.: Web Document Clustering: A Feasibility Demonstration. In: SIGIR'98, Melbourne, Australia.

Indirect Association: Mining Higher Order Dependencies in Data *

Pang-Ning Tan[1], Vipin Kumar[1], and Jaideep Srivastava[1]

Department of Computer Science,
University of Minnesota,
200 Union Street SE,
Minneapolis, MN 55455.
{ptan,kumar,srivasta}@cs.umn.edu

Abstract. This paper introduces a novel pattern called indirect association and examines its utility in various application domains. Existing algorithms for mining associations, such as Apriori, will only discover itemsets that have support above a user-defined threshold. Any itemsets with support below the minimum support requirement are filtered out. We believe that an infrequent pair of items can be useful if the items are related indirectly via some other set of items. In this paper, we propose an algorithm for deriving indirectly associated itempairs and demonstrate the potential application of these patterns in the retail, textual and stock market domains.

1 Introduction

In recent years, there has been considerable interest in extracting association rules from large databases [1]. Conceptually, an association rule indicates that the presence of a set of items in a transaction often implies the presence of other items in the same transaction. The problem of mining association rules is often decomposed into two subproblems : (1) discover all itemsets having support above a user-defined threshold, and (2) generate rules from these frequent itemsets. Under this formulation, any itemsets that fail the support threshold condition are considered to be uninteresting. However, we believe that some of the infrequent itemsets may provide useful insight about the data. Consider a pair of items, $\{a, b\}$, that seldom occurs together in the same transaction. If both items are *highly dependent* on the presence of another itemset, Y, then a and b are said to be indirectly associated via Y (Figure 1).

There are many potential applications for indirect associations. For market basket data, this method can be used to perform competitive analysis. For example, a and b may represent products of competing brands, such as Reebok and Nike. Suppose Reebok marketers are interested in expanding their current market share by attracting Nike customers through direct marketing campaigns. However, instead of promoting to every Nike customers, such a campaign can

---

* supported by NSF ACI-9982274 and AHPCRC Contract No. DAAH04-95-C-0008.

D.A. Zighed, J. Komorowski, and J. Żytkow (Eds.): PKDD 2000, LNAI 1910, pp. 632–637, 2000.
© Springer-Verlag Berlin Heidelberg 2000

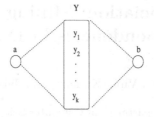

Fig. 1. Indirect Association between a and b via a mediating itemset Y.

be made more effective, in terms of cost-benefit and lift analysis [6], by selecting a smaller target group whose buying behavior resemble that of Reebok customers. Indirect association provides an approach to characterize the group by identifying the set of items that are often bought by both groups of customers.

In the text domain, indirect association often corresponds to synonyms, antonyms or words that are used in the different contexts of another word. As an example, the words *coal* and *data* can be indirectly associated via *mining*. If a user queries on the word *mining*, the collection of documents returned often contains a mixture of both mining contexts. However, with indirect association, one can potentially identify explicitly the different ways in which the queried word appears in the corpus of text documents. Similarly, for stock market data, indirect association can help to identify the different set of events influencing the movement of a stock price.

The importance of indirect relationship between attributes of a dataset has been acknowledged by several authors [5,4]. However, there has not been any direct attempts to explicitly derive such patterns. For example, in [5], Melamed observed that indirectly-associated words tend to reduce the accuracy of automated document translation systems by polluting the lexicon translation tables. Das et al. [4] introduced the notion of external similarity measure between attributes of a database relation. Essentially, external similarity is a measure of proximity between two attributes using the values in other columns, called the probe attributes. The notion of probe attribute is similar to our idea of mediator for indirect association. However, in [4], the role of a probe attribute is minimal; it is used only as far as determining the similarity between two attributes. On the other hand, a mediator is central to the concept of indirect association. Furthermore, probe attributes are chosen according to domain knowledge or constraints specified by a user [4]; whereas mediators are automatically derived from the observation data, as will be described in a later section.

2   Problem Formulation

Let $I = \{i_1, i_2, \cdots, i_d\}$ denotes a set of binary literals (called items) and T is the set of all transactions, $T = \{T_j \mid \forall j : T_j \subseteq I\}$. We will use upper case letters to represent itemsets (or sets of itemsets) and lower-case letters for individual items. Also, let $sup(X)$ denotes the support of an itemset, X.

Definition 1. *An itempair* $\{a, b\}$ *is indirectly associated via a mediator set* Y *if the following conditions hold :*

1. $sup(a, b) < t_s$ *(Itempair Support Condition)*
2. *There exists a non-empty set* Y *such that* $\forall Y_i \in Y$:
   a) $sup(a, Y_i) \geq t_f, sup(b, Y_i) \geq t_f$ *(Mediator Support Condition).*
   b) $d(a, Y_i) \geq t_d,\ d(b, Y_i) \geq t_d$ *where* $d(p, Q)$ *is a measure of the dependence between* p *and* Q *(Dependence Condition).*

Condition 1 is needed because an indirect association is significant only if both items rarely occur together in the same transaction. Otherwise, it makes more sense to characterize the pair in terms of their direct association. An alternative to this condition is to test for independence between the two items. However, it is often the case that independent or negatively correlated itempairs tend to have low support values. Therefore, Condition 1 will effectively consider only itempairs that are slightly or negatively correlated.

Condition 2(a) can be used to guarantee the statistical significance of the mediator set. In particular, for market basket data, the support of an itemset justifies the feasibility of promoting the items together. Support also has a nice downward closure property which allows us to prune the combinatorial search space of the problem.

Condition 2(b) ensures that only items that are highly dependent on both a and b are used to form the mediator set. Over the years, many measures have been proposed to represent the degree of dependence between attributes of a dataset. One such measure is Pearson's linear correlation coefficient, ϕ. For binary variables, it can be shown that within certain range of support values [1], the correlation coefficient $\phi_{x,y}$ can be expressed in terms of the interest factor [3,2], $I(x, y) \equiv \frac{P(x,y)}{P(x)P(y)}$, and support, i.e. :

$$\phi_{x,y} \approx \sqrt{I(x, y) \times \sup(x, y)} .$$

We will use the right-hand side of this expression, called the IS measure, as the dependence measure in Condition 2(b). This measure is desirable because it takes into account both the interestingness and support aspects of a pattern. However, our general framework can accommodate other measures, such as Piatetsky-Shapiro's rule-interest, J-measure and Gini index, which have been shown to be equally good at capturing statistical correlation [7].

3   Algorithm

An algorithm for mining indirect association is given in Table 1. Initially, an itempair support matrix S is constructed by scanning the entire database (step 2). Next, S will be used to prune the itempair space (step 3) based on the following criteria : (1) If the support of a is below t_f or a does not belong to any

---

[1] when $P(x) \ll 1$, $P(y) \ll 1$ and $\frac{P(x,y)}{P(x)P(y)} \gg 1$.

frequent itempairs, then the mediator set for a will always be empty. (2) Any itempairs that violate Condition 1 will be removed. There are two main phases in the FindMediator step : candidate generation and pruning of the mediator. Basically, it is assumed that a lattice of frequent itemsets, FI, has been generated using standard algorithm such as Apriori. During candidate generation, it will find all candidate mediators, $Y_i \subseteq I - \{a, b\}$, such that $\{a\} \cup Y_i \in FI$ and $\{b\} \cup Y_i \in FI$. The IS measure for each Y_i is then computed. Finally, during the pruning phase, candidates that fail Condition 2(b) are removed.

4   Experimental Results

To demonstrate the utility of indirect associations, experiments were carried out using datasets from three application domains : text, retail and stock market data. Table 2 shows a summary of the datasets used along with the thresholds chosen for our experiments.

Reuters-21578 Distribution 1.0 Newswire Articles

This dataset contains a collection of financial and commodity news articles that appeared on Reuters newswire in 1987.[2] Articles from both categories are preprocessed by removing stopwords and stemming each word to its root form. Table 3 shows some of the indirectly associated stemmed words from both collections of news articles. Most of the indirect associations represent the different contexts in which a term may appear in the news collection. For example, the indirect association between Soviet and worker refers to two separate news threads about union : one corresponds to news stories about Soviet Union while the other involves articles on labor unions.

Retail Data

The retail data was obtained courtesy of Fingerhut Corp. As expected, most of the indirect associations correspond to pairs of competing items (Table 4). Instead of doing competitive analysis, we are interested in discovering *surprising* patterns. A pattern is surprising if items belonging to competing product categories are directly associated. Hence, the first itempair is not surprising because it relates two products of different sizes via items that do not have a size distinction. Nevertheless, this type of pattern can be useful to determine what products should be bundled together for upsale promotions. The second itempair involves two products with competing design logos (checkered flag versus Nascar drivers). Since each logo has its own matching comforter, sheet, pillow case, etc., it is not surprising that their joint support is low. However, unlike the previous example, this pattern is unexpected because we do not expect checkered flag comforters and Nascar drivers wallpapers (the mediator) to have a large support value. Upon closer examination, we found that the reason their observed support is high is because the product catalog does not offer any checkered flag

---

[2] available at http://www.research.att.com/~lewis.

Table 1. Basic algorithm for mining indirect association between itempairs.

1. let $S = [sup(a, b)]$ denotes the support matrix for all itempairs (a, b).
2. For each transaction $t_i \in T$, UpdateSupportMatrix(t_i, S).
3. prune the itempair space.
4. for each remaining itempair (a, b) :
    4a. $Y \leftarrow$ FindMediator(S, a, b, t_d, t_f)
    4b. if $Y = \emptyset$, go to step 4 else $\{a, b\}$ is indirectly associated via Y.

Table 2. Summary of dataset parameters and results.

Dataset	t_s	t_f	t_d	n	|T|	# Freq pairs	Indirect pairs
Reuters(Finance)	0.15%	1.5%	0.3	2886	2005	10877	99
Reuters(Commodity)	0.15%	1.5%	0.3	3785	2308	6621	34
Retail Data	0.01%	0.1%	0.1	14462	58565	1174	59
S&P-500	0.25%	2.5%	0.2	976	716	13229	262

Table 3. Indirect association for Reuters-21578 Finance (left) and Commodity (right) datasets.

a	b	Y_i	$d(a, Y_i)$	$d(b, Y_i)$	a	b	Y_i	$d(a, Y_i)$	$d(b, Y_i)$
suppli	shortag	monei	0.4940	0.4597	soviet	strike	union	0.6248	0.4149
partner	temporari	trad	0.3932	0.3043	soviet	worker	union	0.6248	0.3615
partner	dealer	trad	0.3932	0.3064	opec	ico	quota	0.4112	0.5539
shortag	growth	forecast	0.3429	0.3502	opec	coffee	quota	0.4112	0.4515
shortag	inflat	forecast	0.3429	0.3241	iran	tax	oil	0.3225	0.3312

Table 4. Indirect Association for retail data

a	b	Y_i	$d(a, Y_i)$	$d(b, Y_i)$
Comforter Queen (Madrid)	Comforter King (Madrid)	Drapes (Madrid)	0.5634	0.4698
		Valance (Madrid)	0.5771	0.4708
		Pillow (Madrid)	0.5840	0.4446
Comforter Twin (Checkered Flag)	Sheets Twin (Nascar Drivers)	Border Wallpaper (Nascar Drivers)	0.3322	0.2517
Comforter Twin (Checkered Flag)	Curtains (Nascar Drivers)	Border Wallpaper (Nascar Drivers)	0.3322	0.2481
Playstn W/Crash2	Playstn Controller	Playstn memory card	0.2784	0.4448

Table 5. Indirect Association for S&P 500 data

a	b	Y_i	$d(a, Y_i)$	$d(b, Y_i)$
ibm-up	yell-up	lsi-up	0.3244	0.2200
		mu-up	0.2507	0.2005
hwp-down	txn-up	gnt-up	0.2002	0.2290
amgn-down	gt-down	digi-down	0.2303	0.2313
		s-down	0.2548	0.2826
oxy-down	ph-down	adsk-down	0.2740	0.2334
axp-down	nke-up	lsi-down	0.2197	0.2093

wallpapers. As a result, many customers who buy checkered flag comforters end up buying Nascar drivers wallpapers.

S&P 500 Stock Market Data

The dataset represents the daily fluctuation of share prices for S&P-500 stocks from Jan. 1994 to Oct. 1996. Each stock is represented by two attributes, X-up and X-down. The value of X-up (or down) is 1 if the closing price for stock X is significantly higher (lower), by at least 2%, than its previous closing price. Indirect associations can be used for event segmentation, where one can determine the set of events that are causing the price of a stock to move up or down. For example, the first indirect association in Table 5 relates IBM-up with YELL-up via LSI-up and MU-up. IBM is a company that provides various customer solution in information technology while YELL (Yellow Corp) is involved in the transportation business. The mediator contains the stocks of two semiconductor companies, LSI Logic and Micron Technology. This pattern indicates that events involving LSI-up and MU-up, can be partitioned into three disjoint sets - one involving IBM-up, another associated which YELL-up; and a third set of events not related to IBM-up nor YELL-up.

5 Conclusions

In summary, the above results show that indirect association can provide meaningful insight into the data. Such knowledge can not be derived from association rules alone because it involves infrequent itempairs and require analysis of higher order dependencies between items. Due to space limitation, details regarding the complexity analysis of our algorithm and threshold selection issues have been omitted. Interested readers can read the expanded version of this paper in [8].

References

1. R. Agrawal, T. Imielinski, and A. Swami. Database mining: a performance perspective. *IEEE Transactions on Knowledge and Data Engineering*, 5:914–925, 1993.
2. T. Brijs, G. Swinnen, K. Vanhoof, and G. Wets. Using association rules for product assortment decisions : A case study. In *Proc. KDD'99*, San Diego, August 1999.
3. S. Brin, R. Motwani, and C. Silverstein. Beyond market baskets: Generalizing association rules to correlations. In *Proc. SIGMOD'97*, Tucson, AZ, 1997.
4. G. Das, H. Mannila, and P. Ronkainen. Similarity of attributes by external probes. In *Proc. KDD'98*, New York, NY, 1998.
5. D. Melamed. Automatic construction of clean broad-coverage translation lexicons. In *2nd Conf. of the Association for Machine Translation in the Americas*, 1996.
6. G. Piatetsky-Shapiro and B. Masand. Estimating campaign benefits and modeling lift. In *Proc. KDD'99*, San Diego, 1999.
7. P.N. Tan and V. Kumar. Interestingness measures for association patterns : A perspective. Technical Report TR00-036, University of Minnesota, 2000.
8. P.N. Tan, V. Kumar, and J. Srivastava. Indirect association : Mining higher order dependencies in data. Technical Report TR00-037, University of Minnesota, 2000.

Discovering Association Rules in Large, Dense Databases

Tudor Teusan[1,3],Gilles Nachouki[2,4],Henri Briand[1,3], and Jacques Philippe[1,2,3]

[1]Ecole Polytechnique de l'Univ. de Nantes, rue Christian Pauc, Nantes Cedex 3, France
tjeusan@ireste.fr
[2]Irin, Faculté des Sciences et de Techniques, dép. Informatique Cedex
nachouki@irin.univ-nantes.fr
[3]PerformanSe, Espace Performance, Atlanpole, BP 703, 44481, Carquefou
[4]IUT Nantes, dép. Informatique, 3, rue Maréchal Joffre, 44041 Nantes Cedex

Abstract. In this paper we propose an approach for mining association rules in large, dense databases. For finding such rules, frequent itemsets must first be discovered. As finding all the frequent itemsets is very time-consuming for dense databases, we propose an algorithm that is able to quickly discover an image of the complete set containing all the frequent itemsets. We define what an image is, and we present a genetic algorithm for discovering such an image. To monitor the discovery process we introduce the notion of *dynamics* of the algorithm. To measure the performances of our frequent itemsets discovery algorithm, we introduce the notion of *efficiency* of the discovery process.

1 Introduction

The problem of discovering association rules within databases was introduced in [1]. In this section we give a succinct presentation of the problem.

Given a table of boolean values (the database), the fields of the table are named *items*, and the records, *transactions*. An *itemset* X is a set of items. X is included in a transaction T, if T contains all the items in X. The support of an itemset X, *sup(X)*, is the number of transactions that contain X. An itemsets is *frequent* if its support is greater than a user specified value, *minsup*. An *association rule* is a X->Y rule with X, Y itemsets, $X \cap Y = \phi$. The support of a rule is defined as $sup(X \cup Y)$ and the confidence, *conf*, as $sup(X \cup Y)/sup(X)$ (the conditional probability of Y given X).

Given two user-specified *minsup* and *minconf* values, the problem of association rules discovery consists in finding all the rules that satisfy $sup(X->Y) \geq minsup$ and $conf(X->Y) \geq minconf$. This is classically done in two steps. Firstly all the frequent itemsets Z are discovered - the support value being stored with each itemset. Secondly each Z is repeatedly divided in two disjoint itemsets X, Y ($X \cup Y = Z$). If conf(X->Y)≥minconf, the rule is a valid one. With this approach the difficulty resides in the first step. An efficient algorithm for solving the second step is presented in [2].

Starting from the classical, two-steps approach, other approaches have been developed: simultaneously considering multiple minimum supports [7], pre-storing information about itemsets, for OLAP-like, ulterior processing [3], simultaneously using multiple constraints in the mining process [4], or searching only optimal rules according to some interestingness metric [5].

D.A. Zighed, J. Komorowski, and J. Zytkow (Eds.): PKDD 2000, LNAI 1910, pp. 638-645, 2000.
© Springer-Verlag Berlin Heidelberg 2000

The characteristics of the database to be mined have a great impact on the performances of a frequent itemsets discovery algorithm. The prototypical data source for association rules mining is the "basket data". "Basket databases" contain only a few items per transaction – they are "rare" databases. However, basket-data is not the only data source for mining association rules. Databases that come from other domains, and even basket-data, can be dense, in the sense that they have [4]:

- many items in each transaction (as compared to the total number of items),
- many frequently occurring items,
- strong correlations between several items

This paper is focalized on mining association rules in dense databases that meet the first criterion. In this context, we present an approach for efficiently discovering frequent itemsets (section 2), provide a genetic algorithm that follows this approach (section 3), and give test results showing the interest of our approach (section 4).

2 Approach for Finding an Image of the Complete Set of Frequent Itemsets

2.1 Defining an Image

As stated in section 1, the difficult problem of finding the association rules within a database consists in generating all the frequent itemsets. The most common solution [1,2] is to start with the frequent itemsets containing only one item and to add them, step by step, more items, until no growth is possible. Each step generates a set of candidate itemsets that are counted in order to select the true frequent ones. On dense databases the main drawback of this approach is the number of candidates that suffers a combinatorial explosion [4], rendering the algorithms very slow and inefficient.

Rather than exhaustively enumerating all the frequent itemsets, we want to provide an image, a "fingerprint" of data. If finding *all* the frequent itemsets is too time-consuming, we are willing to trade the completeness of the result for an approximate set of frequent itemsets, obtained in a significantly shorter time. This approximate set must be a representative image of the set containing *all* the frequent itemsets [10]:

- It should contain a *number* of frequent itemsets that should be close to the *number* of *all* frequent itemsets.
- If the complete set contains different lengths frequent itemset, the approximate set should also contain, roughly in the same proportions, all lengths frequent itemsets.

2.2 Efficiently Discovering an Image

For presenting our approach we begin by enunciating a well-known result: any set of frequent itemsets is downwards closed; any subset of a frequent itemset is also a frequent one. This leads us to defining the notions of *frontier* and *pseudo-frontier*.

Definition 1: The *frontier* is the set of *maximal frequent itemsets*. A *maximal frequent itemset* is a frequent itemset, which is not contained in any other frequent itemset [11]. Such an itemset gets infrequent by adding it any item not already contained in the itemset.

Definition 2:   A *pseudo-frontier* is a set of *potentially maximal frequent itemsets*. A *potentially maximal itemset* is an itemset that, at a certain moment of the frequent itemsets discovery algorithm, is frequent and not contained in an already discovered frequent itemset. A pseudo-frontier is an approximation of the real frontier.

The frontier contains all the information about the frequent itemsets. If it is known determining the whole set of frequent itemsets is trivial: generating all the subsets of the frontier itemsets solves the problem [11].

For efficiently discovering an image, we search an algorithm that converges towards the frontier. At every moment of its execution it must have different approximations of the frontier: pseudo-frontiers. It must obtain these pseudo-frontiers without systematically counting the underlying subitemsets. In an unpredictable (but finite) time the algorithm would discover the true frontier. As our purpose is to discover an image and not the complete set, we must be able to interrupt the algorithm at a certain moment. Once interrupted it must return the itemsets on the current pseudo-frontier, as well as the underlying frequent subitemsets. All these itemsets must form an image in the sense of the two criteria presented in 2.1.

The interruption moment should be decided by regarding the *dynamics* of such an algorithm. *Dynamics* is defined as number of newly discovered frequent itemsets in the time unit. The algorithm should permanently indicate its dynamics. The user, as well as the algorithm itself, should be able to decide to stop and return the discovered image, if the evolution of this parameter shows that the process becomes inefficient.

To measure the effectiveness of the algorithm, we define the *efficiency*, η, of a frequent itemsets discovery algorithm. Given such an algorithm A, a database D, and a threshold minsup, $\eta = |L|/|T|$, where L is the set of frequent itemsets discovered by A, T is the set of all the itemsets counted by A, and $|X|$ denotes the cardinal of set X.

The optimal value for η is greater than 1, and is given by a purely hypothetical algorithm that only counts the frequent itemsets on the frontier. If the exact support of every frequent itemset is required, the optimal η is 1, and is given by an equally hypothetical algorithm that counts only the frequent itemsets. A real algorithm, as APRIORI [2], working in real conditions, usually has $\eta < 1$.

3   A Genetic Algorithm for Discovering an Image

An efficient pseudo-frontier discovery algorithm has to ignore a significant number of frequent, but few-items, itemsets. It must also not count too many infrequent itemsets, in order to quickly obtain a pseudo-frontier – a synthetic "representation" of the frequent itemsets. A genetic algorithm, in which itemsets represent chromosomes, can generate close-to-the-real-frontier itemsets, without considering all their sub-itemsets, hence finding an approximate image of the complete set of frequent itemsets. If the database is dense such an algorithm gives interesting results, both as execution time, and as efficiency. Figure 1 shows the template of the genetic algorithm, which we will use as a base for further detailing.

```
List find_freq_itemsets (Database db) {
  Population pop, raw_pop;
  Hash_table freq_is;
  Hash_tree  infreq_is;
  pop = initial_population();
  while (! stop_condition) {
    raw_pop = generate_next_raw_pop(pop);
    compute_fitness(raw_pop, db, freq_is, infreq_is);
    store_freq_&_infreq_is(raw_pop, freq_is, infreq_is);
    freq_is = expand_infreq_is(freq_is, infreq_is);
    pop = select_new_population(raw_pop);
  }
  return freq_is;
}
```

Fig. 1. The genetic algorithm – template

Some of the topics traditionally related to the design of the genetic algorithms were treated in a classical manner [6], or have straightforward solutions. The *representation* problem has a simple solution: each itemset is represented as an array of bits, each bit standing for a gene. There is a one-to-one mapping between chromosome and itemsets. The *initial population*, generated by `initial_population()`, is filled with chromosomes containing only "0"s. The raw population (fixed size) is generated in `generate_next_raw_pop()` using 1-cutting-point crossover and 1-gene mutation as *genetic operators*. The selection of the next population out of the raw one is done in `select_new_population()` by the classical "roulette" method; a chromosome's chances of survival being proportional to its fitness value.

3.1 Fitness Function

The fitness function is the core of the genetic algorithm. The principles that led us to its formula closely followed the principles and purposes presented in section 2:

- generate as few as possible infrequent itemsets.
- the discovered pseudo-frontier, and the underlying subitemsets must form a set that is rich in every possible length frequent itemsets.
- generate chromosomes that should be close to the real frontier

The first goal can be satisfied if the value of the fitness function for the infrequent chromosomes is much smaller than its value for any frequent one. A minimal chance of survival is given to the infrequent chromosomes.

The last two objectives can be satisfied if we use *a fitness function that encourages the chromosomes to grow in length, as long as, by growing in length, they stay frequent*. Such a function favors the apparition of long itemsets and, implicitly, of all their subitemsets. At the same time, encourages the *evolution* of the frequent itemsets to the point when the growth becomes impossible, that is, to the frontier.

The chromosomes that are infrequent, but near the frontier are equally interesting. If a chromosome grows in length and gets infrequent, but remains near the frontier, we generate its subitemsets that are, most probably, frequent.

Hence, we use a two-branch fitness function: for all the chromosomes that have a support that is less than 80% of minimum support, the fitness value is a fixed, minimal one, so that such chromosomes have a minimal chance of survival. For those chromosomes that have a support greater than 80% of the minimum required support, the fitness function is presented in the following paragraphs.

For synthesizing such a fitness function, we start from the support. The support diminishes as a chromosome grows in length. As the fitness value has to grow, we have to compensate the loss of support. Let $p(X_i)$ be the frequency of the X_i item in the database (ratio between the number of transactions containing X_i and the size of the database). If a chromosome gets from $X_{i1}X_{i2}\ldots X_{ik}$ to $X_{i1}X_{i2}\ldots X_{ik}X_{ik+1}$, an overestimated loss of support is given by $p(X_{ik+1})$. To compensate it, it is enough to multiply the support of the longer chromosome with $1/p(X_{ik+1})$. Such a correction applied for every item within a chromosome – $1/p(X_{i1})*\ldots*1/p(X_{ik+1})$ – insures that the fitness does not diminish with the growth of the chromosome. In reality, the support does not diminish as much as estimated, so the correction determines a fitness function that grows with the length of the chromosome.

Identifying and multiplying the frequencies of the items within a chromosome each time its fitness has to be evaluated is quite time-expensive. Instead of the different $p(X_i)$ we use a mean frequency, α, computed as the geometric mean of the all $1/p(X_i)$. In this way, for $C = X_{i1}\ldots X_{ik}$, $1/p(X_{i1})*\ldots*1/p(X_{ik})$, becomes $\alpha^{\text{length}(C)}$.

As already pointed out, we are interested in chromosomes that *evolve* and not in long chromosomes. Therefor, we seriously penalize those chromosomes that are not able to grow any further. For achieving this we use the *age of a chromosome*, computed as the number of generations the chromosome has survived. Taking all the above in consideration, the fitness function for the frequent chromosomes is:

$$\text{fitness}(C) = \frac{\text{support}(C) \cdot \alpha^{length\,(C)}}{\text{age}(C)}, \quad \alpha = \frac{1}{\sqrt[n]{p(X_1)p(X_2)\ldots p(X_n)}} \tag{1}$$

3.2 Managing the Discovered, Frequent and Infrequent, Itemsets

This step is reflected in the `store_freq_&_infreq_is()` function in the algorithm template. This is an essential step that permits achieving two major objectives.

The frequent chromosomes that have already been discovered should not be recounted. For this purpose we have used a hash-table in which all the frequent are stored. Newly discovered frequent itemsets, as well as their subitemsets, are stored in the hash-table. The age parameter is also handled using the hash-table.

A newly generated itemset that contains an already discovered infrequent subitemset should not be counted. For achieving this we use a data structure on which we are capable of quickly deciding whether the structure contains a subset of an input, tested-if-infrequent, set. Such a structure is a variant of the hash-tree presented in [2], with itemsets stored not only in the leaves, but at all the levels of the hash-tree.

Once a new population is generated, and the support of its chromosomes has to be evaluated (as the first step of the fitness evaluation), we search the chromosomes in the hash-table containing the frequent itemsets. For those chromosomes not found to be frequent, we try to find a subitemset in the hash-tree containing the infrequent itemsets. Only those chromosomes neither found to be frequent, nor to be infrequent, are counted. After being counted, they are stored in their corresponding structure.

An apart hash-table is filled with those chromosomes that represent infrequent, but near-the-frontier itemsets (infrequent itemsets with an absolute support greater than 80% of the minimum support). Each new such itemset has its subitemsets generated and counted in the `expand_infreq_is()` function.

3.4 Stopping Condition

The stopping condition of the genetic algorithm is related to the *dynamics* of the discovery process as defined in 2.1. At a certain moment of its execution, the genetic algorithm starts to discover few new frequent itemsets, and spend a lot of time on counting infrequent ones. Ideally this situation arrives once all the frequent itemsets are discovered. In practice, as our tests have shown, the algorithm becomes inefficient after discovering 80% to 90% of the total number of frequent itemsets, which is enough for providing an image of the complete set of frequent itemsets. Based on our tests, the algorithm should be stopped (by the user or by itself) once its dynamics gets between 10% and 13% of the initial, "moment 0", dynamics.

4 Tests

The genetic algorithm can efficiently discover frequent itemsets only if the database is memory replicated. Traditional approaches as APRIORI[2] or Partition[9] try to minimize the I/O overhead by reducing the number of passes over the database. Our main objective is to minimize the number of count operations, even if this is achieved at the cost of an arbitrary number of passes over the database.

In our tests we established the behavior of the genetic algorithm for dense databases. We have mainly studied two parameters: dynamics - d, and, efficiency - η. To show the interest of finding an image instead of the complete set of frequent itemsets, we have compared the results with those obtained by an optimized version of APRIORI (optimized candidate generation, memory-replication of the database).

For generating test data we have used the generation algorithm proposed in [2], and publicly available at [12]. We present 3 test cases, for a relatively reduced, a moderate and a high number of frequent itemsets. For the first two tests we used the same database, T30I6D100k [2], with minsup set at 10% and 5%. The third case was tested on a T40I6D100k database with minsup at 10%. In all cases N, the total number of items, was 100. The test platform was an SGI Origin 200, 4 R10000 processors running at 180MHz, 1 Gbyte RAM, under an IRIX 6.5 operating system.

For each test we present two graphics. The first graphic shows de dependency execution time – number of discovered frequent itemsets. The total number of itemsets and APRIORI's total time are also indicated. For the first 5 points we show the percentage of discovered frequent itemsets (as compared with APRIORI) and the percentage of APRIORI's total time, as well as the dynamics around each point. The second graphic presents the number of discovered frequent itemsets – long, dark gray bars, the number of counted frequent itemsets – shorter, lighter gray bars and the number of counted infrequent itemsets – white bars. The number of discovered frequent itemsets and the efficiency are indicated above each group of three bars.

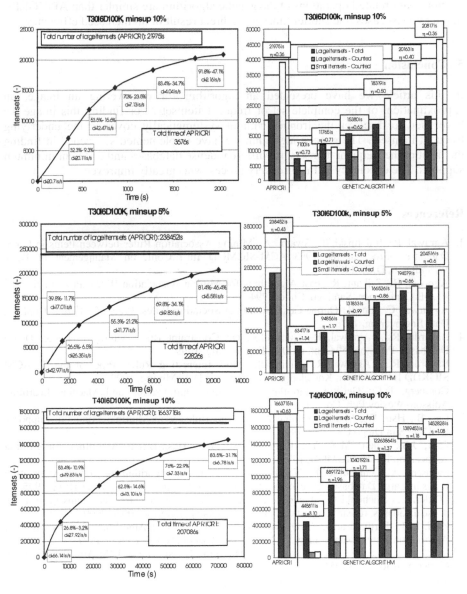

The next to last point figured in each first type graphic roughly corresponds to the stopping point. The corresponding dynamics are 10.43%, 12.98% and 10.58% of the initial dynamics, that is 10%-13% of the initial dynamics. The genetic algorithm has discovered 80% to 90% of the total set of frequent itemsets, in times ranging from 31% to 47% of APRIORI's time. Efficiency graphics show that the genetic process stays significantly more efficient than APRIORI for up to 80%-83% of the complete set. This makes it well suited for finding an image, in better efficiency conditions.

In the first case, the efficiency for discovering more than 90% of the frequent itemsets is the same as APRIORI's. The time is significantly shorter (about 50%) as the not-count-related operations of the genetic algorithm are simpler than APRIORI's. In the two last cases, the shorter times are a direct result of an increased efficiency.

5 Conclusion

In this article we have presented an algorithm for discovering an image, an approximation, of the complete set of frequent itemsets. For finding this image we have tried to discover the frontier without systematically covering the underlying frequent itemsets. The genetic algorithm we have implemented succeeded in finding the most of the frequent itemsets, for different dense databases and different minimum supports. The efficiency of the discovery process was greatly improved.

References

1. Agrawal, R., Imielinski, T., Swami, A.: Mining Association Rules between Items in Large Databases. Proc. of the 1993 ACM-SIGMOD Int'l Conf. on Management of Data, Washington, D.C., 1993
2. Agrawal, R., Srikant, R.: Fast Algorithms for Mining Association Rules. Proc. of the 20th VLDB Conference, Santiago, Chile, 1994
3. Aggarwal, C., Yu, P.: Online Generation of Association Rules. Proc. of 14th Int'l Conf. on Data Engineering, 1998
4. Bayardo, R.J., Agrawal, R., Gunopulos, D.: Constraint Rules Mining in Large, Dense Databases. Proc. of 15th Conf. on Data Engineering, 1999
5. Bayardo, R.J., Agrawal R.: Mining the Most Interesting Rules. Proc. of the 5th ACM SIGKDD Int'l Conf. on Knowledge Discovery and Data Mining, 1999
6. Goldberg, D.E.: Genetic Algorithms in Search, Optimization and Machine Learning, Adisson–Wesley Pub. Co., 1989
7. Liu, B., Hsu, W., Ma, Y.: Mining Association Rules with Multiple Minimum Supports. KDD-99 San Diego CA USA, 1999
8. Park, J.S., Chen, M-S, Yu, P.S.: An Effective Hash-Based Algorithm for Mining Association Rules. SIGMOD, 1995
9. Savasere, A., Omiecinski, E., Navathe, S.: An Efficient Algorithm for Mining Association Rules in Large Databases. Proc. of the 21st VLDB Conference, Zurich, Switzerland, 1995
10. Tudor, J., Nachouki G., A genetic algorithm for discovering frequent itemsets in large, dense databases, research report of Nantes University (IRIN) n° 00.9, 2000
11. Zaki, M.J., Parthasarathy, S., Ogihara, M., Li, W.: Parallel Algorithms for Discovery of Association Rules. Data Mining and Knowledge Discovery 1-4, Kluwer 1997
12. http://www.almaden.ibm.com/cs/quest/syndata.html. Quest Project. IBM Almaden Research Center, San Jose, CA, 95120

Providing Advice to Website Designers Towards Effective Websites Re-organization

Peter Tselios, Agapios Platis, George Vouros

University Of the Aegean, Department of Information and Communication Systems
Karlovassi, 83200, Samos, Greece
{tpe, platis, georgev}@aegean.gr

Abstract. This paper presents a method to help website designers to re-organize sites towards making pages that are "hidden" from site visitors but contain information that is of high importance for them, more accessible to future visitors. Towards this aim we present a method for computing the *probability* of visitors to be in a page and page's *attractability*. The computation of these indicators is based on *transitions* made to pages and on the *information interest* that the site designer assigns to pages. Based on this method, the paper presents *DesignersAdvisor*, which computes pages' *attractability*. *DesignersAdvisor* aims to assist the re-organization of any existing website without posing any additional effort to site visitors and without posing particular requirements concerning the form or the content of web pages.

1. Introduction

Web site designers structure their websites in a way that is comprehensive and more or less intuitive (according to designers' judgment) to the average visitor. However, it is very difficult to organize a site with a large number of pages in such a way that (a) every visitor could easily access information and (b) designers to achieve their communicative goals in the general case.

Our aim is to provide designers with tools that help them monitor visitors' behavior, help them to identify problematic cases in their sites and monitor the progress achieved towards making a "better" site organization. Problematic cases comprise information that website designers intent to present but, somehow, it is not easily accessible by website visitors.

Towards this aim we present a method that takes into account the activity of large number of people who have visited a website, calculates the *probability of visitors to be in a page* and *pages' attractability*. Attractability of a page is a new measure proposed here that depends on the probability of visitors to be in that page and the *information interest* that the site designer assigns to this page. The information interest captures in a simplistic way the communicative intention of website designers.

Based on the proposed method, we have developed *DesignersAdvisor*. This is a prototype tool that utilizes sites' log files in order to provide evidence for the behavior

D.A. Zighed, J. Komorowski, and J. Zytkow (Eds.): PKDD 2000, LNAI 1910, pp. 646–651, 2000.
© Springer-Verlag Berlin Heidelberg 2000

of websites visitors and to compute the attractability of pages based on designers' intentions. *DesignersAdvisor* assists the re-organization of any existing website without requiring any feedback from visitors and without posing particular requirements concerning the form or the content of web pages.

The paper is structured as follows: Section 2 sketches previous work towards re-organizing websites' structure for making information contained therein more accessible. It identifies the key issues of our research. Section 3 presents the proposed framework for modeling visitors, and describes the theoretical basis for the computation of the probabilities for visitors to be in a website page and pages' attractabilities. Section 4 presents *DesignersAdvisor* that implements the framework described in section 3, and provides the results of applying the proposed method to an existing website. Finally, section 5 concludes the paper and presents plans for future work.

2. Related Work - Motivation

Site re-organization aims at either *customizing* a website so as to satisfy the needs and characteristics of an individual user, or at *optimizing (transforming, re-organizing)* the structure of a site so as to make information more accessible and more effective to a large set of users. Furthermore, site re-organization may be content or access based [1]. In the first case re-organization decisions are based on pages' content, while in the second case they are based on requests of past visitors. The aim of this paper is to help designers of websites to structure their sites so as to make pages more accessible to future visitors, taking into account the activity of past visitors. This is an access-based optimization method.

Our desiderata for a method for sites' re-organization are given in figure 1.

• It must not require additional effort from the visitors to provide feedback concerning their interest on pages' or sites organization.
• It must neither add/remove pages in the site automatically, nor alter the structure of the site by adding new pages. Human webmasters must keep the control of the website structure.
• It must suggest web pages that have a high probability to be visited but are not found easily by sites' visitors.
• It must take into account website designers' intents.
• It must facilitate their easy integration into existing websites with the minimal effort from the webmasters.

Fig. 1. Desiderata for sites' re-organization method.

Systems that aim to *transform* the structure of websites in an *access-based manner* may be categorized according to their effect on browsing strategies within a site. According to [13, 14] there are three major browsing strategies:

a.   *Search browsing.* In this case a user, to achieve specific goals, performs the same short navigation sequences relatively infrequently, but does perform long navigational sequences often" [13].   Systems that re-organize websites towards making them more effective to searchers include WebWatcher [7], and PageGather [2] [3]. WebWatcher [7] provides navigational support by predicting links that users may follow on a particular page as a function of their interests. This requires website visitors' feedback on their goal ("what they are looking for") and on whether they have found what they wanted. WebWatcher uses qualitative techniques since users state their interests in broad terms such as "accommodation facilities". On the other hand, PageGather proposes clusters of pages and synthesize index pages automatically. The system supposes that users visit a site with a particular information-seeking goal and therefore, pages that are interesting to them should be *conceptually related*. PageGather follows a statistical approach since it computes the co-occurrence frequencies between pages and creates a similarity matrix for pages.

    b.   *General purpose browsing:* In this case users consult sources that have a high likelihood of interest. As it has been computed in [13], probabilistically, users' behavior is more or less opportunistic. Efforts towards making websites more effective to general-purpose browsers include path traversal approaches such as Footprints [5] and the approach reported in [12]. Footprints computes visitors' access patterns. The basic idea is to visualize the paths that visitors have already followed and provide information concerning ways to proceed within the site. Footprints takes a statistical approach since it counts the number of users following the paths in a site. Similarly, the system reported in [12] computes path traversal patterns. The system computes frequent traversal patterns, i.e., large reference sequences. This is a statistical approach that takes into account the number of occurrences of reference sequences.

    Based on evidence that a large class of website visitors, as already stated, do not repeatedly follow complex navigational patterns, our objective here is to emphasize on individual pages so as to make them more accessible to users. This is a complementary approach to finding path traversal patterns.

    c.   *Serendipitous browsing:* Users behavior in this case is purely random. Long invocation sequences are not repeated. Search engines are utilized to make websites more effective to serendipitous browsers. However, search engines do not assist websites' re-organization towards making them more effective to future visitors.

3. Modeling Visitors and Attractability of Pages

Based on our survey on access-based transformation approaches, it should be noted that most of them are statistical. For general-purpose and serendipitous human browsers that do not repeat navigational patterns and seek for interesting information, it seems that the evolution of their behavior in a website is Markovian. In this case, the transition from one page to another depends only on the information received from the former page. Based on this assumption, a probabilistic approach can provide the necessary information for making predictions on website visitors behavior based on pages' requests. Although this is true in the general case, such a model is not appropriate for sites where the navigational history plays a major role. For instance, this is not appropriate for educational hypermedia where the choice of the next page depends on the set of pages already visited.

Markov chain. Let E be a set of points that represent the space of pages within a site and $X=\{X_n, n \in \mathbb{N}\}$ an E-valued stochastic process on a probability space, whose measure is Pr (with $\mathbb{N}$ with the set of positive integers). X represents visitors' behavior evolution inside the space of web pages. If we assume that the transition from one page to another depends solely on the information contained in the former page, then

X is considered to be a Markov chain. That is, $\Pr\{X_{n+1}=j|X_0, X_1, X_2, ..., X_n\} = \Pr\{X_{n+1}=j| X_n\}$ *a.s.*, where $p(i,j)=\Pr\{X_{n+1}=j|X_n=i\}$ is called the transition probability from i to j. Let α be the initial distribution on E. X is completely defined by the initial distribution function α and the transition matrix p. Therefore, the probability of being in page j at time n is defined to be: $P_j(n)=\Pr(X_n=j)=(\alpha\, p^n)(j)$

The *steady state probability* distribution is given by $\lim\limits_{n\to\infty} P\,(n) = \lim\limits_{n\to\infty} (\alpha\, p^n) = \Pi$.

In this way, the *probability of being in a page i* at an infinite time *n,* is given by

$$\Pi_i = [\ \lim_{n\to\infty} (\alpha\, p^n)\ ](i) \qquad (1)$$

Π_i gives the probability of being in page *i* at a time where the evolution of website visitors behavior has reached a sort of stability. Furthermore, to infer those pages that are of real importance to visitors, based on website designers' intents, we introduce pages' information interest factor.

Attractability: The attractability indicator is defined as a measure combining information on the probabilistic activity of website visitors and information concerning website designers' communication intents. Attractability has been derived from probabilistic performance indicators (e.g. in [10]).

Each web page *i* in a site is associated with an *"information interest factor"* g(*i*) that represents the judgment of site designers on the importance of pages' *information content*. In this case, for instance, index pages are considered to have low information interest. We must notice that it is essential to take the designer's point of view and not to let the system evolve chaotically leaded only by the visitors' requests.

Given that the information interest factor assigned to each page is given by g(*i*) and that a site is observed in a time interval [n, n+N] of length N, where *n* is large enough so that website visitors' behavior has reached a stability (given by steady state probabilities obtained by (1)), we can compute the attractability for each page on [n, n+N] in terms of requests made by websites' visitors for that page. So, we define the *attractability* of a page *i* by:

$$attract_i = g(i)\ \sum_{k=n}^{n+N-1}\!\!\mathbf{Pr}(X_k = i) = N\,g(i)\Pi_i \qquad (2)$$

where $\sum\limits_{k=n}^{n+N-1}\mathbf{Pr}(X_k = i)$ represents the total time spent in page *i* in the time interval [n, n+N] provided by means of (1). In this way, attractability of a page represents the cumulative reward that this page receives in time interval [n, n+N].

Pages accessibility. While it is difficult in the general case to define a mechanism for promoting pages' links [1], we will use the term as a synonym to making pages more accessible. We define a page to be *accessible wrt to its information interest* when the *(Percentage of requests)/(attractability)* ratio of this page is equal to the corresponding ratio of all the pages in the site. Therefore, the proposed method re-organizes websites structure by combining, designers' intentions in terms of pages information interest and data coming from visitors' navigation in the site.

Towards finding pages that should be made more (less) accessible we propose finding (a) pages that have a percentage of visits lower (respectively, higher) than the mean value of the percentage of requests and (b) pages that have attractability greater (respectively, lower) than the mean value of pages attractabilities. Intersecting these two sets we obtain those pages that have low (respectively, high) percentage of visits and high (respectively, low) attractability. This set of pages needs promotion (respectively, needs to be made less accessible).

4   DESIGNERSADVISOR

DesignersAdvisor is a prototype tool that utilizes sites' log files and computes the attractability of pages based on designers' intentions. *DesignersAdvisor* assists the re-organization of any existing website without requiring any feedback from visitors and without posing particular requirements concerning the form or the content of web pages.

The input to the *DesignersAdvisor* is the site's log file. The system preprocesses the log file and cleans it from useless information, such as requests to image files or error entries, and counts the number of transitions from page i to page j during a session. Based on indications provided by others [6,13], we define a session to be a set of requests, where two subsequent requests are made within a time interval of 20 minutes. Using the number of transitions made from page to page, the system calculates the probabilities of visitors to be in each page of the site using (1), the attractability of each page using (2), and indicates the sets of pages that need to be made more/less accessible.

DesignersAdvisor has been applied to the ACAI '99 website log file. The log file contained almost 6000 valid records. These records represented requests to 41 pages made by 576 visitors. Figure 2 provides the information interest factor (which is an integer from 1 to100), the calculated probabilities, percentage of requests and attractabilities for some of the site pages. Let us for instance consider the page "Events". As it is shown in Figure 2, although this page has a small percentage of visits, the probability to be requested is relatively high. Furthermore, the attractability of this page is relatively high, which in conjunction with the number of requests indicates that this page needs to be made more accessible. It must be noticed that results provided by the system may be used (a) as a feedback to designers in order to reconsider their judgment concerning pages' information interest, and (b) to help them indicate the "problematic" cases in their site.

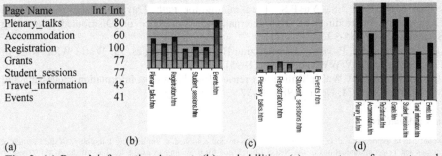

Page Name	Inf. Int.
Plenary_talks	80
Accommodation	60
Registration	100
Grants	77
Student_sessions	77
Travel_information	45
Events	41

(a)   (b)   (c)   (d)

Fig. 2. (a) Pages' information interest, (b) probabilities, (c) percentage of requests and (d) attractabilities for some of the pages.

5. Conclusions – Future Work

This paper presented a simple and effective access-based, probabilistic method, and a tool named *DesignersAdvisor*, to help website designers to re-organize their sites towards making information, which is of high importance and currently hidden, more accessible to future visitors.

Although this paper deals with single pages, the method could be extended to a subset of pages. In that case, the above method can be applied to clusters of pages, and give a clearer view of visitors' behavior in the site. Another issue that is subject to further investigation is the adequate representation of designers' intention. This paper does not provide any guidelines or any method concerning how information interest of pages should be assigned. Last, but not least, an empirical comparison with alternative methods needs to be carried in order to evaluate the provided results.

References

1. Perkowitz M., Etzioni O.: Adaptive Websites: an AI Challenge. In the proceedings of Fifteenth International Joint Conference on Artificial Intelligence (1997)
2. Perkowitz M., Etzioni O.: Adaptive Websites: Automatically synthesizing web pages. Proceedings of the Fifteenth National Conference on Artificial Intelligence, IAAA Press (1998)
3. Perkowitz M., Etzioni O.: Towards Adaptive Websites: Conceptual Framework and Case Study. To Appear in AI Journal.
5. Wexelblat A., Maes P.: Footprints: Visualizing Histories for web browsing. In proceedings of "5th RIAO Conference: Computer-Assisted Information Retrieval on the Internet", Montreal (1997)
6. Paliouras G., Papatheodorou C., Karkaletsis V., Tzitziras P., Spyropoulos C.D.: Learning communities of the ACAI '99 website visitors. In the proceedings of Advanced Course on Artificial Intelligence 1999 (1999)
7. Armstrong R., Freitag T., Joachims T., Mitchell T.: WebWatcher: A learning apprentice for the World Wide Web. In the proceedings of the 1995 AAAI Spring symposium on information gathering form heterogeneous, distributed environments, AAAI Press (1995), 6-12
10. Platis A., Limnios N., Le Du M.: Performability of electrical power systems modeled by non-homogeneous Markov Chains. IEEE Transactions on reliability, vol. 45 No. 4 (1996) 605-610
12. Chen M.S., Park J.S., Yu P.S.: Data mining for Path Traversal Patterns in a Web Environment. Proceedings of. 16th International Conference on Distributed Computing Systems, (1996), 385-392
13. L. Catledge and J. Pitkow, Characterizing Browsing Strategies. In: World Wide Web. In Proceedings of 3rd WWW conference, (1995)
14. Cove, J.F. and B.C. Walsh. Online text retrieval via browsing, Information Processing and Management, Vol. 24, No. 1, (1988) 31-37.

Acknowledgements

We would like to express our sincere thanks to the Software and Knowledge Engineering Laboratory of the Institute of Informatics and Telecommunications of the NRCS Demokritos for providing to us the log files of the ACAI'99 site. It should be indicated that the log files provided had no visitors' personal information. IPs were replaced by small integers.

Clinical Knowledge Discovery in Hospital Information Systems: Two Case Studies

Shusaku Tsumoto

Department of Medicine Informatics, Shimane Medical University, School of Medicine,
89-1 Enya-cho Izumo City, Shimane 693-8501 Japan
E-mail: tsumoto@computer.org

Abstract. Since early 1980's, the rapid growth of hospital information systems stores the large amount of laboratory examinations as databases. Thus, it is highly expected that knowledge discovery and data mining(KDD) methods will find interesting patterns from databases as reuse of stored data and be important for medical research and practice because human beings cannot deal with such a huge amount of data. However, there are still few empirical approaches which discuss the whole data mining process from the viewpoint of medical data.In this paper, KDD process from a hospital information system is presented by using two medical datasets. This empirical study show that preprocessing and data projection are the most time-consuming processes, in which very few data mining researches have not dicussed yet and that application of rule induction methods is much easier than preprocessing.

1 Introduction

Since early 1980's, the rapid growth of hospital information systems (HIS) stores the large amount of laboratory examinations as databases (Van Bemmel, and Musen, 1997). For example, in a university hospital, where more than 1000 patients visit from Monday to Friday, a database system stores more than 1 GB numerical data of laboratory examinations for each year. Furthermore, storage of medical image and other types of data are discussed in medical informatics as research topics on electronic patient records and all the medical data will be stored in hospital information systems within the 21th century. Thus, it is highly expected that data mining methods will find interesting patterns from databases as reuse of stored data and be important for medical research and practice because human beings cannot deal with such a huge amount of data.

In this paper, knowledge discovery and data mining (KDD) process (Fayyad, et. al, 1996) for two medical datasets extracted from a hospital information system is presented. This empirical study show that preprocessing and data proejction are the most time-consuming processes, in which very few data mining researches have not dicussed yet and that application of rule induction methods is much easier than preprocessing.

D.A. Zighed, J. Komorowski, and J. Zytkow (Eds.): PKDD 2000, LNAI 1910, pp. 652–656, 2000.
© Springer-Verlag Berlin Heidelberg 2000

2   KDD Process

2.1   Data Selection

In this paper, we use the following two datasetsfor data mining, which are extracted from two different hospital information systems. One is bacterial test data, which consists of 101,343 records, 254 attributes. This data includes past and present history, physical and laboratory examinations, diagnosis, therapy, a type of infectious disease, detected bacteria, and sensitivities for antibiotics. The other one is a dataset on the side effect of steroid, which consists of 31,119 records, 287 basic attributes. This data includes past and present history, physical and laboratory examinations, diagnosis, therapy, and the type of side effects. The characteristic of the second dataset is that it is a temporal database: although it includes 287 basic attributes, 213 attributes of which have more than 100 temporal records. These datasets are obtained through the first to third steps of KDD process: data selection, data cleaning and data reduction.

In the first step of KDD process, these databases are extracted from two different hospital information systems by simple query process. Table 1 gives results for data selection.

Table 1. Data Selection Results

Datasets	HIS size	Target Data	Time Required
Bacterlal Test	1,275,242(52GB)	361,932(14GB)	2.3 Days
Side-Effect	2,631,549(100GB)	135,749(6GB)	7.3 Days

2.2   Data Cleaning as Preprocessing

After data selection, data cleaning is required since the data obtained from the first step are not clean, including data records not suitable for data analysis. Even though these records are selected by matching the query condition, they do not include enough amount of information. For data cleaning, we define the two-fold cleaning steps: first we select the records which have no missing values in the pre-defined indispensable attributes. Then, we calculate how many attributes in the remaining attributes are used to describe the records in the first step. If the number of attributes used for a case is not sufficient, then this case will removed. For those steps, the indispensable attributes and the threshold for the second selection are given a priori by domain experts. Table 2 summarizes results for data cleaning.

2.3   Data Reduction and Projection as Preprocessing

Data Projection for Bacterial Test Database. From the viewpoint of table processing, data cleaning can be viewed as cleaning steps in the direction of records,

Table 2. Data Cleaning Results

Datasets	Data	Cleaned Data	Time Required
Bacterlal Test	361,932(14GB)	101,343(3GB)	5.2 Days
Side-Effect	135,749(6GB)	31,119(1,5GB)	2.5 Days

that is, in the direction of row. On the other hand, data reduction can be viewed as cleaning steps in the direction of attributes, that is, in the direction of column. Although the data cleaning process is a time-consuming process, it takes much more time to reduct and project data in clinical databases due to the characteristics of biological science, including medicine. The tradition of classification in biology tends to have a large scale of classification systems. Thus, generalization of such over-classified attributes is required to discover rules which are easy for medical experts to interpret. For bacterial test databases, a simple concept hierarchy was used for generalization of values.

Data Reduction for Steroid Side-Effect Database. Since incorporating temporal aspects into databases is still an ongoing research issue in database area (Abiteboul, et. al., 1995), temporal data are stored as a table in hospital information systems (H.I.S.) with time stamps. The characteristics of medical temporal data are as follows(Tsumoto 1999): (1)The Number of Attributes are too many. (2) Irregularity of Temporal Intervals. (3)Missing Values are too many.

In order to deal with medical temporal databases is discussed in (Tsumoto, 1999). Tsumoto introduces extended moving average method, which automatically set up the scale of the temporal interval. Here, we applied this method to this dataset.

2.4   Total Time Required for Data Reduction

In summary, Table 3 gives the total time required for data reduction and projection for each data set, including knowledge acquisition process. The second column gives the type of preprocessing. The third column shows total time required for each process. Finally, the fourth column shows the time required for acquisition of knowledge from domain experts.

Table 3. Total Time Required for Data Reduction

Dataset	Preprocessing	Total Time	Time for Acquisition
Bacterial Test	Projection	15.25 Days	7.0 Days
Side-Effect	Summarization	2.3 Days	0

This table suggests the generalization of values in attributes should be a time-consuming process, especially when domain knowledge is given.

2.5   Rule Induction as Data Mining

After the third step, rule induction based on rough set model (Pawlak, 1991) was applied to two medical datasets. Tsumoto (1998) extends rough-set-based rule induction methods into probabilistic domain.

In this section, we skip this part due to the limitation of space. For further discussion, the readers may refer to (Tsumoto, 1998; Polkowski and Skowron, 1998). The algorithm introduced in (Tsumoto 1999) was implemented on the Sun Spaarc station and was applied to the above two medical databases, the information of which is summarized in Table 4. For rule induction, the thresholds for accuracy and coverage are set to 0.5 and 0.5, respectively.

Table 4. Summary of Data Mining

Data	Size	Attributes	Rules	Computational Time
Bacterial Test	101,343 (3GB)	254	24,335	60 hours (2.5 Days)
Side-Effects	31,119 (1.5GB)	287	14,715	18 hours (0.75 Days)

2.6   Interpretation of Induced Rules

After the data mining step, we obtain many rules to be interpreted by medical experts. Even if the amount of information is very small compared with the original databases, it still takes about one week to evaluate all the induced rules and only small part of rules were found to be interesting or unexpected to medical experts(Table 5).

Table 5. Summary of Induced Rules and Interpretation

	Induced Rules	Interesting Rules
Bacterial-Test	24,335	114 (0.47%)
Side-Effect	14,715	106 (0.72%)

3   Discussion

After the data interpretation phase, about one percent of induced rules are found to be interesting or unexpected to medical experts. In this section, the total KDD process is reviewed with respect to computational time.

Table 6 shows the total time required for KDD process. Each column shows two data sets for each process. Each row includes computational time required for each process. Totally, it takes about one month and three weeks to complete the whole KDD process for bacterial test database and side-effect database, respectively. It is notable that more than 60% of the process is devoted to the three preprocessing processes: data selection, cleaning and reduction. Especially, as for the bacterial test dataset, 22.75 days (79.5%) are used for three processes. It is because domain knowledge should be acquired for generalization of data, as discussed in Section 3. On the other hand, only 4 to 8 percent of total time is spent for data mining process. In the case of bacterial test database, only 2.5 days (8%) is used for rule induction. Therefore, these empirical results suggest that the main KDD processes should be preprocessing rather than discovery of patterns from data.

Table 6. Total Time Required for KDD process

KDD Process	Bacterial Test	Side-Effect
Data Selection	2.3	7.3
Data Cleaning	5.2	2.5
Data Reduction	15.25	2.3
Data Mining	2.5	0.75
Data Interpretation	7.0	7.0
Total Time	32.25	19.85

References

1. Abiteboul, S., Hull, R., and Vianu, V. 1995. Foundations of Databases, Addison-Wesley, NY.
2. Fayyad, U., Piatetsky-Shapiro, G. and Smyth, P. 1996. The KDD Process for Extracting Useful Knowledge from Volumes of data. CACM, 39: 27-34.
3. Pawlak, Z. 1991. Rough Sets, Kluwer Academic Publishers, Dordrecht.
4. Polkowski, L. and Skowron, A. 1998. Rough Sets in Knowledge Discovery Vol.1 and 2, Physica-Verlag, Heidelberg.
5. Tsumoto, S. 1998. Automated extraction of medical expert system rules from clinical databases based on rough set theory, Information Sciences, 112, 67-84.
6. Tsumoto, S. 1999. Rule Discovery in Large Time-Series Medical Databases. In: Proc. 3rd European Conference on Principles of Knowledge Discovery and Data Mining (PKDD), LNAI 1704, Springer Verlag, 23-31.
7. Van Bemmel,J. and Musen, M. A.1997. Handbook of Medical Informatics, Springer-Verlag, NY.

Knowledge Discovery Using Least Squares Support Vector Machine Classifiers: A Direct Marketing Case

S. Viaene[1], B. Baesens[1], T. Van Gestel[2], J.A.K. Suykens[2], D. Van den Poel[3], J. Vanthienen[1], B. De Moor[2], and G. Dedene[1]

[1] K.U.Leuven, Dept. of Applied Economic Sciences,
Naamsestraat 69, B-3000 Leuven, Belgium
{Stijn.Viaene,Bart.Baesens,Jan.Vanthienen,Guido.Dedene}@econ.kuleuven.ac.be
[2] K.U.Leuven, Dept. of Electrical Engineering ESAT-SISTA,
Kardinaal Mercierlaan 94, B-3001 Leuven, Belgium
{Tony.Vangestel,Johan.Suykens,Bart.Demoor}@esat.kuleuven.ac.be
[3] Ghent University, Dept. of Marketing,
Hoveniersberg 24, B-9000 Ghent, Belgium
{Dirk.Vandenpoel}@rug.ac.be

Abstract. The case involves the detection and qualification of the most relevant predictors for repeat-purchase modelling in a direct marketing setting. Analysis is based on a wrapped form of feature selection using a sensitivity based pruning heuristic to guide a greedy, step-wise and backward traversal of the input space. For this purpose, we make use of a powerful and promising least squares version (LS-SVM) for support vector machine classification. The set-up is based upon the standard R(ecency) F(requency) M(onetary) modelling semantics. Results indicate that elimination of redundant/irrelevant features allows to significantly reduce model complexity. The empirical findings also highlight the importance of Frequency and Monetary variables, whilst the Recency variable category seems to be of lesser importance. Results also point to the added value of including non-RFM variables for improving customer profiling.

1 Introduction

The main objective of this paper involves the detection and qualification of the most relevant variables for repeat-purchase modelling in a direct marketing setting. This knowledge is believed to vastly enrich customer profiling and thus contribute directly to more targeted customer contact.

The empirical study focuses on the *purchase incidence*, i.e. the issue whether or not a purchase is made from any product category offered by the direct mailing company. Standard R(ecency) F(requency) M(onetary) modelling semantics underly the discussed purchase incidence model [3]. This binary (buyer vs. non-buyer) classification problem is being tackled in this paper by using least squares support vector machine (LS-SVM) classifiers. LS-SVM's have recently been introduced in the literature [11] and excellent benchmark results have been

D.A. Zighed, J. Komorowski, and J. Żytkow (Eds.): PKDD 2000, LNAI 1910, pp. 657–664, 2000.
© Springer-Verlag Berlin Heidelberg 2000

reported [13]. Having constructed an LS-SVM classifier with all available predictors, we engage in a feature selection experiment. Feature selection has been an active area of research in the data-mining field for many years now. A compact, yet highly accurate model may come in very handy in (on-line) customer profiling systems. Furthermore, elimination of redundant and/or irrelevant features often improves the predictive power of a classifier, in addition to reducing model complexity. On top, by reducing the number of input features, both human understanding and computational performance can often be vastly enhanced.

Section 2 briefly elaborates on some response modelling issues including the description of the data set. In Section 3, we discuss the basic underpinnings of LS-SVM's for binary classification. The feature selection experiment and corresponding results are presented and discussed in Section 4.

2   The Response Modelling Case for Direct Marketing

2.1   Response Modelling and RFM

For mail-order response modelling, several alternative problem formulations have been proposed based on the choice of the dependent variable. In purchase incidence modelling [1,12] the main question is whether a customer will purchase during the next mailing period. Other authors have investigated related problems dealing with both the purchase incidence and the amount of purchase in a joint model [7]. A third alternative perspective for response modelling is to model interpurchase time through survival analysis or (split-)hazard rate models [4]. The purchase incidence model in our experiment uses the traditionally discussed (R)ecency, (F)requency and (M)onetary variables as the main predictor categories. In addition, some extra historical customer profiling variables have been included in the data set. This choice is motivated by the fact that most previous research cites them as being most predictive and because they are internally available at very low cost.

Cullinan [3] is generally considered as being the pioneer of RFM modelling in direct marketing. Since then, the literature has accumulated so many uses of these three variable categories, that there is overwhelming evidence both from academically reviewed studies as well as from practitioners' experience that the RFM variables are the most important set of predictors for modelling mail-order repeat purchasing [1,6]. However, when browsing the vast amount of literature, it becomes evident that only very limited attention has been devoted to selecting the right set of variables (and their operationalisations) for inclusion into the model of mail-order repeat buying.

2.2   The Data Set

We obtained Belgian data on past purchase behaviour at the order-line level, i.e. we know when a customer purchased what quantity of a particular product at what price as part of what order. The total sample size amounts to 5000 customers, of which 37.9% represented buyers. The (R)ecency, (F)requency and

Table 1. A listing of all features (both RFM and non-RFM) included in the direct marketing case.

Recency	Frequency	Monetary	Other
RecYearR	FrYearR	MonHistR	ProdclaT
RecYearN	FrYearN	MonHistN	ProdclaM
RecHistR	FrHistR	MonYearR	GenCust
RecHistN	FrHistN	MonYearN	GenInfo
		Ln(MonHistR)	Ndays
		Ln(MonHistN)	IncrHist
		Ln(MonYearR)	IncrYear
		Ln(MonYearN)	RetMerch
			RetPerc

(M)onetary variables have then been modelled as described in detail in [12]. Here, we briefly cover the basic semantics of the variables included in Table 1.

We used two time horizons for all RFM variables. The Hist horizon refers to the fact that the variable is measured between the period 1 July 1993 until 30 June 1997. The Year horizon refers to the fact that the variable is measured over the last year. All RFM variables have been modelled both with and without the occurrence of returned merchandise, indicated by R and N, respectively. Taking into account both time horizons (Year versus Hist) and inclusion versus exclusion of returned items (R versus N), we arrive at a 2 × 2 design in which each RFM variable is operationalised in 4 ways. The Recency variable is operationalised as the number of days since the last purchase. The Monetary variable is modelled as the total accumulated monetary amount of spending by a customer. Additionally, we include the natural log transformation (*Ln*) of all monetary variables as a means to reduce the skewness of the data distribution. The Frequency variable measures the number of purchase occasions in a certain time period. Apart from the RFM variables, we also included 9 other customer-profiling features, which have also been discussed in detail in [12]. The ProdclaT respectively ProdclaM variables represent the Total respectively Mean forward-looking weighted productindex. The weighting procedure represents the 'forward-looking' nature of a product category purchase, derived from another sample of data. The GenCust and GenInfo variables model the customer/company interaction on the subject of information requests and complaints. The length of the customer relationship is quantified by means of the Ndays variable. The IncrHist and IncrYear variables measure the increased spending frequency over the entire customer history and over the last year, respectively. The RetMerch variable is a binary variable indicating whether the customer has ever returned an item, that was previously ordered from the mail-order company. The RetPerc variable measures the total monetary amount of returned orders divided by the total amount of spending.

Notice that all missing values were handled by the mean imputation procedure [8] and that all predictor variables were normalized to zero mean and unit variance prior to their inclusion in the model.

3   Least Squares SVM Classification

3.1   The LS-SVM Classifier

Given a Training set of N data points $\{y_k, x_k\}_{k=1}^{N}$, where $x_k \in \Re^n$ is the k-th input pattern and $y_k \in \{-1, 1\}$ is the k-th output pattern, Vapnik's SVM classifier formulation [2,9,14] is modified by Suykens [11] into the following LS-SVM formulation:

$$\min_{w,b,e} \mathcal{J}(w, e) = \frac{1}{2} w^T w + \gamma \frac{1}{2} \sum_{i=1}^{N} e_i^2 \tag{1}$$

subject to the equality constraints

$$y_i \left[w^T \varphi(x_i) + b \right] = 1 - e_i, \quad i = 1, ..., N. \tag{2}$$

This formulation now consists of equality instead of inequality constraints and takes into account a squared error with regularization term similar to ridge regression. The solution is obtained after constructing the Lagrangian $\mathcal{L}(w, b, e; \alpha) =$

$$\mathcal{J}(w, e) - \sum_{i=1}^{N} \alpha_i \{ y_i [w^T \varphi(x_i) + b] - 1 + e_i \} \tag{3}$$

where α_i are the Lagrange multipliers. After taking the conditions for optimality, one obtains the following linear system [11]:

$$\begin{bmatrix} 0 & Y^T \\ Y & \Omega + \gamma^{-1} I \end{bmatrix} \begin{bmatrix} b \\ \alpha \end{bmatrix} = \begin{bmatrix} 0 \\ 1 \end{bmatrix} \tag{4}$$

where $Z = [\varphi(x_1)^T y_1; ...; \varphi(x_N)^T y_N]$, $Y = [y_1; ...; y_N]$, $1 = [1; ...; 1]$, $\alpha = [\alpha_1; ...; \alpha_N]$, $\Omega = ZZ^T$ and Mercer's condition [11] is applied within the Ω matrix

$$\begin{aligned} \Omega_{ij} &= y_i y_j \, \varphi(x_i)^T \varphi(x_j) \\ &= y_i y_j \, K(x_i, x_j). \end{aligned} \tag{5}$$

For the kernel function $K(\cdot, \cdot)$ one typically has the following choices: $K(x, x_i) = x_i^T x$ (linear kernel), $K(x, x_i) = (x_i^T x + 1)^d$ (polynomial kernel of degree d), $K(x, x_i) = \exp\{-\|x - x_i\|_2^2 / \sigma^2\}$ (RBF kernel), $K(x, x_i) = \tanh(\kappa \, x_i^T x + \theta)$ (MLP kernel), where d, σ, κ and θ are constants. Notice that the Mercer condition holds for all σ and d values in the RBF and the polynomial case, but not for all possible choices of κ and θ in the MLP case. The LS-SVM classifier is then constructed as follows:

$$y(x) = \text{sign} \left(\sum_{i=1}^{N} \alpha_i y_i K(x, x_i) + b \right). \tag{6}$$

Note that the matrix in (4) is of dimension $(N+1) \times (N+1)$. For large values of N, this matrix cannot easily be stored, such that an iterative solution method for solving it is needed. A Hestenes-Stiefel conjugate gradient algorithm is suggested in [10] to overcome this problem. Basically, the latter rests upon a transformation of the matrix in (4) to a positive definite form [10].

3.2 Calibrating the RBF LS-SVM Classifier

All classifiers were trained using RBF kernels. Estimation of the generalisation ability of the RBF LS-SVM classifier is then realised by the following experimental set-up [13]:

1. Set aside $\frac{2}{3}$ of the data for the training/validation set and the remaining $\frac{1}{3}$ for testing.
2. Perform 10-fold cross validation on the training/validation data for each (σ, γ) combination from the initial candidate tuning sets Σ and Γ typically chosen as follows :
   $\Sigma = \{0.5, 5, 10, 15, 25, 50, 100, 250, 500\} \cdot \sqrt{n}$,
   $\Gamma = \{0.01, 0.5, 1, 10, 50, 100, 500, 1000\} \cdot \frac{1}{N}$.
   The square root $\sqrt{n}$ of the number of inputs n is introduced since $\|x - x_i\|_2^2$ in the RBF kernel is proportional to n and the factor $1/N$ is introduced such that the misclassification term $\gamma \sum_{i=1}^{N} e_i^2$ is normalized with the size of the data set.
3. Choose optimal (σ, γ) from the initial candidate tuning sets Σ and Γ by looking at the best cross validation performance for each (σ, γ) combination.
4. Refine Σ and Γ iteratively by means of a grid search mechanism in order to further optimize the tuning parameters (σ, γ). In our experiments, we repeated this step three times.
5. Construct the LS-SVM classifier using the total training/validation set for the optimal choice of the tuned hyperparameters (σ, γ).
6. Assess the generalization ability by means of the independent test set.

4 The Feature Selection Experiment

Feature selection effectively starts at the moment the LS-SVM classifier has been constructed on the full set of n available predictors. The feature selection procedure is based upon a (greedy) best-first heuristic, guiding a backward search mechanism through the feature space [5]. The mechanics of the implemented heuristic for assessing the sensitivity of the classifier to a certain input feature are quite straightforward. We apply a strategy of constant substitution in which a feature is perturbed to its mean while all other features keep their values and compute the impact of this operation on the performance of the obtained LS-SVM classifier without re-estimation of the LS-SVM parameters α_k and b. This assessment is done using the separate Pruning set, in order to obtain an unbiased estimate of the change in classification accuracy of the constructed classifier. Fig. 1 provides a concise overview of the different steps of the experimental procedure.

Starting with a full feature set F_1, all n inputs are pruned sequentially, i.e. one by one. The first feature f_k to be removed, is determined at the end of *Step 1* (task (4)). After having removed this feature from F_1, the reduced feature set $F_2 = F_1 \setminus \{f_k\}$ is used for subsequent feature removal. At this moment, an iteration of identical *Steps i* is started, in which, in a first phase, the LS-SVM parameters α_k and b are re-estimated on the Training set (task (1) of *Step i*),

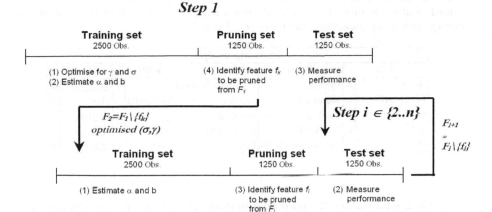

Fig. 1. Experimental set-up consisting of a first step for constructing an optimised LS-SVM classifier on a full feature set and of a subsequent iteration of pruning *Steps i*.

Table 2. Empirical assessment of the RBF LS-SVM classifier trained on the full feature set, i.e. *Full model*, vis-a-vis the RBF LS-SVM classifier trained on the reduced feature set, i.e. *Reduced Model*. *Majority* stands for the majority prediction error.

Results	Full Model	Reduced Model	Majority
Training (2500 Obs.)	77.36%	76.04%	62%
Pruning (1250 Obs.)	76.72%	77.20%	62%
Test (1250 Obs.)	73.92%	73.52%	62%
Features	25	9	0

however without re-calibrating for σ and γ[1], and generalisation ability of the classifier is quantified on the independent Test set (task (2) of *Step i*). Again, feature sensitivities of the resulting classification model (without re-estimation of α_k and b) are assessed on the Pruning set to identify the feature to which the classifier is least sensitive when perturbed to its mean (task (3) of *Step i*). This feature is then pruned from the remaining feature subset and disregarded for further analysis. The pruning procedure is thereupon resumed with a reduced feature set (*Step i + 1*), until all input features are eventually removed. Once all features have been pruned, the prefered reduced model is then determined by means of the highest Pruning set performance.

---

[1] Notice that the originally optimised γ and σ values obtained in task (1) of *Step 1* remain unchanged during the entire feature selection phase. Experimental evaluation showed that this implementation heuristic significantly speeds up the pruning procedure, without having a detrimental effect on the predictive performance of the reduced feature set.

Table 3. Order of feature removal using the pruning procedure presented in section 4. Each feature is qualified by its category with r, f, m, o respectively standing for recency, frequency, monetary and other (cf. Table 1).

Pruning Steps									
1-5		**6-10**		**11-15**		**16-20**		**21-25**	
RetPerc	o	ProdclaM	o	RecHistN	r	FrYearN	f	MonYearR	m
Ln(MonHistN)	m	MonHistR	m	IncrHist	o	Ln(MonHistR)	m	MonYearN	m
RecHistR	r	IncrYear	o	RecYearR	r	MonHistN	m	GenInfo	o
Ndays	o	Ln(MonYearR)	m	RecYearN	r	GenCust	o	FrHistR	f
ProdclaT	o	Ln(MonYearN)	m	FrHistN	f	RetMerch	o	FrYearR	f

Table 2 summarises the empirical findings of the pruning procedure for the RFM case. We contrasted the full model results with those of a binary logistic regression and concluded that the RBF LS-SVM classifier outperformed the latter significantly. Observe how the suggested feature selection method allows to significantly reduce the model complexity (from 25 to 9 features) without any significant degradation of the generalisation behaviour on the independent Test set. The Test set performance amounts to 73.92% for the full model and 73.52% for the reduced model.

The order of feature removal as depicted in Table 3, provides further insight into the relative importance of the predictor categories (cf. Table 1). The reduced model consists of the 9 features that are underlined in Table 3. This reduced set of predictors consists of Frequency, Monetary value and other (non-RFM) variables. It is especially important to note that the reduced model includes information on returned merchandise. Furthermore, notice the absence of the Recency component in the reduced feature set. Inspection of the order of removal of features, while further pruning this reduced feature set, highlights the importance of the Frequency variables. More specifically, the last two variables to be removed belong to this predictor category. Remark that a feature set consisting of only these two features, still yields a percentage correctly classified of 72% on the Test set, which might be considered quite satisfactory.

5 Conclusion

In this paper, we applied an LS-SVM based feature selection wrapper to a real-life direct marketing case involving the modeling of repeat-purchase behaviour based on the well-known R(ecency) F(requency) M(onetary) framework. The sensitivity based, step-wise feature selection method constructed as a wrapper around the LS-SVM classifier allows to significantly reduce model complexity without degrading predictive performance. The empirical findings highlight the role of Frequency and Monetary variables in the reduced model, whilst the Recency variable category seems to be of lesser importance within the RFM model. Results also point to the beneficial effect of including non-RFM customer profiling variables for improving predictive accuracy.

Acknowledgements. This work was partly carried out at the Leuven Institute for Research in Information Systems (L.I.R.I.S.) of the Dept. of Applied Economic Sciences of the K.U.Leuven in the framework of the KBC Insurance Research Chair. This work was partially carried out at the ESAT laboratory and the Interdisciplinary Center of Neural Networks ICNN of the KULeuven and supported by grants and projects from the Flemish Gov. (Res. Counc. KULeuven: GOA-Mefisto; FWO-Flanders: res. projects and comm (ICCoS & ANMMM); IWT: STWW Eureka); the Belgium Gov. (IUAP-IV/02, IV-24); the Eur. Comm. (TMR, ERNSI). S. Viaene, holder of the KBC Research Chair, and B. Baesens are both Research Assistants of L.I.R.I.S. T. Van Gestel is a research assistant, J. Suykens is a postdoctoral researcher, B. De Moor is a senior research associate with the Fund for Scientific Research Flanders (FWO-Flanders), resp. D. Van den Poel is an assistant professor at the department of marketing at Ghent University. J. Vanthienen and G. Dedene are Senior Research Associates of L.I.R.I.S.

References

1. Bauer, A.: A direct mail customer purchase model. *Journal of Direct Marketing*, 2(3):16–24, 1988.
2. Cristianini, N. and Shawe-Taylor, J.: *An Introduction to Support Vector Machines*. Cambridge University Press, 2000.
3. Cullinan, G. J.: *Picking them by their batting averages' recency-frequency-monetary method of controlling circulation*. Manual release 2103. Direct Mail/Marketing Association. N.Y., 1977.
4. Dekimpe, M. G. and Degraeve, Z.: The attrition of volunteers. *European Journal of Operations Research*, 98:37–51, 1997.
5. John, G., Kohavi, R. and Pfleger, K.: Irrelevant features and the subest selection problem. *Machine Learning: proceedings of the Eleventh International Conference, San Francisco*, pages 121–129, 1994.
6. Kestnbaum, R. D.: Quantitative database methods. *The Direct Marketing Handbook*, pages 588–597, 1992.
7. Levin, N. and Zahavi, J.: Continuous predictive modeling: a comparative analysis. *Journal of Interactive Marketing*, 12(2):5–22, 1998.
8. Little, R.J.A.: Regression with missing x's: a review. *Journal of the American Statistical Association*, 87(420):1227–1230, 1992.
9. Schölkopf, B., Burges, C., and Smola, A.: *Advances in Kernel Methods - Support Vector Learning*. MIT Press, 1998.
10. Suykens, J. A. K., Lukas, L., Van Dooren, P., De Moor, B. and Vandewalle, J.: Least squares support vector machine classifiers: a large scale algorithm. *ECCTD'99 European Conf. on Circuits Theory and Design, Stresa, Italy*, pages 839–842, 1999.
11. Suykens, J. A. K. and Vandewalle, J.: Least squares support vector machine classifiers. *Neural Processing Letters*, 9(3):293–300, 1999.
12. Van den Poel, D. : *Response Modeling for Database Marketing using Binary Classification*. Phd. Dissertation. K.U. Leuven, 1999.
13. Van Gestel, T., Suykens, J.A.K., Baesens, B., Viaene, S., Vanthienen, J., Dedene, G., De Moor, B. and Vandewalle, J.: Benchmarking least squares support vector machine classifiers. *CTEO, Technical Report 0037, K.U. Leuven, Belgium*, 2000.
14. Vapnik, V.: *Statistical learning theory*. John Wiley, New-York, 1998.

Lightweight Document Clustering

Sholom M. Weiss, Brian F. White, and Chidanand V. Apte

IBM T.J. Watson Research Center

P.O. Box 218, Yorktown Heights, NY 10598, USA

sholom@us.ibm.com, bfwhite@us.ibm.com, apte@us.ibm.com

Abstract. A lightweight document clustering method is described that operates in high dimensions, processes tens of thousands of documents and groups them into several thousand clusters, or by varying a single parameter, into a few dozen clusters. The method uses a reduced indexing view of the original documents, where only the k best keywords of each document are indexed. An efficient procedure for clustering is specified in two parts (a) compute k most similar documents for each document in the collection and (b) group the documents into clusters using these similarity scores. The method has been evaluated on a database of over 50,000 customer service problem reports that are reduced to 3,000 clusters and 5,000 exemplar documents. Results demonstrate efficient clustering performance with excellent group similarity measures.

1 Introduction

The objective of document clustering is to group similar documents together, assigning them to the same implicit topic. Why is document clustering of interest? The original motivation was to improve the effectiveness of information retrieval. Standard information retrieval techniques, such as nearest neighbor methods using cosine distance, can be very efficient when combined with an inverted list of word to document mappings. These same techniques for information retrieval perform a variant of dynamic clustering, matching a query or a full document to their most similar neighbors in the document database. Thus standard information retrieval techniques are efficient and dynamically find similarity among documents, reducing the value for information retrieval purposes of finding static clusters of large numbers of similar documents [Sparck-Jones, 1997].

The advent of the web has renewed interest in clustering documents in the context of information retrieval. Instead of pre-clustering all documents in a database, the results of a query search can be clustered, with documents appearing in multiple clusters. Instead of presenting a user with a linear list of related documents, the documents can be grouped in a small number of clusters, perhaps ten, and the user has an overview of different documents that have been found in the search and their relationship within similar groups of documents. One approach to this type of visualization and presentation is described in [Zamir *et al.*, 1997]. Here again though, the direct retrieval and linear list remains effective, especially when the user is given a "more like this" option that finds a subgroup of documents representing the cluster of interest to the user.

Document clustering can be of great value for tasks other than immediate information retrieval. Among these task are

D.A. Zighed, J. Komorowski, and J. Zytkow (Eds.): PKDD 2000, LNAI 1910, pp. 665–672, 2000.
© Springer-Verlag Berlin Heidelberg 2000

- summarization and label assignment, or
- dimension reduction and duplication elimination.

Let's look at these concepts by way of a help-desk example, where users submit problems or queries online to the vendor of a product. Each submission can be considered a document. By clustering the documents, the vendor can obtain a overview of the types of problems the customers are having, for example a computer vendor might discover that printer problems comprise a large percentage of customer complaints. If the clusters form natural problem types, they may be assigned labels or topics. New user problems may then be assigned a label and sent to the problem queue for appropriate response. Any number of methods can be used for document categorization once the appropriate clusters have been identified. Typically, the number of clusters or categories number no more than a few hundred and often less than 100.

Not all users of a product report unique problems to the help-desk. It can be expected that most problem reports are repeat problems, with many users experiencing the same difficulty. Given enough users who report the same problem, an FAQ, Frequently Asked Questions report, may be created. To reduce the number of documents in the database of problem reports, redundancies in the documents must be detected. Unlike the summary of problem types, many problems will be similar but still have distinctions that are critical. Thus, while the number of clusters needed to eliminate duplication of problem reports can be expected to be much smaller than the total number of problems reports, the number of clusters is necessarily relatively large, much larger than needed for summarization of problem types.

In this paper, we describe a new lightweight procedure for clustering documents. It is intended to operate in high dimensions with tens of thousands of documents and is capable of clustering a database into the moderate number of clusters need for summarization and label assignment or the very large number of clusters needed for the elimination of duplication.

2   Document Clustering Techniques

The classical k-means technique [Hartigan and Wong, 1979] can be applied to document clustering. Its weaknesses are well known. The number of clusters k must be specified prior to application. The summary statistic is a mean of the values for each cluster. The individual members of the the cluster can have a high variance and the mean may not be a good summary of the cluster members. As the number of clusters grow, for example to thousands of clusters, k-means clustering becomes untenable, approaching the $O(n^2)$ comparisons where n is the number of documents. However, for relatively few clusters and a reduced set of pre-selected words, k-means can do well[Vaithyanathan and Dom, 1999].

More recent attention has been given to hierarchical agglomerative methods [Griffiths et al., 1997]. The documents are recursively merged bottom up, yielding a decision tree of recursively partitioned clusters. The distance measures used to find similarity vary from single-link to more computationally expensive ones, but they are closely tied to nearest-neighbor distance. The algorithm works by recursively merging the single best pair of documents or clusters, making the computational costs prohibitive for document collections numbering in the tens of thousands.

To cluster very large numbers of documents, possibly with a large number of clusters, some compromises must be made to reduce the number of indexed words and the number of expected comparisons. In [Larsen and Aone, 1999], indexing of each document is reduced to the 25 highest scoring TF-IDF words (term frequency and inverse document frequency [Salton and Buckley, 1997], and then k-means is applied recursively, for k=9. While efficient, this approach has the classical weaknesses associated with k-means document clustering. A hierarchical technique that also works in steps with a small, fixed number of clusters is described in [Cutting *et al.*, 1992].

We will describe a new lightweight procedure that operates efficiently in high dimensions and is effective in directly producing clusters that have objective similarity. Unlike k-means clustering, the number of clusters is dynamically determined, and similarity is based on nearest-neighbor distance, not mean feature distance. Thus, the new document clustering method maintains the key advantage of hierarchical clustering techniques, their compatibility with information retrieval methods, yet performance does not rapidly degrade for large numbers of documents.

3   Methods and Procedures

3.1   Data Preparation

Clustering algorithms process documents in a transformed state, where the documents are represented as a collection of terms or words. A vector representation is used: in the simplest format, each element of the vector is the presence or absence of a word. The same vector format is used for each document; the vector is a space taken over the complete set of words in all documents. Clearly, a single document has a sparse vector over the set of all words. Some processing may take place to stem words to their essential root and to transform the presence or absence of a word to a score, such as TF-IDF, that is a predictive distance measure. In addition weakly predictive words, stopwords, are removed. These same processes can be used to reduce indexing further by measuring for a document's vector only the top k-words in a document and setting all remaining vector entries to zero.

An alternative approach to selecting a subset of features for a document, described in [Weiss *et al.*, 2000], assumes that documents are carefully composed and have effective titles. Title words are always indexed along with the k most frequent words in the document and any human-assigned key words.

Not all words are of the same predictive value and many approaches have been tried to select a subset of words that are most predictive. The main concept is to reduce the number of overall words that are considered, which reduces the representational and computational tasks of the clustering algorithm. Reduced indexing can be effective in these goals when performed prior to clustering. The clustering algorithm accepts as input the transformed data, much like any information retrieval system, and works with a vector representation that is a transformation of the original documents.

3.2   Clustering Methods

Our method uses a reduced indexing view of the original documents, where only the k best keywords of each document are indexed. That reduces a document's vector size and the computation time for distance measures for a clustering method. Our procedure for clustering is specified in two parts (a) compute k most similar documents (typically the top 10) for each document in the collection and (b) group the documents into clusters using these similarly scores. To be fully efficient, both procedures must be computationally efficient. Finding and scoring the k most similar documents for each document will be specified as a mathematical algorithm that processes fixed scalar vectors. The procedure is simple, a repetitive series of loops that accesses a fixed portion of memory, leading to efficient computation. The second procedure uses the scores for the k most similar documents in clustering the document. Unlike the other algorithms described earlier, the second clustering step does not perform a "best-match first-out" merging. It merges documents and clusters based on a "first-in first-out" basis.

Table 3.2 describes the data structures needed to process the algorithms. Each of these lists can be represented as a simple linear vector. Table 3.2 describes the steps for the computation of the k most similar documents, typically the top 10, for each document in the collection. Similarity or distance is measured by a simple additive count of words found in both documents that are compared plus their inverse document frequency. This differs from the standard TF-IDF formula in that term frequency is measured in binary terms, i.e. 0 or 1 for presence or absence. In addition the values are not normalized, just the sum is used. In a comparative study, we show that TF-IDF has slightly stronger predictive value, but the simpler function has numerous advantages in terms of interpretability, simple additive computation, and elimination of storage of term frequencies. The steps in Table 3.2 can readily be modified to use TF-IDF scoring.

The remaining task is to group the documents into clusters using these similarly scores. We describe a a single pass algorithm for clustering, with at most k*n comparisons of similarity, where n is the number of documents.

For each document D_i, the scoring algorithm produces a set of k documents, $\{D_j\}$, where j varies from 1 to k. Given the scores of the top-k matches of each document D_i, Table 3.2 describes the actions that may be taken for each matched pair during cluster formation. Documents are examined in a pairwise fashion proceeding with the first document and its top-k matches. Matches below a pre-set minimum score threshold are ignored. Clusters are formed by the document

doclist: The words (terms) in each document. A series of numbers; documents are separated by zeros. *example:* Sequence = 10 44 98 0 24 ... The first document has words 10, 44 and 98. The second document has words 24...

wordlist: The documents in which a word is found. A series of consecutive numbers pointing to specific document numbers.

word(c): A pointer to wordlist indicates the starting location of the documents for word c. To process all documents for word c, access word(c) through word(c+1)-1. *example:* word(1)=1, word(2)=4; wordlist={18 22 64 16 ...} Word 1 appears in the documents listed in locations 1, 2, and 3 in wordlist. The documents are 18, 22, and 64.

pv(c): predictive values of word c = 1+idf, where idf is 1/(number of documents where word c appears)

Table 1. *Definitions for Top-k Scoring Algorithm*

1. Get the next document's words (from doclist), and set all document scores to zero.
2. Get the next word, w, for current document. If no words remain, store the k documents with the highest scores and continue with step (i).
3. For all documents having word w (from wordlist), add to their scores and continue with step (ii).

Table 2. *Steps for Top-k Scoring Algorithm*

pairs not yet in clusters. Clusters are merged when the matched pair appear in separate clusters. As we shall see in Section 4, not allowing merging yields a very large number clusters whose members are highly similar. The single setting of the minimum score has a strong effect on the number of clusters; a high value produces a relatively large number of clusters and a zero value produces a relatively small number of clusters. Similarly, a high minimum score may leave some documents unclustered, while a low value clusters all documents. As an alternative to merging, it may be preferable to repeat the same document in multiple clusters. We do not report results on this form of duplication, typically done for smaller numbers of documents, but the procedure provides an option for duplicating documents across clusters.

3.3   Measures for Evaluation of Clustering Results

How can we objectively evaluate clustering performance? Very often, the objective measure is related to the clustering technique. For example, k-means clustering can measure overall distance from the mean. Techniques that are based on nearest neighbor distance, such as most information retrieval techniques, can measure distance from the nearest neighbor or the average distance from other cluster members.

For our clustering algorithm, distance is measured mostly in terms of counts of words present in documents. A natural measure of cluster performance is

1. If score for D_i and D_j is less than Minimum Score, next pair.
2. If D_i and D_j are already in the same cluster, next pair.
3. If D_i is in a cluster and D_j isn't, add D_j to the D_i cluster, next pair.
4. Cluster Merge Step: if both D_i and D_j are in separate clusters:
   (a) If action plan is "no merging", next pair.
   (b) If action plan is "repeat documents", repeat D_j in all the D_i clusters, next pair.
   (c) Merge the D_i cluster with D_j cluster, next pair.

Table 3. *Actions for Clustering Document Pairs*

the average number of indexed words per cluster, i.e. the local dictionary size. Analogous measures of cluster "cohesion," that count the number common words among documents in a cluster, have been used to evaluate performance[Zamir *et al.*, 1997]. The average is computed by weighing the number of documents in the cluster as in equation 1, where N is the total number number of documents, m is the number of clusters, $Size_k$ is the number of documents in the k-th cluster, and $LDict_k$ is the number of indexed words in the k-th cluster.

$$AverageDictionarySize = \sum_{k=1}^{m} \frac{Size_k}{N} \cdot LDict_k \qquad (1)$$

Results of clustering are compared to documents randomly assigned to the same size clusters. Clearly, the average dictionary size for computed clusters should be much smaller than those for randomly assigned clusters of the same number of documents.

3.4   Summarizing Clustering Results

The same measure of evaluation can be used to find exemplar documents for a cluster. The local dictionary of a document cluster can be used as a virtual document that is matched to the members of the cluster. The top-k matched documents can be considered a ranked list of exemplar documents for the cluster.

Selecting exemplar documents from a cluster is a form of summary of the cluster. The technique for selecting the exemplars is based on matching the cluster's dictionary of words to its constituent documents. The words themselves can provide another mode of summary for a cluster. The highest frequency words in the local dictionary of a cluster often can distinguish a cluster from others. If only a few words are extracted, they may be considered a label for the cluster.

4   Results

To evaluate the performance of the clustering algorithms, we obtained 51,110 documents, taken from reports from customers having IBM AS/400 computer systems. These documents were constructed in real-time by customer service representatives who record their phone dialog with customers encountering problems with their systems.

The documents were indexed with a total of 21,682 words in a global dictionary computed from all the documents. Table 4 summarizes the results for clustering the document collection in terms of the number of clusters, the average cluster size, the ratio of the local dictionary size to random assignment, the percentage of unclustered documents, the minimum score for matching document pairs, and whether merging was used. The first row in the table indicates that 49 clusters were found with an average size of 1027 documents. A random cluster's dictionary was on average 1.4 times larger than the generated cluster; and 1.5% of the documents were not clustered. These results were obtained by using a minimum score of 1 and cluster merging was allowed. All results are for finding the top-10 document matches.

Cnum	AveSize	RndRatio	Unclust %	MinScore	Merge
49	1027.3	1.4	1.5	1	yes
86	579.6	1.4	2.5	2	yes
410	105.5	1.5	16.2	3	yes
3250	15.5	1.8	1.5	1	no
3346	14.9	1.8	2.5	2	no
3789	11.4	1.9	16.2	3	no

Table 4. *Results for Clustering Help-Desk Problems*

A single clustering run, one row in Table 4 currently takes 15 minutes on a 375 MHz RS6000 running AIX. The code is written in Java.

Exemplar documents were selected for each of the 3250 clusters found in the fourth entry of the table. For some large clusters, two or three exemplars were selected for a total of 5,000 exemplar documents. Using the same scoring scheme, each of the exemplars was matched to the original 51,110 documents. 98.4% of the documents matched at least one of of the exemplars, having at least one indexed word in common. 60.7% of the documents matched an exemplar of their assigned cluster, rather than an exemplar of an alternative cluster.

5   Discussion

The lightweight document clustering algorithms achieves our stated objectives. The process is efficient in high dimensions, both for large document collections and for large numbers of clusters. No compromises are made to partition the clustering process into smaller sub-problems. All documents are clustered in one stage.

These clustering algorithms have many desirable properties. Unlike k-means clustering, the number of clusters is dynamically assigned. A single parameter, the minimum score threshold, effectively controls whether a large number of clusters or a much smaller number is chosen. By disallowing merging of clusters, we are able to obtain a very large number of clusters.

The success of clustering can be measured in terms of an objective function. In our case, we are using the local dictionary size of the cluster. In all instances, we see that clustering is far better than random assignment. As expected, the greater the number of clusters, the better the performance when measured by dictionary size. In the help-desk application, it is important to remove duplication, while still maintaining a large number of exemplar documents. The help-desk clusters have strong similarity for their documents, suggesting that they can be readily summarized by a one or two documents. For the largest number of clusters, dictionary size is nearly half that for random document assignment, far better than for smaller number of clusters.

The help-desk application is characterized by a large number of specialized indexed words for computer systems. Future applications will determine the generality of this approach. There is ample room for enhancements to the computer implementation that will lead to faster performance and a capability to run on far larger document collections.

References

[Cutting et al., 1992] D. Cutting, D. Karger, J. Pedersen, and J. Tukey. Scatter/Gather: a Cluster-based Approach to Browsing Large Document collections. In *Proceedings of the 15th ACM SIGIR*. ACM, 1992.

[Griffiths et al., 1997] A. Griffiths, H. Luckhurst, and P. Willett. Using interdocument similarity information in document retrieval systems. In P. Sparck-Jones, K. and. Willet, editor, *Readings in Information Retrieval*, pages 365–373. Morgan Kaufmann, 1997.

[Hartigan and Wong, 1979] J. Hartigan and M Wong. A k-means clustering algorithm. *Applied Statitsics*, 1979.

[Larsen and Aone, 1999] B. Larsen and C. Aone. Fast and Effective Text Mining Using Linear-time Document Clustering. In *Proceedings of the 5th International Conference on Knowledge Discovery ad Data Mining*, pages 16–22. ACM, 1999.

[Salton and Buckley, 1997] G. Salton and C. Buckley. Term-weighting approaches in automatic text retrieval. In P. Sparck-Jones, K. and. Willet, editor, *Readings in Information Retrieval*, pages 323–328. Morgan Kaufmann, 1997.

[Sparck-Jones, 1997] P. Sparck-Jones, K. and. Willet. Chapter 6 - techniques. In P. Sparck-Jones, K. and. Willet, editor, *Readings in Information Retrieval*, pages 305–312. Morgan Kaufmann, 1997.

[Vaithyanathan and Dom, 1999] S. Vaithyanathan and B. Dom. Model Selection in Unsupervised Learning with Applications to Document Clustering. In *Proceedings International Conference on Machine Learning*, 1999.

[Weiss et al., 2000] S. Weiss, B. White, C. Apté, and F. Damerau. Lightweight document matching for help-desk applications. *IEEE Intelligent Systems*, page in press, 2000.

[Zamir et al., 1997] O. Zamir, O. Etzioni, O. Madani, and R. Karp. Fast and Intuitive Clustering of Web Documents. In *Proceedings of the 3rd International Conference on Knowledge Discovery ad Data Mining*. Morgan Kaufmann, 1997.

Automatic Category Structure Generation and Categorization of Chinese Text Documents

Hsin-Chang Yang and Chung-Hong Lee

Department of Information Management, Chang Jung University,
Tainan, 711, Taiwan
{hcyang, leechung}@mail.cju.edu.tw

Abstract. Recently knowledge discovery and data mining in unstructured or semi-structured texts(*text mining*) has been attracted lots of attention from both commercial and research fields. One aspect of text mining is on automatic text categorization, which assigns a text document to some predefined category according to the correlation between the document and the category. Traditionally the categories are arranged in hierarchical manner to achieve effective searching and indexing as well as easy comprehension for human. The determination of categories and their hierarchical structures were most done by human experts. In this work, we developed an approach to automatically generate categories and reveal the hierarchical structure among them. We also used the generated structure to categorize text documents. The document collection is trained by a self-organizing map to form two feature maps. We then analyzed the two maps to obtain the categories and the structure among them. Although the corpus contains documents written in Chinese, the proposed approach can be applied to documents written in any language and such documents can be transformed into a list of separated terms.

1 Introduction

Text categorization concerns about assigning a text document to some predefined category. When a set of documents are well categorized, both storage and retrieval of these documents can be effectively achieved. A primary characteristic of text categorization is that a category reveal the theme of those documents under this category, that is, these documents form a natural cluster of similar context. Thus text categorization provides some knowledge about the document collection. An interesting argument about text categorization is that before we can acquiring knowledge through text categorization, we need some knowledge to correctly categorize documents. One of the key knowledge we need to perform text categorization is the generation of categories and the structure among categories. Traditionally such knowledge is provided by human experts or some semi-automatic mechanisms which incorporate human knowledge as well as computing techniques, for example, natural language processing. Fully automatic category generation is difficult due to two reasons. First, we need to select some important words that can be assigned as category terms(or labels). We

D.A. Zighed, J. Komorowski, and J. Żytkow (Eds.): PKDD 2000, LNAI 1910, pp. 673–678, 2000.
© Springer-Verlag Berlin Heidelberg 2000

use these words to represent categories and provide indexing information for the categorized documents. A proper selection of a term should represent the general theme of the documents under the corresponding category. Such selections were always done by human linguistic experts because we need an insight of the underlying semantic structure of a language. Such insight is hard to automate. Certain techniques such as word frequency counts may help, but it is the human experts to decide which terms are most discriminative and representative. Second, the categories were always arranged in a tree-like hierarchical structure. This hierarchy reveals the relationships among categories. A category associated to higher level nodes of the hierarchy represents a more general theme than those associated to lower level nodes. The parent category in the tree should represent the common theme of its child categories. Thus the retrieval of documents of a particular interest can be effectively achieved through such hierarchy. The hierarchy must be constructed carefully so that irrelevant categories may not be the children of the same parent category. A thorough investigation about the semantic relations among category terms must be conducted to establish a well-organized hierarchy. Therefore, most of text categorization systems focus on categorizing documents according to some human-specified category terms and hierarchy, rather than on generating category terms and hierarchy.

In this work, we provide a method which can automatically generate category terms and establish category structure. Traditionally, category terms are selected according to the popularity of words in the majority of documents. This can be done by human engineering, statistical training, or a combination of the two. In this work, we reverse the text categorization process to obtain the category terms. First we should cluster the documents. The document collection is trained by the self-organizing maps (SOM) [1] algorithm to generate two feature maps, namely the document cluster map and the word cluster map. A neuron in these two maps represents a document cluster and a word cluster respectively. Through self-organizing process the distribution of neurons in the map reveals the similarities among clusters. We select category terms according to such similarities. To generate the category terms, dominating neurons in the document cluster map are first selected as centroids of some *super-clusters*, which each represents a general category. The words associated to the corresponding neurons in word cluster map are then used to select category terms. The structure of categories may also be revealed by examining the correlations among neurons in the two maps.

The corpus used to train the maps consists of documents written in Chinese. We decide to use Chinese corpus by two reasons. First, over a quarter of population on earth use Chinese as their native language. However, techniques for mining Chinese documents are relatively less than for documents written in English. Second, traditional text mining (and other related fields) techniques developed based on English corpora are not quite suitable for processing Chinese documents. Nowadays, demands for Chinese-based text mining techniques arise rapidly by many commercial fields. A difficult problem in developing Chinese-based text mining techniques is that research on lexical analysis of Chinese documents is still in its infancy. Besides, there is still no attempt in discovering knowledge from such a large repository so far. Therefore methodologies developed for English documents play an inevitable role in developing a model for knowledge discovery in Chinese documents. In this work, traditional term-based representation scheme in information retrieval field is adopted for document encoding. The same method developed in our work can naturally extend to English documents because these documents can always be represented by a list of terms.

2   Related Works

Text categorization or classification systems usually categorize documents according to some predefined category hierarchy. An example is the works by CMU text learning group[2] where they used Yahoo hierarchy to categorize documents. Another approach is to automatically generate the category terms as well as hierarchy. McCallum and Nigam [3] used a bootstrapping process to generate new terms from a set of human-provided keywords. Human intervention is still required in their work. Rauber and Merkl [4] used the self-organizing map to cluster text documents. They labeled each neuron with a set of keywords that were selected from the input vectors mapped to this neuron. Those keywords that contribute less quantization error were selected. These keywords together form the label of a cluster, rather than a single term for a category. Moreover, the hierarchical structure among these keywords were not revealed.

3   Automatic Category Hierarchy Generation

To obtain the category hierarchy, we first cluster documents by SOM using the same data set and method in [5] to generate two cluster maps, namely the document cluster map and the word cluster map. A neuron in the document cluster map represents a cluster of documents. Documents associated with neighboring neurons contain words that often co-occur in these documents. Thus we may form a super-cluster by combining neighboring neurons. To form a super-cluster, we first define distance between two clusters: $D(i,j) = ||\mathbf{G}_i - \mathbf{G}_j||$, where i and j are the neuron indices of the two clusters and $\mathbf{G}_i$ is the two-dimensional grid location of neuron i. $||\mathbf{G}_i - \mathbf{G}_j||$ measures the Euclidean distance between the two coordinates $\mathbf{G}_i$ and $\mathbf{G}_j$. We also define the dissimilarity between two clusters: $\mathcal{D}(i,j) = ||\mathbf{w}_i - \mathbf{w}_j||$, where $\mathbf{w}_i$ is the synaptic weight vector of neuron i. We may compute the supporting cluster similarity $\mathcal{S}_i$ for a neuron i from its neighboring neurons by

$$S(i,j) = \frac{\text{doc}(i)\text{doc}(j)}{F(D(i,j)\mathcal{D}(i,j))}$$
$$\mathcal{S}_i = \sum_{j \in B_i} S(i,j), \tag{1}$$

where $\text{doc}(i)$ is the number of documents associated to neuron i in the document cluster map and B_i is the set of neuron index in the neighborhood of neuron *i*. The function $F : \mathbf{R}^+ \rightarrow \mathbf{R}^+$ is a monotonically increasing function. A dominating neuron is the neuron which has locally maximal supporting cluster similarity. We may select dominating neuron by the following algorithm:

1. Find the neuron with the largest supporting cluster similarity. Selecting this neuron as dominating neuron.
2. Eliminate its neighbor neurons so that they will not be considered as dominating neurons.
3. If there is no neuron left or the number of dominating neurons exceeds a predetermined value, stop. Otherwise goto step 1.

A dominating neuron is the centroid of a super-cluster, which contains several clusters. The *i*th cluster (neuron) belongs to the *k*th super-cluster if $\mathcal{D}(i, k) = \min_l \mathcal{D}(i, l)$, where l is a super-cluster. A super-cluster may be thought as a category which contains several sub-categories. Let C_k denote the set of neurons that belong to the *k*th super-cluster. The category terms are selected from words in the word cluster map that associated to neurons in C_k. For all neurons $j \in C_k$, we select the n^*th word as the category term if

$$\sum_{j \in C_k} w_{j_{n^*}} = \max_{1 \leq n \leq N} \sum_{j \in C_k} w_{j_n}. \tag{2}$$

Eq. 2 selects the term that is the most important to a super-cluster since the weight of a synapse in a neuron reflects the willingness that the neuron wants to learn the corresponding input data, i.e. a word in our case.

The terms selected by Eq. 2 form the root nodes of the category hierarchies. The number of generated hierarchies is the same as the number of super-clusters. To find the children of these root nodes, we may apply the above process to each super-cluster. A set of sub-category will be obtained for each hierarchy. These sub-categories form the new super-clusters that are the children of the root node of the hierarchy. The category structure can then be revealed by recursively applying the same category generation process to each new-found super-cluster. We decrease the size of neighborhood in selecting dominating neurons when we try to find the sub-categories. In each category hierarchy, each leaf node represents an individual neuron in the trained map.

4   Automatic Text Categorization

A text document can be categorized into the developed category hierarchy as follows. An incoming document T with document vector $\mathbf{x}_T$ is compared to all neurons in the trained map to find its document cluster. The neuron with synaptic weight vector which is the closest to $\mathbf{x}_T$ is selected. The incoming document is categorized into the category where the neuron been associated to as a leaf node in one of the category hierarchies. Through this way, we can see that the task of text categorization has been done naturally by the category generation process.

5   Experimental Results

We applied our method on the Chinese news articles posted daily in the web by CNA(Central News Agency). In the preliminary experiments a corpus was constructed by randomly selecting 100 news articles posted in Aug. 1, 2, and 3, 1996. A word extraction process was applied to the corpus to extract Chinese words. Total 1475 words were extracted. To reduce the dimensionality of the feature vectors we discarded those words which occur only once in a document. This reduced the number of words to 563. We also constructed a self-organizing map which contains 64 neurons in 8×8 grid format. The number of neurons was determined experimentally such that a better clustering can be achieved. Each neuron contains 563 synapses. The initial training gain was set to 0.4 and the maximal training time was set to 100. After training we labeled the map by documents and words respectively, and obtained the document cluster map and the word cluster map. We applied the category generation process to the document cluster map to obtain the category hierarchies. Fig. 1 depicts the result after the first application of the generation process. Each circle represents a neuron in the maps. The neuron indices are shown at the upper-left corner of each neuron. The size of a circle depicts the supporting similarity of the corresponding neuron obtained by Eq. 1. A dominating neuron for a super-cluster is marked by a cross. The 2-tuple (x, y) beneath each circle shows that the corresponding neuron(cluster) belongs to the super-cluster x where y is its supporting similarity. For example, the 35th neuron has a 2-tuple $(1, 3.0)$ and a cross on it, which shows that this neuron is a dominating neuron for super-cluster 1. The supporting similarity for this neuron is 3.0. Neurons belonging to the same super-cluster locate closely in Fig. 1. The category terms for super-clusters are shown beneath the 2-tuple. Fig. 2 depicts the final category hierarchy. Each tree depicts a category hierarchy where the number in the root node depicts the super-cluster found in Fig. 1. Each node in a tree represents a cluster in the document cluster map. The parent node of some child nodes represents a super-cluster. The number enclosed in every leaf node is the neuron index of its associated cluster in the document cluster map. The root node of each tree is the super-cluster found in the first application of category generation process.

6   Conclusions

In this paper, we present a method to automatically generate category terms and hierarchies. The documents were first transformed to a set of feature vectors. The vectors were used as input to train the self-organizing map. Two maps, namely the word cluster map and the document cluster map, were obtained by labeling the neurons in the map with words and documents respectively. An automatic category generation process was applied to the document cluster map to find some dominating neurons that are centroids of some super-clusters. The category terms of super-clusters were also determined. The same process were applied recursively to each super-clusters to reveal the structure of the categories. Our method used neither human-provided terms nor predefined category structure. Text categorization can easily be achieved in our method.

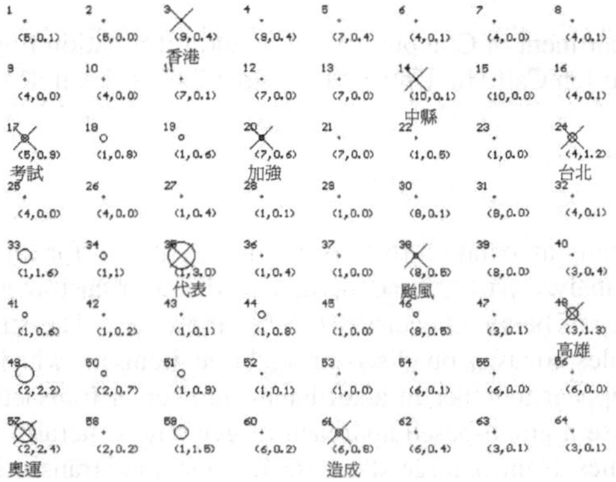

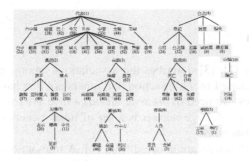

Fig. 1. The supporting similarity for each neuron in the first application of category generation process.

Fig. 2. The category hierarchies of the training documents.

References

1. Kohonen, T.:Self-Organizing Maps. Springer-Verlag, Berlin Heidelberg New York (1997)
2. Grobelnik, M., Mladenić, D.:Efficient Text Categorization. Proceedings of Text Mining Workshop on ECML-98, Chemnitz, Germany (1998)
3. McCallum, A., Nigam, K.:Text Classification by Bootstrapping with Keywords, EM and Shrinkage. Proceedings of ACL '99 Workshop for Unsupervised Learning in Natural Language Processing (1999) 52-58
4. Rauber, A., Merkl, D.:Using Self-Organizing Maps to Organize Document Archives and to Characterize Subject Matter: How to Make a Map Tell the News of the World. Proceedings of 10th International Conference on Database and Expert Systems Applications. Florence, Italy (1999) 302-311
5. Lee, C. H., Yang, H. C.:A Web Text Mining Approach Based on Self-Organizing Map. Proceedings of ACM CIKM'99 2nd Workshop on Web Information and Data Management. Kansas City, Missouri, USA (1999) 59-62

Mining Generalized Multiple-level Association Rules

Show-Jane Yen

Department of Computer Science and Information Engineering
Fu Jen Catholic University, Taipei 242, Taiwan, R.O.C.
sjyen@csie.fju.edu.tw

Abstract. Mining association rules is an important task for knowledge discovery. We can analyze past transaction data to discover customer behaviors such that the quality of business decision can be improved. The strategy of mining association rules focuses on discovering large itemsets which are groups of items which appear together in a sufficient number of transactions. In this paper, we propose a graph-based approach to generate generalized multiple-level association rules from a large database of customer transactions, which describes the associations among items in any concept level. This approach is to scan the database once to construct an association graph, and then traverse the graph to generate large itemsets.

1 Introduction

An *association rule* [1], [3] describes the associations among items in which when some items are purchased in a transaction, the others are purchased, too. In order to find association rules, we need to discover all *large itemsets* from a large database of customer transactions. A large itemset is a set of items which appears often enough within the same transactions.

The following definitions are adopted from [1]. A transaction t *supports* an item x if x is in t. A transaction t supports an itemset X if t supports every item in X. The *support for an itemset* is defined as the ratio of the total number of transactions which support this itemset to the total number of transactions in the database. To make the discussion easier, occasionally, we also let the total number of transactions which support the itemset denote the support for the itemset. Hence, a large itemset is an itemset whose support is no less than a certain user-specified *minimum support*. An itemset of length k is called a k-itemset and a large itemset of length k a large k-itemset.

After discovering all large itemsets, the association rules can be generated as follows: If the large itemset $Y=I_1I_2...I_k$, $k \geq 2$, all rules that reference items from the set $\{I_1, I_2, ..., I_k\}$ can be generated. The antecedent of each of these rules is a subset X of Y, and the consequent Y-X. The confidence of $X \Rightarrow Y\text{-}X$ in database D is the probability that when itemset X occurs in a transaction in D, itemset Y-X also occurs in the same transaction. That is, the ratio of the support for itemset Y to the support for itemset X. A generated rule is an association rule if its confidence achieves a certain user-specified minimum confidence.

D.A. Zighed, J. Komorowski, and J. Zytkow (Eds.): PKDD 2000, LNAI 1910, pp. 679-684, 2000.
© Springer-Verlag Berlin Heidelberg 2000

A *multiple-level association rule* [2] is an association rule, in which all items are described by a set of relevant attributes. Each attribute represents a certain concept, and these relevant attributes form a set of multiple-level concepts. For example, in Table 1, food items can be described by the relevant attributes "category", "content" and "brand," and attribute "category" represents the first-level concept (i.e., the highest level concept), attribute "content" the second-level concept and attribute "brand" the third-level concept. There is a set of domain values for an attribute. Each item in the database contains a domain value for each relevant attribute. For example, if the "category", "content" and "brand" of an item have the domain values "bread," "wheat" and "Wonder", respectively, then this item is described as "Wonder wheat bread" in the database.

Table 1. Item description

Category	Content	Brand
bread	wheat	Wonder
milk	chocolate	Dairyland
....	....	....
milk	2%	Firemost

From the items in the database, we can derive other items at different concept levels. The domain values of the attribute at the first (i.e., the highest) concept level are the items at the first concept level. An item at the kth concept level can be formed by combining a domain value of the attribute at the kth concept level with an item at the (k-1)th concept level. Hence, item "bread" is at the first concept level, item "wheat bread" at the second concept level, and item " Wonder wheat bread " at the third concept level. A *multiple-level association rule* [2] is an association rule which describes the associations among items at the same concept level. For each concept level, both minimum support and minimum confidence are specified.

One may relax the restriction of mining associations among items at the same concept level to allow the associations among items at any concept level. A *generalized multiple-level association pattern* is a large itemset in which each item may be at any concept level, and the support for the large itemset need to achieve the minimum support specified at the lowest concept level among the concept levels of all items in the large itemset. Besides, the support for each item in a generalized multiple-level association pattern needs to achieve the minimum support specified at its concept level. For an item I to be in a generalized multiple-level association pattern, the items at the corresponding higher concept levels of the item I need to be large at their corresponding concept levels. This is to avoid the generation of many meaningless combinations formed by the items at the corresponding lower concept level of the non-large items. For example, if "bread" is not a large item, item "wheat bread" which is at the corresponding lower concept level of item "bread" need not be further examined.

A *generalized multiple-level association rule* is an association rule which describes the associations among items at any concept level, whose confidence achieves the minimum confidence specified at the lowest concept level among the concept levels of all items in the corresponding generalized multiple-level association pattern. This paper focuses on the association pattern generation, because after generating the association patterns, the association rules can be generated from the corresponding association patterns.

2 Mining Generalized Multiple-level Association Patterns

In this section, we present the data mining algorithm GMLAPG (Generalized Multiple-Level Association Pattern Generation) to generate generalized multiple-level association patterns, which includes four phases: Large item generation phase, Numbering phase, Association graph construction phase and Association pattern generation phase.

2.1 Large Item Generation

Because each item in the database contains domain values of the relevant attributes, for an attribute, each domain value is arbitrarily given a unique number. Besides, each item in a transaction is denoted according to its domain values.

In the large item generation phase, GMLAPG scans the database to find all large items at every concept level and build the *bit vector* for each large item. The length of each bit vector is the number of transactions in the database. If an item appears in the ith transaction, the ith bit of the bit vector associated with this item is set to 1. Otherwise, the ith bit of the bit vector is set to 0. The bit vector associated with item i is denoted as BV_i. The number of 1's in BV_i is equal to the number of transactions which support the item i, that is, the support for the item i.

Example 1. Consider the database TDB in Table 2 in which there are three concept levels defined, and the items at each concept level are numbered. For example, in Table 2, item "211" can be the item "Wonder wheat bread", where the first number "2" represents the domain value "bread" of the attribute "category" at level-1, the second number "1" for the domain value "wheat" of the attribute "content " at level-2, and the third number "1" for the domain value "Wonder" of the attribute "brand" at level-3. Assume that the minimum supports $\mathfrak{S}_1$, $\mathfrak{S}_2$ and $\mathfrak{S}_3$ specified at the concept levels 1, 2 and 3 are 4, 3 and 3 transactions, respectively.

In the large item generation phase, GMLAPG finds all large items for every concept level from Table 2. For the first level, only the level-1 items in the transactions are considered. After the large item generation phase, the found level-1 large items are "1**" and "2**", because their supports are above the minimum support, and the associated bit vectors are (1111100) and (1110110), respectively, where the notation "*" represents any item. The level-2 large items are 11*, 12*, 21* and 22*, and the level-3 large items are 111, 211 and 221.

Table 2. A database TDB of transactions

TID	Itemset
100	{111, 121, 211, 221}
200	{111, 211, 222, 323}
300	{112, 122, 221, 411}
400	{111, 121}
500	{111, 122, 211, 221, 413}
600	{211, 323, 524}
700	{323, 411, 524, 713}

2.2 A Numbering Method

Property 1: The support for the itemset $\{i_1, i_2, ..., i_k\}$ is the number of 1's in $BVi_1 \wedge BVi_2 \wedge ... \wedge BVi_k$, where the notation "$\wedge$" is a logical AND operation.

Lemma 1. The support for an itemset X that contains both an item x_i and the item χ_i at the corresponding higher concept level of item x_i will be the same as the support for the itemset X-χ_i.

From Lemma 1, we want to discover all generalized multiple-level association patterns which do not contain both an item x and the item which is at the corresponding higher concept level of item x. For this purpose, all large items for each concept level need to be numbered. In the numbering phase, GMLAPG first apply the *TAXC algorithm* to construct taxonomies for all large items. Each level-1 large item constitutes the root node of a taxonomy. For a node which contains a level-k large item Z, each child node of this node contains a level-$k+1$ large item whose corresponding level-k item is the level-k large item Z.

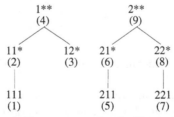

Fig. 1. Two taxonomies for Example 1

For Example 1, two taxonomies with root nodes which contain items 1** and 2**, respectively, are created by applying the algorithm TAXC. Node 1** has two child nodes which contain items 11* and 12*, respectively, because the two items are level-2 large items and their corresponding level-1 item is 1**. The two taxonomies for Example 1 are shown in Fig 1. For a taxonomy, we call the corresponding higher level items of an item x the ancestors of the item x, and the corresponding lower level items of an item y the descendants of the item y.

After constructing the taxonomies, all large items are numbered by applying the *PON method* on the taxonomies. For each taxonomy, PON numbers each item at the taxonomy according to the following order: for each item at the taxonomy, after all descendants of the item are numbered, PON numbers this item immediately, and all items are numbered increasingly. After all items at a taxonomy are numbered, PON numbers items at another taxonomy. For example, all items in the taxonomies in Fig 1 are numbered, where the number within the parentheses below each item is the number of the item.

Lemma 2. If the numbering method PON is adopted to number items, and for every two items i and j (N(i) < N(j)), item ϑ is an ancestor of item i but not an ancestor of item j, then N(ϑ) < N(j), where N(x) denotes the number of large item x after applying the PON method.

2.3 Generalized Association Graph Construction

In the association graph construction phase, GMLAPG applies the GAGC (Generalized Association Graph Construction) algorithm to construct a generalized association graph to be traversed. For every two large items i and j (i < j), if item j is not an ancestor of item i and the number of 1's in $BV_i \wedge BV_j$ achieves the user-specified minimum support, a directed edge from item i to item j is created. Also, itemset (i, j) is a large 2-itemset. In the following, we use the number of an item to represent this item.

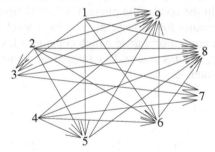

Fig. 2. The generalized association graph for Example 1

After applying the GAGC algorithm in the association graph construction phase, the generalized association graph for Example 1 is constructed in Fig 2, where there are no edges between an item and its ancestors.

2.4 Generalized Multiple-level Association Pattern Generation

In the association pattern generation phase, the algorithm LGDE (Large itemset Generation by Direct Extension) is proposed to generate large k-itemsets (k>2). For each large k-itemset (k $\geq$ 2), the last item of the k-itemset is used to extend the large itemset into k+1-itemsets.

Lemma 3. For a large itemset $(i_1, i_2, ..., i_k)$, if there is no directed edge from item i_k to an item v, then itemset $(i_1, i_2, ..., i_k, v)$ cannot be a large itemset.

Suppose $(i_1, i_2, ..., i_k)$ is a large k-itemset. If there is no directed edge from item i_k to an item v, then the itemset need not be extended into k+1-itemset, because $(i_1, i_2, ..., i_k, v)$ must not be a large itemset according to Lemma 3. If there is a directed edge from item i_k to an item u, then the itemset $(i_1, i_2, ..., i_k)$ is extended into k+1-itemset $(i_1, i_2, ..., i_k, u)$. The itemset $(i_1, i_2, ..., i_k, u)$ is a large k+1-itemset if the number of 1's in $BVi_1 \wedge BVi_2 \wedge ... \wedge BVi_k \wedge BV_u$ achieves the minimum support. If no large k-itemsets can be generated, the algorithm LGDE terminates.

For example, there is a directed edge from the last item 3 of the itemset (2, 3) to item 7 in Fig 2. Hence, the 2-itemset (2, 3) is extended into 3-itemset (2, 3, 7). The number of 1's in $BV_2 \wedge BV_3 \wedge BV_7$ is 3. Hence, the 3-itemset (2, 3, 7) is a large 3-itemset, since the number of 1's in its bit vector is no less than the minimum support threshold.

Theorem 1. If the numbering method PON is adopted to number items and the algorithm GAGC is applied to construct a generalized association graph, then any itemset generated by traversing the generalized association graph (i.e., performing LGDE algorithm) will not contain both an item and its ancestor.

3   Conclusion and Future Work

We propose a graph-based approach to discover generalized multiple-level association patterns. Because the GMLAPG algorithm needs only one database scan and performs logical AND operations, this algorithm is efficient when it is compared with other similar problems [1], [2], [3], [4] which need multiple passes over the database.

For our approach, the related information may not fit in the main memory when the size of the database is very large. In the future, we shall consider this problem by reducing the memory space requirement. Also, we shall apply our approach on different applications, such as document retrieval and resource discovery in the world-wide web environment.

References

1. Agrawal R., Srikant R.: Fast Algorithm for Mining Association Rules. Proceedings of the International Conference on Very Large Data Bases (1994) 487–499
2. Han J., Fu Y.: Mining Multiple-Level Association Rules in Large Databases. IEEE Transactions on Knowledge and Data Engineering (1999) 798–805
3. Park J.S., Chen M.S., Yu P.S.: An Effective Hash-Based Algorithm for Mining Association Rules. Proceedings of ACM SIGMOD, Vol. 24, No. 2 (1995) 175–186
4. Srikant R., Agrawal R.: Mining Generalized Association Rules. Proceedings of the International Conference on Very Large Data Bases (1995) 407–419

An Efficient Approach to Discovering
Sequential Patterns in Large Databases

Show-Jane Yen and Chung-Wen Cho

Department of Computer Science and Information Engineering
Fu Jen Catholic University, Taipei 242, Taiwan, ROC
{sjyen,Cho}@csie.fju.edu.tw

Abstract. *Mining sequential patterns* is to discover sequential purchasing behaviors of most customers from a large amount of customer transactions. The
previous approaches for mining sequential patterns need to repeatedly scan the
large database, and take a large amount of computation time to find frequent
sequences, which are very time consuming. In this paper, we present an algorithm *SSLP* to find sequential patterns, which can significantly reduce the number of the database scans. The experimental results show that our algorithms are
more efficient than the other algorithms.

1 Introduction

Because the capacity of the storage is getting larger, large amount of data can be
stored in the database. Potential useful information may be embedded in the large
databases. Hence, how to discover the useful information exists in such databases is
becoming the popular field in the computer science. The purpose of *data mining* [1],
[3], [4], [5] is to discover the useful information from the large databases, such that
the quality of decision making can be improved.

A *transaction database* consists of a set of *transactions*. A transaction typically
consists of the transaction identifier, the customer identifier (the buyer), the transaction date (or transaction time), and the items purchased in this transaction. Mining
sequential patterns [1], [2], [6] is to find the sequential purchasing behavior of most
customers from a large transaction database. For example, there is a sequential pattern
<{TERMINATOR, TERMINATOR 2}{TRUE AND FALSE}{JUNIOR}> 70% is dis-
covered from the transaction database in a video rental store, which means that sev-
enty percent of the customers rent *TRUE AND FALSE* after renting both
TERMINATOR and *TERMNATOR 2*, and then they rent *JUNIOR* after enjoying *TRUE
AND FALSE*. The manager can use this information to recommend the new customers
to rent *TRUE AND FALSE* and *JUNIOR* when they rent *TERMINATOR* and
TERMINATOR 2.

The definitions about mining sequential patterns are presented as follows: An
itemset is a non-empty set of items. An itemset X is *contained* in a transaction T, if
$X \subseteq T$, and X is a sub-itemset of T.

D.A. Zighed, J. Komorowski, and J. Zytkow (Eds.): PKDD 2000, LNAI 1910, pp. 685–690, 2000.
© Springer-Verlag Berlin Heidelberg 2000

A sequence is an ordered list of the itemsets. A sequence s is denoted as $<s_1, s_2, ...,$ $s_n>$, where s_i is an itemset. A sequence $<a_1, a_2, ..., a_n>$ is *contained* in another sequence $<b_1, b_2, ..., b_m>$ if there exist integers $i_1< i_2< ...< i_n$, $1¡Øi_k ¡Øm$ such that $a_1\subseteq bi_1, ..., a_n\subseteq bi_n$, then $<a_1, a_2, ..., a_n>$ is a *subsequence* of sequence $<b_1, b_2, ...,$ $b_m>$. A *maximum sequence* is a sequence that is not contained in any other sequence.

A *customer sequence* is the list of all the transactions of a customer, which is ordered by increasing transaction-time. A customer sequence c *supports* a sequence s if s is contained in c. The *support* for a sequence s is the number of customer sequences that supports s. If the support for a sequence s satisfies the user-specified minimum support threshold, then s is called *frequent sequence*. Otherwise, s is a *non-frequent sequence*. The *length* of a sequence s is the number of itemsets in the sequence. A sequence of length k is called a *k-sequence*, and a frequent sequence of length k a *frequent k-sequence*. In general, before generating the frequent sequences, we need to generate the *candidate sequences*, and scan the database to count the support for each candidate sequenc to decide if it is a frequent sequence. A candidate sequence of length k is called a *candidate k-sequence*.

For the previous approaches [1], [2] they need to generate a large number of candidates to be counted and make multiple passes over the large database, which are very time consuming. In this paper, we present an algorithm *SSLP* to discover all the sequential patterns. This algorithm can reduce the number of database scans. Moreover, it can effectively decrease the number of candidates to be counted.

The rest of this paper is organized as follows: Section 2 presents the algorithm to discover sequential patterns. The performance evaluation and the experimental results are shown in section 3.

2   Mining Sequential Patterns

We refer to [1] to decompose our algorithms into the same five phases. However, The major work of mining sequential patterns is to find all the frequent sequences. Hence, in the Sequence Phase, we present an efficient algorithm *SSLP* to generate all the frequent sequences from a transformed transaction database.

2.1   Algorithm SSLP (Segmental Smallest and Largest Position)

Let L_k be the set of the k-frequent sequence$s$, and C_k be the set of the candidate k-sequences.

Definition 1. *Let a sequence* $s = <s_1, s_2, ..., s_n>$, *and a customer sequence* $c = <c_1,$ $c_2, ..., c_m>$. *If there exists* $i_1<i_2<...<i_n ¡Al ¡Øi_k ¡Øm$, *such that* $s_1\subseteq ci_1,...,s_n\subseteq ci_n$, *then the position of* s *in* c *is* i_n. *Otherwise, the position is 0.*

For example, consider the customer sequence 3 in Table 1, the position of 2-sequence $<\{A\}\{B\}>$ is 3.

Definition 2. *The position pair of a sequence* s *corresponding to a customer sequence* c *is denoted as (F,L), where* F *and* L *are the minimum value and the maximum value among all the positions of* s *in* c, *respectively. If* s *is not contained in* c, *then the position pair of* s *in* c *is (0,0). If the position of* s *in* c *is* p, *then the position pair of* s *is* (p,p), *where* $1 \leq \emptyset p \leq \emptyset$ *the number of itemsets in* c.

For example, consider the customer sequence 3 in Table 1, the position pair of 2-sequence <{A}{B}> is (3,3).

Table 1. The transformed transaction database (*DB*)

CID	Transformed customer sequence
1	<{A,B,D}{A}{B}>
2	<{A}{B}{A,B,C,D}>
3	<{A,B,D}{A}{B}{C}>
4	<{A,C}{B,C}{C}>

SSLP partitions *DB* into some segments. For each segment, *SSLP* only needs to record the position pair of each candidate sequence corresponding to each customer sequence in the segment. A *segment* consists of *n* customer sequences, where *n* is the *length* of the segment. The *i-segment support* for a sequence *s* is the number of the customer sequences which support *s* in the ith segment. *SSLP* scans the database pass by pass. For the $(k/2)^{th}$ (k=2,4,...) pass, *SSLP* scans *DB* from the first segment to the mth segment, and generates L_k and L_{k-1}. For the ith segment scan, *SSLP* computes the *i*-segment support for each candidate sequence.

Assume that *SSLP* has scanned the ith segment. The *current support* CS*s*(*i*) of a sequence *s* = 1-segment support of *s* +...+ *i*-segment support of *s*. The *maximum support* of *s* is a value that the support of *s* is never greater than this value. Let the maximum support of *a* = <a_1, ..., a_k> be denoted as maxS*a*(*i*), and the *i*-segment support of 1-sequence a_j be S_{ij}. The initial maximum support maxS*a*(0) of *a* is defined as :

$$\text{maxS}_a(0) = \sum_{i=1}^{i=m} min\{Si_1,...,Si_k\}, \text{ m is the number of the segments in the database} \quad (1)$$

After scanning the ith segment, the maximum support of the *k*-sequence *a* can be obtained by the following expression :

$$\text{maxS}_a(i) = \text{maxS}_a(i\text{-}1) \text{-} (min\{Si_1,...,Si_k\} \text{-} i\text{-segment support of } a) \quad (2)$$

For example, the transaction database *DB* (Table 1) is partitioned into 2 segments, and the minimum support is set to 3. Hence, the customer sequences 1 and 2 are in the first segment, and the customer sequences 3 and 4 are in the second segment. The 1-segment supports and 2-segment supports of all the frequent 1-sequences can be found in the Transformation Phase, which are shown in Table 2. In the following, we use this example throughout this section and describe the three steps for the (1/2)*k* pass (*k*=2,4,...) for *SSLP*.

Step 1. Generate C_k from L_{k-1}, compute the initial maximum support of each candidate k-sequence, and then prune the candidate k-sequences that cannot be the frequent k-sequences from C_k.

We refer to [1] to generate C_k form L_{k-1}. After generating candidate k-sequences C_k, if C_k is not empty, then the candidates in C_k need to be further pruned. Otherwise, *SSLP* terminates. If one of the $(k-1)$-subsequences of the candidate k-sequence s is not in L_{k-1}, then s is pruned from C_k.

Table 2. The segments supports of frequent 1-sequences

Frequent 1-sequence	Segment support		Frequent 1-sequence	Segment support	
	Segment 1	Segment 2		Segment 1	Segment 2
<{A}>	2	2	<{C}>	1	2
<{B}>	2	2	<{D}>	1	1

Lemma 1. *If the maximum support of a sequence is less than the minimum support threshold, then the sequence is not a frequent sequence.*

For example, the initial maximum support of <{A}{B}> is 2+2=4, and the initial maximum support of <{A}{D}> is 1+1=2, which cannot be the frequent sequences.

Step 2. Generate C_{k+1} from C_k, compute the initial maximum supports of the candidate $(k+1)$-sequences, and then prune candidate $(k+1)$-sequences which cannot be the frequent $(k+1)$-sequences from C_{k+1}.

For example, the initial maximum support of candidate 3-sequence <{A}{B}{C}> is 1+2=3.

Step 3. Scan the transaction database *DB* from the first segment to the last segment.

Lemma 2. *If the maximum support of a sequence s is equal to the current support of s, and the current support is no less than the minimum support, then s is a frequent sequence.*

Step 3.1. For the ith ($i_i\grave{U}1$) segment scan, *SSLP* records the position pairs of each candidate k-sequence corresponding to each customer sequence in the ith segment, computes the maximum support and the current support for each candidate k-sequence, and then prunes the candidate k-sequences which cannot be the frequent sequences. Besides, if the pruned candidate k-sequence is a subsequence of a candidate $(k+1)$-sequence s, then s is also pruned from C_{k+1}.

For example, Table 3 shows the related information about some of C_2 after scanning the first segment, where candidate 2-sequence <{C}{A}> can be pruned, because its maximum support is less than the minimum support. Besides, according to Lemma

2, we can find some frequent sequences earlier, and these sequences need not be further examined.

Table 3. The related information for some candidate 2-sequences and candidate 3-sequences

Candidate	Position pairs		Maximum support	Current support
	Customer sequence 1	Customer sequence 2		
<{A}{B}>	(3,3)	(2,3)	4-(2-2)=4	2
<{A}{C}>	(0,0)	(3,3)	3-(1-1)=3	1
<{C}{A}>	(0,0)	(0,0)	3-(1-0)=2	0
<{A}{B}{C}>	0	3	3-(1-1)=3	1

Step 3.2. Find the i-segment support of each candidate $(k+1)$-sequence by computing the simple positions according to Definition 3, compute the maximum support and the current support for each candidate $(k+1)$-sequence, and then prune the candidate $(k+1)$-sequences which cannot be the frequent sequences according to Lemma 1.

Definition 3. *Let a sequence* $s = <a_1,...,a_k>$, *and the two* $(k-1)$-*subsequences of s are* $s_1 = <a_1, ..., a_{k-2}, a_{k-1}>$ *and* $s_2 = <a_1, ..., a_{k-2}, a_k>$. *Assume that the position pairs of* s_1 *and* s_2 *corresponding to the customer sequence c are* (f_1,l_1) *and* (f_2,l_2), *respectively. The simple position of s corresponding to the customer sequence c is denoted as* $Sim_s(c)$. *If* $f_1<l_2$ *and* $f_1 \neq 0, f_2 \neq 0$ *then* $Sim_s(c)=l_2$ *else* $Sim_s(c)=0$.

For example, the position pair of <{A}{B}> corresponding to customer sequence 4 is (2,2), and the position pair of <{A}{C}> is (2,3). Hence, the simple position of <{A}{B}{C}> corresponding to customer sequence is 3.

Lemma 3. *A customer sequence c supports a sequence s if and only if the simple position of s corresponding to c is not equal to 0.*

According to Lemma 3, if the simple position of a candidate $(k+1)$-sequence s corresponding to a customer sequence c is not equal to 0, then c support s. Hence, the supports of all the candidate $(k+1)$-sequences can be found by computing the simple positions. Thus, we do not have to scan the transaction database DB for counting the support of s.

SSLP repeats step 3.1 and step 3.2 until all the segments in the transaction database DB are scanned, and the candidate sequences which are remained in C_k and C_{k+1} are the frequent sequences. In our example, *SSLP* finally generates L_2 and L_3:

$L_2 = \{<\{A\}\{B\}>,<\{A\}\{C\}>,<\{B\}\{A\}>,<\{B\}\{C\}>\}, L_3 = \{<\{A\}\{B\}\{C\}>\}$.

3. Experimental Results

We refer to [1] to generate three synthetic transaction databases, and evaluate the performance of *SSLP* by comparing this algorithm with *Aprioriall*.

Suppose the maximum length of the generated candidate sequences by *Aprioriall* is q. Hence, the maximum length of the generated frequent sequences is no greater than q. *Aprioriall* has to scan the given transaction database $q-1$ times, but *SSLP* needs only to scan the database $\left\lceil q-\frac{1}{2} \right\rceil$ times. However, if q is even, then the maximum length of the candidate sequences generated by *SSLP* may be $q+1$. If C_{q+1} is not empty, *SSLP* need to compute the supports for the candidate $(q+1)$-sequences, even though there is no frequent $(q+1)$-sequences generated. The relative execution times for *Aprioriall* and *SSLP* over the minimum support ranging from 20.2% to 19%, in which the segment length is 250. The execution time of *SSLP* is between 1.6 times and 2.6 times as fast as that of *Aprioriall*. We notice that when the minimum support is 20%, the performance gap slightly decreases because the maximum length of the candidate sequences generated by *Aprioriall* is even. *SSLP* outperforms *Aprioriall* ranging from 1.6 to 2.6, and the performance gap increases as the minimum support decreases because the number of database scans increases for *Aprioriall*.

References

1. Agrawal R., et al.: Mining Sequential Patterns. Proceedings of International Conference on Data Engineering. (1995) 3–14
2. Agrawal R., et al.: Mining Sequential Patterns: Generalizations and Performance Improvements. Proceedings of the Fifth Int'l Conference on Extending Database Technology. Avignon, France (1996) 3–17
3. Agrawal R.: Johannes Gehrke, Dimitrios Gunopulos, Prabhakar Raghavan : Automatic Subspace Clustering of High Dimensional Data for Data Minig Applications. Proceedings of the ACM SIGMOD Int'l Conference on Management of Data. Seattle, Washington (1998) 94–105
4. Agrawal R., Bayardo R. J., Srikant R.: Mining-based Interactive Management of Text Databases. IBM Research Report RJ10153. (1999)
5. Savasere A., Omiecinski E., Navathe S.: An Efficient Algorithm for Mining Association Rules in Large Databases. Proceedings of 21st VLDB Conference. Zurich, Swizerland (1995) 432–444
6. Yen S.J., Chen A.L.P.: An Efficient Approach to Discovering Knowledge from Large Databases. Conference on Parallel and Distributed Information Systems. (1996) 8–18.

Using Background Knowledge as a Bias to Control the Rule Discovery Process

Ning Zhong[1], Juzhen Dong[1], and Setsuo Ohsuga[2]

[1] Dept. of Information Eng., Maebashi Institute of Technology, Japan
[2] Dept. of Infor. and Computer Science, Waseda University, Japan

Abstract. This paper investigates a way of using background knowledge in the rule discovery process. This technique is based on Generalization Distribution Table (GDT for short), in which the probabilistic relationships between concepts and instances over discrete domains are represented. We describe how to use background knowledge as a bias to adjust the prior distribution so that the better knowledge can be discovered.

1 Introduction

Over the last two decades, many researchers have investigated inductive methods [3] for learning *if-then* rules and concepts from instances. According to the value of information, these methods can be divided into two types. The first type is based on the *formal* value of information; that is, the real meaning of data is not considered in the learning process. ID3 and Prism are the typical methods of this type [5, 1]. Although *if-then* rules can be discovered by using the methods, it is difficult to use background knowledge in the learning process. The other type of inductive methods is based on the *semantic* value of information; that is, the real meaning of data must be considered by using some background knowledge in the learning process. Dblearn is a typical method belonging to this type [3]. It can discover rules by means of background knowledge represented by concept hierarchies, but if there is no background knowledge, it can do nothing. The question is *"how can both the formal value and the semantic value be considered in a discovery system?"*. Unfortunately, so far there is not any inductive learning method that can consider both of the formal value and the semantic value of information at the same time. It is clear that an ideal rule discovery system should have such feature, that is, on one hand, background knowledge can be used flexibly in the discovery process; on the other hand, if no background knowledge is available, it can also work on the formal value of data.

In [7, 8], we proposed a new methodology called GDT (Generalization Distribution Table), for learning *classification* rules in data with uncertainty and incompleteness. The main features of the GDT are

- It can predict unseen instances and represent explicitly the uncertainty of a rule including the prediction of possible instances in the strength of the rule.

---

[3] The discussion in this paper is limited to *attribute value learning*, which is a major type of inductive learning.

D.A. Zighed, J. Komorowski, and J. Zytkow (Eds.): PKDD 2000, LNAI 1910, pp. 691–698, 2000.
© Springer-Verlag Berlin Heidelberg 2000

- It can flexibly select biases for search control, and background knowledge can be used as a bias to control the creation of a GDT and the rule induction process.

In [9], we discussed the first feature of the GDT. This paper investigates the second feature of the GDT, that is, a way of using background knowledge in the rule discovery process. This paper is organized as follows: First Section 2 is a brief review of the basic concepts of the GDT. Section 3 discusses how to adjust the prior distribution by background knowledge. Section 4 gives a real world example to show the effects of usage of background knowledge. Finally in Section 6 we summarize our work and point out the future research direction.

2   Generalization Distribution Table (GDT)

The central idea of our methodology is to use a variant of transition matrix, which is called *Generalization Distribution Table (GDT)*, as a hypothesis search space for generalization, in which the probabilistic relationships between concepts and instances over discrete domains are represented [7, 8]. A GDT consists of three components:

The first is *possible instances*, which are denoted in the top row of a GDT, are all possible combinations of attribute values in a database. The number of the possible instances is $\prod_{i=1}^{m} n_i$, where m is the number of attributes, n_i is the number of different attribute values in each attribute i.

The second is *possible generalizations* for instances, which are denoted in the left column of a GDT, are all possible cases of generalization for all possible instances. "$*$", which specifies a wild card, denotes the generalization for instances[4]. For example, the generalization $*b_0c_0$ means the attribute a is unimportant for describing a concept. In other words, if $a = \{a_0, a_1\}$ and both $a_0b_0c_0$ and $a_1b_0c_0$ can describe a concept, attribute a does not seem to be important, that is, from $\{b_0c_0\}$, the concept can be described, and so we use the generalization $*b_0c_0$ to describe the concept, and let $*b_0c_0$ represent the set $\{a_0b_0c_0, a_1b_0c_0\}$. The number of the possible generalizations is $\prod_{i=1}^{m}(n_i + 1) - \prod_{i=1}^{m} n_i - 1$.

The third is *probabilistic relationships* between the possible instances and the possible generalizations, which are represented in the elements G_{ij} of a GDT, are the probabilistic distribution for describing the strength of the relationship between every possible instance and every possible generalization. The prior distribution is equiprobable, if any prior background knowledge is not used. Thus, it is defined by the Eq. (1), and $\sum_j G_{ij} = 1$:

$$G_{ij} = p(PI_j | PG_i)$$
$$= \begin{cases} \dfrac{1}{N_{PG_i}} & \text{if } PI_j \in PG_i \\ \\ 0 & \text{otherwise} \end{cases} \tag{1}$$

---

[4] For simplicity, the wild card will be omitted in some places in this dissertation.

where PI_j is the jth possible instance, PG_i is the ith possible generalization, and N_{PG_i} is the number of the possible instances satisfying the ith possible generalization, that is,

$$N_{PG_i} = \prod_{k \in \{l|\ PG[l]=*\}} n_k \qquad (2)$$

where $PG_i[l]$ is the value of the kth attribute in the possible generalization PG_i, $PG[l] = *$ means that PG_i doesn't contain attribute l.

Furthermore, for convenience, letting $E = \prod_{k=1}^{m} n_k$, Eq. (1) can be changed into the following form:

$$G_{ij} = p(PI_j|PG_i) = \begin{cases} \dfrac{\displaystyle\prod_{k \in \{l|\ PG[l] \neq *\}} n_k}{E} & \text{if } PI_j \in PG_i \\ 0 & \text{otherwise} \end{cases} \qquad (3)$$

because of

$$\frac{1}{N_{PG_i}} = \frac{1}{\displaystyle\prod_{k \in \{l|\ PG[l]=*\}} n_k} \cdot = \frac{\displaystyle\prod_{k \in \{l|\ PG[l] \neq *\}} n_k}{\displaystyle\prod_{k=1}^{m} n_k}$$

$$= \frac{\displaystyle\prod_{k \in \{l|\ PG[l] \neq *\}} n_k}{E}.$$

Since E is a constant for a given database, the prior distribution $p(PI_j|PG_i)$ is directly proportional to the product of the numbers of values of all attributes contained in PG_i.

Thus, in our approach, the basic process of hypothesis generation is to generalize the instances observed in a database by searching and revising the GDT. Here, two kinds of attributes need to be distinguished: *condition* attributes and *decision* attributes (sometimes called class attributes) in a database. Condition attributes as possible instances are used to create the GDT, but the decision attributes are not. The decision attributes are normally used to decide which concept (class) should be described in a rule. Usually a single decision attribute is all that are required.

3   Adjusting the Prior Distribution by Background Knowledge

One of the main features of the GDT methodology is that biases can be selected flexibly for search control, and background knowledge can be used as a bias to control the creation of a GDT and the rule discovery process. This section explains how to use background knowledge as a bias to adjust the prior distribution for learning much better knowledge.

As stated in Section 2, when no prior background knowledge as a bias is available, as default, the occurrence of all possible instances is equiprobable, and the prior distribution of a GDT is shown in Eq. (1). However, the prior distribution can be adjusted by background knowledge, and will be un-equiprobable after the adjustment. Generally speaking, the background knowledge can be given in

$$a_{i_1 j_1} \Rightarrow a_{i_2 j_2}, \quad Q,$$

where $a_{i_1 j_1}$ is the $j_1 th$ value of attribute i_1, and $a_{i_2 j_2}$ is the $j_2 th$ value of attribute i_2. $a_{i_1 j_1}$ is called the *premise* of the background knowledge, $a_{i_2 j_2}$ is called the *conclusion* of the background knowledge, and Q is called the *strength* of the background knowledge. It means that Q is the probability of occurrence of $a_{i_2 j_2}$ when $a_{i_1 j_1}$ occurs. $Q = 0$ means that "$a_{i_1 j_1}$ and $a_{i_2 j_2}$ never occur together"; $Q = 1$ means that "$a_{i_1 j_1}$ and $a_{i_2 j_2}$ always occur in the same time"; while $Q = 1/n_{i_2}$ means that the occurrence of $a_{i_2 j_2}$ is the same as the case of without background knowledge, where n_{i_2} is the number of values of attribute i_2. For each instance PI (or each generalization PG), let $PI[i]$ (or $PG[i]$) denote the entry of PI (or PG) corresponding to attribute i. For each generalization PG such that $PG[i_1] = a_{i_1 j_1}$ and $PG[i_2] = *$, the prior distribution between the PG and related instances will be adjusted. The probability of occurrence of attribute value $a_{i_2 j_2}$ is changed from $1/n_{i_2}$ to Q by background knowledge, so that, for each of the other values in attribute i_2, the probability of its occurrence is changed from $1/n_{i_2}$ to $(1-Q)/(n_{i_2}-1)$. Let the adjusted prior distribution be denoted by p_{bk}. The prior distribution adjusted by the background knowledge "$a_{i_1 j_1} \Rightarrow a_{i_2 j_2}, \quad Q$" is

$$
p_{bk}(PI|PG)
$$
$$
= \begin{cases} p(PI|PG) \times Q \times n_{i_2} & \text{if } PG[i_1] = a_{i_1 j_1}, PG[i_2] = *, \\ & \quad PI[i_2] = a_{i_2 j_2} \\ p(PI|PG) \times \dfrac{1-Q}{n_{i_2}-1} \times n_{i_2} & \text{if } PG[i_1] = a_{i_1 j_1}, PG[i_2] = *, \\ & \quad \exists j (1 \leq j \leq n_{i_2}, j \neq j_2) \; PI[i_2] = a_{i_2 j} \\ p(PI|PG) & otherwise \end{cases} \quad (4)
$$

where coefficients of $p(PI|PG)$, $Q \times n_{i_2}$, $\frac{1-Q}{n_{i_2}-1} \times n_{i_2}$, and 1 are called *adjusting factor* (*AF* for short) with respect to the background knowledge "$a_{i_1 j_1} \Rightarrow a_{i_2 j_2}, \quad Q$". They explicitly represent the influence of a piece of background knowledge to the prior distribution. Hence, the adjusted prior distribution can be denoted by

$$p_{bk}(PI|PG) = p(PI|PG) \times AF(PI|PG), \quad (5)$$

and the AF is

$$AF(PI|PG) = \begin{cases} Q \times n_{i_2} & \text{if } PG[i_1] = a_{i_1 j_1}, PG[i_2] = *, \\ & \quad PI[i_2] = a_{i_2 j_2} \\ \dfrac{1-Q}{n_{i_2} - 1} \times n_{i_2} & \text{if } PG[i_1] = a_{i_1 j_1}, PG[i_2] = *, \\ & \quad \exists j(1 \leq j \leq n_{i_2}, j \neq j_2) \ PI[i_2] = a_{i_2 j} \\ 1 & otherwise. \end{cases} \quad (6)$$

So far, we have explained how the prior distribution is influenced by only one piece of background knowledge. We then consider the case that there are several pieces of background knowledge such that for each i $(1 \leq i \leq m)$ and each j $(1 \leq j \leq n_i)$, there is at most only one piece of background knowledge with a_{ij} as its conclusion.

Let S be the set of all pieces of background knowledge to be considered. For each generalization PG, let

$B[S, PG] = \{i \in \{1, \ldots, m\}|$
$\exists i_1(1 \leq i_1 \leq m) \ \exists j_1(1 \leq j_1 \leq n_{i_1}) \ \exists j(1 \leq j \leq n_i)$
$[(\text{there is a piece of background knowledge in } S \text{ with } a_{i_1 j_1} \text{ as its premise}$
$\quad \text{and with } a_{ij} \text{ as its conclusion}) \ \& \ PG[i_1] = a_{i_1 j_1} \ \& \ PG[i] = *] \ \}$,

and for each $i \in B[S, PG]$, let

$J[S, PG, i] = \{j \in \{1, \ldots n_i\}| \ \exists i_1(1 \leq i_1 \leq m) \ \exists j_1(1 \leq j_1 \leq n_{i_1})$
$[(\text{there is a piece of background knowledge in } S \text{ with } a_{i_1 j_1} \text{ as its premise}$
$\quad \text{and with } a_{ij} \text{ as its conclusion}) \ \& \ PG[i_1] = a_{i_1 j_1} \ \& \ PG[i] = *] \ \}$.

Then, we must use the following *adjusting factors* AF_S with respect to all pieces of background knowledge,

$$AF_S(PI|PG) = \prod_{i=1}^{m} AF_i(PI|PG) \quad (7)$$

where

$$AF_i(PI|PG) \quad (8)$$

$$= \begin{cases} Q_{ij} \times n_i & \text{if } i \in B[S, PG], j \in J[S, PG, i], \text{ and } PI[i] = a_{ij} \\ \dfrac{1 - \displaystyle\sum_{j \in J[S, PG, i]} Q_{ij}}{n_i - |J[S, PG, i]|} \times n_i & \text{if } i \in B[S, PG], \\ & \quad \forall j(j \in J[S, PG, i])[PI[i] \neq a_{ij}] \\ 1 & \text{otherwise} \end{cases}$$

where for each i $(1 \leq i \leq m)$ and each j $(1 \leq j \leq n_i)$, Q_{ij} denotes the strength of the background knowledge (in S) with a_{ij} as its conclusion.

Although Q can be any value from 0 to 1 in principle, giving an exact value of Q is difficult, and the more the background knowledge, the more difficult to calculate the prior distribution. Hence, in practice, if "$a_{i_1j_1} \Rightarrow a_{i_2j_2}$" with higher possibility, we treat that Q is 1, that is, $a_{i_2j_2}$ occurs but other values of attribute i_2 do not, when $a_{i_1j_1}$ occurs. In contrast, if "$a_{i_1j_1} \Rightarrow a_{i_2j_2}$" with lower possibility, we treat that Q is 0, that is, $a_{i_2j_2}$ does not occur but other values of attribute i_2 occur in equiprobable, when $a_{i_1j_1}$ occurs. Furthermore, if several pieces of background knowledge with higher possibility, and the conclusions of them belong to the same attribute i_2, all of the attribute values (conclusions) are treated as occurrence in equiprobable, but other values in attribute i_2 are treated as no occurrence.

4   An Application

This section describes a result of an experiment in which background knowledge is used in the learning process to discover rules from a meningitis database [2]. The database was collected at the Medical Research Institute, Tokyo Medical and Dental University. It has 140 instances, each of which is described by 38 attributes that can be categorized into present history, physical examination, laboratory examination, diagnosis, therapy, clinical course, final status, risk factor etc. The task is to find important factors for diagnosis (bacteria and virus, or their more detail classifications) and predicting prognosis. A more detailed explanation on this database could be found at http://www.kdel.info.eng.osaka-cu.ac.jp/SIGKBS.

For each of the decision attributes: DIAG2, DIAG, CULTURE, C_COURSE, and COURSE(Grouped), we run our rule discovery system named GDT-RS, which is a way of implementation of the GDT methodology by combining the GDT with rough sets [10], on it twice: using background knowledge and without using background knowledge, to acquire the rules respectively. For the discretization of continuous attributes, an automatic discretization [4] is used.

4.1   Background Knowledge Given By a Medical Doctor

The experience of a medical doctor can be used as background knowledge:

> If the brain wave (EEG-WAVE) is normal, the focus of brain wave (EEG_FOCUS) is never abnormal;
> If the number of white blood cells (WBCs) is high, the inflammation protein (CRP) is also high.

In the following list, a part of the background knowledge given by a medical doctor is described:

– Never occurring together.
 $EEG\_WAVE(normal) \Leftrightarrow EEG\_FOCUS(+)$
 $CSF\_CELL(low)\ \ \ \ \Leftrightarrow Cell\_Poly(high)$
 $CSF\_CELL(low)\ \ \ \ \Leftrightarrow Cell\_Mono(high)$

– Occurring with lower possibility.

$WBC(low) \Rightarrow CRP(high)$
$WBC(low) \Rightarrow ESR(high)$
$WBC(low) \Rightarrow CSF\_CELL(high)$
$WBC(low) \Rightarrow Cell\_Poly(high)$

– Occurring with higher possibility.

$WBC(high) \quad \Rightarrow CRP(high)$
$WBC(high) \quad \Rightarrow ESR(high)$
$WBC(high) \quad \Rightarrow CSF\_CELL(high)$
$EEG\_FOCUS(+) \Rightarrow FOCAL(+)$
$EEG\_WAVE(+) \Rightarrow EEG\_FOCUS(+)$

Here "high" in brackets denoted in the background knowledge means that the value is greater than the maximal value in the normal values range; and "low" means that the value is less than the minimal value in the normal values range.

4.2   Comparing the Results

The effects of usage of the background knowledge in GDT-RS are as follows:

First, some candidates of rules, which are deleted due to lower strengths during rule discovery without background knowledge, are selected. For example, $rule_1$ is deleted when no background knowledge is used, but after using the background knowledge stated above, it is reserved because its strength increased 4 times.

$rule_1$:

$$ONSET(acute) \wedge ESR(\leq 5) \wedge CSF\_CELL(> 10) \wedge CULTURE(-) \rightarrow VIRUS(E).$$

Without using background knowledge, the strength S of $rule_1$ is 30*(384/E). In the background knowledge given above, there are two clauses related to this rule:

• Never occurring together
$CSF\_CELL(low) \Leftrightarrow Cell\_Poly(high)$
$CSF\_CELL(low) \Leftrightarrow Cell\_Mono(high).$

By using automatic discretization to continuous attributes $Cell\_Poly$ and $Cell\_Mono$, the attribute values in each of attributes are divided into two groups: high and low. Since the high groups of $Cell\_Poly$ and $Cell\_Mono$ do never occur when $CSF\_CELL(low)$ occurs, the product of the numbers of attribute values is decreased to $E/4$, and the strength S is increased to $S = 30 * (384/E) * 4$.

Second, using background knowledge also causes some rules to be replaced by others. For example, the rule

$rule_2$:

$$DIAG(VIRUS(E)) \wedge LOC[4, 7) \rightarrow EEG\_abnormal, \quad S = 30/E$$

can be discovered without background knowledge, but after using the background knowledge stated above, it is replaced by

$rule_{2'}$:

$EEG\_FOCUS(+) \wedge LOC[4, 7) \rightarrow EEG\_abnormal, \quad S = (10/E) * 4.$

The reason is that both of them contain the same instances, but the strength of $rule_{2'}$ becomes larger than that of $rule_2$.

The result has been evaluated by a medical doctor. According to his opinions, both $rule_1$ and $rule_{2'}$ are reasonable, and $rule_{2'}$ is much better than $rule_2$.

Although the similar results can be obtained from the meningitis database by using GDT-RS and C4.5 if such background knowledge is not used, it is difficult that such background knowledge is used in C4.5 [6].

5   Conclusion

We presented a new way of using background knowledge in the rule discovery process. The theoretical and experimental results show that our methodology is very flexible. That is, on one hand, background knowledge can be used flexibly in the discovery process so that the better knowledge can be discovered; on the other hand, if no background knowledge is available, it can also work on the formal value of data.

Our future work includes cooperatively using such background knowledge and different types of background knowledge as biases in more different aspects of the rule discovery process to learn much better knowledge.

References

1. J. Cendrowska, "PRISM: An Algorithm for Inducing Modular Rules", *International Journal of Man-Machine Studies*, Vol.27, (1987) 349-370.
2. Dong, J.Z., Zhong, N., and Ohsuga, S. "Rule Discovery from the Meningitis Database by GDT-RS" (Special Panel Discussion Session on Knowledge Discovery from a Meningitis Database) *Proc. the 12th Annual Conference of JSAI* (1998) 83-84.
3. J. Han, Y. Cai, and N. Cercone, "Data-Driven Discovery of Quantitative Rules in Relational Databases", *IEEE Trans. Knowl. Data Eng.*, Vol.5, No.1 (1993) 29-40.
4. S.H. Nguyen and H.S. Nguyen, "Discretization Methods in Data Mining", L. Polkowski and A. Skowron (eds.) *Rough Sets in Knowledge Discovery*, Vol.1, Physica-Verlag (1998) 451-482.
5. J.R. Quinlan, "Induction of Decision Trees", *Machine Learning*, Vol.1 (1986) 81-106.
6. J.R. Quinlan, *C4.5: Programs for Machine Learning*, Morgan Kaufmann (1993).
7. N. Zhong and S. Ohsuga, "Using Generalization Distribution Tables as a Hypotheses Search Space for Generalization". *Proc. 4th Inter. Workshop on Rough Sets, Fuzzy Sets, and Machine Discovery (RSFD-96)* (1996) 396-403.
8. N. Zhong, J.Z. Dong, and S. Ohsuga, "Discovering Rules in the Environment with Noise and Incompleteness", *Proc. 10th Inter. Florida AI Research Symposium (FLAIRS-97)* edited in the *Special Track on Uncertainty in AI* (1997) 186-191.
9. N. Zhong, J.Z. Dong, and S. Ohsuga, "Data Mining: A Probabilistic Rough Set Approach", L. Polkowski and A. Skowron (eds.) *Rough Sets in Knowledge Discovery*, Vol.2, In Studies in Fuzziness and Soft Computing series, Vol.19, Physica-Verlag (1998) 127-146.
10. N. Zhong, J.Z. Dong, and S. Ohsuga, "GDT-RS: A Probabilistic Rough Induction System", Bulletin of International Rough Set Society, Vol.3, No.4 (1999) 133-146.

Author Index

Acid, Sylvia 309
Alonso, Carlos J. 299
Apte, Chidanand V. 665
Apte, Chid 575

Baesens, B. 657
Barber, Brock 316
Batistakis, Y. 265
Bauer, Valerij 464
Bensusan, Hilan 325
Bernadet, Maurice 24
Blockeel, Hendrik 1
Bobrowski, Leon 331
Boström, Henrik 299
Boulicaut, Jean-François 75
Brazdil, Pavel B. 126
Breunig, Markus M. 232
Brewer, Sam 337
Briand, Henri 483, 638
Bunelle, Pascal 581
Bykowski, Arthur 75

Carvalho, Deborah R. 345
Cau, Denis 96
Chauchat, Jean-Hugues 181
Cho, Chung-Wen 685
Chung, Fu-Lai 510
Ciampi, Antonio 353, 359
Clech, Jérémy 359
Cole, Richard 367
Conruyt, Noël 409
Contreras-Arevalo, Edgar E. 524
Cristofor, Dana 619
Cristofor, Laurentiu 619

Das, Gautam 201
De Campos, Luis M. 309
Dedene, G. 657
De Moor, D. 657
Diatta, Jean 409
Djeraba, Chabane 375
Dong, Juzhen 691
Dong, Guozhu 191
Dougarjapov, Aldar 289
Drossos, Nikos 381
Druet, Christophe 86
Dury, Laurent 388

Džeroski, Sašo 54, 446

Eklund, Peter 367
Ernst, Damien 86

Feelders, A.J. 395
Feschet, Fabien 599
Flach, Peter 255
Forman, George 243
Freitas, Alex A. 345

Gamberger, Dragan 34
Geurts, Pierre 136, 401
Giraud-Carrier, Christophe 325
Grosser, David 409
Guillet, Fabrice 483
Güntzer, Ulrich 159
Gyenesei, Attila 416

Halkidi, Maria 265
Hamilton, Howard J. 316, 432
Han, Eui-Hong (Sam) 424
Hein, Sebastian 277
Hilario, Melanie 106
Hilderman, Robert J. 432
Hipp, Jochen 159
Ho, Tu Bao 211, 458
Holeňa, Martin 440
Hovig, Eivind 470
Hristovski, Dimitar 446
Hsu, Meichun 243
Huang, Weiyun 452
Hvidsten, Torgeir R. 470

Inokuchi, Akihiro 13

Jacques, Yannick 581
Jenssen, Tor-Kristian 470

Kahle, Hans-Peter 277
Kalles, Dimitris 381
Kalousis, Alexandros 106
Kargupta, Hillol 452
Karypis, George 424
Kavalec, Martin 626
Kawasaki, Saori 458
Keane, John A. 542
Keller, Jörg 464

Khabaza, Tom   337
Kindermann, Jörg   558
Knobbe, Arno J.   1
Komorowski, Jan   470
Krętowski, Marek   331
Kriegel , Hans-Peter   232
Kryszkiewicz, Marzena   476
Kumar, Vipin   632
Kuntz, Pascale   483
Kwedlo, Wojciech   464

Lægreid, Astrid   470
Lallich, Stéphane   221
Lanquillon, Carsten   490
Laur, P.A.   498
Lausen, Georg   277, 289
Lavrač, Nada   34, 255
Lechevallier, Yves   353
Lee, Chung-Hong   673
Leherte, Laurence   388
Lehn, Rémi   483
Li, Jinyan   191
Liu, Bing   504
Ludl, Marcus-Christopher   148
Lui, Chung-Leung   510

Ma, Yiming   504
Mannila, Heikki   201
Marshall, Adele   516
Martínez-Trinidad, José F.   524
Masseglia, F.   498, 530
McClean, Sally   516
Millard, Peter   516
Moal, Frédéric   536
Morzy, Tadeusz   65
Motoda, Hiroshi   13
Munteanu, Paul   96
Muyeba, Maybin K.   542

Nachouki, Gilles   638
Nakhaeizadeh, Gholamreza   159
Nguyen, Ngoc Binh   211, 458
Nicoloyannis, Nicolas   593, 599
Nock, Richard   44

Ogihara, Mitsunori   566
Ohsuga, Setsuo   691
Okada, Takashi   550
Oyama, Mayumi   550

Paass, Gerhard   558
Papagelis, Athanasios   381

Park, Byung-Hoon   452
Parthasarathy, Srinivasan   566
Perner, Petra   575
Peterlin, Borut   446
Philippe, Jacques   638
Platis, Agapios   646
Poncelet, P.   498, 530
Puuronen, Seppo   116

Rakotomalala, Ricco   181, 221
Ramamohanarao, Kotagiri   191
Ramstein, Gerard   581
Ras, Zbigniew W.   587
Rigotti, Christophe   75
Ritschard, Gilbert   593
Robardet, Céline   599
Robert, Didier   181
Rodríguez, Juan J.   299
Rossi, Fabrice   605
Rozic-Hristovski, Anamarija   446

Salleb, Ansaf   613
Sander, Jörg   232
Sandvik, Arne K.   470
Savnik, Iztok   277, 289
Sebban, Marc   44
Shapcott, Mary   516
Siebes, Arno.   1
Simovici, Dan A.   619
Sivakumar, Krishnamoorthy   452
Soares, Carlos   126
Spiecker, Heinrich   277
Srivastava, Jaideep   632
Stumme, Gerd   367
Suykens, J.A.K.   657
Suzuki, Einoshin   169
Svátek, Vojtěch   626

Tan, Pang-Ning   632
Teisseire, M.   530
Teusan, Tudor   638
Tjeldvoll, Dyre   470
Todorovski, Ljupčo   54, 255
Tselios, Peter   646
Tsumoto, Shusaku   652
Tsymbal, Alexey   116
Turmeaux, Teddy   536

Van Den Poel, D.   657
Van Der Wallen, Daniël   1
Van Gestel, T.   657
Vanthienen, J.   657

Vautrain, Frédérick   605
Vazirgiannis, M.   265
Velasco-Sánchez, Miriam   524
Vercauteren, Daniel P.   388
Viaene, Stijn   657
Vouros, George   646
Vrain, Christel   536, 613

Wang, Shuren   452
Washio, Takashi   13
Wehenkel, Louis   86, 401
Weiss, Sholom M.   665
White, Brian F.   665

Widmer, Gerhard   148
Wieczorkowska, Alicja   587
Wojciechowski, Marek   65
Wong, Ching Kian   504

Yang, Hsin-Chang   673
Yen, Show-Jane   679, 685

Zakrzewicz, Maciej   65
Zhang, Bin   243
Zhong, Ning   691
Zighed, Djamel A.   359
Żytkow, Jan M.   169